13th Aceh International Workshop and Expo on Sustainable Tsunami Disaster Recovery (The 13th AIWEST-DR 2021)

Held Online

E3S Web of Conferences Volume 340 (2022)

Banda Aceh, Indonesia
26-27 October 2021

Editors:

Alfi Rahman
Yunita Idris
Haekal Azief Haridhi
Muksin
Ella Meilianda
Tursina Musa
Ibnu Rusydy
Anawat Suppasri
Erick Mas
Aaron Opdyke

ISBN: 978-1-7138-4282-8

Printed from e-media with permission by:

Curran Associates, Inc.
57 Morehouse Lane
Red Hook, NY 12571

Some format issues inherent in the e-media version may also appear in this print version.

For additional information, please contact EDP Sciences – Web of Conferences at the address below.

EDP Sciences – Web of Conferences
17, Avenue du Hoggar
Parc d'Activité de Courtabœuf
BP 112
F-91944 Les Ulis Cedex A
France

Phone: +33 (0) 1 69 18 75 75
Fax: +33 (0) 1 69 28 84 91

contact-edps@webofconferences.org

Additional copies of this publication are available from:

Curran Associates, Inc.
57 Morehouse Lane
Red Hook, NY 12571 USA
Phone: 845-758-0400
Fax: 845-758-2633
Email: curran@proceedings.com
Web: www.proceedings.com

TABLE OF CONTENTS

AIWEST - DR 2021
13th Aceh International Workshop and Expo on Sustainable Tsunami Disaster Recovery

The 13th Aceh International Workshop and Expo on Sustainable Tsunami Disaster Recovery - 2021

26-27
October 2021

"Post-the 2004 Indian Ocean Tsunami Challenges: Strengthening Community Resilience from Compound Disasters in Pandemic Situations"

Opening Remarks

Prof. Samsul Rizal
Rector of Universitas Syiah Kuala, Banda Aceh

Moderator 1

Suryane Sulistiana Susanti, PhD
TDMRC Universitas Syiah Kuala

Moderator 2

Dr. Aaron Opdyke
The University of Sydney

Keynote Speakers

Prof. Fumihiko Imamura
Director International Research Institute of Disaster Science (IRIDeS), Tohoku University, Japan

Prof. Rajib Shaw
Professor at Graduate School of Media and Governance, Keio University, Japan

Dr. dr. Safrizal Rahman, M.Kes., Sp.OT
Vice Head of Covid-19 Task Force Universitas Syiah Kuala

Prof. Helene Joffe
Professor of Psychology University College London

Registration: bit.ly/AIWEST-DR-Reg

TOPICS:

Prof. Fumihiko Imamura
"10 Years of the Great East Japan Earthquake and Tsunami in COVID-19 pandemic: cultural, information, disaster and social issues"

Prof. Rajib Shaw
"Governance, technology and citizen behavior in pandemic: Lessons from COVID-19"

Dr. dr. Safrizal Rahman, M.Kes., Sp.OT
Role of Higher Education Institution in Disaster Management during Pandemic era: A Case of Universitas Syiah Kuala

Prof. Helene Joffe
Public responses to COVID-19: Psychosocial Aspects

Organized by: USK, TDMRC, IRIDeS

Supported by: UCL, EPICentre, UPNM

edp sciences

E3S Web of Conferences

Statement of Peer review

In submitting conference proceedings to E3S Web of Conferences, I certify to the Publisher that I adhere to the **Policy on Publishing Integrity** of the journal in order to safeguard good scientific practice in publishing.

1. All articles have been subjected to peer review administered by the proceedings editors.
2. Reviews have been conducted by expert referees, who have been requested to provide unbiased and constructive comments aimed, whenever possible, at improving the work.
3. Proceedings editors have taken all reasonable steps to ensure the quality of the materials they publish and their decision to accept or reject a paper for publication has been based only on the merits of the work and the relevance to the journal.

Title, date and place of the conference

Title of Conference : The 13th of Aceh International Workshop and Expo on Sustainable Tsunami Disaster Recovery (The 13th AIWEST-DR 2021)
Date : 26 - 27 October 2021
Place : Banda Aceh (Virtually)

Proceedings editor(s):

1. Dr. Alfi Rahman
2. Dr. Yunita Idris
3. Dr. Haekal Azief Haridhi
4. Assoc. Prof. Muksin
5. Dr. Ella Meilianda
6. Tursina Musa
7. Ibnu Rusydy
8. Assoc. Prof. Anawat Suppasri
9. Assoc. Prof. Erick Mas
10. Dr. Aaron Opdyke

Date and editor's signature

Dr. Alfi Rahman   Dr. Yunita Idris   Dr. Haekal A Haridhi   A/Prof. Muksin   Dr. Ella Meilianda

Tursina Musa, ST, MT.   Ibnu Rusydy, S.Si., M.Sc.   A/Prof. Anawat Suppasri   A/Prof. Erick Mas   Dr. Aaron Opdyke

E3S Web of Conferences **340**, 01001 (2022)
The 13th AIWEST-DR 2021
https://doi.org/10.1051/e3sconf/202234001001

Indonesia paleotsunami database as an effort to reduce the tsunami disasters in Indonesia

Ihda Ibtihaj[1], *Supriyanto* Suparno[2*], *Muhammad* Rizqy Septyandy[1], *Gamma* Abdul Jabbar[1], *Tri* Rani Puji Astuti[1]

[1]Geology Study Program, University of Indonesia, Pondok Cina, Beji, Depok, 16424, Indonesia
[2]Geophysics Study Program, University of Indonesia, Pondok Cina, Beji, Depok, 16424, Indonesia

Abstract. Indonesia's tectonic activity is one of the most active in the world. These conditions make Indonesia prone to earthquakes, tsunamis, and volcanic eruptions. Tsunamis are large ocean waves that are formed through several processes such as underwater earthquakes, underwater landslides, volcanic activity, and asteroid impacts. Indonesia has been hit by various tsunami events. However, not all characteristic tsunami events are well known and recorded. Knowing the characteristics of a tsunami occurrence is very important to understand the frequency and intensity of tsunamis in the present. This research create a WebGIS database for paleotsunami in Indonesia. Tsunami incident data comes from literature related to tsunami events such as books, journals, reports, final projects, and others. Making WebGIS in this study will be developed using website programming, namely HTML, CSS, and JavaScript. The research results obtained 302 tsunami data. The period of tsunami in this research is from Late Miocene to the last tsunami in 2018. The WebGIS created presents a database of tsunami events in Indonesia, a dashboard of Indonesian tsunami statistics, and a tsunami learning educational feature. It is hoped that the results of this database will serve as a means of information for future paleotsunami research needs and as a means of education regarding the potential for tsunamis in Indonesia.

1 Introduction

Indonesia is a country located at the confluence of four major plates, namely Eurasia, Indo-Australia, the Philippines, and the Pacific [1]. Complex plate tectonic conditions make Indonesia one of the regions in the world with the most active tectonic activity [2]. As a result of these conditions, Indonesia is prone to earthquakes, tsunamis, and volcanic eruptions.

Tsunamis are the most powerful ocean wave motions. The mechanism for the formation of a tsunami is through processes that produce shock waves, such as underwater earthquakes, underwater landslides, volcanic activity, and asteroid impacts [3]. The occurrence of a tsunami is a disaster because it has caused damage, loss of life, and property loss as well as damage to ecosystems [4].

When a tsunami occurs, the waves will carry material from the deep seafloor in the form of clay. When approaching the coast, the tsunami also erodes the beach base, which generally consists of sand. So there is a mixture of beach sand material and deep sea mud. The larger the tsunami that occurs, the tsunami inundation zone (inundation) and the material it carries will spread farther inland. After the tsunami energy weakens, the seawater returns to the sea (backwash) at a slower speed so that the sand material carried does not come back but is deposited on land. Deposition brought by the tsunami will cover the coastal plain and the existing soil or vegetation [5].

Tsunami deposits are deposits carried by the tsunami, which are left on the coast during the inundation phase or returned to the sea during the backwash phase. These deposits are used to identify past tsunami events (paleotsunami) [5]. Paleotsunami is a tsunami that occurs before historical records or there are no written observations [6]. Paleotsunami deposits can be identified using paleotsunami proxies.

The potential for a tsunami disaster is very likely to occur in Indonesia. At least there have been various series of tsunami events that have caused many casualties, both caused by underwater earthquakes and underwater volcanic eruptions. However, in most of the tsunami events that have hit Indonesia, the mechanism and character of the tsunami are not well known [2]. Historical information and data on tsunami events in an area can be used as a reference for understanding the frequency and intensity of tsunami disasters in the present [5, 7]. Therefore, it is necessary to create a historical database of tsunami events in Indonesia as an effort to reduce tsunami risk.

Indonesia Paleotsunami Database will be made based on WebGIS. WebGIS is a Geographic Information System (GIS) that uses the internet network as the main access to distribute spatial data and information as well as GIS analysis [8]. The use of WebGIS was chosen because this method can be easily

* Corresponding author: supriyanto@sci.ui.ac.id

accessed by the general public. The results of this database is expected to be a means of information for future paleotsunami research needs as well as a means of education regarding the potential for tsunamis in Indonesia.

2 Methods

2.1 Paleotsunami proxy

Paleotsunami proxies are various criteria used to identify deposits caused by tsunami processes [9]. The use of proxies in paleotsunami research has not become a standard thing and is still being developed today. The most common paleotsunami proxy and has been widely used in paleotsunami research is the proxy developed by Goff et al. (2011) [10]. Paleotsunami proxy presented in Table 1.

Table 1. Paleotsunami Proxy

Proxy Number	Proxy
1	Grain size
2	Fining upwards sediment
3	Sediment units that differ/contrast with other layers.
4	sub-units that characterize tsunami runup and backwash.
5	Contact unconformity or erosional.
6	Contains a rip-up clast of reworked material.
7	The loading structure at the bottom of the deposit
8	Microscale features may be visible on thin sections
9	Measurement of anisotropy magnetic susceptibility (AMS) combined with grain size analysis.
10	Heavy mineral laminates
11	Increased concentration of elements sodium, sulfur, chlorine, calcium, strontium, magnesium, titanium, zirconium
12	Possible contamination by heavy metals and metalloids
13	Geochemical evidence
14	Shells of marine organisms and rich layers of shells.
15	Crushed shells, wood, and various other materials are often found "inserted" in the layers of deposits.
16	It may be associated with vascular plant remains, paleosoil, or buried bone remains.
17	Associated with an increase of diatoms living in marine environments – brackish water
18	Found a collection of foraminifera shells (or other fossils
19	Terrestrial pollen concentrations decrease on and/or coastal pollen concentrations increase.
20	The existence of archaeological sites that are overlaid or overlain by layers of sediment.
21	There is a collection of archaeological middens
22	Archaeological sites (buildings) that have been damaged by water waves.
23	Archaeological sites that have been re-deposited in inappropriate locations.
24	Replication, layers of coastal archaeological remain and piles of shells
25	Local folklore about the tsunami.
26	Paleogeomorphological acquisitions.
27	Paleogeomorphology at the time of inundation
28	Local or regional tsunamigenic knowledge.
29	Local and regional knowledge of ancient environments
30	Similar coastal deposits were found regionally indicating regional inundation.

2.2 Data collection and processes

The data used in this research include scientific journals, papers, research reports, final projects, or other publications related to the history of the tsunami incident in Indonesia. Data collected on tsunami events in Indonesia can come from valid geoscience, archeological or folklore perspective. Tsunami event data that has been collected then matriculated. Data matriculation processing uses Microsoft Excel 2016.

The data resulting from the matriculation will then be processed into GeoJSON data (Geospatial JavaScript Object Notation). Data tsunami GeoJSON will be data in the form of a point (*point*) indicated that at the location of the point is never the tsunami. In addition, this data contains important information about the occurrence of a tsunami at that location.

Making WebGIS in this research developed using website programming. The programs used are HTML (Hypertext Markup Language), CSS (Cascading Style Sheets), and JavaScript. WebGIS present important information content related to paleotsunami in Indonesia. The attributes used to present content in WebGIS presented in table 2.

Table 2. Database Attributes

No	Attributes
1	Province / Regional Location
2	Location of the tsunami incident
3	The latitude and longitude of the tsunami/paleotsunami point
4	Location status (there is an unknown deposit or deposit)
5	Physical characteristics (tsunami candidate deposition conditions, or other enabling conditions)
6	Classification of the time of occurrence • *Historical* (1800 AD - present) • *Prehistorical* (years <1800 AD)
7	Time of occurrence
8	Validity • Fair (1-4 criteria of proxy) • Moderate (5-8 criteria of proxy) • *Excellent* (> 9 criteria of proxy)
9	Nature of evidence
10	Description of nature of evidence

11	Dating Technique
12	Dating Method
13	Paleotsunami point elevation
14	Inundation zone boundaries
15	The thickness of the sediment/sediment formed
16	The cause of the tsunami
17	Tsunami source location and coordinates
18	Paleotsunami proxy used
19	Reference

2.3 Data analysis

The analysis conducted in this research is quantitative analysis. The mechanism that caused the tsunami that hit Indonesia was recorded with the frequency of occurrence. Paleotsunami proxies used for identification are also recorded in frequency of use in paleotsunami identification studies in Indonesia. The results of the analysis become data on the frequency of mechanisms that cause tsunami events in Indonesia and the frequency of paleotsunami proxies that are widely used in Indonesia.

3 Results and discussion

3.1 Database processing results

The tsunami incident data that was collected and entered into the database amounted to 302 data. Based on this research database tsunami period which is have been collected, oldest recorded tsunami events are paleotsunami deposits thought to be Late Miocene based on the foraminifera fossils contained [5]. Meanwhile, the most recent incident was the tsunami in 2018 which hit Lombok, West Nusa Tenggara and Palu, Central Sulawesi [12]. Data from the literature that is entered into the matriculation is information according to the required attributes. The results of the database matriculation are presented in Figure 1.

Fig. 1. Database matriculation

The distribution of tsunami data in Indonesia, almost all of Indonesia's coasts have been hit by a tsunami. The West Coast of Sumatra Island is the area that has experienced the most tsunamis. The southern regions of Java, North Sulawesi and the Maluku Islands have also experienced many tsunamis. The East Coast of Kalimantan Island is the area least affected by the tsunami. This shows that almost all regions in Indonesia have the potential for a significant tsunami disaster.

3.2 Causes of the tsunami in Indonesia

The mechanism that caused the tsunami in Indonesia has been grouped to determine the dominant cause. The results of data grouping show that the causes of tsunamis in Indonesia have several mechanisms, namely earthquakes, volcanism, landslides and combined mechanisms. the recapitulation of the causes of the tsunami in Indonesia is presented in Figure 2.

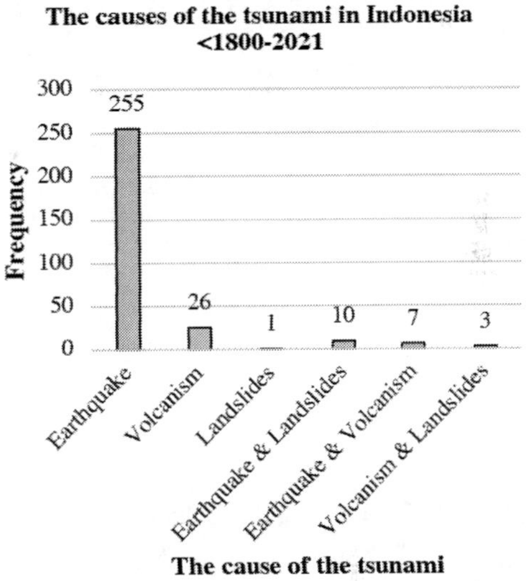

Fig. 2. Chart of the causes of the tsunami in Indonesia

Earthquakes are the mechanism that causes most tsunamis in Indonesia. Based on the database, the occurrence of tsunamis caused by earthquakes mostly occurs in areas associated with the location of active faults. Several active fault zones that are the source of earthquakes that cause tsunamis include the Sunda Megathrust, Banda Megathrust, Flores Rise Fault, North Sulawesi Megathrust, Papua Megathrust, Sangihe Megathrust, and local faults [11, 12].

Coastal areas adjacent to the presence of active faults and subduction zones are areas that have the most potential for tsunamis caused by earthquakes. This is because the area has high seismic activity. So, it causes a lot of earthquakes [2]. Based on the database for the western region of Sumatra, southern Java, the Banda arc, and the Maluku Sea, these are the dominant areas where earthquakes trigger tsunamis. The area is a subduction zone with high seismic activity that causes many earthquakes.

Volcanism activities that occur in Indonesia have been the cause of the tsunami. The eruption of Mount Krakatau on August 26, 1883, was a volcanic activity that caused tsunami waves that impacted Lampung and Banten [13, 14, 15, 16]. The Lampung and Banten regions were also affected by the tsunami on December 22, 2018, due to volcanism from Mount Anak Krakatau.

 https://doi.org/10.1051/e3sconf/202234001001

Volcanic activity from Mount Krakatau and Anak Krakatau is one of the pieces of evidences from the past and present that volcanism has the potential to cause tsunamis.

Underwater avalanches are the mechanism that once caused a tsunami in Indonesia. Tsunamis caused by submarine avalanches are rare according to the database. Although rare, tsunamis caused by avalanches are also a risk of disaster.

3.3 Paleotsunami proxy applied in Indonesia

The Indonesian paleotsunami proxy in this study is a collection of proxies grouped based on the created database. The grouping of proxies is done to find out which proxies are dominant and effectively used in Indonesia. The most widely used proxy is proxy number 28 while the least used is proxy number 23 (figure 3). Based on the grouping of proxy data from the literature used by paleotsunami research in Indonesia, only 28 of the 30 proxies are available. Proxies number 24 and 27 have not been implemented in Indonesia.

Proxy number 28 is the most widely used proxy. This is because the tsunami data was identified dominantly using contextual data. The contextual data used are such as tsunami catalogs or historical records. The use of contextual data as a proxy shows that paleotsunami research in Indonesia needs further development. The application of other proxies that identify field data needs to be developed so that concrete evidence of a tsunami can be found in deposits in the field.

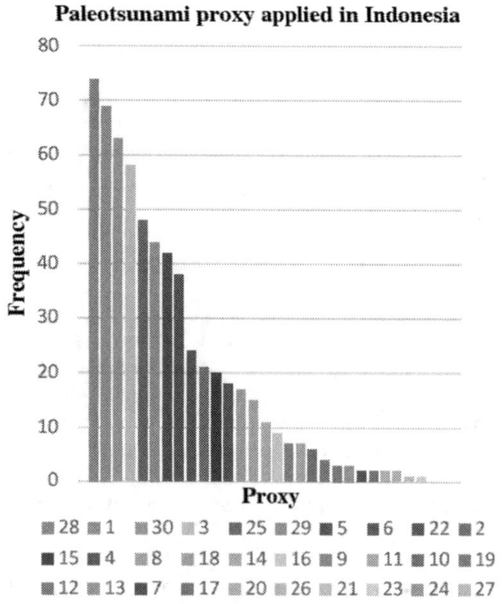

Fig. 3. Chart of paleotsunami proxy applied in Indonesia

Proxy number 24 is the identification of coastal archeological replication and shells due to inundation. Based on the database, paleotsunami research in Indonesia has not done too much identification of paleotsunami with archaeological data. One of the paleotsunami studies that identified paleotsunami deposits with archaeological evidence was the research conducted by Daly et al. (2019) [17]. The research carried out was to identify the presence of ceramic shards on the North Coast of Aceh which indicated its presence on the coast due to the tsunami process. Collaboration of geological data with archaeological data needs to be developed in paleotsunami research. This is done so that proxy number 24 can be applied in Indonesia.

Proxy number 27 is the identification of paleotsunami based on paleogeomorphology when the inundation condition was not due to a storm but a tsunami. When the tsunami waves come to the mainland, the waves hit the coastal morphology so that there are morphological changes [18]. These morphological changes were identified to indicate a tsunami. The application of this proxy requires special research in the form of coastal paleogeomorphological research to reconstruct the morphological condition of the coast which is then correlated with the presence of sediment. Paleotsunami research that applies paleogeomorphological analysis in Indonesia is still minimal, so that proxy number 27 has not been used. The research related to tsunamis and morphological changes was carried out by Maramai & Tinti (1997) after the 1994 Banyuwangi tsunami [18]. However, the 1994 Banyuwangi tsunami was a tsunami that occurred in the present and was recorded. So, the development of paleogeomorphological proxies needs to be done to be applied in Indonesia. This development will have a positive impact because evidence of paleotsunami can increase and the quality of the data will improve.

3.4 Validity of Indonesian tsunami data

The validity of the data in this study was determined based on the number of proxies used in identifying tsunami events. Validity is Excellent if the number of proxies applied for the identification of tsunami events is 9 proxies. Validity is Moderate if the number of proxies applied for the identification of tsunami events is 5-8 proxies. Validity is Fair if the number of proxies applied for tsunami event identification is 1-4 proxies [19]. The results show that the validity of the tsunami incident data in Indonesia is in three categories, namely Excellent, Moderate and Fair. The number of data validity is Excellent as many as 19, Moderate as many as 53 and Enough as much as 230 (figure 4). The validity of this data reflects the quality of tsunami event data based on a large amount of literature and proxies used to identify tsunami events. The more literature and proxies used, the better the validity of the data. The amount of literature but the proxies used are the same and do not make the number of proxies used increase, so the validity of the data does not increase.

Excellent data validity can be said as accurate tsunami event information. Accurate in this study is the characteristics of the tsunami event can be known in detail. This detailed information can be used as a reference for further study for further research on the occurrence of tsunamis. Moderate data validity means that the information presented regarding the characteristics of the tsunami is quite good but not as detailed as the excellent validity. Data that has fair validity does not mean that the data is wrong, but the lack of literature that explains in detail the characteristics of the tsunami occurrence and deposition.

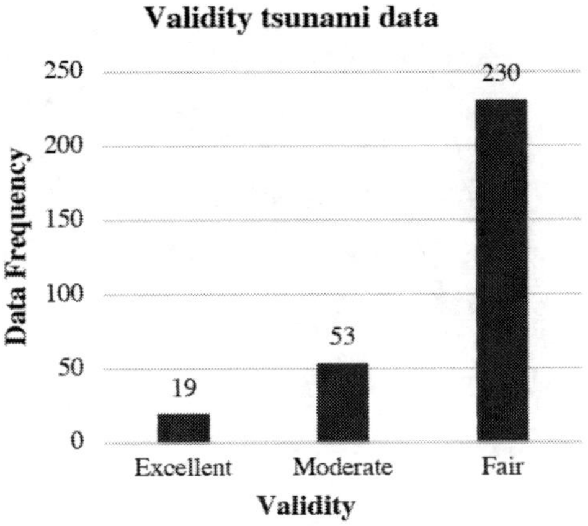

Fig. 4. Chart of the validity tsunami data in Indonesia

Most of the tsunami events in Indonesia have fair data validity. Tsunamis that occur in historical and prehistorical periods are both classified as fair validity. The validity of this fair data is due to the lack of literature that records in detail the tsunami incident. The lack of literature can be due to several reasons, such as the lack of infrastructure that records and reports tsunami events, research on paleotsunami deposits is still limited, the lack of previous community knowledge about tsunamis so that there are no records or just local folklore that lacks detail explaining the tsunami incident.

Improving the validity of tsunami event data in Indonesia can be done by developing paleotsunami research. Paleotsunami research can strengthen data on already recorded tsunami events or find evidence of new paleotsunami events. The development of paleotsunami research can increase the number of proxies used to identify tsunami events in Indonesia so that the validity of the data can be better.

3.5 Geological paleotsunami proxy

Geological proxies are criteria for the identification of suspected paleotsunami deposits based on geological aspects [10]. Geological proxies are the best proxies for identifying tsunami events. When a tsunami occurs, the tsunami waves will carry material from the sea and form sediment on land. The sedimentation results from tsunami waves will become tsunami deposits or known as tsunamiites [4]. A well-preserved tsunami deposit will be strong evidence that a tsunami has occurred, both for identification now and in the future. The presence of tsunami deposits will make geological proxies workable.

The use of geological proxies will make it easier to identify deposits caused by tsunamis because tsunami deposits have distinctive sediment characteristics. The quality of data that uses geological proxies is better than other data that does not apply geological proxies. Based on the database, tsunami events identified by geological proxies have Moderate - Excellent data validity.

Geological proxies that are widely used in paleotsunami research in Indonesia are proxy number 1, namely grain size, and proxy number 3, namely the presence of contrast layers. This proxy is widely used because the identification of grain size is quite easy to do, and the grain size will be correlated with the contrast layer. Differences in grain size in a stratigraphic cross-section of sediment samples for paleotsunami identification will be seen macroscopically, then layers that contrast with the general layer can also be seen macroscopically using this criterion the suspected paleotsunami layer can be identified. Most of the paleotsunami research in Indonesia based on the database has implemented the 1st and 3rd proxies and is supported by other proxies.

However, not all tsunami events in Indonesia have been identified with geological proxies. This is due to the lack of research related to paleotsunami deposits in more detail in several regions in Indonesia. Information related to tsunami events is still based on historical records or anthropological evidence. In the future, the development of geological proxies is needed to identify paleotsunami deposits. The development of paleotsunami research using geological proxies in Indonesia can improve data quality and validity in the database for the better.

3.6 WebGIS Indonesia paleotsunami database

The WebGIS created in this research is a website-based map (WebMap) database of tsunami events in Indonesia. The information presented is data on tsunami events in Indonesia and their characteristics. All information in this WebGIS is free to access at no cost and the data in this WebGIS is open source. The widgets presented in WebGIS are the basic menus in WebGIS. The main features of this WebGIS are layer selection, attribute detail information, and data download. As for this WebGIS will be included supporting features to make the data presented more informative and easily accessible. WebGIS Indonesia Tsunami Event Database can be accessed via the link https://trc-ui.geosinyal.id/

3.6.1 WebGIS view

The initial view of the WebGIS Indonesia Paleotsunami Database will display a Map of Indonesia and data on tsunami events in Indonesia marked with a blue wave symbol. The design of this WebGIS display is made informative and attractive so that the general public can easily access and understand the information in WebGIS. In general, the basic appearance of WebGIS consists of a menu bar at the top and a widget (figure 5).

The menu bar contains information about WebGIS. In the menu bar you can get information about WebGIS such as about, disclaimer, map legend, download dashboard feature data and tsunami study feature. The data presented in WebGIS is open source, including the overall tsunami incident data, proxies used and reference sources.

Widgets in WebGIS are basic features in a Web-based Geographic Information System (figure 6). These widgets include a layer widget to set the data layer that is cool to display. Then, the widget zooms in and out to zoom in and out on the map. Then, a search widget to search for a specific location, a home widget to return to the home screen, and a my location widget to find out where the user is.

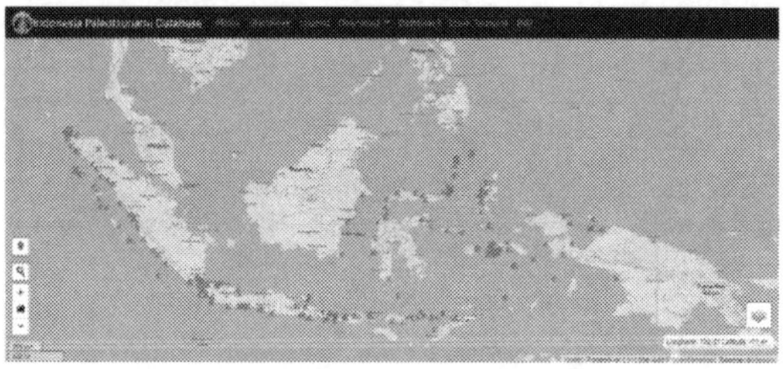

Fig. 5. User interface WebGIS view

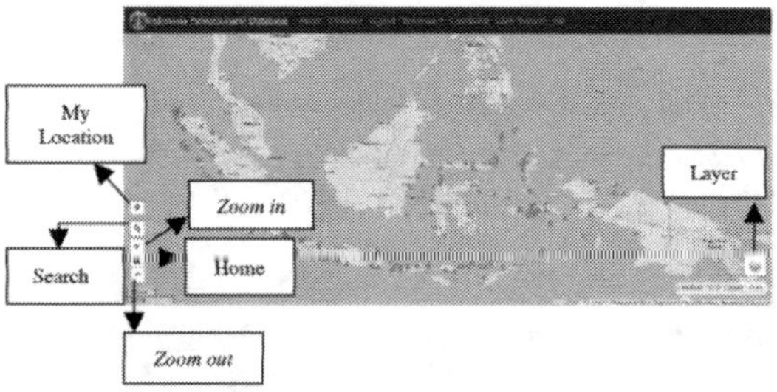

Fig. 6. Widget on WebGIS

3.6.2 WebGIS layers

Layer in WebGIS is divided into 2, namely the main layer and additional layers. The main layer is the main data for tsunami events including the "All Data", "Classification", "Validity", "Nature of Evidence", "Deposit Status", and "Tsunami Cause" layers. Meanwhile, the additional layer is a layer that contains supporting data including the "Source Tsunami", "Fault" layer, "Subduction" layer, and " Indonesia's Volcano" layer. The data presented can be selected based on the available layers. The map will only display data according to the layer that is activated. The display of the layer settings that you want to display is presented in figure 7.

3.6.3 Features detailed Information characteristics of Tsunami

The detailed information feature of tsunami characteristics is a feature of quick access to the attributes of the tsunami database. This feature can be accessed by clicking on the symbol of the desired tsunami event to know in more detail its characteristics. Detailed information will appear after the symbol is clicked. Information will be presented in the form of a pop-up dialog box containing the attributes of the selected tsunami event. The pop-up dialog box display is shown in figure 8.

Fig. 7. Layers available in WebGIS

Fig. 8. The pop-up box displays the tsunami event attributes

3.6.4 Indonesia paleotsunami database dashboard

The Indonesian Tsunami Database Dashboard is a supporting facility in WebGIS. The dashboard provides statistical infographics and tables of tsunami events in Indonesia (figure 9). The infographic on the dashboard displays statistical data from the data layer in WebGIS. The tsunami occurrence table displays tsunami data along with the characteristics of tsunamis that have occurred in Indonesia. This facility can be accessed by clicking the "Dashboard" menu in the WebGIS menu bar.

3.6.5 Learn tsunami features on WebGIS

The tsunami study feature is a supporting feature in WebGIS. The feature can be accessed by clicking the "Learn Tsunami" menu on the menu bar in WebGIS. This feature provides educational information about tsunamis. The educational information presented is in the form of articles on a website view that can be read by the accessor (figure 10). The education delivered was about tsunamis, tsunami mechanisms, paleotsunami,

 https://doi.org/10.1051/e3sconf/202234001001

and paleotsunami proxies. The existence of this feature is a supporting tool for educating the general public who access WebGIS to better understand the data presented in WebGIS and increase their knowledge about tsunamis.

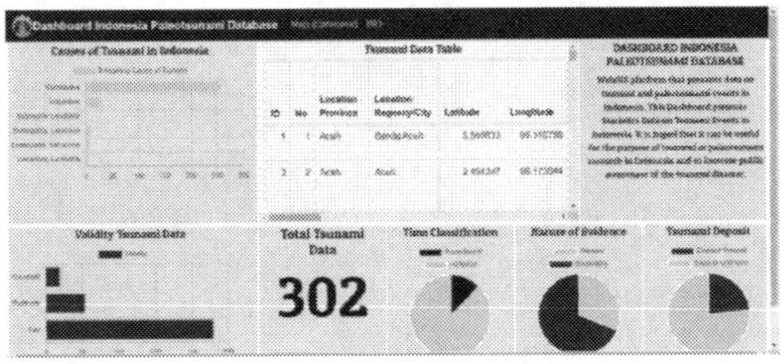

Fig. 9. The display of Indonesia Paleotsunami Database Dashboard

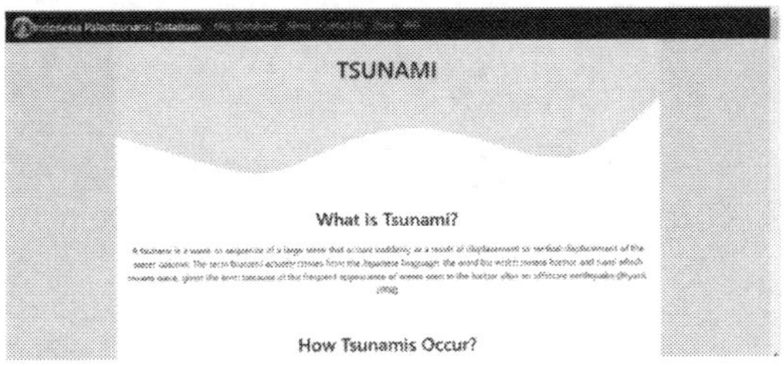

Fig. 10. The display of the learn tsunami feature

4 Conclusion

Data on tsunami events entered into the database amounted to 302 data. Tsunami events in Indonesia were caused by 255 earthquakes, 26 volcanism, 1 landslide, 10 earthquakes - landslides, 7 earthquakes – volcanism, and 3 volcanism - landslides. There are 28 proxies used to identify tsunami events in Indonesia out of 30 available proxies. Proxy number 28 is the dominant proxy used. The application of geological proxies can improve data quality. The validity of the tsunami data in Indonesia was "Excellent" as many as 19, "Moderate" as 53, and "Fair" as 230. WebGIS Indonesia Tsunami Event Database can be accessed via the link https://trc-ui.geosinyal.id/. Information that can be accessed in WebGIS is the database of tsunami events and paleotsunami deposits in Indonesia, the Indonesian tsunami statistical data dashboard, and the educational feature for studying tsunamis. The database in this research will continue to be updated in line with the development of tsunami research in Indonesia

Acknowledgment. The authors would like to thank Program Study Geology and Geophysics Universitas Indonesia for Supporting this research. The authors also would like to thank The Ministry of Research and Technology for providing the Basic Research for Higher Education Grant to fund this research.

References

1. P.C. Nugroho, S.E. Pinuji, A.N. Ichwana, A. Nugraha, S. Wiguna, et al, *IRBI (Indeks Resiko Bencana Indonesia) Tahun 2018*, BNPB, (2018)
2. H. Latief, N.T. Puspito, F. Immamura, Tsunami Catalog and Zones in Indonesia. J. Nat. Disaster Sci., **22**, 25–43 (2000)
3. E. Bryant, *Tsunami The Underrated Hazard*, Springer, (2008)
4. T. Shiki, Y. Tsuji, K. Minoura, T. Yamazaki, *Tsunamiites - Features and Implications*, Elsevier, (2008)
5. Y. Zaim, Daryono, I.N. Sukanta, Y. Rizal, Aswan, R.O. Wahyu, et al., *Jejak Tsunami Masa Lalu diantara Pangandaran dan Cilacap*. Earthquake and Tsunami Center Deputy for Geophysics, (2018)
6. Intergovernmental Oceanographic Commission, *Tsunami Glosary*, UNESCO, (2019)
7. P. Dunbar, H. McCullough, Global tsunami deposits database, J. Nat. Hazards, **63**, 267–278 (2012)
8. R. Abdalla, M. Esmail, *WebGIS for Disaster Management and Emergency Response*, Springer, (2019)
9. Chagué-Goff, C., Schneider, J. L., Goff, J. R., Dominey-Howes, D., & Strotz, L. (2011). Expanding the proxy toolkit to help identify past events - Lessons from the 2004 Indian Ocean Tsunami and the 2009 South Pacific Tsunami. *Earth-Science Rev*. (2011)
10. J. Goff, C. Chagué-Goff, S. Nichol, B. Jaffe, D. Dominey-Howes, Progress in palaeotsunami research, Sediment. Geol., **243–244**, 70–88 (2011)
11. R. Harris, J. Major, Waves of destruction in the East Indies: The Wichmann catalogue of earthquakes and tsunami in the Indonesian region from 1538 to 1877, Geol. Soc. Spec. Publ., **441**, 9–46 (2016)
12. R. Triyono, T. Prasetya, S. D. Anugrah, A. Sudrajat, U. Setiyono, et al., *Katalog Tsunami Indonesia Per-Wilayah Tahun 416-2018*, BMKG, (2019)
13. K. Iida, D. C. Cox, G. Pararas-Carayannis, Preliminary catalog of tsunamis occurring in the Pacific Ocean, Hawaii Inst. of Geophys., (1967)
14. B.W. Mutaqin, F. Lavigne, D.S. Hadmoko, M.N. Ngalawani, Volcanic Eruption-Induced Tsunami in Indonesia: A Review, IOP Conf. Ser. Earth and Environ. Sci., **256** (2019)
15. P. S. Putra, E. Yulianto, Stratigrafi Endapan tsunami Krakatau 1883 di Daerah Limus, Pantai Barat Teluk Semangko, Lampung. J. Lingkungan Dan Bencana Geol., **7** (2016)
16. P. S. Putra, E. Yulianto, Karakteristik Endapan Tsunami Krakatau 1883 Di Daerah Tarahan, Lampung. J. RIS. Geol. dan Pertam., **27** (2017)
17. P. Daly, K. Sieh, T.Y. Seng, E.E. McKinnon, A.C. Parnell et al., *Archaeological evidence that a late 14th-century tsunami devastated the coast of northern Sumatra and redirected history*, in

Proceedings of the National Academy of Sciences, May 2019, America (2019)

18. A. Maramai, S. Tinti, The 3 June 1994 Java Tsunami: A post-event survey of the coastal effects, J. *Nat.l Hazards*, **15**, 31–49 (1997)
19. J. Goff, S. Nichol, D. Kennedy, Development of a palaeotsunami database for New Zealand, J. Nat. Hazards, **54**, 193–208 (2008)

After the Indian Ocean Tsunami (IOT): Natural beach recovery, Meulaboh, Sumatra, Indonesia

Lucy Buck[1*], *Charlie* Bristow[1], *Ella* Meilianda[2]

[1]Earth and Planetary Science Department, Birbeck, University of London, Malet Street, London, WC1E7HX, England
[2]Civil Engineering Department, Engineering Faculty, Universitas Syiah Kuala, Jalan Tgk. Abdur Rauf, Kopelma Darussalam, Banda Aceh, 23111, Indonesia

Abstract. Ground-penetrating radar (GPR) offers an efficient and non-invasive method of identifying and characterising subsurface features. It has previously been used to investigate both tsunami deposits and marine erosion surfaces from tsunamis as well as the structure of the structure of prograding beaches. The present study investigates beach deposits at Meulaboh, western coast of Aceh Province in Sumatra Island of Indonesia, to estimate the volume of sediment that has been deposited since the 2004 Indian Ocean Tsunami, using the GPR with an antenna of 200 MHz. Two profiles perpendicular to the coastline were collected, one 93 m long and the other 30 m long, to capture the internal profile of beach ridge deposition. From the GPR measurement the amount of 1,190,191,716 tons of sediment redeposited along the 1092 m coastline since the 2004 tsunami, with a prograding length of 73 m per year. As beaches provide a good form of tsunami protection the rapid beach recovery and the return of a large amount of sediment helps provide much needed coastal protection to the area.

1 Introduction

On 26th December 2004 a 9.3 earthquake struck off the coast of Sumatra. The epicentre was located 80 km west of the coast of northern Sumatra (95°51'0 W, 3°25'0 N) [9] at a depth of 25-30 km [6]. The resulting tsunami was the most devastating tsunami in recorded history and caused more than 292,000 fatalities in 12 countries [9] with over 1 million left homeless [6].

In Sumatra, the resulting Indian Ocean Tsunami (IOT) consisted of three main waves and a retreat of the sea was observed 10 minutes after the earthquake, the first wave arrived 3 minutes later, the second and largest wave arrived within 5 minutes of the first wave, there are few observations for the third wave however the return flow was achieved 5 minutes after the third wave [21]. A 51 m run up was recorded in Lho Nga [22].

Meulaboh is a city 181 km south of Banda Aceh, on the west side of the Island of Sumatra. The study site is 14 km north of the city on a long, straight part of the NW-SE trending coast, midway between the Bhee river and the headland. The tsunami damaged almost all structures, including the airport, roads, both jetties and a hospital, along the coast and flooded up to 2km inland [13], reaching wave heights of 9-14m and reached 2-2.5km inland [14]. The coast is relatively low-lying and the beach is backed by a sequence of ridges and swales with the tsunami deposits locally preserved in the marshy swales. Previous work by Monecke et al [14] identified tsunami deposits from the 2004 event as well as tsunamis dating to 1290-1400 and 780-990 here. The oldest tsunami deposits are found in the swale furthest from the current beach, approximately 1,800m from the coastline [14].

The tsunami caused widespread erosion along the coast of Sumatra, including the removal of all depositional features such as beaches, low sand dunes and wetlands [15]. However, within weeks deposition features started to reform. The post-tsunami coast undergoes the same coastal processes as the pre tsunami coast. As a results new coastal features formed within the same constraints and the present coast closely resembles the pre tsunami coast [15]. Due to this Liew et al. [15] suggest that sedimentary deposits are a better indication of past tsunamis than coastal morphology.

Ground-penetrating radar (GPR) offers an efficient and non-invasive method of identifying and characterising subsurface features [10, 18]. It has previously been used to investigate both tsunami deposits [5] and marine erosion surfaces from tsunamis [23] as well as the structure of the structure of prograding beaches [2. 4, 17, 19-20, 24].

In this study we use GPR to investigate the beach deposits at Meulaboh and estimate the volume of sediment that has been deposited since the 2004 Indian Ocean Tsunami.

2 Methods

A GPR survey was conducted in September 2019. Two profiles were collected, one 93 m long and the other 30 m long. The profiles were perpendicular to the coast and the locations can be seen in figure 1.3. An antenna frequency of 200 MHz was used with a step size of 0.5

* Corresponding author: lucy.buck.15@ucl.ac.uk

m and a separation of 1m. Topographic data was collected using levelling techniques.

The data was processed using EKKO Project 5. The data underwent a Dewow, SEC2 Gain and a FK migration using velocity of 0.07/s. The sediment volume that has built up since the 2004 tsunami was evaluated using polygons with the lower boundary being the erosional layer created by the tsunami and the upper boundary being the surface. The area of this polygon was used to calculate the area of sediment that had accumulated since the tsunami. To calculate the volume the length of the erosional ridge along this part of the coast was used.

Fig.1. 1) Sattellite image showing Aceh and the location of Banda Aceh, the location of image B is shown by the red box. 2) The coast north of Meulaboh showing the erosional ridge in yellow, the location of image C is shown by the red box. 3) The location of the two GPR profiles used in this survey. Line A starts just beyond the erosional ridge, line B ends by the sand cliff at the high tide line. (Source: Google Earth).

3 Result

The GPR data for line A can be seen in figure 2 and is shown with both a depth and time scale. The GPR profile shows several reflections with a strong amplitude. There are multiple inclined reflections, in orange, that dip from left to right, that is towards the sea. In addition, there are concave reflections that are between 15 m and 40 m wide and, with one exception, start at the surface and reach 9 m deep across the entire the profile. The inclined reflections that dip towards the sea are interpreted as beach progrades with each reflection marking a former position of the beach.

Due to its correspondence with the ridge that marks the scarp formed by the 2004 IOT the reflection marked in blue is interpreted to be the tsunami erosion surface. The concave reflections, shown in brown, are interpreted as erosion surfaces either formed by earlier coastal erosion, or the river that runs parallel to the coast behind the beach. Prior to the tsunami the coconut palm trees extended much further toward the sea at the time of study the trees were over 15 years old, the age of the tsunami, and were only present inland from the ridge, further reinforcing the idea that the ridge and erosional surface were created by the tsunami.

If the erosional surface is extrapolated to the sea the volume of sand above that line represents the sediment that has been added to the beach since the 2004 tsunami. Along this single profile this has been calculated to be 329 m^3. The beach ridge has been measured to be 1902 m long. If the sediment volume in the GPR is multiplied along the length of the beach ridge then 625,758 m^3 or 1,190,191,716 tons of sediment, has been returned to the beach since the tsunami, assuming a bulk density of 1.6 g/cm3 for beach sand. To put this in context, an Olympic swimming pool has a volume of 2,500 m^3, so that is equivalent to 250 Olympic swimming pools of sand. This amount of sand would take 38,263 standard 25 tons dumper trucks to move. Using the ridge as the erosional limit the beach has grown in length by 177 m between 2004 and 2019 giving a horizontal rate of progradation of 7.3 m per year.

4 Discussion

The rapid recovery of the beach north of Meulaboh documented here is consistent with observations of post tsunami beach recovery after the 2011 Tohoku Tsunami in Japan [8] and at Lampu Uk, the bay immediately north to Leupung 60% of the sediment lost due to the tsunami was regained within 6 months [16]. This rapid beach recovery has also been seen in Thailand after the 2004 IOT [3,7,11]. The implication is that sediment eroded from the beach by the tsunami has remained close to the coast so that it was available to be transported back onshore under normal wave conditions.

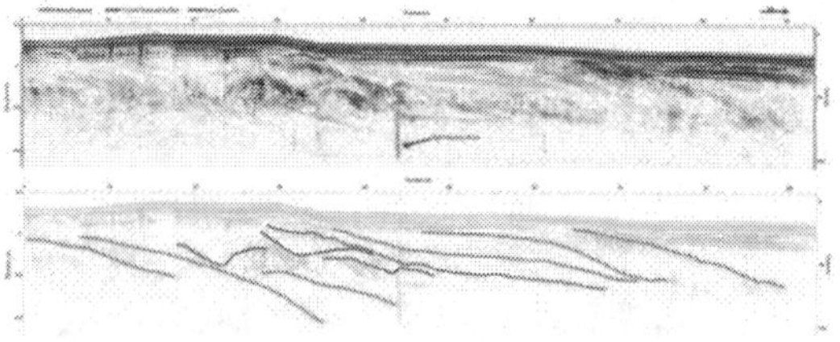

Fig. 2. GPR data from line A, this is the longer profile that was taken further inland. The top image shows the uninterpretated but processed data, the below image shows the interpretations used here. The tsunami erosional surface is shown in blue, the beach progrades are shown in orange and the paleochannels in brown. The data is presented with an elevation scale on the left and a velocity scale on the right. For all relevant processing a velocity of 0.07 m/s was used. The sea is to the right of the figure.

The natural restoration of the beach in less than 20 years has implications for shoreline management on tsunami prone coasts. Construction of artificial sea defenses immediately after a tsunami event, and artificial beach nourishment may not be required if the beach will restore itself [12]. It would also be advisable to restrict development within the areas that have recovered to provide a buffer against potential tsunami erosion in the future [11, 15].

5 Conclusion

Meilianda et al. found that at Lampu Uk, the bay immediately north to Leupung, 60% of the sediment lost due to the tsunami was regained within 6 months [16]. The data from Meulaboh suggests that this rapid sedimentation has continued down the coast. As beaches provide a good form of tsunami protection the rapid beach recovery and the return of a large amount of sediment helps provide much needed coastal protection to the area.

Acknowledgements. This research was completed with funding from the London NERC DTP studentship NERC Grant Number NE/L002485/1.

References

1. Athukorala, P. and Resosudarmo, B. (2005) The Indian Ocean Tsunami: Economic Impact, Disaster Management, and Lessons. Asian Economic Papers, 1-39, 4(1)
2. Bristow, C. and Pucillo, K. (2006) Quantifying rates of coastal progradation from sediment volume using GPR and OSL: The Holocene fill of Guichen Bay, south-east South Australia. Sedimentology, 769-788, 53(4)
3. Choowong, M., Phantuwongraj, S., Charoentitirat, T., Chutakositkanon, V., Yumuang, S. and Charusiri, P (2009) Beach recovery after 2004 Indian Ocean tsunami from Phang-nga, Thailand. Geomorphology, 134-142, 104(3-4)
4. Engels, S, and Roberts, M. C. (2005) The Architecture of Prograding Sandy-Gravel Beach Ridges Formed During the Last Holocene Highstand: Southwestern British Columbia, Canada. Journal of Sedimentary Research, 75 (6): 1052–1064.
5. Gouramanis, C., Switzer, A., Polivka, P., Bristow, C., Jankaew, K., Dat, P., Pile, J., Rubin, C., Yingsin, L., Ildefonso, S. and Jol, H. (2015) Ground penetrating radar examination of thin tsunami beds — A case study from Phra Thong Island , Thailand. Sedimentary Geology. 149-165, 329
6. Grilli, S. T., Ioualalen, M., Asavanant, J., Shi, F., Kirby, J. T., and Watt, P. (2007) Source Constraints and Model Simulation of the December 26, 2004, Indian Ocean Tsunami. Journal of Waterway, Port, Coastal and Ocean Engineering, (2007), 414-428, 133(6)
7. Grzelak, K., Kotwicki, L. and Szczucinski, W. (2009) Monitoring of Sandy Beach Meiofaunal Assemblages and Sediments after the 2004 Tsunami in Thailand. Polish J. of Environ. Stud. Vol. 18, No. 1, 43-51
8. Hoang, V. C., Tanaka, H. and Mitobe, Y. (2019)Morphological recovery of beach severely damaged by the 2011 great east Japan tsunami. Esturine, Coastal and Shelf Science. 226
9. Ioualalen, M., Asavanant, J., Kaewbanjak, N., Grilli, S. T., J. T. and Watts, P. (2007) Modeling the 26 December 2004 Indian Ocean tsunami: Case study of impact in Thailand. Journal Of Geophysical Research, Vol. 112, C07024
10. Jol, H and Bristow, C. (2003) An introduction to ground penetrating radar (GPR) in sediments. Geological Society London Special Publications
11. Kendall, M. A., Aryuthaka, C., Chimonides, J., Daungnamon, D., Hills, J., Jittanoon, C., Komwachirapitak, P., Kongkaew, V., Mittermeyr, A., Monthum, Y., Nimsantijaroen, S., Paterson, G. L. J., Foster-Smith, R., Foster-Smith, J. and Thongsin, N. (2009) Post-Tsunami Recovery of Shallow Water Biota and Habitats on Thailand's Andaman Coast. Pol. J. Environ. Stud. 18(1):69–75
12. Knight, J. and Goff, J. R. (2016) Coastal science for post-tsunami reconstruction. Holocene, 1334-1340, 26(8)
13. Lee, V. J., Low, E. Ng, Y. Y. and Teo, C. (2005) Disaster Relief and Initial Response to the Earthquake and Tsunami in Meulaboh, Indonesia. Annals Academy of Medicine. Vol. 34 No. 9
14. Monecke, K., Finger, W., Klarer, D., Kongko, W., McAdoo, B., Moore, A. and Sudrajat, S. (2008) A 1,000-year sediment record of tsunami recurrence in northern Sumatra. Nature, 1232-1234, 455(7217)
15. Liew, C. S., Gupta, A., Wong, P. P. and Kwoh, L. K. (2009) Recovery from a large tsunami mapped over time: The Aceh coast, Sumatra
16. Meilianda, E., Dohmen-Janssen, C., Maathuis, B., Hulscher, S and Mulder, J. (2010) Short-term morphological responses and developments of Banda Aceh coast, Sumatra Island, Indonesia after the tsunami on 26 December 2004. Marine Geology, 96-109, 275(1-4)
17. Moore, A., Nishimura, Y., Gelfenbaum, G., Kamataki, T. and Triyono, R. (2004) Sedimentary deposits of the 26 December 2004 tsunami on the northwest coast of Aceh , Indonesia. Earth Planets Space. 253-258, 58.
18. Neal, A. (2004) Ground-penetrating radar and its use in sedimentology : principles , problems and progress. Earth-Science Reviews,, 261–330, 66
19. Neal, A. and Roberts, C. L. (2000) Applications of ground-penetrating radar (GPR) to sedimentological, geomorphological and geoarchaeological studies in coastal environments. Geological Society, London, Special Publications, 175, 139-171
20. Neal, A., Pontee, N.,Pye, K and Richards, J. (2002) Internal structure of mixed-sand-and-gravel beach deposits revealed using ground-penetrating radar. Sedimentology. 789-804, 49(4)
21. Paris, R., Wassmer, P., Sartohadi, J., Lavigne, F., Barthomeuf, B., Desgages, E., Grancher, D., Baumert, P., Vautier, F., Brunstein, D. and Gomez, C. (2009) Geomorphology Tsunamis as geomorphic crises : Lessons from the December 26 , 2004 tsunami in Lhok Nga , West Banda Aceh (Sumatra , Indonesia). Geomorphology. 59-72, 104 (1-2)
22. Paris, R., Wassmer, P., Sartohadi, J., Lavigne, F., Barthomeuf, B., Desgages, E., Grancher, D., Baumert, P., Vautier, F., Brunstein, D. and Gomez,

C. (2010) Boulder and fine sediment transport and deposition by the 2004 tsunami in Lhok Nga (western Banda Aceh , Sumatra , Indonesia): A coupled offshore – onshore model. Marine Geology, 43-54, 268(1-4)

23. Switzer, A., Bristow, C. and Jones, B. (2006) Investigation of large-scale washover of a small barrier system on the southeast Australian coast using ground penetrating radar. Sedimentary Geology. 145-156. 183.
24. Tamura, T., Murakami, F., Nanayama., F., Watanabe, K. and Saito, Y. (2008) Ground-penetrating radar profiles of Holocene raised-beach deposits in the Kujukuri strand plain, Pacific coast of eastern Japan. Marine Geology. 11-27, 248 (1-2)

Landslide triggered tsunami modelling: A study in Anak Krakatoa collapse

Siti Fatimah[1*], *Abd* Hafidz[1], *Muhammad* Syahmi Zikri[1], *Nine* Safira[1], *Tito* Latif Indra[2], and *Supriyanto*[1]

[1]Department of Geosciences, Universitas Indonesia, Depok, Indonesia
[2]Department of Geography, Universitas Indonesia, Depok, Indonesia

Abstract. The Anak Krakatoa volcano in Sunda Strait is a tsunami threat to the southern part of Sumatra Island and the west part of Java Island as the eruptions and landslides it generates may trigger a tsunami. As the coasts of West Java are densely populated areas, if a tsunami occurs, then the loss and casualties would be massive. Therefore, a hazard assessment in the area is necessary which includes a simulation of possible tsunami occurring in the region. We simulated the 2018 tsunami in Sunda Strait triggered by the collapse of the Anak Krakatoa flank using the landslide parameters inferred from previous studies simulating that the 2018 tsunami event. The water wave propagation in this simulation demonstrates a tsunami that travels rather fast, where the tsunami reaches the Panaitan Island in 20 minutes and has reached the mainland around 30 minutes. The simulated landslide created a water wave amplitude as high as 60 m in the nearby islands of Krakatoa Archipelago, down to less than 10 m in the mainland of Java Island. This result relatively correlates with the run-up height data measured in the field by previous studies in 2019 and 2020. The shape of the coastline also determines how the water waves affect the area, which should be an essential factor in the hazard assessment.

1 Introduction

The subduction zone alongside the west of Sumatra all the way to the south coast of Java had made the islands in those areas vulnerable to tsunamis, either caused by tectonic earthquakes, volcanic eruptions, or other causes of submarine landslides. The Krakatoa Archipelago, a volcanic island group in the Sunda Strait, threatens the adjacent islands including Java and Sumatra with the possible tsunami caused by its volcanic activities. The Krakatoa volcanic eruption on 26 August 1883 is recorded as the deadliest volcanic explosion creating a tsunami and causing at least 36000 death tolls [1]. The emerging of the Anak Krakatoa volcano after the eruption becomes another tsunami threat in the area. The collapse of the Anak Krakatoa volcano flank in 2018 had caused a landslide that triggers the latest tsunami in Sunda Strait, resulting in hundreds of fatalities.

Not far from the threatening volcano, the west coast of Java Island is a densely populated area and is well-known for its tourist resorts. As consequence, the area is not only vulnerable but also at a high risk of a tsunami with a possible immense number of losses and casualties. That being the case, the simulation of a possible landslide-triggered tsunami in the area is important as the information of arrival time and the estimated wave height would improve the early warning system and hazard assessment in the area. In terms of the hazard it poses to people, landslides and volcano flank collapse like in the case of Krakatoa represent an important tsunami source [2].

2 Geological setting

The Krakatoa volcanic complex is part of an NNE-SSW trending lineament of a quaternary volcanic edifice that lies approximately perpendicular to the Java Trench [3]. After the 1883 eruption of Krakatoa, which is one of the largest volcanic explosions in history, the northern portion of Krakatoa Island disappeared and was replaced by a caldera [4]. In 1928, Anak Krakatoa (the Child of Krakatoa) emerged on the submarine caldera formed during the 1883 eruption [5]. The caldera, together with three surrounding islands are remnants of the 1883 Krakatoa eruption, creating a complex called Krakatoa Archipelago. The archipelago consists of Anak Krakatoa Volcano, Sertung Island, Panjang Island, and Rakata Island (Fig. **1**).

The cone of Anak Krakatoa was built up by the recurrent lava flows and pyroclastic fall deposits up to 350 m high before it collapses in 2018 [5]. Accompanied by a series of volcano eruptions from 22 to 26 December 2018, the collapse of the volcano flank generated a tsunami in Sunda Straits on 22 December 2018 [6]. The eruption started on 23 October 2007 with minor eruptions that created ash clouds and Plumes and continued with fluctuating eruptive activity until October 2018 when the volcano became more active with strombolian and explosive events. This eruptive phase continued for 175 days until 22 December 2018, when the activity suddenly evolved into a sector

* Corresponding author : fatimahsiti@ui.ac.id

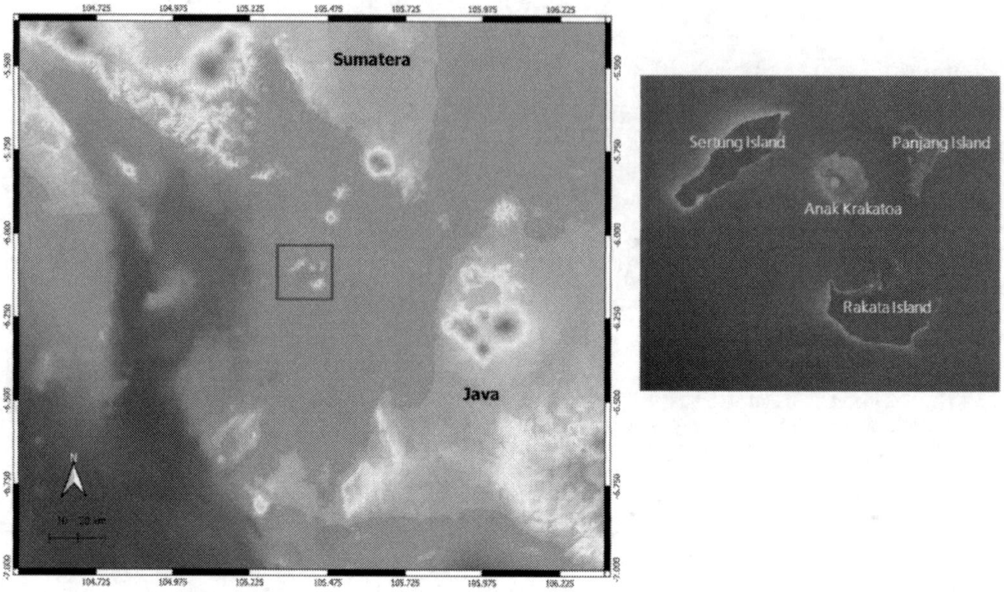

Fig. 1. The Anak Krakatoa volcano is a part of a volcanic island group (Krakatoa archipelago) along with Sertung Island, Panjang Island, and the remnants of Rakata cone of the old Krakatoa Island.

proposed that the collapse plane must extend below sea level to be able to drastically change the shape and volume of the island as observed from the photographs and radar images. He stated that the subaerial area of the island was estimated to be reduced by 49% due to the eruption.

The Anak Krakatoa flank collapse had been simulated in 2012, six years before the 2018 event, by Giachetti [9]. The study presents the simulation of the Anak Krakatoa landslide and tsunami propagation. The study infers that the highest wave amplitude is directed towards the southwest and that the presence of the Sertung and Rakata islands causes the wave height to be reduced in the northwest and southwest of the landslide scenario location.

3 Methodology

Numerical modeling is a fit approach to forecast wave propagation and inundation on land [10]. Different approaches of modeling have been presented for the 2018 Sunda Strait tsunami [8], [5]. In this research, we used a nonlinear Boussinesq equation in the wave propagation model provided by Geowave - a simulation software developed and managed in parts by Applied Engineering, Inc. The tsunami propagation model was performed to model water waves while accommodating the dispersive behavior of the wave during the generation, propagation [10], and inundation of tsunami waves [5]. Paris [5] compared shallow water simulation and Boussinesq simulation to assess the potential dispersive effects in Sunda Strait. His research shows that even though the first wave from both simulations is similar in the near field, there are dispersive effects in the far field where the period of the first water wave increases during the propagation. The Boussinesq model does not constrain horizontal velocities to have a constant value wave model thus it could give us accurate vertical tsunami run-up and horizontal tsunami inundation in one simulation [11].

The bathymetric data of Krakatoa and Banten used in modeling was obtained from the DEMNAS (http://tides.big.go.id/DEMNAS/index.html), a part of the Indonesian Geospatial Information Agency owned by the government. It has a spatial resolution of 0.27-arcsecond, using vertical datum EGM2008. It was created from several sources including IFSAR (5m resolution), TERRASAR-X (5m resolution), and ALOS PALSAR (11.25m resolution).

The characteristics of a tsunami greatly depend on the source parameter, hence it is an essential factor in tsunami simulation. For a tsunami generated by subaqueous landslides, we need to be mindful of the volume, the dynamics of sliding masses, and water depth. The inadequacy in the information of landslide geometry is the challenging part of this modeling as it allows multiple possibilities of resulting models. We used parameters inferred from previous studies, including a landslide simulation by Paris [5] and a collapse scenario presented by Grilli [8]. The landslide simulation performed by Paris presented that the height of the Anak Krakatoa subaerial edifice was reduced from approximately 350 m before its collapse to about 120 m during its flank landslide which causes the 2018 tsunami in Sunda Strait. We defined the geometry of the landslide plane from a landslide simulation result performed by Paris [5] and a model of the Anak Krakatoa collapse scenario constructed from high-resolution satellite imagery and aerial photography by Grilli [8]. The slope is set to be 8° [10, 5] with the total estimated volume of our collapse model is about 0.18 km^3 which is in between the estimation made by [5] and

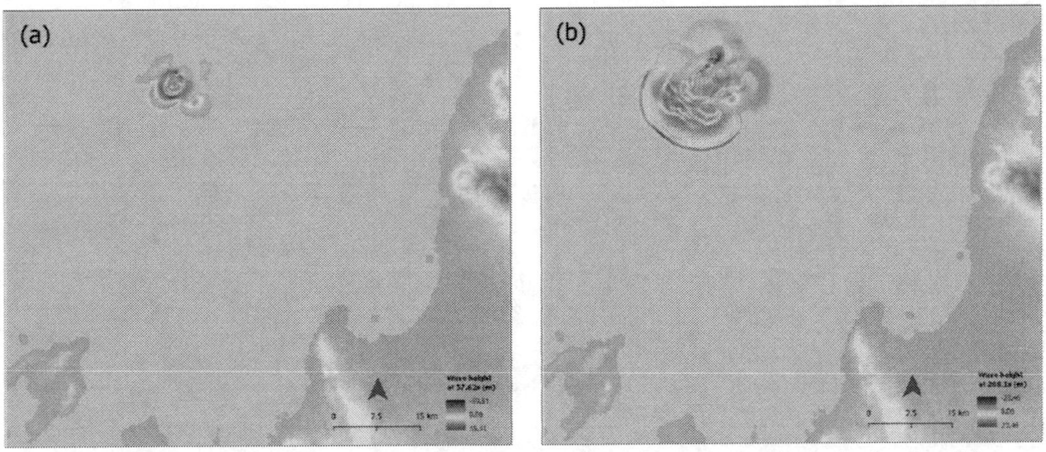

Fig. 2. The landslide at the southwest rim of Anak Krakatoa triggered water waves traveling in the direction of the mass movement (a) with maximum wave height of almost 60 m at time 58 s (b) which decreases down to around 23.5 m less than 5 minutes after the trigger.

[8]. We assumed an initial depth of 25 m [8] and a density of 1500 kg/m^3 [9].

Previous studies [5, 6, 12, 13] had provided field measurement data of inundation and run-up heights caused by the 2018 tsunami in Sunda Straits. In this study, we used two measurement data [6, 12] which we reckon would complement each other. The field measurement is mainly conducted in the Tanjung Lesung peninsula and around Marina Jambu where tsunami traces were identified. The observation suggests that Tanjung Lesung is the area with the highest run-up height along the coast of West Java in respect to the 2018 event tsunami.

4 Discussion

The debris avalanche of the landslide is directed to the southwest, thus displacing water in that direction and creating water waves as high as 60 m. In less than 1 minute, the first wave crest hit Rakata Island and Sertung Island in the southeast and the west of the landslide scar (**Fig**(a)). The wave with the highest amplitude travels in the same direction of mass movement and the opening between islands make it possible for the wave to go way through without a hitch. On the other hand, the presence of islands around the Anak Krakatoa volcano serves as barriers for the water which attenuates the wave traveling to the north and east of Anak Krakatoa (**Fig**(b)) wherein we see much lower amplitude.

The waves going to the southwest are the fastest to travel and Panaitan Island in the south became the first to be hit by the water wave. It might be associated with the bathymetry between Panaitan Island and the Krakatoa archipelago. The tsunami wavefront hit the island with a maximum amplitude of about 10.8 m, 20 minutes and 10 seconds after the landslide occurs (**Fig. 2**(a)). The crest of the leading wave then arrives at Tanjung Lesung beach in 5 minutes, having a water surface elevation of almost 10 m (**Fig. 2**(b)). Several minutes later, the waves enter Lada Bay with much lower wave of about only 2 m high. Directly after, at 1786 s after the trigger, the tsunami more than 13 m high hit Ujung Kulon, the most western part of Java Island (**Fig. 2**(c)). At 30 and a half minutes, the water waves reach Carita and Marina Jambu (**Fig. 2**(d)) with moderate wave height between 3 to 5 m and were followed by other parts of the mainland. The wave have gotten completely into the west coast of Java around 36 minutes after the landslide occurs.

The waves approaching the mainland, generally have lower amplitude and travel at a slower speed as a result of the shallower bathymetry. It is since a tsunami act as a shallow-water wave, where the traveling speed depends only on the depth of the ocean. Considering the landslide scar is oriented to the southwest, the wave is expected to have the highest crest in that direction. Therefore, we expect that the area in the southern part of the mainland would be subjected to higher water waves, which is consistent with our simulation.

We compare the arrival time of our simulation with the results from Grilli [9] and Paris [5] for two locations covered by those studies, Tanjung Lesung and Marina Jambu. From Table 1. It is noted that our tsunami simulation gives faster water wave propagation than two others but there is only 1-2 s difference. There is no recorded arrival time at Tanjung Lesung to be compared, but the data from Marina Jambu recorded the arrival time to be 33 minutes [5] which only differ one second compared to the simulation in this study. This signifies our simulation result, though more data to be compared is needed.

Table 1. The travel time of water waves computed in this study, compared with previous study

Travel time (s)			Location
Computed	Giachetti et al. (2012)	Paris et al. (2019)	
26	28	29	Tanjung Lesung
32	33	34	Marina Jambu

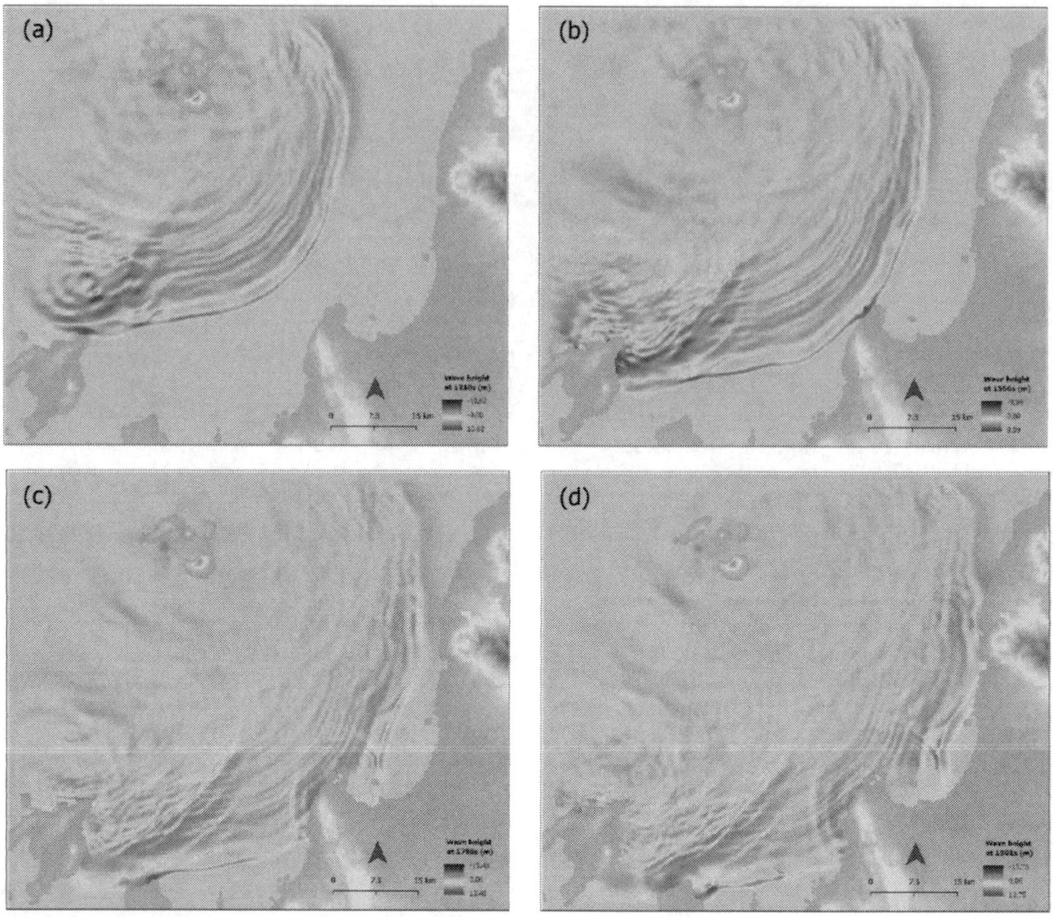

Fig. 2. The water surface elevation of tsunami modeling (a) at time 1210 s (20 minutes) as the wave hit Panaitan Island with height of 10.8 m. (b) After about 5 minutes (at 1556 s), the wave first reaches the mainland at Tanjung Lesung beach with maximum height of almost 10 m.

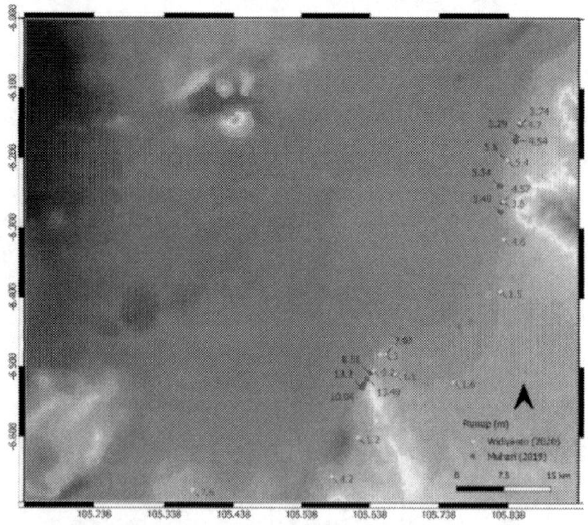

Fig. 3. The run-up height measurement data at several sites obtained from field surveys conducted by Widiyanto [11] (green bulltes) and Muhari [6] (red bullets)

A narrow bay such as the Lada Bay tends to amplify tsunami height, but both the simulation result and observation of run-up data in the field showed that the water wave here is not as high as expected, which we assume it was due to the presence of the Tanjung Lesung peninsula acting as a barrier for the Bay. As shown in **Fig. 3**, the run-up measurement from Widiyanto [12] recorded a low run-up height and scarce tsunami trace in this bay, meaning that the area is less affected by the 2018 tsunami. On the contrary, Tanjung Lesung peninsula had experienced such severe damage [6] with the highest run-up height measured in the research area. Based on field observation by Muhari [6], the run-up height in the peninsula is between 7 – 13.5 m with an average height of 10.6 m, while the highest run-up measured by Widiyanto [12] at the west coast of Tanjung Lesung gives 9.2 m. These field data agree with our simulation result whose water height is as high as 10 m in Tanjung Lesung but only 2 m at Lada Bay. The observation data of run-up heights around Marina and Carita are between 3-5 m, which is also very similar to our results. The comparison we mentioned might only be done at sites where the tsunami traces can be observed and measured. However, as the measured run-up values at most area are in agreement with the water waves height of our simulation, we could also assume that the expected tsunami water height and arrival time at other places would conform with our result as well as. That being said, we could later use the simulation result and its parameter to aid in the work of tsunami hazard assessment and evacuation plan.

Both the run-up height observation and wave simulation suggest that Tanjung Lesung peninsula and Ujung Kulon are likely to be heavily affected by a tsunami. The fact that Tanjung Lesung is one of the tourist destinations in the region, become alarming as it is often packed with tourists. In the 2018 tsunami, the area is severely damaged, including the forest and beach areas. Further work is needed to determine the area with high risks, as the height if water wave does not always directly related to run-up height and inundation distance.

5 Conclusion

The landslide scenario in this research produces water waves as high as 60 m soon after the landslide, with the highest wave traveling to the south, echoing the direction of the debris movement. The water wave direction is, indeed, bound by the orientation of the landslide plane and also the presence of barriers, in this case, are the islands surrounding Anak Krakatoa. After coming out of the Krakatoa complex, the water wave travels radially in all directions. The highest wave crest, however, is aiming to the south to the Panaitan Island. The first wave then hit this Island with a height of more than 10 m.

In the mainland, the area first hit by the wave is the Tanjung Lesung peninsula, then followed by Ujung Kulon and the other area in the west coast of Java in the north of Tanjung Lesung. Tanjung Peninsula and Ujung Kulon both are hit by average height of 10 m water waves, the highest among other parts of the coast. Both areas have relatively deeper bathymetry than Marina Jambu or Carita Beach in the north, whose wave amplitude is much lower, which might contribute to the the height of water wave. The run-up measurement in the field conducted by Muhari [6] and Widiyanto [11] also show very similar pattern and are in agreement with each other, though the exact value of water wave and run-up height might not be comparable. That being the case, our simulation result is in agreement with the run-up record measured based on the tsunami traces in the field. However, we need to analyze further into the inundation record as the wave energy might be converted into run-up height or inundation distance depending on the topography/bathymetry of the coastlines.

References

1. B. H. Choi, E. Pelinovsky, K. O. Kim and J. S. Lee, *Natural hazards and earth System Sciences,* 321-332 (2003)
2. T. Zengaffinen, F. Lovholt, G. K. Pedersen and A. Muhari, *Pure and Applied Geophysics*, 2493-2516 (2020)
3. S. Nishimura, H. Harjono and S. Suparka, *GeoJournal,* **2**, 28 (1992)
4. N. Nomanbhoy and K. Satake, *Geophys. Res. Lett.,* **22**, 4, 509-512 (1995)
5. A. Paris, P. Heinrich, R. Paris and S. Abadie, *Pure Appl. Geophys.*, **177**, 571-590 (2019)
6. A. Muhari, M. Heidarzadeh, H. Susmoro, H. D. Nugroho, E. Kriswati, Supartoyo, A. B. Wijanarto, F. Imamura and T. Arikawa, *Pure Appl. Geophys.* **176**, 5219-5233 (2019)
7. T. R. Walter, M. H. Haghighi, F. M. Schneider, D. Coppola, M. Motagh, J. Saul, A. Babeyko, T. Dahm, V. R. Troll, F. Tilmann, S. Heimann, S. Valade, R. TRiyono, R. Khomarudin, N. Kartadinata, M. Laiolo, F. Massimetti and P. Gaebler, *Nature Communications,* **10**, 4339 (2019)
8. S. T. Grilli, D. R. Tappin, S. Carey, S. F. L. Watt, S. N. Ward, A. R. Grilli, S. L. Engwell, C. Zhang, J. T. Kirby, L. Schambach and M. Muin, *Sci Rep,* **9**, 11946 (2019)
9. T. Giachetti, R. Paris, K. Kelfoun and B. Ontowirjo, *Geological Society London,* 79-90 (2012)
10. I. E. Mulia, S. Watada, T. Ho, K. Satake, Y. Wang and A. Aditiya, *Geophys. Res. Lett.*, **47**, 14 (2020)
11. P. Watts, S. Grilli, J. Kirby, G. Fryer and D. Tappin, *Nat. Hazards Earrth Syst. Sci.,* **3**, 5, 391-402 (2003)
12. W. Widiyanto, S.-C. Hsiao, W.-B. Chen, P. B. I. R. T. Santoso and W.-C. Lian, *Nat. Hazards Earth Syst. Sci.*, 20, 933-946 (2020)
13. Syamsidik, Benazir, M. Luthfi, A. Suppasri and C. L.K., *Nat. Hazards Earth Syst. Sci.,* **20**, 2, 549-565 (2020)
14. J. Mori, W. D. Mooney, Afnimar, S. Kurniawan, A. I. Anaya and S. Widiyantoro, *Seismological Research Letters,* 201, (2007)

https://doi.org/10.1051/e3sconf/202234001004

Mineralogy, geochemistry, and genesis of glauconite mineral from paleotsunami deposit in Lebak, Banten, Indonesia

Rasis Abi Tiyana[1*], *Supriyanto*[2], *Tri* Rani Puji Astuti[1], *Gamma* Abdul Jabbar[1], and *Muhammad* Rizqy Septyandy[1]

[1]Geology Study Program, University of Indonesia, Pondok Cina, Beji, Depok, 16424, Indonesia
[2]Geophysics Study Program, University of Indonesia, Pondok Cina, Beji, Depok, 16424, Indonesia

Abstract. Bagedur coast in Lebak regency, Banten province, Indonesia is part of lowland-swampy swale area that consist of various sediment deposit. Based on its sediment characteristics and geomorphology profile, Bagedur coast is assumed as depositional environment of paleotsunami deposit. Hand Auger drilling method is used to collecting loose sediment sample from eight observation stations. The depth of hand drilling in BG 01 station to BG 08 station is consecutively as follows: 115 cm, 100 cm, 80 cm, 700 cm, 400 cm, 450 cm, 160 cm, and 143 cm. This research is carried out to understand the diagenetic process of glauconite mineral in presumption paleotsunami deposit. The diagenetic process of glauconite mineral is identified based on its mineralogical and geochemical characteristics by petrography and X-Ray Fluorescence (XRF) methods. Observation samples is taken from green sand sediment bed in hand drilling core. Petrographic analysis under polarized microscope shows various glauconite mineral which recognized as matrix, oval pellets, replacement of precursor mineral, and bioclastic shell infillings. XRF analysis result show that glauconite in observation site are classified as glauconitic-illite type. The diagenetic stage is interpreted as early stage of diagenetic (nascent) which characterized by low Fe_2O_3 and K_2O content relative to Al_2O_3 content.

1 Introduction

Bagedur coast and its surroundings are an area composed by various sedimentary deposits which some of them comes from the marine environment deposited in alluvium units. Geomorphologically, this area is categorized as a lowland which is characterized by geomorphological units of coastal embankment, ridge, and swale. Based on the composition of the sedimentary material and geomorphological characteristics, this area is assumed to preserve paleotsunami deposits [16]. In addition to the presence of swamps which formed in low-land areas, sediment layers that are presumed as paleotsunami deposits can be identified through the presence of glauconite minerals in the hand drilling core.

Sediment layers which contain glauconite will tend to show a greenish colour in the hand drilling core. Glauconite can be used as stratigraphic marker of marine deposits because the formation of glauconite will take place specifically in the marine environment [15]. The depositional environment of glauconite is strongly influenced by the presence of a semi-confined microenvironment. Semi-confined microenvironment is a partially isolated depositional environment with a semi-closed system which able to provide ionic exchange to occur at medium rate [14]. In that condition, glauconite pellets will perform ion exchange from its surrounding environment to increase its maturation to become fully-evolved glauconite.

Glauconite has macroscopic characteristics in the form of spherical peloids in sediments, sandstones, and carbonates with a green oval appearance [15]. Glauconite will undergo morphological deformation through several stages of evolution. There are four stages of glauconite evolution which explain its diagenesis process, they are: nascent, slightly-evolved, evolved, and highly-evolved. The evolution stages of glauconite will indicate the time of glauconite grains spent in seawater-sediment interface zone, which later can be used as indication of sedimentation rate in those facies [5]. The presence of glauconite in hand drilling core taken from the swamp in the swale area can lead us to a presumption of the presence of paleotsunami deposits at those areas.

In this research, the authors conduct sediment sampling around Bagedur Coast, Lebak, Banten on alluvium deposits, especially in swampy areas by Hand-Auger drilling method Figure 1. This research is carried out because there are still few studies about the diagenesis of glauconite minerals in the sediment layers of presumed tsunami deposit. Glauconite samples were taken from the green sediment layers on the hand drilling core. The sample will then be analysed for its mineralogy and geochemical properties using petrography and X-Ray Fluorescence (XRF) methods. Based on the results from those methods, the mineralogical and geochemical characteristics of glauconite could be recognized, thus it can be used to

* Corresponding author: rasisabitiyand@gmail.com

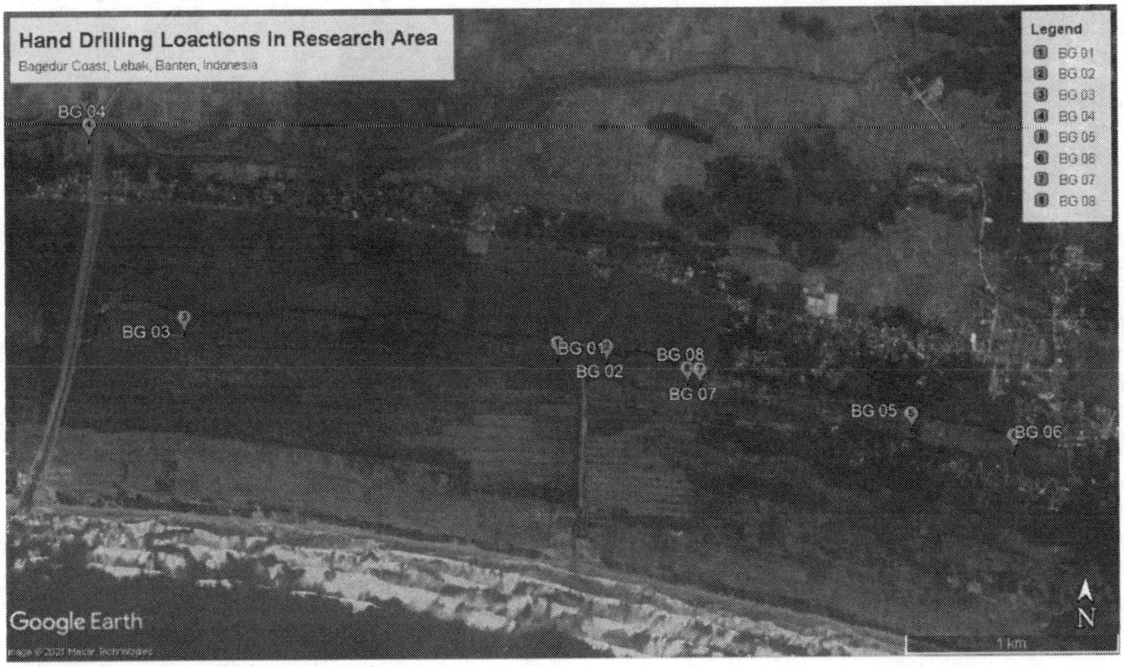

Fig. 1. Research location

explain the diagenesis and depositional process of glauconite at the research area.

2. Geological settings

Based on the division of physiographic zones of West Java, the research location is included in the Dome and Mountains section of the Central Depression Zone in the Bayah Mountains. Bayah Mountains is located at south of Banten, extending from Ujung Kulon in the west to Pelabuhan Ratu in the east. Bayah Mountains is formed at Eocene age which consists of a strongly folded area at the southern part of old andesite rocks and the northern part at the central area of the Bogor zone. Bayah area and its surroundings are part of the Sundaland margin of West Java and Oligocene magmatic arc [20].

The geomorphology of the Banten region is generally divided into three groups, namely plain morphology, moderately sloping hills (low-moderate waves) and steep hills. Lowland morphology is generally found at the northern and southern parts. The plains area is formed by an area that has an altitude of less than 50 m above sea level and a coastal area with altitude of 0-1 m above sea level. Sampling locations in Bagedur Coast has a geomorphological characteristic of swale in lowlands area. Sediment sampling is focused on swamps zone with loose soil sediment in order to gain greater depth from hand drilling penetration. The location with these geomorphological features is formed with low endogenous and exogenous energy, especially in the process of sediment accumulation in the swamp environment. These conditions will help us to identify sediment deposition anomalies that may occur, such as the occurrence of sediment layers with coarse grains among sediment layers with finer grains in a swamp environment that has low depositional energy.

The research area includes Alluvium (Qa) and Coastal Step (Qc) deposits which formed at Holocene age on the Regional Geological Map of Cikarang on 1:100,000 scale [19]. Those two units are composed of sediments of gravel, sand, silt, clay, silt, and pumice gravel. Furthermore, biogenic compositions were also found in the form of shell fragments and carbonate sediments.

Based on previous researches, there are two presumed paleotsunami deposits in the Banten, which took place at Wanasalam Coast, Wanasalam Region, Banten [21], and Lebak Beaches, Banten [16]. The paleotsunami deposits in those area was located in lowland geomorphology units in the form of swamps, rice fields, and plantations around the coastal area. Sediment deposits that thought to be formed due to the past tsunami process at the coast of Wanasalam, Banten, have characteristics of very fine grain size, poor grain sorting, low organic material content, and relatively higher carbonate material [21]. Meanwhile, the characteristics of the sediments in paleotsunami deposits at the Lebak Beaches show erosional contacts with a finer grain size pattern (fining upward) with poor grain sorting [16]. Foraminifera microfossils were also found in the sediment layer. The paleotsunami deposits which found in the Lebak beach is interpreted to have a relationship to a large-scale earthquake that hit the southern part of Java with a magnitude of more than 9 Mw which occurs every 600 years [11].

https://doi.org/10.1051/e3sconf/202234001004

3 Methods

Sampling method for collecting loose sediment sample in the field is carried out by Hand Auger drilling method Figure 2. Hand Auger drilling instrument consists of several steel rods which can be installed at the end of each other rods. The rods are used as both drill bits and drill core container that will be penetrated into the ground and then rotated in order to collect sediment sample in the borehole. Hand Auger steels can be stacked consecutively until it reaches certain depth in the borehole. After we drilled sediment layers out from the soil, we then move it to pipes which used as sample container.

Fig. 2 Hand-Auger drilling method.

There are 8 sampling locations in the research area. The depths of hand drilling from BG 01 station to BG 08 are consecutive as follows: 115 cm, 100 cm, 80 cm, 700 cm, 400 cm, 450 cm, 160 cm, and 143 cm. Among sediment layers in the drilling core, there are several layers of green sand sediment deposit at certain depths which assumed to contain glauconite minerals.

The green sand sediment layers which contain glauconite minerals then are analysed by petrography and XRF methods to recognize their geochemical properties. Petrography analysis is carried out using Leica DM750 microscope with magnification of 10x to 40x. Sample from microscope objective lenses then captured using camera to perform point counting analysis by making grids towards captured sample image.

XRF analysis is carried out by several steps of preparation. The first step is to pound the sample until it becomes smooth, then put it in a bottle with a volume of 50 ml. To ascertain the weight of the sample, the sample is weighed on a scale until it shows the desired value. The sample then filled with distillation water and be put into an ultrasonic dispenser for 30 minutes. After that, the sample will be put into a centrifuge at 1000 rpm for 4 minutes and reprocessed for 30 minutes at a speed of 3000 rpm. The results from the second centrifuge process will be put into sample glass.

4 Results and discussion

4.1 Stratigraphy of Bagedur coast

Loose sediment deposit layer that composes the drill core has varied characteristics. There are some layers of soil and peat soil with brown to blackish grey colour with high water content so that it looks muddy in the drilling core. The grain size of loose sediment in the drill core has a range between clay to medium sand.

There are other sediment components in the sediment layer at the drilling core in the form of wood fragments, shell fragments, and a layer of green sand sediment. Wood fragments were found in peat soil and sediment layers with grain sizes ranging from silt to medium sand at the shallow depth (0-10 cm) and the greater depth (540 cm) (Figure. 3 a). Shell fragments were found at BG 04 and BG 06 stations (Figure 3 b). Shell fragments at station BG 04 were found at depths of 370-400 cm, 430-470 cm, 510-540 cm, and 620-680 cm, while shell fragments at station BG 06 were found at depths of 155-190 cm, 190-240 cm, 245-280 cm, 326-390 cm, and 390-440 cm.

Among all sediment layers in the drilling core, several layers of green sediment were found (Figure. 3 c). Those layers indicate the properties of glauconite in hand specimens. Layers of green sand sediment were found at several stations with different depths, they are BG 02 at the depth of 25-35 cm and 60-65 cm, BG 05 at depth 218-228 cm and 238-240 cm, BG 06 at depth 142-145 cm, BG 07 at depth 93-130 cm, and at depth BG 08 0-2 cm and 83-90 cm. The thickness of the green sand sediment layer varies at different stations with a range between 3-37 cm.

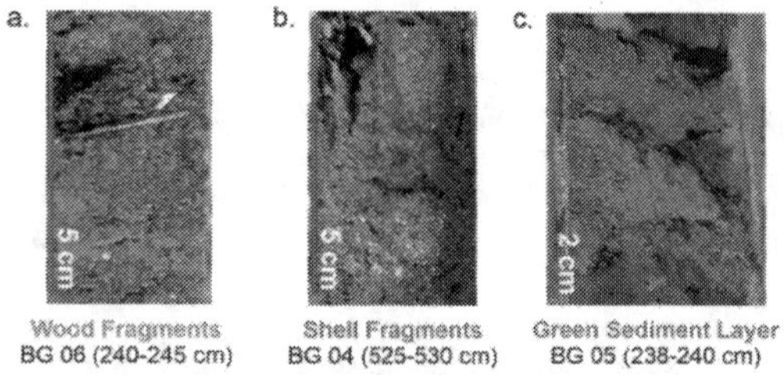

Fig. 3. Sediment components in hand drilling core

Based on the profile of stratigraphic log which taken from swamp environment, stratigraphic correlations are carried out between stations to determine the presence of presumed paleotsunami deposit at the research area. Correlation to indicate paleostunami layers were carried out between green sediment layers which assumed to contain glauconite minerals in it. The correlation result show that presumed paleotsunami layer were found between various layers of green sand sediment at stations BG 02, BG 08, BG 07, BG 05, and BG 06

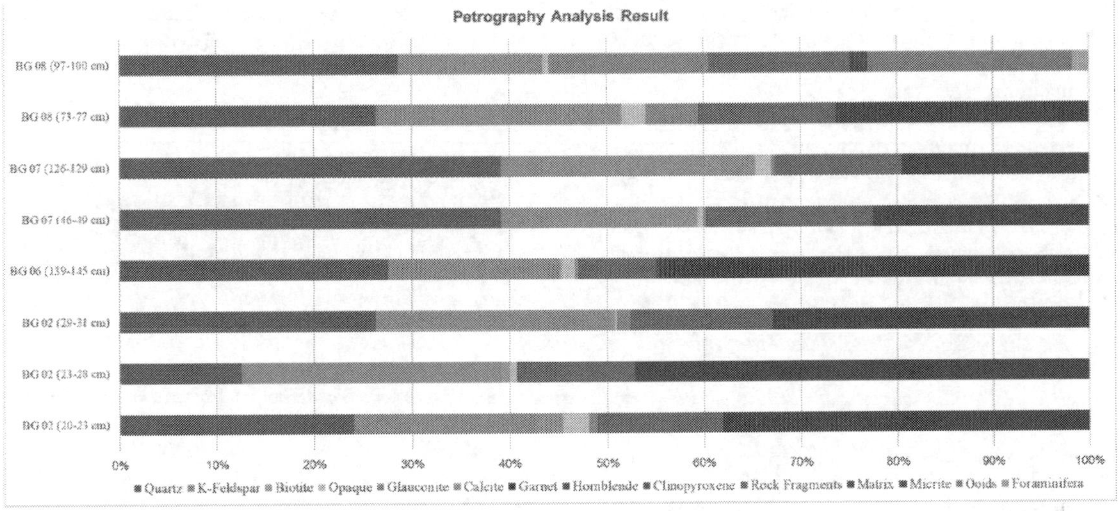

Fig.4. The composition of petrography samples in research area

4.2 Petrography analysis

Petrography images were identified using the point counting method at 10x and 40x magnifications through parallel and crossed polarization to determine the optical properties and percentage of the minerals and other components (Figure 5). The average number of points counting in each sample is 4417 points from 4 fields of view. Minerals and other constituent components found in petrographic samples include: quartz, k-feldspar, calcite, biotite, opaque, glauconite, garnet, hornblende, clinopyroxene, shell fragments, ooids, foraminifera, rock fragments, and matrix. The result of mineral compositions percentage from point counting method is shown on Figure 4.

Fig. 5. Point counting in petrography analysis

From the observations, glauconite minerals were found in several samples with various optical properties. Glauconite minerals were found in samples BG 02 at a depth of 20-23 cm, BG 02 at depth of 29-31 cm, BG 07 at depth of 46-49 cm, BG 07 at depth of 126-129 cm, BG 08 at depth of 73-77 cm, and BG 08 at depth 97-100 cm with an average grain size of <0.02 mm and yellowish brown to dark green.

The lowest percentage of glauconite recorded in a thin section has the smallest value of 0.2% in the BG 07 sample at a depth of 46-49 cm and the largest percentage value is 16.17% in the BG 08 sample at a depth of 97-100 cm. Glauconite is found in various forms. They were appeared as a matrix that fills the empty space between minerals (Figure 6 a), rounded oval pellets, replacement of quartz and feldspar minerals, and infillings in shell fragments. The space on the shell fragments was previously a place for the accumulation of sediment and detrital minerals that experienced a change in chemical composition to glauconite minerals, while the outer shell component became a semi-permeable barrier that formed a semi-enclosed microenvironment for ion exchange with seawater and surrounding sediments.

4.2.1 Evolution stage of glauconite

The petrographic properties from thin sections observations can aid us in making an interpretation about the maturity of glauconite pellets, the type of substrate, and the formation mechanism of glauconite (autochthonous or allochthonous). The Petrographic properties in the aspects of mineral colour, grain shape, and replacement of precursor minerals by glauconite are the most common parameters used in determining the evolutional stages of glauconite [6].

The replacement of glauconite in precursor minerals, especially in fragments of quartz and k-feldspar minerals that appear as rock fragments, are seen in samples BG 02 at the depth of 23-28 cm and BG 06 at 139-145 cm. Glauconite replacement in sample BG 02 (23-28 cm) indicates that the glauconite is entering an early stage of evolution (nascent) (Figure 6 b). This stage is indicated by the replacement of glauconite in a rounded oval shape with a bright yellow-green colour which still leaves a small amount of trace from the precursor minerals (K-feldspar). Replacement of glauconite at slightly evolved stage is shown in sample BG 06 139-145 cm (Figure 6 c). At this stage, glauconite replacement will tend to show a dark greenish colour with fewer trace of precursor minerals than the nascent stage evolution [6].

The colour of glauconite minerals that undergo the early stage of evolution (nascent) will tend to show dark yellow to light green colour. The colour of mineral body will be darker green as they processed to the higher stage of evolution [9]. Petrography sample that shows higher colour of evolution stages can be seen in sample BG 07 (126-129 cm) (Figure 6 d). In addition to the dark green colour, glauconite minerals that have undergone a high evolution stage are also characterized by rounded oval shape, this feature is shown in sample BG 08 (97-100 cm) (Figure 6 e).

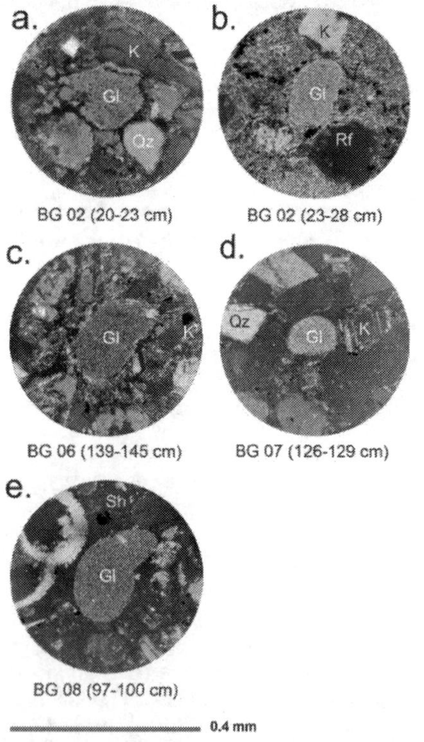

Fig. 6 Variations of glauconite form in petrography samples (a. Matrix; b and c. replacement of k-feldspar minerals; d. and e. rounded oval)

4.2.2 Substrates of glauconite based on petrography characteristics

Through petrography analysis, we recognized two types of substrates for glauconite evolution processes to take place by its shape characteristics, they are: faecal pellets and bioclastic. Semi-rounded faecal pellets substrate was seen on sample BG 07 (126-129 cm) with a size of 0.1 mm (Figure 7 a), and bioclastic substrate can be seen on sample BG 08 (97-100 cm) (Figure 7 b). Some of glauconites show an empty pattern on its mineral body which is assumed as the trace of shell fragment that previously served as bioclastic substrate for glauconite evolution process (Figure 7 c). Bioclastic substrates will completely disappeared at the early stage of evolution process due to the effect of dissolution process [7].

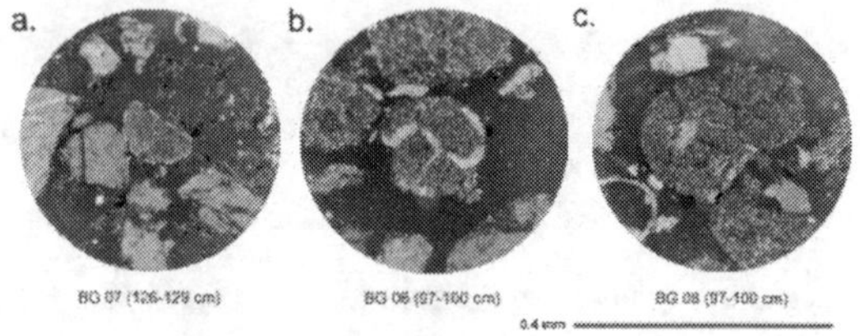

Fig. 7. Variations of glauconite substrates in petrography samples (a. Fecal pellets; b and c. replacement of k-feldspar minerals; d. and e. rounded oval)

4.2.3 Mechanism of glauconite formation

Glauconite mineral found in the research location is assumed to be the result of transportation from the marine environment to the coastal environment by the tsunami waves. Based on that condition, the formation mechanism process of glauconite can be classified as an allochthonous process. Glauconite minerals which experienced a transportation process from its place of origin (allochthonous) will show petrographic characteristics with good sorting with other components, mineral habits in the form of granules, round grain shape, similarity of grain size with other detrital materials, and lack of fracture features in mineral body [1].

Grain sorting in petrographic samples which contain glauconite mineral shows moderate sorting of grain components. The shape of the glauconite mineral in the thin sections appears as a granular form with a round grain shape without any fracture features on the grain surface. The grain size comparison between glauconite minerals with crystals and other detrital materials in thin sections shows a similar size range that not indicates a significant difference, which is between 0.04-0.2 mm.

4.3 XRF analysis

XRF analysis was carried out on eight samples from stations BG 05, BG 06, and BG 08. Each sample is composed by various compounds with up to 17 different compounds found in one sample. SiO_2 became the most dominant compounds in all samples, with the lowest percentage of SiO_2 is recorded in sample BG 06 (221-222 cm) by 40.7%, and the largest percentage is shown in sample BG 08 (41-43 cm) with values up to 75.2%. This result is also match with the result of petrography analysis, where the most abundant mineral found in each sample is quartz with chemical formula of SiO_2.

Observations on the chemical composition of each sample are focused on potassium (K_2O), iron (Fe_2O_3), magnesium (MgO), and aluminium (Al_2O_3). This is because the content of potassium, iron, and aluminium

https://doi.org/10.1051/e3sconf/202234001004

can be used to determine the stage of maturity in the stage of glauconite evolution [15].

4.3.1 Cross-plot correlations

Based on XRF data interpretations, it can be seen that there is a repeating pattern in most of the samples. The pattern shows that the composition of SiO_2 is the most abundant compound in all of the samples, followed by Al_2O_3, Fe_2O_3, MgO, and K_2O compounds. Quantitative correlation cross-plots between the existing chemical compositions were carried out to determine the effect of the composition of one compound on the composition of other compounds. The correlation is represented in a polynomial regression graph of four pairs of compounds that are compared with each other, they are: Al_2O_3 vs Fe_2O_3, K_2O vs Al_2O_3, K_2O vs Fe_2O_3, and K_2O vs MgO.

The comparison between the pairs of compounds above was carried out to determine the relationship of chemical compound exchanges that occurred during the glauconite evolution process. Chemical compounds that have a major influence in determining the level of glauconite evolution are Al_2O_3, K_2O, and Fe_2O_3. Along with the increase in K_2O and Fe_2O_3 and decrease in Al_2O_3 compounds during the maturation process, MgO compounds will be released during the process of sediment infillings and seawater intrusion into the semi-confined microenvironment [15]. MgO depletion pattern can also be analysed towards its relationship with K_2O compounds.

By plotting the composition of chemical compounds between samples on a polynomial regression graph, the value of r^2 will be obtained which represents the strength or weakness of the influence between two chemical compounds that are compared between each other. Based on the cross-plot results, the correlation between compounds in all graphs shows an intermediate relationship, with a range of r^2 values from 0.2968-0.5384. The polynomial correlation shows a pattern of increasing and decreasing certain compounds from different samples.

The correlation pattern in all of the cross-plot graphs shows a mixed pattern between compared samples. The relationship between Al_2O_3 and Fe_2O_3 shows a positive correlation from BG 08 stations in the west to BG 05 and 06 stations in the east with r^2 value of 0.4839 (Figure 8 a).

Changes in the correlation pattern of K_2O and Al_2O_3 compounds were shown in the samples from BG 08 and 06 with BG 05 (Figure 8 b). The initial pattern showed a positive correlation between samples at BG 05 stations (237-238 cm), BG 06, and BG 08 which then showed a negative correlation. to the sample BG 05 (226-228 cm). The r^2 value of this correlation is 0.5384.

The correlation between K_2O and Fe_2O_3 from the analysed samples shows a change in trend which is indicated by the curve pattern (Figure 8 c). In this correlation, it can be seen that there is a change from a negative correlation between BG 05 and BG 08 which then becomes a positive correlation in BG 06. There is a decrease in Fe_2O_3 content and an increase in K_2O content between sample BG 05 at the depth of 226-228 cm and 237-238 cm. The r^2 value of this correlation is 0.4918.

The correlation between K_2O and MgO also shows a similar pattern to the correlation between K_2O and Fe_2O_3 compounds (Figure 7 d). A negative correlation characterized the decrease in compound composition simultaneously at samples from BG 05 and BG 08, which then turned into a positive correlation at samples from BG 06. The r^2 of this correlation is 0.5132.

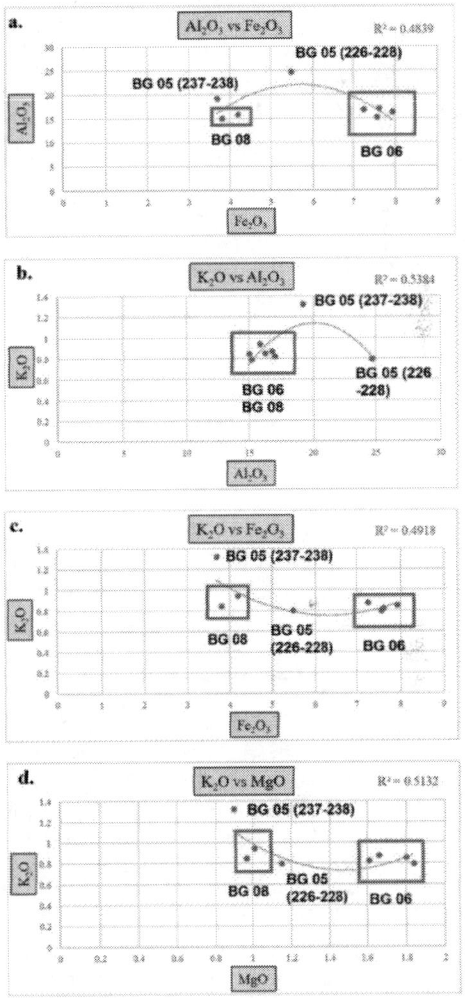

Fig. 8. Cross-plots of chemical compounds in XRF Analysis (a. Al_2O_3 vs Fe_2O_3; b. K_2O vs Al_2O_3; c. K_2O vs Fe_2O_3; d. K_2O vs MgO)

4.3.2 Geochemical interpretations

The increase in the composition of K_2O and Fe_2O_3 occurs when the burial process take place at seawater-sediment interface zone in the semi-confined microenvironment where the supply of both compounds predominantly comes from water and marine sediments [10]. The K ion in the K_2O compound also act as interlayer cation that binds the tetrahedral and octahedral sheets on the glauconite mineral structure [2]. This pattern can be seen in the correlation graph between K_2O and Fe_2O_3 between samples from BG 05 (226-228 cm) and BG 06.

The decrease in Al_2O_3 content is due to the substitution process of Al_2O_3 compounds with Fe_2O_3 in the octahedral layer of glauconite minerals [12]. As a result of this substitution process, the composition of the Al_2O_3 compound will decrease as the Fe_2O_3 became more abundant. This substitution process will also affect the crystal structure of the glauconite mineral, where based on the chemical formula of the glauconite mineral $[K_{0.6-0.85}(Fe^{3+}, Al, Mg, Fe^{2+})(OH)_2(Si_{4-x}Al_x)O_{10}$, the Fe^{3+} content is described as having relatively more abundant composition than Al. The decreasing pattern of this compound is shown in the correlation graph between Al_2O_3 with Fe_2O_3 and K_2O. The negative correlation on the Al_2O_3 vs Fe_2O_3 graph is shown by the relationship between sample BG 05 (226-228 cm) and sample BG 06, while a negative correlation on the Fe_2O_3 and K_2O graph is shown between sample BG 05 (237-238 cm) and sample BG 05 (226-228 cm).

Table 1. Comparison between glauconite chemical content in Bagedur coast with [14].

Compounds	Usual composition (Odin and Fullagar, 1988)	Composition in XRF samples	Criteria Evaluation
SiO_2	47.5 – 50 wt%	42 – 75.2 wt%	Fulfilled
Al_2O_3	3.5 – 11 wt%	15 – 27.7 wt%	Fulfilled
Fe_2O_3	19 – 25 wt%	3.68 – 7.93 wt%	Not Fulfilled
MgO	2 – 5 wt%	0.899 – 1.84 wt%	Not Fulfilled
K_2O	3 – 9 wt%	0.791 – 1.32 wt%	Not Fulfilled

Odin and Fullagar classified typical abundances of major cations which present in glauconitic green grains [14] (Table 1). Based on the analysis result, there were 3 compounds that did not meet the criteria range to be classified as glauconite green grains.

The composition of K_2O from all samples are 0.9-1.32 wt%, which not meet the minimum value of 3 wt% in the classification of Odin and Fullagar [14]. The low value of K_2O is assumed to be caused by the high presence of substrates, especially for the bioclastic substrates. The formation of smectite minerals in the early stages of the evolution of glauconite minerals is thought to be the cause of the low MgO content. Smectite is formed through the process of replacing sediment grains with Mg and Fe ions which decreases the composition of both ions along with the development of smectite minerals [3].

For Fe_2O_3, the average percentage of the 8 XRF samples was 5.94 wt%, which again did not meet the minimum criteria of Fe_2O_3 content by 19 wt%. The low content of Fe_2O_3 is mainly influenced by the presence of an oxic depositional environment [14]. The existence of an oxic depositional environment is caused by the supply of fresh water which causes a decrease in the mobility of Fe ions, thus making the sample have a low Fe content [2].

4.3.3 Type of glauconite mineral

There are two types of glauconite minerals which distinguished by the composition of K_2O and Fe_2O_3, they are illitic minerals (Fe-illite) and glauconite mica (true glauconite) [15]. Based on the cross-plot results, the samples were categorized as illitic minerals with the composition of K_2O between 0.796-1.32%, and Fe_2O_3 between 3.68-7.93% (Figure 9).

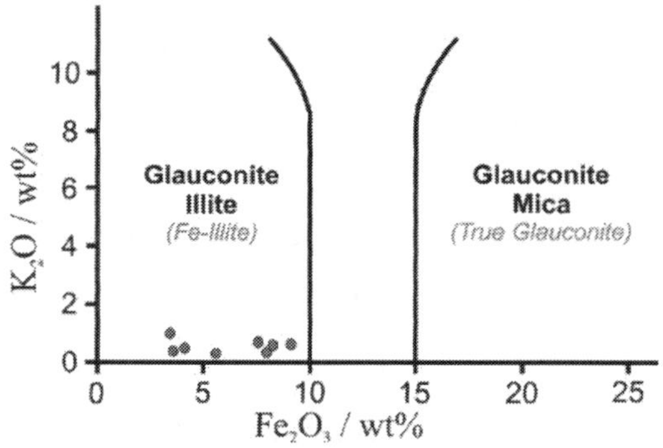

Fig. 9 Types of glauconite based on K_2O and Fe_2O_3 cross-plots

4.3.4 Substrate of glauconite based on chemical composition

The bioclastic substrate is assumed as dominant substrate in diagenesis of the glauconite pellets based on XRF analysis result. Bioclastic substrates are characterized by high Al_2O_3 content [18], where it was found that the Al_2O_3 content from all samples shows a relatively high composition with the value ranging from 15-24.7 wt%. The content of CaO at sample BG 06 from all depths also shows a relatively high composition with a range between 22.9-25.5 wt%, where this result is also in line with the finding of bioclastic components in the form of shell fragments in the stratigraphic logs at the same depths.

The bioclastic substrate is also characterized by low content in Fe and K. The low content of Fe and K of bioclastic substrates in all samples relative to the fecal pellets is because fecal pellets have a higher porosity and organic carbon content so that they can support the enrichment of Fe in glauconite pellets [17].

Bioclastic substrates such as Foraminifera, gastropods, Bryozoa, and Ostracods are porous, but the diameter of the pores are too small (<5 m), which makes seawater supply and diffusion of seafloor sediments that carry K and Fe ions into the substrate is restricted [4].

The low supply of seawater and seafloor sediment that enters through the pores of microorganisms makes Fe_2O_3 and K_2O levels in XRF samples are low.

4.3.5 Evolution of glauconite grains

The composition of K_2O, Fe_2O_3, and Al_2O_3 can be used to determine the evolution stages of glauconite minerals. The stages of glauconite evolution were divided into four parts: nascent, slightly evolved, evolved, and highly evolved [14]. The composition of K_2O, Fe_2O_3 and Al_2O_3 in all samples can be seen in Table 3. Analysis result shows that all samples undergone an early stage of evolution (nascent) which is characterized by the percentage of chemical composition of K_2O in less than 4%, Fe_2O_3 content between 8-12%, and Al_2O_3 content between 16-20% (Fig. 10).

Table 2. Content of major glauconite chemical compounds in samples from Bagedur coast

Content (wt%) / Samples	K_2O	Fe_2O_3	Al_2O_3
BG 05 (226-228 cm)	0.796	5.49	24.7
BG 05 (237-238 cm)	1.32	3.68	19.2
BG 06 (221-222 cm)	0.791	7.57	15.2
BG 06 (240-242 cm)	0.849	7.93	16.3
BG 06 (249-251 cm)	0.871	7.26	16.8
BG 06 (436-437 cm)	0.819	7.62	17
BG 08 (0-2 cm)	0.94	4.2	15.9
BG 08 (41-43 cm)	0.844	3.8	15

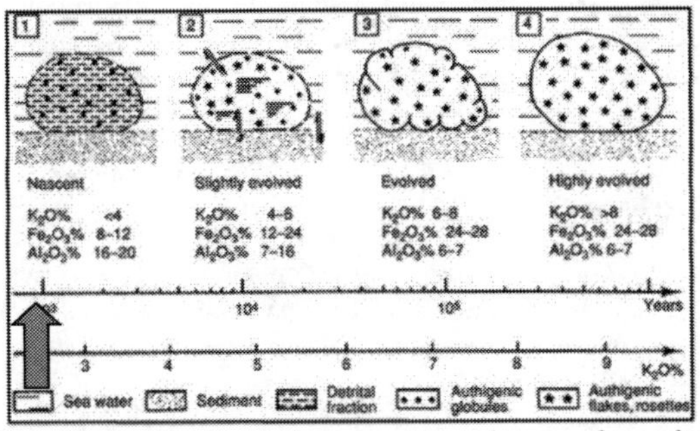

Fig. 10 Major chemical composition plot on glauconite evolution diagram by [15]

The evolution of glauconite pellets was significantly controlled by the sedimentation rate and the physical characteristics of the substrate in the form of porosity and permeability [19] . The condition of the porous and permeable substrate supports the enrichment of K and Fe compounds that enter from the environment around the substrate, and a low sedimentation rate is also needed to enable stable condition for glauconite maturation to not disturbed by the accumulation of large amounts of terrigenous material.

4.4 Diagenesis process of glauconite

The diagenesis of glauconite minerals at the research area follows the "Verdissement" theory [15] (Figure 11). Glauconite formed from oxidation process of organic material and the exchange of compounds between the substrate and the surrounding environment in a semi-confined microenvironment. The process of glauconite evolution occurs in the seawater-sediment interface in seafloor. Oxidation of organic material supported by the redistribution of sediment by bottom-current winnowing. Sub-oxic and semi-confined microenvironment conditions are then formed through the process of decomposition of organic material from faecal pellets and microfossil shells substrates [4].

There is an infilling of detrital material from loose sediments and mineral debris into the pores of the substrate which will change the composition of chemical compounds in advanced process due to contact with seawater. Fe-Al smectite then formed by rapid crystal growth process in detrital phyllosilicates minerals in response of changes in chemical [15, 13]. The duration of Fe-Al smectite formation is limited by the rapid sedimentation rate, where the presence of Fe-Al smectite will then be replaced by a further evolutionary process into Fe-illite or glauconite mica.

Chemical composition change in semi-confined microenvironment cause the dissolution of the substrate until it be removed completely [7]. The dissolution of the bioclast substrate at the research area was encountered in samples BG 07 (126-129 cm) and BG 08 (97-100 cm) which left glauconite in an empty space in the shell.

Ions from the surrounding environment of the substrate will affect the chemical composition of the glauconite. Al and K ions are predominantly supplied by seawater, while Fe ions come from the diffusion process from sediment deposits at the seafloor [4]. Dissolution of the substrate will release chemical compounds, such as the faecal pellets that supplies Fe and K ions [2].

Fe-Al smectite will then turned into Fe-Illite or glauconite mica by increasing the number of Fe ions through a chemical diffusion process. At the research area, glauconite shows a tendency to grow as illitic minerals (Fe-illite) based on the analysis of Fe_2O_3 content which has a low value (<10 wt%). The process of glauconite evolution at the research area is ended at the nascent evolutionary stage.

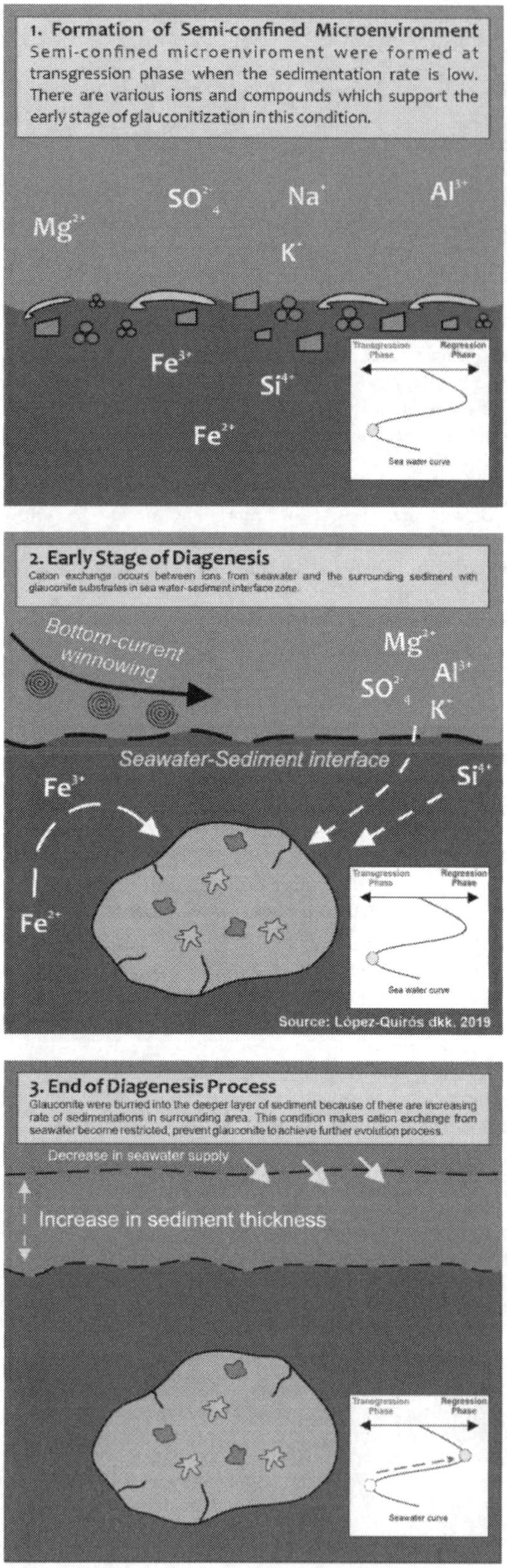

Fig. 11 Diagenesis process of glauconite in Bagedur Coast

5. Conclusions

Based on the results of geochemical and mineralogical analysis using petrography and XRF methods, we got several conclusions about the characteristics of glauconite in research area, they are:

1. The types of sediment in research area are composed by soil, peat soil, clay, silt, very fine sand to medium sand, and a layer of green sand sediment. There are also other components in the form of wood fragments, detrital minerals, and shell fragments between sediment layers at several sampling stations with varying depths.
2. Glauconite were found in green sand sediment layers at 5 stations: BG 02, BG 07, BG 08, BG 05, and BG 06. Glauconite has petrographical properties of greenish yellow to dark green colour with various shapes, appearing as a matrix, oval-shaped pellets, replacement of precursor minerals, and infillings of voids in shell fragments.
3. The geochemical properties of glauconite indicates that they are entering the early stage of evolution (nascent). Glauconite shows characteristics as an illitic mineral. The composition of primary chemical compounds is same in all samples. They have relatively higher Al_2O_3 content than Fe_2O_3 and K_2O.
4. The formation of glauconite in research area follows Verdissement mechanism (Odin and Matter, 1981), where glauconite form a semi-confined microenvironment through their interactions with the substrate, seawater and surrounding sediments. Glauconite then undergoes an evolutionary process through the exchange of chemical compounds and stopped at the nascent stage.

The limitation of this research is we only focus about the analysis of glauconite diagenesis in paleotsunami deposit. For further interpretation about the tsunami source, radiometric dating method should be conducted to answer those question.

References

1. Amorosi, A. Detecting compositional, spatial, and temporal attributes of glaucony: a tool for provenance research. Sedimentary Geology, **109**(1-2), 135-153 (1996)
2. Amorosi, A., Sammartino, I., & Tateo, F. Evolution patterns of glaucony maturity: a mineralogical and geochemical approach. Deep Sea Research Part II: Topical Studies in Oceanography, **54**(11-13), 1364-1374 (2007)
3. Amorosi, A., Guidi, R., Mas, R., & Falanga, E. Glaucony from the Cretaceous of the Sierra de Guadarrama (Central Spain) and its application in a sequence-stratigraphic context. International Journal of Earth Sciences, **101**(2), 415-427 (2012)
4. Banerjee, S., Chattoraj, S. L., Saraswati, P. K., Dasgupta, S., & Sarkar, U. Substrate control on formation and maturation of glauconites in the Middle Eocene Harudi Formation, western Kutch,

India. Marine and Petroleum Geology, **30**(1), 144-160 (2012)

5. Banerjee, S., Bansal, U., & Thorat, A. V. A review on palaeogeographic implications and temporal variation in glaucony composition. Journal of Palaeogeography, **5**(1), 43-71 (2016)
6. Bansal, U., Banerjee, S., Ruidas, D. K., & Pande, K. Origin and geochemical characterization of the glauconites in the Upper Cretaceous Lameta Formation, Narmada Basin, central India. Journal of Palaeogeography, **7**(2), 99-116 (2018)
7. Bornhold, B. D., & Giresse, P. Glauconitic sediments on the continental shelf off Vancouver Island, British Columbia, Canada. Journal of Sedimentary Research, **55**(5), 653-664 (1985)
8. Huggett, J. M., & Cuadros, J. Glauconite formation in lacustrine/palaeosol sediments, Isle of Wight (Hampshire Basin), UK. Clay Minerals, **45**(1), 35-49. (2010)
9. Huggett, J. M., Petroclays, H., Time, I., Space, I., Dating, A., & Illit, F. Minerals: Glauconites and Green Clays. Reference Module in Earth Systems and Environmental Sciences; Elsevier: Amsterdam, The Netherlands (2013)
10. López-Quirós, A., Escutia, C., Sánchez-Navas, A., Nieto, F., Garcia-Casco, A., Martín-Algarra, A., ... & Salabarnada, A. Glaucony authigenesis, maturity and alteration in the Weddell Sea: An indicator of paleoenvironmental conditions before the onset of Antarctic glaciation. Scientific reports, **9**(1), 1-12 (2019)
11. McCaffrey, R. Global frequency of magnitude 9 earthquakes. Geology, **36**(3), 263-266 (2008)
12. McConchie, D. M., Ward, J. B., McCann, V. H., & Lewis, D. W. A Mössbauer investigation of glauconite and its geological significance. Clays and Clay Minerals, **27**(5), 339-348 (1979)
13. Meunier, A., & El Albani, A. The glauconite–Fe-illit–Fe-smektit problem: a critical review. Terra Nova, **19**(2), 95-104 (2007)
14. Odin, G. S., & Fullagar, P. D. Chapter C4 geological significance of the glaucony facies. In Developments in sedimentology (Vol. **45**, pp. 295-332). Elsevier (1988)
15. Odin, G. S., & Matter, A. De glauconiarum origine. Sedimentology, **28**(5), 611-641. (1981)
16. Putra, P. S., Yulianto, E., Praptisih, P., Supriatna, N., Trisuksmono, D., Amar, A., ... & Griffin, J. STUDI PALEOTSUNAMI DI SELATAN JAWA PALEOTSUNAMI STUDY IN THE SOUTH COAST OF JAWA. PROSIDING GEOTEKNOLOGI LIPI (2015)
17. Pryor, W. A. Biogenic sedimentation and alteration of argillaceous sediments in shallow marine environments. Geological Society of America Bulletin, **86**(9), 1244-1254 (1975)
18. Smaill, J. B. Geochemical variations in glauconitic minerals: application as a potassium fertiliser resource. (2015)
19. Sudana, D. & Santosa, S. Geology of the Cikarang Quadrangle, Java: Pusat Penelitian dan Pengembangan Geologi, Bandung, **13** (1992)
20. Sujanto, F. X. dan Sumantri, Y. R. Preliminary Study on the Tertiary Depositional Patterns of Java. Bulletin of Scientific Contribution, Volume **13**, Nomor 3. Desember 2015: 182-191. (1977)
21. Zulkarnain, D. A., Amijaya, H., & Yulianto, E. Karakteristik Endapan Paleotsunami Di Pesisir Wanasalam, Daerah Wanasalam, Banten. In Proceeding, Seminar Nasional Kebumian Ke-10 Peran Penelitian Ilmu Kebumian Dalam Pembangunan Infrastruktur Di Indonesia 13–14 September 2017; Graha Sabha Pramana (2017)

Investigating characteristics of tsunami hazards for west coast of Aceh Besar district, Indonesia

Ibrahim[1,2,3], *Syamsidik*[1,2,4*], *Azmeri*[1,4], *Muttaqin* Hasan[1,4], and *Tursina*[2]

[1]Doctoral Program, School of Engineering, Universitas Syiah Kuala (USK), Jl. Syech Abdurrauf No. 7, Darussalam, Banda Aceh, 23111, Indonesia
[2]Tsunami and Disaster Mitigation Research Center (TDMRC), Universitas Syiah Kuala, Jl. Prof. Dr. Ibrahim Hasan, Banda Aceh, 23233, Indonesia
[3]Civil Engineering Department, Politeknik Negeri Lhokseumawe, Jl. Banda Aceh-Medan, Km. 280,3, Lhokseumawe, 24301, Indonesia
[4]Dept. of Civil Engineering, Universitas Syiah Kuala, Jl. Syech Abdurrauf No. 7, Darussalam, Banda Aceh, 23111, Indonesia

Abstract. The seismic activities of a long section of the Sunda Plate revealed the vulnerability of the Indian Ocean to tsunami along the west coast of Sumatra, including on the west coast of Aceh Province. This megathrust caused the great tsunamigenic earthquake of 2004 and can be a possible candidate for rupture in the future and will affect or destroy everything, especially buildings along the coast. The west coast of Aceh Besar is one of the worst areas hit by tsunami waves and can cause tsunamis in the future. The objectives of this study are to determine the height, distance, and area of the tsunami inundation on the west coast of Aceh Besar District based on three earthquake magnitudes scenarios Mw 8.0, 8.5, and 9.2. The COMCOT (Cornell Multigrid Coupled Tsunami Model) tsunami model was used to obtain the height, area, and distance of the tsunami inundation. It was found that the largest tsunami inundation was caused by the tsunami generated by the magnitude Mw 9.2 earthquake. The maximum tsunami run up reached up to 20 meters with an inundation distance of 6 km from the coastline and a tsunami inundation area of 40 km^2 distributed along the west coast of Aceh Besar District.

1 Introduction

The seismic activities along Sunda Plate revealed the vulnerability of the Indian Ocean to tsunami along the west coast of Sumatra. That includes the area along the west coast of Aceh Province. This megathrust is the source of the great tsunamigenic earthquakes of 2004. The tsunami in the area can still be a possible candidate for rupture in the future and will affect or destroy everything especially building along the coast. In addition to the Mw 9.1 – 9.3 2004 earthquake [1], other large earthquakes have also occurred along with the Sunda plate with a magnitude of Mw 7.7 to Mw 8.9 [2]. The return period of an earthquake with a magnitude of 7.5 to 9.3 is 55 to 500 years [3].

The earthquake and tsunami that struck Aceh in December 2004 are the worst natural disaster in living memory [4].The west coast area of Aceh Besar District, which is part of the west coast of Aceh, was one of the areas most severely hit. . This area was hit by a tsunami as high as 15 to 30 meters which destroyed many buildings and caused many casualties [5,6]. After 16 years of the tsunami, this area was rebuilt with various uses such as residential areas, tourism, industries, and public facilities. The population rate in this area reaches 11.05% [7]. Generally, this area is located or concentrated on the coast. With the history of previous earthquake events and the potential for tsunami events in the future, research is needed to investigate the potential hazards of tsunami waves in the context of tsunami wave mitigation.

Several studies have been conducted in this area. Research on tsunami estimated times of arrival (ETAs) for cities in Aceh has been done [8]. They found the ETAs for several cities on the west coast are between 25 -78 minutes based on the 2004 tsunami simulation. Another research for this location focus on sediment deposits, coastal erosion, and coastal change [9–17]. The west coast area is also included in the area with a high level of tsunami hazard [18]. Research on the potensial earthquake along west of Sumatra subduction zone had been carry out, return period of earthquake surrounding this area are 50 to 200 years with magnitude from Mw 8.0 to 9.2 [19]. However, research on the potential tsunami hazard in the future for the west coast of Aceh Besar has not been carried out. This research is needed to provide sufficient scientific approaches to mitigate tsunami and minimize the casualty in the future.

This study aims to investigate the tsunami hazard on the west coast of Aceh Besar District. The tsunami hazard determined in this research is height, distance, and area of the tsunami inundation based on three earthquake magnitudes scenarios, i.e., Mw 8.0, 8.5, and

*Corresponding author: syamsidik@tdmrc.org

9.2. The scenario earthquake with Mw 9.2 similar to 2004 andaman tsunami and used to validate the model. The COMCOT (Cornell Multigrid Coupled Tsunami Model) tsunami model was used to obtain the height, area, and distance of the tsunami inundation. Results of this study are expected to be a reference in the preparation of tsunami disaster mitigation for this area.

The research is located on the west coast of Aceh Besar District, as shown in Fig 1. The west coast of Aceh Besar consists of 3 sub-districts, namely Lhoknga, Leupung, and Lhoong. Sloping beaches dominate the beach, and some of them have steep beaches with hills. The Lhoknga area is a sloping area inhabited by many residents, tourist areas, and industry (cement factory). During the 2004 tsunami, most of this area was washed away by the tsunami and many casualties. Leupung and Lhoong are located in the southern part of the Lhoknga area and also densely populated areas. Some of these areas are flat, and the coastline is closed to the hills.

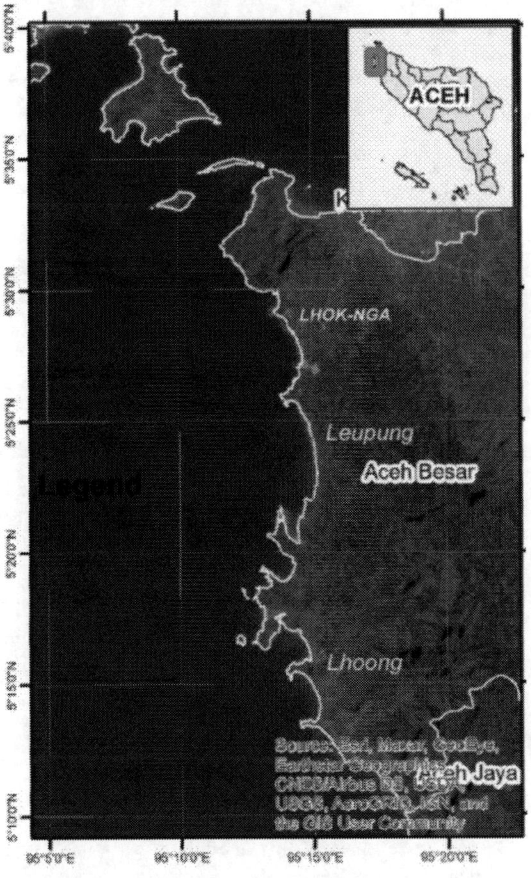

Fig 1. Research location from Google Earth.

2 Methods

Tsunami waves may change according to their position from the shoreline and the depth of the water. In deep water, the wave height to length (H/L) ratio is significant, and in shallow water, the wave height to water depth (H/d) ratio is relatively high [20]. When a tsunami wave reaches land, the terms used to describe a tsunami wave are flow depth (depth of water), tsunami height (the elevation of water above the reference level), runup height (the maximum elevation the wave reaches at the maximum inundation), inundation distance (maximum horizontal intrusion), and inundation limit (inland limit of wetting) [21,22]. A schematic explanation of the tsunami terms is shown in Fig 2.

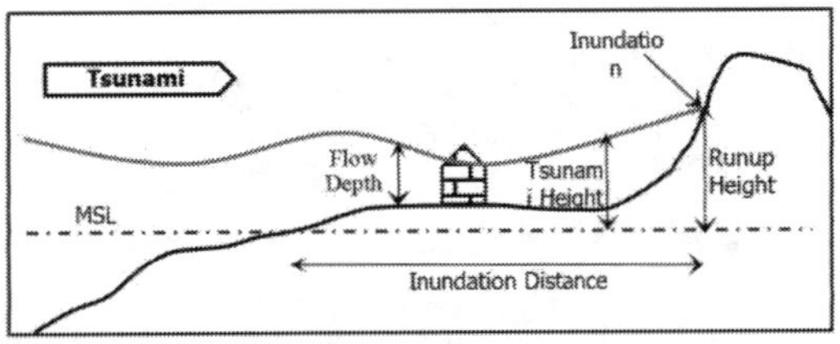

Fig 2. Schematic explanation of some commonly used tsunami terms (refer to IOC UNESCO [23]).

The investigation of tsunami inundation was conducted using numerical simulation using the Cornell Multi-grid Coupled Tsunami model (COMCOT). This numerical model consists of two stages, namely:

1. Tsunami wave generator;
2. Propagation of tsunami waves from the epicenter to the mainland (inundation) using a nested grid system.

The equations used in COMCOT are linear and non-linear equations of the shallow water equation for both spherical and cartesian coordinate systems [24]. For tsunamis in the deep sea, the height of the tsunami is less than the water depth, so the shallow water linear equation for spherical coordinates was used as follows:

$$\frac{\partial \eta}{\partial t} + \frac{1}{R\cos\varphi}\left\{\frac{\partial P}{\partial \psi} + \frac{\partial \eta}{\partial \psi}(\cos\varphi\, Q)\right\} = 0 \quad (1)$$

$$\frac{\partial P}{\partial t} + \frac{gh}{R\cos\varphi}\frac{\partial P}{\partial \psi} - fQ = 0 \quad (2)$$

$$\frac{\partial Q}{\partial t} + \frac{gh}{R}\frac{\partial \eta}{\partial \psi} + fP = 0 \quad (3)$$

$$f = \Omega \sin\varphi \quad (4)$$

When the tsunami waves reach the coastal area, the equation applied is a non-linear equation. In COMCOT, the shallow water non-linear equation for the spherical coordinate system was used as follows:

$$\frac{\partial \eta}{\partial t} + \frac{1}{R\cos\varphi}\left\{\frac{\partial P}{\partial \psi} + \frac{\partial}{\partial \psi}(\cos\varphi\, Q)\right\} = 0 \quad (5)$$

$$\frac{\partial P}{\partial t} + \frac{1}{R\cos\varphi}\frac{\partial}{\partial \psi}\left\{\frac{P^2}{H}\right\} + \frac{1}{R}\frac{\partial}{\partial \varphi}\left\{\frac{PQ}{H}\right\} + \frac{gH}{R\cos\varphi}\frac{\partial \eta}{\partial \psi} - fQ + F_x = 0 \quad (6)$$

$$\frac{\partial Q}{\partial t} + \frac{1}{R\cos\varphi}\frac{\partial}{\partial \psi}\left\{\frac{PQ}{H}\right\} + \frac{1}{R}\frac{\partial}{\partial \varphi}\left\{\frac{Q^2}{H}\right\} + \frac{gH}{R}\frac{\partial \eta}{\partial \psi} + fP + F_y = 0 \quad (7)$$

$$F_x = \frac{gn^2}{H^{\frac{7}{3}}}P(P^2 + Q^2)^{1/2} \quad (8)$$

$$F_y = \frac{gn^2}{H^{\frac{7}{3}}}Q(P^2 + Q^2)^{1/2} \quad (9)$$

In equations (1)–(7), η represents the water surface elevation; (P,Q) is the volume fluxes in X (West-East) direction and Y (South-North) direction; (φ,ψ) is the latitude and longitude of the Earth; R is the radius of the Earth; g is the gravitational acceleration; h is the water depth. Term f represents the Coriolis force coefficient due to the rotation of the Earth; Ω is the rotation rate of the Earth; H is the total water depth, and H = η + h; (F_x, F_y) represents the bottom friction in X and Y directions;

n is the Manning's roughness coefficient, and g is the gravitational acceleration

The validation was done using the Aida method [25–27]. Validation between observations and computations was carried out using variables tsunami height and inundation distance. The equations used in the Aida coefficient are as follows.

$$K_i = \frac{x_i}{y_i} \quad (10)$$

$$\text{Log } K = \frac{1}{n}\sum_{i=1}^{n} \log K_i \quad (11)$$

$$\text{Log } k = \sqrt{\frac{1}{n}\sum_{i=1}^{n} (\log K_i)^2 - (\log K)^2} \quad (12)$$

The *K* and *k* are Aida coefficient where *K* represents geometric mean and *k* represent geometric standard deviation. Variable x_i and y_i are the observed and computed variables. The results were classified as a good agreement if $0.8 \leq K < 1.2$ and $k \leq 1.60$. The smaller the geometric standard deviation (*k*), the better result.

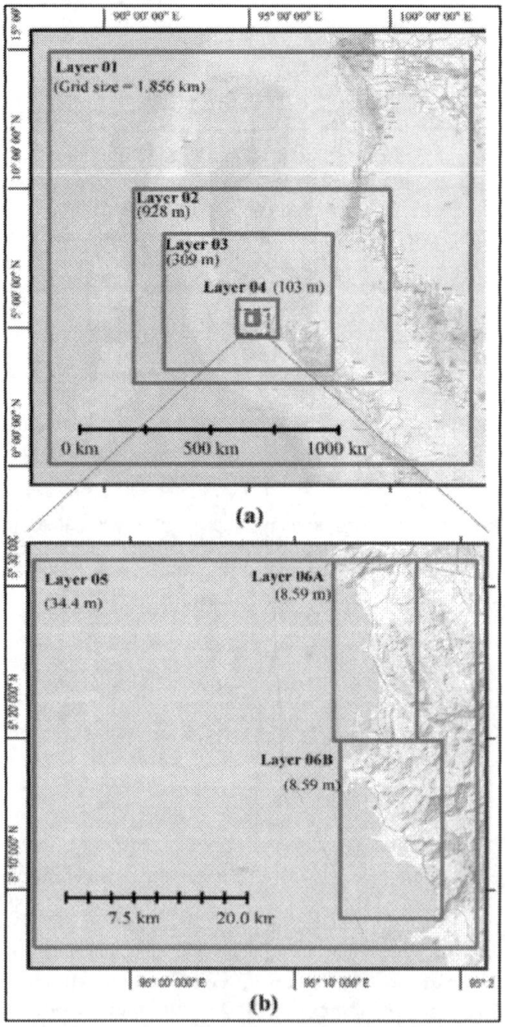

Fig 3. Nested grids for COMCOT model simulation domain: (a) Layers 1–4 and (b) Layer 5, 6A, and 6B

3 Model Setup

3.1 Input Data and Grid Setup

The numerical simulations were carried out in several stages. The first stage was generating an initial tsunami waveform around the source. In the tsunami wave generator stage, the required input data were bathymetry of the Indian Ocean up to the study area as shown in layer 01 (Fig 3(a)). This data was obtained from GEBCO (General Bathymetric Chart of the Ocean) with a resolution of 1-minute grid size (or 1,856 km) [28]. The source of the wave generator in this model comes from earthquakes based on three earthquake magnitudes scenarios of -Mw 8.0, 8.5, and 9.2 in accordance with historical seismic activity for study areas. Earthquake parameters Mw 9.2 are similar to the 2004 tsunami in the Andaman Sea [1].

The second stage in tsunami modeling was the propagation of tsunami waves from the epicenter to the mainland (inundation). More detailed bathymetric, topographic, and land use data (manning values) were required in this stage. Bathymetry data were obtained from BATNAS (National Bathymetry) with a resolution of 6 arc second and DEMNAS (National Digital Elevation Model (DEM) data with a spatial resolution of 0.27-arcsecond or 8.35 m which can be accessed through http://tides.big.go.id [29]. Land use data can be accessed through http://appgis.menlhk.go.id/[30]. The land use at the study area consisted of forest/plantation (n=0.035), housing/build-up area (n=0.040), open land/rice field (n=0.02), and ponds (n=0.017).

Table 1. Rupture parameters.

Scenario		**1**	**2**		**3**					
Mw		**8.0**	**8.5**		**9.2**					
L (Km)		165	125	180	20 0	12 5	18 0	14 5	12 5	38 0
W (km)		75	75	75	15 0	15 0	15 0	15 0	15 0	15 0
Dis (m)		8	8	8	14	13	15	7	7	7
Strike (°)		340	335	340	32 3	33 5	34 0	34 0	34 5	7
Dip (°)		15	15	15	15	15	15	15	15	15
Rake (°)		90	90	90	90	90	90	90	90	90
Epicenter	**Long.**	93	93	93	94	93	93	92	92	92
	lat	6	4	6	3	4	6	7	8	11
Total L (km)		165	305		1155					
Depth, D (Km)		10	10		10					

The last stage of this modeling was validation. Validation was carried out using the Aida method for tsunami height and inundation distance variables. The observed tsunami height variable was obtained from NOAA [14]. While the data on the inundation limit inundation distance were digitized from aerial photos before the tsunami (June 2004) and after the tsunami (January 2005) [31].

A nested grid scheme was used to simulate the tsunami wave from tsunami sources to the study area. The nested grid is divided into seven layers. A combination of bathymetric and topographic data from

GEBCO, BATNAS, and DEMNAS was used in the grid. Layer 1 used a larger grid size (1.856 km) and a smaller grid on layers 6A and 6B (8.59 m). The map of the simulation layer is shown in Fig. 3.

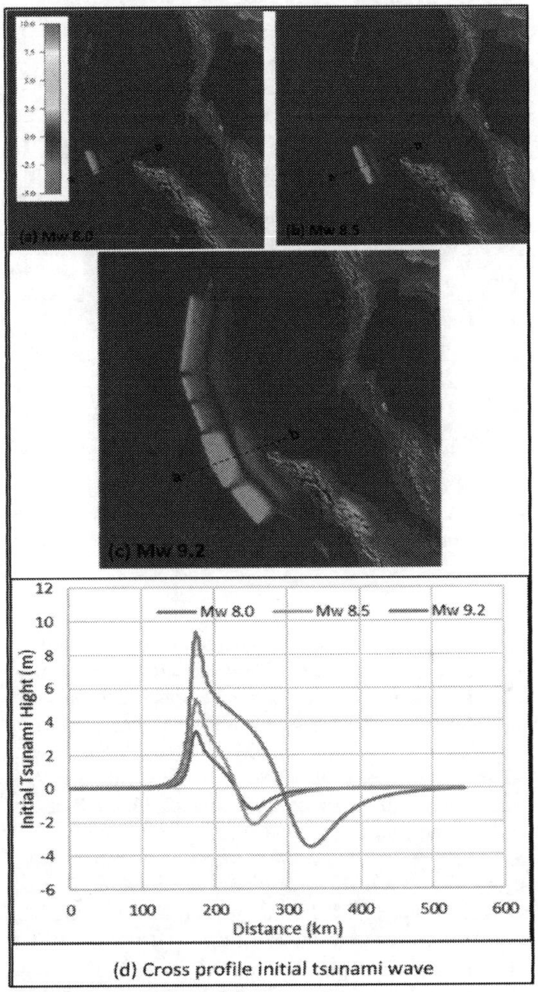

Fig 4. Initial tsunami height with earthquake magnitude Mw 8.0 (a), Mw 8.5 (b) Mw 9.2(c), and cross profile of initial tsunami wave for different magnitude (d)

4 Results and Discussion

4.1 Initial Condition

The wave generation was based on three earthquake magnitudes: Mw 8.0, Mw 8.5, and Mw 9.2. Possibility return period for this earthquake are from 50 to 200 yaers[19]. For scenario Mw 9.2, the rapture parameters were based on the 2004 earthquake conditions [1]. The rupture parameters for Mw 8.0 and Mw 8.5 scenarios were calculated using the scaling law proposed by Wells and Coppersmith [32]. The location of the wave generator was along the Sumatra Andaman subduction zone, and the location of the epicenter was predicted to be at the location of the earthquake in 2004. The rapture parameters for this study are shown in Table 1.

Initial sea survace movement cause by an earthquake is calculated using elastic finite fault plan theory [24]. Initial sea surface movement for the various scenarios is shown in Fig 4. The green color represents positive waves, and the magenta represents negative waves. The initial wave for scenario Mw 8.0 is shown in Fig 4(a). It can be seen that the tsunami generator is a single fault with a maximum wave of +3.40 m and a minimum of -1.21 m. For the scenario of Mw 8.5, the maximum positive and minimum waves are +5.30 m and -2.15 m with two rupture segments, as shown in Fig 4(b). The highest wave at the epicenter occurs in Mw 9.2 scenario with the maximum positive wave of +9.378 m and the minimum wave of -3.507 m with six rupture segments (Fig 4(c)). The cross-section profile of tsunami wave shape at the epicenter for all scenarios is shown in Fig 4(d).

4.2 Validation

Validation was carried out on the tsunami waves reaching land or inundation. Validation was carried out for the tsunami height and inundation distance variables based on the Mw 9.2 earthquake magnitude scenario (2004 tsunami). There are 172 observation points for the 2004 tsunami height distributed along the west coast of Aceh Besar District conducted by NOAA [14]. The results of tsunami inundation modeling are shown in Fig 8 (a), (b), and (c). From the validation results, using the Aida method, it was found that the results of the tsunami modeling in this study were a good agreement with the observed data with a value of $K = 1{,}139$ and a value of $k = 1{,}585$ (Table 2).

Table 2. Validation of tsunami height and inundation distance.

Aida Coefficient	Required range	Result			
		Tsunami Height (NOAA) [14]		Inundation Distance	
Geometric mean (K)	K ($0.8 \le K \le 1.2$)	1.139	*Good Agreement*	1.036	*Good Agreement*
Geometric standard deviation (k)	k ($k \le 1.6$)	1.585	*Good Agreement*	1.130	*Good Agreement*

Validation was also carried out for the variable inundation distance. The inundation limit line was obtained based on the comparison of aerial photos before the tsunami (June 2004) and aerial photos after the tsunami (January 2005), as shown in Fig 5 [31]. The post-tsunami images (Fig 5(b), (c), and (d)) show that the area inundated was characterized by brown land color (dead vegetation). The limit of this inundation area was used in the validation process. In this study, the location of the inundation measurement was determined as many as 30 points located along the coast. This location was used as a measurement point for the inundation distance between the digitized inundation limit from aerial photographs and simulation (Fig 6). The inundation distance varies from 500 m (Leupung area) to 6000 m (Lhoknga area) depending on topographical conditions. Using Aida method, validation for inundation distance variable has also a good agreement with $K = 1{,}036$ and $k = 1{,}130$. (Table 2).

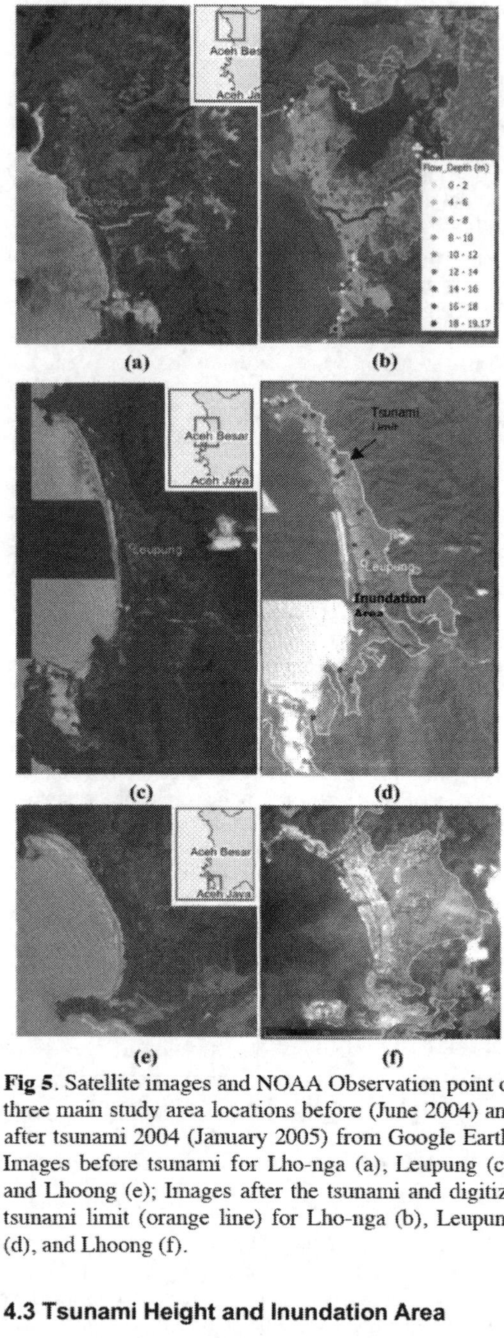

Fig 5. Satellite images and NOAA Observation point of three main study area locations before (June 2004) and after tsunami 2004 (January 2005) from Google Earth. Images before tsunami for Lho-nga (a), Leupung (c), and Lhoong (e); Images after the tsunami and digitize tsunami limit (orange line) for Lho-nga (b), Leupung (d), and Lhoong (f).

4.3 Tsunami Height and Inundation Area

The modeling results for three earthquake scenarios are shown in Fig 8. The image shows the tsunami height distribution (green to red) with a Google Earth map background. The farthest inundation limit and highest tsunami height occurred due to the tsunami with Mw 9.2, as shown in Fig 8 (a), (c), and (e). Maximum inundation distance occurs in Lhoknga Subdistrict with an inundation distance of 6 km with a maximum tsunami height of 13 meters. In the Leupung Subdistrict, it can be seen that this area experienced the highest tsunami hit (red color), up to 20 meters, but the inundation distance is not too far because there are hills near the coast.

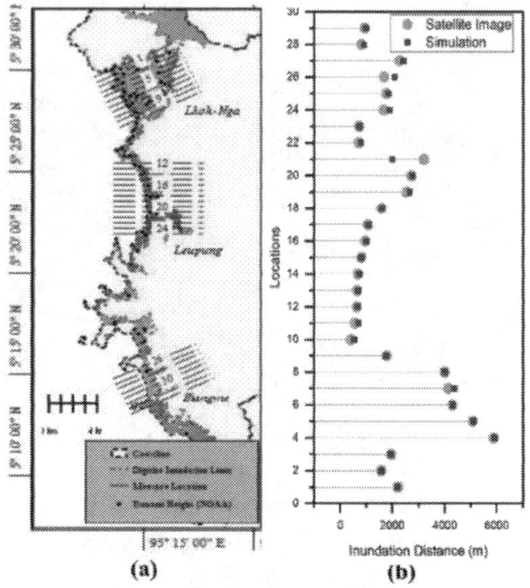

Fig 6. (a) Locations of measure inundation limit and (b) comparing inundation distance (b) from satellite image after the tsunami and result from COMCOT.

Fig 9 shows a map of the inundation area based on the flow depth categories. The total area of tsunami inundation generated by the Mw 9.2 earthquake is 40.91 km² or the largest inundation area compared to the other two scenarios. The tsunami inundation area caused by the scenarios Mw 8.5 and Mw 8.0 earthquakes was 15.40 km² and 5.72 km². In terms of inundation area, it can be seen that the ratio of the tsunami inundation area with the increase in earthquake magnitude from Mw 8.0, Mw 8.5 to Mw 9.2 is approximately three times. Fig 7 shows the area of flow depth for the three earthquake scenarios. From the figure, it can be seen that the distribution area for each tsunami flow depth due to Mw 9.2 is more evenly distributed compared to the distribution area caused by Mw 8.5 and Mw 8.0. For the Mw 9.2 scenario, the largest area of ± 10 km² occurs at a flow depth of 2-4 m. For the Mw 8.5 and Mw 8.0 scenarios, the largest area occurs at a flow depth of 0-2 meters.

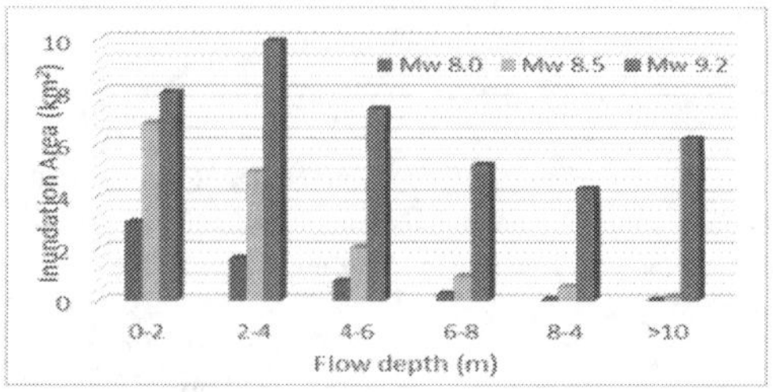

Fig 7. Areas of tsunami inundation with different flow depth and earthquake magnitude

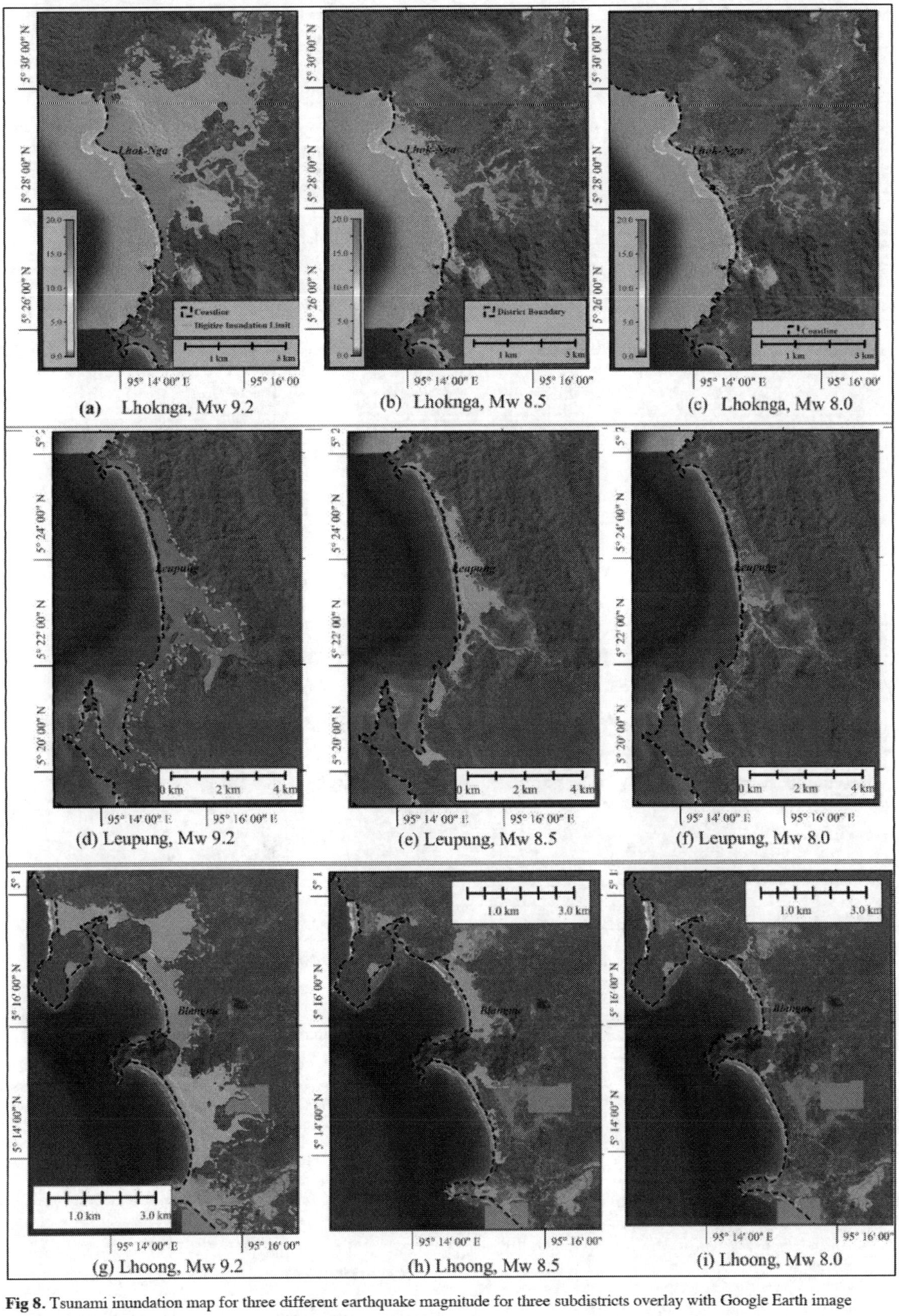

Fig 8. Tsunami inundation map for three different earthquake magnitude for three subdistricts overlay with Google Earth image

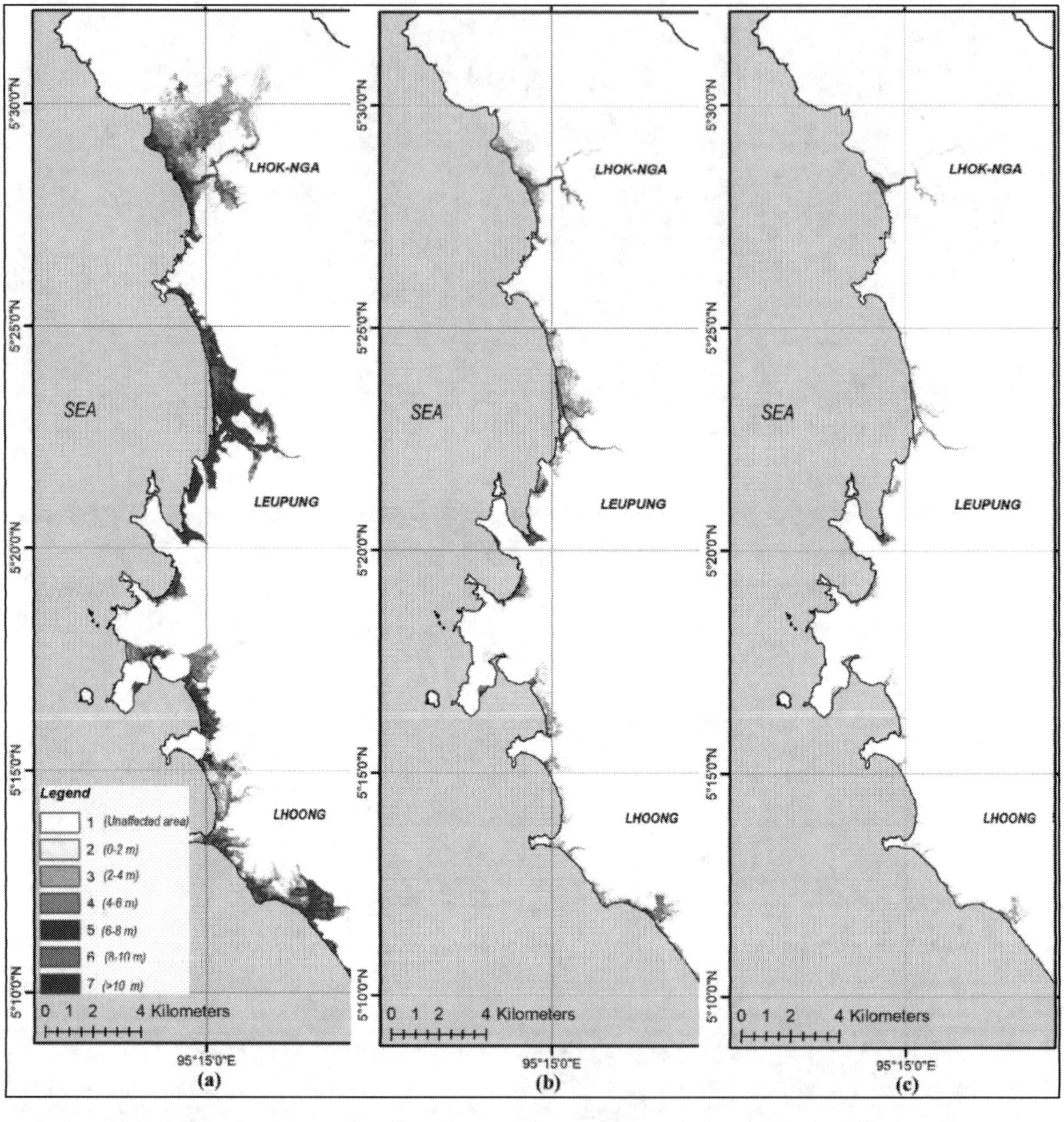

Fig 9. Flow depth and inundation area for scenario earthquake magnitudes-Mw 9.2(a), 8.5 (b), and 8.0 (c)

5 Conclusion

The tsunami caused by the earthquake magnitude Mw 9.2 was found to have the most significant impact on the West coast of Aceh Besar District. From the results of our simulations, this research concludes that:

- The highest tsunami was in the sub-district of Leupung, with a tsunami height up to 20 m.
- The largest inundation distance is in the Lhoknga sub-district, with an inundation distance of 6 km
- The tsunami inundation area reached 40 km^2 distributed along the west coast of Aceh Besar District.

With such results, projections of the development of the area should consider historical evidence of tsunami as in the 2004 Indian Ocean tsunami and also smaller intensity tsunamis as revealed in this research.

Acknowledgments. Authors thank to Partnership Enhanced Engagement in Research (PEER) of the United States Agency for International Development (USAID) and NASEM for providing research grant No. 5-395 title: Incorporating climate change induced sea-level rise information into coastal cities' preparedness toward coastal hazards. The grant has enabled authors to collect some data and attend a writeshop to publish this article.

References

1. S. Koshimura, T. Oie, H. Yanagisawa, and F.

Imamura, Coast. Eng. J. **51**, 243 (2009)
2. A. Muhammad, K. Goda, and N. Alexander, Front. Built Environ. **2**, 33 (2016)
3. W. Sengara, H. Latief, and M. S. B. Kusuma, Geomech. Geoengin. An Int. J. **5**, 57 (2010)
4. A. Azmeri, C. Mutiawati, N. Al-Huda, and H. Mufiaty, Int. J. Disaster Manag. **1**, 35 (2017)
5. Y. Tsuji, Y. Tanioka, H. Matsutomi, Y. Nishimura, T. Kamataki, Y. Murakami, T. Sakakiyama, A. Moore, G. Gelfenbaum, S. Nugroho, B. Waluyo, I. Sukanta, R. Triyono, and Y. Namegaya, J. Disaster Res. **1**, 103 (2006)
6. I. Nistor and K. Xie, in *Indian Ocean Tsunami*, edited by Tad S Murty, U. Aswathanarayana, and N. Nirupama, 1st ed. (CRC Press, London, 2007), pp. 185–208
7. Badan Pusat Statistik (BPS), *Kabupaten Aceh Besar Dalam Angka 2019* (Aceh Besar, 2019)
8. Syamsidik, T. M. Rasyif, and S. Kato, Int. J. Disaster Risk Reduct. **14**, 403 (2015)
9. L. Li, Q. Qiu, and Z. Huang, Nat. Hazards **64**, 1549 (2012)
10. R. Paris, P. Wassmer, J. Sartohadi, F. Lavigne, B. Barthomeuf, E. Desgages, D. Grancher, P. Baumert, F. Vautier, D. Brunstein, and C. Gomez, Geomorphology **104**, 59 (2009)
11. B. Ontowirjo, R. Paris, and A. Mano, Nat. Hazards **65**, 1967 (2013)
12. K. Satake, Y. Nishimura, P. S. Putra, A. R. Gusman, H. Sunendar, Y. Fujii, Y. Tanioka, H. Latief, and E. Yulianto, Pure Appl. Geophys. **170**, 1567 (2013)
13. A. R. Gusman, Y. Tanioka, and T. Takahashi, Earth, Planets Sp. **64**, 3 (2012)
14. (n.d.)
15. T. Siregar Novita, Progr. Stud. Kehutan. USU (2016)
16. Syamsidik, M. Al'ala, H. M. Fritz, M. Fahmi, and T. M. Hafli, Nat. Hazards Earth Syst. Sci. **19**, 1265 (2019)
17. T. Rasyif, S. Syamsidik, M. Al'ala, and M. Fahmi, J. Coast. Conserv. **20**, (2016)
18. N. Horspool, I. Pranantyo, J. Griffin, H. Latief, D. Natawidjaja, W. Kongko, A. Cipta, B. Bustaman, S. Anugrah, and H. Thio, Nat. Hazards Earth Syst. Sci., 14, 3105–3122, 2014 **2**, (2014)
19. A. Jihad, U. Muksin, Syamsidik, A. Suppasri, M. Ramli, and V. H. Banyunegoro, Int. J. Disaster Risk Reduct. **51**, 101800 (2020)
20. E. Bryant, *Tsunami: The Underrated Hazard (Third Edition)*, Third Edit (Springer International Publishing, Kuala Lumpur, 2014)
21. USGS, (n.d.)
22. Intergovernmental Oceanographic Commission, (2019)
23. UNESCO-IOC, 1 (2012)
24. X. Wang, Cornell Univ. **6**, 1 (2009)
25. I. AIDA, J. Phys. Earth **26**, 57 (1978)
26. Tursina, Syamsidik, S. Kato, and M. Afifuddin, Int. J. Disaster Risk Reduct. **55**, 102084 (2021)
27. R. Komata, IOP Conf. Ser. Earth Environ. Sci. **326**, (2019)
28. International Hydrographic Organization (IHO) and (IOC) Intergovernmental Oceanographic Commission (IOC), UNESCO (n.d.)
29. Pusat Jaring Kontrol Geodesi dan Geodinamika, (n.d.)
30. Direktorat Jenderal Planologi Kehutanan, (n.d.)
31. Google Earth (n.d.)
32. D. Wells and K. Coppersmith, Bull. Seismol. Soc. Am. **84**, 974 (1994)

Assessing slope failure of soil erodibility problem by soil dispersive identification

Reza Pahlevi Munirwan[1*], and *Munirwansyah* Munirwansyah[1]

[1]Department, of Civil Engineering, Universitas Syiah Kuala, Jl. Tgk. Syech Abdur Rauf No. 7 Banda Aceh, Aceh 23111, Indonesia

Abstract. Dispersive soils are becoming a common building material. Due to their susceptibility to internal erosion and leakage, dispersive soils should only be used in combination with precise engineering measures to avoid catastrophic failures. Dispersive soils stabilization is critical and has been investigated in several studies conducted throughout the world. Erosion is a significant issue in structures built on sloping contours. As was the case with St. 670+250 Lipat Kajang road in Aceh Singkil. Soil erosion happens as a result of water's dispersion and transport force. Dispersive soil is one of the factors that contribute to an increase in the soil erodibility index. The objective of this research is to develop a method for enhancing the soil's dispersive qualities. In this investigation, specimens were prepared in three different soil mix plans (10%, 20%, and 30%) and then tested using a pinhole. The quantity of erodibility that happens in Sta. 670 + 250 Lipat Kajang - Aceh Singkil Regency is based on the results of testing the soil's dispersive properties because the soil in this area is highly dispersive.

1 Introduction

Soil instability must be considered in road-building plans, even more so if the road is built on a sloped topography. One of the issues that can contribute to slope soil instability is the dispersive nature of the soil [1]. As a tropical country, Indonesia has a high risk of geological disasters and landslides each year, particularly during the rainy season [2] and [3]. The Indonesian government has made measures to mitigate the risk of road damage, including enhancing road geometry, reinforcing slopes, and rehabilitating eroded soil [4] and [5]. However, this appears insufficient to address road slope degradation, as field evidence indicates that cracks and landslides caused by erosion continue to occur on the surface of road slopes [6]. Therefore, soil dispersive contributing to slope erodibility should also be evaluated as a possible cause of road collapse [7].

Dispersive soils are a type of problematic soil frequently encountered in engineering projects throughout several countries and geographical areas [1]. Alluvial soils formed by the deposition of mudrocks in aquatic environments are well-known to be dispersive. Still, soils formed by the weathering of metamorphic and igneous rocks are less dispersive and are typically non-dispersive [8]. In dispersive soils, repulsive forces between clay particles are greater than clay particles attractive forces, resulting in a poor and unstable soil structure. These soils are prone to erosion. Even exposure to a modest flow can readily result in particle separation from the soil structure and erosion of the soil [9].

Due to the high cost of repairing significant earth constructions, substantial effort has been directed to various techniques of soil treatment [5] [10] and [11]. The simplest engineering solution for avoiding difficulties related to dispersive soils is to replace them with another soil [1]. However, this technology has limitations in terms of project economics. In addition, in many circumstances, soil remediation has been regarded as a more practical choice.

Fig. 1. Lipat Kajang Road – Aceh Singkil Regency Destruction

This study aims to ascertain whether or not the soil along the slope of the Lipat Kajang road segment in Aceh Singkil is dispersive (figure 1). Additionally, it is vital to determine the best ways to stabilize or improve the soil's dispersive qualities on road slopes.

* Corresponding author: r.munirwan@unsyiah.ac.id

https://doi.org/10.1051/e3sconf/202234001006

2 Study Literature

Soil erosion is a natural disaster and physical phenomenon that can cause significant harm to both humans and the environment [1]. It severely impacts farmland, civil engineering structures, and cutting subsurface pipelines or cables exposed. Furthermore, soil erosion worsened structural engineering difficulties such as landslides, ground subsidence, and bank failures, culminating in disastrous and permanent losses. Due to soil erosion, the most common structural collapses are piping collapses and slope erosion due to rainfall [12].

From [13], an investigation of several soil indices as indications of excessive soil erodibility in Anambra State, south-eastern Nigeria, examines the topic through the perspective of soil indices. The study discovered soil characteristics by investigating soil samples collected from several study regions and their varying degrees of prediction and contributing to soil erodibility within the work area. The study of soil data indicates that the soils in the survey area are very erodible and susceptible to erosivity at dispersion rates (DR) ranging around 0.429-0.865% assuming that soils with a dispersion rate higher than 0.15% are naturally erodible. In addition, soils with a low organic matter concentration are more prone to erosive soil conditions.

The research from [14] primary purpose was to investigate the erosion tendency of compacted biochar amended soil (BAS) derived from four unique biochars. Biochar was manufactured in-house and combined with soil at 5% and 10% concentration of the weight. Eighty-one pinhole erosion trials were performed in total to determine the rate of erosion of bare soil and BAS at three various compaction stages using the same amount of compaction energy. It was discovered that the erosion rate decreased as the water content increased gradually for BAS, which was ascribed primarily to the particles changing direction from flocculated to scattered along the compaction curve. The application of biochar to soil decreased erosion in the dry condition and increased wet state erosion.

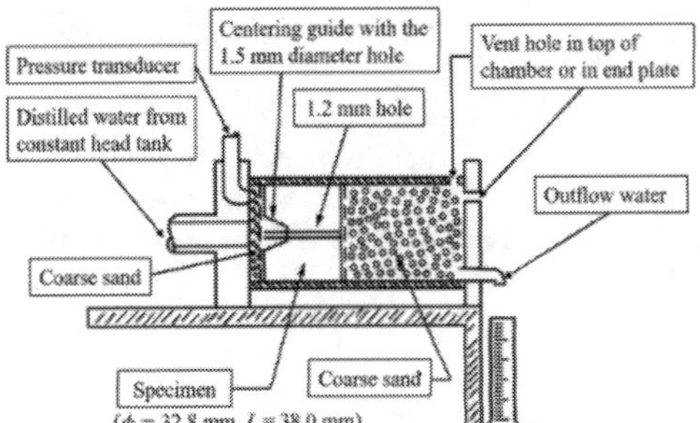

Fig. 2. Diagrammatic representation of the pinhole test equipment [12]

[15] conducted a review of the literature on soil erosion trends, conflicts, and perspectives in the context of South African. Additionally, the feasibility of employing the link between aggregate erodibility and stability as an indicator of soil erosion susceptibility was studied. This connection is frequently utilized in place of time-consuming and costly local soil erosion investigations and models. However, there are conflicting reports on its potential to provide rapid data regarding the soil's vulnerability to erosion. Nevertheless, a review of the South African and international references revealed that aggregate stability is a commonly used physical measure of soil erodibility. Nonetheless, there is no consensus regarding the most appropriate aggregate stability measures to utilize.

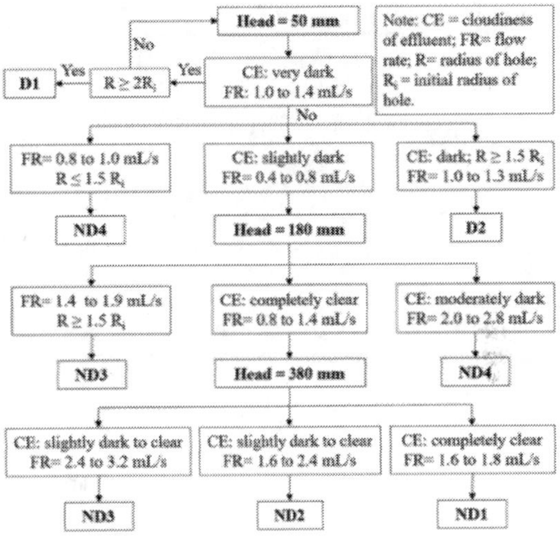

Fig. 3. Procedure for conducting pinhole tests and classification scheme [12]

[16] provides a comprehensive overview of current knowledge and experiences on core soil filtration, especially dispersive soils. The engineering challenges related to dispersive soils are also explored, and fundamental discoveries from past research on protective filters. It is worth noting that the current review considers both traditional, more empirical filter design criteria based on particle sizes and the current, more conceptual state-of-the-art filter specifications. That is based on constriction sizes, with the discussion of their respective pros and contras. This review's knowledge should help investigate, plan, and operate associated geoenvironmental and geotechnical projects.

3 Methods

Numerous studies have been undertaken to determine the erodibility of soils using various instruments and testing methods, including the crumb test, pinhole test, jet erosion test, hole erosion test, flow pump test, and flume erosion test. Among these tests, the pinhole test is widely recognized as the most trustworthy for classifying and identifying soil erodibility. It directly assesses the erodibility and dispersibility of the soil due to the water flowing through a small hole in the middle of the soil sample (figure 2). The advantage of this method is the small sample size required, the ease with which the sample is prepared, and the straightforward testing technique. The procedure for conducting pinhole tests and classification schemes can be seen in figure 3.

Table 1. Mixing presentation of soil specimens

Specimens Code	Lipat Kajang Soil (%)	Gle Geunteng Soil (%)
LK	100	-
LG_{10}	90	10
LG_{20}	80	20
LG_{30}	70	30

This research approach was designed by reviewing the subject matter and literature studies, and other comparable references. The soil sampled for this study is from the slopes of the Lipat Kajang road, which connects Lipat Kajang Sta. 670+250 to Aceh Singkil Regency. Additionally, soil stabilization measures will be made using clay Gle Geunteng Village in the Peukan Bada District of Aceh Besar. This research sampled is disturbed soil. The soil specimens of the mixing presentation are shown in table 1.

Table 2. Result of flow rate and hole size after tests from different water level experiments.

Sample	Diff. Water Level	Time	Flow Rate	Hole size after tests
	(mm)	(second)	(ml/s)	(mm)
LK	50	5	1.20	4.0 – 4.5
	50	5	1.36	3.5 – 4.0
	50	5	1.50	3.5 – 4.0
LG_{10}	50	10	1.13	1.5 – 2.0
	50	10	1.11	1.5 – 2.0
	50	10	1.08	2.0 – 2.5
LG_{20}	50	10	0.87	1.0 – 1.5
	50	10	0.90	1.0 – 1.5
	50	10	0.92	1.5 – 2.0
LG_{30}	180	5	1.50	2.5 – 3.0
	180	5	1.40	3.5 – 4.0
	180	5	1.43	3.5 – 4.0

The purpose of this research is to conduct geotechnical investigations on the erodibility of soils, utilizing a pinhole test to determine soil dispersive qualities. The test object, specifically the original soil combined with varying percentages of clay, will be modeled. The specimens were made according to the various clay addition plans (10%, 20%, 30%) and then molded into test specimens with a diameter (d) of 32.5 mm and a length (L) of 37.5 mm.

Testing with pinhole instruments at 50 mm, 180 mm, 380 mm, and 1020 mm water height differences. Results of flow rate and hole size after tests from different water level experiments can be seen in table 2. The investigation was conducted in ascending order. The results collected during the laboratory identification of the dispersive qualities of soil using pinholes will be classed as D1 (highly dispersive), D2 (dispersive), ND4 (possibly dispersive), ND3 (possibly dispersive), ND2 (not dispersive), and ND1 (not dispersive). After determining the soil's dispersive qualities on the natural soil, specimens were modeled by combining the soil sample (Lipat Kajang + Glee Geunteng) with the specified percentage of the mixture.

4 Results

Soils with high clay content are generally more resistant to erosion than soils with low clay content. AASHTO classifies Lipat Kajang soil as group A-4(2), sandy, loamy silt, yellowish gray, whereas Glee Geunteng soil is classified as group A-7-6(31), reddish-brown silty clay. Based on the soil physical properties, the Lipat Kajang soil is more likely to be dispersive when compared to the Glee Geunteng soil. This is because the Lipat Kajang soil contains less clay than the Glee Geunteng soil. Therefore, for initial identification, the Lipat Kajang soil is identified as dispersive soil. Meanwhile, Glee Geunteng is categorized as non-dispersive soils.

The findings of the pinhole test used to determine the soil's dispersive qualities are classified based on the difference in water level test time, flow rate, water turbidity, and hole diameter. The dispersive nature of the soil is one of the variables that contribute to the increase in the soil erodibility index. This is because dispersive soil is quickly scattered, particularly during the wet season.

Field evidence indicates that on the slopes of the Lipat Kajang road in Aceh Singkil Regency, small holes run with cloudy and hazy water caused by distributed soil particles. If it occurs on a large scale, it can result in a landslide, which is extremely dangerous for road users. Landslides on the slopes of the Lipat Kajang road in Aceh Singkil Regency are generally caused by the soil's incapacity to endure the destructive force of water, as the soil in the area is classified as type D1, a very dispersed soil classification. The result of dispersive soil classification can be seen in table 3.

5 Conclusions

Based on the findings of the tests, it can be stated that the lower the clay content of the soil, the greater the likelihood that the soil will be dispersive in nature. In Aceh Singkil Regency, a landslide occurred on the slopes of the Lipat Kajang road Sta. 670+250, which occurred because the soil is very dispersive. Based on the findings of the pinhole test, it has been classified as D1. A mixture of 30 percent Glee Geunteng clay can improve the soil dispersive qualities on the slopes of the Lipat Kajang Sta. 670+250 Aceh Singkil District, bringing them from category D1 to ND3 which is not dispersive soil.

Table 3. The classification result of dispersive soil

Sample	End of test water visual		Classification
	side	top	
LK	Very Cloudy	Cloudy	D1
	Very Cloudy	Cloudy	D1
	Very Cloudy	Cloudy	D1
LG_{10}	Cloudy	Cloudy	D2
	Cloudy	Cloudy	D2
	Cloudy	Cloudy	D2
LG_{20}	Medium Cloudy	Cloudy	ND4
	Medium Cloudy	Cloudy	ND4
	Medium Cloudy	Cloudy	ND4
LG_{30}	Slightly Cloudy	Medium Cloudy	ND3
	Medium Cloudy	Medium Cloudy	ND3
	Medium Cloudy	Medium Cloudy	ND3

References

1. A. Masoodi, M. R. M. Tabatabai, A. Noorzad, and A. Samadi, J. Hydrol. Eng. **24**, (2019)
2. Munirwansyah, M. Irsyam, R. P. Munirwan, H. Yunita, and M. Zulfan Usrina, IOP Conf. Ser. Mater. Sci. Eng. **352**, (2018)
3. M. Sungkar, M. Munirwansyah, R. P. Munirwan, and D. Safrina, IOP Conf. Ser. Mater. Sci. Eng. (2020)
4. M. Munirwansyah, R. P. Munirwan, and H. Yunita, Int. J. Adv. Sci. Eng. Inf. Technol. **8**, 870 (2018)
5. R. P. Munirwan, M. Munirwansyah, M. Marwan, P. J. Ramadhansyah, and V. Kamchoom, IOP Conf. Ser. Mater. Sci. Eng. (2020)
6. B. Bin Xu and W. Si, Matec Web Conf. (2016)
7. A. Maharaj, L. Van Rooy, and P. Paige-Green, J. South African Inst. Civ. Eng. **57**, 31 (2015)
8. S. Moravej, G. Habibagahi, E. Nikooee, and A. Niazi, Geoderma **315**, 130 (2018)
9. A. Bernatek-Jakiel and J. Poesen, Earth-Science Rev. **185**, 1107 (2018)
10. D. Sundary, R. P. Munirwan, and N. Al-Huda, J. Phys. Conf. Ser. (2021)
11. M. Munirwansyah, R. P. Munirwan, M. Sungkar, and F. Fachrurrazi, AIP Conf. Proc. **020029**, (2019)
12. B. H. Dinh, A. D. Nguyen, S. Y. Jang, and Y. S. Kim, Int. J. Geo-Engineering **12**, (2021)
13. R. U. Ayadiuno and D. C. Ndulue, Int. J. Mod. Agric. **10**, 3451 (2021)
14. H. Kumar, S. Prakash, S. Bordoloi, S. Sreedeep, P. Lin, G. Mei, A. Garg, and A. K. Sarmah, Sci. Total Environ. **672**, 698 (2019)
15. A. D. Nciizah and I. I. C. Wakindiki, Arch. Agron. Soil Sci. 37 (2014)
16. A. H. Vakili, M. R. Selamat, P. Mohajeri, and H. Moayedi, Geotech. Geol. Eng. (2018)

The synergy between government institutions in clean water preparedness for earthquake and tsunami disaster management in Banda Aceh City

M. Arifin[1*], *M.* Dirhamsyah[2], and *Nizamuddin* [3]

[1] Departement of Disaster Science, Universitas Syiah Kuala, Banda Aceh, 23111, Indonesia
[2] Departement of Mechanical Engineering, Faculty of Engineering, Universitas Syiah Kuala, Banda Aceh City 23111, Indonesia
[3] Departement of Information Science, Faculty of Math and Science, Universitas Syiah Kuala, Banda Aceh City 23111, Indonesia

Abstract. Synergy is coordination and cooperation between groups of people in order to achieve an efficient goal. An example of a synergy is the synergy between government agencies to improve access to clean water in relation to disaster management. This study aims to describe the synergy between Indonesia's government agencies to improve access to clean water in relation to disaster management, determine the factors that affect the synergy, and recommend a synergy model. The study was conducted through a qualitative approach. Data were collected through in-depth interview with the heads of government agencies that had a duty to improve access to clean water, observation, and literature review. The data analysis technique was triangulation technique. The results of this study was that the synergy between the government agencies had not run well due to several inhibiting factors that were inefficient planning, overlapping work division, and no evaluation.

1 Introduction

There are so many catastrophic events that occur in Indonesia each year, and the number is increasing in the last few years. The number of catastrophic events in 2016, 2017, 2018 was 2,306, 2,391 and 2,572 respectively. From 2007 to 2018, earthquakes and tsunamis are two disasters that caused the highest number of casualties [1]. Natural disasters such as floods, volcanic eruptions, earthquakes, and tsunamis cause damage to the communication system, electricity supply, water supply, and in this crisis situation it is difficult to provide emergency response services [2].

Disasters can cause damage to the clean water distribution network system affecting the supply of clean water for millions of people. Like the earthquake that occurred in Northridge California in 1994 which resulted in 500,000 people losing clean water services [3]. For example, the earthquake that occurred in Northridge California in 1994 caused 500,000 people losing access to clean water. The earthquake also caused a huge fire due to the failure of clean water distribution by the fire department [4].

On March 11, 2011 at 2:46 p.m. local time, Japan was rocked by an earthquake with a magnitude of 9.0 on the Richter Scale and caused a tsunami on the East coast of Japan, which disrupted water supply for 2,300,000 households from Tohoku to Kanto due to damage to the clean water distribution network [5]. Meanwhile, the earthquake that occurred on April 14, 2016 in the vicinity of Kumamoto City with a magnitude of 6.4 on the Richter Scale caused 85,000 households to not get clean water due to the cessation of clean water distribution due to 69 of the 96 water sources in Kumamoto City experiencing turbidity that cannot be controlled received [6].

The tsunami in Banda Aceh in 2004 caused almost all river water to be processed before used, including the distribution pipes and well water in Lampaseh area. Moreover, it was found a high population of Coliform bacteria in the Krueng Raya Muara River as many as 3100 colonies / 100 ml. Only one of the two water treatment plants in Banda Aceh is still operating, about 75% of the distribution pipe network is damaged, and river water is also damaged. Also states in general the negative impacts of the earthquake and tsunami disaster on water infrastructures were that (1) there was damage to the intake towers, the water treatment plants, and distribution pipes, (2) the damage occured in areas near the shoreline, (3) the level of damage was up to 90%, and (4) there was a problem in operational costs [7].

After the tsunami in Banda Aceh City occurred, the people there had difficulty accessing clean water. Some people in Gano Village, Syiah Kuala District, Banda Aceh City buy bottled drinking water/gallon water which is expensive, the cost is about 20% or 25% of their income, the water is used for bathing and washing [8]. The disaster caused power failures so that the tap did not turn on. Some people had access to well water, but it was very limited to 3 or 5 households because the wells were located in coastal areas. To use the wells, it was not free. The owner charged a price [8]. More than 60% of Banda Aceh City was destroyed. Some parts of the land disappeared, and the water infrastructures water were damaged so that only 20% of clean water was available after the tsunami. Groundwater that was contaminated by sea water was a big problem because it damaged the infrastructure of the Regional Drinking Water Company of Banda Aceh City[9].

In Lombok on August 5 2018 at 19.46 local time measuring 6.9 on the Richter Scale an earthquake occurred where the epicenter was 27 km northeast of

* Corresponding author: arifin90@mhs.unsyiah.ac.id

North Lombok with a depth of 10 km [10]. According to the National Disaster Management Agency, the earthquake caused 506 people to die, 1,406 people injured, and 430,946 people to relocate. It also caused damage to the infrastructures, including water infrastructures [5].

Similar to the earthquake and tsunami that occurred in Palu, the earthquake also caused the people difficulty accessing clean water. Based on the interview with Darwis who was a UNICEF Indonesia representative in Palu and worked to improve access to Water, Sanitation, and Hygiene (WASH) on February 9, 2019, at 16.00 Western Indonesian Time (WIB), the clean water supply was very limited. In places of refuge, there was no clean water managed by the Regional Drinking Water Company and the boreholes owned by the society were damaged. Most residents whose houses were affected by the disaster in Palu City would choose to stay in their relatives' houses if the houses were safer because in the places of refuge the facilities were inadequate such as lack of access to clean water and a limited number of public toilets [11]. The earthquake that occurred in Palu measuring 7,5 on the Richter scale caused the 270-meter-long water pipes meters from the total length of 28,079 that brought water to Tanamodindi Village to break and leak. It caused the society of Tanamodindi Village to get difficulty accessing clean water [12].

The earthquake in Lombok in 2019 also caused damage to the irrigation networks by 45% or 97 units, roads by 28% or 61 units, drinking water supply networks and wastewater treatment installations by 15% or 32 units, and the damage caused the society in Lombok get difficulty to access clean water [13]. Banda Aceh is a disaster-prone region especially earthquakes and tsunamis because Banda Aceh is bordered by Samudra Hindia where the Eurasian plate and the Indo-Australian plate meet. Banda Aceh is also prone to submarine earthquakes and inland earthquakes because there is the Great Sumatran Fault adjacent to Banda Aceh City [14].

The government of Banda Aceh City has an important role in disaster management. In accordance with the mandate of the Constitution Number 24 of 2007 concerning Disaster Management, the government of Banda Aceh City needs to mitigate disasters so that obstacles or problems that occur such as in 2004 can be overcome. Disaster preparedness is needed because the tsunami occurred in Aceh has a recurrence period that can occur once every 600 years [15]. Based on the prediction of the researchers from the Tsunami Disaster Mitigation Research Center (TDMRC), there will be an earthquake measuring 7.0 on the Richter scale which has the potential to cause a tsunami [16]. To deal with the possibility of a tsunami, the government of Banda Aceh City should prepare a system that enables disaster victims to access clean water. This is part of disaster mitigation which if performed well, it can help to improve access to clean water in post-disaster period.

The Regulation of the Head of Indonesian National Board for Disaster Management Number 4 of 2008 concerning the Guidelines for Preparing Disaster Management Plans states that the government can carry out disaster management planning in the form of disaster preparedness carried out by the Regional Disaster Management Agency in coordination with various agencies in pre-disaster, during disaster, and post-disaster period. The Indonesian National Board for Disaster Management has also issued the Guidelines regarding Clean Water, Sanitation and Hygiene Assistance in Disaster Emergencies as the basis or reference for cross-sector related disaster management processes in meeting the needs of clean water for disaster victims.

Based on the data collected by the researchers through reading the constitutions, the mayor regulations, research journals, and having discussions with the research supervisors, government agencies that are responsible for improving access to clean water in Banda Aceh City were the Regional Drinking Water Company of Tirta Daroy of Banda Aceh City, the Department of Public Works and Spatial Planning of Banda Aceh City, the Development Planning Agency at Sub-National Level of Banda Aceh City, and the Public Health Office of Banda Aceh City, and the Regional Disaster Management Agency of Banda Aceh City.

Moreover, the Regional Drinking Water Company of Tirta Daroy of Banda Aceh City, the Department of Public Works and Spatial Planning of Banda Aceh City, and the Development Planning Agency at Sub-National Level of Banda Aceh City coordinate the work of the procurement and management of clean water infrastructures in Banda Aceh City. The Regional Disaster Management Agency of Banda Aceh City is responsible for coordinating the implementation of disaster management activities with other institutions related human resources and logistics. On the other hand, the Public Health Office of Banda Aceh City is responsible for maintaining the quality of clean water in Banda Aceh.

In conclusion, each institution has its own role. This is what is important in this study, to ask several questions, namely: How is the synergy between government agencies in preparing clean water for earthquake and tsunami disaster management in Banda Aceh City? clean water for earthquake and tsunami disaster management in Banda Aceh City?, and How to create a model of synergy between government agencies in disaster management efforts, one of which is the logistics of clean water supply in Banda Aceh City?. So that this study can describe in more detail how the synergy between the government agencies to improve access to clean water in relation to earthquake and tsunami disaster management in Banda Aceh City.

2 Research Method

This study used qualitative approach. The researchers collected data through interview, observation, and literature review. The interview was conducted with five government agencies responsible for improving access to clean water in Banda Aceh City that were the head of the Sub-Division of Public Works, Spatial Planning and Defense of the Development Planning Agency at Sub-National Level of Banda Aceh City, the head of the Emergency and Logistics Division of the Regional

Disaster Management Agency of Banda Aceh City, the head of the Environmental Sanitation and Water Supply Division of the Department of Public Works and Spatial Planning of Banda Aceh City, the head of Planning and Technical Development of the Regional Drinking Water Company of Tirta Daroy of Banda Aceh City, and the executor of the Food and Beverage Processing Site Program of the Public Health Office of Banda Aceh City.

The researchers conducted an observation at the government agencies that were responsible for improving access to clean water. The researchers observed how the Regional Drinking Water Company of Tirta Daroy of Banda Aceh City processed, produced, and distributed clean water. The researchers also visited the logistics warehouse of the Regional Disaster Management Agency of Banda Aceh City. The researchers collected data from research journals, books, magazines published by the government agencies, the constitutions, and other supporting documents related to clean water supply.

After collecting data, the researcher conducted data reduction that was the activities of simplying, classifying, and discarding unrelevant data. The data was classified into 3 categories that were the roles of each government agency, the synergy pattern between the government agencies, the factors that affected the synergy between the government agencies. The criteria to describe the supporting and inhibiting factors were communication, awareness of the importance of coordination, participant competence, agreement, commitment, incentives, continuity of planning [17], evaluation [18], work division [19], responsibility, mutual contribution, and effort [20].

The next stage was that the researchers presented the data about the roles of each government agency, the synergy pattern between the government agencies, the factors that affected the synergy between the government agencies. The researchers also created a synergy model between the government agencies to improve access to clean water. The final stage was that the researchers drew a conclusion of how the synergy between the government agencies to improve access to clean water, the synergy pattern between the government agencies, and the factors that affected the synergy between the government agencies. The researchers created a synergy model between the government agencies to improve access to clean water based on logistics management system and disaster management equipment from the National Disaster Management Agency.

3 Results and discussion

3.1 The government agencies as executors of the policy on improving access to clean water

Based on the collected data, not all of the responsible government agencies executed the policy on improving access to clean water. The Development Planning Agency at Sub-National Level of Banda Aceh City, the Department of Public Works and Spatial Planning of Banda Aceh City, and the Regional Drinking Water Company of Tirta Daroy of Banda Aceh City were the government agencies as executors of the policy on improving access to clean water. On the other hand, the Regional Disaster Management Agency of Banda Aceh City and the Public Health Office of Banda Aceh City were related agencies, but they did not take part as executors of the policy on improving access to clean water. This is illustrated in the figure 1.

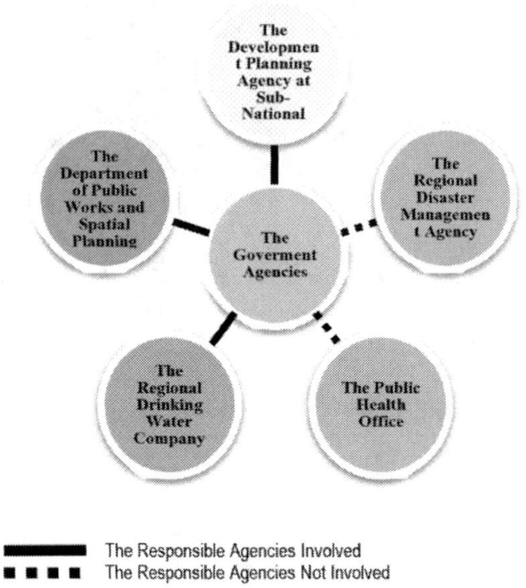

Fig. 1. The government agencies as executors of the policy on improving access to clean water.

3.2 The synergy between the executors

Based on the collected data, here are the roles of each executor. The Development Planning Agency at Sub-National Level of Banda Aceh City played a role in funding and budgeting for the project, the Department of Public Works and Spatial Planning of Banda Aceh City played a role in building the Drinking Water Supply System, and the Regional Drinking Water Company of Tirta Daroy of Banda Aceh City played a role in managing the clean water supply in Banda Aceh City. This synergy is illustrated in the figure 2.

The Public Health Office of Banda Aceh City played a role in monitoring the quality of clean water. The Public Health Office of Banda Aceh City did not inform the results of the work to the Regional Drinking Water Company of Tirta Daroy of Banda Aceh City. On the other hand, the Regional Drinking Water Company of Tirta Daroy of Banda Aceh City was the agency responsible for managing clean water supply in Banda Aceh City. The Public Health Office of Banda Aceh City should have submitted the report on monitoring the quality of clean water.

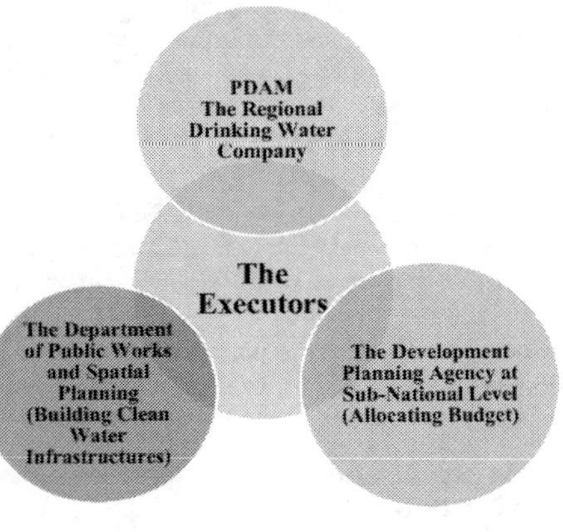

Fig. 2. The synergy between the executors.

The Public Health Office of Banda Aceh City carried out its duty by only involving the Community Health Center in each sub-district in Banda Aceh City. The report on monitoring the quality of clean water published on the Health Profile Book of Banda Aceh City and was not submitted to the Regional Drinking Water Company of Tirta Daroy of Banda Aceh City. The Regional Disaster Management Agency of Banda Aceh City responsible for coordinating the implementation of disaster management activities with other institutions had not yet carried out integrated disaster management because in the strategic plan of the Regional Disaster Management Agency of Banda Aceh City for 2018-2020 there was no clean water plan in earthquake and tsunami disaster management.

3.3 Factors affecting the synergy between the government agencies

3.3.1 The supporting factors

These supporting factors were concluded based on the analysis of the criteria mentioned above. In the aspect of communication, the Development Planning Agency at Sub-National Level of Banda Aceh City, the Department of Public Works and Spatial Planning of Banda Aceh City, and the Regional Drinking Water Company of Tirta Daroy of Banda Aceh City established an effective communication. For example, the Development Planning Agency at Sub-National Level of Banda Aceh City held a development plan meeting with the Regional Work Units of Banda Aceh City.

The results of the meeting became the guidelines for the Development Planning Agency at Sub-National Level of Banda Aceh City in carrying out its duties one of which was the construction of Drinking Water Supply System buildings. This is in accordance with what is stated by (Hasibuan & Hasibuan, 2016) that in communication science there is a systematic effort to firmly formulate the principles, and on these principles information is conveyed in the form of opinions and attitudes.

In the aspect of awareness of the importance of coordination, the Development Planning Agency at Sub-National Level of Banda Aceh City, the Department of Public Works and Spatial Planning of Banda Aceh City, and the Regional Drinking Water Company of Tirta Daroy of Banda Aceh City had awareness of the importance of coordination. This is concluded based on the interview with the head of the three institutions.

Each institution stated the importance of coordination, especially in the construction of Drinking Water Supply System buildings in Banda Aceh City. This is in accordance with the opinion expressed by Hasibuan and Hasibuan (2016) regarding the factors that affect coordination that are the awareness of each institution with other institutions to adjust theirselves to one another or their duties with other institutions so that these institutions do not work independently.

In the aspect of participant competency, each agency involved people who were competent in the construction of Drinking Water Supply System buildings. In the aspect of agreement, commitment, and incentives, the Development Planning Agency at Sub-National Level of Banda Aceh City, the Department of Public Works and Spatial Planning of Banda Aceh City, and the Regional Drinking Water Company of Tirta Daroy of Banda Aceh City kept to the agreement and showed a commitment. Also, everyone involved in the construction of Drinking Water Supply System buildings got incentives.

3.3.2 The inhibiting factors

These inhibiting factors were concluded based on the analysis of the criteria mentioned above. In the aspect of continuity planning, the program was not well planned because the progeam was based on a proposal from the Regional Drinking Water Company of Tirta Daroy of Banda Aceh City which was accommodated by the Department of Public Works and Spatial Planning of Banda Aceh City which then the proposal was submited to the Development Planning Agency at Sub-National Level of Banda Aceh City. The Department of Public Works and Spatial Planning of Banda Aceh City was the one who responsible to do the planning and assessment of the construction of Drinking Water Supply System buildings, and The Department of Public Works and Spatial Planning of Banda Aceh City should not only accept the proposal from the Regional Drinking Water Company of Banda Aceh City.

The Department of Public Works and Spatial Planning should do checking in the field to ensure that the proposal of the Regional Drinking Water Company of Tirta Daroy was suitable or not. In the aspect of work division, it was unclear due to the overlapping tasks between the Department of Public Works and Spatial Planning of Banda Aceh City and the Regional Drinking Water Company of Tirta Daroy of Banda Aceh City regarding the pipeline network planning of the Regional Drinking Water Company.

The Department of Public Works and Spatial Planning of Banda Aceh City was the one who should carry out the planning of the construction of the Drinking Water Supply System buildings. However, the task was carried out by the Regional Drinking Water Company of Tirta Daroy with a Detail Engineering Design (DED). So, it is clear that the Department of Public Works and Spatial Planning of Banda Aceh City did not carry out their duties properly.

In the aspect of evaluation, there was no evaluation of the agreement. Evaluation is very important to form an opinion of the quality of the program. From the evaluation, it can be seen which programs have been carried out and which programs have not been completed. This conclusion was obtained based on the interview in the field. The Department of Public Works and Spatial Planning of Banda Aceh City, the executor of the construction of the Drinking Water Supply System buildings, did not have the data such as how long the pipes had been built. In fact, this is the basic data as the guidelines for the next Drinking Water Supply System development. It can be concluded that the Department of Public Works and Spatial Planning of Banda Aceh City did not evaluate its work.

This indicated that the data on the construction of the Drinking Water Supply System buildings was unclear which then caused unclear construction planning of Drinking Water Supply System building in Banda Aceh City. The construction of Drinking Water Supply System buildings usually ruined the other infrastructures such as roads. Such problem occurred because the Department of Public Works and Spatial Planning of Banda Aceh City, the executor of the construction of Drinking Water Supply System buildings, did not have valid data on the construction of Drinking Water Supply System buildings. The Department of Public Works and Spatial Planning of Banda Aceh City also built roads, bridges, drainage, and irrigation which were not synchronous between each sector. This can be concluded that in the field there was often damage to pipelines on the basis of other infrastructure development.

The inhibiting factors described above are in accordance with what is stated by Soewarno (2006) that the unclear formulation of duties, authorities, and responsibilities of each institution, the work procedure are not understood by the parties who sometimes raise doubts between them.

3.4 The synergy model of clean water access improvement

To realize the synergy between the government agencies to improve access to clean water in Banda Aceh City, it is necessary to create a synergy model so that the Development Planning Agency at Sub-National Level of Banda Aceh City, the Department of Public Works and Spatial Planning of Banda Aceh City, the Regional Drinking Water Company of Tirta Daroy of Banda Aceh City, the Regional Disaster Management Agency of Banda Aceh City, and the Public Health Office of Banda Aceh City can be together and involved in clean water access improvement.

This synergy model was created based on several findings in the field and based on the supporting and inhibiting factors affecting the synergy of the government agencies.

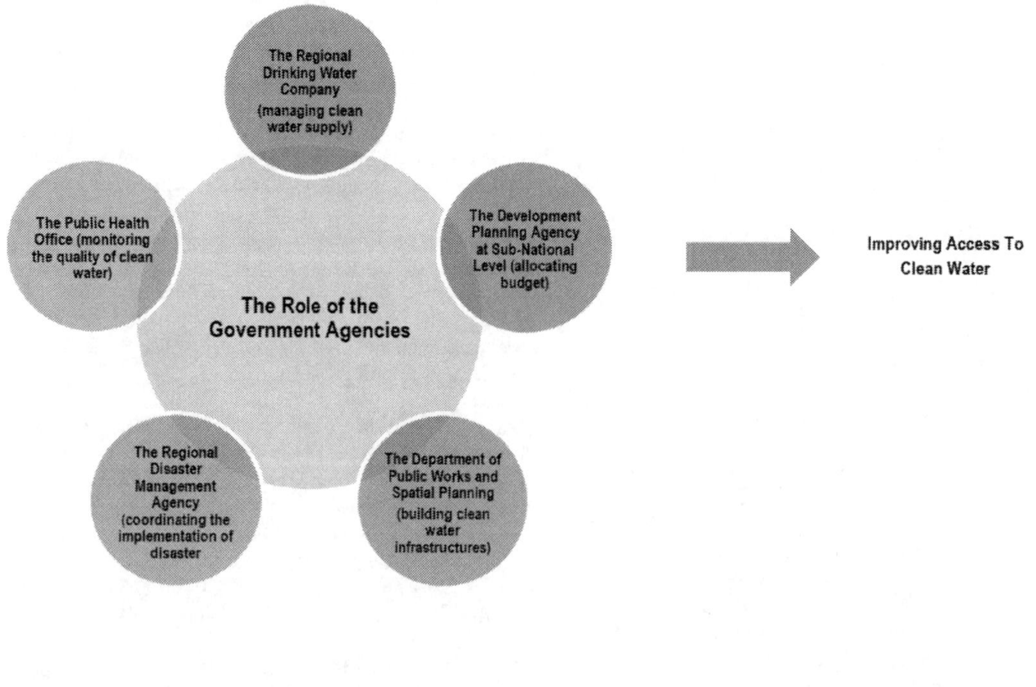

Fig. 3. The synergy model of clean water access improvement

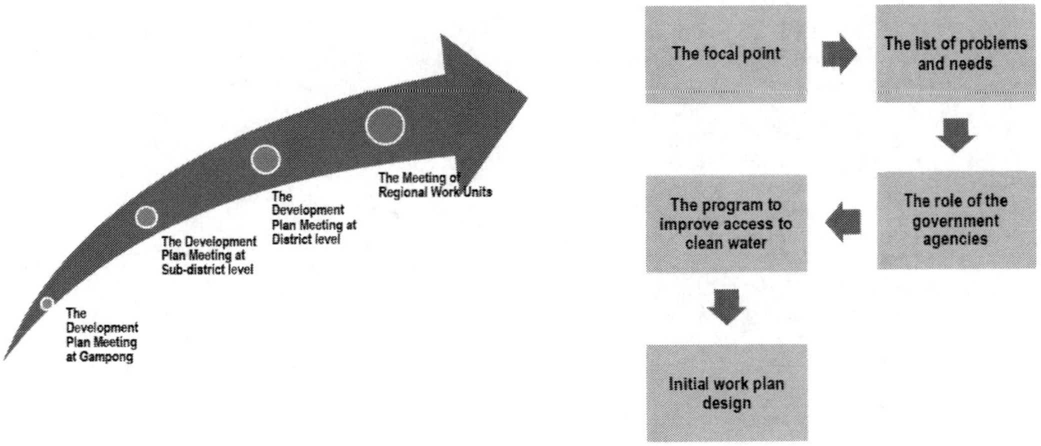

Fig. 4. Clean water access improvement planning into the regional medium term development plan

The focal point is the Regional Disaster Management Agency of Banda Aceh City, the agency responsible for disaster management. The Regional Disaster Management Agency of Banda Aceh City should coordinate the program of improving access to clean water with other agencies. The Regional Disaster Management Agency of Banda Aceh City must be the one who designs the policy, and it also must be more active in carrying out its duties. The Regional Disaster Management Agency of Banda Aceh City must conduct meetings with the Department of Public Works and Spatial Planning of Banda Aceh City, the Regional Drinking Water Company of Tirta Daroy of Banda Aceh City, and the Public Health Office of Banda Aceh City. The data obtained from the meetings should be written on a document that lists problems and needs related to the program of improving access to clean water.

The list of problems and needs will provide information on the potential number of disaster victims, how many liters of clean water is needed, how many water storage tanks are needed. After an earthquake and tsunami, the building of Drinking Water Supply System is ruined, so raw water cannot be processed into clean water. As a result, it is necessary to provide cars to distribute clean water and equipment to distill water.

Each agency has its role in improving access to clean water. The Regional Disaster Management Agency of Banda Aceh City has a role to assess the number of disaster victims, how much clean water is needed, how many water storage tanks are needed, how many water tank cars are needed, and how many distillers are needed. The Department of Public Works and Spatial Planning of Banda Aceh City has a role to provide needs from the data of the Regional Disaster Management Agency of Banda Aceh City and repair the building of of Drinking Water Supply System. The Regional Drinking Water Company of Tirta Daroy of Banda Aceh City has a role to ensure the availability of water in water storage tanks and equipment for distilling water, distribute water, and ensure that the clean water supply is enough. The Public Health Office of Banda Aceh City has a role to monitor water quality.

4 Conclusion

Not all agencies related to clean water access improvement are involved in implementing the policy. Therefore, it is necessary to create a synergy model between government agencies in clean water access improvement, and the synergy model is included in the Disaster Management Plan implemented by the Regional Disaster Management Agency of Banda Aceh City together with the Department of Public Works and Spatial Planning of Banda Aceh City, the Regional Drinking Water Company of Tirta Daroy of Banda Aceh City, and the Public Health Office of Banda Aceh City. Furthermore, the Disaster Management Plan is included in the Strategic Plan of each government agency involved. Furthermore, the Disaster Management Plan is recommended to the Development Planning Agency at Sub-National Level of Banda Aceh City to be included in the Regional Medium Term Development Plan.

References

1. BNPB, Kepala Pus. Data Inf. Dan Humas BNPB (2018)
2. H. A. Eiselt and V. Marianov, Eur. J. Oper. Res. **216**, 563 (2012)
3. L. McReynolds and R. L. Simmons, Journal-American Water Work. Assoc. **87**, 65 (1995)
4. C. R. Scawthorn, Earthq. Spectra **27**, 419 (2011)
5. A. R. Hidayat, R. Triatmadja, and I. Supraba, IOP Conf. Ser. Earth Environ. Sci. **426**, (2020)
6. B. P. Wham, S. Dashti, K. Franke, R. Kayen, and N. K. Oettle, Lowl. Technol. Int. **19**, 151 (2017)
7. T. W. Ignasius D.A. Sutapa, LIMNOTEK Perair. Darat Trop. Di Indones. **21**, 135 (2014)

https://doi.org/10.1051/e3sconf/202234001007

8. J. Song, M. Han, T. Kim, and J. Song, Desalination **248**, 233 (2009)
9. A. Rahman, S. Iwasaki, S. A. Sutton, A. Sakurai, and Parmakope, *Urban Water Management Issues and Challenges After the 2004 Indian Ocean Tsunami Recovery: Lessons Learned from Banda Aceh City, Indonesia* (Springer Singapore, 2019)
10. A. Ganas, V. Tsironi, and S. Valkaniotis, (2018)
11. Y. Darma, B. Sulistyantara, and Yonvitner, IOP Conf. Ser. Earth Environ. Sci. **501**, (2020)
12. T. Anasiru, S. Tahir, and Anawardiayah, Siimo Eng. **4**, 1 (2020)
13. M. S. Yadnya, A. Tahmid, A. Zainuddin, B. Kanata, T. Zubaidah, and P. Paniran, J. Gema Ngabdi **2**, 63 (2020)
14. Gusmeri, Fadhil, and R. S. Oktari, *Strategi Mitigasi Bencana Tsunami Dan Banjir Rob Yang Diperparah Oleh Kenaikan Muka Air Laut Akibat Perubahan Iklim Di Kota Banda Aceh* (2019)
15. D. H. Natawidjaja, J. Ris. Geol. Dan Pertamb. **25**, 49 (2015)
16. I. Rusydy, U. Muksin, Mulkal, Y. Idris, M. N. Akram, and Syamsidik, in *AIP Conf. Proc.* (AIP Publishing LLC, 2018), p. 20012
17. H. Soewarno, Penerbit PT. Toko Gunung Agung, Jakarta (2006)
18. Manullang, *Dasar - Dasar Manajemen* (Ghalia Indonesia, Yogyakarta, 2008)
19. M. S. P. Hasibuan and H. M. S. P. Hasibuan, *Manajemen Sumber Daya Manusia* (Bumi Aksara, 2016)
20. M. West, Cetakan Kelima. Penerjemah: Srikandi Waluyo. Yogyakarta: Penerbit Kanisius (2002)

An update on reservoir operating rules following changes on the effective capacity due to land erosion

Azmeri Azmeri[1*], *Devi* Sundary[1], *Cut* Mutiawati[1], *Hafnidar* A. Rani[2]

[1]Civil Engineering Department, Universitas Syiah Kuala, Abdur-Rauf street No. 7 Darussalam Banda Aceh, 23123, Indonesia
[2]Civil Engineering Department, Universitas Muhammadiyah Aceh, Muhammadiyah street No. 91 Batoh Banda Aceh 23245, Indonesia

Abstract. Reservoirs play a role in decreasing the temporal variability of river flow by regulating the reservoir water release for several downstream water needs. The Keuliling Reservoir was built to meet the needs of irrigation water. However, the dry season in Aceh Besar Regency significantly impacts irrigation areas because water from the Keuliling Reservoir cannot meet the demands. During the dry season, which is also the beginning of the second rice planting season (for gadu rice variety), the water source of the Keuliling Reservoir significantly decreases. The rate of land erosion leads to sedimentation, which may cause a reduction in the effective reservoir capacity. A decline in reservoir function due to sedimentation can affect the reservoir's performance. The reservoir operating rules need to be adjusted due to a decline in capacity caused by sedimentation. This study aims to develop updated reservoir operating rules that can be used as an operational guideline for the Keuliling Reservoir. The study used a non-linear program to optimize the reservoir operation, which required objective functions and constraints. Optimization will result in reservoir operations that can supply water according to the target rice field area in normal and wet years. The irrigation area is 578 hectares for Group I (Keuliling Upstream Irrigation Area) and 1053 ha for Group I (Keuliling Downstream Irrigation Area). However, in the dry year of Group I (DI Keuliling Hulu), with an irrigation area of 578 hectares, the rice planting pattern (one rice planting season) begins with the planting season in October I. In addition, in Group II (DI Keuliling Hilir), with an irrigation area of 781 ha, the rice planting pattern (one rice planting season) starts the planting season in October II. The reduced inflows during dry years have decreased the target irrigation area by 113 ha. This Keuliling Reservoir operating manual can determine the reservoir water release to balance it with needs and storage targets.

1 Introduction

Water is an essential resource for humanity but at times becomes scarce and requires a struggle to obtain, especially in dry areas [1]. Water is primarily used for consumption; thus, its availability needs to be well-planned to prevent conflicts among water users.

Reservoirs play a critical role in ensuring the availability of water and meeting various water demands. Reservoirs can reduce the temporal variability of flows by adjusting the release. However, water release controlled from reservoirs has a downstream detrimental effect if it neglects the ecological river flows [2].

Decision-makers often face difficulties in selecting the appropriate system for reservoir water management. The decisions taken can be controversial because of different water users of the implementation of the alternatives provided [3]. Therefore, optimization of water resource management is carried out to evaluate various modifications of reservoir operations. An optimum reservoir operating system has a vital role in meeting downstream water demands [4]. Improvements to the reservoir operating system are also required if variations in hydrological conditions, such as climate change and reservoir storage capacity alterations due to sedimentation, occur [5].

The dry season in Aceh Besar Regency significantly impacts irrigation areas because water from the Keuliling Reservoir cannot meet the demands. During the dry season, which is also the beginning of the second rice planting season (for gadu rice variety), the water source of the Keuliling Reservoir significantly decreases. The rate of land erosion leads to sedimentation, which may cause a reduction in the effective reservoir capacity. A decline in reservoir function due to sedimentation can affect the reservoir's performance.

Previous studies suggest that reservoir operations need to be updated due to sedimentation problems [6, 7]. The adjustments are required to control sedimentations and operations and, thus, produce a good performance. The reservoir operating rules need to be adjusted due to a decline in capacity caused by sedimentation. This study aims to develop updated reservoir operating rules that can be used as an operational guideline for the Keuliling Reservoir.

*Corresponding author: azmeri@unsyiah.ac.id

2 Reservoir operation formula

The development of the reservoir operation model is done by identifying reservoir inflow and outflow. The reservoir inflow is a long-term time-series data, while the reservoir outflow is the reservoir evaporation in the forms of irrigation needs, ecological river flows, spillway discharge [8].

The reservoir operation modeling formulation is the conservation of water balance for time intervals, as the following formula:

$$V_t - V_{t-1} = \Delta V = I_t - R_t - E_t - L_t \tag{1}$$

where:
V_{t-1} : The water volume stored in the t-1 time (m^3);
V_t : The water volume stored in the t time (m^3);
I_t : The *inflow* discharge in t time (m^3/s in t time (m^3/s);
Rt : The release of water from the reservoir (m^3/s);
E_t : The evaporation in t time (m^3); and
L_t : The other water loss in t time (m^3).

Bathymetri process at Waduk Keuliling is carried out at Waduk Keuliling Area. The depth measurement of the Keuliling reservoir was performed using the ultrasonic wave reflection method which is perpendicularly emitted to the bottom of the water surface using the transducer (an echo sounder tool) and then the wave will be reflected back to the transducer [9].

3 Modeling of Keuliling reservoir operation optimization

The following are the parameters used in the stages of modeling the operation of the Keuliling Reservoir:

1. Flow Discharge

The river discharge data were from five channels leading to the Keuliling Reservoir. The discharge data were from the rain data undergoing rainfall-runoff analysis from 1995 to 2015. Then, semi-monthly discharges were 20%, 50%, and 80 % for the wet, normal, and dry years.

2. Irrigation water need and the ecological river flows

a. Data showed that the area of the irrigation served by the Keuliling Reservoir was 2,131 Ha (Balai Wilayah Sungai Sumatera I 2010).
b. The minimum river flow that must be available at all times as ecological river flows to maintain the life of river biota is 5% of the Alue Keuliling River discharge.

3. Parameters of Keuliling Reservoir Operation Optimization

The optimization method requires an objective function, a constraint function, and a decision variable to obtain an optimal reservoir operation pattern. Each function is explained in the following sections:

a. Objective function

The optimization target is the objective function of the Keuliling Reservoir operation system. This objective function is to maximize the release of irrigation water to supply water for the agricultural area where the rainfall does not meet the water needs for crops [8].

b. Constraint function

The constraint function applies to storage reservoirs. At each stage of the operation, the reservoir storage volume is limited by the maximum and minimum operation volume.

$$V_{min} \leq V_t \leq V_{max} \rightarrow V_t = V_1 \text{ s/d } V_{24} \tag{2}$$

In addition, there is an obstacle in the release of water from the reservoir (*Rt*) that must be sufficient for the capacity of river, maintenance flow, and irrigation water demand. *Vmin* and *Vmax* are reservoir life storage limits.

$$R_1 \text{ s/d } R_{24} \geq MF \text{ (Maintenance Flow)} \tag{3}$$

$$R_1 \text{ s/d } R_{24} \geq CR \text{ (Capacity of River)} \tag{4}$$

$$R_1 \text{ s/d } R_{24} \geq Demands \tag{5}$$

The reservoir water level in the 25th semi-month or at the beginning of semi-month for the operation of the following year must be greater than or equal to the water level at the beginning of the operating year.

$$TMA_{semi\text{-}month\text{-}25} \geq TMA_{t=1} \tag{5}$$

c. Decision variable

The amount of reservoir water release limited by constrained conditions is a decision variable in the operation of the reservoir for semi-monthly needs in the operating period of the wet year (20%), normal year (50%), and dry year (80%).

The study used a non-linear program to optimize the reservoir operation, which required objective functions and constraints. The release of water from the reservoir intended to satisfy irrigation water demands is the decision variable [2]. The study observed a connection between land erosion, changes in effective capacity, and adjustments in reservoir operating rules. The rules are helpful as an operational guideline during dry, normal, and wet years.

4 Results and discussion

After conducting bathymetry measurement of the Keuliling Reservoir inundation, a re-analysis was carried out on the changes in the reservoir capacity. If compared to the results of the 2015 measurements, a decrease in reservoir capacity due to sedimentation was observed. The situation required adjustments to the reservoir operating rules. An optimal operating condition was obtained by analyzing the optimal operating rules; thereby, maintaining the reservoir's function to control flood and supply irrigation water. Optimization will result in reservoir operations that can supply water according to the target rice field area in normal and wet years. The irrigation area is 578 hectares for Group I (Keuliling Upstream Irrigation Area) and 1053 ha for Group I (Keuliling Downstream Irrigation Area). However, in the dry year of Group I (DI Keuliling Hulu), with an irrigation area of 578 ha, the rice planting pattern (one rice planting season) begins with the planting season in October I. In addition, in Group II (DI Keuliling Hilir), with an irrigation area of 781 ha, the rice planting pattern (one rice planting season) starts the planting season in October II. The reduced inflows during dry years have decreased the target irrigation area by 113 ha. In dry years, the initial water level is set at +45.8 m (elevation of the total reservoir capacity) to optimize the total reservoir release. The variable used is a semi-monthly reservoir release for total demand, including for irrigation and maintenance. The constraints in dry years include a water level above +40 m at the end of the operating year and a semi-monthly reservoir release greater than the total demand. Another constraint is a semi-monthly volume larger than the minimum volume and less than the maximum reservoir volume. The irrigation area in year 2013 and the target of area after study sediment that can be irrigated for each group are given in Table 1.

Table 1. Irrigation area year 2013 and exiting area

Periode	Irrigation Area year 2013 (Ha)			Existing Irrigation Area (Ha)		
	Group I	Group II	Group III	Group I	Group II	Group III
Wet	578	1053	500	578	1053	809 (+309)
Normal	578	1053	500	578	1053	390 (-110)
Dry	578	894	500	578	781 (-113)	0

A plan for reservoir operating rules for wet, normal, and dry years, as well as the actual operating rule of the Keuliling Reservoir, was prepared after updating the reservoir operating rules (Figure 1). The actual operating rule is based on the daily recorded reservoir gate opening intakes.

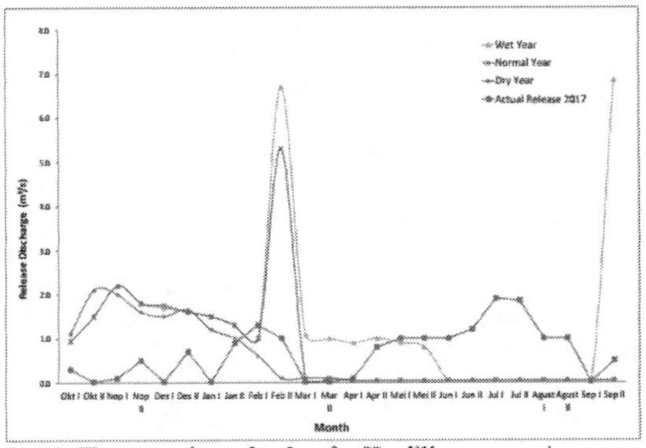

Fig. 1. The operating rule plan for Keuliling reservoir on wet, normal, and dry years

There are significant differences between the actual and planned operating rules. The differences happened due to: A decline in the reservoir capacity caused by sedimentation, and on actual release, a large amount of water supply was observed in the third planting season (June – September). However, the third planting season was not included in the normal reservoir operating rule.

5 Conclusion

This study succeeded in providing an update on the operating pattern of the Keuliling Reservoir due to changes in the reservoir capacity. A decline in the reservoir capacity is caused by sedimentation and actual release. Optimization will result in reservoir operations that can supply water according to the target rice field area in normal and wet years. The irrigation area is 578 hectares (ha) for Group I (Keuliling Upstream Irrigation Area) and 1053 ha for Group II (Keuliling Downstream Irrigation Area). The operation of reservoirs in dry years with water deficit has affected the water supply for irrigation. However, in the dry year of Group I (DI Keuliling Hulu), with an irrigation area of 578 ha, the rice planting pattern (one rice planting season) begins with the planting season in October I. In addition, in Group II (DI Keuliling Hilir), with an irrigation area of 781 ha, the rice planting pattern (one rice planting season) starts the planting season in October II. The reduced inflows during dry years have decreased the target irrigation area by 113 ha.

References

1. T. Bohmelt, T. Bernauer, H. Buhaug, N. P. Gleditsch, T. Tribaldos, and G. Wischnath, Global Environmental Change 29 (2014) 337–348.
2. H. Wang, E. D.Brill, R. S.Ranjithan, A. Sankarasubramanian, Advances in Water Resources 78 (2015) 9–21.
3. C. S. Lee, Chemosphere 87 (2012) 608–613.
4. A. Azmeri, I.K.Hadihardaja, N. Shaskia, K.S.Admaja, Proceedings of the 3rd International Conference on Construction and Building Engineering (ICONBUILD) (2017) AIP Conf. Proc. 1903, 100001-1–100001-9.
5. D. A. Wulandari, Master thesis. Diponegoro University. Semarang (2013).
6. N.M. Khan, M.S. Babel, R. Clemente, T. Tingsanchali, Water Resources Management 26 (2012) 11
7. D. A. Wulandari, D. Legono, S. Darsono, International Journal of Science and Engineering. Vol. 6(1)4 (2014)16-23
8. V. Tinoco, P. Willems, G. Wyseure, and F. Cisneros, Journal of Hydrology: Regional Studies 5 (2016) 213–225
9. A. Armido, A. Azmeri, E. Fatimah, N. Nurbaiti, S.N. Yolanda. 2020. The sedimentation datasets of Keuliling reservoir. Data in Brief. 32 106181.

Improvement of geotechnical properties of clayey soil with saw dust ash stabilization

Khaizal Jamaluddin[1], and *Reza* Pahlevi Munirwan[1*]

[1]Department, of Civil Engineering, Universitas Syiah Kuala, Jl. Tgk. Syech Abdur Rauf No. 7 Banda Aceh, Aceh 23111, Indonesia

Abstract. Rapid urbanization and construction are driving up the consumption of cement and aggregate used in building construction. However, the manufacture of cement and concrete boosts carbon dioxide emissions and depletes the supply of natural aggregates. As a result, alternative materials for building construction are required. A replacement material, saw dust ash (SDA), is obtained from woodfired power stations, pulp, paper mills, and other wood-burning factories. However, the SDA manufacturing sector necessitates enormous tracts of land in rural locations. As a result, some wood waste is sprinkled throughout the open land, while others are burned in incinerators, contributing to air pollution, a significant environmental issue. The purpose of this study is to provide a brief overview of the practical usage of SDA as a substitute material in the building sector, especially for soil stabilization. The article provides an overview of SDA use in the construction sector and examines development trends such as strength and durability. The findings indicate that implementing SDA into soi the soil stabilization sector benefits both sustainability and zero-waste technology. Environmental concerns about using SDA are also discussed, as are methods for minimizing the dangerous influence on the surrounding environment for future disaster risk reduction due to soil failure.

1 Introduction

The shortage of good quality soil suitable for construction and the recent emphasis on environmentally friendly mitigation strategies have prompted geotechnical engineering to explore and perfect systems for reusing excavated soils [1] and [2]. Preserving and improving existing pavements requires a sizable portion of transportation ministries' expenditures in every nation worldwide [3]. Approaches for lowering construction costs and extending the life of pavements can assist in better maintaining the transport system on a shoestring budget [4].

Roads constructed in the modern era are supposed to have a high standard of protection and comfort for their users [5]. Enhancement of soil can be accomplished through stabilization, modification, or combining the two [6]. While soil modification is accomplished by adding various admixtures such as lime or cement to alter the soil's index properties, soil stabilization is accomplished by treating the soil to increase its durability and strength to propose it is suitable for construction [7]. Soil stabilization is the process of modifying or adjusting the qualities of the soil to meet specific engineering requirements [8]. Compaction and the use of admixtures are two methods for soil stabilization. Lime and cement are two often used stabilizers for modifying the characteristics of soils.

Recent research suggests that solid wastes such as rice husk ash and fly ash can be used to stabilize soils in place of or in addition to cement or lime [9]. Saw dust is a by-product of the wood and lumber sawing sectors. Saw dust had limited cementitious value on its own. Still, when combined with water, it chemically interacts to generate cementitious compounds, which contribute to the increase of soil compressibility and strength parameter [10]. Silica is the prime component of saw dust ash (SDA), and it is responsible for the ash's reaction. Building technology is constantly evolving to accommodate changing transportation trends, building materials, and sub-grade conditions. The majority of pavement failures are due to the presence of poor subgrade conditions [1].

The industrial output of a wood manufacturing plant produces a variety of by-products. One of the majors is saw dust ash, produced when residue from wood chips, bark, sawdust, and cuts is used as an energy source [11]. The yearly production of saw dust ash is around 10,000 tons, but with increased production, the amount is approximately double. The created saw dust ash keeps gathering the mineral nutrients included in the waste wood, excluding nitrogen molecules, which are primarily discharged into the gaseous state during ignition. The nutrient mineral components present in solid compounds persist reasonably steady throughout the heat processing of combustion [12]. Specific components are in similar quantities as they were in the wood pulp composition, which is necessary for

* Corresponding author: r.munirwan@unsyiah.ac.id

developing effective ways and technologies for utilizing the elements' chemo dynamic cycle in soil systems [13]. Nitrogen components from the wood are destroyed during the firing process and discharged into the environment as waste gases, primarily as oxides.

This research aims to briefly re-examine the effect of saw dust ash on soil stabilization and its reaction mechanism using extensive physicochemical and micro-analysis. Additionally, the effects of ionic exchange on the mineralogy and microstructure of saw dust ash-treated soil are demonstrated convincingly. Finally, based on micro mechanisms and physicochemical observations, the strength behavior of SDA-treated soil is explored. Environmental friendly issue and future disaster risk reduction due to soil failure may be explained by using saw dust ash for soil stabilization and construction.

2 Statement of problem

The construction industry needs alternative and substitute alternatives for cement because of its price and pollution-prone manufacturing process. In addition, the cement industry accounts for around 7% of global CO emissions, with around 650–920 kg CO2 released during the preparation of one ton of cement [14]. Thus, a product capable of reducing cement needs and the associated CO emission is an urgent requirement in the modern-day.

Fig. 1. Waste wood produced during manufacturing processing

An option to address the over-consumption of cement is to reduce the built environment's reliance on cement by including domestically available by-products as reinforcing cementitious and other low carbon components into the concrete manufacturing process [14]. Agricultural waste ashes, for example, rice husk ash (RHA) and palm oil fuel ash (POFA), have recently been investigated their use as construction material. Due to the high silica concentration, RHA becomes pozzolanic and can be used as a supplemental cementitious material in eco-concrete applications. Meanwhile, POFA has demonstrated significant potential as a cement substitute in concrete technology. Other agricultural waste ashes, such as coffee husk ash, sugarcane bagasse ash, corn cob ash, and saw dust ash, can be used as additions in building materials.

Agricultural wastes are typically incorporated into building in the form of ash. However, concerns about sustainable development and zero-waste innovation have prompted particular ashes as cement substitutes [15]. Currently, more than 70% of SDA is disposed of in the environment in a different structure. When some types of wood, such as chips and barks, are ignited, a substance termed SDA is formed. SDA is confined to specific specified dimensions necessary for optimal harvest development. SDA is produced globally through ignition in woodfired power plants, sawmills, paper mills, and other timber manufacturing processes.

Fig. 2. Burning of saw dust and saw dust ash [5]

Between 1992 and 2010, the FAO stated that roughly 98.2 x 10^7 m^3 SDA was produced worldwide on an annual average basis [12]. China generated 15.3 x 10^7 m^3 of SDA during this period, followed by Brazil with 14 x 10^7 m^3, the United States with 13 x 10^7 m^3, the Federation of Russian with 7.9 x 10^7 m^3, and France 7.7 x 10^7 m^3. Wood waste generated during industrial processing can be seen in Fig 1.

Apart from all other ashes, SDA is developing into a more vibrant and nutrient-dense mass stream. SDA is primarily composed of destroyed wood and wastes from the wood industry; it is easy biomass and can also be referred to as a limitless fuel; hence, it is desirable for large-scale biomass ignition plants [18]. However, employing SDA as a substitute for conventional building materials has limitations. As a result, regulations on using SDA as a substitute for cement or lime in construction materials are necessary. Fig. 2 illustrates the process of combusting saw dust and the resulting saw dust ash.

Table 1. Chemical composition of wood ash

Compound	Composition (%)	
	[16]	[17]
Aluminum Oxide (Al2O3)	23.6	14.72
Silicon Dioxide (SiO2)	27	25.8
Potassium Oxide (K2O)	6.3	9.55
Calcium Oxide (CaO)	21.5	29.8
Ferric Oxide (Fe2O3)	2.7	0.95
Magnesium Oxide (MgO)	11	5.25
Sodium Oxide (Na2O)	7.9	7.5
Loss on ignition, LOI	-	2.7

3 Saw dust ash properties

SDA comes in various chemical components, depending on the region in which it is produced and the combustion procedure used. Still, the principal oxides found in industrial SDA are mixed oxides, as illustrated in Table 1.

As illustrated in Fig. 3, the XRD data from [7] verified the presence of numerous main crystalline phases. Calcite - $CaCO_3$ is the primary crystalline component, followed by quartz - SiO_2 and potassium and calcium carbonate - $K_2Ca(CO_3)_2$.

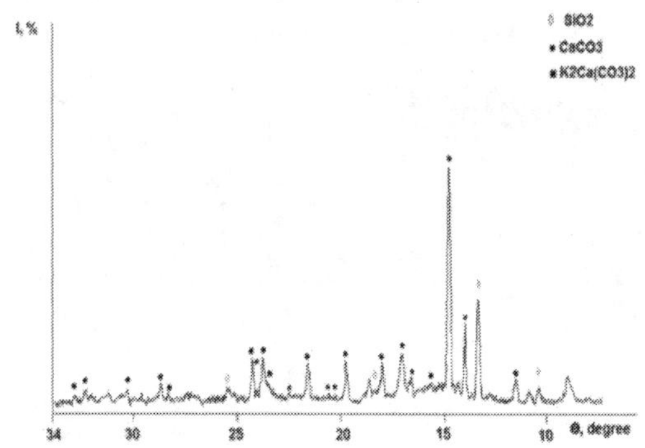

Fig. 3. Typical XRD pattern wood ash [7]

Fig. 4 illustrate SEM analysis of SDA as angular, brittle, and elongated granules with jagged edges of SDA from [11].

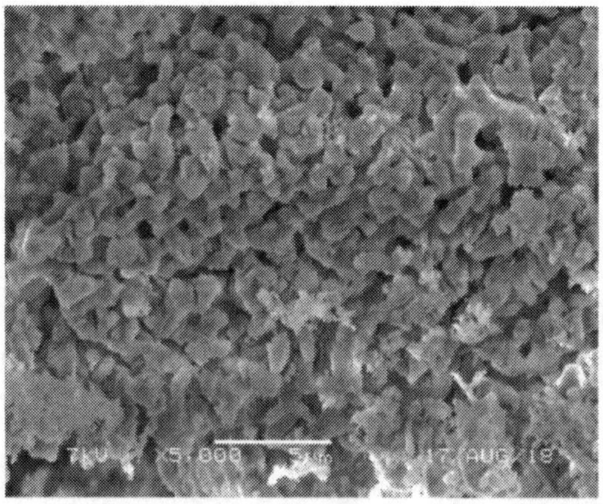

Fig. 4. Typical SEM of wood ash [11]

4 Saw dust ash application

Some research for applications of SDA utilization in road construction has been conducted in few past years. However, most of the SDA application is moderately effective to improve the physical and mechanical parameter of soil.

A comprehensive experimental investigation from [5] was conducted to evaluate SDA's soil improvement potential through CBR and UCS tests. The experimental analysis demonstrated that the inclusion of SDA increases the CBR and UCS significantly. Additionally, the CBR values obtained are within the suggested limits of Asphalt Institute for highway sub-base and sub-grade. Thereby, it is concluded from this study that SDA, an industrial waste, is a cost-effective stabilizing agent for road construction in clayey fills; however, its efficiency can be enhanced by integrating it with other bonding materials like lime, and thus becomes a viable alternative use of industrial waste to reduce road construction costs, especially in rural communities.

From [18], lateritic soil was treated with SDA by dry weight and compacted using the British Standard light compaction method to determine the chemical compatibility with landfill leachate municipal solid waste (MSW) in comparison to tap water. On treated soil specimens, index, compaction, and compatibility tests were conducted utilizing long-term hydraulic conductivity. The test findings indicate that hydraulic conductivity increased in the presence of permeation liquid. While hydraulic conductivity increased overall when test specimens were penetrated with MSW leachate, it dropped for samples treated for 4% SDA but improved for all other treated samples compared to natural lateritic soil.

The effectiveness of wood ash as a stabilizing agent for soft clay has been examined by [17]. It is discovered that stabilizing existing soil improves the engineering features of the soil, specifically the UCS, shear strength parameters, workability, and compaction and compressibility properties. As a result, laboratory experiments on these parameters were conducted on a fixed amount of wood ash. Chemical analysis of wood ash reveals roughly 30% calcium oxide, which causes it to act similarly to a pozzolanic material. Additionally, the test result indicates that the soil might be made lighter by increasing its strength, moisture content, and compressibility with the addition of ash.

[11] examined the strength, stiffness, and microstructural response of marine-deposited clays in Cyprus modified with cement and wood-ash as a cement substitute. Portland cement (7, 10, and 13% by weight) was substituted with varying amounts of wood ash (5 and 10% by weight) to produce concrete with two varied dry densities (1400 and 1600 kg/m^3) and three different curing times (7, 28, and 60 days). UCS, porosity, direct shear, and pulse velocity testing were conducted. Additionally, SEM, XRD, and EDX studies were applied to characterize the microstructures of clay–wood–ash–cement mixes. The results indicated that substituting 5% wood ash for cement resulted in higher performance. Further analysis of the microstructures of wood-ash–cement–clay blends revealed the creation of a porous matrix with stable linkages. Additionally, blends' porosity and strength qualities exhibit unique connections with their porosity and binder contents, as proven by other green substitute materials and soils.

The purpose of the research from [16] was to identify the optimal wood ash proportion for stabilizing black cotton soil. The laboratory study is designed to ascertain the soil sample's geotechnical properties. Wood ash is added in progressive mass percentages of 5%, 10%, 15%, and 20% per soil dry unit weight. Additionally, stabilized soils are cured for seven days for Atterberg limit and compaction tests and soaking for four days for CBR tests. Wood ash reduces the plasticity index, OMC,

https://doi.org/10.1051/e3sconf/202234001009

and swelling, according to an experimental study. On the other hand, adding wood ash can enhance CBR, maximum dry density, and UCS. The efficient wood ash ratio for stabilizing black cotton soil is between 10% and 15% of the soil mass, with a preference for results between 13% and 15%.

5 Conclusions

SDA has the potential to be used as a substitute for soil stabilization in soil stabilization applications. SDA is capable of being used in large quantities. Numerous researches have examined a few key characteristics and demonstrated comparable strengths. Considerable progress has been made toward zero-waste technologies. SDA has been used in various amounts in the past for soil stabilization.

This brief review focused on the effect of SDA as a partial substitute for soil stabilization. SDA is regarded as additional source material or additive that increases the reactivity of binders. SDA-incorporated soil stabilization with various ashes, synthetic compounds, or filaments performed better than untreated soil. This behavior is mainly explained by the synergistic effects of SDA's pozzolanic and filler properties on the microstructure, resulting in a denser microstructure. As a result, by consolidating SDA in soil stabilization, the porous nature of the primary waste materials can be decreased.

References

1. R. P. Munirwan, M. Munirwansyah, M. Marwan, P. J. Ramadhansyah, and V. Kamchoom, IOP Conf. Ser. Mater. Sci. Eng. (2020)
2. M. Zhang, J. Yi, and D. Feng, Sci. Prog. **103**, 1 (2020)
3. M. K. Atahu, F. Saathoff, and A. Gebissa, J. Rock Mech. Geotech. Eng. **11**, 337 (2019)
4. C. C. Ikeagwuani and D. C. Nwonu, J. Rock Mech. Geotech. Eng. **11**, 423 (2019)
5. W. A. Butt, K. Gupta, and J. N. Jha, Int. J. Geo-Engineering (2016)
6. D. Sundary, R. P. Munirwan, and N. Al-Huda, J. Phys. Conf. Ser. (2021)
7. E. Serafimova, M. Mladenov, I. Mihailova, and Y. Pelovski, J. Univ. Chem. Technol. Metall. **46**, 30 (2011)
8. D. S. Vijayan and D. Parthiban, Environ. Technol. Innov. **20**, 101108 (2020)
9. M. Uduebor, O. J. Oyedepo, and O. B. Dorcas, J. Geotech. Transp. Eng. **5**, 55 (2019)
10. B. Bunyamin, R. P. Munirwan, M. Ridha, and N. Hendrifa, IOP Conf. Ser. Mater. Sci. Eng. (2021)
11. A. Ekinci, M. Hanafi, and E. Aydin, Minerals (2020)
12. K. Tamanna, S. N. Raman, M. Jamil, and R. Hamid, Constr. Build. Mater. **237**, (2020)
13. N. Ijaz, F. Dai, and Z. ur Rehman, J. Environ. Manage. **262**, 110285 (2020)
14. D. Tavakoli, M. Hashempour, and A. Heidar, Pertanika J. Sci. Technol. 499 (2018)
15. Z. Sabzi, J. Environ. Friendly Mater. **2**, 33 (2018)
16. S. S. Sefene, Geotech. Geol. Eng. (2020)
17. B. D. Nath, G. Sarkar, S. Siddiqua, M. Rokunuzzaman, and M. R. Islam, Adv. Civ. Eng. (2018)
18. K. J. Osinubi, A. O. Eberemu, J. R. Oluremi, and S. T. Ijimdiya, Proc. Inst. Civ. Eng. - Waste Resour. Manag. 1 (2017)

Earthquake behavior in Sumatra territory based on time and distance between consecutive large event

Samsul Anwar[1*]

[1]Universitas Syiah Kuala, Department of Statistics, Banda Aceh 23111, Indonesia

Abstract. The purpose of this study is to understand the behavior of earthquakes with a specific magnitude (M) in the Sumatra territory based on their duration of time and the distance between two earthquakes that happen consecutively. We explore 1,641 data sets of duration and distances with a specific magnitude that were generated from 458 earthquakes with M $\geq$ 5.0 during 2009-2017. This study revealed that the occurrence of the later (second) earthquake at any location (referred to in this paper as a recurring earthquake) with a bigger or equal magnitude as the previous one depends on duration especially for magnitude under 5.8. The comparison analyses showed that the recurring earthquake events tended to have similar average distances. However, the length of time between two earthquakes is different for lower magnitudes (under 6.0), whereas between an earthquake and a recurring earthquake with a magnitude above 6.0, the length of time tends to be similar.

1 Introduction

Indonesia, a country located in between two continents, Asia and Australia, is squeezed by four large tectonic plates of the Earth: Indies, Australia, Eurasian, and Pacific plates. The Indies-Australian plate moves to the north, pounding the Eurasian plate, moving at a speed of 50-70 mm/year [1]. The pounding zones between these two plates are along the sea trench of Sumatra, Java, Bali, and Lombok islands. The movement of collisions and the shifting of these large plates respond mechanically. These plate movements have caused many earthquakes to occur in the Indonesia archipelago, one of them in the fault of the Sumatra territory. According to NOAA's National Geophysical Data Center, at least 235 tsunamis have occurred in the last four centuries and Indonesia is one of the countries that is most exposed to earthquakes in the world [2].

The oblique tugging between the Indies-Australian and Eurasian plates had a significant influence on tectonic and volcanic conditions in northern Sumatra [3]. It caused the Sumatra fracture zone to become reactive when the depth of the earthquake center was not more than 50 kilometers. In general, the Sumatra territory is divided into two sources of tectonic earthquakes, those that are in subduction zones along the sea, and those in the zones along the land with a fracture system of the Sumatra territory [4]. There are several massive earthquakes that have happened in this fractured area, including the earthquake in Padang in 2009, Mentawai in 2010 and the earthquake and tsunami Aceh in 2004. The earthquake is one of the tsunamigenic events that caused a tsunami. In Indonesia, tsunamis in the northern part of the equator tend to happen less frequently compared to the southern part of the equator [5].

The earthquake known as the Sumatra-Andaman earthquake happened on December 26, 2004 [6,7], and it is one of the earthquakes whose source is the subduction of oceanic plates into the Earth along the continental banks of western Sumatra, which extends to the south of Java, Bali, and Lombok islands [3]. This 2004 earthquake with a magnitude of 9.1 - 9.3 [6, 8–10], was also felt in Bangladesh, India, Malaysia, Myanmar, Thailand, Si ngapore, Sri Lanka and the Maldives [7] and caused a tsunami wave with a peak of 51 meters (167.3 ft) [11]. It destroyed everything not only in the southern coastal area of Sumatra island, but also in the other areas along the coastline of the Indian Ocean, and even reached some countries on the coast of the Africa continent. The number of victims caused by this earthquake and the tsunami was approximately 230,000 people in 14 countries [2]. The Sumatra-Andaman earthquake is the third biggest earthquake that has ever been recorded with a seismograph and had the longest duration in history, about 8.3 to 10 minutes [12]. The earthquake caused the entire planet to vibrate 1 centimeter (0.4 inches) [13]. In addition, the Sumatra-Andaman earthquake also triggered the formation of a new fault pattern, specifically around the Sumatra Island which could potentially cause other earthquakes with large magnitudes. It can be shown that an earthquake with a large magnitude (M $>$ 5.5) might occur without being preceded by a smaller earthquake. Although the probability is relatively small, an earthquake with a magnitude $>$ 7.6 can possibly occur before an earthquake with a magnitude between 5.0 and 7.6

* Corresponding author: samsul.anwar@unsyiah.ac.id

[14]. Another study showed that the Aceh region, the nearest area from the epicentre of the Sumatra-Andaman earthquake, has a high seismicity level [15].

In order to minimize the risk of disasters caused by earthquakes in the future, the mitigation effort needs to be prepared as much as possible. Knowledge of the past can help reduce losses in the future by identifying and defining earthquake and tsunami hazards [8]. This includes the study of earthquake behavior, which is the first step in understanding the phenomena of earthquakes. The behavior can be studied through the research of hazard seismicity. The base seismic hazard assessment is based on the statistical distribution of the recurrence interval [16]. It is assumed that the times between large consecutive earthquakes (inter-event times or recurrence intervals) follow a certain statistical distribution [17]. The duration and magnitude play an increasingly important role in earthquake prediction models [18].

The objective of this study is to understand the earthquake behavior in the Sumatra territory based on the duration and distance between two earthquake events with particular magnitudes. The duration data is analysed using a survival analysis approach in order to understand the hazard seismicity. Furthermore, for each set of earthquakes with a certain magnitude, the duration, as well as the distance between earthquakes was compared by using Analysis of Variance (Anova) or the Kruskal-Wallis test, when applicable. Together, these two analyses give more comprehensive information about the earthquake behavior in the Sumatra territory.

2 Materials and methods

2.1 Data and settings

This study analyses the earthquake dataset catalog of National Meteorology, Climatology, Geophysics Agency (BMKG) from 2009 to 2017 (http://dataonline.bmkg.go.id), with latitudes between -9.66 and 5.57 degrees, longitudes between 94.08 and 105.54 degrees, which is the area northern Sumatra, off the West Coast of northern Sumatra, southern Sumatra, and southwest of Sumatra, Indonesia. The data set used in this study is the length of time (in days) between two earthquakes, where the later (second) earthquake (termed as recurring earthquake) has a bigger or equal magnitude (M) to the previous one. A recurring earthquake differs from an aftershock because an aftershock happens in the same location as the original earthquake. This paper uses recurring earthquakes as another event in the Sumatra territory. The distance between these two earthquakes' epicentres is measured in kilometers (km). The magnitudes to be analysed in this study are as follows: $M \geq 5.0$, $M \geq 5.1$, $M \geq 5.2$, $M \geq 5.3$, $M \geq 5.4$, $M \geq 5.5$, $M \geq 5.6$, $M \geq 5.7$, $M \geq 5.8$, $M \geq 5.9$, $M \geq 6.0$, $M \geq 6.1$, $M \geq 6.2$ and $M \geq 6.3$. The higher magnitudes ($M \geq 6.4$ and so on) are not evaluated due to small sample size (<5). As an illustration, if the first earthquake has a magnitude of 5.0, then the data set is the length of time from this earthquake to the event of the next earthquake with $M \geq 5.0$.

2.2 Methods

This duration data of a recurring earthquake event is a set of survival data. Hence, the data can be analysed by using the survival analysis method. Specifically, the hazard rate is used to help understand how the earthquake behavior in the Sumatra territory is related to the duration of the recurring earthquake events with the specific magnitudes, as stated previously. There are eight probability distributions used to test the duration data set namely the Exponential, Gamma, Gumbel, Logistic, Lognormal, Normal, Normal Family and Weibull distributions.

Furthermore, the length of time as well as the distance between two earthquakes of different magnitudes evaluated are compared using the Analysis of Variance (Anova) or Kruskal-Wallis test, depending on the normality and homogeneity variances assumption. A p-value less than 0.10 was considered significant. These analyses would give a better understanding of earthquake behavior, whether or not different magnitudes would require different durations and distances between the earthquake and the recurring earthquake events to happen. The Shapiro-Wilk test is employed to analyse the distance data set to test its normality assumption. As for the duration data, it is checked in a part of survival analysis. Additionally, Levene's test is performed to check the homogeneity variances assumption. Analyzing data in this study used software R version 3.5.0.

Performing the analysis in this study is divided into three steps:

1. Estimating the probability distribution of the duration of the recurring earthquake events with a specific magnitude using the Maximum Likelihood Estimator (MLE) method that is integrated in the GAMLSS package. The chosen probability distribution is the one having the smallest AIC value.
2. Analyzing the earthquake behavior through the hazard rate according to the probability distribution estimated in the first step.
3. Comparing magnitudes evaluated to each other in order to investigate whether or not the recurring earthquake happened for a similar duration and at a similar distance.

2.3 Literature background

In order to support the methodology used in this study, some brief background theories are provided.

2.3.1 Survival analysis

The lifetime of a unit is the length of time, T, from the first time the unit works until the time it fails. The time variable T is defined as a random variable because it cannot be predicted precisely. There are four important functions in the survival analysis, namely: probability density function (p.d.f.), cumulative density function (c.d.f.), survival function, and hazard function. One of the most important functions in the survival analysis is a hazard function ($h(t)$). $h(t)$ is a function that provides the instantaneous failed rate of a unit that already has lifetime at time t ($0 < t < \infty$), or could be written as follows [19]:

$$h(t) = \lim_{\Delta \to 0} \frac{F(t+\Delta)-F(t)}{\Delta P\{T>t\}} = \frac{f(t)}{S(t)} \quad (1)$$

where the $f(t)$ is a p.d.f., $F(t)$ is a c.d.f. and $S(t)$ is a survival function with the following forms:

$$f(t) = P\{T = t\} \quad (2)$$

$$F(t) = P\{T \le t\} = \int_0^t f(x)dx \quad (3)$$

$$S(t) = 1 - \{T \le t\} = P\{T > t\} = \int_t^\infty f(x)dx \quad (4)$$

Accordingly, knowing the p.d.f. and c.d.f. would lead to the hazard function based on Equation (1) which is also related to Equation (2) and (4).

2.3.2 Maximum likelihood estimator (MLE)

Let θ denote the parameter set of the distribution function $F(t)$. Since the likelihood function for a failure observation t is $L(t) = f(t;\, \theta)$, then the overall likelihood function is written as:

$$L(\theta) = \prod_{i=1}^{n} L_i(\theta) \quad (5)$$

The parameter is determined by maximizing the $L(\theta)$ or easier in logarithmic form, i.e.

$$\ln[L(\theta)] = \sum_{i=1}^{n} \ln[L_i(\theta)] \quad (6)$$

The maximizing problem can be solved by doing differentiation on $\ln[L(\theta)]$ against each parameter in the probability distribution [20].

2.3.3 Akaike information criterion (AIC)

The information criterion is appropriate for model selection when the candidate models are either nested or non-nested [20]. The AIC is formulated as:

$$AIC = -2\ln(L) + 2m \quad (7)$$

where L is likelihood function and m is the number of parameters in the model. Model with the smallest AIC value is better than others [21].

2.3.4 Probability distributions in survival analysis

There are eight probability distributions tested in this study, namely: Exponential, Gamma, Gumbel, Logistic, Lognormal, Normal, Normal Family, and Weibull distributions. Table 1 provides the p.d.f. and c.d.f. of each probability distribution tested [22].

Table 1. PDF and CDF of the examined probability distributions

Probability Distribution	p.d.f.	c.d.f.
Exponential, EX(μ)	$\frac{1}{\mu} exp\left(\frac{-y}{\mu}\right)$	$1 - exp\left(\frac{-y}{\mu}\right)$
Gamma, GA(μ,σ)	$\frac{y^{\frac{1}{\sigma^2}-1} e^{\frac{-y}{\sigma^2\mu}}}{(\sigma^2\mu)^{\frac{1}{\sigma^2}}\Gamma\left(\frac{1}{\sigma^2}\right)}$	$\frac{\gamma(\sigma^{-2}, y\mu^{-1}\sigma^{-2})}{\Gamma(\sigma^{-2})}$, where $\gamma(a,x) = \int_0^x t^{a-1}e^{-t}dt$
Gumbel, GU(μ,σ)	$\frac{1}{\sigma} e^{\left[\left(\frac{y-\mu}{\sigma}\right)-e^{\left(\frac{y-\mu}{\sigma}\right)}\right]}$	$1 - e^{\left[-e^{\left(\frac{y-\mu}{\sigma}\right)}\right]}$
Logistics, LG(μ,σ)	$\frac{1}{\sigma}\left\{e^{\left[-\left(\frac{y-\mu}{\sigma}\right)\right]}\right\}\left\{1 + e^{\left[-\left(\frac{y-\mu}{\sigma}\right)\right]}\right\}^{-2}$	$\left\{1 + e^{\left[-\left(\frac{y-\mu}{\sigma}\right)\right]}\right\}^{-1}$
Log Normal, LOGNO (μ,σ)	$\frac{1}{\sqrt{2\pi\sigma^2}}\frac{1}{y} e^{\left[-\frac{[\log(y)-\mu]^2}{2\sigma^2}\right]}$	$\Phi\left[\frac{(\log(y)-\mu)}{\sigma}\right]$
Normal, NO (μ,σ)	$\frac{1}{\sqrt{2\pi}\sigma} e^{\left[-\frac{(y-\mu)^2}{2\sigma^2}\right]}$	$\Phi\left[\frac{(y-\mu)}{\sigma}\right]$
Normal Family, NOF(μ,σ,v)	$\frac{1}{\sqrt{2\pi}\sigma\mu^{\frac{v}{2}}} e^{\left[-\frac{(y-\mu)^2}{2\sigma^2\mu^v}\right]}$	$\Phi\left[\frac{(y-\mu)}{\sigma\mu^{\frac{v}{2}}}\right]$
Weibull, WEI(μ,σ)	$\frac{\sigma y^{\sigma-1}}{\mu^\sigma} e^{\left[-\left(\frac{y}{\mu}\right)^\sigma\right]}$	$1 - e^{\left[-\left(\frac{y}{\mu}\right)^\sigma\right]}$

2.3.5 Kruskal-wallis test

Given data set $t_{11}, t_{12}, t_{13}, \ldots, t_{1n_1}$ are samples from population 1 and $t_{k1}, t_{k2}, t_{k3}, \ldots, t_{kn_k}$ are samples from population k. Let R_{ij} be the rank of the data point t_{ij} and $R_i = \sum_{j=1}^{n_i} R_{ij}$ be the sum of ranks in the ith sample and $N = \sum_{i=1}^{k} n_i$. Thus, the statistics H of Kruskal-Wallis test is defined as:

$$H = \frac{12}{N(N+1)} \sum_{i=1}^{k} \frac{R_i^2}{n_i} - 3(N+1) \quad (8)$$

The distribution of statistics H is approximated by Chi-square distribution with $k-1$ degree of freedom [23].

2.3.6 Mann-whitney u test

Given data sets from two groups, says X and Y, with n_X and R_X as the number of samples and the sum of ranks assigned to the first group respectively, then n_Y and R_Y are the number of samples and the sum of ranks assigned to the second group respectively, the Mann-Whitney U test for each group is defined as:

$$U_X = n_X n_Y + \left(\left(n_X (n_X + 1) \right) / 2 \right) - R_X \quad (9)$$

$$U_Y = n_X n_Y + \left(\left(n_Y (n_Y + 1) \right) / 2 \right) - R_Y \quad (10)$$

In terms of median, two groups are considered to have different medians if, by consulting the Mann and Whitney tables, the probability corresponding to the $\min(U_X, U_Y)$ is smaller than the probability of the predetermined α threshold [24].

3 Results and discussion

3.1 Data characteristics

During 9 years (2009 - 2017), there were 46,335 earthquakes that occurred in Indonesia with a magnitude of between 1.5 and 7.8 and with a depth of hypocenters of between 1 to 750 meters. 7,970 (17.20%) of those earthquakes happened in the Sumatra territory between the latitudes of -9.66 to 5.57, and longitudes of 94.08 to 105.54. In agreement with a previous study [25], this paper shows that almost all segments in the northern part of Sumatra are in active condition. Among those 7,970 earthquakes, 503 (6.31%) of them have M ≥ 5.0. Table 2 provides the number of earthquake events in Indonesia and Sumatra territory for M ≥ 1 and M ≥ 5.

Table 2. Number of earthquake events in Indonesia (2009 – 2017)

Area (Magnitude)	Years								
	2009	2010	2011	2012	2013	2014	2015	2016	2017
Indonesia (M ≥ 1.0)	3,676	5,999	4,451	6,624	4,130	4,668	5,073	5,625	6,089
Sumatra (M ≥ 1.0)	827	1363	944	1085	814	751	696	699	791
Sumatra (M ≥ 5.0)	107	82	69	48	45	38	34	36	44

Table 3. Descriptive statistics of the recurring earthquake events at all evaluated magnitudes

Magnitude (≥)	Frequency	Duration time (days)				Distance (km)			
		Min	Median	Mean	Max	Min	Median	Mean	Max
5.0	458	0.50	3.00	6.82	70.00	2.22	434.20	529.70	1775.00
5.1	331	0.50	4.50	9.21	77.00	2.22	448.50	542.50	1775.00
5.2	246	0.50	6.00	11.69	90.00	2.22	470.80	535.60	1634.00
5.3	178	0.50	8.00	15.76	145.00	2.23	397.50	492.60	1501.00
5.4	122	0.50	11.00	20.58	165.00	2.23	355.10	459.30	1494.00
5.5	86	0.50	13.00	24.44	176.00	2.23	432.10	517.30	1494.00
5.6	71	0.50	16.00	29.66	176.00	2.23	490.90	524.30	1408.00
5.7	43	0.50	27.00	36.29	162.00	2.23	490.90	515.80	1228.00
5.8	28	0.50	32.00	50.04	165.00	2.23	329.00	444.60	1033.00
5.9	24	0.50	33.00	58.09	165.00	2.23	329.00	434.00	1045.00
6.0	19	0.50	48.50	61.39	173.00	2.23	232.00	415.00	1045.00
6.1	15	0.50	32.50	55.75	169.00	2.23	518.50	509.00	1033.00
6.2	11	0.50	32.50	58.95	231.00	2.23	774.00	583.90	1033.00
6.3	9	0.50	33.00	44.81	123.00	2.23	242.70	390.30	803.30

However, this study only analysed data from 458 (91.05%) earthquake events with M ≥ 5.0. There was too small a sample size for events with magnitudes above 6.3, so they were not analysed. Note that the duration and distance between epicenters can only be calculated between two events. Accordingly, the number of data

https://doi.org/10.1051/e3sconf/202234001010

points in the data set of duration and distance can be formulated as $n-1$, where n is the number of earthquake events. Furthermore, earthquakes with M ≥ 5.0 can be reclassified into several new categories with bigger magnitude. For example, from this data set, earthquake events can be subdivided into specific M ≥ 5.1 intervals and then the duration and distance between the first to the next earthquake with equal or bigger magnitude can be calculated. As mentioned before, in this study earthquake events are evaluated at M ≥ 5.0, M ≥ 5.1, ... and M ≥ 6.3. Using this logic, a total of 1,641 new data points of duration and distances are generated from 458 earthquakes with M ≥ 5.0. Some descriptive statistics of duration and distance between recurring earthquake events are provided in Table 3.

According to Table 3, we can see the most frequent recurring earthquakes that happened in the Sumatra territory have a magnitude of M ≥ 5.0, and this magnitude earthquake has happened 458 times between 2009 - 2017. Consequently, there are 457 durations and distances or 27.85% of total data used in this study. The least frequent magnitude is nine times for M ≥ 6.3. In general, it could also be seen that the intensity of the recurring earthquake tends to be smaller exponentially along with the growth of the earthquake magnitude. Fig. 1 shows the distribution of duration for all recurring earthquake events in time sequence order of each magnitude evaluated during 2009 - 2017.

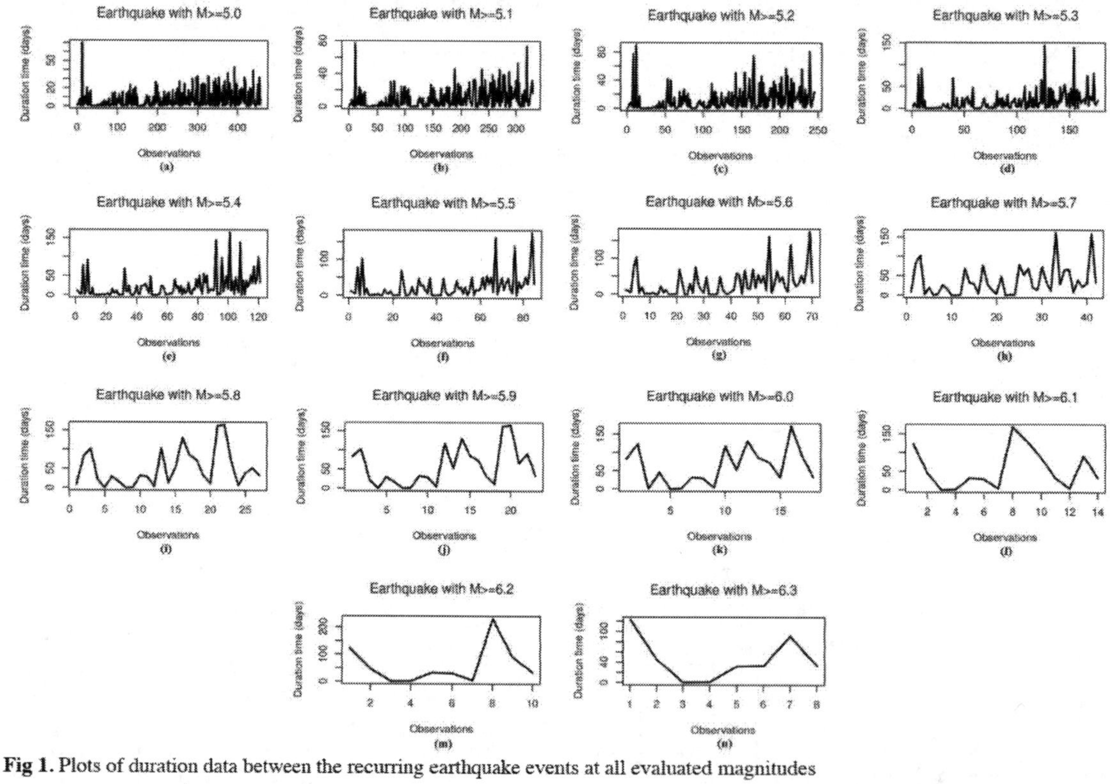

Fig 1. Plots of duration data between the recurring earthquake events at all evaluated magnitudes

According to Figure 1, it is known that the duration of the recurring earthquake events with an increasing magnitude measured tends to be longer. It indicates that earthquakes with a bigger magnitude would occur less often when compared to earthquakes with smaller magnitude. Moreover, the larger the magnitude of earthquakes, the less frequently they occur. Note that, in the survival analysis, data points with zero value could not be analysed. Therefore, the occurrence of the recurring earthquakes on the same day as the previous one in this study would be given a small value. In the data set, there are 45 days when the recurring earthquakes happened on the same day with a varying number of recurring earthquake events. On average, there are 2 recurring earthquake events per day. Accordingly, by assuming the recurring earthquakes happen in a uniform time, the recurring earthquake duration on the same day then will be given as 0.5 day (12 hours) instead of 0.

From Table 3, it is also known that the mean of the duration of the recurring earthquake tends to be higher as the magnitude is greater. The mean time between an earthquake and a recurring earthquake with M ≥ 5.0 is 6.82 days and has a median of 3 days, with a maximum of 70 days since the first earthquake. The minimum duration of a recurring earthquake is 0.5 days, which means it happens on the same day as the first earthquake. Furthermore, the mean and median time between the earthquake and a recurring earthquake with M ≥ 6.3 increases to 44.81 and 33 days respectively, with the minimum and maximum time as 0.5 days and 123 days after the first earthquake with M ≥ 6.3. Another interesting fact is that the recurring earthquakes with a maximum magnitude evaluated (6.3) might occur on the same day as a previous earthquake with a similar magnitude. The earthquake's magnitude varies from one earthquake to the next and the earthquake does not necessarily repeat at regular intervals [8]. Slightly different, the mean and median of distances between the recurring earthquakes seem not to have constant increasing or decreasing patterns. However, the maximum distance between recurring earthquakes tends to be smaller in accordance with the increase of its magnitude.

3.2 Estimating probability distribution of duration data

The maximum likelihood estimator (MLE) is used to estimate the probability distribution for the duration data of the recurring earthquake events with a specific magnitude. The GAMLSS package provides the MLE with various probability distributions that can be tested. In this study, there are eight probability distributions to be evaluated, namely: Exponential, Gamma, Gumbel, Logistic, Lognormal, Normal, Normal Family and Weibull distributions. Three of these distributions (the Gamma, Lognormal, and Weibull) are frequently used for earthquakes, because they share three properties commonly observed for earthquake inter-event times [26]. The probability distribution which has the highest suitability level for the distribution of the data set is the one that has the smallest Akaike Information Criterion (AIC) value among all probability distributions tested. Table 4 provides AIC values for all probability distributions evaluated of the recurring earthquake events with M ≥ 5.0, M ≥ 5.1, … and M ≥ 6.3.

Table 4. AIC values of the examined probability distributions for duration data between the recurring earthquake events at all evaluated magnitudes

Magnitude (M)	Probability Distribution							
	EXP	**GA**	**GU**	**LG**	**LOGNO**	**NO**	**NOF**	**WEI**
M ≥ 5.0	2671.12	2654.56	3656.16	3148.21	2616.73*	3241.44	3243.44	2646.21
M ≥ 5.1	2127.11	2101.63	2816.82	2481.11	2087.65*	2546.45	2548.45	2096.05
M ≥ 5.2	1696.86	1666.60	2226.88	1978.01	1654.84*	2038.78	2040.78	1660.81
M ≥ 5.3	1332.14	1300.85	1769.43	1542.19	1295.80*	1603.65	1605.65	1296.23
M ≥ 5.4	975.87	943.09	1266.46	1127.68	940.90	1167.29	1169.29	939.99*
M ≥ 5.5	715.37	695.70	910.56	815.68	699.13	842.26	844.26	694.75*
M ≥ 5.6	616.59	599.51*	755.84	695.56	606.78	708.56	710.56	600.15
M ≥ 5.7	387.68	384.88*	454.84	426.75	393.61	431.16	433.16	385.71
M ≥ 5.8	267.29*	268.16	299.58	288.96	276.31	289.10	291.10	268.64
M ≥ 5.9	234.85*	235.72	255.90	250.33	245.05	249.25	251.25	236.41
M ≥ 6.0	186.22*	186.62	200.32	196.73	195.14	195.72	197.72	187.61
M ≥ 6.1	142.58*	142.76	158.93	155.49	146.97	154.80	156.80	143.34
M ≥ 6.2	103.53	102.65*	121.88	115.99	104.92	116.94	118.94	103.01
M ≥ 6.3	78.84*	79.66	88.04	85.76	82.91	85.58	87.58	80.18

* indicates the smallest AIC value

According to the selection criteria, it is known that there are four probability distributions that have the smallest AIC value among all magnitudes evaluated. Those probability distributions are Lognormal, Weibull, Gumbel, and Exponential. Therefore, the duration of recurring earthquake events among all magnitudes evaluated could be estimated to follow one of those four probability distributions. Based on information provided at Table 4, it is known that the recurring earthquake events for M ≥ 5.0, M ≥ 5.1, M ≥ 5.2 and M ≥ 5.3 are approximated to follow Lognormal distribution, the recurring earthquake events for M ≥ 5.4 and M ≥ 5.5 are approximated to follow the Weibull distribution, the recurring earthquake events for M ≥ 5.6, M ≥ 5.7 and M ≥ 6.2 are approximated to follow the Gumbel distribution, while the rest follow the Exponential distribution.

3.3 Hazard rate of the recurring earthquake events

According to the results of the estimation probability distribution with the smallest AIC criteria, the recurring earthquake data set for each magnitude evaluated in this study approximately follows one of the following probability distributions: Lognormal, Weibull, Gumbel, and Exponential distribution. However, the Lognormal distribution is more appropriate than the Gaussian (Normal) or Weibull distributions for data of actual repeat time/difference between repeat times [27]. Another study showed that the Lognormal model provided only a slightly better fit than the Gaussian (Normal) or Weibull models by choosing a set of models without any physical or theoretical basis [28]. The parameters of those distributions used in this study are estimated through MLE and shown in Table 5.

Table 5. Parameter estimations of the selected probability distributions at all evaluated magnitudes

Magnitude (M)	Probability Distribution Estimated	Parameter	
		μ	σ
M ≥ 5.0	Lognormal, LOGNO (μ, σ)	1.166	0.2734
M ≥ 5.1	Lognormal, LOGNO (μ, σ)	1.380	0.3582
M ≥ 5.2	Lognormal, LOGNO (μ, σ)	1.541	0.4091
M ≥ 5.3	Lognormal, LOGNO (μ, σ)	1.761	0.4690
M ≥ 5.4	Weibull, WEI (μ, σ)	2.745	-0.3967
M ≥ 5.5	Weibull, WEI (μ, σ)	2.950	-0.3722
M ≥ 5.6	Gamma, GA (μ, σ)	3.390	0.2840
M ≥ 5.7	Gamma, GA (μ, σ)	3.591	0.1924
M ≥ 5.8	Exponential, EX (μ)	3.913	-
M ≥ 5.9	Exponential, EX (μ)	4.062	-
M ≥ 6.0	Exponential, EX (μ)	4.117	-
M ≥ 6.1	Exponential, EX (μ)	4.021	-
M ≥ 6.2	Gamma, GA (μ, σ)	4.077	0.2909
M ≥ 6.3	Exponential, EX (μ)	3.802	-

Once the parameters of those distributions are known, the p.d.f. and c.d.f. of the distributions are able to be estimated through the equations in Table 1. Furthermore, the hazard rate is also derivable from p.d.f. and c.d.f. due to its relation as described in Equation (1). The earthquake behavior could be observed through the hazard rate of the recurring earthquake duration. Along with the hazard definition, the hazard rate in this study implies the instantaneous rate of recurring earthquake events with a specific magnitude evaluated after not having any event with this magnitude within t days. From another perspective, the hazard rate indicates the number of failure events over a specific time period. In this study, the hazard rate shows the number of recurring earthquake events per day at the evaluated time t days. Figure 2 shows the comparison of hazard rates for all magnitudes evaluated. The X-axis represents the duration (days), while Y-axis represents the hazard rate.

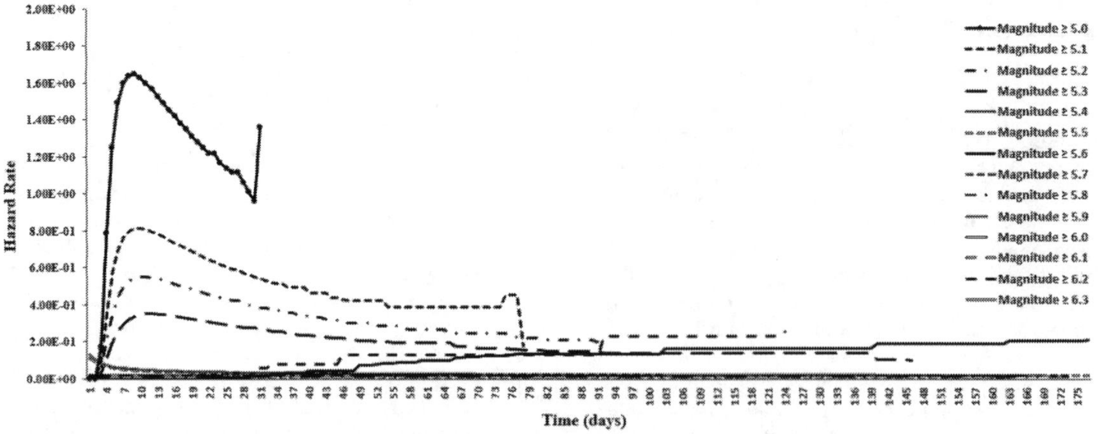

Fig. 2. Hazard rates comparison for recurring earthquake events at all evaluated magnitudes

As shown in Fig. 2, the recurring earthquakes for each magnitude evaluated have a different hazard rate. There are four main trends of the hazard rate behavior, those which increase until a certain time t and then decrease for the rest of the time, those which decrease over time t, those which increase over time t and those which are constant with time t. This result implies that, in general, the hazard rate is dependent on time t. In other words, the number of possible earthquake events per day is different within all times t for a specific magnitude evaluated. Note that the length of the hazard rates over the time t are different for each magnitude evaluated due to the fact that those magnitudes have different maximum time to happen as shown in Table 3. The longer time period, the longer the length of the hazard rate.

Specifically, the hazard rates for recurring earthquakes with magnitudes bigger than 5.0 but smaller than 5.4 would likely increase significantly until around time ten days after the initial earthquake and then decrease for the following time t to the maximum time that it could possibly happen, which also can be seen in Figure 2. Among those magnitudes, lower magnitudes tend to have a bigger hazard rate (i.e. would more likely to happen). Also, recurring earthquakes for $M \geq 5.4$ and $M \geq 5.5$ tend to have decreasing hazard rate trends. However, the rates are relatively small compared to the previous magnitudes evaluated. The following two bigger magnitudes ($M \geq 5.6$ and $M \geq 5.7$) have contrast behavior which increases with time t at the maximum hazard rate of 0.20 for $M \geq 5.6$ and 0.008 for $M \geq 5.7$. These results confirm that the earthquake phenomenon in the Sumatra territory could happen anytime with different hazard rates for magnitude under 5.8. While for recurring earthquakes with bigger magnitudes, hazard rates are constant for any point in time t except for $M \geq$ 6.2.

These results imply that the hazard rate for these specific magnitudes under 5.8 are dependent on time, while bigger magnitudes are independent of time. Some studies reported time-dependent seismic hazard assessment in several regions [29–31], while some other researchers investigated both time-dependent and time-independent seismic hazard assessments in some areas like California [32] and Greece [33]. Furthermore, an earthquake with bigger magnitude tends to occur at a lower rate compared to earthquakes with smaller magnitudes. However, for a massive magnitude such as Sumatra-Andaman earthquake in 2004, it will take at least 300 to 400 years before an earthquake with similar magnitude will occur again [7].

3.4 Comparison analysis of duration and distance between recurring earthquake events

In this section, the duration and distance between recurring earthquake events for all magnitudes evaluated are compared by using Analysis of Variance (Anova) or Kruskal-Wallis test when applicable. The first step in this analysis is to check whether or not the data set meets the normality assumption by using the Shapiro-Wilk test and homogeneity variances by Levene's test. Shapiro-Wilk test is generally more powerful than the Kolmogorov-Smirnov test [34,35]. Shapiro-Wilk test can be used when the sample size is between 3 and 5,000 [36]. The null hypothesis in the test states that the data set follows normal distribution. In this study, the null hypothesis will likely reject when the p-value is smaller than 0.10. It is important to note that, from the previous section we know that the duration data set for all magnitudes evaluated are not normally distributed. Hence, the Shapiro-Wilk test is only employed for the distance data set. The results of the Shapiro-Wilk test of distance data set for all magnitudes evaluated are presented in Table 6.

Table 6. Shapiro-Wilk test for distance between the recurring earthquake events at all evaluated magnitudes

Magnitude (M)	Distance (km)		
	Shapiro-Wilk Test		Decision
	Statictics	p-value	
M ≥ 5.0	0.921	<0.001*	Not Normal
M ≥ 5.1	0.924	<0.001*	Not Normal
M ≥ 5.2	0.933	<0.001*	Not Normal
M ≥ 5.3	0.917	<0.001*	Not Normal
M ≥ 5.4	0.908	<0.001*	Not Normal
M ≥ 5.5	0.932	<0.001*	Not Normal
M ≥ 5.6	0.947	0.005*	Not Normal
M ≥ 5.7	0.946	0.047*	Not Normal
M ≥ 5.8	0.919	0.038*	Not Normal
M ≥ 5.9	0.919	0.065*	Not Normal
M ≥ 6.0	0.849	0.008*	Not Normal
M ≥ 6.1	0.842	0.017*	Not Normal
M ≥ 6.2	0.862	0.081*	Not Normal
M ≥ 6.3	0.821	0.048*	Not Normal

* indicates the test is significant

Similar to duration data in Table 4, Table 6 shows that all distance data sets do not meet the normality assumption due to the p-value of the Shapiro-Wilk test that is less than 0.10. Moreover, Levene's test is used to check whether the variances of k samples are equal. Levene's test is powerful and robust to non-normality and has become a very popular tool for checking the homogeneity of variances [37]. The null hypothesis of Levene's test states that the variances of k samples are equal (homogenous) and will likely reject when the p-value is smaller than 0.10. The Levene statistics for duration and distance data set are 1.778 and 52.649 respectively. The null hypothesis for both variables is rejected due to small p-values. The p-values for duration and distance are 0.041 and <0.001 respectively. Those values are smaller than 0.10 which means that the variances of duration and distance data on 14 magnitudes evaluated are not equal.

Accordingly, the nonparametric Kruskal-Wallis test is used as an alternative method. Kruskal-Wallis test is a preferred procedure in many situations to compare more than two independent samples [38]. The Kruskal-Wallis test can be used as an alternate method to Anova [39]. In this case, the null hypothesis of the test states that all magnitudes evaluated have a similar mean rank instead of a mean value. Rejecting the null hypothesis means that there are at least two magnitudes evaluated that have different mean rank in duration or distance data. In this study, the null hypothesis will likely be rejected when the p-value is smaller than 0.10. Table 7 presents the result of the Kruskal-Wallis test.

Table 7. Kruskal-Wallis test for duration and distance between recurring earthquake events

Statistics	Duration time	Distance
H	193.244	7.763
p-value	<0.001*	0.933

* indicates the test is significant

According to Table 7, only duration data has a different mean rank between each magnitude evaluated due to a small p-value (<0.001). While for distance data, there is no statistical evidence that different magnitudes evaluated have different mean ranks of distance. This result confirms that recurring earthquake events with a random magnitude will occur in similar mean distance to other events with specific magnitudes. On average, the recurring earthquake events with any larger magnitude to the previous one is not necessary to happen in the nearer or farther area from the epicentre of the first earthquake. This finding is in line with the descriptive statistics in Table 3 where the mean distance between magnitudes is close to each other, averaging about 492.42 kilometers.

Furthermore, nonparametric Mann-Whitney U test is employed for the duration data in order to know which magnitudes have different median duration. The Mann-Whitney U test is a statistical test that can be used to compare two independent groups that do not require large normally distributed samples [24]. In this study, Mann-Whitney U test compares two magnitudes to prove whether or not these magnitudes have a different duration. In this study, a pair with a p-value less than 0.10 is considered to have different medians of time between an earthquake and recurring earthquake. According to combination theory, all possible combinations of r from m total categories is determined by the following equation:

$$\binom{m}{r} = \frac{m!}{(m-r)!r!} \quad (11)$$

where $n! = n \cdot (n-1) \cdot (n-2) \cdot \ldots 3 \cdot 2 \cdot 1$ [40]. Therefore, there are 91 possible combinations of 2 from 14 magnitudes evaluated using Equation (11). For convenience and simplicity, the p-values of the Mann-Whitney U test for those 91 possible combinations are presented in a matrix table form with a size of 14 × 14 (Table 8).

Table 8. p-value of the Mann-Whitney U tests for all possible comparative analysis

Mag.	M ≥ 5.0	M ≥ 5.1	M ≥ 5.2	M ≥ 5.3	M ≥ 5.4	M ≥ 5.5	M ≥ 5.6	M ≥ 5.7	M ≥ 5.8	M ≥ 5.9	M ≥ 6.0	M ≥ 6.1	M ≥ 6.2	M ≥ 6.3
M ≥ 5.0	-	0.029*	0.001*	<0.001*	<0.001*	<0.001*	<0.001*	<0.001*	<0.001*	<0.001*	<0.001*	<0.001*	0.002*	0.005*
M ≥ 5.1	0.029*	-	0.203	0.006*	0.001*	<0.001*	<0.001*	<0.001*	<0.001*	<0.001*	<0.001*	<0.001*	0.004*	0.008*
M ≥ 5.2	0.001*	0.203	-	0.135	0.029*	0.002*	<0.001*	<0.001*	<0.001*	<0.001*	<0.001*	0.001*	0.009*	0.016*
M ≥ 5.3	<0.001*	0.006*	0.135	-	0.376	0.062*	0.012*	<0.001*	<0.001*	<0.001*	<0.001*	0.003*	0.024*	0.036*
M ≥ 5.4	<0.001*	0.001*	0.029*	0.376	-	0.330	0.098*	0.007*	<0.001*	<0.001*	0.001*	0.010*	0.052*	0.071*
M ≥ 5.5	<0.001*	<0.001*	0.002*	0.062*	0.330	-	0.432	0.055*	0.004*	0.002*	0.003*	0.036*	0.121	0.149
M ≥ 5.6	<0.001*	<0.001*	<0.001*	0.012*	0.098*	0.432	-	0.248	0.028*	0.010*	0.014*	0.089*	0.232	0.276
M ≥ 5.7	<0.001*	<0.001*	<0.001*	<0.001*	0.007*	0.055*	0.248	-	0.203	0.076*	0.059*	0.222	0.436	0.466
M ≥ 5.8	<0.001*	<0.001*	<0.001*	<0.001*	<0.001*	0.004*	0.028*	0.203	-	0.612	0.458	0.836	1.000	1.000
M ≥ 5.9	<0.001*	<0.001*	<0.001*	<0.001*	<0.001*	0.002*	0.010*	0.076*	0.612	-	0.762	0.975	0.829	0.804
M ≥ 6.0	<0.001*	<0.001*	<0.001*	<0.001*	0.001*	0.003*	0.014*	0.059*	0.458	0.762	-	0.775	0.665	0.597
M ≥ 6.1	<0.001*	<0.001*	0.001*	0.003*	0.010*	0.036*	0.089*	0.222	0.836	0.975	0.775	-	0.906	0.891
M ≥ 6.2	0.002*	0.004*	0.009*	0.024*	0.052*	0.121	0.232	0.436	1.000	0.829	0.665	0.906	-	1.000
M ≥ 6.3	0.005*	0.008*	0.016*	0.036*	0.071*	0.149	0.276	0.466	1.000	0.804	0.597	0.891	1.000	-

* indicates the test is significant

This matrix is symmetric due to the fact that comparison for example between M ≥ 5.0 and M ≥ 5.5 is similar to comparison between M ≥ 5.5 and M ≥ 5.0, and so forth. The diagonal does not have a p-value since the Mann-Whitney U test only compares two different groups. Table 8 shows that the comparison between the magnitudes of up to M ≥ 5.9 each other will conclude a similar decision (they have different median), except for sequential magnitude comparison such as between M ≥ 5.9 and M ≥ 5.8 or between M ≥ 5.8 and M ≥ 5.7. Those pairs with small p-values have a different median of duration. Additionally, the comparison between any magnitudes lower than M ≥ 5.5 (M ≥ 5.0, M ≥ 5.1, …, M ≥ 5.4) and the rest (M ≥ 5.5, M ≥ 5.6 , …, M ≥ 6.3) are also significant as marked with sign (*). In contrast, the comparison between recurring earthquakes with bigger magnitudes (M ≥ 6.0) are not significant due to larger p-values (>0.10). This finding proves that recurring earthquakes with bigger magnitude (M ≥ 6.0) do not necessarily happen for a different length of time. As presented in Table 3, the median duration between recurring earthquakes with lower magnitudes have an increasing trend, which supports the Mann-Whitney U test. Therefore, it can be concluded that the recurring earthquakes for lower magnitude (under 6.0) will happen for a different duration. The lower the magnitude, the faster the recurring earthquake event will happen, while recurring earthquakes for bigger magnitudes (M ≥ 6.0) tend to happen for similar durations.

4 Conclusions

In the present study, the earthquake behavior in the Sumatra territory was assessed based on the duration and distance between an earthquake and the next (later) earthquake event (termed as recurring earthquake) which had a magnitude greater than or equal to the previous one. The survival analysis approach revealed that recurring earthquakes with specific magnitude evaluated have different hazard rates following their probability distributions. Earthquakes with magnitude under 5.8 could happen anytime with different rates, while the hazard rates are constant for larger magnitudes except for those events with $M \geq 6.2$. Furthermore, the duration as well as the distance between the recurring earthquakes at any magnitude evaluated are compared with each other and it was found that the recurring earthquake events tend to happen at a similar average distance for any magnitude. However, the duration between recurring earthquake events is different particularly for lower magnitude under 6.0. This research has significant implications for earthquake researchers and the Indonesian government in understanding the phenomenon of earthquakes in Sumatra territory.

References

1. D. H. Natawidjaja, *Gempabumi Dan Tsunami Di Sumatra Dan Upaya Untuk Mengembangkan Lingkungan Hidup Yang Aman Dari Bencana Alam* (Jakarta, 2007)
2. R. Paris, F. Lavigne, P. Wassmer, and J. Sartohadi, Mar. Geol. **238**, 93 (2007)
3. J. Jatnika, A. D. Nugraha, and Wandono, J. Meteorol. Dan Geofis. **16**, 113 (2016)
4. L. D. Rifai and I. P. Pudja, J. Meteorol. Dan Geofis. **11**, 147 (2010)
5. S. Anwar, J. Lingkung. Dan Bencana Geol. **12**, 33 (2021)
6. T. Lay, H. Kanamori, C. J. Ammon, M. Nettles, S. N. Ward, R. C. Aster, S. L. Beck, S. L. Bilek, M. R. Brudzinski, R. Butler, H. R. DeShon, G. Ekström, K. Satake, and S. Sipkin, Science (80-.). **308**, 1127 (2005)
7. F. Løvholt, H. Bungum, C. B. Harbitz, S. Glimsdal, C. D. Lindholm, and G. Pedersen, Nat. Hazards Earth Syst. Sci. **6**, 979 (2006)
8. K. Satake and B. F. Atwater, Annu. Rev. Earth Planet. Sci. **35**, 349 (2007)
9. S. Stein and E. A. Okal, Nature **434**, 581 (2005)
10. C. Subarya, M. Chlieh, L. Prawirodirdjo, J.-P. Avouac, Y. Bock, K. Sieh, A. J. Meltzner, D. H. Natawidjaja, and R. McCaffrey, Nature **440**, 46 (2006)
11. R. Paris, M. Cachão, J. Fournier, and O. Voldoire, Géomorphologie Reli. Process. Environ. **16**, 109 (2010)
12. NSF, (2005)
13. M. Walton, (2005)
14. S. Anwar, J. Lingkung. Dan Bencana Geol. **10**, 159 (2019)
15. E. Lusiani, S. Anwar, and M. F. Nugraha, J. Meteorol. Dan Geofis. **19**, 71 (2018)
16. J. J. Clague, C. Yorath, R. Franklin, and B. Turner, *At Risk : Earthquakes and Tsunamis on the West Coast* (Tricouni Press, Vancouver, 2006)
17. Á. González, J. B. Gómez, and A. F. Pacheco, J. Seismol. **10**, 131 (2006)
18. D. Shanker and A. Harbindu, in *13 Th World Conf. Earthq. Eng.* (Vancouver, B.C., Canada, 2004), p. Paper No. 3024
19. S. Zacks, *Introduction to Reliability Analysis : Probability Models and Statistical Methods* (Springer, New York, 1992)
20. R. Jiang, *Introduction to Quality and Reliability Engineering* (Springer, Berlin, Heidelberg, 2015)
21. H. Akaike, IEEE Trans. Autom. Control **19**, 716 (1974)
22. R. Rigby, M. Stasinopoulos, G. Heller, and F. De Bastiani, *Distributions for Modelling Location, Scale and Shape: Using GAMLSS in R* (2017)
23. D. C. Montgomery and G. C. Runger, *Applied Statistics and Probability for Engineers* (John Wiley & Sons, Inc., Hoboken, New Jersey, 2003)
24. N. Nachar, Tutor. Quant. Methods Psychol. **4**, 13 (2008)
25. J. Jatnika, A. D. Nugraha, and W. Wandono, J. Meteorol. Dan Geofis. **16**, (2016)
26. A. J. Michael, Bull. Seismol. Soc. Am. **95**, 1594 (2005)
27. B. C. Papazachos and C. A. Papaioannou, Pure Appl. Geophys. PAGEOPH **140**, 593 (1993)
28. S. P. Nishenko and R. Buland, Bull. Seismol. Soc. Am. **77**, 1382 (1987)
29. C.-H. Chan, Y.-M. Wu, C.-T. Cheng, P.-S. Lin, and Y.-C. Wu, Nat. Hazards Earth Syst. Sci. **13**, 1143 (2013)
30. G. F. Panza, A. Peresan, A. Magrin, F. Vaccari, R. Sabadini, B. Crippa, A. M. Marotta, R. Splendore, R. Barzaghi, A. Borghi, L. Cannizzaro, A. Amodio, and S. Zoffoli, Nat. Hazards **69**, 1179 (2013)
31. G. P. Biasi and S. C. Thompson, Bull. Seismol. Soc. Am. **108**, 39 (2018)
32. M. D. Petersen, T. Cao, K. W. Campbell, and A. D. Frankel, Seismol. Res. Lett. **78**, 99 (2007)
33. C. A. Papaioannou and B. C. Papazachos, Bull. Seismol. Soc. Am. **90**, 22 (2000)

34. Z. Hanusz, J. Tarasinska, and W. Zielinski, REVSTAT-Statistical J. **14**, 89 (2016)
35. N. M. Razali and Y. B. Wah, J. Stat. Model. Anal. **2**, 21 (2011)
36. P. Royston, J. R. Stat. Soc. Ser. C (Applied Stat. **44**, 547 (1995)
37. J. L. Gastwirth, Y. R. Gel, and W. Miao, Stat. Sci. **24**, 343 (2009)
38. A. Vargha and H. D. Delaney, J. Educ. Behav. Stat. **23**, 170 (1998)
39. R. R. Pagano, *Understanding Statistics in the Behavioral Sciences*, Fourth Edi (WestPublishing Company, Eagan, Minnesota, 1994)
40. S. Ross, *A First Course in Probability*, Eighth Edi (Prentice Hall, Inc., New Jersey, 2010)

Understanding of flow liquefaction phenomena in Palu City from shear wave velocity profiles

Adrin Tohari[1*], *Imamal* Muttaqien[2], and *Resi* Wasilatus Syifa[3]

[1]Research Center for Geotechnology, National Research and Innovation Agency (BRIN), Bandung, West Java, 40135, Indonesia
[2]Faculty of Science and Technology, State Islamic University of Sunan Gunung Djati (IUN), Bandung, West Java 40614, Indonesia
[3]Former undergraduate student, Physics Study Program, State Islamic University of Sunan Gunung Djati (UIN), Bandung, West Java 40614, Indonesia

Abstract. On 28 September 2018, a strong earthquake (Mw 7.5) struck Donggala Regency triggering flow liquefaction (locally termed *Nalodo*) phenomena in two residential areas in Palu City. Mitigating such an earthquake collateral hazard requires understanding the geological condition controlling such liquefaction. This paper presents a geophysical study to clarify the geological condition controlling the *Nalodo* phenomena in Palu City. The objectives of this study were to estimate shear wave velocity (V_s) profiles in the *Nalodo* and non-*Nalodo* sites and evaluate liquefaction potential based on the shear wave velocity profiles. The study involved a series of array microtremor surveys in two *Nalodo* sites and three non-*Nalodo* sites in Palu City. The results show that V_s generally increases with depth but varies from one site to another. Based on the V_s profiles, a thick soil layer with V_s of less than 175 m/s is present from the ground surface in the Balaroa and Petobo sites. Contrary, this soil layer is detected at a deeper depth in the non-*Nalodo* sites. Based on the liquefaction potential analysis, the near-surface soil layers in the Balaroa and Petobo sites are clearly liquefiable for all the PGA values used in this study. Although deeper soil layers can also be liquefied, no liquefaction occurred in non-*Nalodo* sites because a thick medium soil layer exists near the surface. Thus, a thick soil layer with V_s of less than 175 m/s near the ground surface controls the *Nalodo* phenomena in Palu City. Further studies to clarify other factors controlling flow liquefaction are recommended.

1 Introduction

The 28 September 2018 strong earthquake (M_w 7.5) triggered by the movement of the Palu-Koro Fault zone also caused massive flow liquefaction phenomena (locally termed *Nalodo*) in highly populated areas in the Balaroa dan Petobo sub-district in Palu City (Figure 1). The flow liquefaction phenomena have annihilated residential areas and claimed many lives in two affected areas. Since future earthquake events have still potentially triggered flow liquefaction, understanding the sub-surface condition leading to flow liquefaction is requisite to establish proper effort to mitigate such liquefaction.

According to [1, 2], a flow liquefaction failure commonly takes place in gently sloping areas with a slope of more than 5% and is induced by the presence of a thin sand layer below the dense soil layer [2]. However, the flow liquefaction phenomenon in Palu City only took place locally in Balaroa and Petobo areas. This discrepancy in flow liquefaction vulnerability is probably associated with the sub-surface geological condition that varies from one region to another.

The understanding of liquefaction phenomena is generally obtained from sub-surface geotechnical investigations such as geotechnical drilling with standard penetration tests (*SPT*) and cone penetration

Fig. 1. Flow liquefaction disaster phenomena in Balaroa (above), and Petobo areas (below) due to the 28 September 2018 earthquake (M_w 7.5).

* Corresponding author: adrin.tohari@gmail.com

tests (CPT). Although these geotechnical methods provide detailed geotechnical data essential for liquefaction analysis, such as soil types, soil strength, and groundwater level, these methods are time-consuming and costly.

Many recent studies have applied geophysical methods to study flow liquefaction phenomena in Palu City [3], [4], [5] [6]. While all these studies have provided information on the sub-surface geology and hydrological condition of the flow liquefaction areas, none has attempted to evaluate the spatial variation of sub-surface geological conditions of the flow liquefaction and no flow liquefaction areas. Thus, a geophysical study that provides a substantial understanding of the geological condition controlling flow liquefaction phenomena is favorable.

This paper presents the results of a geophysical investigation using the array microtremor method to provide an understanding of the geological condition controlling flow liquefaction phenomena in Palu City. The objectives of the study were to (1) estimate shear wave velocity (V_s) profiles and (2) evaluate liquefaction potential based on the shear wave velocity profiles in flow liquefaction and no flow liquefaction sites.

2 Geological condition and tectonic setting of Palu City

2.1 Geological condition

According to the geomorphological classification [7], Palu City can be classified into three geomorphological classes: fluvial and structural denudational origins. The fluvial origins dominate the center part of Palu City and consist of alluvial fan, river beds, and flood plains with a slope inclination of 0 to 5%. Meanwhile, the structural denudational origins consist of mainly rolling and hilly topography and hilly to mountainous ridges with a slope inclination of higher than 5%. These structural denudational units extend to the north and east Palu regions. Field observations conducted after the earthquake showed that all liquefaction manifestations occurred in the fluvial geomorphological units.

Geologically, Palu City is mainly characterized by two different types of rock formations, i.e., alluvial and coastal deposits (Qap) and Celebes Molasse (QTms) [8]. The alluvial and coastal deposits are the youngest sediments in this area and are likely Holocene in age. They consist of gravel, sand, silt, and coral limestones deposited in rivers, deltas, and shallow seas.

According to [9], the alluvial deposits can be divided into young and old alluvial fan deposits, flood, alluvial, and old river channel deposits. The alluvial fan deposits are found on the hilly lowlands stretching on the east and west sides of the valley. According to Figure 2, flow liquefaction phenomena in the Balaroa and Petobo areas involved the alluvial and old alluvial fan deposits.

On the other hand, the Celebes Molasses consists of sandstones, conglomerates, mudstones, marls, and coral limestones, which are poorly consolidated and originated from older formations [8]. They are found on both ridges at lower altitudes, unconformably overlying

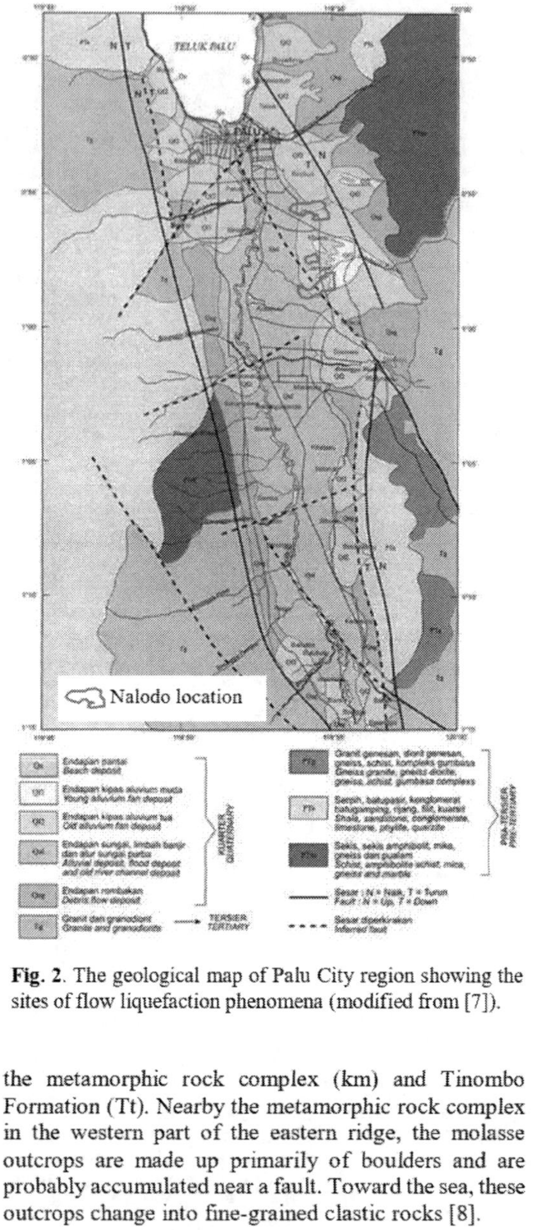

Fig. 2. The geological map of Palu City region showing the sites of flow liquefaction phenomena (modified from [7]).

the metamorphic rock complex (km) and Tinombo Formation (Tt). Nearby the metamorphic rock complex in the western part of the eastern ridge, the molasse outcrops are made up primarily of boulders and are probably accumulated near a fault. Toward the sea, these outcrops change into fine-grained clastic rocks [8].

2.2 Tectonic setting

Tectonically, the geological structure in the Palu City area is primarily controlled by the main active fault, namely Palu-Koro Fault, with left-lateral movement and an NNW–SSE trend [10-14]. This strike-slip fault is recognizable within the basin, crosses the lower portion of old alluvial fans, and is characterized by the intra-basin ridge in the slightly uplifted ground surface from the adjacent surfaces [15]. The 28 September 2018 earthquake (M_w 7.5) revealed the main Palu-Koro fault line, traceable from Sigi Regency, Palu City, and Palu Bay [14, 16]. The fault movement yielded a maximum slip of 6±0.5 m within Palu City, with an average of up to 4.7 m in the northern and southern parts of the city [16].

https://doi.org/10.1051/e3sconf/202234001011

3 Methods

3.1 Microtremor data acquisition and processing

A series of array microtremor measurements were conducted using four McSEIS MT-NEO accelerometers to estimate shear wave velocity (V_s) profiles in 5 sites, as shown in Figure 3 and summarized in Table 1. At each site, one microtremor sensor was placed at the center of the circle with a radius, r. The other three were placed on the circle with the shape of a regular triangle. The measurement duration and the sampling frequency were 45 minutes and 100 Hz, respectively. Sequential measurements were carried out five times by changing the array radius; r = 5, 10, 20, 40, and 80 m.

The measurement data was transformed from the time domain into the frequency domain. Then, each dispersion curve (phase velocity vs. frequency) was calculated using the SPAC method [18]. After processing each array individually, the dispersion curve for the 5, 10, 20, 40, 80 m arrays were combined to form a composite dispersion curve. Then, inversion analyses were undertaken on the combined dispersion curve to estimate a shear wave velocity (V_s) profile for each site. Then, based on the V_s profile at each site, the average shear wave velocity to a depth of 30 meters (V_s30) was calculated according to Equation 1 to determine the site class.

$$V_s30 = \frac{30}{\Sigma_i N \frac{h_i}{V_i}} \tag{1}$$

, where N is the number of soil layers, h_i is the thickness of each soil layer, and V_i is the shear wave velocity for each soil layer.

Table 1. Array microtremor measurement sites.

Site	Coordinates		Remarks
Balaroa	0°54.425'S	119°50.541'E	F*
Boyaoge	0°54.895'S	119°51.189'E	NF**
South Tatura	0°55.015'S	119°52.492'E	NF**
North Birobuli	0°55.096'S	119°54.044'E	NF**
Petobo	0°56.623'S	119°54.961'E	F*

* Flow liquefaction ** No flow liquefaction

3.2 Liquefaction potential analysis

Analysis of liquefaction potential with the shear wave velocity method was conducted in 3 (three) stages: (1) calculation of cyclic stress ratio (*CSR*), (2) calculation of cyclic resistance ratio (*CRR*), and (3) calculation of the factor of safety against liquefaction (*FS*).

The calculation of *CSR* was conducted using a simplified equation of [19] as shown in Equation 2, where a_{max} is the maximum earthquake acceleration at the ground surface (*PGA*), g is the acceleration of gravity, σ_v is the initial total vertical stress at depth z, σ'_v is the initial effective vertical stress, and r_d is the shear stress reduction factor that accounts for the dynamic response of the soil profile, according to [20].

$$CSR = 0.65 \times \frac{a_{max}}{g} \frac{\sigma_v}{\sigma'_v} \times r_d \tag{2}$$

The 28 September 2018 earthquake (M_w 7.5) produced a *PGA* of 0.34g, according to [21]. Meanwhile, according to [22], the *PGA* value in all study sites is in the range of 0.4 to 0.6g. This current study applied *PGA* values of 0.33 to 0.6g for calculating *CSR* by considering the site coefficient based on the site class [22].

Fig. 3. Array microtremor measurement sites: (a) Balaroa, (b) Boyaoge, (c) South Tatura, (d) North Birobuli, and (e) Petobo.

Meanwhile, the cyclic resistance ratio (*CRR*) is calculated based on the shear wave velocity [23], as shown in Equation 3. V_{S1} is the stress-corrected shear-wave velocity (m/s), and V^*_{S1} is the limiting upper value of V_{S1} for cyclic liquefaction occurrence, which varies between 200 and 215 m/s, depending on the fines content of the soil, and *MSF* is a magnitude scaling factor. V_{S1} is given by Equation 4 [24], where *Pa* is the reference stress of 100 kPa. The magnitude scaling factor is given by Equation 5 [25], where M_w is the earthquake moment magnitude.

$$CRR = \left[0.022\left(\frac{V_{S1}}{100}\right)^2 + 2.8\left(\frac{1}{V^*_{S1}-V_{S1}} - \frac{1}{V^*_{S1}}\right)\right] MSF \quad (3)$$

$$V_{S1} = V_S\left(\frac{P_a}{\sigma'_v}\right)^{0.25} \quad (4)$$

$$MSF = \left(\frac{M_w}{7.5}\right)^{-2.56} \quad (5)$$

Finally, the factor of safety against liquefaction (*FS*) was determined using Equation 6 [23]:

$$FS = CRR/CSR \quad (6)$$

By convention, when *FS* is less than 1, liquefaction is most likely to occur. On the other hand, liquefaction is unlikely to occur when *FS* is higher and equal to 1.

4 Results and discussions

4.1 Profil V_s and site class

Figure 4 shows the shear wave velocity (V_s) profile for each site obtained from the inversion process of each dispersion curve. It is evident from this figure that V_s values generally increase with depth but vary from one site to another. For example, in the Balaroa site, the V_s profile indicates that a bedrock layer is likely to exist from a depth of deeper than 90 m. The bedrock layer is much deeper in other sites than the Balaroa site. According to Figure 4, all sites are classified into site class D (S_D), with the values of V_s30 varying between 230 and 280 m/s.

The soil layer with V_s values of less than 175 m/s about 7 m thick from the ground surface is present in the Balaroa and Petobo sites. This soil layer of less than 3 m thick is also detected at depths of 2-3 m in the North Birobuli site. Meanwhile, the soil layer is detected deeper in the Boyaoge site than in the Balaroa and Petobo areas. It is also apparent from Figure 4 that this soil layer seems not to exist in the South Tatura site at a shallow depth. This evidence indicates that the presence of this thick soil layer of low V_s near the ground surface is one of the controlling factors of flow liquefaction phenomena in Palu City during the 28 September 2018 earthquake. In other words, the absence of this soil layer near the ground surface will lead to no liquefaction phenomena. The results of this study are apparently in good agreement with the sub-surface geotechnical drilling data conducted by JICA [26], which show that flow liquefaction occurrence in Balaroa and Petobo areas involved loosely-packed sand, with *N-SPT* values of less than 10.

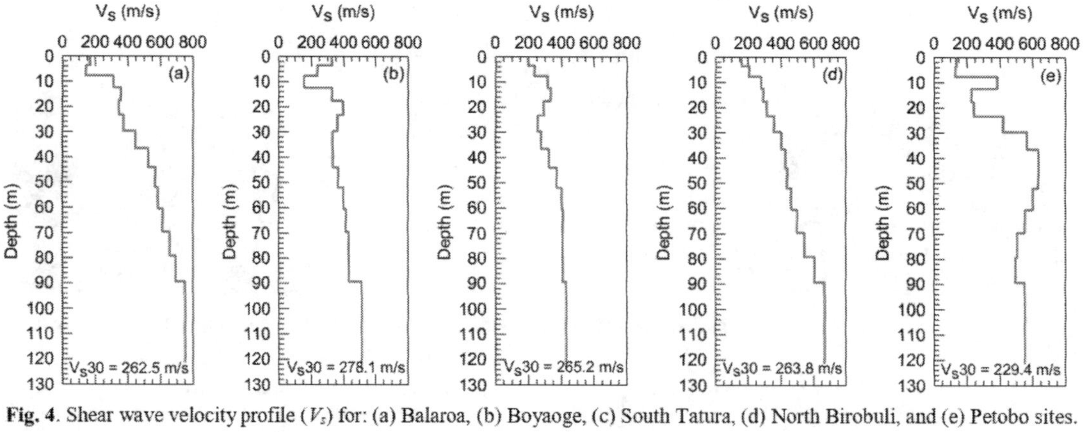

Fig. 4. Shear wave velocity profile (V_s) for: (a) Balaroa, (b) Boyaoge, (c) South Tatura, (d) North Birobuli, and (e) Petobo sites.

4.2 Liquefaction potential

The results of liquefaction potential analysis in all sites up to the depth of 30 m are shown in Figure 5. As seen in this figure, the thick soil layers of low shear wave velocity (V_s < 175 m/s) in Balaroa and Petobo sites up to the depth of up to 7 m are liquefiable for all the prescribed PGA values. It is also worth noting that this soil layer at depths of 2-3 m in the North Birobuli site is also liquefiable for all the prescribed PGA values. However, only sand boiling, rather than flow liquefaction, occurred at this site during the 28 September 2018 earthquake.

As also seen in Figure 5, soil layers at deeper depths in the Boyaoge, South Tatura, and Petobo sites are also liquefiable. However, no liquefaction manifestation was observed in Boyaoge and South Tatura sites during the earthquake. This discrepancy is probably associated with the existence of thick, hard soil layers near the ground surface, preventing liquefaction from occurring in these two sites.

5 Discussions

The results of this current study point out the role of a thick, soft to medium soil layer near the ground surface

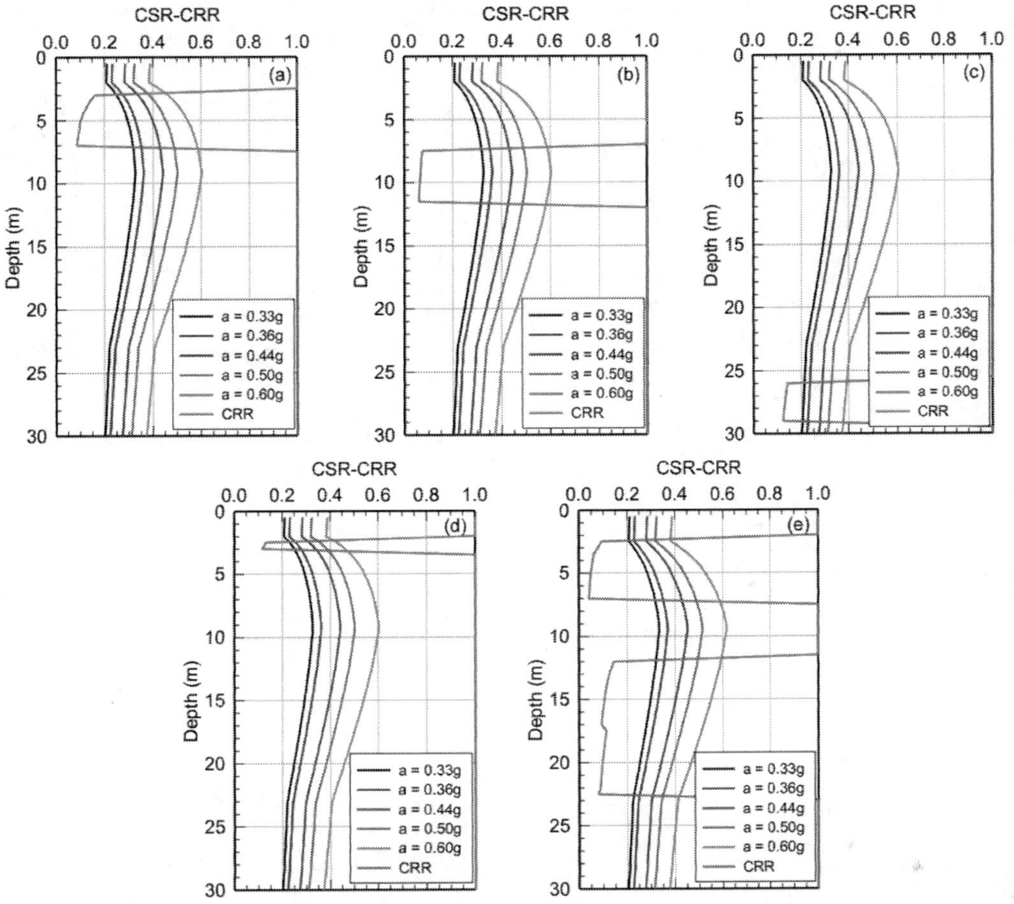

Fig. 5. Graph of CSR versus CRR for each site: (a) Balaroa, (b) Boyaoge, (c) South Tatura, (d) North Birobuli, and (e) Petobo.

as one of governing factors in the flow liquefaction occurrence in Balaroa and Petobo. However, although thick loosely-packed sediments may almost exist in Palu City, other factors may have a critical role in controlling flow liquefaction because flow liquefaction phenomena only occurred in Balaroa and Petobo areas. Based on microtremor measurements in the Balaroa, [4] found that the subsurface topographical condition in the flow liquefaction area seems to play an essential role in the occurrence of flow liquefaction at this site. Another geophysical study by [6] also indicated that the subsurface topographical conditions governed the flow liquefaction phenomena in Balaroa and Petobo areas. Thus, these studies imply the significance of a detailed subsurface study, using a combination of geophysical and geotechnical methods to establish a better understanding of subsurface geological and topographical conditions required for the potential flow liquefaction risk assessment in Palu City. Finally, massive flow liquefaction in Balaroa and Petobo may also occur in other coastal cities, built on sedimentary basins, prone to strong earthquake shaking.

6 Conclusions

Based on the analysis of all shear wave velocity profiles, the existence of a thick soil layer with V_s values less than 175 m/s near the ground surface controls the flow liquefaction phenomena in Palu City. Although deeper soil layers may liquefy, the presence of a thick, densely-packed soil layer near the ground surface may hinder liquefaction at the ground surface. While this current study has shown how near-surface soil layers control the flow liquefaction phenomena in Palu City, further studies aiming to clarify other factors controlling flow liquefaction are recommended.

Acknowledgment. The microtremor array measurements were made possible by the 2019 INSINAS research grant from the Ministry of Research, Technology, and Higher Education. The first author would also like to thank the Japan International Cooperation Agency (JICA) for providing a microtremor equipment grant in 2017.

References

1. T.L. Youd, Civil Eng. 48, 4, 47 (1978)
2. T.L. Youd, *Geologic effects—liquefaction and associated ground failure*, in Proceedings of the Geologic and Hydrologic Hazards Training Program, 5-30 March 1984, Denver, Colorado (1984)
3. M. Zeffitni, M. Basir-Cyio, S. Napitupulu, Worosuprojo, J. Phys.: Conf. Series. **1434** (2020).

4. A. Cipta, A. Rudyanto, H. Afif, R. Robiana, A. Solikhin, A. Omang, Supartoyo, S. Hidayati, *Unearthing the buried Palu-Koro fault and the pattern of damage caused by the 2018 Sulawesi Earthquake using HVSR inversion* in Characterization of Modern and Historical Seismic–Tsunamic Events, and Their Global–Societal Impacts. Editor / Y. Dilek; Y. Ogawa; Y. Okubo. Geol. Soc. London, Spec. Pub. **501(1)**, 185 (2021)
5. A. Yulianur, T. Saidi, B. Setiawan, Sugianto, M. Rusdi, J. Eng. Sci. Tech. **15**, 5, 2871 (2020)
6. A. Tohari, D.D. Wardhana, M. Hanif, K. Koizumi, E3S Web of Conferences, **331**, 03002 (2021).
7. R.A. van Zuidam, *Aerial photo interpretation terrain analysis and geomorphology mapping*, Smits Publisher, The Haque, Netherlands, 442 p (1985).
8. H. Sumadirdja, T. Suptandar, S. Hardjoprawiro, D. Sudana, *Reconnaissance geological map of the Palu quadrangle, Sulawesi*, Pusat Penelitian dan Pengembangan Geologi, Bandung (1973)
9. A. Soehaimi, M. Firdaus., I. Effendi, Earthquake *Susceptibility Zonation Map of Palu and its surrounding area*, Pusat Penelitian dan Pengembangan Geologi, Bandung (2000) in Indonesian
10. J.A. Katili, Geol Rundsch. **59**, 581 (1970)
11. H.D. Tjia, Bull. Geol. Soc. Malaysia **10**, 73 (1978)
12. O. Bellier, M. Sébrier, D. Seward, T. Beaudouin, M. Villeneuve, E. Putranto, Tectonophys. **413**, 201 (2006)
13. M.R. Daryono, *Paleoseismologi Tropis Indonesia (Dengan Studi Kasus di Sesar Sumatra, Sesar Palukoro-Matano, dan Sesar Lembang)*. Dissertation Doctoral Program. Institut Teknologi Bandung (2016). Unpublished
14. I.M. Watkinson, R. Hall, *Fault systems of the eastern Indonesian triple junction: evaluation of Quaternary activity and implications for seismic hazards Geohazards in Indonesia*, in Earth Science for Disaster Risk Reduction. editor / P. Cummins; I. Meilano, Geol. Soc. London **441**, 71 (2017)
15. A. Patria, P. S. Putra, Geosci. Lett. **7**, 1 (2020)
16. A.I. Abdullah, Abdullah, J. Phys.: Conference Series, **1434**, 012009 (2020)
17. H. Bao, J.P. Ampuero, L. Meng, E.J. Fiedling, C. Liang, C.W.D. Milliner, T. Feng, H. Huang, Nat. Geosci. **12**, 200 (2019)
18. K. Aki, Bull. Earth. Res. Inst. **35**, 3, 415 (1957)
19. H. B. Seed, I.M Idriss, J. Soil Mech. Found. Div., ASCE. **97(SM9)**, 1249 (1971).
20. S.S.C. Liao, R.V. Whitman, J. Geotech. Engrg. **112**, 3, 373 (1986).
21. T. Kiyota, H. Furuichi, R.F. Hidayat, N. Tada, H. Nawir, Soils Found **60**, 722 (2020)
22. Badan Standarisasi Nasional, SNI 1726:2019 (2019)
23. R.D. Andrus, K.H. Stokoe II, J. Geotech. Geoenviron. Eng., ASCE. **126(11)**, 1015 (2000)
24. P.K. Robertson, D.J. Woeller, W.D.L. Finn, Can. Geotech. J. **29**, 4, 686 (1992)
25. T.L Youd, I.M. Idriss, R.D. Andrus, I. Arango, G. Castro, J.T. Christian, R. Dobry, W.D.L. Finn, L.F.Harder Jr., M.E. Hynes, K. Ishihara, J.P. Koester, S. S. C Liao, W.F. Marcuson III, G.R. Martin, J.K Mitchell, Y. Moriwaki, M.S. Power, P.K. Robertson, R.B. Seed, K.H. Stokoe II, J. Geotech. Geoenviron. Eng., ASCE, **127**, 10, 817 (2001).
26. JICA, *Project for Development of Regional Disaster Risk Resilience Plan in Central Sulawesi*, Project Report (2019).

Focal mechanism and source determination of the Padang Lawas earthquake april 30, 2020, using moment tensor inversion

Retno Agung Prasetyo[1*], *Supriyanto* Rohadi[1], *Nelly* Florida Riama[1], *Yusuf* Hadi Perdana[1], *Rahmat* Setyo Yuliatmoko[1], *Lewi* Ristiyono[2] and *Marzuki* Sinambela[2]

[1]Research and Development Centre, Meteorological Climatology and Geophysical Agency (BMKG), Jl. Angkasa I No.2 Kemayoran, Jakarta Pusat, 10610, Indonesia
[2]Meteorological, Climatological And Geophysical Regional I (BBMKG Wilayah I), Medan, 20131, Indonesia

Abstract. The earthquake focal mechanism is one of the essential parameters in understanding the characteristics of an earthquake source. For example, the earthquake mainshock at the Padang Lawas Regency and South Tapanuli Regency on April 30, 2020, caused some damage to houses, schools, and mosques, especially in the South Tapanuli Regency. BMKG released the first information that the earthquake was 1.17N, 99.53E, 24 kilometers northwest of Padang Lawas, North Sumatra province, at a depth of 10 kilometers, Thursday (30/4/2020) at 15.20.25 Local Time with a magnitude of 5.6. The purpose of this study is to determine the focal mechanism of the Padang Lawas earthquake mainshock on April 30, 2021, and determine the fault nodal plane that best suits local tectonic conditions. We used the moment tensor inversion method by using waveform velocity data from the BMKG mini regional station seismic network installed in 2019 and close to the epicenter, namely TTSM, PLSM, LTSM, and RRSI. We also used the velocity model from AK135. In this study, we applied the filter technique using a Butterworth bandpass filter with a lower limit of 0.05 Hz, an upper limit of 0.10 Hz, and a cut frequency of 0.13 Hz. The result of the moment tensor inversion shows the variance of the observation and calculation data is 0.3824. The earthquake source parameters show that the Nodal I plane with strike/dip/slip values is 100.1/88.3/7.5 and the Nodal II plane with strike/dip/slip values is 9.9/82.5/178.3. The value of Seismic Moment (Mo) is 0.4653E+17 Nm or Mw 5. Based on the distribution of aftershocks and the mechanism pattern of the Tapanuli earthquake source on April 30, 2020, it is consistent with the nodal plane II. The mechanism is in the form of a strike-slip fault and corresponds to the movement mechanism of the Sumatran fault system in the Toru segment. Therefore, the tectonic conditions in the southern part of Padang Lawas Regency and South Tapanuli Regency are classified as active. Thus, the government and society need to be aware and make mitigation efforts against the impact of the next earthquake in the future.

1 Introduction

Sumatra is a part of Indonesia that has been frequently hit by significant and destructive earthquakes in the last two decades. The major earthquake in Sumatra was sourced from tectonic activity in the subduction zone in the west of Sumatra under the Indian Ocean and tectonic activity from the Sumatra Fault System (SFZ), which is located on the mainland of the island of Sumatra.

SFZ consists of several segments whose slip rate is known [1]. The SFZ is located extending from Banda Aceh to the Sunda Lampung Strait for 1900 km [2]. The SFZ crosses several cities in Sumatra and can cause a large earthquake in each fault segment.

Several cities in the southern region of North Sumatra, namely Tarutung, Padang Sidempuan, Sipirok, Padang Lawas, and Mandailing Natal, are trajectories of several SFZ segments. This condition affects earthquake activity which often occurs and has an impact on these areas. One of the significant earthquakes that hit this area occurred on April 30, 2020, with a magnitude of 5.6 at 15.20.25 WIB and the epicenter coordinates of 1.17° North, 99.53° East, at a distance of 24 kilometers Northwest of Padanglawas Regency with a hypocenter depth of 10 Km. This earthquake caused damage in the South Tapanuli area, which damaged two places of worship, an elementary school building, and four houses in Aek Libung, Sayur Matinggi, South Tapanuli. There are no scientific publications related to these earthquake focal mechanisms. Is this earthquake-related to the Toru segment or Angkola or Barumun, or another local fault activity?

Analysis of the mechanism of the Padanglawas earthquake source is needed to obtain information on the fault plane following the surrounding tectonic conditions. Furthermore, knowing the fault plane can be used as a reference for mitigation efforts against earthquake hazards in earthquake-affected areas. Map of these research areas and the mainshock location can be seen in Figure 1.

* Corresponding author: prasetyo.agung@bmkg.go.id

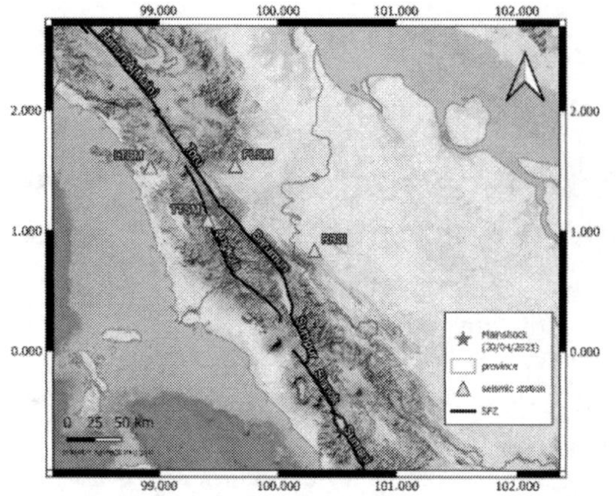

Fig. 1. Location of EQ Mainshock (red star) and BMKG seismic station (yellow triangle) that used, also tectonic fault (black line) layout at this research area.

2 Data

The data used in this study is the waveform recording of the Padanglawas earthquake on April 30, 2020, which was obtained from the BMKG seismic station closest to the epicenter location, namely TTSM, PLSM, LTSM, and RSSI. A complete description of the BMKG seismic station used in this study is shown in Table 1. All of these seismic stations were installed in 2019.

Table 1. Metadata of seismic stations used in the study.

Station	Digitizer	Sensor	Lat (Deg)	Long (Deg)	Altitude (m)
TTSM	Centaur	Trillium-Compact	1.085	99.41	173
PLSM	Centaur	Trillium-Compact	1.540	99.64	88
RRSI	Centaur	Trillium Horizon	0.841	100.3	90
LTSM	Centaur	Trillium-Compact	1.529	98.93	5

Earthquake recording data from several stations closest to the earthquake source used in this study can be seen in Figure 2. The nearest station to the epicenter is TTSM, which is 12 km, and the farthest is RRSI, which is 100 km.

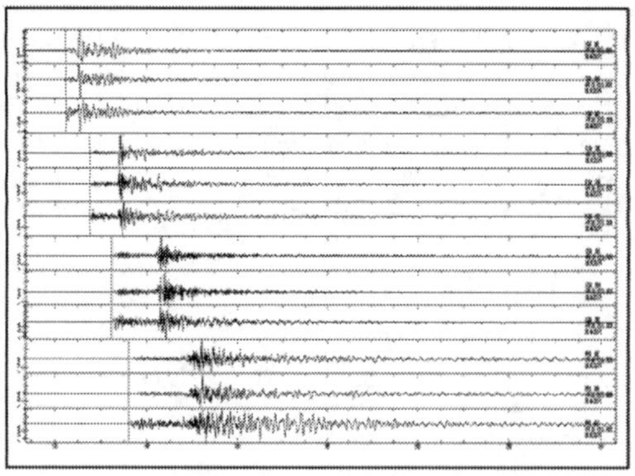

Fig. 2. Waveform seismogram (count unit) at station TTSM, PLSM, LTSM, and RRSI, which is each station is 3 component. Software using SAC [3].

3 Method

The method used in this study is the moment tensor inversion with a double couple approach [4]. We use SAC for data processing, so converting from miniseed data into SAC format is necessary. The velocity structure AK135 was used in this study. The filter uses the Butterworth bandpass filter method with a limit of 0.05 – 0.1 Hz and a cut frequency of 0.13 Hz. We removed the instrument response during pre-processing steps by deconvolution of the response file for each seismometer in the SAC code. The moment tensor inversion process can be seen in Figure 3 below:

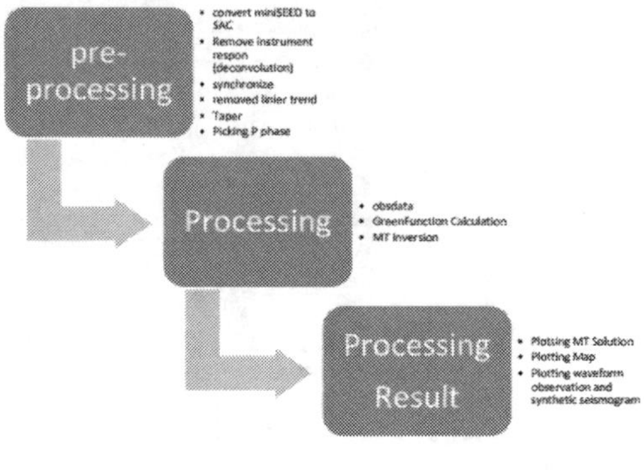

Fig. 3. Diagram processing of the moment tensor inversion

We used moment tensor inversion code from Yagi. This method provides a reliable technique to estimate the moment tensor solutions for moderate earthquakes [5]. In the processing stage, the obsdata, according to Figure 2, contains a description of the data that will be processed into an observation signal which includes the station name data and station components, the seismograph data type used is velocity, the signal duration starts from 5 seconds before P, and the total signal duration is 120 seconds, the period of the sampling time used is 0.5. The type of output signal is displacement.

4 Result and discussion

The result of the moment tensor inversion shows the variance of the observation and calculation data is 0.3824. The earthquake source parameters show that the Nodal I with strike/dip/slip values is 100.11/88.33/7.53 and the Nodal II with strike/dip/slip values is 9.89/82.47/178.31. The value of Seismic Moment (Mo) is 0.4653E+17 Nm or Mw 5.04.

Based on the location of the main earthquake, which is near the end of the Renun segment and Angkola segment, and the area of the dominant aftershock distribution being on the southeast side of the main earthquake epicenter, the nodal plane II is more suitable as a fault plane parameter. Furthermore, these results can be strengthened by analysis of the direction of compression and dilation. The Sumatran fault mechanism is dextral so that the nodal plane II is more suitable.

We also compare the focal mechanism resulting from the moment tensor inversion with the focal mechanism from the calculation of the initial polarity of the P wave using the first motion module [6] in Seiscomp3. The focal mechanism results from the first motion analysis showed NP1: 95/80/2 and NP2: 4/88/170.

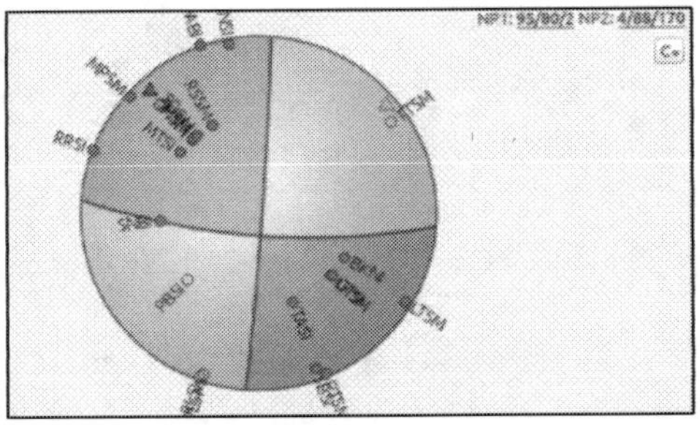

Fig. 4. Output from first motion analysis show the focal mechanism of mainshock April 30, 2020. Compression (white) and Tension (brown).

The results of the comparison of the focal mechanism data processing show a similar focal mechanism pattern between the moment tensor inversion results in Figure 5 and the First motion polarity in Figure 4. Furthermore this technique add strong confidence about the mechanism of the main earthquake on April 30, 2020 which is slightly different when compared to the historical map of the focal mechanism of earthquakes around the Sumatran fault [7].

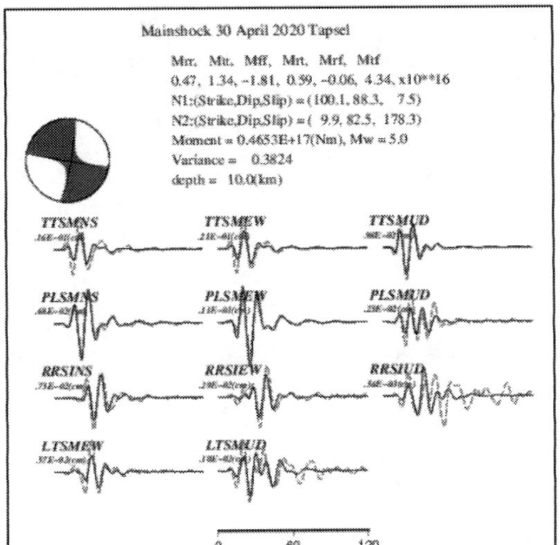

Fig. 5. Output from the moment tensor inversion show the mechanism, source parameter, and the fitting waveform between synthetic (red line) and observation data (black line).

Based on Figure 5, it can be seen that the fitting of observation and calculation data is suitable, especially in the low-frequency signal pattern is reasonable for Magnitude greater than 5 [8]. On the other hand, the signal pattern with a reasonably high amplitude indicates a displacement due to earthquake energy released in seismic waves.

BMKG informed that the source of the Padanglawas earthquake on April 30, 2020, was the result of the Toru Fault activity. The active fault mapping study based on [2] was then explicitly detailed in the Tapanuli area by [1], namely the Angkola Fault and Toru Fault split into several smaller segments. Toru has a slip rate of 9.3±1.8 mm/year, and Angkola has a slip rate of 4-5 mm/year [9]. Therefore, the potential for earthquakes in the study area is still relatively high, and earthquake mitigation efforts must continue to be carried out by the local government.

5 Conclusion

The mainshock of Padanglawas earthquake on April 30, 2020, had Mw 5.0, the moment tensor inversion results obtained NP1: 100.1/88/7.5 and NP2: 9.9/82.5/178.3. Thus, NP2 is more in line with the dextral mechanism of the Sumatran Fault, also supported by the distribution of dominant aftershocks in the southeast direction. Earthquake mitigation efforts need to be carried out in the Toru Fault and Angkola Fault surrounding areas because the potential for earthquakes in this area is relatively high.

References

1. Natawidjaja, D. H. (2018). Major Bifurcations, Slip Rates, and A Creeping Segment of Sumatran Fault Zone in Tarutung-Sarulla-Sipirok-Padangsidempuan, Central Sumatra, Indonesia. Indonesian Journal on Geoscience, 5(2), 137-160.
2. Sieh, K., & Natawidjaja, D. (2000). Neotectonics of the Sumatran fault, Indonesia. Journal of Geophysical Research: Solid Earth, 105(B12), 28295-28326.
3. Helffrich, G., Wookey, J., & Bastow, I. (2013). The seismic analysis code: A primer and user's guide. Cambridge University Press.
4. Kikuchi, M., & Kanamori, H. (1991). Inversion of complex body waves—III. Bulletin of the Seismological Society of America, 81(6), 2335-2350.
5. Yagi, Y., & Nishimura, N. (2011). Moment tensor inversion of near source seismograms. Bulletin of the International Institute of Seismology and Earthquake Engineering, 45, 133-138.
6. Triantafyllis, N., Sokos, E., & Ilias, A. (2016). SCISOLA: REAL-TIME MOMENT TENSOR MONITORING FOR SEISCOMP3. Bulletin of the Geological Society of Greece, 50(2), 1120-1129.
7. Weller, O., Lange, D., Tilmann, F., Natawidjaja, D., Rietbrock, A., Collings, R., & Gregory, L. (2012). The structure of the Sumatran Fault revealed by local seismicity. Geophysical Research Letters, 39(1).
8. Abdel-aal, A. A. K., & Yagi, Y. (2017). Earthquake source characterization, moment tensor solutions,

and stress field of small-moderate earthquakes occurred in the northern Red Sea Triple Junction. Geosciences Journal, 21(2), 235-251.
9. Natawidjaja, D. H. (2018, February). Updating active fault maps and sliprates along the Sumatran Fault Zone, Indonesia. In IOP Conference Series: Earth and Environmental Science (Vol. 118, No. 1, p. 012001). IOP Publishing.

Analysis of land use change using normalized difference vegetation index (NDVI) as input parameter to flood risk assessment of Krueng Kluet watershed in South Aceh, Indonesia

Fitri Zaitun Nurnalisa[1*], *Ella* Meilianda[1], and *Azmeri*[1]

[1]Civil Engineering Department, Engineering Faculty, Syiah Kuala University, Banda Aceh, Indonesia

Abstract. Flood events have been escalating in the last decade at the Kluet River, one of the major rivers in the South Aceh district, Aceh Province, Indonesia. The frequent flood events have made remarkable changes to the river morphology including river meander and delta particularly at the downstream. One of the interplaying parameters contribute to flood events is the land use changes in the watershed. This study aimed to analyze the land use changes of Krueng Kluet Watershed using the Normalized Difference Vegetation Index (NDVI). The satellite image data series were obtained to remotely monitor the land use change over 20 years period (1998-2018). NDVI was used in classifying the vegetation density distribution value of the Krueng Kluet Watershed. The output of the study explained that there were increase in the plantation, bareland, grassland, water and settlement areas, while in contrary the forest area and bush. There was an increase by 8.05%, 0.15%, 0.14%, 0.10%, and 0.07% respectively to the plantation, bareland, grassland, water, settlements/urban areas, during the period of 1998-2018. On the other hand, the forest and bush were significantly decreased by 8.25%, and 0.3, respectively in the same period. Overall, the land use changes have also increased the Curve Number (CN) of Krueng Kluet Watershed from 61.8 in 1998 to 62.1 in 2008 and to 62.5 in 2018. This suggests the land use change as an important parameter to take into account in flood risk assessment of the Krueng Kluet Watershed

1 Introduction

Activities carried out in watersheds influence the ecosystem changes, i.e land use change. The land use change in upstream affects downstream due to water discharge fluctuation, sediment content and other dissolved material content [1].

Flood events have been escalating in the last decade in the Kluet River, one of the major rivers in the South Aceh district of Aceh Province, Indonesia. The report of Aceh Watering Institution, SID (Design Investigation Survey) of The Flood Control and Krueng Kluet River Management in South Aceh District, informed that Krueng Kluet River has several problems including river bank damage. It was effected by the floods transporting the sediments and impacting river capacity reduction [2]. Haba Rakyat also mentioned that River bank erosion due to flood in North Kluet Sub District become higher [3].The frequent flood events have made remarkable changes of the river morphology, in the form of changing course of river meander and formation of delta by sedimentation, particularly at the downstream. The Krueng Kluet Watershed have been subjected to human-induced land use change for a long time, including intensive deforestation for economic purposes. Any adverse changes occur in the watershed may have been correlated to the changes of the river morphology, and will be potential to increase the risk of flood event in future.

Watershed characteristics highly impacting to surface runoff were watershed area and shape, topography, and landuse. For the dense vegetation landuse, the rain stands in the vegetation before percolating to the soil. The condition affects the low surface runoff. While for bareland area or no vegetation, the rain highly percolate to the soil and affect the high surface runoff flowing to the river. The condition improve the high river flow [4]. The conversion from permeable area to impermeable area influences imbalance ecosystem in a watershed and affects the watershed condition. The land use change impacts surface run off time and volume, and land erosion. In bareland area where no vegetation, the rainfall becomes surface runoff and flows into the river. The improvement of surface runoff volume affects flood and damage of the downstream river area [5].

To comprehend the natural processes or human activities affecting the changes of river morphology, it is required to study the river area condition. This study used Geographic Information System (GIS) and remote sensing. The application of both geographic information systems remote sensing become effective and efficient scientific tool and technique in analyzing the changes of river morphology spatially and temporally. Since the

*Corresponding author: ella_meilianda@unsyiah.ac.id

basic problem identified and the process comprehended, the proper handling can be provided to keep the stable river flow and not endanger the surrounding river area [6].

2 Method

2.1 Data collection

The data used in this study were secondary data acquired from satellite images. The data can be summarized as follows.

2.1.1 Land use data

The land use data used was the Land Satellite Data (LANDSAT) acquired from the United States Geological Survey (USGS) website (http://earthexplorer.usgs.gov/) from 1998 to 2018.There are only three set of satellite data of acceptable quality (e.g., no cloud cover) covering the area of this study, i.e. from the satellite images of 1998, 2008 and 2018.

2.1.2 Digital elevation model (DEM) data

Digital Elevation Model (DEM) data is required in this study to obtain topographic data and used in determining watershed border. DEM data can be download from national DEM (DEMNAS) website (http://tides.big.go.id/DEMNAS/).

2.1.3 Precipitation data

The rainfall data collected from 1998 to 2018 to calculate surface flow rate was retrieved from the satellite-based rainfall data, which can be downloaded from here http://giovanni.gsfc.nasa.gov/giovanni/ The data was downloaded for the period of 20 years. The maximum rainfall data is 141.49 mm.

2.2 Analysis technique

The data analysis phases in this study were divided into three phases, which is summarized as follows:

1. Spatial analysis of topographic data watershed boundary determination;
2. Spatial analysis of land use change;
3. Calculation of the peak discharge using hydrograph of SCS synthetic unit.

2.2.1 The determination of watershed using DEM

The analysis of watershed border determination used a hydrology tool in ArcGIS 10.3 software, called "watershed". The placement of the "pour point" is an important step in the watershed delineation process. "pour point" must be located in high flow accumulation area because it is being used to calculate the total water flow contributing to a certain point location..

After all processes finished, it can be continued to the process of river flow area delineation using the "Watershed" tool resulting "Flow Direction" and "Pour Point" as the inputs. To continue "Geoprocessing" process, watershed raster data is converted into polygon by using "Conversion" tool to process to the next spatial analysis phase [7].

2.2.2 The analysis of land use change

Satellite image analysis was started by downloading the land satellite image of 1998, 2008 and 2018. The image cropping was executed based on study area boundary, image improvement, and radiometric correction. Landsat image can be further analysed to classify the land use [8], [9] and [10]. Normalized Difference Vegetation Index (NDVI) in equation (1) can be used to obtain the vegetation density distribution value of the studied watershed.

$$NDVI = \frac{NIR-RED}{NIR+RED} \quad (1)$$

It explains that NDVI is normalized difference water index, NIR is near infrared, and RED is red reflectance

2.2.3 Peak discharge analysis using SCS method

There are several methods to calculate a peak discharge. One of the methods is SCS (Soil Conservation Service) by The Soil Conservation Service (1972), which method computes abstractions from storm rainfall. The hypothesis of the SCS method is that the ratios of the two actuals to the two potential quantities of discharge are equal, that is

$$\frac{Fa}{S} = \frac{Q}{P-Ia} \quad (2)$$

From the continuity principle

$$P = Q + Ia + Fa \quad (3)$$

$$Q = \frac{(P-Ia)^2}{P-Ia+S} \quad (4)$$

$$Ia = 0{,}2\,S \quad (5)$$

On this basis

$$Q = \frac{(P-0{,}2\,.S)^2}{P+0{,}8\,.S} \quad (6)$$

It describes that Q is excess rainfall (mm), P is maximum rainfall (mm), and S is potential maximum infiltration rate (mm);

The variation between rainfall and surface flow (S) correlate with curve number (CN). The equation can be detailed as below:

$$S = (\frac{25400}{CN} - 254) \quad (7)$$

Curve number (CN) is varied from 0 to 100 affected by hydrologic soil group of Antecedent Moisture Content

(AMC). AMC is divided into 3 categories, i.e AMC I (dry), AMC II (normal) and AMC III (saturated) [11]. Based on soil type, AMC, and land use, Watershed CN can be calculated by using the following equation:

$$CN\ DAS = \frac{(\sum_{i=1}^{n} AiCNi)}{(\sum_{i=1}^{n} Ai} \quad (8)$$

It explains that CN is *curve number* and Ai is Waterdhed Area (ha)

AMC II (normal) was categorized by infiltration rate and soil texture. This category was divided into 4 (four) groups. They are Group A (sandy soil deep containing dust and clay elements) with infiltration rate of 8-12 mm/hour, Group B (sandy soil shallower than A, fine to medium texture) with infiltration rate 4-8 mm/hour, Group C (shallow soil containing sufficient clay) with infiltration rate 1-4 mm/hour, and Group D (mostly clay, shallow, with an impermeable layer near the soil surface) with infiltration rate 0-1 mm/hour [12].

The SCS dimensionless Hydrograph is a synthetic unit hydrograph in which the discharge is expressed by the ratio of discharge q to to peak discharge q_p and the time by ratio of time t to the time of rise of the unit hydrograph, T_p. Given the peak discharge and lag time for the duration of excess rainfall, the unit hydrograph can be estimated from the synthetic dimensionless hydrograph for the given basin. Figure 1.a shown such a dimensionless hydrograph, prepared from the unit hydrographs of a variety of watershed. The value of q_p and T_p may be estimated using a simplified model of a triangular unit hydrograph as shown in Figure 1.b, where the time is in hours and the discharge in m³/s [13]. From a review of a large number of unit hydrograps, the Soil Conservation Service suggest the time of recession may be approximated as 1.67 T_p. As the area under the unit hydrograph should be equal to a direct runoff of 1 cm, it can be shown:

$$q_p = \frac{CA}{T_p} \quad (9)$$

$$T_p = \frac{t_r}{2} + t_p \quad (10)$$

$$t_p = 0,6\ x\ T_c \quad (11)$$

where C = constant (2.08), A is the watershed area in square kilometers, and $T_r = 0.6\ T_c$. Where T_c is the time of consentration of the watershed. The concentration time can be calculated by the Kirpich formula (11)

$$T_c = 0,01947 L^{0,77} S^{-0,385} \quad (12)$$

Were L = flow length and S = slope. The coordinate unit hydrograph are given in the Table 1.

Table 1. Coordinate unit hydrograph

t/tp	q/qp	t/tp	q/qp	t/tp	q/qp
0	0	1	1	2.4	0.18
0.1	0.015	1.1	0.98	2.6	0.13
0.2	0.075	1.2	0.92	2.8	0.098
0.3	0.16	1.3	0.84	3	0.075
0.4	0.28	1.4	0.75	3.5	0.036
0.5	0.43	1.5	0.66	4	0.018
0.6	0.6	1.6	0.56	4.5	0.009
0.7	0.77	1.8	0.42	5	0.004
0.8	0.89	2	0.32		
0.9	0.97	2.2	0.24		

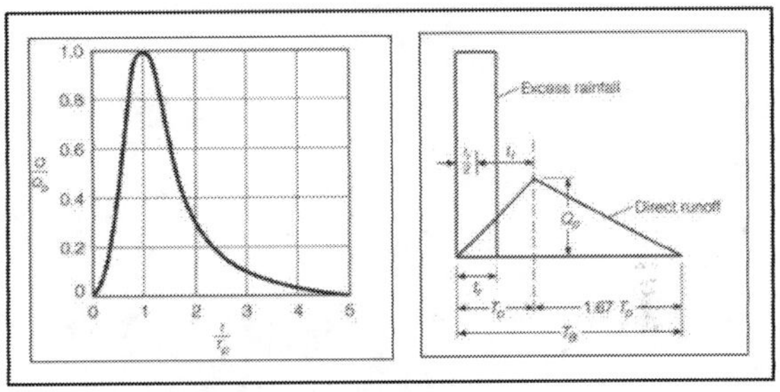

Fig. 1. Soil Service Synthetic unit hydrograph.

3 Result

3.1 Delineation of watershed

The delineation of the watershed in this study was carried out by using ArcGIS 10.5 software. The input data was DEM data downloaded from USGS EarthExplorer. Due to the running process, it was resulted that the total area of Krueng Kluet Watershed is 2,954.31 km². It can be viewed in Figure 2.

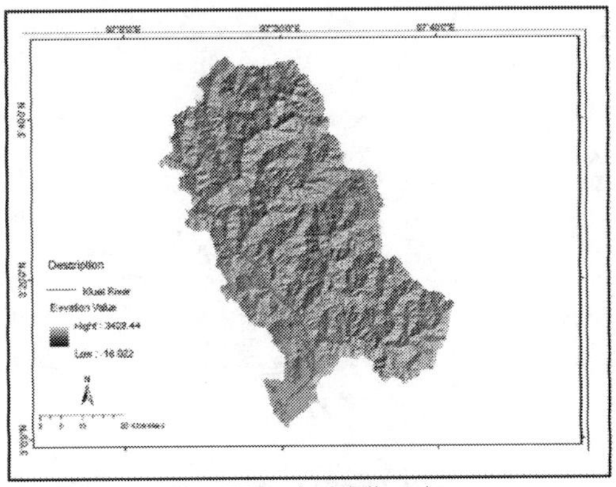

Fig. 2. Krueng Kluet Watershed delineation map.

3.2 Spatial analysis of land use change

3.2.1 *Radiometric correction*

Radiometric correction is carried out to improve image quality due to surface reflection errors or the curvature of the earth and other factors including the direction of sunlight, weather conditions, atmospheric conditions and other factors resulting the more accurate information. The results of the radiometric correction can be seen in the following figure (Figure 3 and 4).

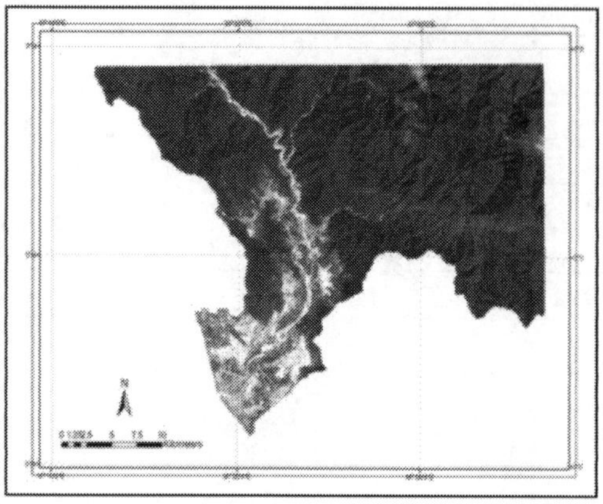

Fig. 3. Before radiometric correction.

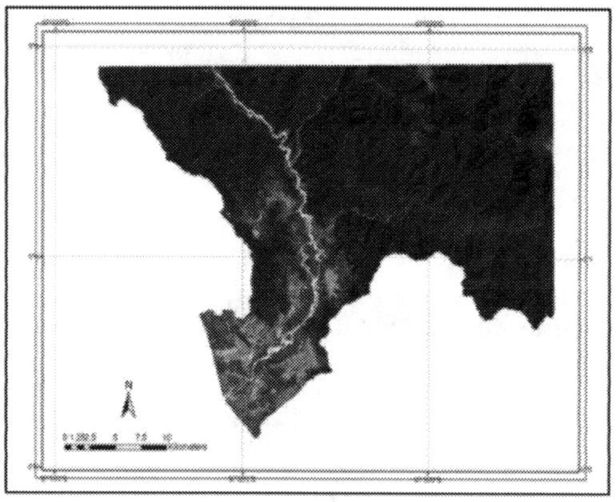

Fig. 4. After radiometric correction.

3.2.2 Normalized Difference Vegetation Index (NDVI)

NDVI can be categorized into several groups. NDVI of -1.0 to 1.0 represents wetland, barren rock areas and non-vegetated objects. NDVI of 0.2 to 0.5 represent the rare vegetation, i.e, bush, grassland and old plants. NDVI of 0.6 to 0.9 represents dense vegetation, i.e, temperate and tropical climate forests, the plants on the peak growth phase. The analysis of NDVI can be summarized into the following figures.

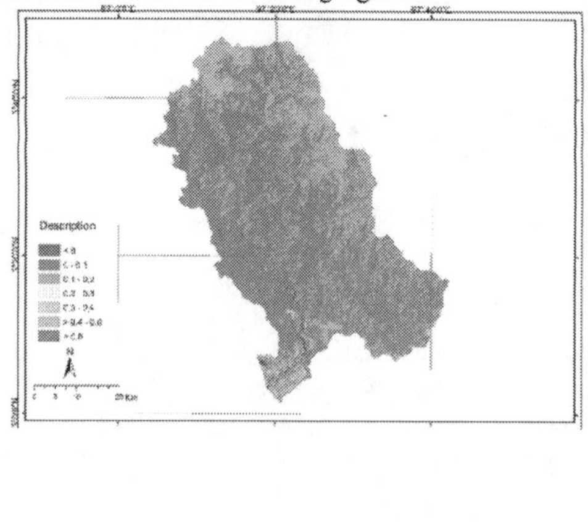

Fig. 5. NDVI of Krueng Kluet Watershed on 1998.

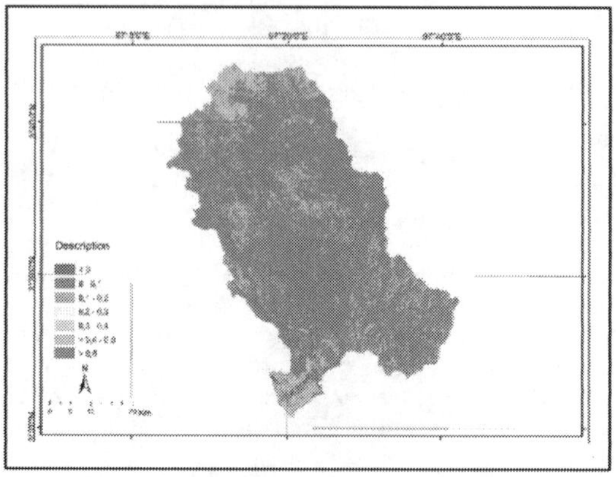

Fig. 6. NDVI of Krueng Kluet Watershed on 2008.

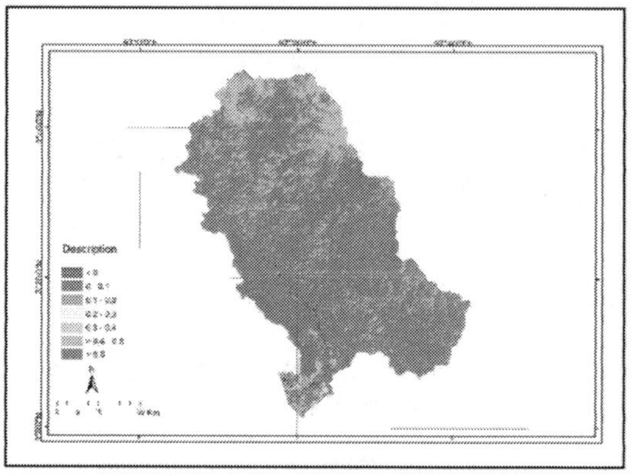

Fig. 7. NDVI of Krueng Kluet Watershed on 2018.

The analysis of land use using NDVI resulted in 7 land use categories as summarized into Table 1. Based on the analysis output, the land use changes during 20 (twenty) years period from 1998 to 2018 can be calculated. Land use change due to classification output during 20 (twenty) years period can be explained in Table 2 and 3.

Table 2. NDVI classification

NDVI	Land Use Type	Total Area (Km^2)		
		1998	2008	2018
< 0	Water	6.36	4.51	9.47
0-0.1	Settlement /urban area	1.68	2.60	3.86
0.1-0.2	Bareland area	2.83	3.43	7.53
0.2-0.3	Grassland	8.25	6.26	12.56
0.3-0.4	Bush	28.35	19.61	19.58
0.4-0.5	Plantation	373.66	501.92	611.68
> 0.5	Forest	2,533.18	2,415.98	2289.56
Total		**2,954.31**	**2,954.31**	**2,954.31**

Table 3. Land use change due to classification output

NDVI	Land Use Type	Total Area (%)		
		1998	2008	2018
< 0	Water	0,22	0,15	0,32
0-0.1	Settlement /urban area	0,06	0,09	0.13
0.1-0.2	Bareland area	0,10	0,12	0,25
0.2-0.3	Grassland	0,28	0,21	0,42
0.3-0.4	Bush	0,96	0,66	0,66
0.4-0.5	Plantation	12,65	16,99	20,70fr
> 0.5	Forest	85,75	81,78	77,50
Total		100,00	100,00	100,00

3.3 Peak discharge calculation using scs method

Peak discharge calculation using hydrograph method requires *Excess Rainfall Hytograf* (ERH) of effective rainfall. ERH calculation by using SCS method requires *Curve Number* (CN) of each observation describing watershed characteristic both land use and soil type.

3.3.1 Curve Number (CN) Factor Classification

The classification of soil is categorized based on FAO Digital Soil Map of the World. CN calculation can be shown in Map.

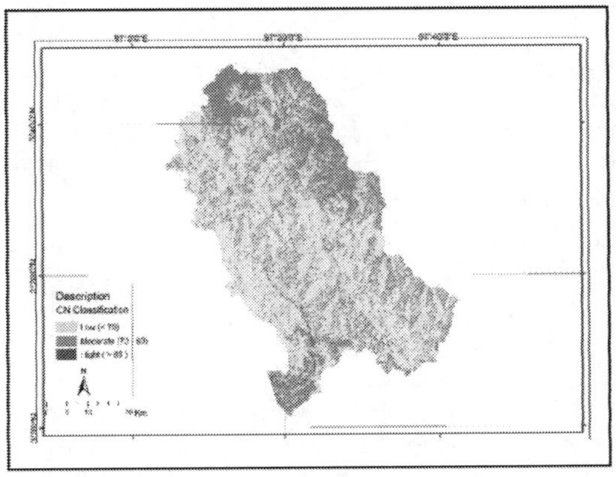

Fig. 8. CN of Krueng Kluet Watershed on 1998.

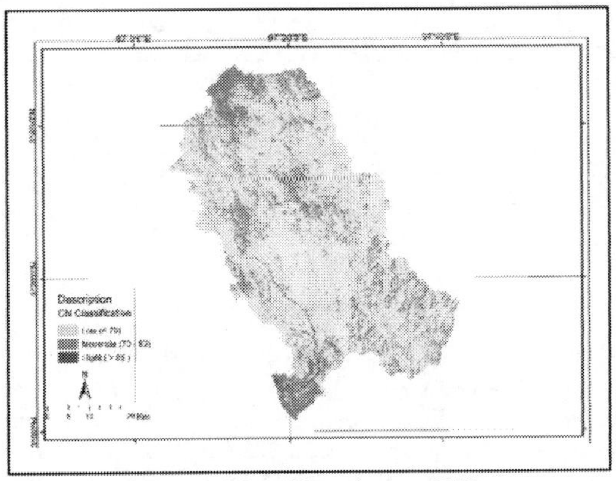

Fig. 9.CN of Krueng Kluet Watershed on 2008.

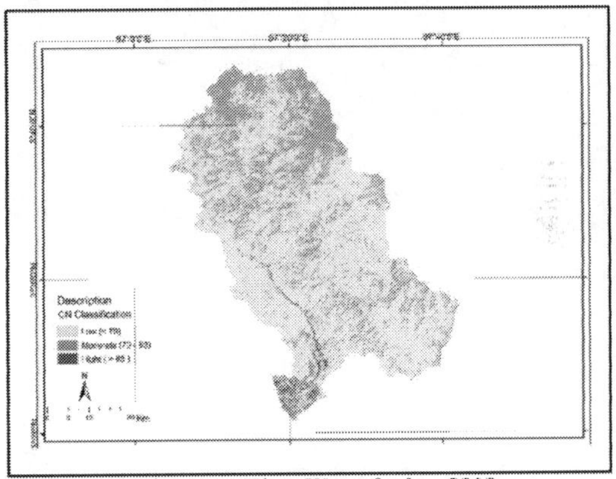

Fig. 10. CN of Krueng Kluet Watershed on 2018.

CN classification maps describe that CN of 1998 is 61.8. It increases in 2008 to 62.1. it also improves in 2008 to become 62.5.

3.3.2 Potential maximum infiltration (S)

The Potential maximum infiltration and surface flow related to *curve number* (CN) as explained in Equation 7. The calculation rainfall and surface flow variation is summarized in the below table.

Table 4. Potential maximum infiltration (S)

1998		2008		2018	
CN	S (mm)	CN	S (mm)	CN	S (mm)
61.8	157.003	62.1	155.018	62.5	152.400

3.3.3 Excess rainfall (Q)

The effective rainfall calculation using SCS method requires maximum daily rainfall from 1998 to 2018 is 141.49 mm in 2016. The total of surface run off volume (Q) is calculated by multiply effective rainfall and Krueng Kluet Watershed Area as 2,954.314 km^2. The output of effective rainfall calculation can be summarized as below table.

Table 5. Excess Rainfall (Q)

Year	P (mm)	S (mm)	Q (m)
1998	141.49	157.003	0.045
2008		155.018	0.046
2018		152.400	0.047

The flood hydrograph is calculated by multiply effective rainfall and unit hydrograph [10]. The maximum peak discharge in 1998 is 3,774.48 m^3/sec, in 2008 is 4,525.81 m^3/sec, and in 2018 is 4,568.31 m^3/sec. The result of flood hydrograph of 1998, 2008, and 2018 are summarized as Figure 11-13.

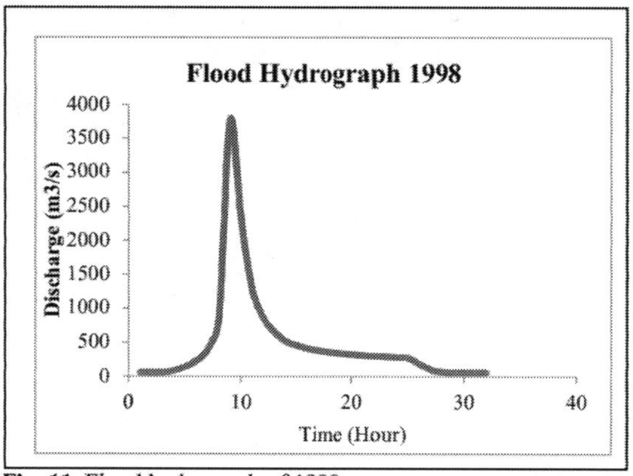

Fig. 11. Flood hydrograph of 1998.

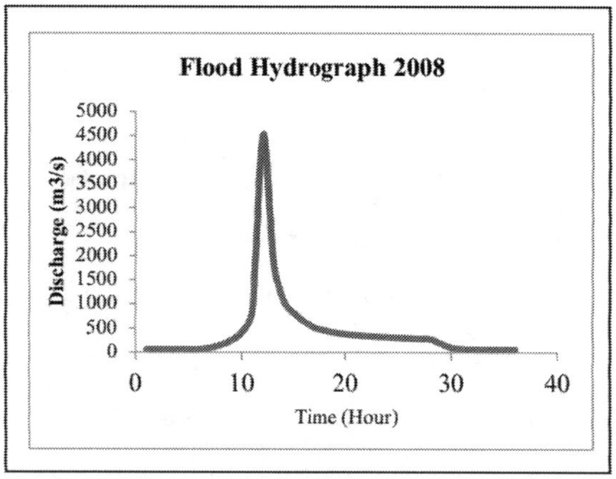

Fig. 12. Flood hydrograph of 2008.

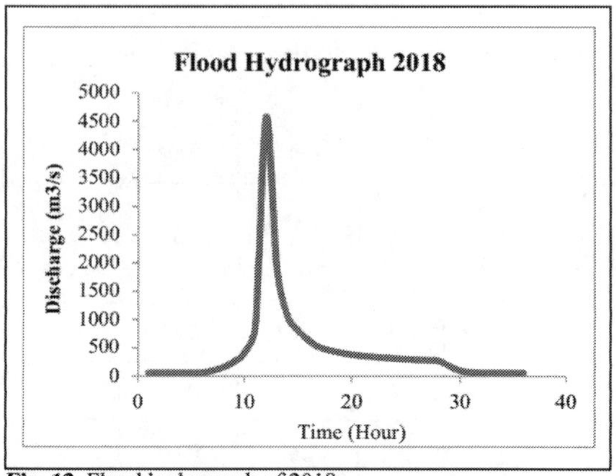

Fig. 13. Flood hydrograph of 2018.

4 Conclusion

The results of the study described that there were significant land use changes mostly of the plantation, settlement and forest areas. Over a 20-years period of observation, we found that there was an increase in the plantation and settlement areas, while in contrary the forest area.

In the period of 1998-2008, there were the increase of plantation area from 12.5% to 16.99%, bareland area from 0.10% to 0.12% and settlements/urban area and from 0.06% to 0.09%. Otherwise, the forest area, bush area, grassland area and water area were significantly decreased i.e. from 85.75% 81.78%, from 0.96% to 0.66%, from 0.28% to 0.21%, and from 0.22% to 0.15%.

In the period of 2008-2018, there were the increase of water area, settlements/urban area, bareland area, grassland area, and plantation i.e. from 0.15% to 0.32%, from 0.09% to 0.13%, 0.12% to 0.25%, from 0.21% to 0.42%, and 16.99% to 20.70% respectively. Whereas the forest are significantly decreased to 4.28%. Bush area was constant and the area was still 0.66%.

This suggests a significant changes of forest, plantation and settlement. The forest decrease from 85.75% in 1998 to 77.50% in 2018 while the plantation and settlement increase from 12.65% in 1998 to 20.70% and from 0.06% in 1998 to 0.13% in 2018. The change of forest and plantation and settlement impact significantly to the Curve Number (CN). The land use changes have also increased the Curve Number (CN) of Krueng Kluet Watershed, i.e, 61.8 in 1998, 62.1 in 2008, and 62.5 in 2018. The maximum peak discharge in 1998 is 3,774.48 m^3/sec, in 2008 is 4,525.81 m^3/sec, and in 2018 is 4,568.31 m^3/sec.

The land use analysis using *Normalized Difference Vegetation Index* (NDVI) of 1998 to 2018 resulted the changes of forest and plantation areas. The forest area decreases while the plantation areas increases. The more forest area decreases, the more CN (curve number) increase. It moderately contribute to the increase of flood discharge in every observation year.

The forest reduction impact flood discharge improvement. It influences sediment supply to the river. The sediment will load in the bottom of the river and affecting the decreasing of river's capacity. The discharge improvement and the river's capacity impact flood risk improvement.

References

1. Suripin, *Soil and Water Resources Conservation* (Andi, Yogyakarta, 2001)
2. A. Watering Institution, *The Final Report of The Flood Control and Krueng Kluet River Management in South Aceh District* (Aceh Government, Banda Aceh, 2015)
3. Haba Rakyat, *Kruet River Erosion become higher, the community move the grave*, (https://habarakyatonline.com/erosi-sungai-kluet-makin-parah-warga-terpaksa-bongkar-kuburan/, accessed 2 January 2019)4. O. Laoh, IPB (2002).
4. O. E. H. Laoh, *Physical, Social, Economic, and Land Use Factors Correlations to Erosion and Sedimentation in Catchment Area (Tondano Case Study, North Sulawesi)*, (IPB, Bogor, 2002).
5. G. P. Verrina, D. D. Anugrah, and Sarino, 2013 **1**, 12 (n.d.)
6. P. K. Langat, L. Kumar, and R. Koech, 2019 92 (n.d.)
7. T. Lilesand and R. W. Kiefer, *Remote Sensing and Image Interpretation* (Gajah Mada University Press, Yogyakarta, 1997)
8. B. Nath, 2014 **2**, (21-27)
9. D. F. Morawitz, T. Blewett M., A. Cohen, and M. Albert, 2006 85 (n.d.)
10. M. M. Aburas, S. H. Abdullah, M. F. Ramli, and Z. Ash'aari, 2015 238 (n.d.)
11. R. H. McCuen, *Hydrologic Analysis and Design* (Pretice-Hall, Inc, Upper Saddle, New Jersey, US, 1989)
12. C. Asdak, *Hydrology and Watershed Management* (Gadjah Mada University Press, Yogyakarta, 2002)
13. V. T. Chow, D. R. Maidment, and L. W, *Applied Hydrology* (Mc Grow-Hill Science Engineering, 1998)

Incorporating dynamics of land use and land cover changes into tsunami numerical modelling for future tsunamis in Banda Aceh

Tursina[1,4], *Syamsidik*[1,2,4,*], *Shigeru* KATO[1,3], and *Mochammad* Afifuddin[1,2]

[1] Doctoral Program, School of Engineering, Universitas Syiah Kuala, Darussalam, Banda Aceh, 23111 Indonesia
[2] Civil Engineering Department, Faculty of Engineering, Universitas Syiah Kuala, Darussalam, Banda Aceh, 23111 Indonesia
[3] Department of Architecture and Civil Engineering, Toyohashi University of Technology, Toyohashi, Aichi 441-8580, Japan
[4] Tsunami and Disaster Mitigation Research Center (TDMRC), Universitas Syiah Kuala, Darussalam, Banda Aceh, 23111 Indonesia

Abstract. This study aims to investigate the tsunami exposure to current land use land cover (LULC) with the LULC predictions for the next 50 years, including the impact of sea-level rise (SLR) in Banda Aceh. This study began with a spatio-temporal dynamic analysis using LULC changes projection. Firstly, Markov Chain was used to simulate the changing trends in land use from 2011 to 2015. The changing trends were used to simulate land use predictions for 2015-2065 using Cellular Automata (CA). There were six main types of LULC classifications, i.e., ponds, built-up areas, mangroves, bare land, urban forests, and water areas. Furthermore, the map resulting from the land use simulation was used as the Manning Coefficients for tsunami simulation using the COMCOT model. The initial tsunami waves were generated based on the 2004 Indian Ocean Tsunami caused by 9.15 Mw earthquake. As a dynamic tsunami hazard approach, a sea-level rise of 0.35 m was considered for the next 50 years. The study results show the built-up area will be affected by the tsunami, about two-thirds of the city's built area. Growth tends to occur in the southern part of the city outside the tsunami hazard zone. But there will also be the growth of built-up areas near the coast. Although much development is observed in the southern part of the city, the coastal area also needs attention because the sea level rise factor can exacerbate the impacts of tsunamis in the future.

1 Introduction

Population in coastal areas grows rapidly from time to time. The primary effect of high population growth is a change in land use and land cover (LULC). Population growth in coastal areas is followed by potential associated coastal hazards. The threat of hazards facing coastal areas becomes more concerning due to the impacts of global warming inducing sea level rise (SLR) [1]. Several studies have proven that SLR can cause flooding [2] and exacerbate the impact of tsunami waves [3], [4].

The 2004 Indian Ocean tsunami caused massive destructions on coastal cities in several countries in the Indian Ocean basin. The worst affected area was Banda Aceh in Indonesia, with around 165,000 people killed. At the time, the tsunami inundated 60% of the city of Banda Aceh, destroying everything in its path due to the extreme wave velocity. Coastal sub-districts directly adjacent to the beach experienced complete damage by the tsunami [5]. During the rehabilitation and reconstruction period between 2005-2009, Banda Aceh underwent massive reconstruction with 90.8% growth in the built-up area [6]. As the capital city of Aceh Province, population growth is mainly an urbanization factor. Urbanization commonly occurs in the main city like Banda Aceh because it is the center of government, economy, and education.

A tsunami is a significant disaster event with low frequent but high impacts events. Imamura et al. [7] explain that future disaster impacts are increasingly complex. Therefore, multi-disciplinary researches are needed to provide comprehensive and sustainable disaster risk reductions [7]. The phenomenon of the tsunami threat is not a static event. There are other coastal hazards that will contribute to more significant impacts on the community. On the other hand, the element at risk is also dynamic along with the changes in the population. Preparing disaster mitigation is only based on static historical data; without taking into account the dynamic factors. However, such an approach provides a limited understanding of longer-term tsunami risk. For this reason, it is necessary to incorporate dynamic hazard components and the dynamics of the elements at risk.

In a previous study, Sleeter et al. [8] observed the human exposure in 2061 by the tsunami with predicting land use in the US-North Pacific without considering the SLR. Another study examined the impact of flooding due to SLR with projected land use in 2030 and 2080 in Bay County, North Florida [9]. Up to the present, only limited studies could be identified evaluating impacts of the future tsunami based on dynamic LULC changes.

The objective of this research is to investigate the impact of land-use exposure in 2065 in tsunami-affected areas in Banda Aceh. The 2065 projection was chosen to highlight the effect of sea-level rise in the next 50-year. The year 2065 has also entered the 3rd period of Banda Aceh's spatial planning regulation after the tsunami recovery. The affected area was analyzed based

* Corresponding author: syamsidik@tdmrc.org

https://doi.org/10.1051/e3sconf/202234001014

on the city administrative boundaries. The results of this research are expected to fill the gap from long-term tsunami mitigation efforts, especially for disaster risk reduction-based spatial planning in Banda Aceh.

2 Study area

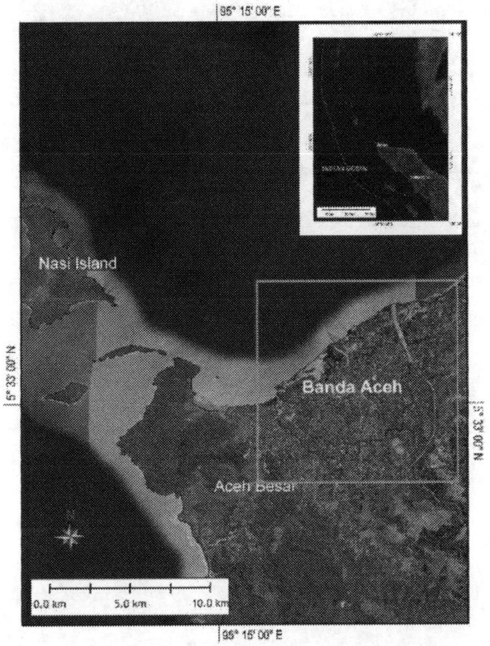

Fig. 1. Banda Aceh's study area (shown by the *blue box*). This map is visualized from World Imagery.

As illustrated in Fig. 1, Banda Aceh is located at the northern extremity of the Sumatra mainland. Banda Aceh covers 60 km^2 and has a 15-km shoreline. Banda Aceh city is flanked on the south and east by the Aceh Besar district. Banda Aceh is exposed to coastal hazards due to its flat inland geography and average height of 0.8 meters above sea level. Based on the study of 100-years SLR projection [10], Banda Aceh will experience coastal flooding and retreat of the shoreline up to 98 m. The city also is situated closely to the Sumatra-Andaman subduction zone, which can produce a large tsunami-like event similar to the 2004 Indian Ocean tsunami. If a tsunami occurs under the sea level rise situation, the tsunami will spread further and exacerbate the impact of human exposure more significantly than the 2004 tsunami [4].

Demographically, the study area includes urban and rural settlements. Meuraxa, Kuta Raja, Kuta Alam, and Syiah Kuala are the four seaside sub-districts of the city. Two of them – Meuraxa and Kuta Raja - were damaged by the 2004 tsunami. As the capital city of Aceh province, Banda Aceh is the center of government and the economy with the highest population density compared to other districts in Aceh. A study revealed that the community preparedness of Banda Aceh toward coastal hazards is a low category. The low category also includes an overall assessment of the city's resilience factors from coastal hazards [11].

3 Method

This research has three main steps: LULC change modeling, tsunami modeling, and land use exposure assessment. The first step is land use projection in 2065 using Markov-Chain and Cellular Automata [12] based on historical data, LULC 2011 and 2015. The second step is tsunami simulation using Cornell Multi Grid Coupled Tsunami Model (COMCOT) with tsunami wave generation from the 2004 earthquake with 9.15 Mw. The tsunami simulation was run on two scenarios, current condition without sea-level rise with LULC 2015 and future condition with 0.35 m of SLR 50 years with LULC 2065. The LULC maps are used for Manning's roughness coefficient on tsunami simulation. By overlaying the tsunami inundation map and the LULC map, we estimate the land use exposure in five land use classes.

3.1 Land use land cover classification

The study area's land use map was created using data from Bappeda (Banda Aceh Development and Planning Agency) for the years 2011 and 2015. These maps were interpreted from visual image analysis produced by Worldview-3 image. Water bodies, built-up areas, bare land, fish ponds, city forests, and mangrove forests were grouped into six key land use types for this study. This land cover classification was performed using ArcGIS software, adjusting to the administrative boundaries of Banda Aceh city. The land category and the descriptions of each land use can be seen in Table 1.

Table 1. Land cover classification [13]

LULC	Description
Waterbody	Lagoon area, rivers, wetland
Built-up area	Settlement area, transportation network, area of industry, government office dan public building area
Bare land	Open land that has no building
Fishpond	Fishpond area
City forest	Evergreen forest covers with large canopy covers
Mangrove	Area of coastal forest or mangrove forest

Land cover maps for 2011, 2015, and 2019 are presented in Fig. 2. To validate the land use simulation, an actual 2019 map was used. This map was digitized from the SPOT 5 satellite image. The current scenario (LULC 2015) and the future (LULC 2065) were used as manning roughness coefficient maps for tsunami simulations. The classification of Manning coefficient are waterbody (0.013), fishpond (0.017), bare land (0.02), mangrove (0.035), built-up area (0.04), and city forest (0.035). Land use is assigned to each grid cell in the research region, and Manning's coefficient is assigned to each land use.

https://doi.org/10.1051/e3sconf/202234001014

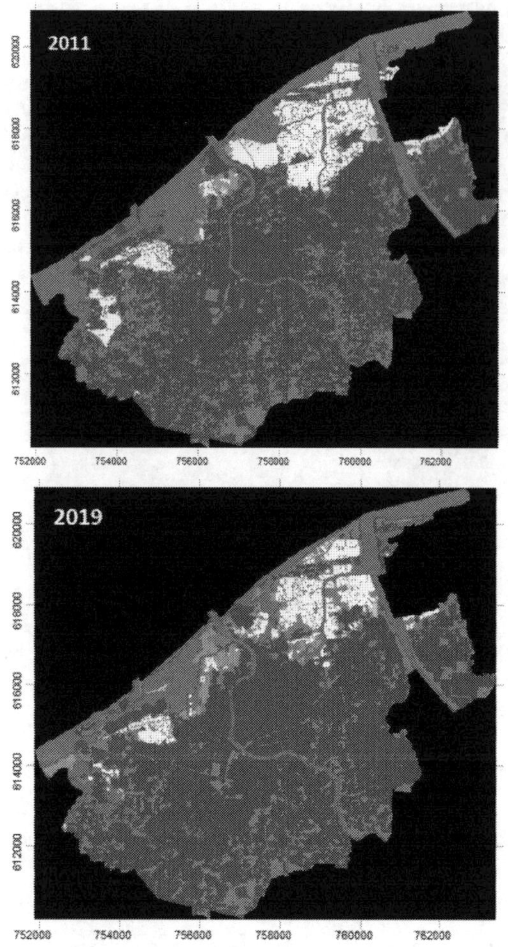

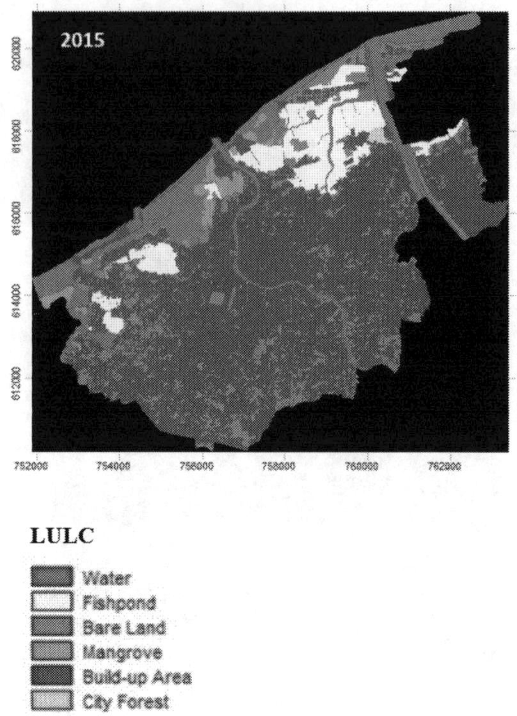

Fig 2. The land use land cover maps, (a) 2011, (b) 2015 and (c) 2019

3.2 Land use modelling

To forecast future land use in 2065, a Markov Chain and Cellular Automata model was used. This model is already part of IDRISI Selva software that is capable of land-use simulation [14]. For this study, the model was developed to produce a spatial projection of change based on historical data (2011-2015). The information was collected from Bappeda Banda Aceh The selection of data for that year was because the tsunami recovery process was considered to have returned to normal condition so that the trend of changing land use can be used to predict land use in the next 50 years. This data was used to determine the transition matrix of the two maps. The use of two actual maps in projection land use has been done by Mansour et al. [15]. Their projection map shows the good agreement of validation. The overall land use projection flowchart can be seen in Fig. 3.

3.2.1 Markov Chain

The Markov chain model analyzes change by calculating the probability transition area over a period of time. A probability matrix was applied to predict land use [16]. The Markov chain is a stochastic model that uses the trend of LULC changes in the preceding year to estimate the probability of land-use change. The model in this research was based on transition probabilities. From 2011 to 2015, the LULC's transition probability matrix, which was used to forecast land use, changes. [17].

Markov Chain uses the following mathematical equations [18]:

$$L_{(t+1)} = P_{ij} \times L_{(t)} \tag{1}$$

$$P_{ij} = \begin{bmatrix} P_{11} & P_{12} & \cdots & P_{1m} \\ P_{21} & P_{12} & \cdots & P_{2m} \\ \cdots & \cdots & \cdots & \cdots \\ \cdots & \cdots & \cdots & \cdots \\ P_{m1} & P_{m2} & \cdots & P_{mm} \end{bmatrix} \tag{2}$$

P_{ij} is a transition matrix from land-use i to land-use j, where $L_{(t+1)}$ and $L_{(t)}$ are land-use at time t+1 and t, respectively. And i and j are land use at first and second times.

3.2.2 Celular Automata-Markov (CA-Markov)

Simulating dynamic spatial phenomena and predicting future land-use change has been done using the CA-Markov model. The combined CA–Markov chain model also takes advantage of the Markov chain expected amount of land-use change and the CA models dynamic explicit spatial simulation. As a result, the CA–Markov model can be used to model spatial land-use change [18].

The equation used in the Cellular Automata module is as follows

$$S(t, t+1) = f(S(t), N) \quad (3)$$

Where S is the limit of limited and discrete cellular states, N denotes the cellular field, t and $t+1$ denote the time difference, and f denotes the transformation rule.

In this study, the actual map years 2011 and 2015 were used as starting points to create a four-year transition matrix. The year 2015 was aimed at projecting condition four years (as in 2019). The actual map 2019 was used for validation of the 2019 model. Map 2015 was set as the starting year to predict land-use change in 2065 by applying the same transition matrix with the number of Cellular Automata iterations was 50.

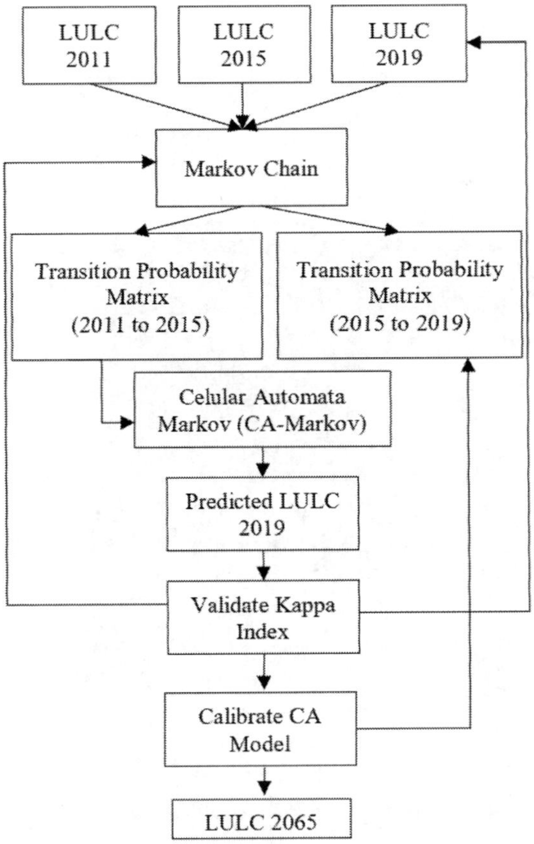

Fig 3. Flowchart of land use projection on Markov and CA Markov model

3.3 Tsunami numerical modelling

Tsunami numerical modeling was run using the Cornell Multi-grid Coupled Tsunami Model (COMCOT). The tsunami simulation includes tsunami generation from the earthquake source to the tsunami propagation and inundation. Some studies have been successfully demonstrated tsunami propagation along the Sumatra-subduction zone using COMCOT [19] [20]. In this study, the earthquake source is data from the 2004 earthquake with an earthquake magnitude of Mw 9.15 with a thrust fault earthquake mechanism. The source of the earthquake as the center of seabed deformation is reproduced by a fault model that provides information on earthquake parameters (Table. 2). The earthquake parameters, including magnitude, depth, fault length, fault width, epicenter coordinates, and angle of movement, have been developed by Koshimura et al. [21]. These have been validated for tsunami heights on land. With a total fault length of 1,155 km and a depth of 10 km, this fault model stretches from the Andamans to the northern half of the island of Simeulu.

Table 2. Fault parameter Mw 9.15 earthquake [21]

Parameter	I	II	III	IV	V	VI
L (km)	200.0	125.0	180.0	145.0	125.0	380.0
W (km)	150.0	150.0	150.0	150.0	150.0	150.0
Dis (m)	14.0	12.6	15.1	7.0	7.0	7.0
Strike (°)	323	335	340	340	345	7
Dip (°)	15	15	15	15	15	15
Rake (°)	90	90	90	90	90	90
Lon. (°)	94.4	93.32	92.87	92.34	91.88	91.9
Lat. (°)	3.03	4.48	5.51	7.14	8.47	11.0

Land use maps of the current and future conditions assigned the Manning coefficient are used as roughness variables in the tsunami numerical simulation using COMCOT. The Manning coefficient does not apply in Layers 1 to 5, and the various Manning coefficient is only used on the innermost layer (Layer 6). Tsunami waves were simulated using an earthquake magnitude Mw 9.15 using six simulation layers with a grid size, 8.5 m x 8.5 m on the smallest layer. The grid settings as shown in Table 3, and the domain computations are shown in Fig. 4.

COMCOT is a two-dimensional horizontal model that uses shallow water equations (SWE) to simulate depth-averaged velocities [22,23]. Leapfrog methods are used to solve the equations. SWEs were switched between linear and nonlinear in the model. The linear SWEs in a spherical coordinate can be seen in Equations (4) to (6).

$$\frac{\partial \eta}{\partial t} + \frac{1}{R\cos\varphi}\left\{\frac{\partial P}{\partial \psi} + \frac{\partial}{\partial \varphi}(\cos\varphi Q)\right\} = 0, \quad (4)$$

$$\frac{\partial P}{\partial t} + \frac{gh}{R\cos\varphi}\frac{\partial \eta}{\partial \psi} - fQ = 0, \quad (5)$$

$$\frac{\partial Q}{\partial t} + \frac{gh}{R}\frac{\partial \eta}{\partial \varphi} + fP = 0, \quad (6)$$

Equations (7) to (13) explain the nonlinear SWEs as follows:

$$\frac{\partial \eta}{\partial t} + \frac{1}{R\cos\varphi}\left\{\frac{\partial P}{\partial \psi} + \frac{\partial}{\partial \varphi}\cos\varphi\, Q\right\} = 0, \tag{7}$$

$$\frac{\partial P}{\partial t} + \frac{1}{R\cos\varphi}\frac{\partial}{\partial \psi}\left\{\frac{P^2}{H}\right\} + \frac{1}{R}\frac{\partial}{\partial \varphi}\left\{\frac{PQ}{H}\right\} + \frac{gH}{R\cos\varphi}\frac{\partial \eta}{\partial \psi} - fQ + F_x = 0, \tag{8}$$

$$\frac{\partial Q}{\partial t} + \frac{1}{R\cos\varphi}\frac{\partial}{\partial \psi}\left\{\frac{PQ}{H}\right\} + \frac{1}{R}\frac{\partial}{\partial \varphi}\left\{\frac{Q^2}{H}\right\} + \frac{gH}{R}\frac{\partial \eta}{\partial \varphi} + fP + F_y = 0, \tag{9}$$

$$f = 2\Omega \sin\varphi, \tag{10}$$

The bottom friction terms using Manning's formula are described as follows:

$$F_x = \frac{gn^2}{H^{7/3}}P(P^2 + Q^2)^{1/2}, \tag{11}$$

$$F_y = \frac{gn^2}{H^{7/3}}Q(P^2 + Q^2)^{1/2}, \tag{12}$$

$$H = \eta + h, \tag{13}$$

Where η is the water surface elevation; (P, Q) denotes the volume fluxes in X (West-East) and Y (South-North) directions, respectively; (φ, ψ) denote the latitude and longitude of the Earth; The Earth's radius is R; the gravitational acceleration is g, and h is the water depth. f is the Coriolis parameter, and Ω is the Earth's angular velocity.

3.4 Land use land cover exposure

Land use land cover (LULC) current condition and future prediction data are overlaid with a simulated tsunami inundation map with an earthquake magnitude of Mw 9.15. The map is specified in the coordinates WGS1984-UTM zone-46N. The LULC in the tsunami-affected area was calculated for each class of land cover. The total affected area and the safe area are also calculated spatially. This process is done with ArcGIS. Furthermore, the results of the current and the future conditions are compared.

Table 3. Simulation layers and types of SWE Equations

Domain	Layer 1	Layer 2	Layer 3	Layer 4	Layer 5	Layer 6
Latitude (°)	0.1 to 14.93	3.0 to 10.0	4.08 to 8.98	5.27 to 6.70	5.5 to 5.69	5.515 to 5.615
Longitude (°)	88.1 to 102.8	91.0 to 100	92.05 to 97.98	94.51 to 95.99	95.14 to 95.39	95.272 to 95.380
Number of Grid	882×890	1080×840	2135×1764	1598×1538	810×616	1400×1296
Grid size (m)	1856	928	309.33	103.11	34.37	8.50
Ratio	-	2	3	3	3	4
Time Step (sec.)	0.2	0.1	0.033	0.011	0.004	0.001
Equations	Linear SWE's					Nonlinear SWE's

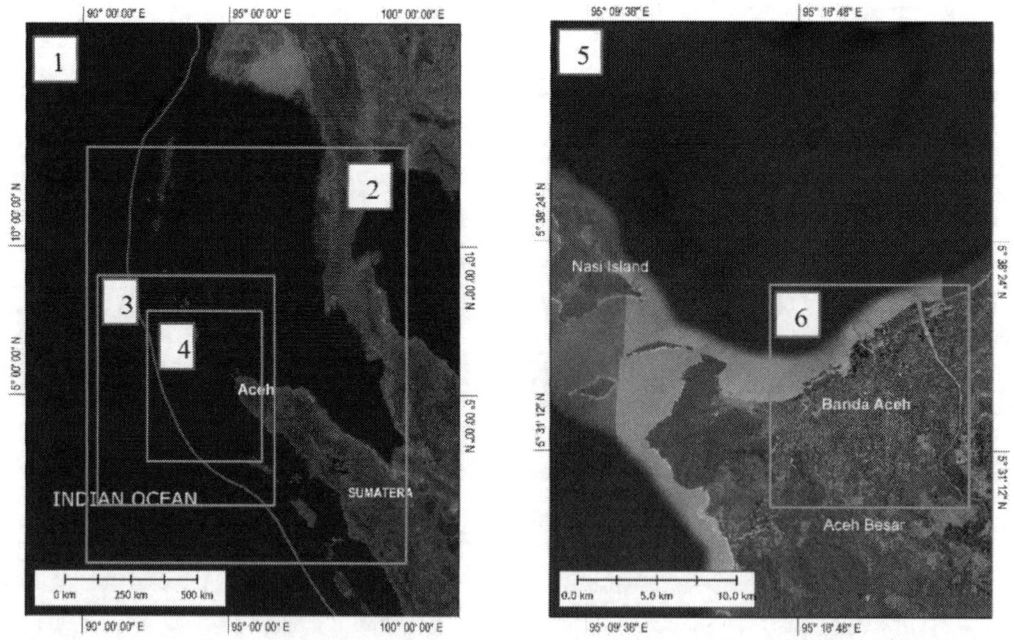

Fig 4. Tsunami simulation domain. The bathymetry from Layer 1 to Layer 5 was provided by GEBCO. JICA updated the data on Layer 6 study area while the non-uniform Manning's Roughness coefficient was used.

https://doi.org/10.1051/e3sconf/202234001014

4 Result and discussion

4.1 Land use projection in 2065

The transition probability matrix between 2011 and 2015 maps is used to predict LULC changes in the next 50 years. The 2015 map is set as the initial year for estimating land use in 2065. Initially, from the transition probability of changes in 2011-2015 (4 years), then land use is calculated for the next four years, namely 2019, to assess the validity of the data. The validation process was carried out using Kappa index statistics to see the suitability of the 2019 projected map and the 2019 actual map. When $0.75 \leq$ Kappa < 1, the maps are in good agreement [24]. Fig. 5 shows the comparison of the area of each land use in the 2019 simulation. The validation result shows the map is in good agreement with the value Kstandar of 0.79, Kno of 0.84, and Klocation of 0.83. These values are feasible for the next analysis.

The 2065 LULC shows that the built-up area of Banda Aceh City will increase significantly. Compared to the 2015 LULC, the bare land area has changed a lot to build-up areas in the central and southern parts of the city. The simulation results demonstrate a quantitative change in the area (Fig. 6) in each land cover type. The largest land cover is the build-up area, with an area of more than 32 km².

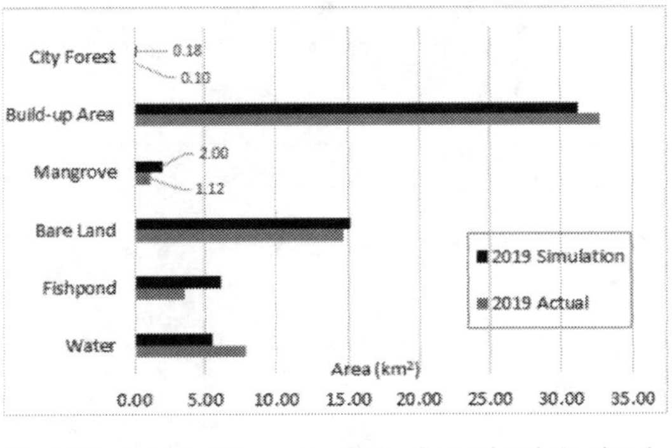

Fig 5. The output of Kappa Analysis of actual and simulated land use map 2019

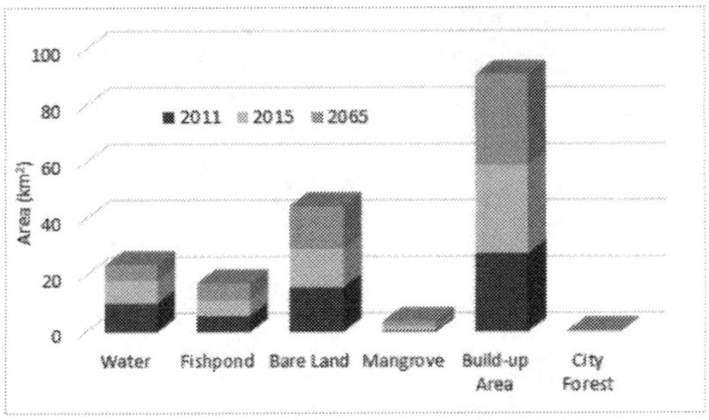

Fig 6. Quantity of previous and predicted LULC change 2011, 2015 and 2065

4.2 Tsunami inundation

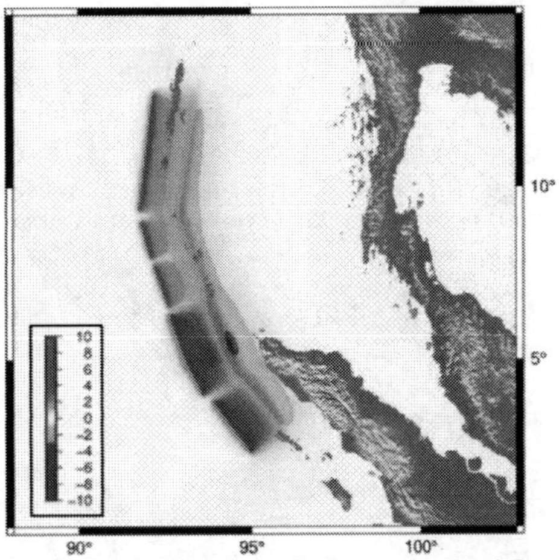

Fig 7. Initial water level change in the computational domain by 9.15 Mw earthquake. The elevation in meter

Tsunami wave propagation was simulated by a numerical model using COMCOT. The initial water level of the 2004 tsunami is shown in Fig. 7. This change in water level occurs due to differences in deformation along the fault zone. Blue color indicates a decrease in water level elevation, while an increase in water level elevation is indicated by red. The decrease in elevation in the western part of the fault close to the west coast of Aceh caused the seawater to retreat shortly after the earthquake. Then, 30 minutes after the earthquake, the tsunami waves came to the coast of Banda Aceh and inundated the 3 km coastal area of the city. In the 2065 future scenarios, the tsunami totally flooded five sub-districts: Meuraxa, Kutaraja, Kuta Alam, Syiah Kuala, and Jaya Baru, as shown in Fig. 8.

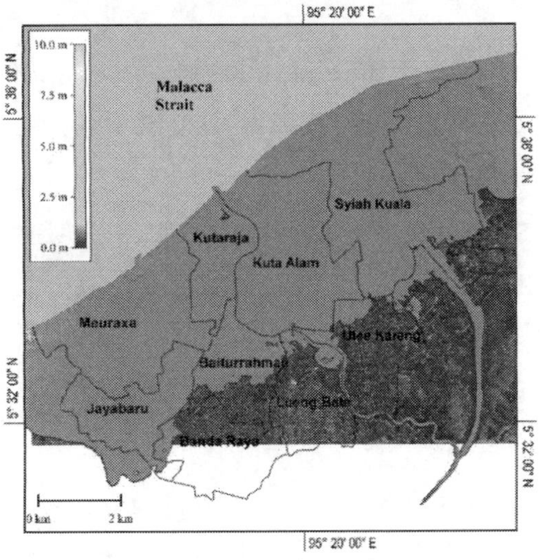

Fig 8. Tsunami inundation of future scenario

4.3 Tsunami exposure on LULC

Simulations were run for both inundation scenarios, including current conditions in 2015 without SLR and future conditions in 2065 with the 50-year SLR incorporated in, as shown in Fig 9. Under current conditions, the inundated area was 41.68 km^2 or 69.5% of the total city area. In the future scenario in 2065, the inundated area increased to 44.70 km^2 or 74.5%. Only 15.30 km^2 of a safe area was left in the southern part of the city. The overlaying of these two scenarios indicates that all of the LULC classes were affected by the tsunami. These conditions occur because the various LULC in Banda Aceh are more concentrated in coastal areas than the southern part of the city.

Figure 10 depicts the land-use area that was flooded by the tsunami. The fish ponds area, in both scenarios, were all affected by the tsunami. In the 2004 tsunami event, fish ponds took a significantly long time to recover to be used again by fishermen. It means that fishermen's livelihoods were disrupted for a long time. The land cover most affected by the tsunami in the scenario of 2065 was the built-up area with more than 20 km^2 or ½ of the tsunami-affected site. The built-up area here includes residential areas, public buildings, roads, offices, and economic regions. The larger the residential area affected, the larger the number of the affected population. Other vital assets located in the affected built-up area will result in economic losses of the city.

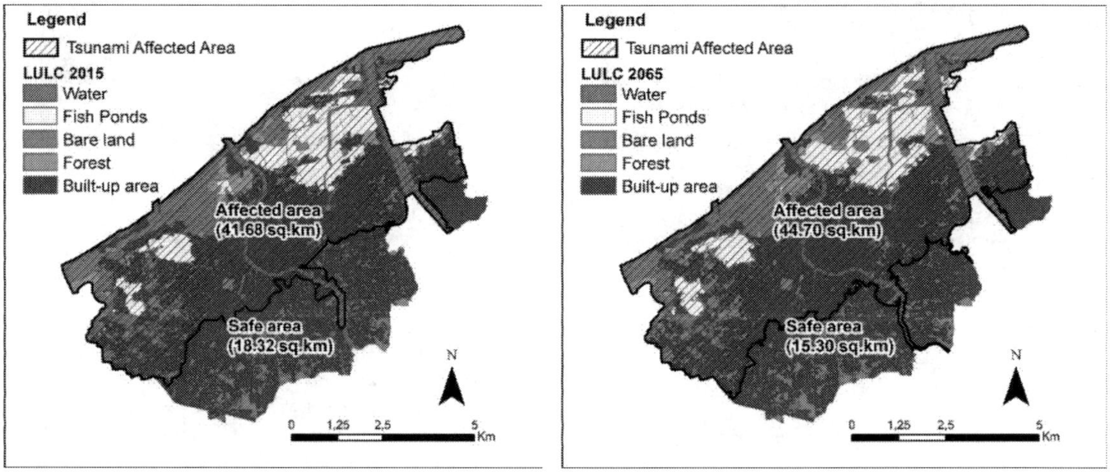

Fig 9. Land use affected by the tsunami inundation in current and future condition

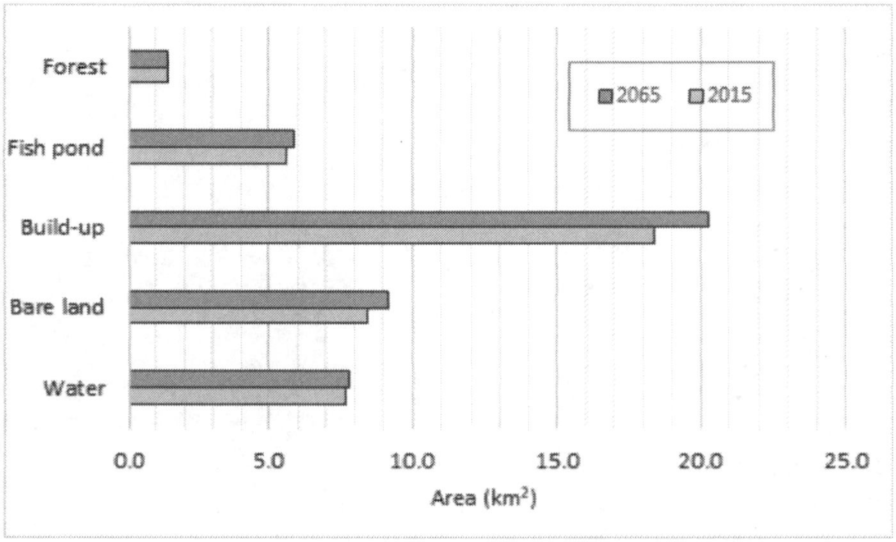

Fig 10. The total area of each type of land cover that was impacted by the tsunami, in a square kilometer.

5 Conclusion

This study examined on how the tsunami might affect Banda Aceh's land use in the future. The dominant built-up area grew in the southern part of the city or the safe zone area. Not only in the safe zone area, but regional growth also occurred in areas affected by the tsunami. The total built-up area affected by the tsunami was 2/3 of the total built-up area of the city. This reveals the high vulnerability of the community to tsunamis in the future. Other types of land cover that were also affected, such as bare land, ponds, and the affected water area, will also increase in future tsunami scenarios. Bare land will experience a negative growth trend because most of it will be converted into built-up areas. Furthermore, the SLR will exacerbate the impact of the tsunami on the city. Thus, the built area, previously a safe zone, will become an affected area in 2065.

The results of this analysis provide important information for policymakers, especially urban spatial planning, which will determine the city's resilience to tsunamis. For example, the tsunami safe zone boundary map used for city spatial reference needs to be updated. The results of this study are also useful as initial information in assessing tsunami risk from the aspect of economic growth in the future. Increasing the built-up area also means increasing population. Therefore, it will also contribute to a larger human exposure. This result gives insight into the needs for sustainable and long-term tsunami risk reduction for Banda Aceh.

Acknowledgments. Authors want to acknowledge Partnership Enhanced Engagement in Research (PEER) of United States Agency for International Development (USAID) and NASEM for providing research grant No. 5-395 title: Incorporating climate change-induced sea-level rise information into coastal cities' preparedness toward coastal hazards. The grant has enabled authors to collect some data and attend a writeshop to publish this article.

References

1. M. Siegert, R. B. Alley, E. Rignot, J. Englander, and R. Corell, One Earth **3**, 691 (2020)
2. V. Gornitz, M. Oppenheimer, R. Kopp, R. Horton, P. Orton, C. Rosenzweig, W. Solecki, and L. Patrick, Urban Clim. **33**, 100654 (2020)
3. L. Li, A. D. Switzer, Y. Wang, C. Chan, Q. Qiu, and R. Weiss, Sci. Adv. 1 (2018)
4. Tursina, Syamsidik, S. Kato, and M. Afifuddin, Int. J. Disaster Risk Reduct. **55**, 102084 (2021)
5. Syamsidik, R. S. Oktari, K. Munadi, S. Arief, and I. Z. Fajri, Nat. Hazards (2017)
6. A. Achmad, S. Hasyim, B. Dahlan, and D. N. Aulia, Appl. Geogr. **62**, 237 (2015)
7. F. Imamura, S. Penmellen, A. Suppasri, and A. Muhari, Prog. Disaster Sci. 1 **1**, 0 (2019)
8. B. M. Sleeter, N. J. Wood, E. Soulard, and T. S. Wilson, Int. J. Disaster Risk Reduct. (2017)
9. J. Song, X. Fu, Y. Gu, Y. Deng, and Z. Peng, (2017)
10. Al'ala and Syamsidik, IOP Conf. Ser. Earth Environ. Sci. 273 (2019)
11. R. S. Oktari, Syamsidik, R. Idroes, H. Sofyan, and K. Munadi, Water (2020)
12. A. A. A. Al-sharif and B. Pradhan, **Arab J Geo**, (2013)
13. M. Tarawally, X. Wenbo, H. Weiming, T. Darlington, and M. Biniyam, Remote Sens. Appl. Soc. Environ. **16**, 100262 (2019)
14. G. B. Ruben, K. Zhang, Z. Dong, and J. Xia, Sustain. **12**, (2020)
15. S. Mansour, M. Al-Belushi, and T. Al-Awadhi, Land Use Policy **91**, (2020)
16. Q. Wang, Q. Guan, J. Lin, H. Luo, Z. Tan, and Y. Ma, Ecol. Indic. **122**, 107231 (2021)
17. M. S. Mondal, N. Sharma, M. Kappas, and P. . Garg, Geocarto Int. (2013)
18. A. A. A. Al-sharif and B. Pradhan, Arab J Geosci (2013)
19. Syamsidik, Tursina, A. Meutia, M. Al'ala, M. Fahmi, and E. Meilianda, J. Earthq. Tsunami **11**, (2017)
20. Syamsidik, T. M. Rasyif, A. Suppasri, M. Fahmi, M. Al'ala, W. Akmal, T. M. Hafli, and A. Fauzia, Int. J. Disaster Risk Reduct. **47**, 101572 (2020)
21. S. Koshimura, T. Oie, H. Yanagisawa, and F. Imamura, Coast. Eng. J. **51**, 243 (2009)
22. X. Wang and P. L. F. Liu, J. Earthq. Tsunami **1**, 273 (2007)
23. X. Wang and P. L. F. Liu, J. Hydraul. Res. **44**, 147 (2006)
24. R. G. P. Jr and L. C. Schneider, Agric. Ecosyst. Environ. **85**, 239 (2001)

Landslide tsunami impact assessment toward human lives using numerical modelling: A case study from Banten, Indonesia

Muhammad Ali[1*], *Teuku* Rasyif[1,2], and *Teuku* Mahlil[1]

[1]Department of Civil Engineering, Faculty of Infrastructure Planning, Universitas Pertamina, Jl. Teuku Nyak Arief, Jakarta Selatan, DKI Jakarta, 12220, Indonesia
[2]Tsunami and Disaster Mitigation Research Center (TDMRC), Universitas Syiah Kuala, Jl. Prof. Dr. Ibrahim Hasan, Gampong Pie, Banda Aceh, 23233 Indonesia

Abstract. After the eruption of Mount Anak Krakatoa, December 22, 2018, which was accompanied by a large landslide into the waters of the Sunda Strait, resulting a tsunami that hit South Coast of Lampung and West of Banten. According [1], 437 peoples were death, 31,942 peoples were injured and 10 peoples were lost due to this disaster. The tsunami impact toward human lives was calculated by using the human instability equation [3] and TUNAMI Two Layer model was applied to simulate the landslide and propagation of tsunami [2]. The simulation results were validated with observations and field measurements data [1]. The result of simulation showed that flow depth of the tsunami reached 0.29-5.83 m with the tsunami velocity about 0.23-7.49 m/s along the shore of Pandeglang Regency. Using Aida Equation for validate simulation results, has K = 0.96733. The results of calculations using the human instability equation show that the value of hazard risk (HR) > 1 because all critical velocity values lower than velocity of simulation. So, it was mean that tsunami waves occurred in Banten especially Pandeglang Regency has high risk impact for the balance of the human body and this condition related with the observation condition that show many victims included people died and injured in this area.

1 Introduction

According to geographic condition, Indonesia located among of three main tectonic plate as known as Indo-Australian, Eurasian, and Pacific. This was made Indonesia often experiences natural disaster such as earthquake and eruption of volcanic. The biggest earthquakes in Indonesia ever happened at December 26, 2004 and also made highest tsunami depth in along of Aceh. In Addition, volcanic activity in Indonesia can be categorized as very high, which is indicated by the number of active volcanoes. There are around 127 active volcanoes that make Indonesia become country with the highest number of volcanoes in the world [4]. History records major volcanic natural disasters have hit Indonesia, one of which was devastating eruption of Mount Anak Krakatau on August 27, 1883. This eruption was referred to largest, deadliest, and destructive volcanic eruption in history that caused large tsunami waves hit almost all the South Coast of Sumatra and West Coast of Java with wave height up to 30 m [5].

On December 22, 2018, Mount Anak Krakatau erupted again which caused landslide of caldera and triggering a tsunami wave in the Sunda Strait. The tsunami wave hit the South Coast of Lampung and West Coast of Banten with the worst affected area is Pandeglang Regency. The number of victims died was 437 people, 31,942 people was injured, and 10 people still missing [1]. The highest tsunami wave height reached 13 m occurred on the West Coast of Banten. Furthermore, Sunda Strait also has potential for tsunami caused by megathrust subduction tectonic [6].

Tsunami waves are classified into tectonic and volcanic tsunamis based on causative factor. Almost 90% of tsunamis in Indonesia occurs due to earthquakes at the seabed [7]. And then, factors that caused volcanic tsunamis are (1) volcanic eruptions cause by earthquakes at sea, (2) underwater volcanic eruptions, (3) pyroclastic flow (ignimbrites and nuees ardentes), (4) caldera collapse, (5) soil and cold lava rock slides, (6) base surge due to shock waves, (7) hot rock avalanches, (8) entry of lava into waters, (9) air waves due to explosions, (10) lava avalanches into the waters [8].

Pandeglang Regency is the name of a district located in Banten Province with coordinated 6° 21' – 7°10 LS and 104° 48' – 106° 11' BT. Pandeglang Regency has administrative boundaries, namely Serang Regency for the North, Indian Ocean for the South, Sunda Strait for the West, and Lebak Regency for the East. This location has total area is around 2,746.81 km² with 35 sub-districts and also has total population in 2019 reached 1,211,909 people with population density of 441 people/km². Topography of Pandeglang Regency form

* Corresponding author: mghofurma@gmail.com

with low elevation for Central and South Regions and high elevation for the North Region [9].

The location of research was conducted in the Pandeglang Regency such as Carita District and Labuan District. Carita District is focused on area of Sambolo Bay, Carita and Sukajadi Village. For Labuan District is focused on Labuan, Teluk, and Cigondang Village. This location chosen because has U shoreline type and fairly steep depth contour type which has high risk potential to produce highest tsunami wave. In addition, the chosen location are dominated by tourism and resident areas.

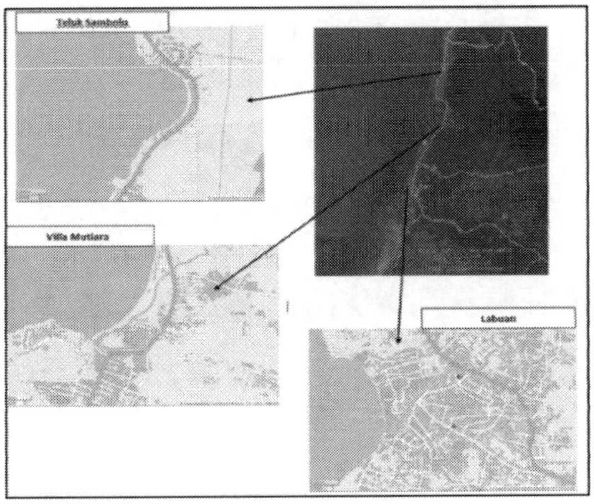

Fig. 1. Location of research

One of the risks of tsunami is inundation that rise to the ground and propagated on the mainland with certain distance. Factor that affected the distance of inundation are sand dunes, cliffs, vegetation in the coastal area, the nature and configuration of the beach, and the most influential are elevation of the beach and speed of wave propagation. The inundation of tsunami wave that hit the mainland can damage infrastructure and also pose a threat to humans due to generate hydrodynamic forces

The impact of tsunami waves that spread to mainland or inundation on human stability need to studied seriously. This study is aimed for investigating impact of the 2018 Sunda Strait tsunami on human lives at Pandeglang coastal area. Process simulation in this study applied numerical modelling to simulate the process of propagation of tsunami waves caused by landslide of Mount Anak Krakatau. To carry out the simulation process, a numerical modelling program namely TUNAMI Two-Layer [2] was used. And then, risk of inundation to humans can be assessed by modelling human instability equations. The output of simulation with TUNAMI Two-Layer and human instability modelling, author can make map classification level threat of danger to humans from hazard ratio value (HR) and expected to be reference parameter in making appropriate mitigation especially in area direct with coastal and tourist resort in Pandeglang.

The hazard ration value has been used in spatial form will indicate areas that have high risk when inundated by inundation, the value will be benchmark for how many people will be affected by brunt of tsunami inundation. The result from this study will be validate with the real condition in the impact area.

Assessment of tsunami impact caused by landslide of caldera of Mount Anak Krakatoa needs to be studied to investigate, educate and provide information about the magnitude of the impact, starting from height and velocity of tsunami wave when they hit the mainland and range of inundation. These are important factors to know hazardous area with different level of risk. Although, there is a very little probability of tsunami with same scenario in the future, as long as Mount Anak Krakatoa is still active, threat of tsunami in Banten especially in Pandeglang Regency is still exist and coupled with risk of megathrust earthquakes which pose a major threat in the future.

Research on assessment of tsunami caused landslide of caldera of Mount Anak Krakatoa to determined impact on humans using mathematical modelling that is human instability equations. However, this research still has limitation by unavailability of statistical data on distribution of affected victims in part of region especially in Carita and Labuan District. Therefore,

2 Data and methods

Data collection method in this research are literature study related to tsunami modelling due to landslide and using secondary data such as bathymetry and topographic data, landslide parameters, weight and height of human. Bathymetric and topographic data used to determine depth and condition of the location so best geometric shape and resolution can be reviewed by data input of simulation. Bathymetric and Topographic data obtained through National Bathymetry (BATNAS) and National Digital Elevation Model (DEMNAS) on https://tanahair.indonesia.go.id/demnas/#/batnas.

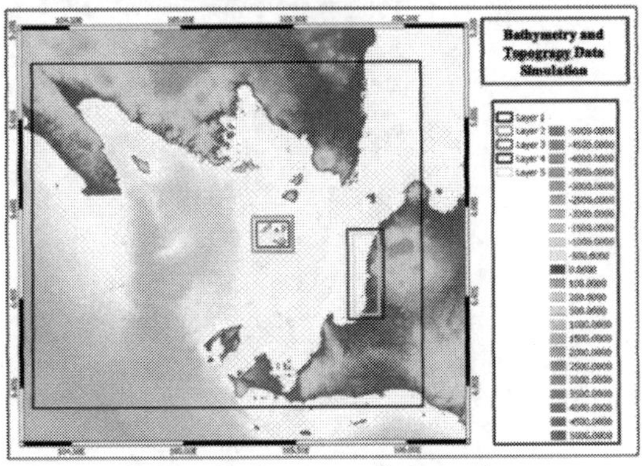

Fig. 2. Layers data of bathymetry and topography.

Grid resolution of both data are 185 m and 8 m. In this research, grid layer will be divided into five layers with grid resolution are 180 m, 60 m, 20 m, 60 m, and 20 m. First layer until third layer are using for simulating area of landslide Anak Krakatau Mountain and propagation of tsunami in sea. Fourth and fifth layer are using for simulating propagation tsunami in mainland. Processing to get layer data from bathymetry and topography using Generic Mapping Tools (GMT), Quantum GIS (QGIS), and Fortran software.

Table 1. Setting Input Data Layer

Location	Grid Size (m)	Coordinate	
		Minimum	Maximum
Layer 1	180	(104.320;-6.903)	(106.078;-5.345)
Layer 2	60	(105.314;-6.201)	(105.495;-6.040)
Layer 3	20	(105.300;-6.180)	(105.478;-6.066)
Layer 4	60	(105.779;-6.500)	(105.903;6.250)
Layer 5	20	(105.750:-6.500)	(105.961;-6.250)

Landslide mechanism data is used to determine movement of landslide and triggering tsunami waves. Landslide mechanism data in this research still adheres by [2] because process of landslide modelling has a fairly complicated with sophisticated computer specification. Landslide parameter data obtained from research data by [2] are maximum height of landslide and sliding time. Maximum height of landslide and sliding time is around 301, 2 m and 10.8 minutes with landslide volume reaching 0.182 km^3. These data are one of important parameters for input in TUNAMI Two-Layers. This modelling will simulate landslide that triggered tsunami by modelling interaction between tsunami waves and subaerial-submarine landslide as top and bottom layers [2]. TUNAMI Two-Layer using governing equations as follows:

a. First layer equation
Using continuity equation as follows:

$$\frac{\partial Z_1}{\partial t}+\frac{\partial Q_{1x}}{\partial x}+\frac{\partial Q_{1y}}{\partial_y}=0 \quad (1)$$

Using momentum equation at x and y directional that write as follows:

$$\frac{\partial Q_{1x}}{\partial t}+\frac{\partial}{\partial x}\left(\frac{Q^2{}_{1x}}{D_1}\right)+\frac{\partial}{\partial y}\left(\frac{Q_{1x}Q_{1y}}{D_1}\right)+gD_1\frac{\partial Z_1}{\partial x}+gD_1\frac{\partial Z_2}{\partial x}+\tau_{1x}=0 \quad (2)$$

$$\frac{\partial Q_{1y}}{\partial t}+\frac{\partial}{\partial x}\left(\frac{Q_{1x}Q_{1y}}{D_1}\right)+\frac{\partial}{\partial y}\left(\frac{Q^2{}_{1y}}{D_1}\right)+gD_1\frac{\partial Z_1}{\partial y}+gD_1\frac{\partial Z_2}{\partial y}+\tau_{1y}=0 \quad (3)$$

b. Second layer equation
Using continuity equation as follows:

$$\frac{\partial Z_2}{\partial t}+\frac{\partial Q_{2x}}{\partial x}+\frac{\partial Q_{2y}}{\partial_y}=0 \quad (4)$$

Using momentum equation at x and y directional that write as follows:

$$\frac{\partial Q_{2x}}{\partial t}+\frac{\partial}{\partial x}\left(\frac{Q^2{}_{2x}}{D_2}\right)+\frac{\partial}{\partial y}\left(\frac{Q_{2x}Q_{2y}}{D_2}\right)+gD_2\frac{\partial Z_2}{\partial x}+gD_2\frac{\rho_1}{\rho_2}\frac{\partial Z_1}{\partial x}+\tau_{2x}=0 \quad (5)$$

$$\frac{\partial Q_{2y}}{\partial t}+\frac{\partial}{\partial x}\left(\frac{Q_{2x}Q_{2y}}{D_2}\right)+\frac{\partial}{\partial y}\left(\frac{Q^2{}_{2y}}{D_2}\right)+gD_2\frac{\partial Z_2}{\partial y}+gD_2\frac{\rho_1}{\rho_2}\frac{\partial Z_1}{\partial y}+\tau_{1y}=0 \quad (6)$$

$$\rho_2=(1-\vartheta)\rho_g+\vartheta\rho_1 \quad (7)$$

Where ρ_1, ρ_g, ρ_2, and ϑ is density of seawater, granular material, sediment, and porosity. In this modelling, fluidized precipitate is hypothesized as immiscible. Index 1 and 2 relates to the upper layer and second layer. Z(x,y,t), i=1,2 is the level of layer at each point (x,y) at time t, where the level value is measured from a given reference level. Q(x,y,t), i=1,2 is the vertically integrated discharge in the x and y directions. $\tau_i(x,y,t)$ is the bottom stress in each layer at each points (x,y) at time t. In this study, parameters were kept fixed as follows: water density = 1,000 kg/m^3, landslide density = 2,000 kg/m^3, and manning coefficient = 0,025 [2].

The simulation results will validate using observational data by [3]. Validation analysis using Aida equation, sum square error (SE), mean square error (MSE), root mean square error (RMSE). Observational data that using in this study located on Marina Jambu tidal gauge with 49 points locations. Validation using Aida equation will result two parameter values of K and k, which is compared with standard of Japan Society of Civil Engineering (JSCE). Based on JSCE, best recommended value of K and k is 0.95< K <1.05 and k < 1.45. Aida equations write as follows.

$$\log K=\frac{1}{n}\sum_{i=1}^{n}\log K_i \quad (8)$$

$$\log k=\sqrt{\frac{1}{n}\sum_{i=1}^{n}(\log K_i)^2-(\log} \quad (9)$$

$$K_i=\frac{x_i}{y_i} \quad (10)$$

Where variable of x_i is observational data and y_i is simulation data. Human instability modelling needed data of human weight and height were used for input data. These data based on research by [3]. Analysis of human instability using five equations that developed by [11][12][3][13][14] with result of this equations is critical velocity. Five human instability equations write as follows:

1) Abt, et al. (1989)

$$DU_C = 0.0929(e^{0.001906HM+1.09})^2 \tag{11}$$

2) Karnoven, et al. (2000)

$$DU_C = 0.004HM + 0.2 \tag{12}$$

3) Ramsbottom, et al. (2004)

$$D(U_C + 0.5) + D = Hazard\ Lev \tag{13}$$

4) Lind and Hartford (2000)

$$DU_C = K_0 M^{0.5} \tag{14}$$

5) Jonkman and Penning-Roswell (2008)
   a. Momentum

$$DU_C = \left(\frac{2HMg\cos\alpha}{C_D\beta\rho}\right)^{0.5} \tag{15}$$

   b. Friction

$$DU_C{}^2 = \frac{2\mu gM}{C_D\beta\rho} \tag{16}$$

This value will be used to calculate HR value at each location. Therefore, it is possible to know level of hazard risk at each location and make easily to facilitate the proper mitigation planning process. Analysis process of human instability in this study using Fortran to get critical velocity value and QGIS for illustrated HR values in spatial model. Equation to get HR value as follows:

$$HR = \frac{U_S}{U_C}, \qquad HR \leq 1 \tag{17}$$

3 Results and discussion

a. Validation

Validation using Aida Equation produces two parameter values namely K and k are 0.967 and 1.691. Based on JSEC standards, value of k does not meet requirements and it is mean that accuracy of simulation data is not good. The value of k does not meet JSCE standard because boundary equation used in TUNAMI Two-Layer not include parameter of building effect, even though it has effect in flow of tsunami wave in the field. According [10], building barriers can increase tsunami wave flow resistance relatively in modelling compared absence of barriers. Therefore, including building parameter in modelling become problem limitation, so that results of wave height tsunami become less accurate.

Then, validation using root mean square error-observations standard deviation ratio (RMSE-SR) and Nash-Sutcliffe effiecency (NSE) produce values of 0.453 and 0.547. If RMSE-SR and NSE value is 0<RMSE_SR ≤0.5 and 0.36<NSE≤0.75, which means that the simulation results have a very good level of accuracy and interpretation of result data is qualified. Moreover, result of simulation also calculate using coefficient of relation (R), the following values are 0.462. This value in the category of sufficient accuracy because in range of 0.25≤R≤0.5. So, based on calculation of Aida equations, RMSE-SR, NSE, and category of sufficient, it can be concluded that the simulation data have interpretation as qualified data and good be used as data reference for further analysis.

b. Initial condition

Simulation result can display as spatial of Mount Anak Krakatau when it has landslide for 10 minutes. In addition, sea water level changes shown when landslide occurs.

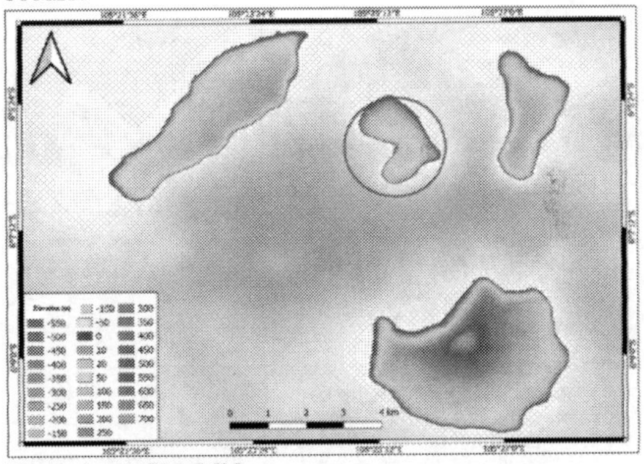

Fig 3. Area of landslide

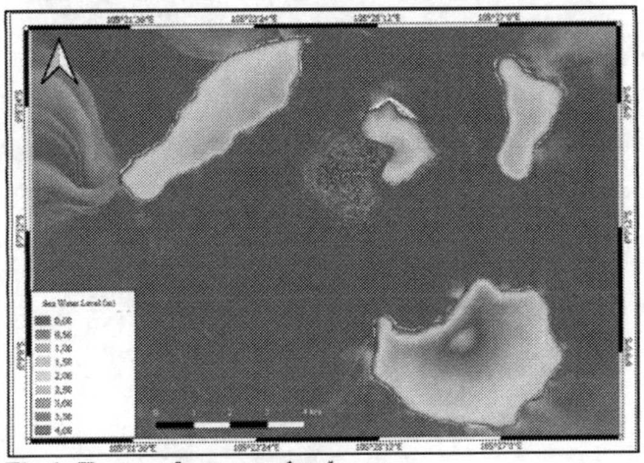

Fig 4. Change of sea water level

c. Wave height of tsunami

Wave height at Carita and Sukajadi Villages are dominated by 4-5 m with inundation range of 200-800 m. The magnitude of dominating wave height is 72.894% of total simulation data.

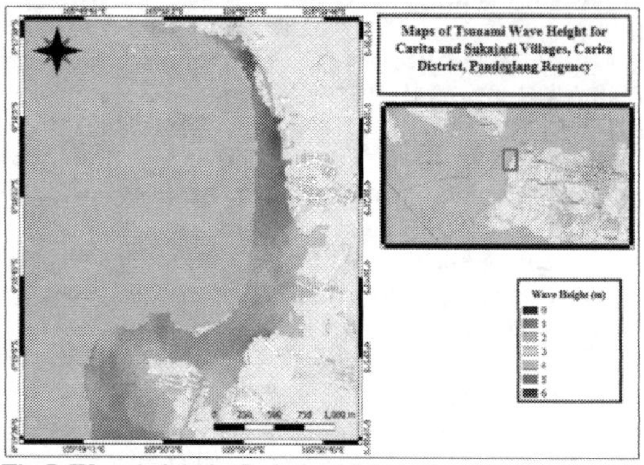

Fig 5. Wave height in Carita and Sukajadi

Wave height at Sambolo Bay is dominated by 3-4 m with inundation range of 400-500 m. The magnitude of dominating wave height is 80.030% of total simulation data.

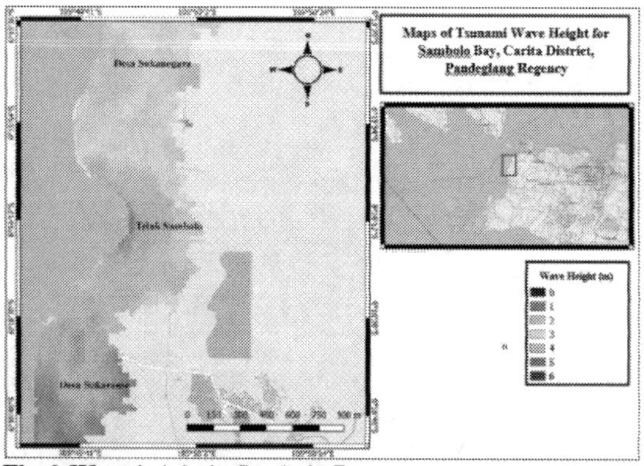

Fig 6. Wave height in Sambolo Bay

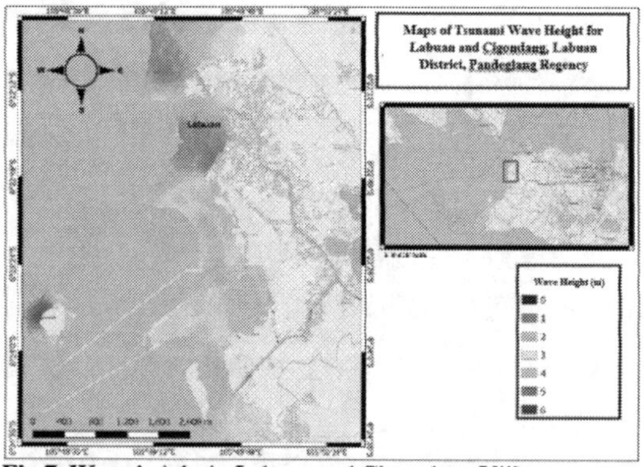

Fig 7. Wave height in Labuan and Cigondang Villages

Wave height at Labuan and Cigondang Villages are dominated by 3-4 m with inundation range of 500-1500 m. The magnitude of dominating wave height is 47.5% of total simulation data.

Based on simulation results of tsunami wave height at three locations, it can be concluded that tsunami waves are dominated by height above 2 m with highest tsunami waves in the coastline area reaching 4-6 m. Tsunami wave that hit the coastline are became highest due to factors from changes in wavelength of tsunami from deep waters to shallow waters, sea depth, and shape of coastline. When tsunami waves reach land and inundated, their height will decrease according to elevation height and level of soil roughness at the scene. Tsunami wave height that exceeds 3 m is classified as a tsunami wave which is very dangerous for both building structures and humans. This is one of the assessment parameters that be used as hazard risk analysis and mitigation planning in tsunami-affected areas [15].

d. Wave velocity of tsunami

Velocity of tsunami waves in Sambolo Bay dominated by 3-4 m/s with percentage of 3-4 m/s and on shoreline has 4 m/s, it means that velocity is very fast and dangerous. Furthermore, tsunami waves in Sambolo also have a very high wave.

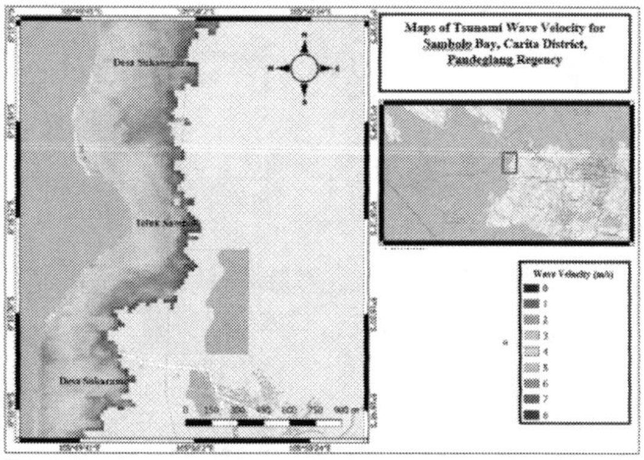

Fig 8. Tsunami wave velocity in Sambolo Bay

Velocity of tsunami waves in Carita and Sukajadi dominated by 2-3 m/s with percentage data of 23.737% and on the shoreline is 4-7 m/s.

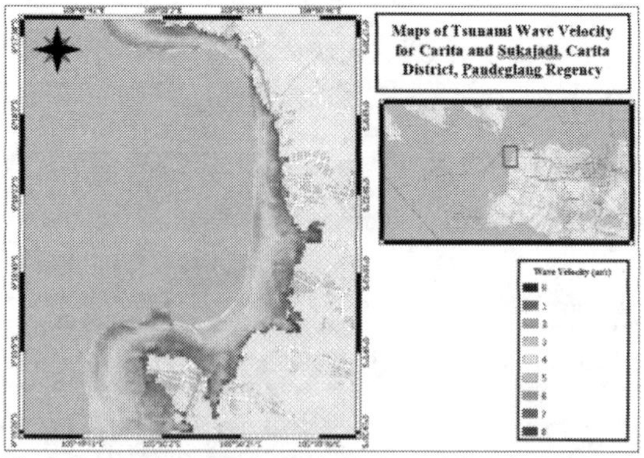

Fig 9. Tsunami wave velocity in Carita and Sukajadi Villages

Then, velocity of tsunami wave in Labuan and Cigondang dominated by 3-4 m/s with percentage

27.601% and on the shoreline of area is 4-6 m/s. This value same to case of Sambolo Bay.

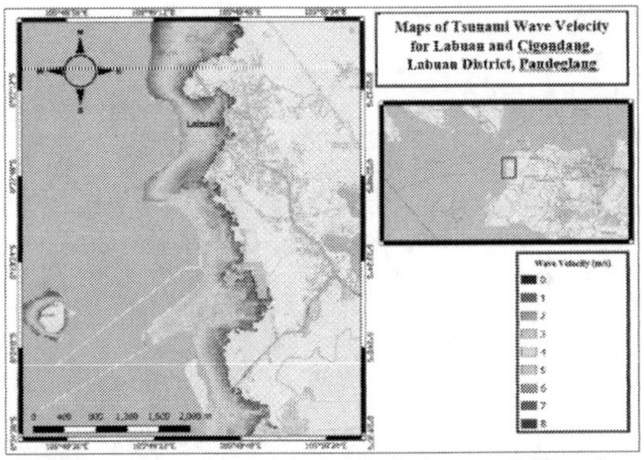

Fig 10. Velocity of tsunami wave in Labuan and Cigondang villages

Based on the simulation result, velocity of tsunami wave in three locations are dominated by more than 2 m/s. this velocity is relatively fast and increases risk of danger because it is accompanied by a fairly high tsunami height. Velocity of wave tsunami in this location became fast because factor of depth of sea and elevation along the shoreline.

e. Time arrival of tsunami

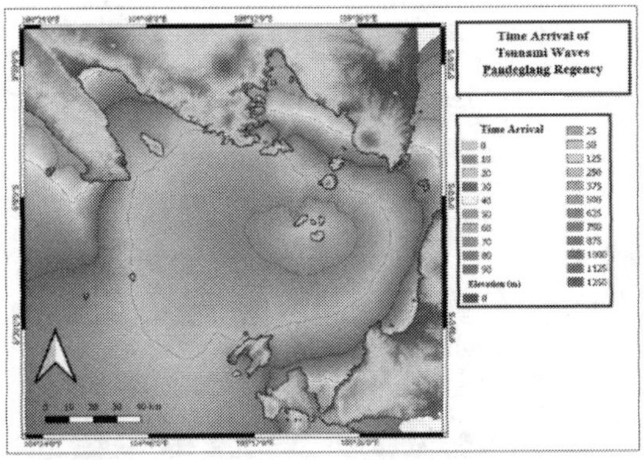

Fig 11. Time arrival of tsunami wave in Pandeglang Regency.

First tsunami wave reached shoreline in Sambolo Bay, Carita-Sukajadi Village, and Labuan-Cigondang Village in time of 36 minutes, 37 minutes, and 43 minutes. It is important to analyze time of tsunami wave propagation to determine mitigation process such as determined evacuation route and locations.

f. Human instability and hazard ratio analysis

The results of the calculation from data sample using five human instability equations show that Karnoven equation resulted best estimated compared to other equations. So, in this study, result from calculation using Karnoven equation would be used to calculate hazard ratio. The Output from calculation of human instability is critical velocity. For Sambolo Bay case has critical velocity for male dominated by 0.050-0.059 m/s with percentage of 44.444-57.658% and 0.060-0.069 m/s with percentage of 42.417-52.327%. This value for case of age 4-12 years and above 12 years. Then, female case has same value from male case, just different in percentage value. So, Sambolo Bay has HR value for any range of age is HR = 1.

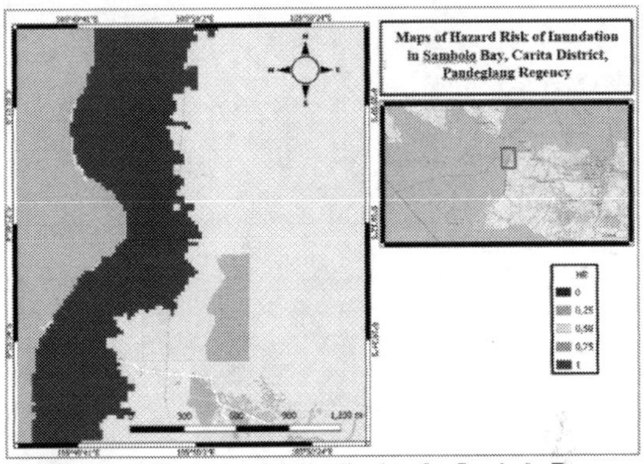

Fig 12. Hazard ratio spatial distribution for Sambolo Bay

For Carita and Sukajadi case, dominated critical velocity for male case are 0.040-0.049 m/s with percentage of 48.181-60.165% and 0.050-0.059 m/s with percentage of 59.361-70.444%. This value for case of age 4-9 years and above 9 years old. For female case, value of critical velocity has same with male case, just different in percentage point. In this location also has HR =1, because velocity of tsunami height larger than critical velocity itself.

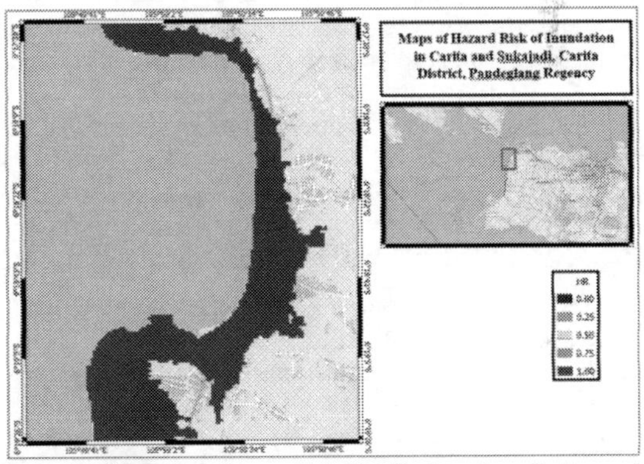

Fig 13. Hazard ratio spatial distribution for Carita and Sukajadi villages

For Labuan and Cigondang case, value of critical velocity for male case dominated by 0.050-0.069 m/s with percentage of 45.900%-54.068%. This value for case of age 4 year until above 80 years old. For female case, value of critical velocity has same with male case just different in percentage distribution. HR value in this location also has HR = 1, same like with other location.

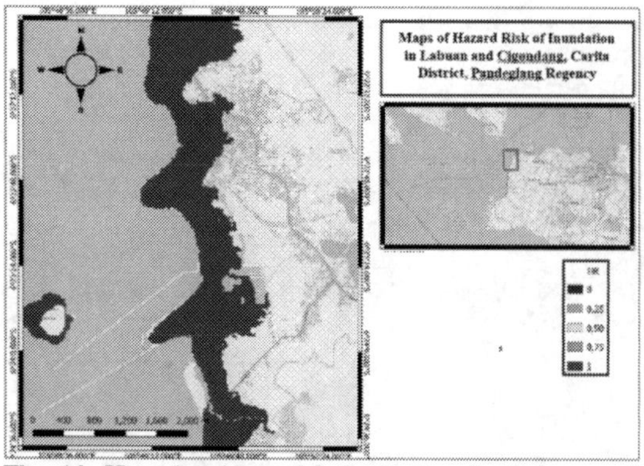

Fig. 14. Hazard ratio spatial distribution for Labuan and Cigondang villages

The result of HR calculations shows HR = 1 for all location in this study, it shows that humans who are in the red zone will not be able to maintain their body balance when exposed to tsunami waves. As the result, humans will be carried away by current of tsunami and caused several possibilities that occurs as surviving without being injured, injured and dead. Based on interim report [16] at December 22, 2018, number of victims in Carita District are 50 people died and 50 people injured. Meanwhile in Labuan District there are 6 people died and 50 people injured. This data show that tsunami occurred in both locations poses a risk to human lives, although data is not detailed and is not final result. However, overall report of final victim data released by [1], the number of victims died was 437 people, 31,942 people was injured, and 10 people still missing. In addition, Carita and Labuan District were two most severely affected by tsunami based on [16]. This indicates that the results of analysis of human instability are in line with actual condition in the field based on risk distribution mapping in Figure 12, Figure 13, and Figure 14.

4 Conclusions

Result in this study showed that tsunami waves caused by landslide of Mount Anak Krakatau had a dominant height wave of above 2 m with velocity over 2 m/s. Meanwhile, critical velocity that can received by humans only under 1 m/s. Therefore, value of hazard risk is HR = 1, which means all area in this study that were inundated by tsunami are unsafe for stability of human body and this analysis are in line with actual condition in the field based on risk distribution mapping. Further study is needed to investigate human risk due to tsunami by applying high resolution data and detail data of victims report including death and injured at any area of study to validation simulation results and human instability equation. Furthermore, using other possibility scenario to enhance mechanism of tsunami disaster in the future and also can preparing step of disaster mitigation.

Acknowledgement. Authors are very grateful to supports received from Dr. Kwanchai Pakoksung, IRIDeS, Tohoku University for providing the TUNAMI Two-Layers model for this study.

References

1. Syamsidik, Benazir, M. Luthfi, A Suppasri, L. K. Comfort. Natural Hazards and Earth System Sciences. 20 (2020) 549-585
2. K. Pakoksung, A. Suppasri, A. Muhari, F. Imamura. Geoscience Letters 7 (2020) 1-20.
3. R. Nakamura, M. Mäll, T. Shibayama. Natural Hazard. 99 (2019) 391-422.
4. Center of Volcanology Survey of Indonesia. Citation from *https://www.vsi.esdm.go.id/index.php/kegiatanpvmbg/kegiatan-diseminasi-informasi/2829-press-releaseaktivitas-vulkanik-g-anak-krakatau-31-desember-2019.* (January, 2019)
5. Sutawijaja, I. The Disaster Prevention Research Institute Annuals. 40 (1997)
6. Maemo, F., & Imamura, F. Indonesia Jurnals Geopysh Res. 116 (2011)
7. Latief, H., Puspito, & Imamura. Tsunami Catalog and Zones in Indonesia. 22 (2000) 25-43
8. Latter, J. H. Bulletin Volcanologique. (1981) 467-490
9. Statistic Center Institution (BPS). *Kabupaten Pandeglang Dalam Angka 2020.* (2020) 1-25
10. Aida, I. J. Phys Earth. (1978) 57-73
11. Abt, S., Wittler, Taylor, & Love. J Am Water Resour As. 25 (1989) 881-890
12. Flooad Hazard Research Centre, Middlesex University Risk, & Policy Analysis. FD2321/TR2 Guidance Document. (2006)
13. Lind, Hartford, & Assaf. J Am Water Resour. (2004) 89-96
14. Jokman & Penning-Roswell. J Am Water Resour. 44 (2008) 5-15
15. Rahmawan. *Studi Potensi Bahaya Tsunami di Selatan Jawa.* (2012)
16. National Board Disaster Management (BNPB). Citation from *https://bpbd.bantenprov.go.id/read/berita/240/LAPORAN-SEMENTARA-TSUNAMI-BANTEN.* (December, 2018).

Current state of the seismic broadband network of Indonesia: Is it possible for the early warning system from a subaerial source-induced tsunami?

Haekal A. Haridhi[1,2,3,*], *Dimas* Sianipar[4], *Syahrul* Purnawan[1,3], *Ichsan* Setiawan[1,3], *Bor-Shouh* Huang[5]

[1]Department of Marine Sciences, Faculty of Marine and Fisheries, Universitas Syiah Kuala, Banda Aceh 23111, Indonesia.
[2]Tsunami and Disaster Mitigation Research Center (TDMRC), Universitas Syiah Kuala, Banda Aceh 23111, Indonesia.
[3]Research Center for Marine Sciences and Fisheries, Universitas Syiah Kuala, Banda Aceh 23111, Indonesia.
[4]Agency for Meteorology, Climatology, and Geophysics (BMKG), Jakarta 10610, Indonesia.
[5]Institute of Earth Science, Academia Sinica, Taipei 115, Taiwan

Abstract. Lesson learned from the Anak Krakatau volcanic eruption followed by the tsunami on 22 December 2018 has generated a huge gap in the readiness of the Indonesian early warning system to detect such an event. Consequently, the official was found it difficult to broadcast the right warning within a considerable time to the local residences to evacuate or take necessary actions, which resulted in the loss of life and properties. There were more than 430 people died, and countless properties were destroyed when the tsunami swept the coastlines of southern Sumatra and western Java. While, this event is among the perfect events to study about the subaerial slide since it was occurred on near-surface with moderate magnitude M5.9 and creates the subaerial slide during the process (flank collapsed), and the broadband seismic stations managed by the Agency for Meteorology, Climatology, and Geophysics (BMKG) are already established and record the signal. Usually, the seismic source from the eruption was not complicated, and the seismic signal that recorded by the BMKG broadband seismic stations should correspond to the subaerial slide processes. Since Indonesia is an archipelago nation, the seismic station managed by the BMKG is distributed on the island with a distance ranging from a couple of kilometers to a tenth of kilometers away between stations. For this reason, this study tries to highlight several important suggestions in order to improve the detection using the signal recorded by the BMKG broadband seismic stations relate to such an event.

1 Introduction

The Krakatau volcanic eruption followed by the tsunami around Sunda strait (Indonesia) that occurred on 22 December 2018 was not the first-ever recorded event; unfortunately, it has repeatedly to occur. In recent history, it occurred in 1883, 1928, 1930, and 1981 [1]. Unfortunately, there were no acquisition devices available in that time and at the region to record the seismic wave energy for the events that were occurred in the 18 – 19 centuries; thus, the mechanism of how the Krakatau volcanic eruption could generate a tsunami is still unclear [2–4]. At the end of 19 century and early 20 centuries, the Agency for Meteorology, Climatology, and Geophysics (BMKG) of Indonesia was established and operated the permanent seismic network distributed in the Indonesia region. With the increasing number of broadband digital seismic stations, it is possible nowadays to study this type of event.

The Anak Krakatau volcanic eruption followed by the tsunami on 22 December 2018 has generated a huge gap in the readiness of the Indonesian early warning system to detect such an event. Consequently, the official was found it difficult to broadcast the right warning within a considerable time to the local residences to evacuate or take necessary actions, which resulted in the loss of life and properties. There were more than 430 people died, and countless properties were destroyed when the tsunami swept the coastlines of southern Sumatra and western Java [5,6]. While this event is among the perfect events to study about the subaerial slide, i.e., it was occurred on a near-surface with low seismic magnitude M5.9 and created the subaerial slide during the process (flank collapsed). At the same time, the broadband seismic stations managed by the Agency for Meteorology, Climatology, and Geophysics (BMKG) are already established and record the signal. Usually, the seismic source from the volcanic eruption was not complicated, and the seismic signal that recorded by the BMKG broadband seismic stations should correspond to the subaerial slide processes. Since Indonesia is an archipelago nation, the seismic station managed by the BMKG is distributed on the island with a distance ranging from a couple of kilometers to a tenth of kilometers away between stations. For this reason, this study tries to highlight several important suggestions in order to improve the detection using the

*Corresponding author: haekal.azief.haridhi@unsyiah.ac.id

signal recorded by the BMKG broadband seismic stations.

This study aims to evaluate the Indonesia early warning system from the existing BMKG broadband seismic stations on the event of the Anak Krakatau volcanic eruption event on 22 December 2018.

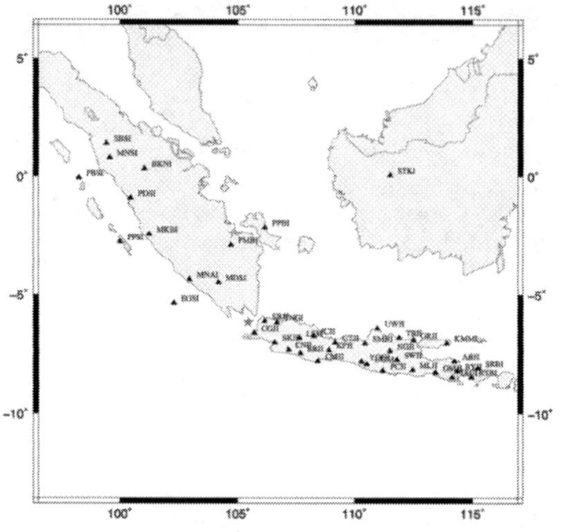

Fig. 1. The spatial distribution of the BMKG broadband seismic stations that recorded the 22 December 2018 Anak Krakatau volcanic eruption. The stations are indicated in the black triangle with text shown the key station name, and the red star indicates the Anak Krakatau event location.

2 Methods

The broadband signals of the Anak Krakatau eruption on 22 December 2018 were obtained upon request to the BMKG. As the data is received, all seismic records were deconvolved with their respective instrument responses to velocity by using the transfer function in SAC [7,8]. With all available stations, as shown in Fig.1, the signal is clearly observed and well recorded, as shown in Fig. 2. Then the manual picking of the event duration on the vertical, radial, and transversal components was done throughout all signals. It is commonly known that the low-frequencies surface waves of a large landslide could be recorded at the stations at long distances, while the high-frequency waves could be recorded with a distance up to a few hundred kilometers [9]. The landslide signals were obviously distinguishable from an earthquake signal; thus, the landslide signal usually observed has a short duration of less than ~100 seconds, has an impulse onset, unclear S-wave onsets. Therefore it is distinguishable between landslide and earthquake events [9,10].

The low-frequencies surface wave recorded at the available stations reflects the series of unloading and reloading of the slide material to the sliding surface from the start of sliding-loading until the sliding process stopped. The high-frequency signal may indicate the scattering effect of the sliding material, such as debris clasts, therefore to estimate the landslide volumes, the low-frequency surface waves are used.

Then, the energy of the signals was calculated following the procedure described in [9], which uses the sum of the squared band-passed velocity signals as an energy parameter, as is seen in equation 1.

$$E = \Sigma_{t=t_0}^{t=t_1} v(t)^2 dt \quad (1)$$

Where v(t) is the 0.05 Hz low-pass filtered velocity record (cm/s) at time t, t_0 is the first arrival, while t_1 is the end of events durations, through the duration of the events indicated as T1 and T2 in Fig. 2b.

Using the detailed studies of landslide and seismic energy done by [9,10], we empirically derive an attenuation relation. A linear function with a constant slope is obtained through the resulted calculated energy from the observation in the function of distance [11]. The regress linear function follow the regress proposed by [9];

$$log_{10}E = \alpha log_{10}R + \beta \quad (2)$$

Where R is the source-station distance (km), α is a constant slope, and β is the intercept. Thus, the energy of the seismic wave at 0 distance or at the source could be predicted using equation 2.

Due to the subaerial nature of landslides, the slide material volume is difficult to estimate. Therefore, we use the detailed study by [9] where, in their study, they compiled five landslide events and measured its associate slide material through airborne LiDAR topographic surveys. They proposed a linear relation that fit between the seismic wave energy and slide volume;

$$E|_{R=1km} = 8.5 \times 10^{-20} V^{2.06} \quad (3)$$

Where V is the volume of the landslides.

Thus, we could predict how large the energy was at 0 distance or at the source. The predicted energy at 0 distance is then compared to the study done by [9] to empirically obtain the subaerial slide volume using equation 3.

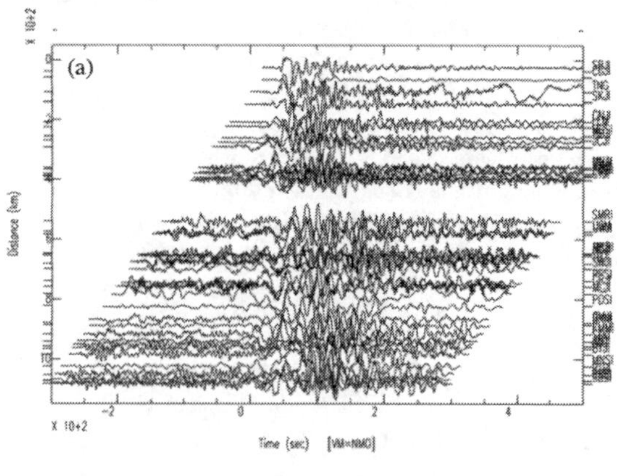

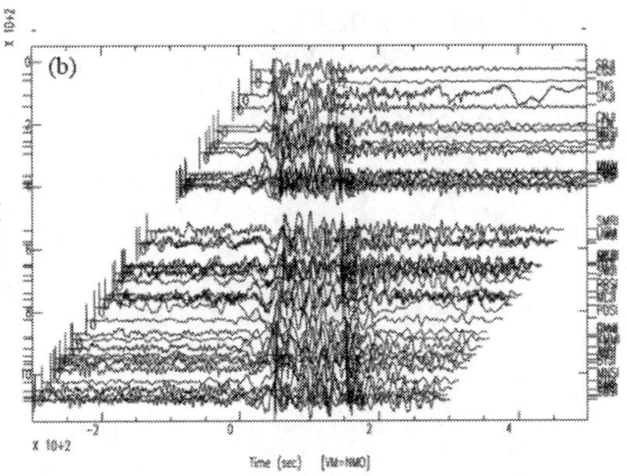

Fig. 2. The normal moveout (NMO) Signal Stacking Subprocess (SSS) with an apparent velocity of 3.2 km/s of the available seismogram. The signals are low-pass filtered to 0.05 Hz, with the x-axis indicate the time in second, the left and right y-axis indicate the station distance and names, respectively. (a) The signal without picking the event duration, (b) with origin time, and picks of T1 and T2 indicate the start and end of the event durations.

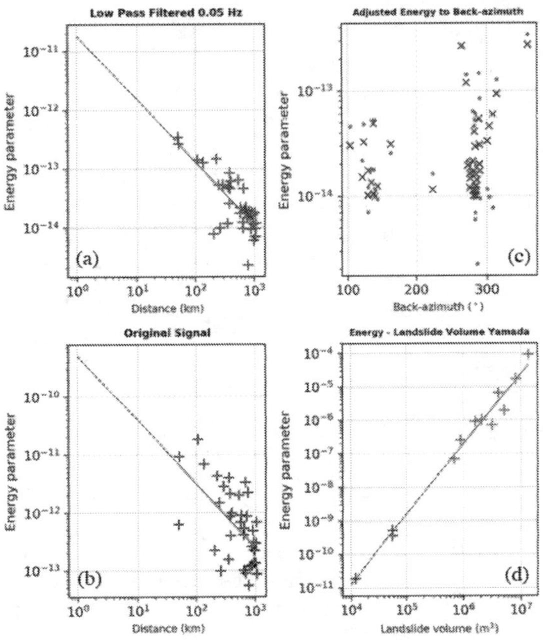

Fig. 3. (a) and (b) are shown the energy parameter against the distance, where the y-axis indicates the energy parameter and the x-axis indicate the distance of the low-pass filtered and original signals, respectively. The plus symbol with the dot is obtained from the observation, and it empirically derives an attenuation relation linear regression is shown in the solid red line, with the dashed blue line as the prediction to the 0 distance. (c) Energy parameter against back-azimuth where the y-axis indicates the energy parameter and x-axis indicate the degree. The green dots symbol indicates the observed energy, while the blue x symbol indicates the normalized energy. (d) The energy parameter against landslide volumes, with the plus green symbol, are the detailed studies of the landslide at Japan done by [9] and predicted linear regression resulted from the observation in this studies is shown in the dashed blue line, and the observations for the low-pass filtered and original signal are shown as a plus symbol with blue and red colors, respectively.

3 Results and discussion

The studies of how large the volumes of the Anak Krakatau flank collapse or subaerial slide on the event of 22 December 2018 has been done using numerous approaches, i.e., Synthetic Aperture Radar (SAR) images analysis [12], inversion of tsunami wave model [13,14], the Anak Krakatau eruption source and derived its slide volume with varying its single-force dip and basal friction [5], and the tsunami run-up studies [6].

However, it is essential to highlight the current state of the seismic broadband network to confirm the subaerial slide volumes resulting from the energy parameter through the information retrieved from the earlier studies. Even studies mentioned earlier have not come to a common term about how large the slide volume was; however, ~0.004 km^3 - ~0.2 km^3 are in good agreement [5,12].

Fig. 3 showed the resulted subaerial landslide is between ~0.0001 - ~0.0006 km^3 obtained in this study. This result indicates that the volumes derived from the energy parameter through the distance of the available stations are ten times lower than the other studies mentioned earlier. Therefore, several factors need to be accounted, i.e., there is no observation within a considerable distance (probably several kilometers) from the source, which is the closest station that recorded this event is located at a distance of 50.2 km away; consequently, some energy already attenuates; the referenced studies for the empirical approach was not reliable to predict subaerial slide volumes; and the coverage azimuth of stations surrounding this event mostly on two angles ~120° and ~280 °, therefore, respective to the maximum slide volume trajectory the recorded signal was not sufficient. It is worth noted that since 2019, BMKG is increasing the number of seismic stations in the Indonesia region.

Although the results indicate that the seismogram records as a promising approach to estimate the landslide or subaerial landslide volumes, however, to make this system available as part of the Early Warning System for landslide as well as the landslide-induced tsunami in Indonesia is still far from reliable. As the matter of fact, the closest to the Early Warning System for such feature is exhibited by Taiwan. Taiwan has developed a near real-time landquake monitoring system (NRLMS) [15], which successful to locate the landslide location, directions, and volume sizes via waveform inversion, with the entire process of location and volume determination taking a few seconds [16], thus allows the stakeholders to have the critical time to announce warnings and or take necessary actions. Further development of the Indonesia Early Warning System should consider the NRLMS. It offers the use of seismogram signal from the available seismic station managed by the BMKG with ease to implement, as well as it is cost-effective and efficient. Indonesia's Early

Warning System should be adopted or it is suggested to developed such system as the NRLMS in Taiwan.

4 Conclusion

From this study, we have shown that the current seismic network has the potential to predict the subaerial landslide volumes, although, in this study, the result indicates it may be ten times larger than the landslide volume predicted in this study (~0.0001 - ~0.0006 km^3). It is considered that the increasing amount of stations will make a reliable slide volume estimation from the seismic wave energy and could be made a possible threshold of a similar event and their slide volume that may generate a tsunami wave.

These important issues need to be addressed for the similar geological features and other tectonic features as well, i.e., faults extension from on land to offshore, such as Sumatra Fault Zone (SFZ), Palu-Koro fault, and others.

However, some additional task needs to be considered: (1) Increase the number of seismic stations at the concerned region with broadband and strong motion instrument available (co-site), (2) Data management and backup system (to reduce missing data from the offline stations), (3) Develop a near real-time monitoring trough seismic waveform inversion, and (4). Enhanced communication infrastructure to quickly broadcast the early warning messages to the concerned society.

Although, at the moment the early warning system in Indonesia to detect a landslide event is unreliable, however, further development should consider the use of the seismogram records and inversion to quickly locate, and determine the landslide volumes.

Not only that, but it is also extremely important to enhance the community resilience through tsunami drills and or disaster preparedness at the concerned region so that the society at those concerned locations has the ability to manage when this event and or any other disaster events occur.

Acknowledgment. We would like to thank the BMKG for the continuous collaboration and made the seismograms available for this study. Sincere thanks are due to the TDMRC for their effort in organizing the 13th Aceh International Workshop and Expo on Sustainable Tsunami Disaster Recovery (AIWEST-DR). This research is funded by the Ministry of Education, Culture, Research, and Technology through Universitas Syiah Kuala, under the Research Institutions and Community Service with contract number: 169/UN11/SPK/PNBP/2021. The writing of this manuscript is supported by the Matching Fund – Kedaireka funded by the Ministry of Education, Culture, Research, and Technology Indonesia awarded to the Research Center for Marine Sciences and Fisheries, Universitas Syiah Kuala. The Generic Mapping Tools is being used to draw the map figure [17].

References

1. NGDC, 2019 (2019)
2. P. W. Francis, J. Volcanol. Geotherm. Res. **25**, 349 (1985)
3. N. Nomanbhoy and K. Satake, Geophys. Res. Lett. **22**, 509 (1995)
4. I. Yokoyama, J. Volcanol. Geotherm. Res. **9**, 359 (1981)
5. L. Ye, H. Kanamori, L. Rivera, T. Lay, Y. Zhou, D. Sianipar, and K. Satake, Sci. Adv. **6**, 1 (2020)
6. Syamsidik, Benazir, M. Luthfi, A. Suppasri, and L. K. Comfort, Nat. Hazards Earth Syst. Sci. **20**, 549 (2020)
7. P. Goldstein, D. Dodge, M. Firpo, and L. Minner, Int. Geophys. **81**, 1613 (2003)
8. P. Goldstein and A. Snoke, Inc. Institutions Seismol. Data Manag. Cent. Electron. Newsl. (2005)
9. M. Yamada, Y. Matsushi, M. Chigira, and J. Mori, Geophys. Res. Lett. **39**, 1 (2012)
10. M. Yamada, H. Kumagai, Y. Matsushi, and T. Matsuzawa, Geophys. Res. Lett. **40**, 2998 (2013)
11. H. Kanamori, J. I. M. Mori, E. Hauksson, T. H. Heaton, K. Hutton, and L. M. Jones, Bull. Seismol. Soc. Am. **83**, 330 (1993)
12. R. Williams, P. Rowley, and M. C. Garthwaite, Geology **47**, 973 (2019)
13. M. Heidarzadeh, T. Ishibe, O. Sandanbata, A. Muhari, and A. B. Wijanarto, Ocean Eng. **195**, 106733 (2020)
14. A. Muhari, M. Heidarzadeh, H. Susmoro, H. D. Nugroho, E. Kriswati, Supartoyo, A. B. Wijanarto, F. Imamura, and T. Arikawa, Pure Appl. Geophys. **176**, 5219 (2019)
15. W. A. Chao, Y. M. Wu, L. Zhao, H. Chen, Y. G. Chen, J. M. Chang, and C. M. Lin, Sci. Rep. **7**, 1 (2017)
16. W. A. Chao, T. R. Wu, K. F. Ma, Y. T. Kuo, Y. M. Wu, L. Zhao, M. J. Chung, H. Wu, and Y. L. Tsai, Seismol. Res. Lett. **89**, 1335 (2018)
17. P. Wessel and W. H. F. Smith, Eos, Trans. Am. Geophys. Union **79**, 579 (1998)

Rock slope kinematic analysis for planar failure: A probabilistic approach

Ibnu Rusydy[1,7*], *Mulkal* Mulkal[2], *Haqul* Baramsyah[2], *Hendri* Ahmadian[3], *Hamzah* Hussin[4], *Nafisah Al*-Huda[1,5], *Marwan* Marwan[6]

[1]Geological Engineering Department, Engineering Faculty, Universitas Syiah Kuala, Indonesia
[2]Mining Engineering Department, Engineering Faculty, Universitas Syiah Kuala, Indonesia
[3]Information Technology Department, Faculty of Science and Technology, UIN Ar-Raniry, Indonesia
[4]Department of Geoscience, Faculty of Earth Science, Universiti Malaysia Kelantan, 17600 Jeli, Kelantan, Malaysia
[5]Civil Engineering Department, Engineering Faculty, Universitas Syiah Kuala, Indonesia
[6]Geophysical Engineering Department, Engineering Faculty, Universitas Syiah Kuala, Indonesia
[7]Tsunami and Disaster Mitigation Research Center (TDMRC), Universitas Syiah Kuala, Indonesia

Abstract. The probabilistic kinematic analysis for rock slope has been conducted on one slope along USAID road in Aceh Province, Indonesia. This research aims to develop the modelled probability density function (PDF) and determine the probabilistic of planar failure occurrence (P_{op}). The geometry of discontinuity planes (dip and dip directions) and slope geometry (slope angle and slope face) were collected from our previous study. One slope with planar failure criteria was selected. The Monte Carlo simulations were performed in generating 100 new random values in 100 time iterations to produce modelled PDFs for the geometry of discontinuity plans based on statistical parameters of field observed data. The probabilistic of each experimental PDFs were computed to produce the probabilistic of planar failure occurrence. The result reveals that the distribution of dip and dip directions for experimental PDFs are considered Beta and Normal distributions. The statistical parameters produced in the model are almost likely similar to observed data. It means the model that was developed are reliable and conscientious. The rule of $\Phi < \beta j < \beta s$ and dip directions (α_j) within ±20° to slope face (α_s) are utilised as the boundaries to calculate the probabilistic of planar failure occurrence (P_{op}) which revealing 0.26.

1 Background

Rock slope failure analysis for mining and civil projects has attracted rock engineers to develop numerous methods to analyse the rock slope stability, including the rock slope kinematic analysis, limit equilibrium, numerical modelling, empirical approach, and rock mass classification [1–5]. In a conventional approach, those methods merely concern a single input parameter value known as deterministic, and it previously applies in deterministic slope kinematic analysis. Nevertheless, the rocks' physical, mechanical, and structural properties vary from point to point and naturally in rock [6]. Thus, the conventional approach is required to be improved obviously to the probability approach considering the uncertainty of geological structural data collected from the surface of rock slope to increase the reliable result in slope design and prevention measure.

The rock slope kinematic analysis approach studies the typology of slope failure without any consideration of the force working on the slope [2,5,7]. The type of potential failures determined in the rock slope kinematic approach was based on the stereography interpretation, which has been successfully conducted by numerous researchers [2,5,7]. However, their approaches are assumed as a deterministic approach picking the mean values from stereography as a single value. The probabilistic rock slope kinematic analysis has been introduced by numerous researchers [8–10], considering the effect of variability of joint orientations. Furthermore, the probability of planar failure due to variability of structural features have been conducted by Rahim et al. [11] and Farhan and Rai [12] and give reliable results. The comprehensive study on how the variability of discontinuity plane drives the planar failure of rock slope is elaborated and scrutinised in this paper. Hence, this study aims to develop the modelled probability density function (PDF) and determine the probabilistic of planar failure occurrence (P_{op}) at USAID highways rock slope in Aceh province, Indonesia.

Overall, this paper discusses the deterministic kinematic analysis, which previously has been conducted by Rusydy et al. [5] for Slope 1, revealing the planar failure potential at the joint set (J2). The data were then re-analysed by performing Monte Carlo simulation to generate a large number of random values based on statistical parameters from the joint set (J2). The probabilistic kinematic analysis was conducted to determine the probabilistic of planar failure occurrence

* Corresponding author: ibnu@unsyiah.ac.id

(P_φ) and how the variability of joint and slope geometries influence the failure.

2 Methodology

This research was conducted at the rock slopes along USAID highways side, build after Indian Ocean Tsunami hit Aceh province. During construction, a multitude of rock slopes have been exposed; hence, some slopes are vulnerable and need to be investigated and analysed. The dip and dip direction of discontinuity planes (e.g. joints, bedding) on the rock slopes were recording directly employing geological compass by Rusydy et al. [2,5,7,13]. The dip and dip direction discontinuity planes data were plotted in stereographic projection to determine the number of planes most likely having a similar direction with the slope face, and the dips of planes are higher than friction angle (Φ) but lower than slope angle. The Alejano et al. [14] suggested method was utilised in this study to determine the basic friction angle.

2.1 Rock slope kinematic analysis

Rock slope kinematic analysis is associated with the rock slope movement process without considering the forces working the rock slope. This approach is introduced by Hoek & Bray [15] and Goodman [16]. The analysis and calculation are stands on the stereography projection, which are projected in the dip (β_j) and dip direction of discontinuity planes (α_j), which is 3D into the 2D model [2,5,7]. Other input data required in the kinematic analysis is friction angle (Φ) determined by employing the tilt testing suggested by Alejano et al. [14].

The Orient software developed by Vollmer [17] was utilised in this research to produce a stereography

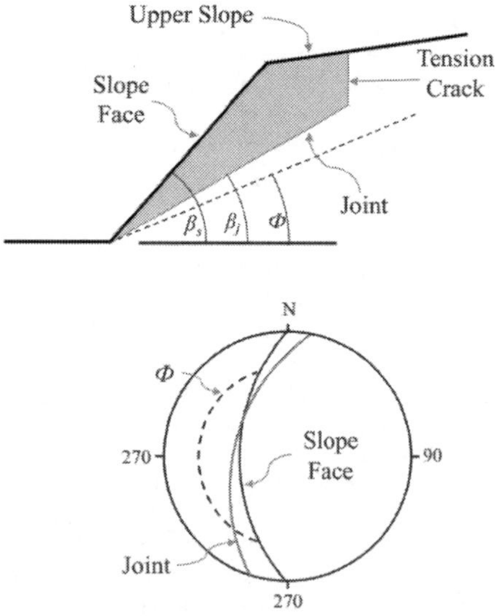

Fig. 1. Geometric circumstances for Plane failure (a) The graphic view of planar failure, (b) stereography projection, after modified from Wyllie & Mah [18].

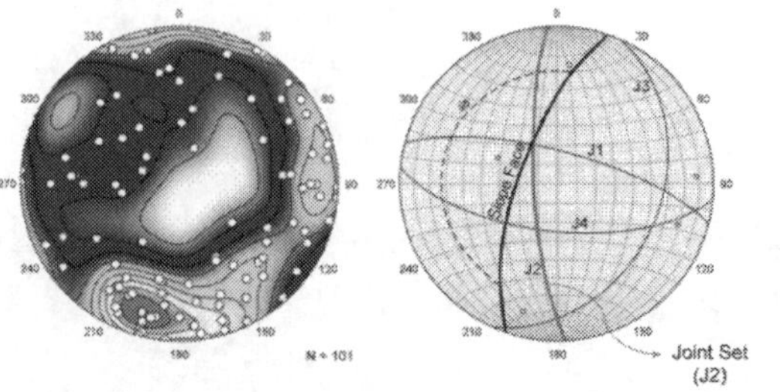

Fig. 2. Stereography projection plot for Slope 1 from Rusydy et al. [5], Joint set (J2) in red line formed the planar failure mode.

projection model (see Fig. 2). The mean of plane orientations (β_j *and* α_j) are determined from Orient software as the joint sets. Those joint sets combine with the orientation of slope (β_s *and* α_s) and friction angle (Φ) are deployed to determine the typology of failures (e.q. planar, wedge, toppling). Using the mean of plane orientation without considering the variability of those orientations is known as the deterministic approach.

Ten rock slopes have been investigated by Rusydy et al. [2] and Rusydy et al. [5], employing the scan-line method. Four from ten slopes are potentially had planar failure refer to deterministic kinematic analysis. Nonetheless, this paper merely discusses one slope as an example to introduce the probabilistic approach in analysing the planar failure probabilities.

Planar failure is a common failure typology in rock slope stability study; nevertheless, it requires numerous geometric circumstances to occur. Wyllie and Mah [18] noted that the planar failure occurs when the dip direction of joint planes (α_j) are within ±20° to the slope face (α_s). The following condition to planar failure to occur is when the dip of joint (βj) is lower than slope angle (βs), yet it must be higher than friction angle (Φ), or in another way it can be written as $\Phi < \beta j < \beta s$. The last circumstances is a present of tension crack in the upper part of the slope, as denoted in Fig. 1. In this study, those circumstances play a crucial role as the boundaries in determining probabilistic of planar failure occurrence except the presence of tension crack due to it is undefined in the field.

2.2 Probabilistic kinematic analysis

Probabilistic analysis is performed when the field data are insufficient or burdensome to assign a single value to calculate the model. Probabilistic describe the degree of belief in the truth of circumstances [19]. Hence, this study revealed probabilistic values between 0 and 1 to express the degree of belief for planar failure to occur. Obregon and Mitri [20] argued that the probability of failure for slope stability could be computed by encountering two models (series and parallel) influenced by numerous parameters. The series model is described as a single line from starting point to the end. If one parameter fails, it could trigger the rock slope failure. While in the parallel system, it requires all failure components to trigger the rock slope failure. The

 https://doi.org/10.1051/e3sconf/202234001017

planar failure in rock slope is more likely similar to the parallel system requiring at least two geometric circumstances to slope to have planar failure as mentioned previously ($\Phi < \beta j < \beta s$) as the first component and the second component is dip direction (α_j) within ±20° to the slope face (α_s), please see Fig. 3. According to Obregon and Mitri [20], the probability of all components are expressed as P[$p_1 \cap p_2 \ldots \cap p_n$]. It means the failure could happen in a parallel system when all components fail, and in probability analysis, they have an intersection relationship.

The probabilistic of failure (*Pf*) for parallel system failure in rock slope can be simplified as denoted in Equation 1. This rule assumes all the parameters (*N*) governing the rock failure are independent and has their probabilities (*Pi*).

$$Pf = \prod_{i=i}^{N}(Pi) \quad (1)$$

Naturally, the values of dip direction (α_j), dip (βj), and friction angle (Φ) vary from site to site, from point to point while the slope angle (βs) and slope face (α_s) have a single value. The variability α_j, βj, and Φ lead to uncertainty; accordingly, the probabilistic kinematic analysis approach is necessary to cope with this variability. The probabilistic density function (PDF) of α_j and βj were generated at the first stages utilising the R programming software, while the Φ is assuming as a single value. The example of PDFs and histograms in this study is denoted in Fig. 4.

After the dip and dip direction PDFs are developed, the kinematic instability is quantified as the probability of pole or plane triggering the failure mechanism. The probabilistic of planar failure occurrence (P_{op}) is written in Equation 2 as suggested by Obregon and Mitri [20].

$$P_{op} = Pr[\Phi \le \beta_j \le \beta_s].Pr[\alpha_s - 20^o \le \alpha_j \le \alpha_s + 20^o] \quad (2)$$

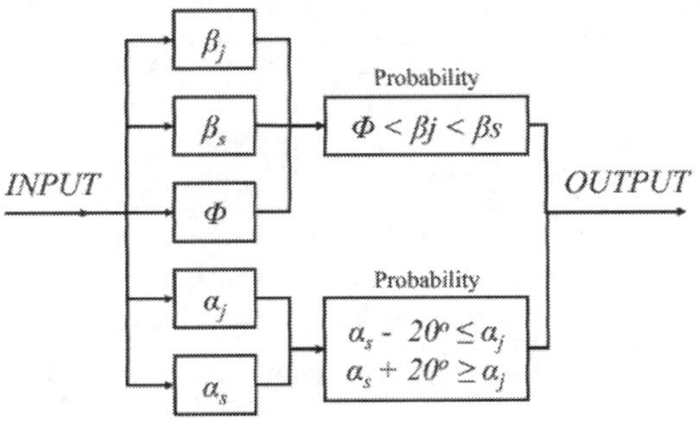

Fig. 3. Parallel system for planar failure in this study

2.3 Monte carlo simulation

The Monte Carlo simulation was run in this study to generate large numbers of dips and dip directions. This simulation is the most common approach in stochastic analysis in that the random values to the model is generated from sampled statistical distribution or frequency distributions like histograms and probability density function (PDF) [21,22].

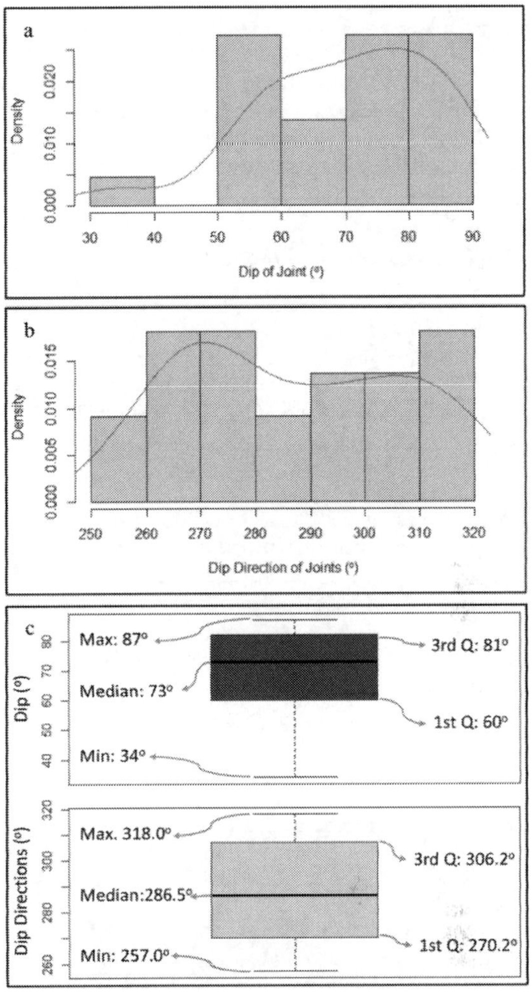

Fig. 4. (a) The Histogram, (b) PDF, and (c) Boxplot of joint geometries for Slope 1 joint set 2 (J2).

The number of observations for the joint set (J2) is merely 22 data, and it is insufficient to perform the probabilistic analysis directly from the PDF of field observed data. Thus, the Monte Carlo simulation was performed to generate large new random values based on statistic characterisation from field observed data (βj, α_j) distributions (see Fig. 3). 100 new data have been generated in 100x iteration for dips and dip directions using R data analysis software. As denoted in Fig. 4, the dip of joints has a Beta distribution typology while the dip direction is recognised as Normal distribution typology. In performing the simulation for Beta distribution, it requires the *Shape1*, and *Shape2* from observed data. The EnvStats package in R programming software developed by Millard [23] was employed to determine the *Shape1* and *Shape2* values from field observed data. Due to the dip direction data distribution is assumed as a normal distribution, it merely requires the mean (μ) and standard deviation (σ) values.

3 Results

The observed data distributions are the key in developing the experimental model utilising the Monte Carlo simulation; accordingly, the result of experimental models varies from 100x of iteration, and probabilistic kinematic analysis for J2 discusses.

3.1 The experimental models

The EnvStats R package was employed for Beta distribution and revealed the *Shape1* value for a dip of joints is 7.753, and *Shape2* is 3.324 from observed data (*J2*). Furthermore, the dip direction was recognised as Normal distribution, having 287.136 of μ and 20.185 of σ. The experimental distribution was developed by employing those statistical parameters.

In developing experimental distribution, 100 times iteration was run for 100 data, and all those simulations are plotted in one histogram and PDF as denoted in Fig. 5. Even though the range of the dip from the observed data is 34 to 87, the Monte Carlo simulation reveals bit higher data than expected. Yet, this simulation result is reliable, and the first quarter (*Q1*) and the third quarter (*Q3*) are similar to observed data. The distribution of dip direction experimental histogram and PDF is considered as normal distribution where the median and mean value are almost likely similar.

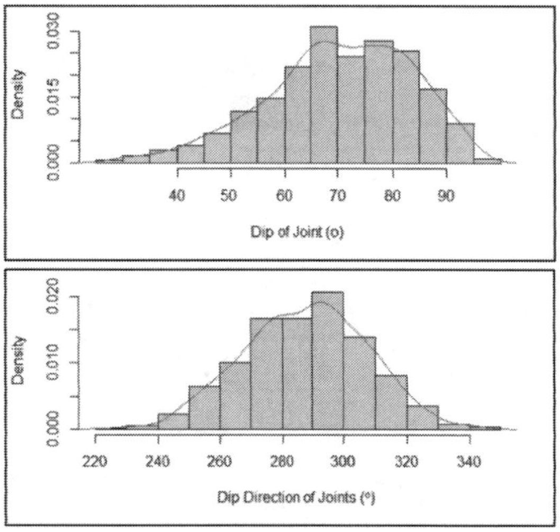

Fig. 5. The Experimental Histogram and PDF joint geometries for Slope 1 joint set 2 (J2) with N=100 and iteration 100 times. The Dip has Beta distribution typology, whilst the Dip direction has Normal distribution.

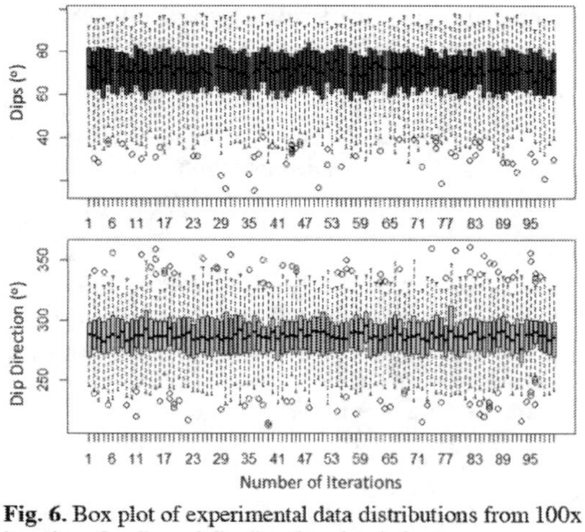

Fig. 6. Box plot of experimental data distributions from 100x iterations for dip and dip direction simulations

Fig. 6 denotes the boxplots of experimental data in a different iteration, the number of iteration as shown in the x-axis. The boxplot reveals the minimum, *Q1*, median, *Q3*, and maximum values sequentially from the bottom to the top. From 100 iteration, the median values of dip vary from 66.41° to 75.51°, while the actual median value from field data is 73°. The field observed median value for dip direction is N 286.5°E, whilst the modelled values fluctuate in 100 times iteration from N 276.9°E to N 295.3°E as shown in Fig. 6.

3.2 Probabilistic kinematic

After all simulated models reveal reliable results, the next stage is computing the probabilistic planar failure occurrence (P_{pf}) employing Equation 2. The geometric circumstances for planar failure are played an essential role as the boundaries in calculating the probabilistic to dip and dip direction in the PDF model.

According to Rusydy et al. [5], the investigated slope has 27° of friction angle (Φ) and the slope angle (βs) is 70°; accordingly, those data are utilised as boundaries in a dip of joint PDF curve to estimate the area between them. The area of 0.47 represents the probability of the first component as denoted in Fig. 7a; obviously, this area is calculated in R programming software. The second component as shown in Fig. 7b, 0.56 probability, act as the probability of dip direction (α_j) toward ±20° to the slope face (α_s). In this research, the slope facing to N 289° E; means any joints having

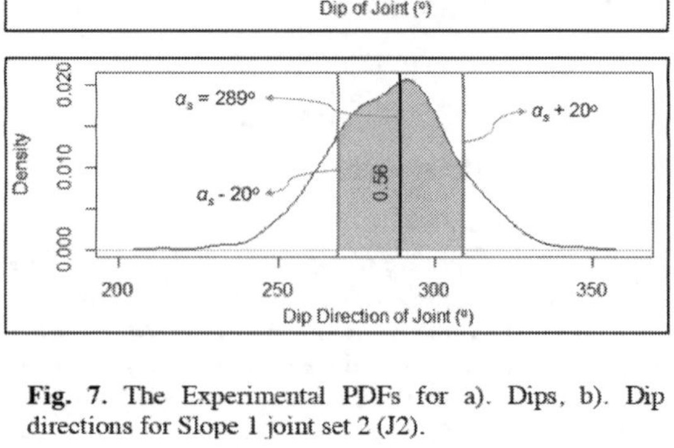

Fig. 7. The Experimental PDFs for a). Dips, b). Dip directions for Slope 1 joint set 2 (J2).

dip direction ≥ N 268° E to ≤ N 309° N are categorised as planar failure potential.

Table 1. Summary of fitted parameters for joint orientations

Data	Geometry	Distributions	Parameters*
Observed Data	Dip	Beta	Shape1 = 7.753
			Shape2 = 3.324
	Dip Direction	Normal	Mean = 287.136
			Sd = 20.185
Experimental Data	Dip	Beta	Shape1 = 7.500
			Shape1 = 3.193
	Dip Direction	Normal	Mean = 286.56
			Sd = 19.787

Note: Sd: Standard Deviation*

The connection between the first component and second component is recognised as an intersection probability relationship. The total probability of two components (0.47 and 0.56) are computed by multiplying those components; hence, the probabilistic of planar failure occurrence (P_e) in this study is 0.26. All parameters are employing in simulation as denoted in Table 1.

4 Conclusion

The slope in this study area has structural geology features from bedding, joints, and minor fault crossing the rock faces. The planar failure mainly occurs due to the dip direction of joints higher or lower than 20° from the slope face, and the dips are higher than the internal friction angle yet lower than the slope angle. Those geometric rules for planar failure obviously can be utilised as the boundaries in calculating the probability in PDF curves either for observed data or experimental/simulated data.

The experimental PDFs were developed based on the statistical parameters from observed data, and they must be scrutinised carefully to produce reliable results. This study identifies two typologies of data distributions, beta and normal which are required *shape1*, *shape2* for beta and mean, the standard deviation for normal distribution.

This merely conducted probabilistic kinematic analysis for planar failure in one rock slope. A multitude of rock slopes needs to be investigated by applying probability analysis to deal with the uncertainty and variability of rock properties. It is suggested to conduct more comprehensive research and cover more rock slopes in the future to mitigate rock slope failure alongside roads or probably in mining areas.

Acknowledgement. The authors would like to express profound gratitude to *Lembaga Penelitian dan Pengabdian Kepada Masyarakat* Universitas Syiah Kuala for providing a research grant through *Penelitian H-Index* scheme 2021 and to Geology and Mining Engineering students who involve in field data acquisition.

References

1. H. Basahel and H. Mitri, Int. J. Min. Sci. Technol. **29**, 357 (2019)
2. I. Rusydy, N. Al-Huda, M. Fahmi, and N. Effendi, Struct. Durab. Heal. Monit. **13**, 379 (2019)
3. I. Rusydy, T. F. Fathani, N. Al-Huda, Sugiarto, K. Iqbal, K. Jamaluddin, and E. Meilianda, Environ. Earth Sci. **80**, 1 (2021)
4. H. Gunawan, N. Al-Huda, M. Sungkar, A. Yulianur, and B. Setiawan, IOP Conf. Ser. Mater. Sci. Eng. **796**, 12044 (2020)
5. I. Rusydy, N. Al-huda, M. Fahmi, N. Effendi, A. Muslim, and M. Lubis, **30**, 93 (2020)
6. M. Abdulai and M. Sharifzadeh, Geotech. Geol. Eng. **37**, 1223 (2019)
7. I. Rusydy, N. Al-Huda, K. Jamaluddin, D. Sundary, and G. S. Nugraha, Ris. Geol. Dan Pertamb. **27**, 145 (2017)
8. H. Park, J. Lee, K. Kim, and J. Um, Eng. Geol. (2015)
9. X. Zhou, J. Chen, Y. Chen, and S. Song, Bull. Eng. Geol. Environ. **76**, 1249 (2017)
10. S. K. Hong and H. J. Park, Econ. Environ. Geol. **52**, 231 (2019)
11. A. F. Abdul Rahim, N. Simon, T. R. Mohamed, A. G. Md. Rafek, A. S. Serasa, Y. Chen, M. Zhang, K. E. Lee, and T. L. Goh, Bull. Geol. Soc. Malaysia **67**, 83 (2019)
12. M. A. Farhan and M. A. Rai, in *Int. Symp. Earth Sci. Technol. 2019* (2020)
13. I. Rusydy and N. Al-Huda, Carbonates and Evaporites **36**, (2021)
14. L. R. Alejano, J. Muralha, R. Ulusay, C. C. Li, I. Pérez-Rey, H. Karakul, P. Chryssanthakis, and Aydan, Rock Mech. Rock Eng. **51**, 3853 (2018)
15. E. Hoek and J. D. Bray, *Rock Slope Engineering* (CRC Press, 1981)
16. R. E. Goodman, *Introduction to Rock Mechanics* (Wiley New York, 1989)
17. F. W. Vollmer, (2017)
18. D. C. Wyllie and C. Mah, *Rock Slope Engineering* (CRC Press, 2004)
19. S. H. Begg, M. B. Welsh, and R. B. Bratvold, (2014)
20. C. Obregon and H. Mitri, Int. J. Min. Sci. Technol. **29**, 629 (2019)
21. M. Sari, C. Karpuz, and C. Ayday, Comput. Geosci. **36**, 959 (2010)
22. Y. Feng, J. P. Harrison, and N. Bozorgzadeh, Rock Mech. Rock Eng. **52**, 5071 (2019)
23. S. P. Millard, *EnvStats: An R Package for Environmental Statistics* (Springer Science & Business Media, 2013)

A preliminary result of microtremor study to identify the subsurface condition in the Aceh Tenggara region

Andrean V. H. Simanjuntak[1,3], *Umar* Muksin[2,3*]

[1]Badan Meteorologi Klimatologi dan Geofisika (BMKG), 22123Aceh Besar, Aceh, Indonesia
[2]Department of Physics of Universitas Syiah Kuala (USK), 22122 Banda Aceh, Aceh, Indonesia
[3]Tsunami and Disaster Mitigation Research Center (TDMRC), Universitas Syiah Kuala (USK), 23111 Banda Aceh, Aceh, Indonesia

Abstract. The Aceh Tenggara region is located in an tectonically active region and it can be categorized as the prone-area. Historically, there were many major earthquakes occured the active fault called the Tripa segment that has generated massive damages and losses. The study of the seismic vulnerability around the highly populated city in the Aceh Tenggara region, Kutacane City, as the center of economy and office activities must be conducted to prepare the suitable earthquake mitigation program. The microtremor have been recorded at 25 sites with one km grid interval between all points. The record lengths have 30 minutes duration at each site, and they were measured with the Taurus Seismometer in miniseed format. The data were analyzed by using the HVSRpy in order to obtain the Horizontal to Vertical Spectral Ratio for each site by means of the geometric-mean and the multiple azimuths methods. Both methods provided a similar result at both focuses with dominant frequency of 0.8 Hz at the AT01 site and 4.8 Hz at the AT14 location. The preliminary result shows that the higher dominant frequency were recorded along the southern part of the Kutacane city which is associated with the Alas formation.

1 Introduction

The Aceh Tenggara region is surrounded by active faults and categorized as the seismic prone area as shown in Figure 1. The Tripa fault is the longest segment along the Sumatran fault in Aceh that has generated earthquakes with a magnitude of M 6.8 in 1990 and with M 7.2 in 1996 [1, 2]. Another major earthquake also occurred in the northern part of the Tripa fault on July 2, 2013 with M 6.2 which is strongly felt by the local people in the Aceh Tenggara region [2, 3, 4].

All major earthquakes generated the intensity of between 4 – 6 MMI (Modified Mercalli Intensity). The intensity was influenced by the subsurface condition from the basin structure in the Aceh Tenggara region [5, 6, 7, 8]. The seismic surface waveform will be trapped in long period and amplified up to 2.0 times greater than the normal recording in the basin area [9, 10, 11]. On other hand, geomorphological studies show that the highest risk level occurred in the basin with alluvial sediment [12, 3].

Furthermore, Irwandi et al. [12] updated the seismic hazard for the Southeast Aceh region to the probability figure exceeding 2% and 10% for 50 years to produce the bedrock PGA (Peak Ground Acceleration) map. The basin structure may amplify the PGA due to the young alluvium (Qh) structure. The amplification results clearly show that the PGA can reach 0.9g and 1.5g for 10% and 2% which are likely to be exceeded for 50 years with ratio reaching 1.6 times in the more densely populated areas in the Kutacane basin.

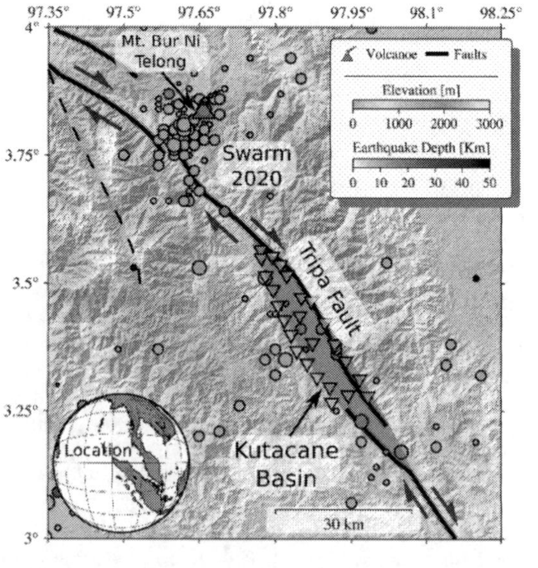

Fig. 1. The tectonic system in the study area with an active fault namely Tripa fault. There were swarm activities in 2020 in the northern part of study area.

Here, we conducted a microtremor study to figure out the subsurface condition and its impact represented by the vulnerability level. Microtremor research has been widely studied through the HVSR method was used to investigate the site characteristics for earthquake mitigation for example, the soil condition after the earthquake [5], the site and soil amplification effect

* Corresponding author: muksin.umar@unsyiah.ac.id

https://doi.org/10.1051/e3sconf/202234001018

[11], microzonation [9, 10] and the shallow sedimentary properties [12].

Theoretically, microtremor utilizes the ratio of horizontal to vertical components of ambient noise (microtremor) to study the site condition. The site condition was represented by the dominant frequency (f0), the amplification factor (A), and the seismic vulnerability index (Kg) of recording microtremors.

In this paper, we applied the HVSR method for two points as the preliminary study. The two points show the different parameters which were influenced by lateral heterogeneity. The future study is to continue the data analysis for more points that are believed to be essential in term of learning the seismic vulnerability for disaster mitigation program in the Aceh Tenggara region.

2 Data and method

2.1 Data observation and location

Microtremor were recorded in the Kutacane city which is part of the Aceh Tenggara region. The Kutacane city was chosen mainly because the city located along the active fault, stands on the basin structure, and densely populated. The microtremor records with a total of 25 locations, as shown in Figure 3, were measured by using a short period Taurus seismometer with a sampling rate of 100 sps (sample per second) and recording time between 30 - 45 minutes for all observation points.

The local geology of study area with the regional scale of 1:150.000 is shown in Figure 2. The Aceh Tenggara region is part of Medan sheet on the right area while left area is included in the Tapaktuan sheet. The rock structure of the study area is composed by Alas (Ppa) and Bohorok formation (Pub) that is formed since Paleozoic age by high pressure from the subduction activity [10, 11].

The Alas Formation (Ppa) located on the southern part is more compact than the northern part due to the high topographical contour. The Alas formation consists of limestone rock (Murl) while the younger formation of Alluvium (Qh) filled with the clastic sediment that can be found on the northern part. This formation comprises of the fractured sedimentary rock of pull-apart basin caused by tectonic activities along the Sumatran fault [10]. The alluvium sediment known as the Kutacane graben has been formed since the Cenozoic to the Neogene period. The graben is categorized as the product of an active fault namely Tripa fault. The Tripa fault moves in the right-lateral direction with slip rate of ~20cm/yr.

The pull-apart basin in the Aceh Tenggara proves to be a very interesting structural surface with a swarm activity in the northern part (See Figure 1), the dextral earthquake in the central part and located close to the Mount Sinabung in the southern part. As a densely populated area, the Kutacane city region can be categorized as a disaster prone-area that needs a comprehensive study to highlight the potential hazard especially caused by the earthquake.

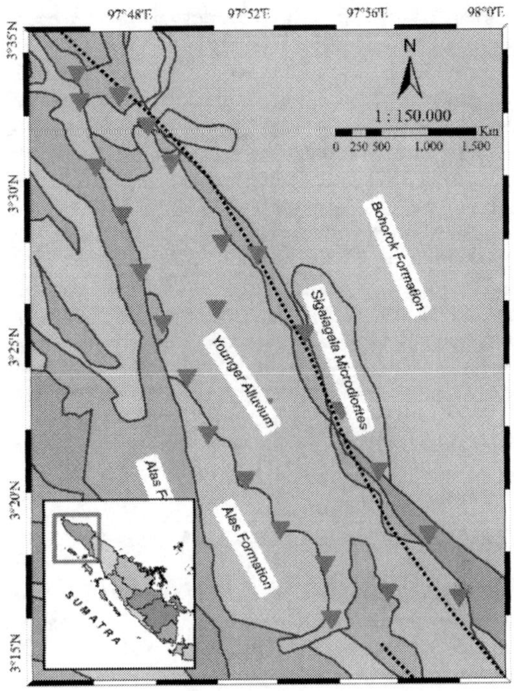

Fig. 2. Map of study area and local geology in the Southeast Aceh. Total of observation points were 25 sites that covered the basin structure. The basin structure in the Southeast Aceh consists of young alluvium sediment (Qh).

The earthquake in the Aceh Tenggara region is dominantly triggered by the subduction from Indo-Australia-and-Eurasia convergence. The shaking from the earthquake is likely amplified when it travels through the graben and the basin structure in Kutacane. A routine observation from the Meteorological, Climatological, and Geophysical Agency [Badan Meteorology Klimatologi dan Geofisika (BMKG)] shows an average intensity of moderate earthquake in the range of 4 – 7 MMI [12, 13]. For that reason, we try to conduct a microtremor study to highlight the frequency and the amplification characteristics in the Southeast Aceh by using the geometric-mean and the multiple azimuth methods. This research aims to provide a preliminary of microtremor parameters that can be used to study the earthquake mitigation plan and preparation in the Southeast Aceh region

2.2 Horizontal Vertical Spectral Ratio (HVSR)

The method of H/V ratio compares the Fourier spectrum of horizontal to vertical component which represented the amplification of a site Nakamura [14]. By using similar approach, Stanko et al. [15] determined the H/V values based on the maximum spectrum of H/V value of 1 was found on the site in hard rock structure.

On the other hand, the amplification parameter in the stations that stand on the soft rock structure has the maximum greater than 1. From the amplification, a further analysis can be formulated as a transfer function of H/V spectral ratio [16]. Then, the time domain must be changed into the frequency domain with the Fast

Fourier Transform (FFT) to find out the specific frequency for upper and lower frequency corner. The Bandpass Butterworth filter with an appropriate corner frequency can be used automatically to get a typical range micro frequency [17].

Here, two horizontal components are combined with the vertical component with the geometric-mean and the multiple azimuths method. The processing procedure was following the recommendations of [18, 9] namely, that if a single HVSR trend is determined, then the lognormal median curve can be calculated using the geometric-mean. Furthermore, a statistical modelling of f0 can be calculated and the multiple-azimuth can be used [20, 21]. In this paper we used a python program namely hvsrpy, which is an open-source package for HVSR analysis [20, 21]. The hvsrpy was built to highlight the new HVSR processing techniques that is proposed by [18, 19, 20, 22].

3 Result and discussion

For the HVSR analysis, we employ the settings of the time tab window selection process which includes the sub-tabs of window length, cosine taper width, and time-domain filter as in [18, 19].

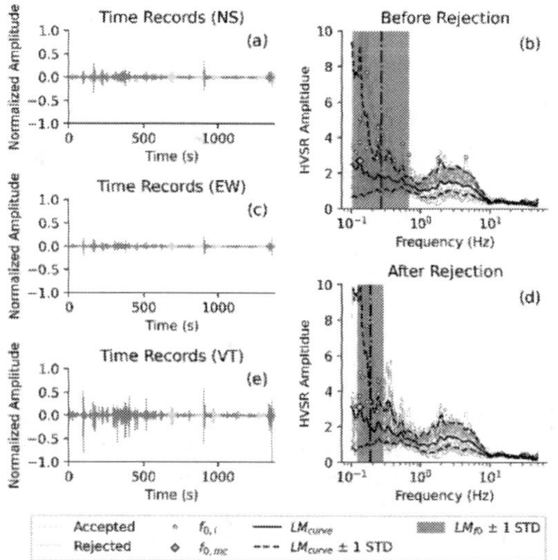

Fig. 3. Example results from the hvsr recording at point one (AT01) which had been calculated with the geometric-mean method. The seismic waveform is illustrated in the three geographical components (a) in the north-south (NS) component, (c) east-west (EW) component, and (e) vertical (VT) component. with rejected and accepted windows in the Frequency-domain.

In other hand, SESAME (2004) noted that the HVSR processing will be better with the time windows in a minimum ten times of 1/f0. In this study, transient effects were determined by using anti-triggering correlation of Short-Term Average (STA) and the Long-Term Average (LTA) at 0 – 30 seconds with a window of 60 seconds and a cosine taper of ± 10% as shown in Figure 3, while the azimuth angle in Figure 4.

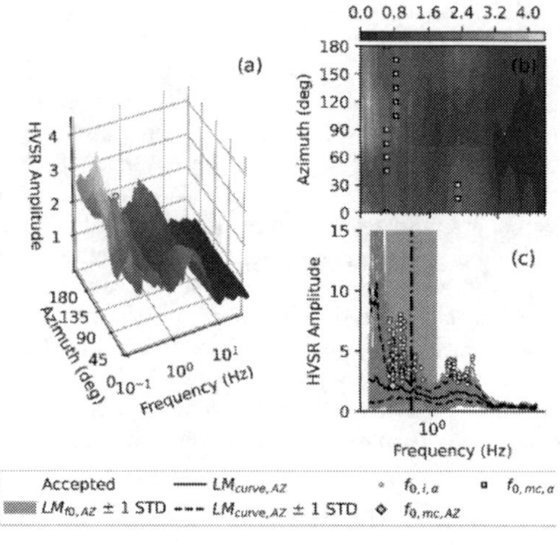

Fig. 4. Example results from the seismic waveform at point one (AT01). (a) The median curve for all azimuth angles in the frequency-azimuth-amplitude space. (b) The spectrogram between the lognormal median trend and all azimuth angles. (c) The HVSR trend from all azimuth angle.

We also analysed another point in the left part of the study area at AT14 as shown in figure 5 and 6 for the comparison values.

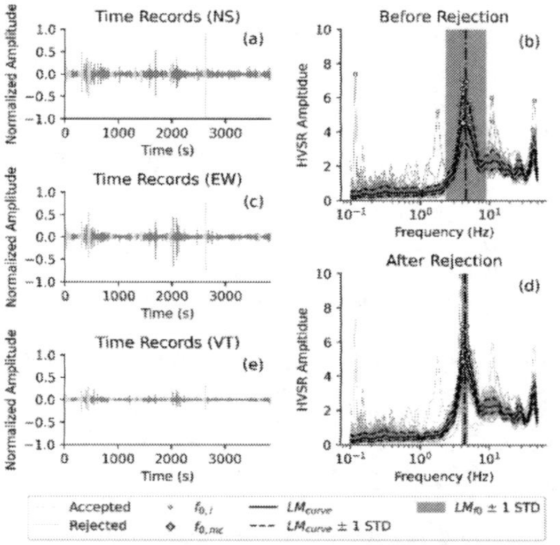

Fig. 5. Example results from the seismic waveform at another point (AT14). The seismic waveform is presented in the three geographical components (a) north-south, (c) east-west, and (e) vertical with the windows rejected in the frequency domain.

The Konno-Ohmachi (1998) smoothing coefficient of 40 is applied to smooth the seismic waveform data at a frequency range of 0.1 – 50 Hz. Then, the geometric-mean processing is used for the H/V results in the lognormal frequency-amplitude (see Figure 3). Point one (AT01) is assumed with soft alluvial sediment formation that has frequency of ~0.8 Hz. Meanwhile, another method that used in this paper is the multiple azimuths (Figure 4). We also applied the Konno-Ohmachi smoothing coefficient of 40 at the frequency range of 0.1 – 50 Hz. The similar technique is used for H/V values in the lognormal frequency-amplitude distribution as shown in Figure 5 and 6.

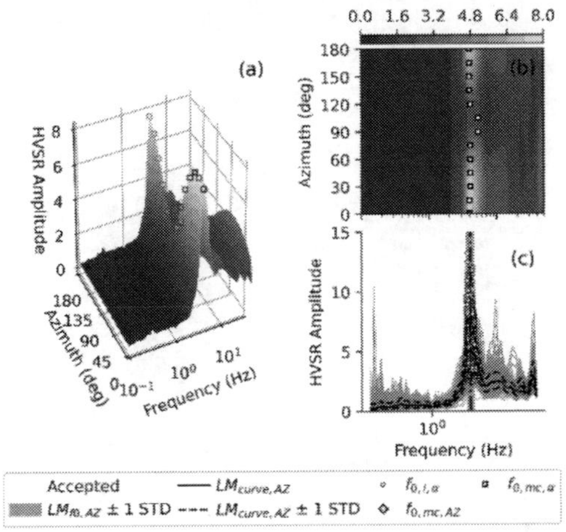

Fig. 6. The results example from the seismic waveform at another point (AT14. (a) The median trend for all azimuth angle in the frequency-azimuth-amplitude domain. (b) The spectrogram between the lognormal median curve and all azimuth angle. (c) The HVSR curves from all azimuth angle.

Point AT14 shows a better result with a lognormal distribution and a greater convergence. It can be assumed that the recording quality has a low noise and the subsurface may be more compact than those of the AT01 point. The AT14 shows a convergence of representative frequency at 4.8 Hz from the geometric mean. Meanwhile, the frequency from the multiple-azimuths method also produces similar values with the frequency range of 4.8 – 5.0 Hz. Both methods have a low uncertainty meaning that the selected window is optimum along the whole trace.

4 Conclusion

We perform a preliminary analysis for two points at AT01 point and AT14 point. The analysis was carried out with the help of the geometric-mean and the multiple azimuths. Both methods noticed that the more compact material are located on the eastern part of the study area with a ~4.8 Hz representative frequency whilst the soft material rests in the northern part with a ~0.8 Hz representative frequency. The convergence results of low uncertainty can be found if the trace has a low noise and greater length of recording time. This preliminary study is important to be applied for all point measurements.

References

1. Pasari, S., Simanjuntak, A. V., Mehta, A., Sharma, Y. Pure and Applied Geophysics, 1-18. (2021).
2. Nurana, I., Simanjuntak, A. V. H., Umar, M., Kuncoro, D. C., Syamsidik, S., & Asnawi, Y. (2021). Spatial Temporal Condition of Recent Seismicity In The Northern Part of Sumatra. Elkawnie: Journal of Islamic Science and Technology, 7(1), 131-145.
3. Muksin, U., Bauer, K., Muzli, M., Ryberg, T., Nurdin, I., Masturiyono, M., Weber, M. Journal of Asian Earth Sciences, 171, 20-27. (2019)
4. Muksin, U., Rusydy, I., Erbas, K., & Ismail, N. (2018, April). Investigation of Aceh Segment and Seulimeum Fault by using seismological data; A preliminary result. In Journal of Physics: conference series (Vol. 1011, No. 1, p. 012031). IOP Publishing.
5. Simanjuntak, A. V., Muksin, U., Setiawan, Y. In IOP Conference Series: Earth and Environmental Science (Vol. 273, No. 1, p. 012021). IOP Publishing. (2019, June).
6. Qadariah, Q., Simanjuntak, A. V., Umar, M. Journal of Aceh Physics Society, 7(3), 127-132. (2018).
7. Rusydy, I., Idris, Y., Muksin, U., Cummins, P., & Akram, M. N. (2020). Shallow crustal earthquake models, damage, and loss predictions in Banda Aceh, Indonesia. Geoenvironmental Disasters, 7(1), 1-16.Yunita Idris
8. Simanjuntak, A. V., Muksin, U., & Rahmayani, F. (2018, May). Microtremor survey to investigate seismic vulnerability around the Seulimum Fault, Aceh Besar-Indonesia. In IOP Conference Series: Materials Science and Engineering (Vol. 352, No. 1, p. 012046). IOP Publishing.Andre hvsr
9. Simanjuntak, A. V. H., Asnawi, Y., Umar, M., Rizal, S., & Syukri, M. (2020). A Microtremor Survey to Identify Seismic Vulnerability Around Banda Aceh Using HVSR Analysis. Elkawnie: Journal of Islamic Science and Technology, 6(2), 342-358.Asrillah
10. Irwandi, I., Muksin, U., & Simanjuntak, A. V. (2021). Probabilistic seismic hazard map analysis for Aceh Tenggara district and microzonation for Kutacane city. In IOP Conference Series: Earth and Environmental Science (Vol. 630, No. 1, p. 012001). IOP Publishing.
11. Simanjuntak, A. V., Muksin, U., & Sipayung, R. M. (2018, December). Earthquake relocation using HypoDDMethod to investigate active fault system in Southeast Aceh. In Journal of Physics: Conference Series (Vol. 1116, No. 3, p. 032033). IOP Publishing.
12. Julius, A. M., & Widana, I. D. K. K. In IOP Conference Series: Earth and Environmental Science (Vol. 708, No. 1, p. 012106). IOP Publishing. (2021, April)
13. Julius, A. M., Pribadi, S., & Muzli, M. In IOP Conference Series: Earth and Environmental Science (Vol. 132, No. 1, p. 012012). IOP Publishing. (2018, March).
14. Yutaka Nakamura; What Is the Nakamura Method?. Seismological Research Letters 2019;; 90 (4): 1437–1443. doi: https://doi.org/10.1785/0220180376
15. Stanko, D., Markušić, S., Strelec, S., & Gazdek, M. (2016). Seismic response and vulnerability of historical Trakošćan Castle, Croatia using HVSR method. Environmental Earth Sciences, 75(5), 368.

16. Tarabusi, G., & Caputo, R. (2017). The use of HVSR measurements for investigating buried tectonic structures: the Mirandola anticline, Northern Italy, as a case study. International journal of earth sciences, 106(1), 341-353.
17. Bignardi, S. (2017). The uncertainty of estimating the thickness of soft sediments with the HVSR method: A computational point of view on weak lateral variations. Journal of Applied Geophysics, 145, 28-38.
18. Cheng, T., Cox, B. R., Vantassel, J. P., & Manuel, L. (2020). A statistical approach to account for azimuthal variability in single-station HVSR measurements. Geophysical Journal International, 223(2), 1040-1053.
19. Cox, B. R., Cheng, T., Vantassel, J. P., & Manuel, L. (2020). A statistical representation and frequency-domain window-rejection algorithm for single-station HVSR measurements. Geophysical Journal International, 221(3), 2170-2183.
20. Vantassel, J. P., & Cox, B. R. (2021). SWinvert: a workflow for performing rigorous 1-D surface wave inversions. Geophysical Journal International, 224(2), 1141-1156.
21. Vantassel, J. P., & Cox, B. R. (2021). A procedure for developing uncertainty-consistent Vs profiles from inversion of surface wave dispersion data. Soil Dynamics and Earthquake Engineering, 145, 106622.
22. Asnawi, Y.; Simanjuntak, A.; Muksin, U.; Okubo, M.; Putri, S.I.; Rizal, S.; Syukri, M., (2022). Soil classification in a seismically active environment based on shear wave velocity and HVSR data. Global J. Environ. Sci. Manage., 8(3): 1-18.

Development of fragility curve based on incremental dynamic analysis curve using ground motion Aceh earthquake

Zu Irfan[1*], *Abdullah*[2], and *Moch.* Afifuddin[2]

[1]PhD Candidate, Doctor of Engineering Study Program, Syiah Kuala University, Jl. Syech Abdurrauf, Banda Aceh, 23111, Indonesia
[2]Civil Engineering Department, Syiah Kuala University, Jl. Syech Abdurrauf, Banda Aceh, 23111, Indonesia

Abstract. History records that Aceh has been hit by earthquakes several times, including the largest recorded, on December 26, 2004 with a magnitude of 9.3 SR. In connection with the history of disasters that have occurred, there is a need for safety and disaster preparedness. It is very important to conduct a feasibility study of the existing public buildings used in disaster preparedness for vulnerability against earthquakes. The type of building damage caused can predicted by referring to the fragility curve. This research conducted on evaluating the seismic performance of existing buildings in Banda Aceh City by developing a fragility curve based on Incremental Dynamic Analysis (IDA). The existing buildings that are the main focus of this study include the Tsunami Vertical Evacuation (TVE) of Lambung, TVE Alu Deah Teungoh, TVE TDMRC, SDN 48 Banda Aceh, SMPN 11 Banda Aceh, Baiturrahim Ulee Lheu Mosque, and the Subulussalam Punge Mosque. The IDA method is applied to predict and estimate the performance of the structure using several ground motions. The resulting set of IDA curves were further analyzed to form the fragility curve. The fragility curve becomes the basis for assessing the seismic performance of building structures rationally.

1 Introduction

Aceh province is located on a megathrust and ring of fire, with the potential for a large earthquake to occur. History records that several large earthquakes have occurred in Aceh. However, in 2004 an earthquake measuring 9.3 on the Richter Scale devastated the city of Banda Aceh which was followed by a tsunami. In connection with the history of disasters that have occurred, there is a need for safety and disaster preparedness, especially in dealing with building damage. Feasibility studies of existing buildings against earthquakes are very important to do because existing buildings are buildings used in disaster preparedness. The type of building damage caused by the earthquake can be seen and predicted by referring to the fragility curve. The fragility curve is a curve that shows how big the probability of the building's vulnerability when exposed to an earthquake [1].

Research related to the fragility curve has been carried out by Irvansyah [2] who studied the fragility curve of school buildings using the IDA method and Aulia [3] with the pushover method. Therefore, it is very important to be developed the fragility curve considering that Banda Aceh City is prone to disasters, especially earthquakes. The fragility curve will show the probability of damage to the building structure with various discrete damage conditions based on Hazus [4], namely light damage, moderate damage, extensive damage, and complete damage. Based on the above hypothesis, a research will be conducted on evaluating the seismic performance of the existing buildings in Banda Aceh City by developing a fragility curve based on incremental dynamic analysis. The method of Incremental Dynamic Analysis (IDA) is applied to predict and estimate the performance of the structure using several ground motions. The resulting set of IDA curves were further analyzed to form the fragility curve of the building structure. The fragility curve becomes the basis for assessing the seismic performance of building structures rationally.

2 Data and method

2.1 Research object

This study only focuses on assessing the condition of the existing building which is carried out by survey or direct assessment. Existing buildings are buildings that are in the utilization stage. This study examines existing buildings scattered in Meuraxa District, Banda Aceh City. The distribution of these buildings can be seen in Figure 1 as follows:

* Corresponding author: zuirfan.eng@gmail.com

https://doi.org/10.1051/e3sconf/202234002001

Fig 1. Map of the distribution of existing buildings that are the object of research

Existing buildings that are the object of research in this study include the Tsunami Vertical Evacuation (TVE) of Lambung, TVE Alu Deah Teungoh, TVE TDMRC, SDN 48 Banda Aceh, SMPN 11 Banda Aceh, Baiturrahim Ulee Lheu Mosque, and the Subulussalam Punge Mosque. Existing buildings that are the object of research can be seen in Figure 2 to 8.

Fig 2. TVE Lambung

Fig 3. TVE Alu Deah Teungoh

Fig 4. TVE TDMRC

Fig 5. SDN 48 Banda Aceh

Fig 6. SMPN 11 Banda Aceh

Fig 7. Baiturrahim Ulee Lheu Mosque

Fig 8. Subulussalam Punge Mosque

2.2 Ground motion records

Ground motion records taken are earthquakes that have occurred and caused severe damage to building structures. In addition to earthquake data from the Aceh area, this study also used several records of earthquake data that had occurred outside the Aceh area. The ground motion records used is the earthquake that causes significant damage and losses. The more earthquake data used, the more the characteristics of the structure will be shown. In this study, 22 earthquake records will be used which can be seen in Table 1.

 https://doi.org/10.1051/e3sconf/202234002001

Table 1. List of Ground Motion Records used.

Gempa	Magnitude
Chi-Chi (Taiwan)	7,3
Friuli (Italia)	6,5
Hollister (USA)	5,6
Imperial Valley (USA)	6,5
Kobe (Japan)	6,9
Kocaeli (Turkey)	7,4
Landers (USA)	7,3
Loma Prieta (USA)	6,9
Northridge (USA)	6,7
Trinidad (USA)	6,7
Andaman-East 2004	9,3
Andaman-North 2004	9,3
Simeulu 1-East 2012	8,5
Simeulu 1-North 2012	8,5
Simeulu 2-East 2012	8,1
Simeulu 2-North 2012	8,1
Bener Meriah-East 2013	6,1
Bener Meriah-North 2013	6,1
Pidie Jaya-East 2016	6,5
Pidie Jaya-North 2016	6,5
Banda Aceh-East 2020	5,5
Banda Aceh-North 2020	5,5

Source: www.peer.berkeley.edu [5] dan BMKG [6]

2.3 Modelling structure

Structural modeling was carried out on 7 existing buildings in Ulee Lheue. The structural system that will be modeled into SAP2000 is in the form of a space frame. SAP2000 is software from Computer and Structures, Inc. SAP 2000 has an integrated system to model, analyze, design and optimize a structure. Loads that are inputted in the SAP2000 program are loads that are reviewed in structural planning in the form of dead loads, live loads and earthquake loads. To obtain the ultimate load from the loads that may occur on the structure, a combination of factored loads is carried out that follows the rules in SNI 1726-2019 [7]. The results of the modeling of the building structure can be seen in Figure 9 to 15.

Fig 9. 3D Model of TVE Lambung

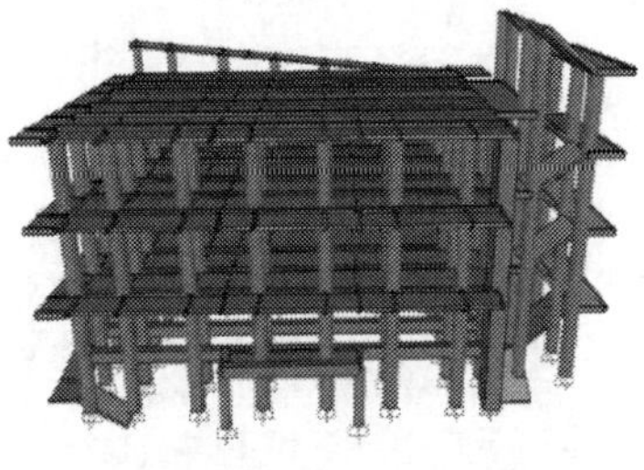

Fig 10. 3D Model of TVE Alu Deah Teungoh

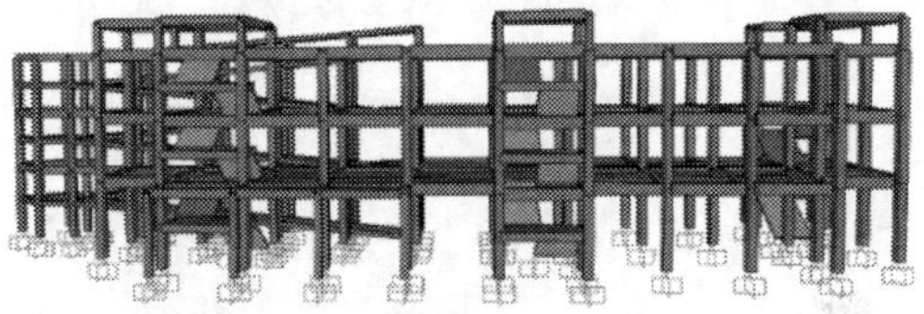

Fig 11. 3D Model of TVE TDMRC

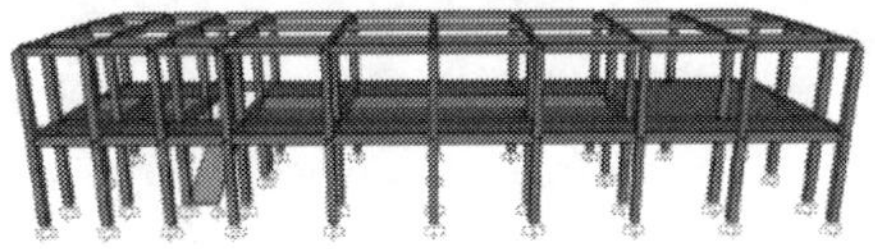

Fig 12. 3D Model of SDN 48 Banda Aceh

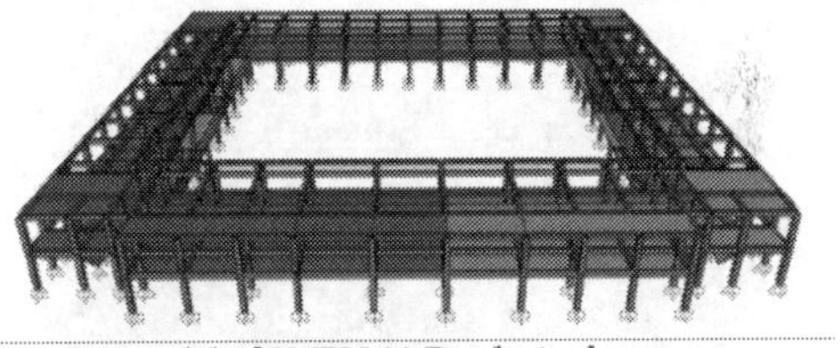

Fig 13. 3D Model of SMPN 11 Banda Aceh

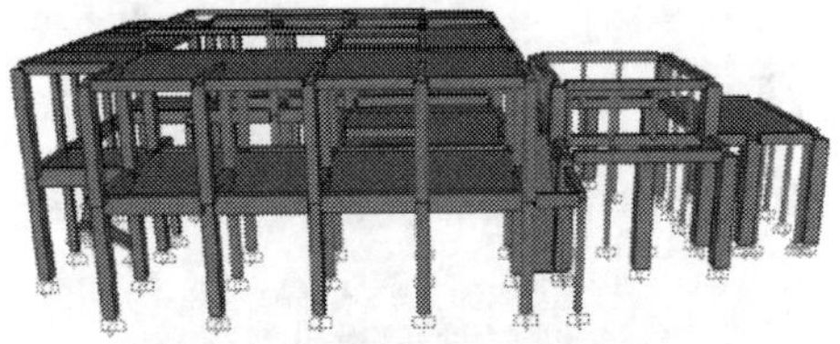

Fig 14. 3D Model of Baiturrahim Ulee Lheu Mosque

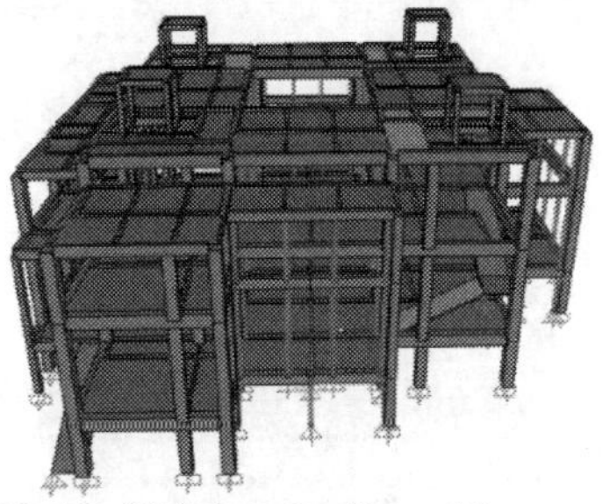

Fig 15. 3D Model of Subulussalam Punge Mosque

2.4 Incremental dynamic analysis

According to Tiwari and Kasnale [8], incremental dynamic analysis (IDA) is an analysis to study the whole earthquake with different intensities, which is applied to the model of the building structure until it collapses. IDA is done by performing a non-linear time history analysis. In non-linear time history analysis, the structural model is subjected to one or more earthquake records (accelerogram) which is multiplied by a scale factor (scalar multiplication) for various levels of earthquake intensity that are expected to occur. 'force' the structure to work over the entire range of behaviour; from the elastic phase to the plastic phase (inelastic) until it finally reaches global dynamic instability or the structure collapses.

After the structural modelling is completed, then the next step is to input the earthquake history data used. Then run a dynamic analysis of time history in stages with increasing earthquake intensity scale. In general, the analysis carried out in this study uses the addition of a scale factor of 0.5; 1 and 2. These additions are used to determine the limits of repeated load increases. The results of the analysis are displacements in the x and y directions. This displacement value is used to determine the value of the drift ratio which is useful in making the IDA curve.

2.5 Fragility Curve

Based on the results of structural analysis using the IDA method, IDA curves with various variations are obtained according to the number of earthquake records used as earthquake loads. The IDA curve can be further analyzed probabilistically by using the fragility function and expressed in the form of a fragility curve. In this study, the value of limit states follows the research of Kim and Shinozuka which is summarized in Pitiliakis [9]. The fragility curve shows the probability of damage to a structure when it receives an earthquake load with a certain intensity at its service limit. Based on the analytical method, the function is derived by Sadradin [10] through a log-normal distribution approach as follows:

$$P = \Phi\left(\frac{\ln(X) - \lambda}{\beta_{RU}}\right) \quad (1)$$

where;

P = probability of structural damage,
Φ = The standard normal function,
βRU = The total uncertainty of the structure,
x = Ground motion parameter, PGA (g),
λ = average of ln (x).

3 Result and discussion

3.1 Results of IDA curve analysis

In making the IDA curve, an Intensity measure in the form of Peak Ground Acceleration (PGA) is needed for the x-axis. Meanwhile, the damage measure used is the drift ratio for the y-axis. The analysis was stopped when the drift ratio was found to be more than 7-8%, because it had crossed the largest limit state value. The results of the IDA curve analysis for earthquakes in the x and y directions can be seen in Figures 16 to 29.

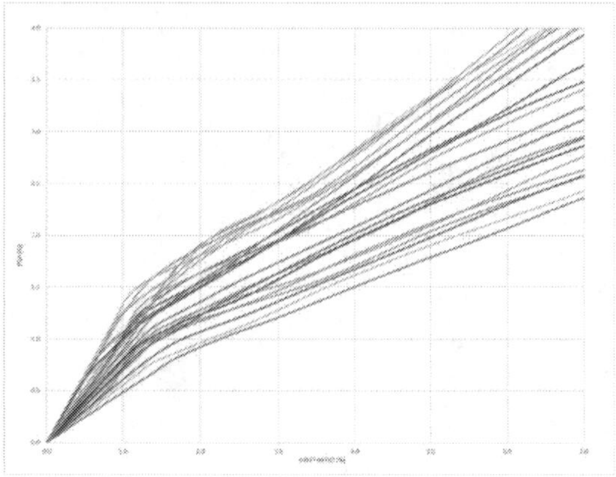

Fig 16. IDA Curve of TVE Lambung x-direction

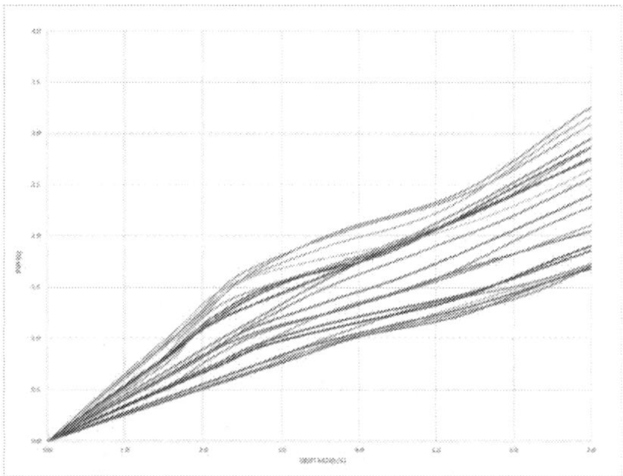

Fig 17. IDA Curve of TVE Lambung y-direction

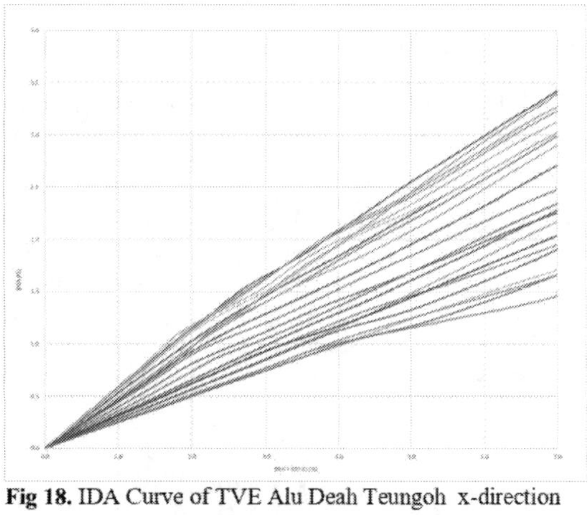

Fig 18. IDA Curve of TVE Alu Deah Teungoh x-direction

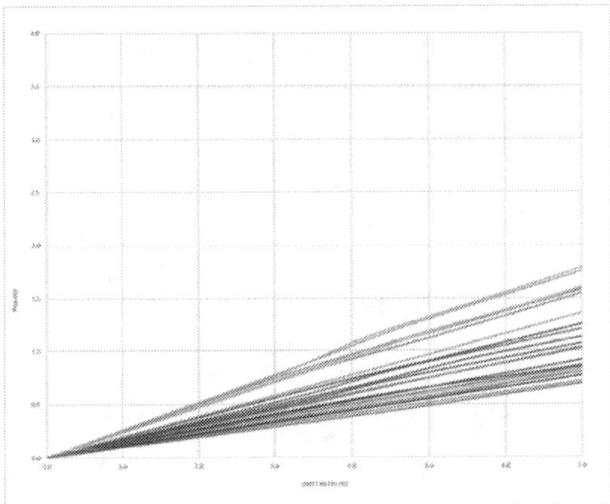

Fig 19. IDA Curve of TVE Alu Deah Teungoh y-direction

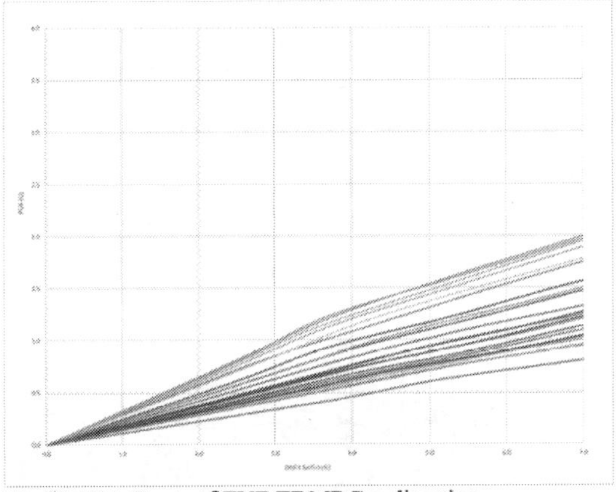

Fig 20. IDA Curve of TVE TDMRC x-direction

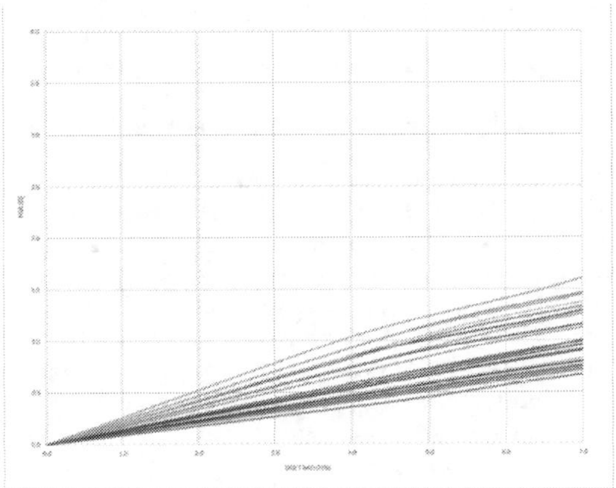

Fig 21. IDA Curve of TVE TDMRC y-direction

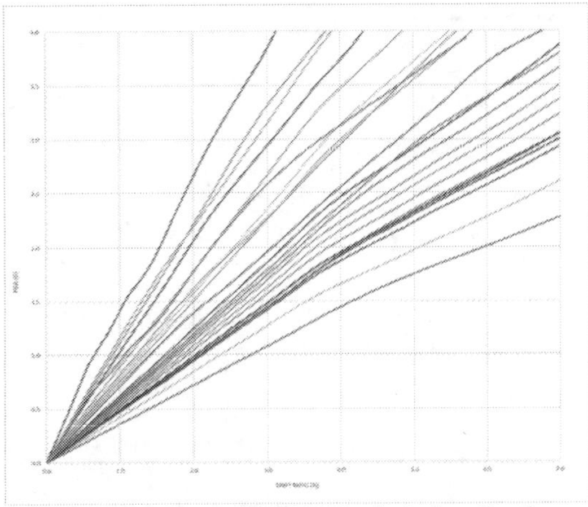

Fig 22. IDA Curve of SDN 48 Banda Aceh x-direction

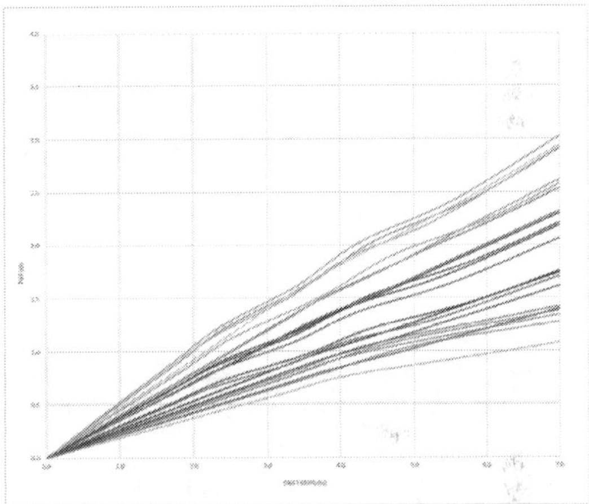

Fig 23. IDA Curve of SDN 48 Banda Aceh y-direction

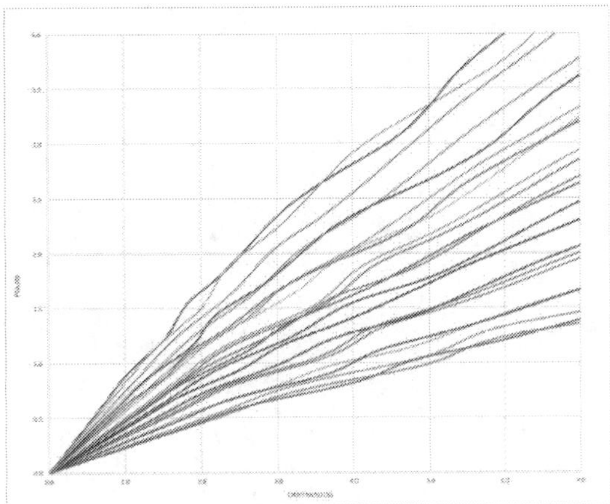

Fig 24. IDA Curve of SMPN 11 Banda Aceh x-direction

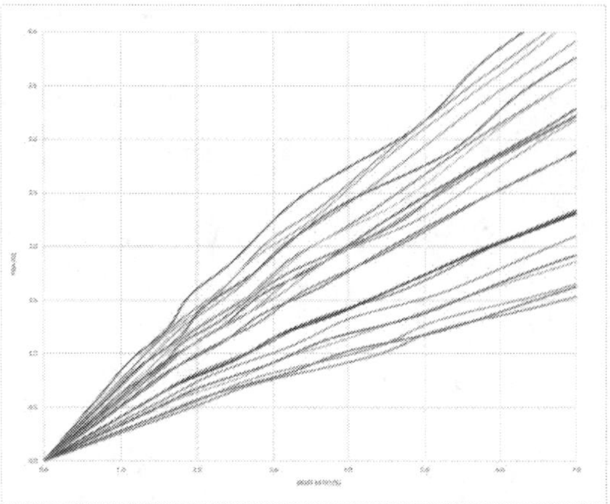

Fig 25. IDA Curve of SMPN 11 Banda Aceh y-direction

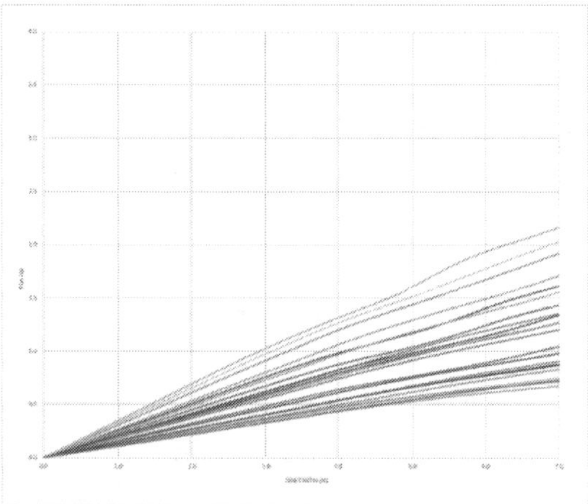

Fig 26. IDA Curve of Ulee Lheu Mosque x-direction

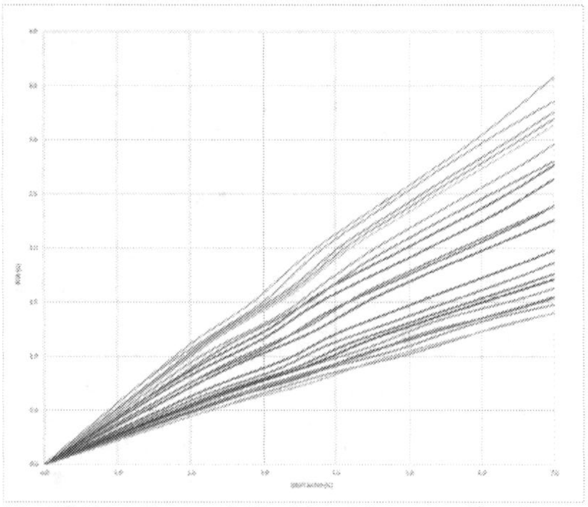

Fig 27. IDA Curve of Ulee Lheu Mosque y-direction

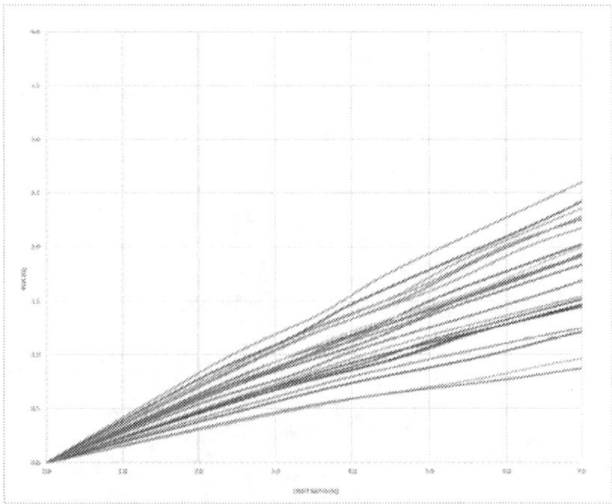

Fig 28. IDA Curve of Punge Mosque x-direction

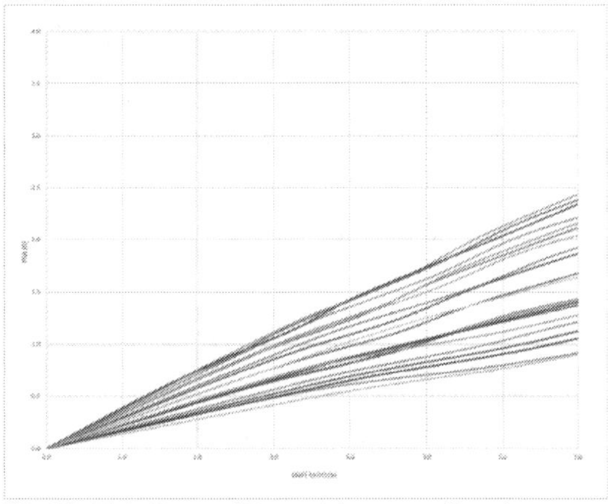

Fig 29. IDA Curve of Punge Mosque y-direction

3.2 Results of fragility curve analysis

The probability equation for the fragility curve is a function of the cumulative standard normal distribution. The variables needed are the ground motion parameter (PGA), the average of the ground motion parameters according to the limit state, and the total uncertainty. Based on these results, the complete fragility probability value is obtained at various limit states. The fragility probability value is plotted on the y-axis (ordinate) along with the ground motion parameter on the x-axis (abscissa). With this curve, the evaluation of the probability of existing building structures being damaged due to variations in the intensity of earthquake loads during their service life can be done quantitatively and rationally. The results of the fragility curves in the x and y directions can be seen in Figure 30 to 43.

https://doi.org/10.1051/e3sconf/202234002001

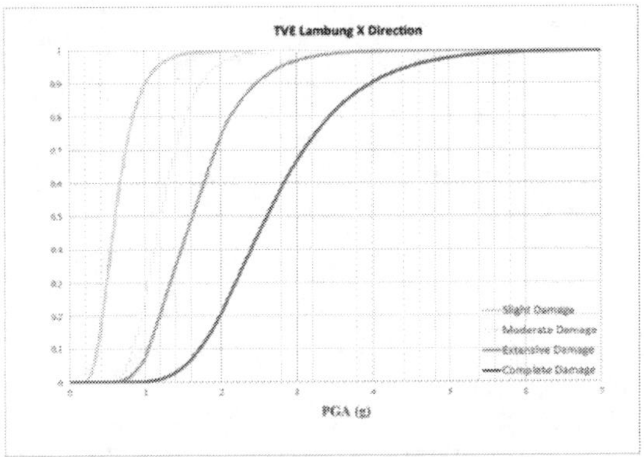

Fig 30 . Fragility Curve of TVE Lambung x-direction

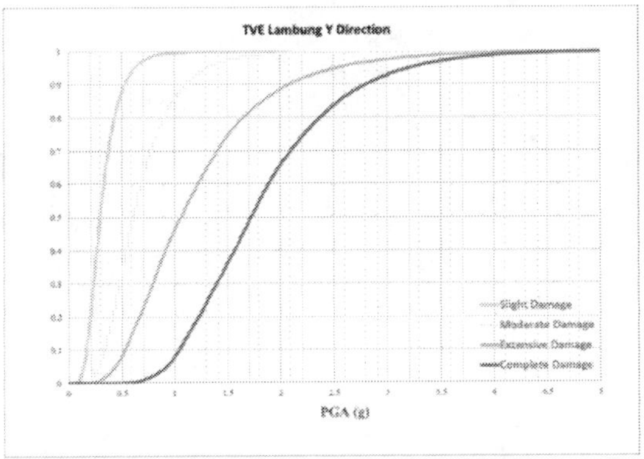

Fig 31. Fragility Curve of TVE Lambung y-direction

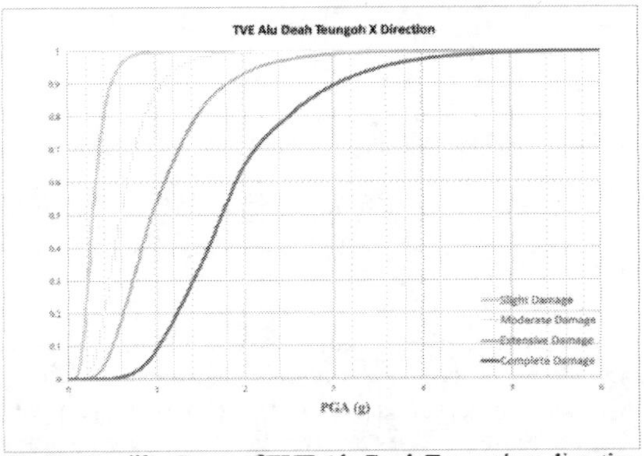

Fig 32. Fragility Curve of TVE Alu Deah Teungoh x-direction

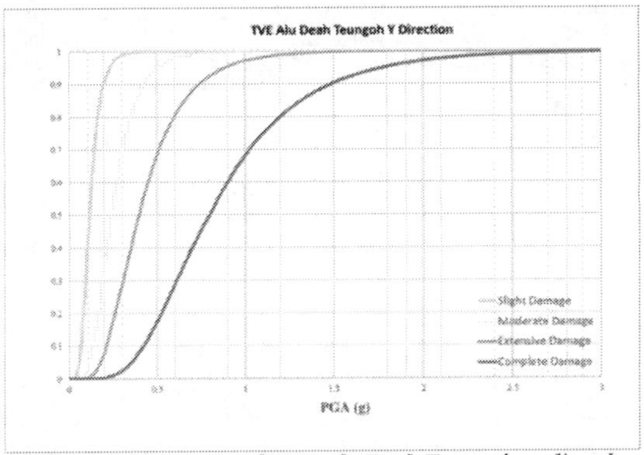

Fig 33. Fragility Curve of TVE Alu Deah Teungoh y-direction

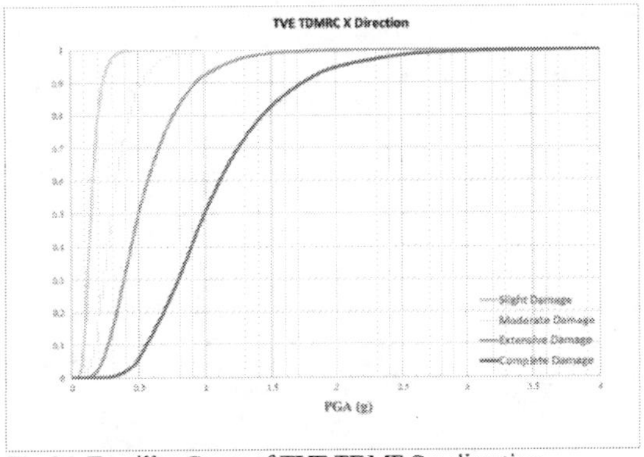

Fig 34. Fragility Curve of TVE TDMRC x-direction

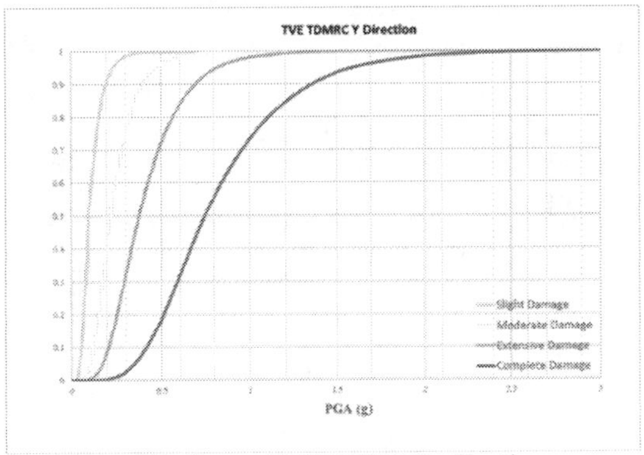

Fig 35. Fragility Curve of TVE TDMRC y-direction

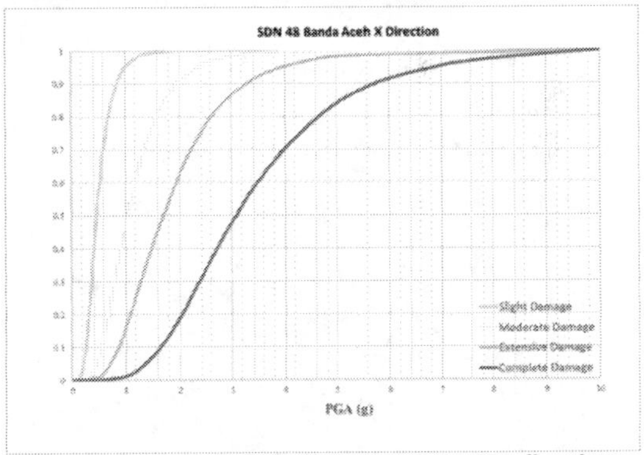

Fig 36. Fragility Curve of SDN 48 Banda Aceh x-direction

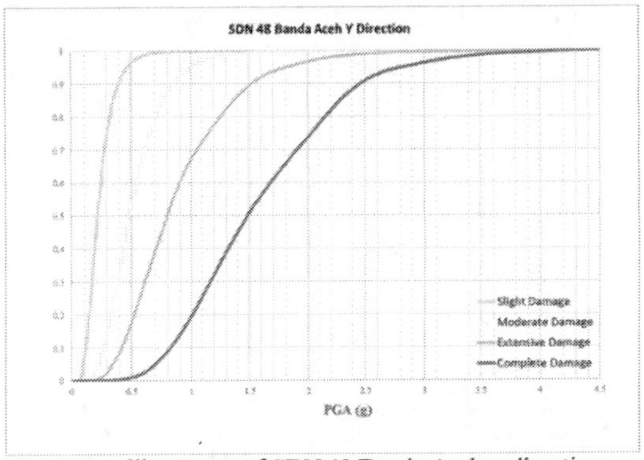

Fig 37. Fragility Curve of SDN 48 Banda Aceh y-direction

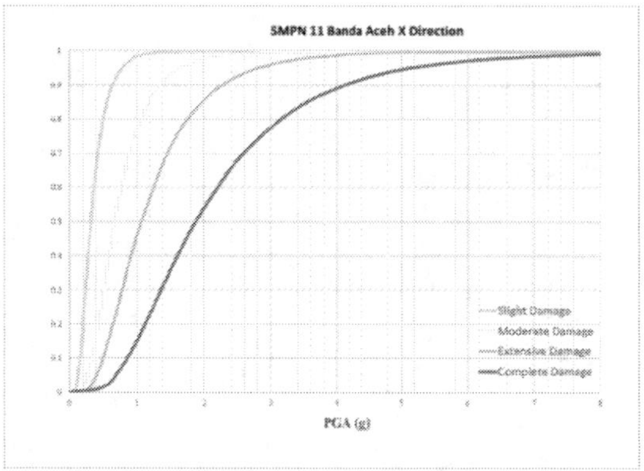

Fig 38. Fragility Curve of SMPN 11 Banda Aceh x-direction

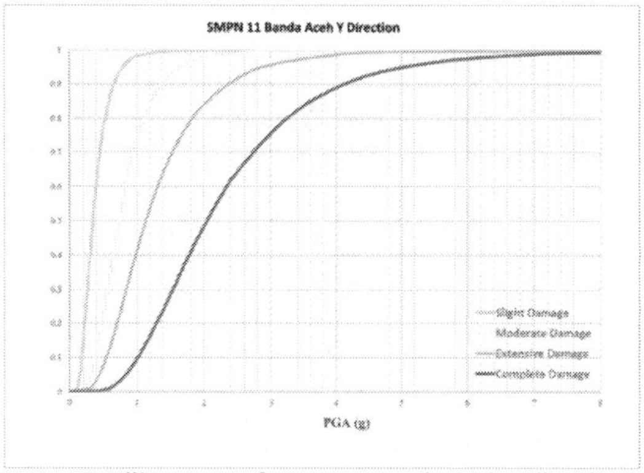

Fig 39. Fragility Curve of SMPN 11 Banda Aceh y-direction

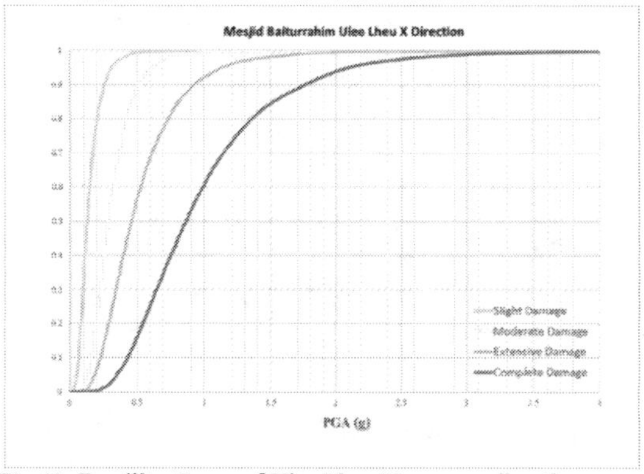

Fig 40. Fragility Curve of Ulee Lheu Mosque x-direction

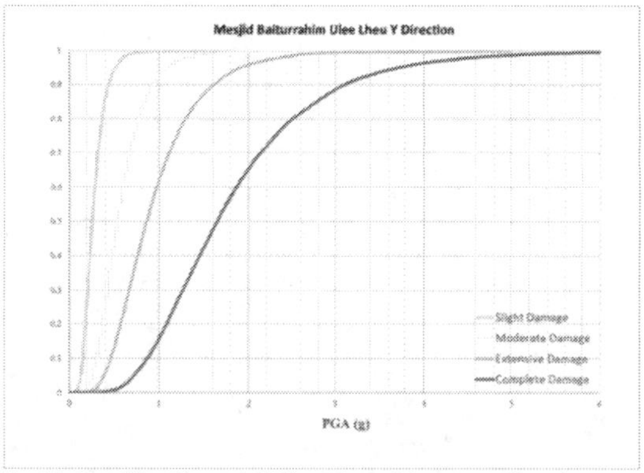

Fig 41. Fragility Curve of Ulee Lheue Mosque y-direction

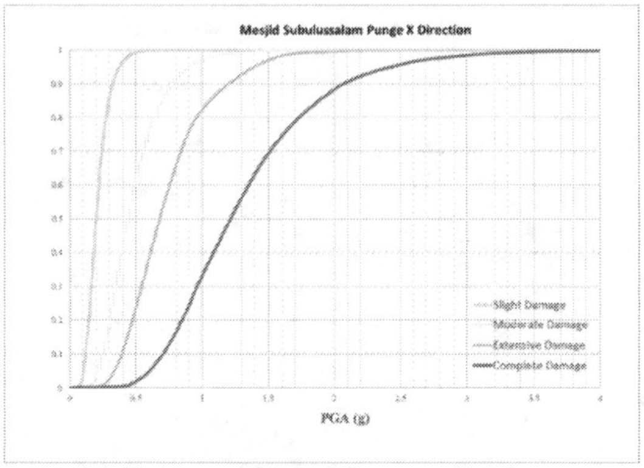

Fig 42. Fragility Curve of Punge Mosque x-direction

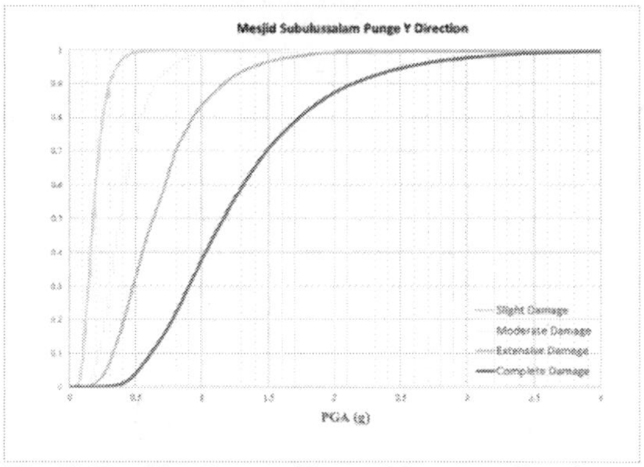

Fig 43. Fragility Curve of Punge Mosque y-direction

Based on the fragility curve formed, the building structure has a 100% probability of damage which can be seen in Table 2. From Table 2 below, it can be seen that the building structures for TVE, schools, and mosques have various probability values. This is influenced by the shape of the building, the number of floors of the building, the loading, and also the dimensions of the structure used.

The shape of the building is very influential for the probability of the x and y directions, we can see this in the Subulussalam Mosque which is almost symmetrical. The probability values in the x direction and y direction are the same and close to .. Unlike the case with the Baiturrahim Mosque which has an asymmetrical shape, the probability values for the x and y directions have different values.

Tabel 2. Fragility curve Results for each building

Existing Building	Earthquake Direction	*Peak Ground Acceleration* (g)			
		***Slight Damage* (100 %)**	***Moderate Damage* (100 %)**	***Extensive Damage* (100 %)**	***Complete Damage* (100 %)**
TVE Lambung	x	2.1	2.7	4.5	7.0
	y	1.3	2.6	4.4	5.0
TVE Alu Deah Teungoh	x	1.1	2.3	4.0	6.0
	y	0.4	1.0	1.8	3.0
TVE TDMRC	x	0.4	1.1	2.1	4.0
	y	0.5	0.8	1.6	3.2
SDN 48 Banda Aceh	x	1.9	3.8	7.6	10
	y	0.9	1.7	3.2	4.2
SMPN 11 Banda Aceh	x	2.0	3	4.8	8.0
	y	1.6	2.6	4.7	8.0
Baiturrahim Ulee Lheu Mosque	x	0.6	1.0	2.1	3.4
	y	0.9	1.8	3.2	5.7
Subulussalam Punge Mosque	x	0.7	1.4	2.2	3.7
	y	0.7	1.4	2.3	3.8

4 Conclusion

The results of the fragility curve analysis show that the building structure is more susceptible to the y direction than the x direction. The vulnerability values of TVE Lambung, TVE Alu Deah Teungoh, TVE TDRMC, SDN 48, SMPN 11, Mesjid Ulee Lheu, and Mesjid Punge each have a probability sequentially, namely 100% slight damage in earthquakes with a PGA of 1, 3; 0.4; 0.5; 0.9; 1.6; 0.9; 0.7, moderate damage at PGA 2.6; 1; 0.8; 1.7; 2.6; 1.8; 1.4, extensive damage on PGA 4.4; 1.8; 1.6; 3.2 ; 4.7 ; 3.2 ; 2,3, and complete damage during an earthquake with a PGA of 5; 3; 3.2 ; 4.2 ; 8 ; 5.7 ; 3.8. These results indicate that it is necessary to carry out periodic maintenance of the existing building in order to become a strong building and ready for use if a large-scale earthquake occurs.

For the existing building structure, it is recommended that structural mitigation efforts related to improving the condition and strengthening of the building structure, especially in the column section, are recommended. One of the efforts to strengthen the flexural column is done by adding reinforcement method. In addition, other column retrofitting efforts can be in the form of column jacketing. Jacketing material in the form of a sheath that can protect the concrete against damage. This sheath material can be metal, steel, rubber, concrete, composites. This method is commonly used for column reinforcement repair.

Acknowledgement. This research was conducted independently without external funding. The authors are grateful to Mr. Rifqi Irvansyah for the assistance provided in the experimental work.

References

1. National Institute of Building Science (NIBS). 2002. *HAZUS (Hazard US), (1999): Earthquake Loss Estimation*. National Institute of Building Sciences, NIBS: Washington D.C.
2. Irvansyah, Rifqi, 2019. Evaluasi Kinerja Seismik pada Tipikal Bangunan Sekolah di Kota Banda Aceh dengan Pengembangan Kurva Kerapuhan Berbasis Incremental Dynamic Analysis. Universitas Syiah Kuala, Indonesia.
3. Aulia, Fajarul, 2019. Evaluasi Kinerja Seismik pada Tipikal Bangunan Sekolah di Kota Banda Aceh dengan Pengembangan Kurva Kerapuhan Berbasis Pushover Analysis.. Universitas Syiah Kuala, Indonesia.
4. Federal Emergency Management Agency, 2001. *Earthquake Loss Estimation Methodology: Hazus MH 2.1*. Department of Homeland Security FEMA Mitigation Division: Washington D.C.
5. Pacific Earthquake Engineering Research Center. Berkeley, University of California, USA. http://peer.berkeley.edu/peer_ground_motion_data base. Accesed on 25 Agustus 2020.
6. Badan Meteorologi Klimatologi dan Geofisika (BMKG) Data Base. Accesed on 27 September 2020.

7. Badan Standarisasi Nasional, 2019. Standar Nasional Indonesia 1726-2019: Tata Cara Perencanaan Ketahanan Gempa untuk Struktur Bangunan Gedung dan Non Gedung. Badan Standarisasi Nasional, Jakarta.
8. Tiwari, V. dan Kasnale, A., 2017. Incremental Dynamic Analysis of RC Frames. *IRJET*, **4**(6): 697-700.
9. Pitilakis, K., et al, 2005. SYNER-G : Typology Definiton and Fragility Functions for Physical Elements at Seismic Risk. Springer: Netherlands.
10. Sadraddin, H., 2015. *Fragility Assesment of High-Rise Reinforced Concrete Buildings.* Tesis, Civil and Construction Engineering Western Michigan University, Kalamazoo.

Development of fragility curve due to the effect of tsunami force on existing building in Banda Aceh city

Zu Irfan[1,*], *Abdullah*[2], *Moch.* Afifuddin[2]

[1]PhD Candidate, Doctor of Engineering Study Program, Syiah Kuala University, Jl. Syech Abdurrauf, Banda Aceh, 23111, Indonesia
[2]Civil Engineering Department, Syiah Kuala University, Jl. Syech Abdurrauf, Banda Aceh, 23111, Indonesia

Abstract. Aceh is located on the Indian Ocean Coast which is very prone to Tsunamis, this is due to frequent large earthquakes due to Megatrust faults, such as those that occurred during the earthquake on December 26, 2004. Based on the 2012 Aceh earthquake, the Acehnese prefer horizontal evacuation to vertical evacuation [1]. In this case, many people in Banda Aceh do not believe in the strength of TVE buildings (distrust building). From this research, it it is possible for the Acehnese will do the same thing in the future when the tsunami occurs. Horizontal evacuation puts them at risk of traffic accidents so that they cannot reach a safe area in a short time, whereas through vertical evacuation they can reach the TVE of a building in a short time due to its proximity to settlements. When the earthquake and tsunami occured Banda Aceh in 2004, many buildings were destroyed due to the earthquake load and also the Tsunami load. For this problem, a vulnerability analysis of buildings that can be used as TVE buildings can be carried out. If an earthquake magnitude more than 7.0 Mw can cause a tsunami, the TVE building should be counted on the structural strength caused by the Tsunami waves. In this research, a hydrodynamic analysis will be carried out due to air pressure and collision which is a function of velocity and depth of flow and also the influence of drag and drift variables on the building.

1 Introduction

Aceh is located on the Indian Ocean Coast which is very vulnerable to Tsunami hazards, this is because large earthquakes often occur due to megatrust faults, such as what happened during the earthquake in 26 December 2004. Therefore, it is necessary to assess the vulnerability of the building to forces due to earthquake and tsunami loads. The Indian Ocean coast is very vulnerable to Tsunamis [2], and Sumatra Island is a seismic gap area which has the potential for large-scale earthquakes [3].

Y. Goto et al. found that during the earthquake that occurred on April 11, 2012, in Banda Aceh, although in 2012 it did not cause a Tsunami, 78% of refugees were stuck in traffic. Only about 22% of those evacuated from nearshore areas reached their destination within 20 minutes after the earthquake (this only increased to 43% at 40 minutes) [4]. The survey also found that vertical evacuation reduced evacuation time: 71% of those evacuated to a tall building or mosque reached this destination within 20 minutes (this rose to about 91% at 40 minutes), while only ~15% of those who evacuated horizontally reach their destination within 20 minutes (this rises to only 35% at 40 minutes) [4].

Jamie W. McCaughey found that during the 2012 Aceh earthquake, many people preferred to carry out

From the study it was found that those who evacuated horizontally to inland destinations during the 2012 earthquake in Aceh province or intend to do in the

* Corresponding author: zuirfan.eng@gmail.com

horizontal evacuations rather than using vertical evacuation buildings that had been built such as escape buildings, this was because many people in Banda Aceh lacked confidence in the power of TVE buildings (distrust buildings) [1].
future think that escape buildings are unsafe and they think they can reach their inland destinations in time. However horizontal evacuation puts them at risk of traffic accidents, that they cannot reach safe inland destinations in a short time, but if they can reach escape buildings in a short time because they are close to settlements. The research also found several alternative designs that can increase the use of escape buildings, namely escape buildings are on a horizontal evacuation route to the interior (away from the coast), it is also mentioned that escape buildings that can substantially increase community intention to evacuate to escape buldings are, if it is a place of worship/mosque building.

Based on this research, many mosque buildings located on horizontal evacuation routes to the interior are also used as places of refuge, so that in the future mosques can be used as buildings that can function as buildings for Tsunami evacuation or later in this study called Tsunami Vertical Evacuation Buildings. TVE Buildings).

2 Study area

This study only focuses on assessing the condition of the existing building which is carried out by survey or direct assessment. Existing buildings are buildings that are in the utilization stage. This study examines existing

Fig. 1. Map of the distribution of existing buildings that are the object of research

buildings scattered in Meuraxa District, Banda Aceh City. Existing buildings that are the object of research in this study include the Tsunami Vertical Evacuation (TVE) of Lambung, TVE Alu Deah Teungoh, TVE TDMRC, SDN 48 Banda Aceh, SMPN 11 Banda Aceh, Baiturrahim Ulee Lheu Mosque, and Subulussalam Punge Mosque.

3 Methodology

3.1 Tsunami loads

The calculation of the tsunami loads is carried out by simulating buildings with tsunami flow heights as high as 2 meters, 4 meters, 6 meters, 7 meters, and 8 meters. This height is taken randomly so that it can show the drift ratio at each level. The equation used in calculating the tsunami load is using the standard from FEMA P646 [5]. The tsunami forces that are reviewed are hydrodynamic forces, impulsive forces, and floating debris impact forces.

Hydrodynamic forces are used when water flows around a building structure. so it can be concluded that this force was applied to all structures affected by the tsunami. hydrodinamic forces can be computed using the following equation

$$F_h = \frac{1}{2}\, \rho_s C_d B (hu^2)_{max} \qquad (1)$$

where ρ_s is the fluid density including sediment (1100 kg/m^3 = 2.13 slugs/ft^3), Cd is the drag coefficient, B is the breadth of the structure in the plane normal to the direction of, h is flow depth, and u is flow velocity at the location of the structure. The drag coefficient may be conservatively taken as Cd = 2.0; the actual value is shape-, orientation-, and size-dependent.

The combination hu^2 represents the momentum flux per unit mass per unit width. The value (hu^2)max can be roughly estimated based on information in the inundation map, using

$$(hu^2)_{max} = gR^2\left(0{,}125 - 0{,}235\frac{z}{R} + 0{,}11\left(\frac{z}{R}\right)^2\right) \qquad (2)$$

where g is the acceleration due to gravity, R is the design runup elevation taken as 1.3 times the maximum runup elevation, R*, and z is the ground elevation at the base of the structure.

While the impulsive force will act on the component that is subjected to the front end of the tsunami wave which usually occurs in buildings that have been previously flooded which can be calculated by the equation:

$$F_s = 1{,}5\, F_d \qquad (3)$$

At the time of the tsunami there was large debris that was carried away by the flow of water, resulting in a debris impact force on the structure of the building. The large debris impact force can cause damage to the building, so it needs to be calculated using the equation:

$$F_i = 1{,}3\, u_{max} \sqrt{k\, m_d(1+c)} \quad (4)$$

where 1.3 is the importance Coefficient for Risk Category IV structures that is specified by ASCE 7 Chapter 5 for debris impacts,
umax is the maximum flow velocity carrying the debris at the site, can be computed using:

$$u_{maks} = \sqrt{\left(2\, g\, R\, \left(1 - \frac{z}{R}\right)\right)} \quad (5)$$

and *c* is a hydrodynamic mass coefficient which represents the effect of fluid in motion with the debris. In this study, lumber of wood log with the mass and the effective stiffness of the debris equal to 450 kg and $2.4 \cdot 10^6$ N/m respectively is considered as the waterborne debris, this value approximately proposed by FEMA P646 [5].
These tsunami forces will be used in a combination of tsunami loading based on FEMA P646 using the equation:

$$U = 1{,}2\ D + 1{,}0\ TS + 1{,}0\ LREF + 0{,}25\ L \quad (6)$$
$$U = 0{,}9\ D + 1{,}0\ TS \quad (7)$$

3.2 Fragility curve

The fragility curve describes the probability that it exceeds the specified level of damage in a tsunami event with a certain intensity [6]. In this study, the damage model based on HAZUS that defines four structural damage states according to the drift ratio based on Medina (2019)[7], can be seen in Table 1.

Table 1. Damage States Limits

Structural Damage States	Drift Damage Index HAZUS – 1–3 stories
Slight	<0.20%
Moderate	0.20%–0.50%
Extensive	0.50%–1.20%
Complete	1.20%–2.80%

The results of nonlinear numerical analysis of inundation depth and drift ratio can be used to obtain a fragility curve which is expressed as a two-parameter lognormal distribution function. The following equation will give the cumulative probability of exceeding the damage level conditioned on the tsunami intensity measure.

$$P_{(DSi\,|\,IM)} = \Phi\left(\frac{\ln(IM) - \ln(\overline{IM_i})}{\beta}\right) \quad (8)$$

where is Φ a standard normal cumulative distribution function, IM is a measure of tsunami intensity expressed in inundation depth (in units of m), IMi is the median value (in units of m) for each level of damage and β is the uncertainty calculated through the log standard deviation parameter in each level of damage. The value of uncertainty can be calculated using the equation.

$$\beta = \sqrt{\beta_c^2 + \beta_D^2} \quad (9)$$

where, c is the uncertainty of the structural capacity which taken as 0.15 (based on the HAZUS MH-Technical Manual)[8] and D is the uncertainty of the structural demand which is considered by calculating the logarithmic dispersion at each level of damage. The results of the calculation of the standard deviation of the total uncertainty in each damage condition.

4 Result and discussion

The tsunami fragility curve requires the value of the inundation depth in meters (m) according to the limit state. The value of inundation depth at each damage limit for each height can be seen in Table 2-15

Table 2. TVE Lambung x direction

Structural Damage States	1st stories	2nd stories	3rd stories
Slight (<0,2%)	18	22	22
Moderate (0,2%-0,5%)	25	32	32
Extensive (0,5%-1,2%)	33	42	42
Complete (1,2%-2,8%)	42	54	54

Table 3. TVE Lambung y direction

Structural Damage States	1st stories	2nd stories	3rd stories
Slight (<0,2%)	0,2	6	14
Moderate (0,2%-0,5%)	0,5	11	26
Extensive (0,5%-1,2%)	1,2	17	40
Complete (1,2%-2,8%)	6,4	26	61

Table 4. TVE Alu Deah Teungoh x direction

Structural Damage States	1st stories	2nd stories	3rd stories
Slight (<0,2%)	20	24	24
Moderate (0,2%-0,5%)	29	35	35
Extensive (0,5%-1,2%)	37	46	46
Complete (1,2%-2,8%)	47	60	60

Table 5. TVE Alu Deah Teungoh y direction

Structural Damage States	1st stories	2nd stories	3rd stories
Slight (<0,2%)	0,1	3,5	5
Moderate (0,2%-0,5%)	1,3	6,5	12
Extensive (0,5%-1,2%)	5,5	11	20
Complete (1,2%-2,8%)	11,2	31	31

Table 6. TVE TDMRC x direction

Structural Damage States	1st stories	2nd stories	3rd stories
Slight (<0,2%)	3,1	7,2	14
Moderate (0,2%-0,5%)	5,1	14,5	19,5
Extensive (0,5%-1,2%)	9,8	23	23,5
Complete (1,2%-2,8%)	16	35,5	29

Table 7. TVE TDMRC y direction

Structural Damage State	1st stories	2nd stories	3rd stories
Slight (<0,2%)	0,4	3,0	4,9
Moderate (0,2%-0,5%)	1,45	4,8	7,9
Extensive (0,5%-1,2%)	2,65	5,9	12
Complete (1,2%-2,8%)	4,61	8,46	17,8

Table 8. SDN 48 Banda Aceh x direction

Structural Damage States	1st stories	2nd stories
Slight (<0,2%)	2	8
Moderate (0,2%-0,5%)	7	12,5
Extensive (0,5%-1,2%)	14	16,5
Complete (1,2%-2,8%)	24	21,2

Table 9. SDN 48 Banda Aceh y direction

Structural Damage States	1st stories	2nd stories
Slight (<0,2%)	0,1	1
Moderate (0,2%-0,5%)	0,2	2
Extensive (0,5%-1,2%)	0,5	3
Complete (1,2%-2,8%)	1	5

Table 10. SMPN 11 Banda Aceh x direction

Structural Damage States	1st stories	2nd stories
Slight (<0,2%)	12	14
Moderate (0,2%-0,5%)	20	20
Extensive (0,5%-1,2%)	25	26
Complete (1,2%-2,8%)	32	33

Table 11. SMPN 11 Banda Aceh y direction

Structural Damage States	1st stories	2nd stories
Slight (<0,2%)	0,05	0,1
Moderate (0,2%-0,5%)	0,1	0,3
Extensive (0,5%-1,2%)	0,3	1
Complete (1,2%-2,8%)	1	2,1

Table 12. Mesjid Baiturrahim Ulee Lheu x direction

Structural Damage States	1st stories	2nd stories
Slight (<0,2%)	0,01	0,02
Moderate (0,2%-0,5%)	0,17	0,2
Extensive (0,5%-1,2%)	0,39	0,4
Complete (1,2%-2,8%)	0,95	0,95

Table 13 Mesjid Baiturrahim Ulee Lheu y direction

Structural Damage States	1st stories	2nd stories
Slight (<0,2%)	22	23
Moderate (0,2%-0,5%)	33	35
Extensive (0,5%-1,2%)	43,9	45
Complete (1,2%-2,8%)	56,2	58,5

Table 14. Mesjid Subulussalam Punge x direction

Structural Damage States	Base	1st stories	2nd stories
Slight (<0,2%)	5	10	16
Moderate (0,2%-0,5%)	12,5	24	34
Extensive (0,5%-1,2%)	22	39	54
Complete (1,2%-2,8%)	35,5	62	84

Table 15. Mesjid Subulussalam Punge y direction

Structural Damage States	Base	1st stories	2nd stories
Slight (<0,2%)	0,05	0,1	8
Moderate (0,2%-0,5%)	0,3	0,6	17
Extensive (0,5%-1,2%)	0,73	1,3	29
Complete (1,2%-2,8%)	1,65	3,18	46

Based on the displacement results for each tsunami height, it can be seen that the structure responds differently to each tsunami inundation that is used as a load. However, generally they have the same tendency, displacement in the y-direction is larger than the x-direction. But, it depends on the condition of the building's location towards tsunami.
The probability value on the fragility curve is plotted on the y-axis (ordinate) and the inundation depth on the x-axis (absis). so that the results of the brittle curve are obtained at various damage limits, it can be seen in Fig 2-15.

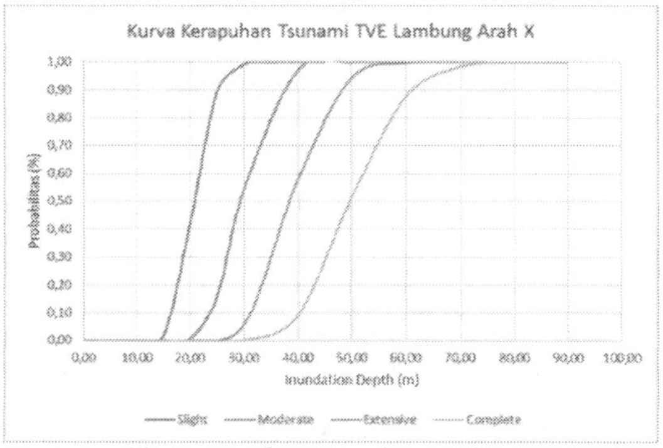

Fig. 2. Fragility Curve of TVE Lambung x-direction

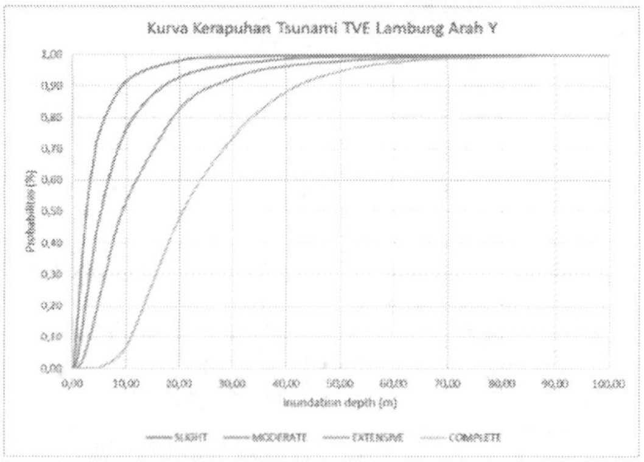

Fig. 3. Fragility Curve of TVE Lambung y-direction

 https://doi.org/10.1051/e3sconf/202234002002

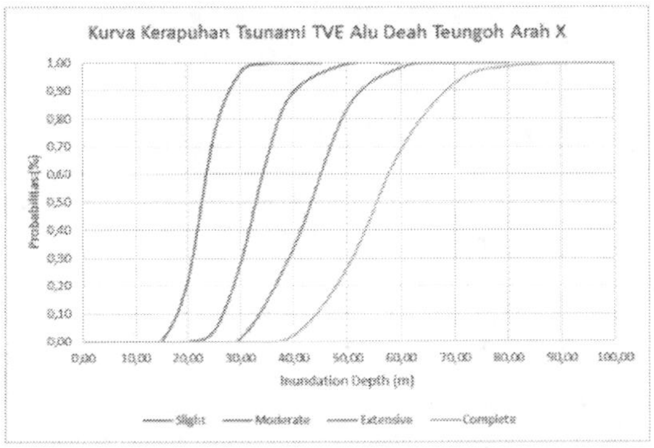

Fig. 4. Fragility Curve of TVE Alu Deah Teungoh x-direction

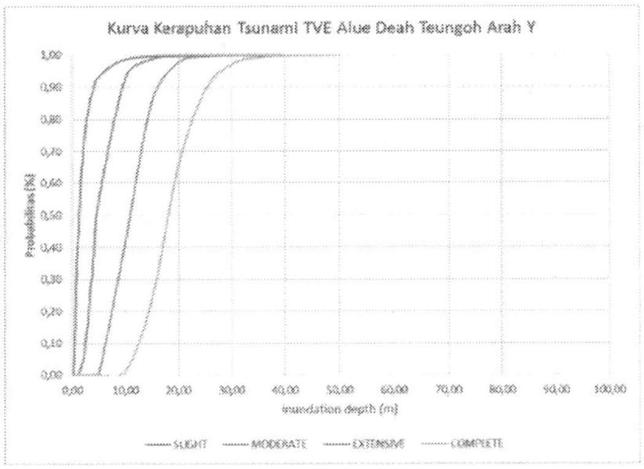

Fig. 5. Fragility Curve of TVE Alu Deah Teungoh y-direction

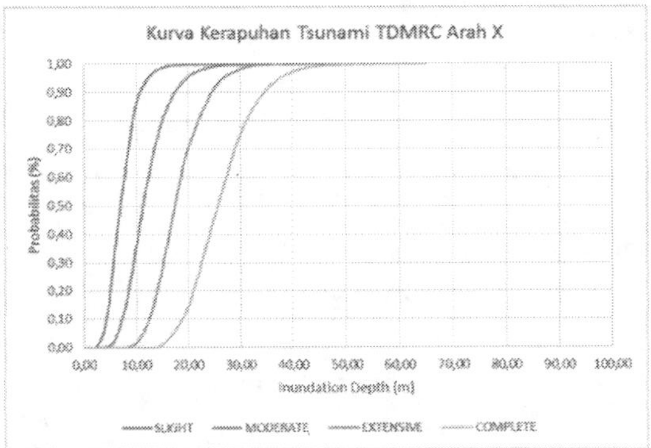

Fig. 6. Fragility Curve of TVE TDMRC x-direction

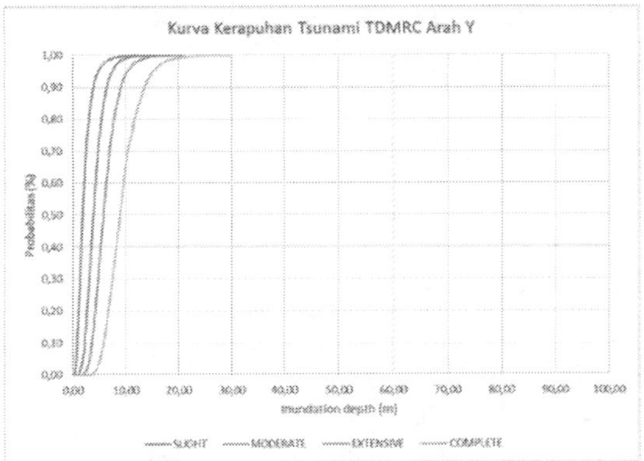

Fig. 7. Fragility Curve of TVE TDMRC y-direction

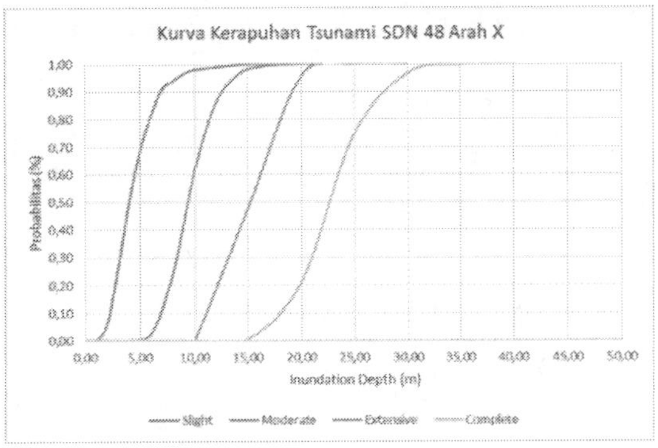

Fig. 8. Fragility Curve of SDN 48 Banda Aceh x-direction

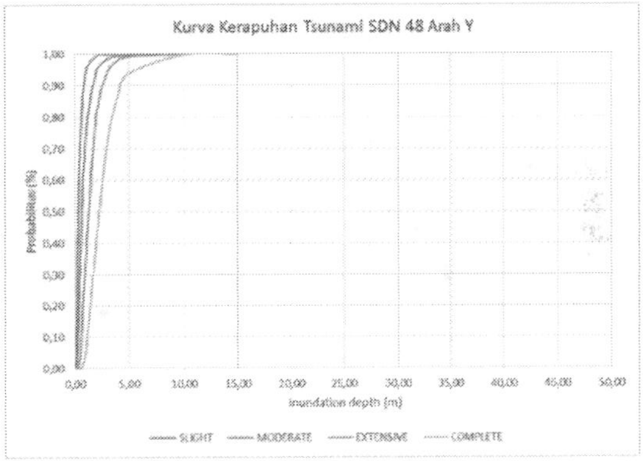

Fig. 9. Fragility Curve of SDN 48 Banda Aceh y-direction

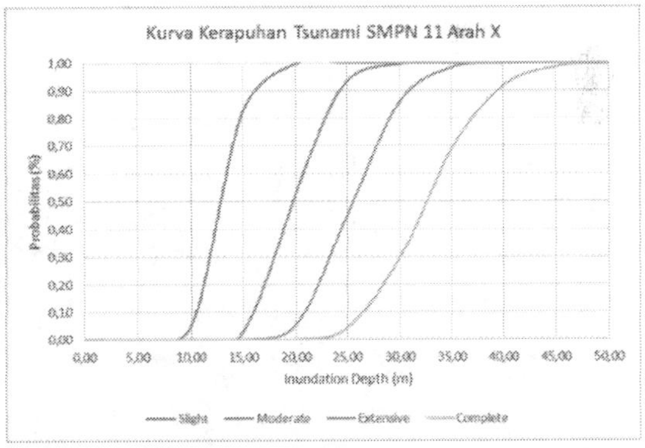

Fig. 10. Fragility Curve of SMPN 11 Banda Aceh x-direction

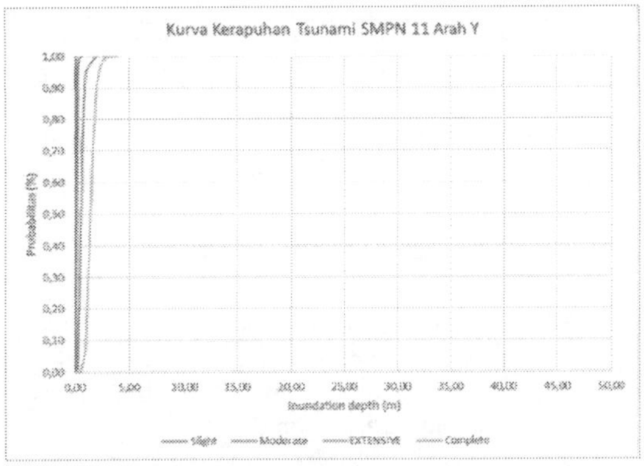

Fig. 11. Fragility Curve of SMPN 11 Banda Aceh y-direction

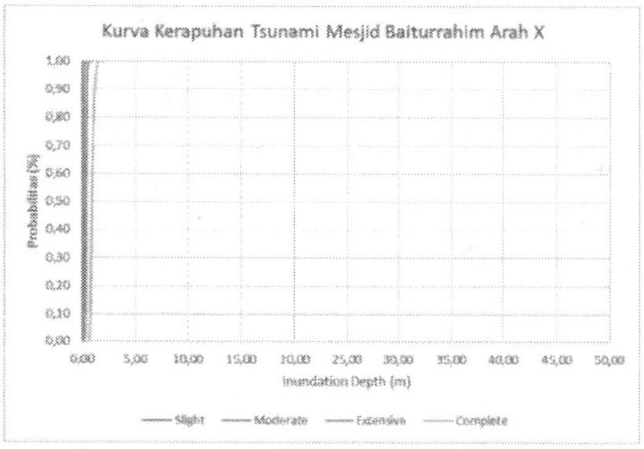

Fig. 12. Fragility Curve of Ulee Lheu Mosque x-direction

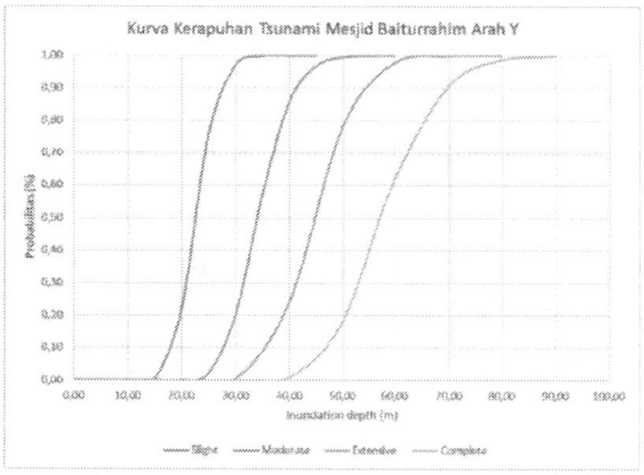

Fig. 13. Fragility Curve of Ulee Lheu Mosque y-direction

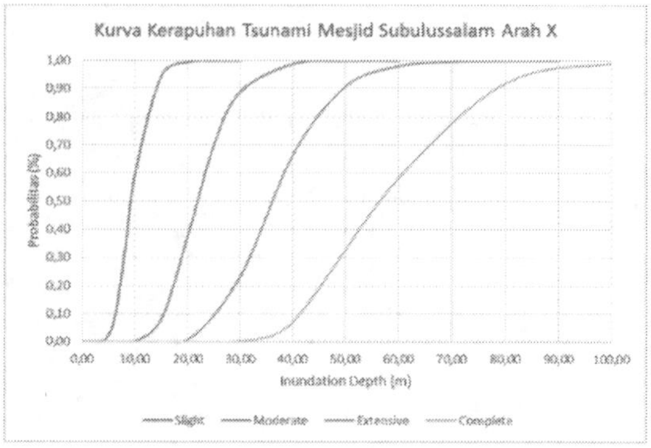

Fig. 14. Fragility Curve of Punge Mosque x-direction

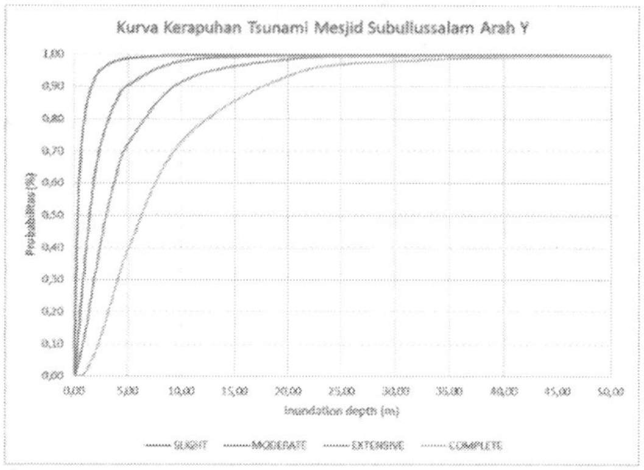

Fig. 15. Fragility Curve of Punge Mosque y-direction

Based on the fragility curve formed, the building structure has a 100% probability of damage which can be seen in Table 16.

Table 16. Fragility curves for each building

Existing Building	Tsunami Direction	*Inundation Depth* (m)			
		Slight Damage **(100 %)**	***Moderate Damage*** **(100 %)**	***Extensive Damage*** **(100 %)**	***Complete Damage*** **(100 %)**
TVE Lambung	x	30	50	60	80
	y	30	60	70	100
TVE Alu Deah Teungoh	x	35	60	80	100
	y	30	30	33	45
TVE TDMRC	x	23	37	43	57
	y	10	12	15	24
SDN 48 Banda Aceh	x	15	17	22	33
	y	2	5	7	10
SMPN 11 Banda Aceh	x	20	30	36	47
	y	0,1	0,5	2	3
Baiturrahim Ulee Lheu Mosque	x	0,1	0,5	1	2
	y	32	50	64	85
Subulussalam Punge Mosque	x	20	40,5	70	100
	y	10	20	25	45

https://doi.org/10.1051/e3sconf/202234002002

5 Conclusion

The results of the damage probability analysis based on the fragility curve can be used to predict the level of damage to buildings in Meuraxa District with reinforced concrete structures due to tsunami in the future.

Acknowledgement. This research was conducted independently without external funding. The authors are grateful to Amirah Fasya for the assistance provided in the experimental work.

References

1. Jamie W. Mc Caughey et. all. (2017), Trust and distrust of tsunami vertical evacuation buildings: Extending protection motivation theory to examine choices under social influence, International Journal of Disaster Risk Reduction 24.
2. Roshan, A.D. et.all (2016) "Tsunami Hazard Assessment of Indian Coast" Natural Hazards 82(2) DOI: 10.1007/s11069-016-2216-1.
3. Delfebriyadi, 2010, "Rekayasa Gempa Teknik Sipil", CV. Ferila., Padang.
4. Goto, Y. et.all (2012), "Tsunami evacuation simulation for disaster education and city planning" Paper: Dr7-1-5086; 2012/1/24.
5. Federal Emergency Management Agency. 2012. Guidelines for Design of Structures for Vertical Evacuation from Tsunami Second Edition. Department of Homeland Security FEMA Mitigation Division: Washington D.C.
6. Karafagka, S., Fotopoulou, S., and Pitilakis, K. 2018. "Analytical tsunami fragility curves for seaport RC buildings and steel light frame warehouses", Soil Dynamics and Earthquake Engineering. 112 (2018): 118-137.
7. Medina, S. et. all. "Tsunami analytical fragility curves for the Colombian Pacific coast: A reinforced concrete building example", Engineering Structures. 196 (2019) 109309.
8. Federal Emergency Management Agency, 2004, Direct physical damage-general building stock, HAZUS-MH Technical manual, Chapter 5, Washington D.C.

A comparative analysis of seismic and tsunami fragility curves for school buildings in Banda Aceh coastal area

Rifqi Irvansyah[1*], *Ibnu* Rusydy[1 2], *Andrean V H* Simanjuntak[3]

[1] Tsunami and Disaster Mitigation Research Center (TDMRC), Universitas Syiah Kuala (USK), Jl. Hamzah Fanshuri, Darussalam, Banda Aceh, 23111, Indonesia
[2] Department of Geological Engineering, Faculty of Engineering, Universitas Syiah Kuala, Jl. Syeh Abdurrauf No. 7, Banda Aceh, 23111, Indonesia
[3] BMKG, stage of Aceh Besar, Mata Ie, Aceh Besar, 23352, Indonesia

Abstract. The potential for a large disaster in Banda Aceh City requires that the school buildings to be highly resilient to the risks that may arise. One of the efforts to evaluate the performance of the building structure is by developing a fragility curve that shows the probability of damage to school buildings. For seismic susceptibility, the fragility curve is formed using the Incremental Dynamic Analysis method which uses historical earthquake data. As for tsunami susceptibility, the fragility curve is formed by reviewing the inundation heights that have occurred based on the data that have been available. The aim of this study is to elaborate the seismic and tsunami susceptibility of the typical school buildings that located in three coastal districts in Banda Aceh City. The summary results of this study indicate that it is necessary to carry out periodic maintenance of school buildings in the coastal area of Banda Aceh city so that they become strong buildings and are ready to be used if a large-scale disaster occurs.

1 Introduction

History records that in 2004 a 9.1 magnitude earthquake, followed by a tsunami, devastated the city of Banda Aceh. The impact of the earthquake and tsunami caused damage to the structures and infrastructure, public facilities, and other buildings.

Casualties and damages on educational facilities suffered significant losses. It was noted that 45,000 students and 1,870 teachers died, and 2,065 educational facilities were damaged [1]. Learning from the history of disasters, it is important to ensure the planning, construction, and operation of educational facilities fulfill the earthquake and building standards.

At date, Indonesian building codes regarding earthquake resistant and tsunami loads for building are following the SNI 1726-2019 [2] and SNI 1727-2020 [3]. Since, the feasibility study of school buildings against earthquake and tsunami hazards is crucial, therefore, it is important to ensure that the school building structures are planned to be safe against earthquakes and tsunamis, specifically in coastal areas.

The earthquake and tsunami risk of school buildings in Banda Aceh had been assessed and reported by Gentile [4]. According to Gentile et al. [4], the seismic fragility of school buildings post-2012 is decreased compared to the pre-2012 structures. The relative seismic risk of the considered buildings is particularly similar, while the relative tsunami risk shows a strong dependence with the distance from the coast.

The purpose of this study is to develop the analytical fragility curve for the typical school's building in Banda Aceh. The fragility curve will show the probability of damage of the building structure in various damage level, namely slight damage, moderate damage, extensive damage, and complete damage. By referring to the fragility curve, an assessment on the vulnerability of buildings due to earthquakes and tsunamis can be predicted. The seismic fragility curve is formed using the Incremental Dynamic Analysis (IDA) method [5] and producing an IDA curve. This IDA Curve will be analyzed further to develop a seismic fragility curve. While the tsunami fragility curve is formed by calculating the cumulative probability of a measure of the tsunami intensity, which is the height of the tsunami inundation [6] and producing capacity curve. The capacity curve will be analyzed further to develop a tsunami fragility curve.

In this study an analytical fragility curve shows the probability of damage to school buildings in 3 coastal districts of Banda Aceh City based on the analysis of earthquake and tsunami loading. With the probability of damage to the building, the results can mitigate disasters in the future. The potential liquefaction impact on the buildings will also be assessed based on the calculated liquefaction map of Banda Aceh.

* Corresponding author: rifqirvansyah@gmail.com

https://doi.org/10.1051/e3sconf/202234002003

2 Methodology

2.1 Research object

The object of this study is school buildings located in 3 coastal districts in Banda Aceh City. Field surveys were carried out on 51 buildings from 34 schools. The location of the schools is plotted in Fig 1.

According to the school building data obtained from the field surveys, 3 schools building was chosen representing the 3 coastal districts. These buildings were selected since it could represent the most dominant building form, the most dominant number of floors, the distance to the closest fault, and the closest distance to the beach. The three buildings are :

1. Schools in Syiah Kuala District are represented by SDN 15 Banda Aceh with the following data:
   - Construction type: Reinforced Concrete (RC)
   - Number of Floors: 2 stories
   - Stories height: 4 m (1st floor), 3.5 m (2nd floor)
   - Structural dimensions:
     Columns: rectangular column with 300 x 400 mm
     Beams: rectangular beams with 300 x 400 mm and 450 x 550 mm.

2. Schools in Kutaraja District are represented by SDN 70 Banda Aceh with the following data:
   - Construction type: Reinforced Concrete (RC)
   - Number of Floors: 2 stories
   - Stories height: 3,5 m
   - Structural dimensions:
     Columns: rectangular column with 300x300 mm
     Beams: rectangular beams with 250 x 400 mm and 300 x 500 mm.

3. Schools in Meuraxa District are represented by SDN 7 Banda Aceh with the following data:
   - Construction type: Reinforced Concrete (RC)
   - Number of Floors: 2 stories
   - Stories height: 3 m
   - Structural dimensions:
     Columns: rectangular column with 300 x 300 mm
     Beams: rectangular beams with 300 x 400 mm and 450 x 600 mm.

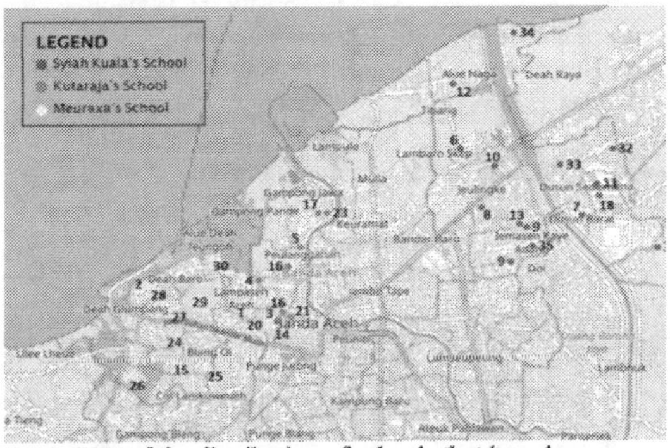

Fig 1. Map of the distribution of schools that have been surveyed

2.2 Modeling structure

The selected buildings were modelled by using the structural analysis software of SAP2000. Then, finite element (FE) analysis was performed to analysed each model against the earthquake and the tsunami loading. Fig 2 to Fig 4 shows the school building and the FE model of each representative building, respectively.

Fig 2 (a). Syiah Kuala's school building

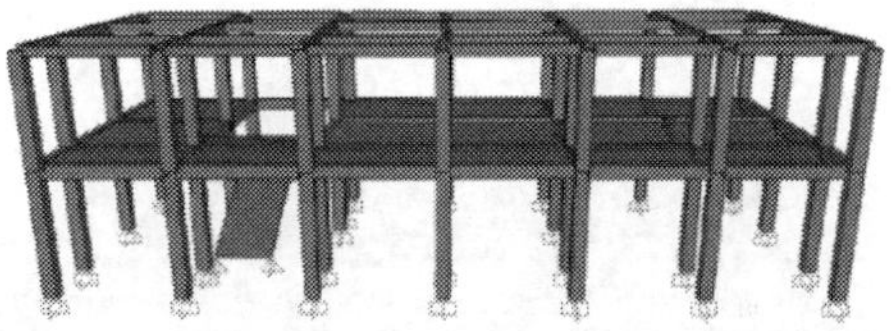
Fig 2 (b). Syiah Kuala's school 3D modelling

Fig 3 (a). Kutaraja's school building

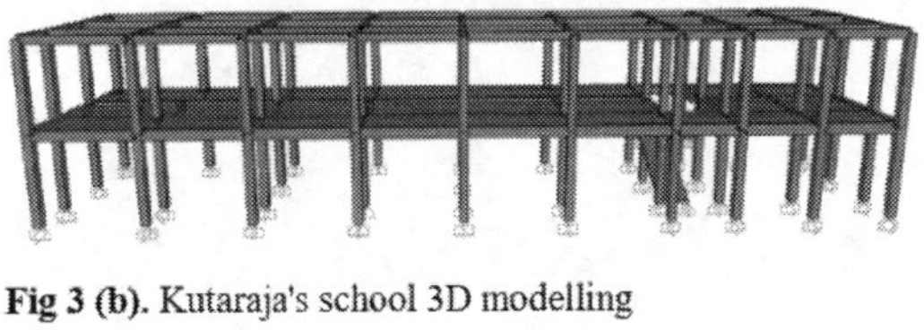
Fig 3 (b). Kutaraja's school 3D modelling

https://doi.org/10.1051/e3sconf/202234002003

Fig 4 (a). Meuraxa's school building

Fig 4 (b). Meuraxa's school 3D modelling

2.3 Response spectrum

The earthquake load indicated in the SNI earthquake is in the form of a spectrum response following the 2017 Earthquake Source and Hazard Map [7]. The rules for determining the spectrum response refer to the SNI 1726-2019 regulations. Based on the calculation of SNI 1726-2019, the spectrum response in the three review districts is as follows:

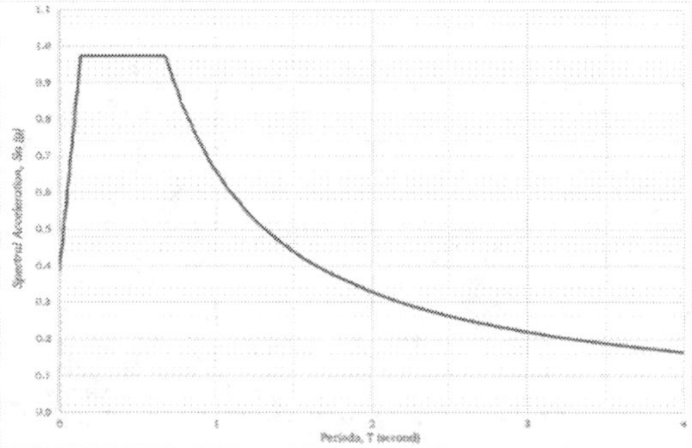

Fig 5. Spectrum response of Syiah Kuala's school

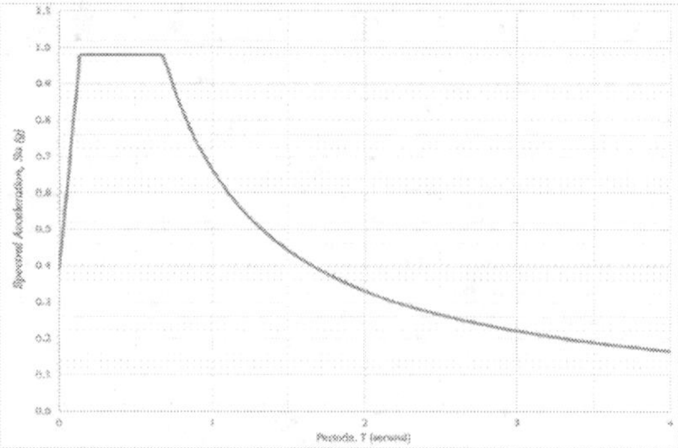

Fig 6. Spectrum response of Kutaraja's school

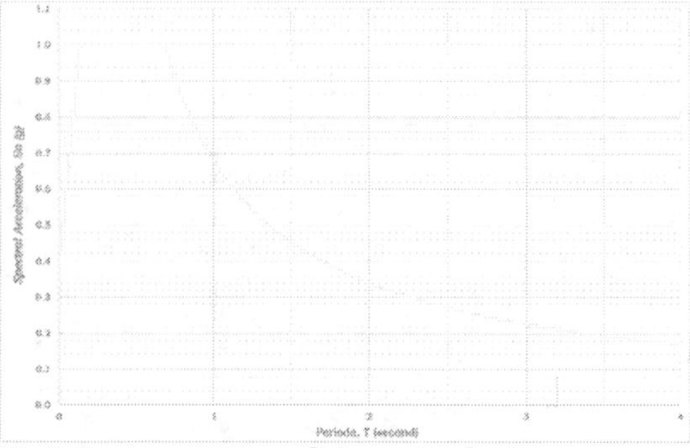

Fig 7. Spectrum response of Meuraxa's school

2.4 Ground motion records

Earthquake data taken are earthquakes that have occurred in Aceh and caused severe damage to structures. This study used 12 ground motion records which can be seen in Table 1.

Table 1. Ground motion records used

Ground Motion Record	Mw
Andaman-East 2004	9,1
Andaman-North 2004	9,1
Simeulu 1-East 2012	8,5
Simeulu 1-North 2012	8,5
Simeulu 2-East 2012	8,1
Simeulu 2-North 2012	8,1
Bener Meriah-East 2013	6,1
Bener Meriah-North 2013	6,1
Pidie Jaya-East 2016	6,5
Pidie Jaya-North 2016	6,5
Banda Aceh-East 2020	5,5
Banda Aceh-North 2020	5,5

Source : BMKG database [8]

The ground motion data in a waveform in ASCII format was converted into an accelerogram using the DADiSP/SE 6.7 application. Then the accelerogram is converted into a response spectrum using seismosignal software. Furthermore, the response spectrum of various earthquakes is matched to the response spectrum of the target using seismomatch software by adjusting the scale factor to obtain a spectrum that matches the design response spectrum. The scale factor calculation uses the spectral acceleration value of the ground motion records response with the spectral acceleration value of the target response spectrum.

A frequency content modification method called spectral matching is used to adjust ground motion records to the target response spectrum. To adjust the ground motion record with the target response spectra using a scale factor. The calculation of the scale factor (SF) is based on the following Kalkan and Chopra formula [9]:

$$SF = \frac{\sum_{i=1}^{n} \bar{A}\ Ai}{\sum_{i=1}^{n} Ai\ Ai} \quad (1)$$

with;

SF = scale factor, Ā = Spectral acceleration target, Ai = Spectral acceleration scaled

The accelerogram that has been scaled is inputted into SAP2000 as a load with a scale factor = g I/R, where g = acceleration due to gravity (9.81 m/s^2), I is the main earthquake factor, which is 1.5 as a building risk categorized as number IV (important building for disaster mitigation), and R is a reduction coefficient factor of 8.

2.5 Tsunami loading

Tsunami loading is determined based on regulations from SNI 1727-2020. The tsunami forces reviewed are hydrostatic, hydrodynamic, and floating debris collision, which can be seen in table 2. These forces are inputted to the building according to the direction of the tsunami and the height of the inundation that occurs.

Tabel 2. The formula of the tsunami force based on SNI 1727-2020

Type of Forces	Tsunami Consideration	Formula
Hydrostatic	Hydrostatic forces occur when slowly moving water hits a member of the structure.	$F_h = \frac{1}{2}\gamma_s b h^2{}_{max}$ (2)
Hydrodynamic	Hydrodynamic forces are used when water flows around the building structure	$F_d = \frac{1}{2}\rho_s I_{TSU} C_d b(h_e u^2)$ (3)
Floating debris impact	Impact forces originating from debris drift can cause damage to buildings.	$F_i = 1{,}47\ C_0\ I_{tsu}$ [KN] (4)

with;

γ_s = weight density of hydrostatic fluid (kg/m^3)
b = width of the column under force (m)
h_{max} = maximum inundation depth (m)
ρ_s = mass density of hydrodynamic fluid (kg/m^3)
I_{tsu} = priority factor for tsunami force
C_d = drag coefficient based on quasi-steady force
h_e = inundation height of individual elements (m)
u = tsunami flow velocity (m/det)
C_o = coefficient of orientation (of debris)

The calculation of the tsunami load is carried out by simulating buildings with a tsunami flow height that has occurred, such as the 2004 tsunami. The data was obtained based on previous research, as Iemura [10] and Tursina [11]. Iemura's research [10] shows the distribution of the Tsunami Pole points, which shows the height of the tsunami inundation. Meanwhile, Tursina's research [11] is based on the 2004 tsunami flow depth simulation in Banda Aceh using NOAA survey data and the Tsunami Pole. The value of the inundation depth that occurs at the object point can be seen based on Tursina [11], but the data is in the form of a range of values so that the value to be used is difficult to determine. For this reason, a comparison is made with Iemura [10] research, which shows the value of inundation depth based on the nearest tsunami pole. The value of the tsunami inundation that occurred at the coordinates of the building under study can be seen in table 3.

Tabel 3. Data of tsunami inundation 2004

School Building	Tsunami Inundation based on:	
	The Nearest Tsunami Poles - Iemura [10]	Tsunami flow depth simulation – Tursina [11]
Syiah Kuala	3.8 m	2 - 4 m
Kutaraja	7 m	4 - 6 m
Meuraxa	7 m	4 - 6 m

The height of the tsunami inundation is taken in stages starting from 1 m, 2 m, 3 m, and so on until the highest tsunami height occurs. The result of this height difference shows the drift ratio at each level. The tsunami loading will be carried out in combination with the tsunami loading based on SNI 1727-2020 using the equation:

$$1{,}2\ D + F_{TSU} + 0{,}5\ L + 0{,}2\ S \quad (5)$$

$$0{,}9\ D + F_{TSU} \quad (6)$$

with;

D = Dead load
L = Live load
S = Snow load
F_{TSU} = Tsunami Force

In terms of the combination of tsunami loading, it does not take into account the seismic damage at the initial state of capacity reduction. The Seismic and tsunami analysis are two different independent analysis.

2.6 Incremental dynamic analysis

According to Tiwari and Kasnale [12], incremental dynamic analysis (IDA) is an analysis to study the whole earthquake with different intensities, which is applied to the building structure model until it collapses. IDA is done by performing a non-linear time history analysis. In a non-linear time history analysis, the structure is assigned one or more accelerograms multiplied by a scale factor, which are performed at various levels of earthquake intensity that are expected to occur. The level of additional earthquake intensity is increased until the structure experiences an elastic phase to a plastic phase (inelastic) until it finally reaches global dynamic instability or the structure collapses. The IDA method will be used in a combination of earthquake loading based on SNI 1726-2019 using the equation:

$$1{,}2D + Ev + Emh + L \quad (7)$$

$$0{,}9D - Ev + Emh \quad (8)$$

with;

Ev = Effect of vertical seismic force
Emh = Effect of horizontal seismic force

After the structural modeling is completed, the next step is to input the earthquake history data used. Then run a dynamic analysis of time history gradually with an increasing earthquake intensity scale. The analysis was carried out using the addition of a scale factor of 0.5; 1 and 2. These additions are used to determine the limits of repeated load increases. The results of the analysis are displacements in the x and y directions. This displacement value is used to determine the value of the drift ratio, which is useful in making the IDA curve.

2.7 Seismic fragility curve

Based on the structural analysis results using the IDA method, IDA curves with various variations are obtained according to the number of earthquake records used as earthquake loads. The IDA curve can be analyzed further probabilistically by using the fragility function and expressed in a fragility curve. In this study, the value of limit states follows the research of Kim and Shinozuka, which is summarized in Pitiliakis [13].

Tabel 4. Limit states IDA method

Damage states	Drift Ratio
Slight	0.7%
Moderate	1.5%
Extensive	2.5%
Complete	5.0%

The fragility curve shows the probability of damage to a structure when it receives an earthquake load with a certain intensity at its service limit. Based on the analytical method, the function is derived by Sadradin [14] through the log-normal distribution approach as follows:

$$P = \Phi\left(\frac{\ln(X) - \lambda}{\beta_{RU}}\right) \quad (9)$$

with;

P = Probability of structural damage,
Φ = Standard normal function of cumulative distribution,
β_{RU} = The total uncertainty of the structure,
x = Ground motion parameter, PGA (g),
λ = mean of ln (x).

2.8 Tsunami fragility curve

To construct a tsunami fragility curve, data on tsunami inundations that have occurred is needed. The data was obtained based on previous research, as Iemura [10] and Tursina [11]. The tsunami fragility curve depicts the probability of a specified level of damage in a tsunami event with a certain tsunami inundation intensity. In this study, the fragility curve was calculated using the cumulative distribution of the damage index and the log-normal function. The level of damage reviewed is seen based on the value of the drift ratio that occurs; this provision is determined by what is summarized by Medina [15], whose limit states can be seen in Table 5.

Table 5. Damage states limits of tsunami fragility function

Damage States	Drift Damage Index HAZUS – 1–3 stories
Slight	< 0.20%
Moderate	0.20% – 0.50%
Extensive	0.50% – 1.20%
Complete	1.20% – 2.80%

The results of non-linear numerical analysis of the tsunami inundation depth and the drift ratio obtained are used to obtain a tsunami fragility curve expressed as a two-parameter log-normal distribution function. The tsunami fragility curve equation used is based on Karafagka's research [6] which provides an assessment of the cumulative probability of exceeding the level of damage conditioned on the measure of tsunami intensity.

$$P_{(DSi\,|\,IM)} = \Phi\left(\frac{\ln(IM) - \ln(\overline{IM_i})}{\beta}\right) \quad (10)$$

Where Φ is a standard normal cumulative distribution function, IM is a measure of tsunami intensity expressed in inundation depth (in units of m), IMi is the median value (in units of m) for each level of damage. It is the uncertainty calculated via the log standard deviation parameter in each damage level. The uncertainty value can be calculated using the following equation:

$$\beta = \sqrt{\beta_C^2 + \beta_D^2} \quad (11)$$

Where β_C is the structural capacity uncertainty taken as 0.15 (based on the HAZUS MH Technical Manual) [16] and D is the structural review uncertainty taken into account by calculating the logarithmic dispersion at each level of damage. The results of the calculation of the standard deviation of the total uncertainty in each damage condition.

3 Result and discussion

3.1 Results of IDA curve analysis

In making the IDA curve, an Intensity measure in Peak Ground Acceleration (PGA) is needed for the x-axis. Meanwhile, the damage measure used is the drift ratio for the y-axis. The analysis was stopped when the drift

https://doi.org/10.1051/e3sconf/202234002003

ratio was more than 7-8% because it had crossed the largest limit state value. The results of the IDA curve analysis for earthquakes in the x and y directions can be seen in Fig 8.

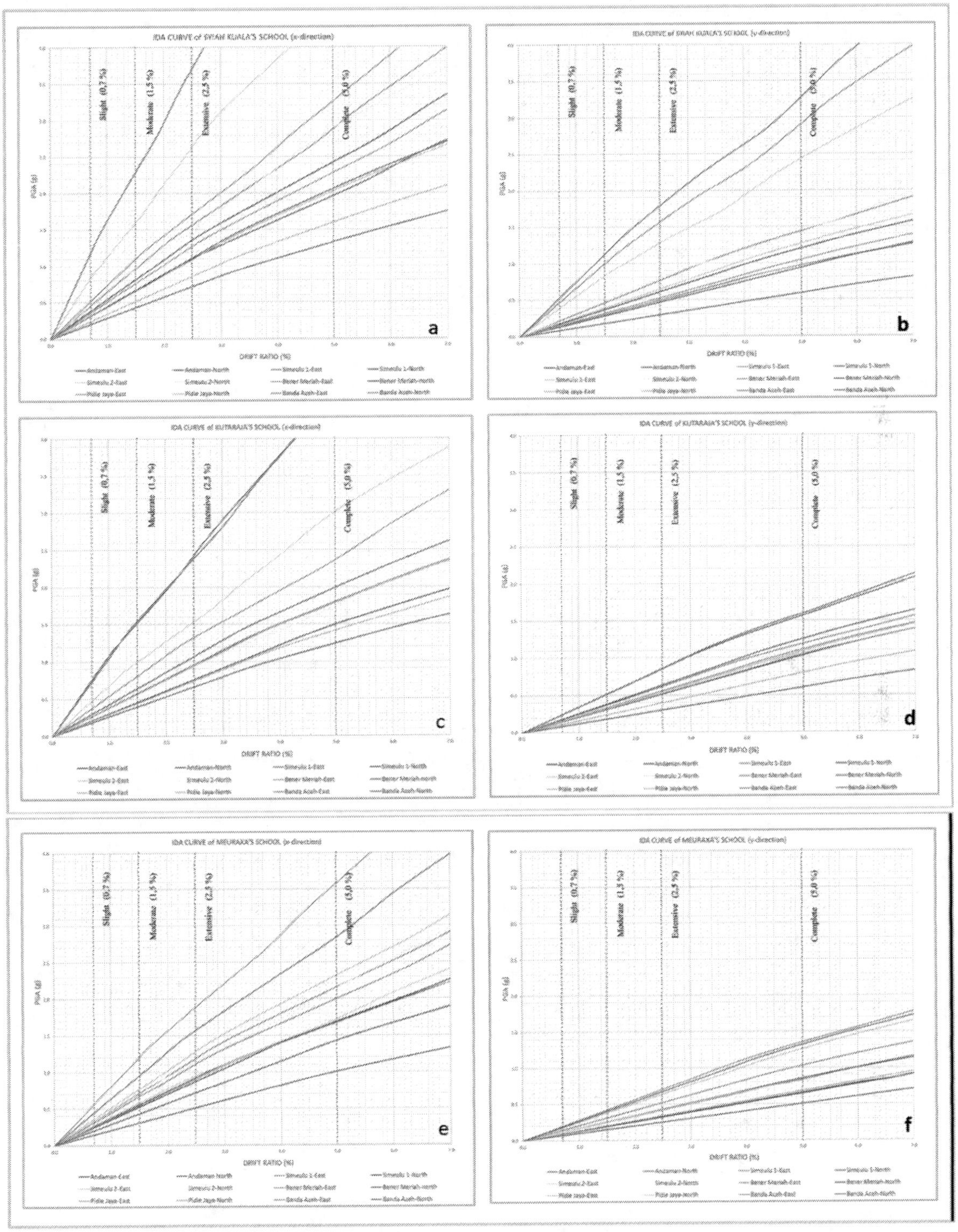

Fig 8. IDA curve of 3 school buildings with two direction

https://doi.org/10.1051/e3sconf/202234002003

3.2 Result of tsunami capacity curve analysis

The drift ratio resulting from differences in the modeled tsunami inundation is used to determine the structure's capacity limit. This value is adjusted to the inundation height to form a capacity curve. This curve is needed to form a tsunami fragility curve which requires a limit value of the drift ratio that occurs with the depth of the modeled tsunami inundation. The value of inundation depth at each damage limit for each height can be seen in Fig 9.

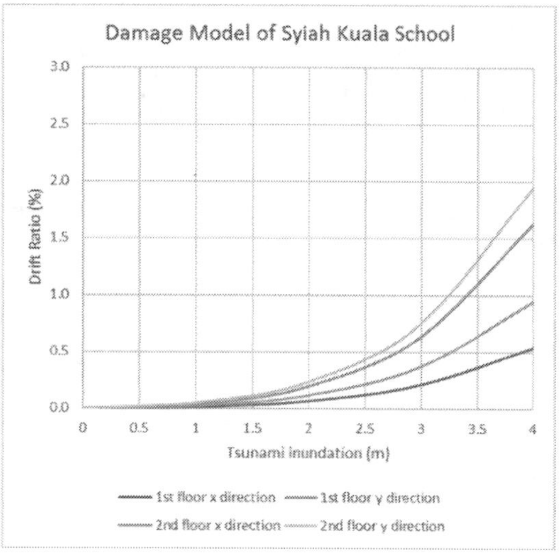

Fig 9 (a). Tsunami damage state of Syiah Kuala's school

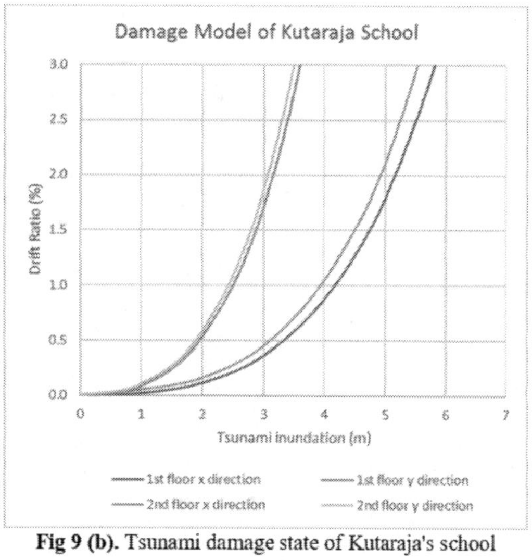

Fig 9 (b). Tsunami damage state of Kutaraja's school

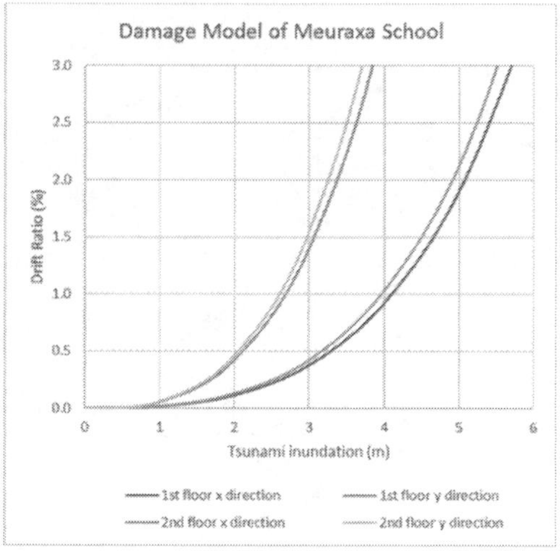

Fig 9 (c). Tsunami damage state of Meuraxa's school

3.3 Results of seismic fragility curve analysis

The variables needed to form the seismic fragility curve are the ground motion parameter (PGA), the average of the ground motion parameters according to the limit state, and the total uncertainty. Based on these results, the complete fragility probability value is obtained at various limit states. The fragility probability value is plotted on the y-axis (ordinate) and the ground motion parameter on the x-axis (abscissa). With this curve, the evaluation of the probability of the building structure being damaged due to variations in the intensity of the earthquake load during its service life can be carried out quantitatively and rationally. The results of the fragility curve in the x and y directions can be seen in Fig 10.

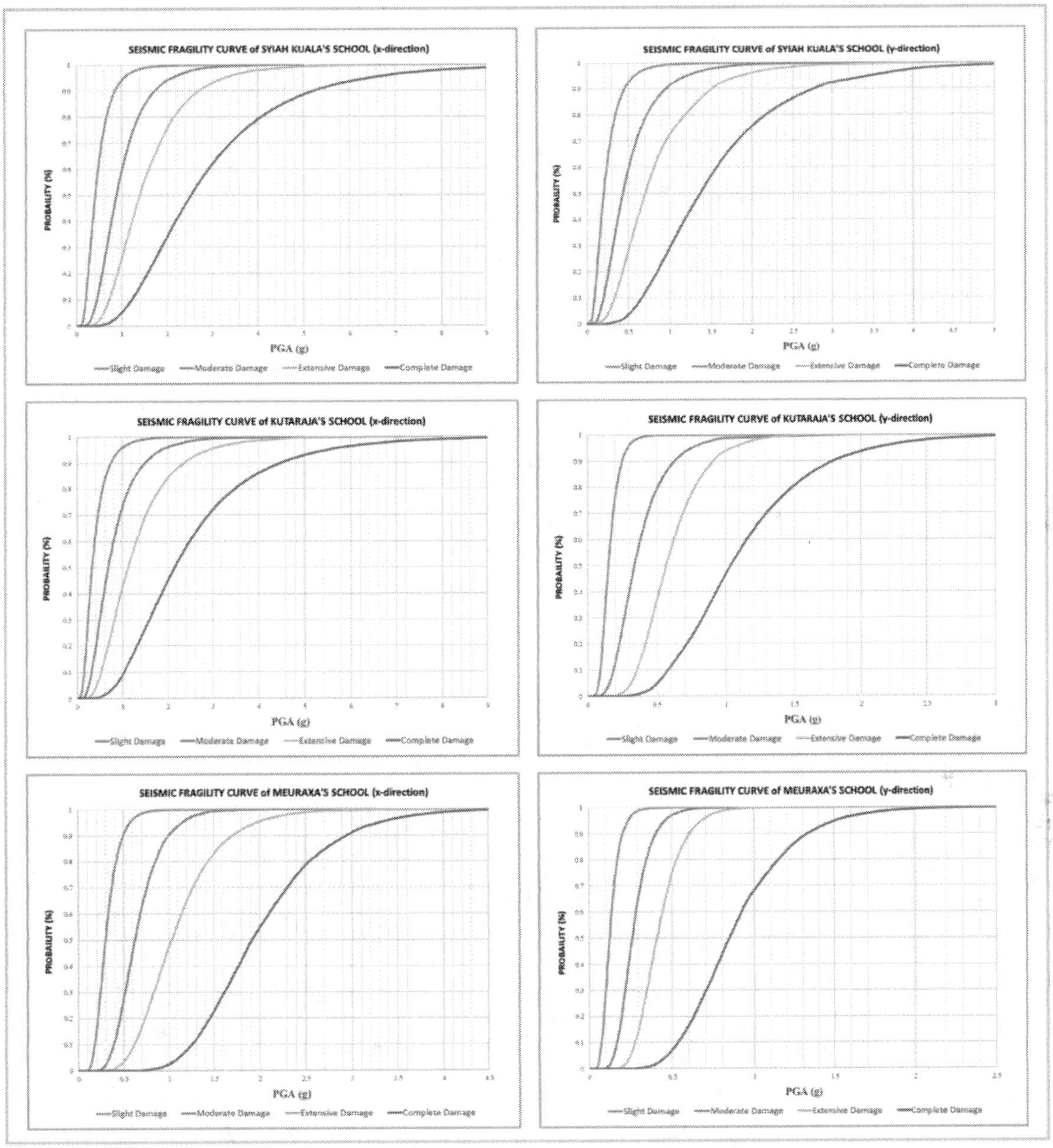

Fig 10. Seismic fragility curve

With the fragility curve, the evaluation of post-earthquake structural failure is easier to predict. The trick is to assume the magnitude of the earthquake that will occur in the future. For example, it is assumed that in the future, there will be an earthquake with a PGA value of 0.75 g, then the possibility of damage to the school building structure can be seen in Table 6 as follows:

Table 6. Probability of damage due to an earthquake with PGA prediction 0.75 g

Building School with PGA value 0.75 g		***Probability of Damage (%)***			
		Slight	***Moderate***	***Extensive***	***Complete***
Syiah Kuala	x	86	38	11	2
	y	98	83	55	14
Kutaraja	x	91	55	24	4
	y	100	95	79	24
Meuraxa	x	99	70	21	1
	y	100	100	97	35

3.4 Results of tsunami fragility curve analysis

The results of displacement and drift ratio show that for each modeled tsunami height, it is seen that the building structure deforms and responds differently. But in general, it has the same tendency; the displacement in the y-direction is bigger than in the x-direction. However, it depends on the condition of the direction of the tsunami and the height of the inundation that occurs. On the tsunami fragility curve, the probability value is plotted on the y-axis (ordinate) and the inundation depth on the x-axis (abscissa). The results of the tsunami fragility curve formed at various damage limits can be seen in Fig 11.

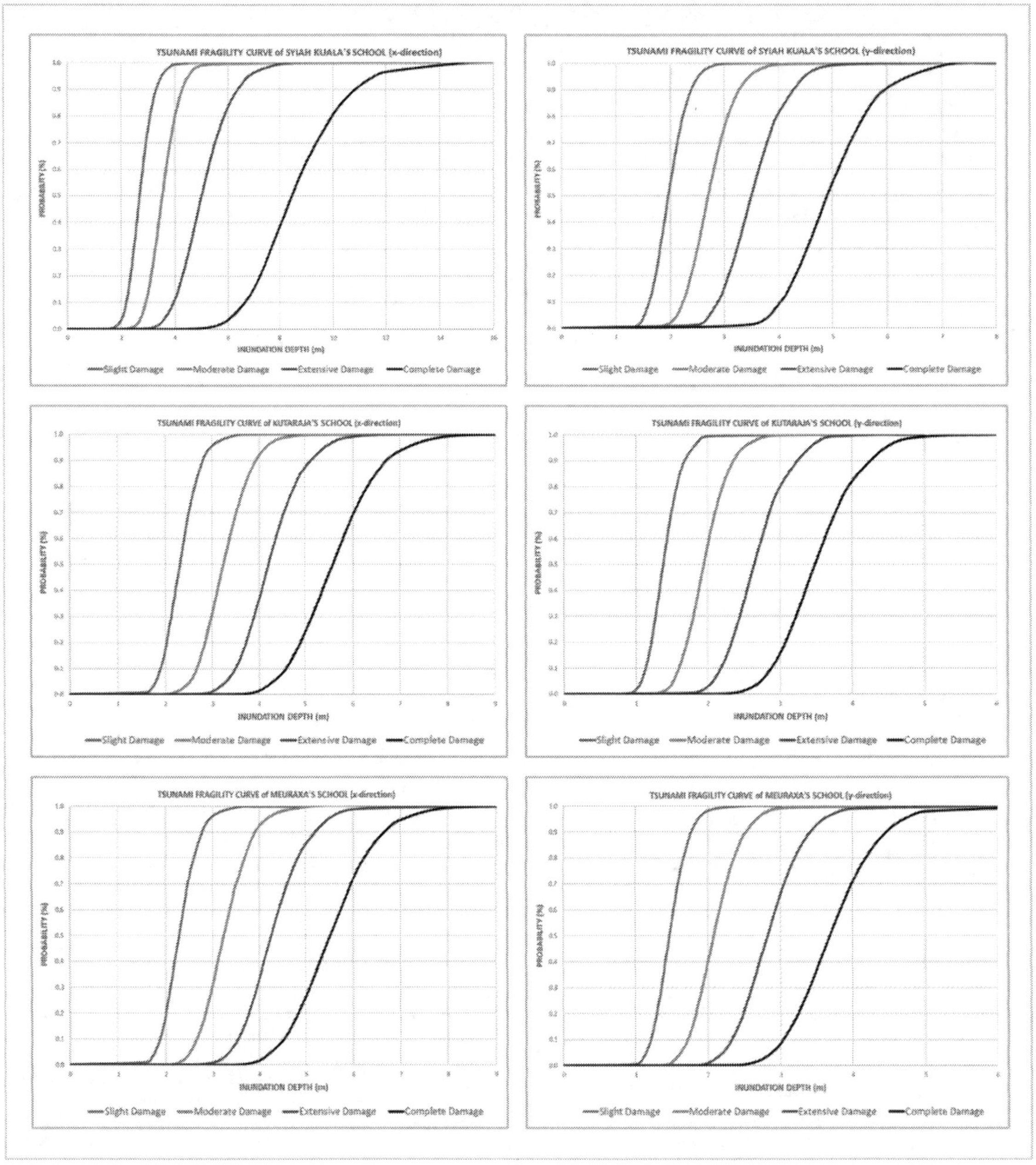

Fig 11. Tsunami fragility curve

The tsunami fragility curve can also predict the damage that occurs by assuming the height of the tsunami inundation that will occur in the future. For example, it is assumed that in the future, there will be a tsunami with an inundation depth of 3 m, then the possibility of damage to the school building structure can be seen in Table 7 as follows:

Table 7. Probability of damage due to a tsunami with a predicted inundation depth of 3 m

Building School with inundation depth 3 m		*Probability of Damage (%)*			
		Slight	*Moderate*	*Extensive*	*Complete*
Syiah Kuala	x	72	18	1	0
	y	100	78	16	2
Kutaraja	x	95	31	2	0
	y	100	100	80	15
Meuraxa	x	96	32	2	0
	y	100	100	68	9

3.5 The result of liquefaction vulnerability

In addition to vulnerability due to the earthquake and tsunami, it is also necessary to review the vulnerability to potential liquefaction. The potential for liquefaction in Banda Aceh City can be seen in Rusydy's research [17] which describes a map of the potential for liquefaction based on a groundwater level map and a map of the vulnerability of the soil to liquefaction. Liquefaction is assumed to cause additional damage to buildings in the light, medium, high, and very high vulnerability zones. The distribution of school buildings on the liquefaction vulnerability map can be seen in Fig 12.

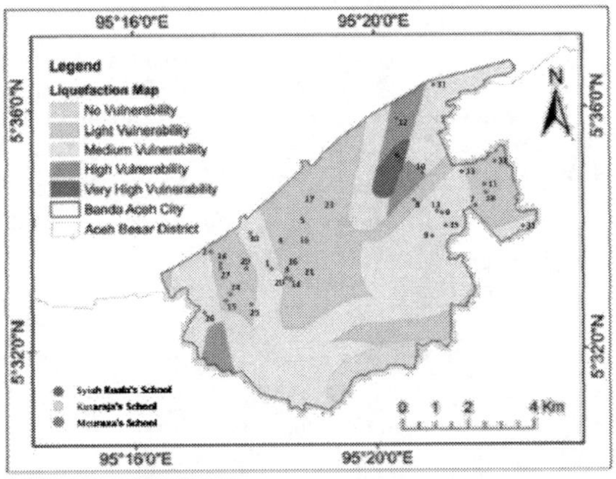

Fig 12. The distribution of school buildings on the liquefaction vulnerability map [17].

Based on the map above, it can be seen that generally, school buildings in the coastal area of Banda Aceh City have a light vulnerability, which means they are still in the safe stage. However, this potential vulnerability cannot be ignored because several schools are included in the medium vulnerability area, so further action is needed for disaster mitigation efforts.

4 Conclusion

The seismic fragility curve analysis results show that the building structure is more susceptible to the y-direction than the x-direction. The vulnerability values studied each has a 100% probability of damage, as shown in Table 8. For vulnerability due to tsunami, the building structure has a 100% probability of damage, as shown in Table 9.

Table 8. The result of seismic fragility curve

Building School		*Peak Ground Acceleration* (g)			
		Slight	*Moderate*	*Extensive*	*Complete*
Syiah Kuala	x	2.1	3.4	5.5	11.0
	y	1.3	2.2	3.4	6.0
Kutaraja	x	2.0	3.4	5.7	9.5
	y	0.5	1.3	1.7	3.1
Meuraxa	x	1.0	1.6	3.0	4.5
	y	0.4	0.7	1.0	2.2

Table 9. The result of tsunami fragility curve

Building School		*Inundation Depth (m)*			
		Slight	*Moderate*	*Extensive*	*Complete*
Syiah Kuala	x	4.5	6.0	8.3	15.0
	y	3.0	4.2	5.6	7.3
Kutaraja	x	3.6	5.2	6.5	8.5
	y	2.2	3.1	3.9	5.4
Meuraxa	x	3.5	5.0	6.4	8.2
	y	2.2	3.2	4.0	6.2

In addition to earthquake and tsunami hazards, liquefaction vulnerability also needs to be taken care of on disaster mitigation. Based on the liquefaction potential map, school buildings in coastal sub-districts have vulnerabilities, which can be seen in Table 10.

Table 10. Potential of liquefaction

Building School	Potential Liquefaction
Syiah Kuala	Light - Medium vulnerability
Kutaraja	Light vulnerability
Meuraxa	Light - Medium vulnerability

The results of the probabilistic analysis of damage based on the fragility curve can be used to predict the level of damage to buildings in the coastal district of Banda Aceh. For probabilistic earthquake susceptibility, the PGA earthquake magnitude is the main parameter that must be considered. In contrast, tsunami susceptibility can be seen based on the height of inundation that occurs in buildings. As well as earthquake and tsunami vulnerability, the liquefaction potential must also be considered because liquefaction also affects ground movement activity. The summary results of this study indicate that it is necessary to carry out periodic maintenance of school buildings in the coastal area of Banda Aceh city so that they become strong buildings and are ready to be used if a large-scale disaster occurs.

Acknowledgment. This research was conducted and performed under AAUI-MAIPARK Research Grant 2020 and funded by PT Reasuransi Maipark Indonesia. The authors are highly thankful to Dwi Agam Riantara and Farah Fadillah for their help in the data collection.

References

1. Bappenas and International Community, Indonesia: Preliminary Damage and Loss Assessment; The December 26, 2004, Natural Disaster, (Bappenas, Indonesia, 2006)
2. B. S. Nasional, Standar Nasional Indonesia 1726-2019: Tata Cara Perencanaan Ketahanan Gempa untuk Struktur Bangunan Gedung dan Non Gedung (BSN, Jakarta, 2019)
3. B. S. Nasional, Standar Nasional Indonesia 1727-2020: Beban Desain Minimum dan Kriteria Terkait untuk Perancangan Bangunan Gedung dan Struktur Lain (BSN, Jakarta, 2020)
4. R. Gentile, et al., From Rapid Visual Survey to Multi-Hazard Risk Prioritisation and Numerical Fragility of School Buildings, NHESS, **19**: 1365–1386 (2019)
5. D. Vamvatsikos, and C. A. Cornell, Incremental Dynamic Analysis, EESD, **27(3)**: 97 – 125 (2002)
6. S. Karafagka, S. Fotopoulou, and K. Pitilakis, Analytical Tsunami Fragility Curves for Seaport RC Buildings and Steel Light-Frame Warehouses, SDEE, **112**: 118-137 (2018)
7. Pusat Studi Gempa Nasional, Peta Sumber dan Bahaya Gempa Indonesia Tahun 2017 (Kementerian PUPR, Bandung, 2017)
8. Badan Meteorologi Klimatologi dan Geofisika (BMKG) Data Base, Accesed on 25 April 2021.
9. E. Kalkan, dan A. K. Chopra, Practical Guidelines to Select and Scale Earthquake Records for Nonlinear Response History Analysis of Structures (Earthquake Engineering Research Institute, London, 2010)
10. H. Iemura, et al., Information Dissemination for Reality-Based Tsunami Disaster Education, Presented on *The 14th World Conference on Earthquake Engineering* (Beijing, China, 12-17 October 2008)
11. Tursina, and Syamsidik, Reconstruction of the 2004 Tsunami Inundation Map in Banda Aceh Through Numerical Model and Its Validation with Post-Tsunami Survey Data, IOP Conf. Series: Earth and Environmental Science, **273** (2019)
12. V. Tiwari, and A. Kasnale, Incremental Dynamic Analysis of RC Frames, IRJET, **4**(6): 697-700 (2017)
13. K. Pitilakis, et al., SYNER-G : Typology Definition and Fragility Functions for Physical Elements at Seismic Risk, (Springer, Netherlands, 2005)
14. H. Sadraddin, Fragility Assessment of High-Rise Reinforced Concrete Buildings, (Civil and Construction Engineering Western Michigan University, Kalamazoo, 2015)
15. S. Medina, et al., Tsunami Analytical Fragility Curves for The Colombian Pacific Coast: A Reinforced Concrete Building Example, ES, **196** (2019) 109309.
16. FEMA, Direct Physical Damage-General Building Stock, (HAZUS-MH Technical manual, Chapter 5, Washington D.C., 2004)
17. I, Rusydy, et al., Shallow Crustal Earthquake Models, Damage, and Loss Predictions in Banda Aceh, Indonesia, GD, **7**: 8 (2020)

The damage effect on the structural capacity of school building columns due to tsunami inundation in Banda Aceh city after 16 years

Bunga Raihanda[1*] and *Abdullah*[2]

[1] Student of Disaster Science Department, Postgraduate Program, Universitas Syiah Kuala, Jl. Hamzah Fanshuri, Darussalam, Banda Aceh, 23111, Indonesia
[2] Department of Civil Engineering, Faculty of Engineering, Universitas Syiah Kuala, Jl. Syeh Abdurrauf No. 7, Banda Aceh, 23111, Indonesia

Abstract. The 2004 Indian Ocean earthquake and tsunami caused catastrophic damage to buildings and infrastructure in Banda Aceh. The impact and damage depended on several factors such as the distance of the area from the coast. It was reported that in Banda Aceh, the tsunami wave inundation reached the city center and affected over a third of the population. Due to priority reasons at the time, although the structure was submerged in tsunami inundation, numerous public facilities including school buildings still in use without any proper repair. This study aims to determine the effect of column damage due to the tsunami inundation on the strength of the building. The school building of SMPN 6 Banda Aceh is selected as the case study. Seven analysis cases were carried out to simulate the degradation of the column cross-section caused by tsunami inundation. The reduction of the column cross-section was determined from 5 to 25% with increments of 5%. The analysis results revealed that the decrease in strength has occurred from the case of 5% degradation and the vulnerability started to increase with the increase of degradation level. In the case of 25% degradation, 90% of the structure's column was at an unsafe level.

1 Introduction

The 2004 Indian Ocean earthquake and tsunami caused catastrophic damage to buildings and infrastructure in Banda Aceh, the capital city of the Aceh province at the northern tip of the island of Sumatera. It was reported that in Banda Aceh, the tsunami wave inundation reached the city center and affected over a third of the population [1]. It includes the deaths of over 45,000 students and 1,870 teachers and damages of 2,065 educational facilities [2].

The impact and damage depended on several factors including the distance of the area from the coast. Numerous school buildings located relatively far from the coast were not completely damaged. Due to priority reasons at the time, such buildings are still in use without any proper repair. However, the building's structure, in particular the column on the first floor, was submerged in tsunami inundation. This condition could decrease the concrete structural strength of the column, which might lead to disrupting the overall strength of the building. This condition affects one of the disaster risks factors [3], namely vulnerability (V).

Thus, to have a better understanding of the disaster risk, as one of the priority actions in the Sendai Framework for Disaster Risk Reduction (SFDRR) [4], further analysis on buildings submerged by tsunami inundation is required. Then, the risk (R) can be reduced by lowering the value of vulnerability (V) or increasing capacity (C).

This study aims to determine the damage level of the building structure submerged by the 2004 Indian Ocean earthquake and tsunami. The school building of SMPN 6 Banda Aceh is selected as the case study since the building was not collapsed after the earthquake and experienced the tsunami inundation.

In this study, finite element analysis was conducted to determine the effect of column damage due to the tsunami inundation on the strength and the capacity of the overall structure. The analysis results, the capacity and the strength of the column, will determine the risk level of the school building against earthquake. At the end, recommendations for repair, retrofitting, and others mitigation efforts will be proposed to reduce the risk and preventing the collapse of the building.

2 Methodology

2.1 Research object

This research is intended for a building located in the earthquake and tsunami-affected area in Banda Aceh City. The building must experience tsunami inundation without significant damage. Thus, structural rehabilitation was not carried out. Based on these

*Corresponding author: bungaraihanda@gmail.com

conditions, the school building of SMPN 6 Banda Aceh was selected as the research object.

2.2 Building assessment

Field surveys were carried out at SMPN 6 Banda Aceh to collect building structure data. During the survey, direct observation was also done to assess the current condition of the building. The data obtained from the survey are as follows,

1. The location coordinate of the building is : 05°34'03" North Latitude - 95°20'37" East Longitude
2. Year of construction : 2001
3. Construction type : Reinforced concrete (RC)
4. Number of floors : 2 stories
5. Stories height : 4 meters
6. Structural dimensions :
   Columns : rectangular column with 300×300 mm and 300×400 mm
   Beams : rectangular beams with 250×400 mm and 300×500 mm
7. Concrete property : (fc') 20,75 MPa
8. Steel strength :
   Longitudinal : fy = 400 MPa, fu = 550 MPa
   Transversal : fy = 240 MPa, fu = 370 MPa

2.3 Visual observation of column damage

Direct observations at the SMPN 6 Banda Aceh building revealed that the concrete portion of the column which was submerged by the tsunami has cracked. Moreover, some of the concrete covers have peeled off and the exposed reinforcement was corroded. These conditions are shown in Fig. 1.

2.4 Finite element model of the structure

The building's structure was modeled by using the structural analysis software SAP2000. The structure was assumed to have a similar design as a reinforced concrete with open structures, in other words, frames with very minimal wall infill. Then, the structure data was input into a model in SAP2000 analysis. The building image is shown in Fig. 2 and the finite element (FE) model of the structure is shown in Fig. 3. In this building, two types of columns are used. The location of the column is illustrated in Fig. 4.

2.5 Load condition

In this study, the applied loading and the combination of factored loads is carried out according to Indonesian Building Code (SNI) as a reference. The load was applied depending on the analysis cases. For the case of the building at the time it was built, the general loading standard from SNI 1727-1989 [5] and the special earthquake loading standard from SNI 1726-2002 [6] were applied. For the case of the existing conditions, the loading was referred to the latest SNI, where general loading standard from SNI 1727-2020 [7] and the earthquake-specific loading from SNI 1726-2019 [8] were applied.

The earthquake loading used in the spectrum response analysis was in the form of spectral plots. The earthquakes loading based on the building code of SNI 1726-2002 and SNI 1726-2019 are shown in Fig. 5 and Fig. 6, respectively.

2.6 Tsunami inundation height

According to the research conducted by Iemura *et al* (2008) [9] in building the Tsunami Height Memorial Poles in Banda Aceh and Aceh Besar, the tsunami inundation at SMPN 6 Banda Aceh is 1.80 m high with a distance to the coast of about 3.85 Km. The tsunami memorial pole at SMPN 6 Banda Aceh is shown in Fig. 7.

(a) Column cracks (b) Concrete cover loss (c) Corroded reinforcement

Fig 1. Column conditions

Fig 2. School building

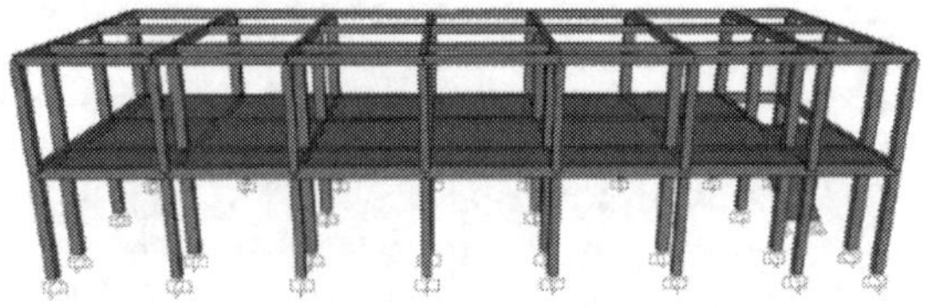

Fig 3. FE 3D model of the structure

 https://doi.org/10.1051/e3sconf/202234002004

Fig 4. Column plan

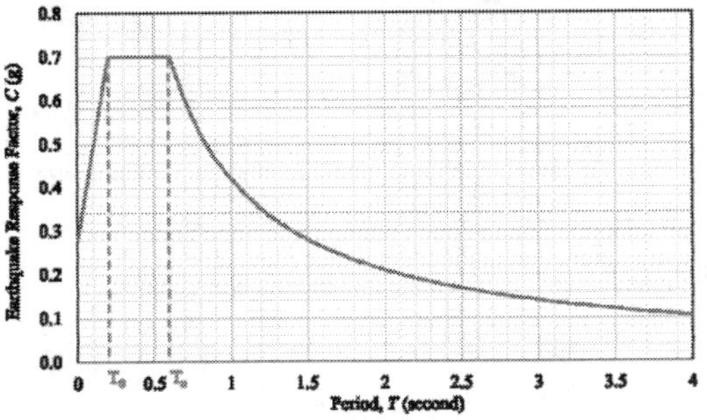

Fig 5. Spectrum of SNI 1726-2002

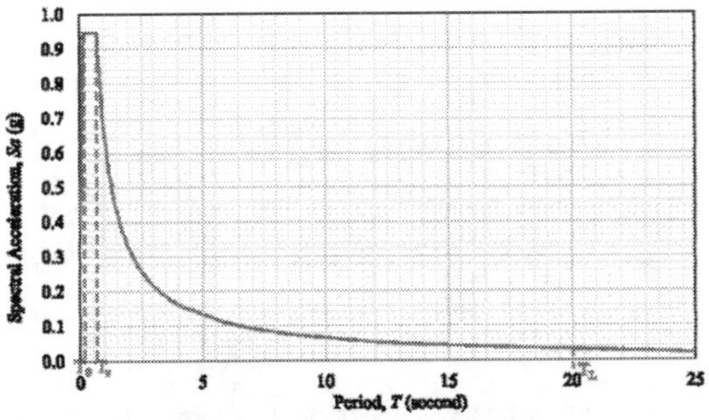

Fig 6. Spectrum of SNI 1726-2019

Fig 7. Tsunami memorial pole in SMPN 6 Banda Aceh

2.7 Analysis cases

In actual condition, the column was built following the cross-section of the design size. Then, since the column was submerged by tsunami inundation and was not repaired, the degradation of column section, cracking, and cover loss is unavoidable. The column cross-section sizes gradually decreased and lead the reinforcement to corrode.

To simulate the effect of degradation of the column cross-section caused by tsunami inundation, seven scenario of analysis cases were carried out as illustrated in Fig. 8. The degradation of the column size was carried out sequentially to evaluate the capacity of the building structure. The reduction of the column cross-section was determined from 5 to 25% with increments of 5%. The degradation percentage of column cross-section were applied on each side of the width and depth.

Case 1 was the condition of the column cross-section following the design size and the applied earthquake loading from SNI 1726-2002 was used. Case 2 was the condition of the column cross-section following the design size but the special earthquake loading from SNI 1726-2019 was used. Case 3 to 6 was analyzed by reducing the cross-sectional size of the column due to concrete cover deterioration caused by tsunami inundation. In case 7 the analysis was carried out with

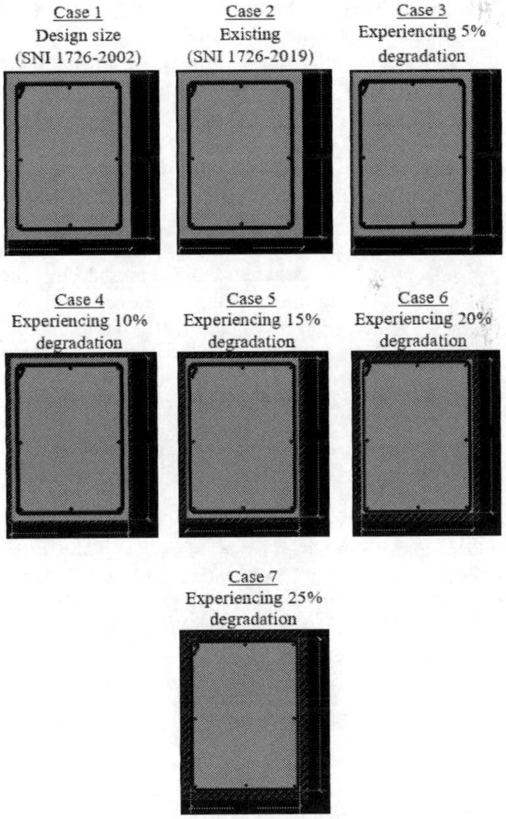

Fig 8. Column degradation included in the analysis

complete loss of the concrete cover, and the degradation of the size of the reinforcement due to corrosion.

Case 7 was carried out to simulate the above conditions. In this case, the column was experienced complete concrete cover loss and the reinforcement started to corrode. Thus, it was required to reduce the cross-sectional areas of the reinforcement in the analysis. The reduction in the reinforcement cross-section due to corrosion of 5.64 μm/yr was used in the analysis referred to the research conducted by Al-Hadi (2016) [10].

3 Result and discussion

3.1 Initial structural capacity

Finite element (FE) analysis was carried out by using SAP2000 software to obtain the structural capacity of the building. Initially, the analysis of Case 1 and Case 2 was carried out to ensure if the structural capacity of the building meets the former and latest Indonesia Building Code (SNI). The visualization of the analysis results of Case 1 and Case 2 are shown in Fig. 9 and Fig. 10, respectively. An indicator is provided in each figure. The grey color indicates the safest condition and the red color indicates the most severe conditions.

In Case 1, the analysis result shows that most of the columns are in a light blue color, indicating that the structure is safe. In Case 2, where the applied earthquake load was updated to the latest SNI, the analysis result shows more columns had turned green, indicating that the capacity of the column was decreased, in other words, the vulnerability level of the column was increased. However, the structure is still within the safe level. According to the results, it can be concluded that differences in structural capacity were observed by updating the earthquake loading standard.

3.2 Analysis results of degradation section

Degradation of concrete columns due to tsunami inundation was simulated by reducing the column cross-section from 5 to 25% with increments of 5%, which represented as Case 3 to Case 7. The visualization of the analysis results for each case is shown in Fig. 11 to Fig. 15, respectively. An indicator is provided in each figure. The grey color indicates the safest condition and the red color indicates the most severe conditions.

Comparing the analysis results of Case 1 to Case 7, it can be observed that the structural capacity is gradually decreased along with the degradation of the column cross-section. In other words, as the capacity of the structure is decreased, the vulnerability level of the structure is increased. The comparison of the number of vulnerability columns of each analysis case is listed in Table 1.

Start from Case 3, the red columns were observed. It indicates that the strength of the column was decreased and become incapable to bear the applied loads. According to the analysis results, when 10% of column cross-section has degraded, as in Case 4, it can be concluded that the risk of structural building collapse has increased. When the column cross-section has degraded for 15-25% as in Case 5-7, the structure is no longer safe and will fail because above 30% of main column of the structure is no longer able bearing the weight of the structure. Deteriorated column caused structural failure where the entire building will collapse because the column is main structural component that play important role bearing the weight of building.

Table 1. Comparison of column vulnerability due to degradation

Case Number	Number of Columns		Number of Vulnerabe Columns		Percentage of Vulnerable Columns (%)	
	K1	K2	K1	K2	K1	K2
Case 1	8	21	0	0	0	0
Case 2	8	21	0	0	0	0
Case 3	8	21	0	1	0	4.8
Case 4	8	21	0	4	0	19.1
Case 5	8	21	0	9	0	42.9
Case 6	8	21	3	17	37.5	80.9
Case 7	8	21	4	19	50.0	90.5

K1 = column 300x300 mm, **K2** = column 300x400 mm.

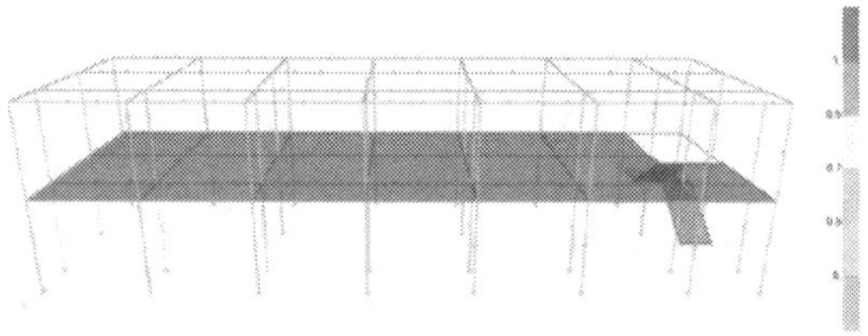

Fig 9. Case 1 - Structural capacity check visualization

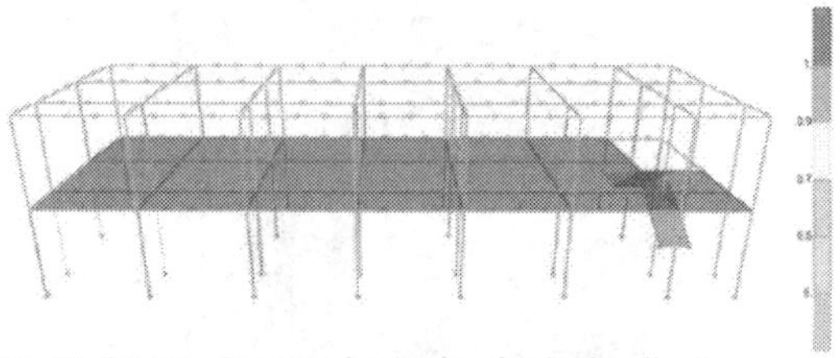

Fig 10. Case 2 - Structural capacity check visualization

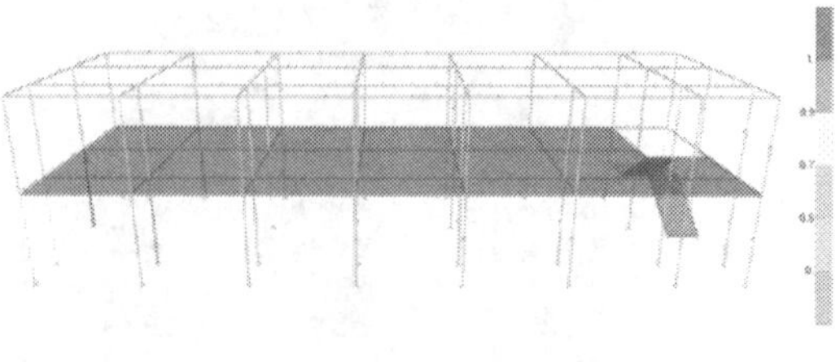

Fig 11. Case 3 - Structural capacity check visualization

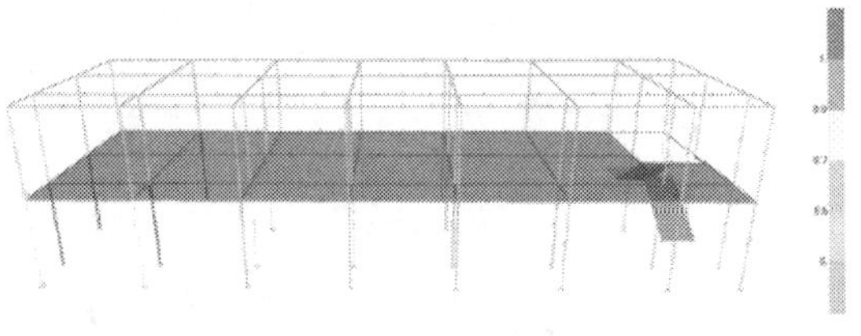

Fig 12. Case 4 - Structural capacity check visualization

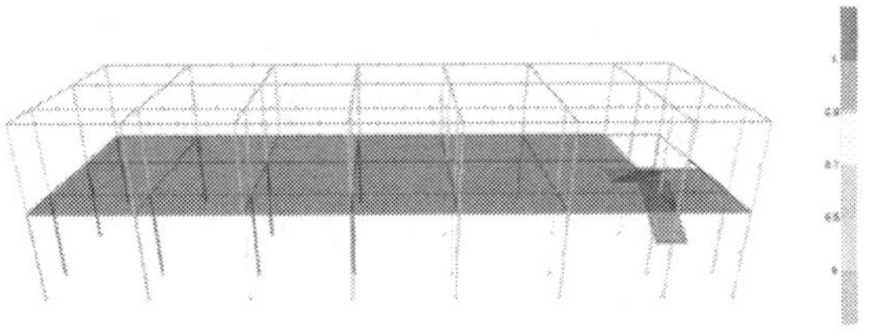

Fig 13. Case 5 - Structural capacity check visualization

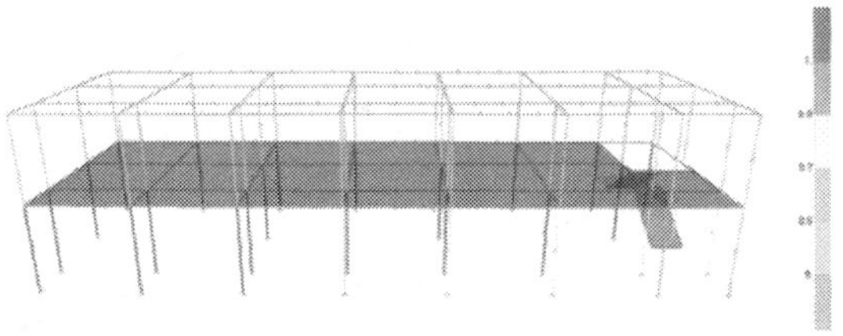

Fig 14. Case 6 - Structural capacity check visualization

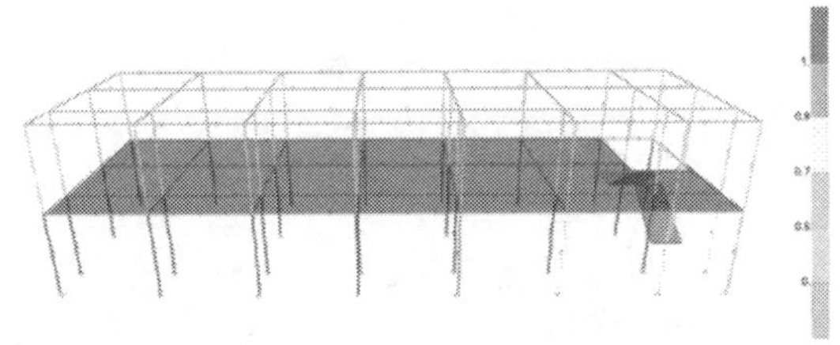

Fig 15. Case 7 - Structural capacity check visualization

3.3 Retrofitting and recommendations

During the field survey and direct observation, it was found that most of the column on the first floor, which experienced 1.8 meter high of tsunami inundation, was degraded. Associating direct observation data and Finite Element (FE) analysis results, the columns of SMPN 6 Banda Aceh structure is classified as heavy damage.

Such conditions of the school buildings are certainly hazardous to the building users, includes students and teachers who are included in the vulnerable group on the disaster risk index. In addition, recognizing that school-age children spend majority of their day hours at school, school buildings need to be protected from any vulnerabilities. Moreover, resilient schools are an effective means for disseminating disaster risk reduction awareness in the communities and can provide emergency shelters immediately after disasters.
As the structural mitigation efforts, it is recommended to evaluate the condition and the function of the school buildings, as well as other public facilities. Particularly, those submerged to the tsunami inundation and unrepaired yet, since similar damage might have occurred. The capacity of the existing structure should be enhanced by applying retrofitting methods.

One of retrofitting methods for flexural columns is by adding reinforcements. This method is effectively applied to vulnerable columns so that it can increase column capacity and reduce direct shear failure.

The other retrofitting method is column jacketing. Jacketing is a method of structural retrofitting and strengthening aims to increase or restore the structural capacity of existing structures by providing wrapping sheets of additional material around the existing member. This additional material can be of several types such as concrete, steel, FRP, and composites.

In the end as the non-structural mitigation strategies, it is expected that the local government and stakeholders could develop strategies and policies, such as regular assessments to evaluate the damaged school buildings, to reduce disaster risk by knowing the vulnerability of school building earlier.

4 Conclusion

The analysis result shows that most of the columns are in a light blue color at the design condition, indicating that the structure was safe. Then, for at the existing building condition, where the applied earthquake load was updated to the latest SNI (Indonesia building code), the analysis result shows more columns had turned green, indicating that the capacity of the column was decreased, in other words, the vulnerability level of the column has increased.

The degradation of concrete columns results for each case shows that the structural capacity is gradually decreased along with the degradation of the column cross-section. The capacity of the structure is decreased, and the vulnerability level of the structure is increased. The first degradation scenario, in Case 3, already shows that the strength of the column was decreased and become incapable to bear the applied loads. The worst condition starts in case 5, where 30% of the main column of the structure became unable to bear the weight of the structure. In other words, the structure is no longer safe and failure might happen at this point. As the main structural component that play important role in bearing the weight of the structure, deteriorated column caused the structural failure and the entire building might collapse.

According to the analysis results, the conditions of the school buildings are certainly hazardous to the building users, including students and teachers who are included in the vulnerable group on the disaster risk index. Therefore, structural and non-structural mitigation strategies should be done especially by the local government to reduce hazards and to increase capacity of the building structures to minimize the disaster risk.

Acknowledgement. This research was conducted independently without external funding. The authors are grateful to Ms. Cut Atika Putri and Mr. Rifqi Irvansyah for the assistance provided in this study.

References

1. Ghobarah, A., Saatcioglu, M. and Nistor, I. 2006. The impact of the 26 December 2004 earthquake and tsunami on structures and infrastructure. Engineering structures, 28(2), pp.312-326.
2. Bappenas and International Community. 2006. Indonesia: Preliminary Damage and Loss Assesment; The December 26, 2004 Natural Disaster. Bappenas: Indonesia.
3. Adiyoso, W. 2018. Manajemen Bencana: Pengantar dan Isu-Isu Strategis. Bumi Aksara.
4. The United Nation Office for Disaster Risk Reduction (UNDRR). 2015. Sendai Framework for Disaster Risk Reduction 2015-2030. The United Nation Office for Disaster Risk Reduction: Geneva, Switzerland.
5. Badan Standarisasi Nasional. 1989. Standar Nasional Indonesia 1727-1989: Tata Cara Perencanaan Pembebanan untuk Rumah dan Gedung. Badan Standarisasi Nasional: Jakarta.
6. Badan Standarisasi Nasional. 2002. Standar Nasional Indonesia 1726-2019: Tata Cara Perencanaan Ketahanan Gempa untuk Struktur Bangunan Gedung dan Non Gedung. Badan Standarisasi Nasional: Jakarta.
7. Badan Standarisasi Nasional. 2019. Standar Nasional Indonesia 1726-2019: Tata Cara Perencanaan Ketahanan Gempa untuk Struktur Bangunan Gedung dan Non Gedung. Badan Standarisasi Nasional: Jakarta.
8. Badan Standarisasi Nasional. 2020. Standar Nasional Indonesia 1727-2020: Beban Desain Minimum dan Kriteria Terkait untuk Perancangan Bangunan Gedung dan Struktur Lain, Badan Standarisasi Nasional: Jakarta.
9. Iemura, H., Pradono, M. H., Sugimoto, M., Takahashi, Y., dan Husen, A. 2012. Tsunami Height Memorial Poles In Banda Aceh And Recommendations For Disaster Prevention, Proceedings of the International Symposium on Engineering Lessons Learned from the 2011 Great East Japan Earthquake, Tokyo, Japan, March 1-4.
10. Al Hadi, H. 2016. Aplikasi Metode Linear Polarization Resistance Untuk Mengukur Laju Korosi Infrastruktur Pasca Sepuluh Tahun Tsunami 2004. ETD Unsyiah.

Strengthening capacity study of reinforced concrete school building structure in Banda Aceh toward earthquakes and tsunami hazards

Putri Istiqamah[1*], *Siti* Humayra[1], and *Huzaim*[2]

[1]Student of Civil Engineering Department, Postgraduate Program, Syiah Kuala University, Jl. Syeh Abdurrauf, No. 7, Banda Aceh, 23111, Indonesia
[2]Civil Engineering Department, Faculty of Engineering, Syiah Kuala University, Jl. Syeh Abdurrauf, No. 7, Banda Aceh, 23111, Indonesia

Abstract. Banda Aceh City was a high level areas of the earthquake and tsunami damage that occurred in 2004. Building damage prevention can be done by planning earthquake-resistant school building structures. The rebuilding needs to be reviewed for its capacity structure because of the high risk building location of earthquake and tsunami hazards. This study aims to determine the behavior and capacity of the school building structure using the earthquake and tsunami scenario in 2004 in Banda Aceh. Capacity analysis by column enlargement of the existing building is one of the research variables. Earthquake load refers to data from the Sumatera-Andaman earthquake on December 26, 2004 and tsunami load refers to FEMA P646. Two typical school buildings were taken from data recapitulation of 100 buildings survey, namely SDN 48 and SDN 8, which represented the overall typical school building in Banda Aceh. This study uses the pushover analysis method on the SAP2000 software. Based on the capacity curve, the value of displacement in the x-y directions in the existing building are greater than the column variation building at SDN 48 and SDN 8. While the shear force value in the existing building are smaller than the column variation building. This shows that with the addition of column dimensions, the cross-sectional area gets bigger and affects the capacity of the building structure.

1 Introduction

Earthquake and tsunami are natural disasters that can cause casualties and buildings damage. Real evidence of this disaster could be seen in 2004, an earthquake measuring 9.3 SR that once rocked Banda Aceh City and was followed by a tsunami wave. The estimated impact of the 2004 earthquake and tsunami was that as many as 45,000 students and 1,870 teachers died, as well as damage to 2,065 educational facilities [2]. School buildings, especially elementary schools, are educational facilities that have a high priority index because they accommodate many of the burdens of life such as students and teachers. Prevention of damage to buildings caused by the earthquake and tsunami can be pursued by planning school building structures that meet the criteria for earthquake-resistant buildings, so that they can be used as temporary shelters in disaster emergency situations. The damage level to buildings is influenced by the strength and intensity of the earthquake, the configuration of the building structure, the stiffness and strength of the building structure, and the quality of building materials. Buildings that suffered light and moderate damage only needed repairs to certain parts, while buildings that suffered heavy damage and total damage required overall repairs to rebuilding. School buildings located in areas affected by the earthquake and tsunami need to be reviewed for their capacity to withstand earthquake and tsunami loads. This is related to efforts to minimize the potential damage caused by the earthquake and tsunami. This study uses earthquake and tsunami loading with earthquake and tsunami scenario in 2004. Capacity analysis by adding column dimensions from the existing building is one of the research variables. This variable will be simulated using parametric studies. The selection of research objects was carried out based on typical building criteria in the form of a top view of the building, number of floors, and building materials, as well as observe the condition of the building and the vulnerabilities found during the field survey. The results of data recapitulation on 100 school buildings and obtained two typical school buildings, namely SD Negeri 48 and SD Negeri 8 which represent the overall typical building in Banda Aceh City. The results of this study were obtained of displacement values, shear forces, and interstory drift to see the behavior of the building structure. The results of the capacity curve obtained from the pushover analysis are evaluated to obtain a comparison of the structural capacity values for the existing building and the column variation building. The capacity curve is converted into a capacity spectrum

* Putri Istiqamah: putriistiqamah13@gmail.com

https://doi.org/10.1051/e3sconf/202234002005

curve that can determine the level of damage to the building. The probability value for the level of building damage is determined using and the developed seismic and tsunami fragility curve, so it can be seen the level of damage that occurs to the building. So that with the addition of column dimensions it can be used as an effort to strengthen the capacity of the building structure. The limitation of the problem in this study is that the research location is in Banda Aceh area, the planning does not take into account the lower structure of the building, and the analysis of earthquake and tsunami loading is carried out separately.

2 Literature review

2.1 Earthquake

Earthquakes are defined as natural vibrations that occur at certain locations and are unsustainable. Earthquakes are usually caused by sudden movements of the earth's crust (plates). The sudden movement occurs because of a source of force as the cause, both from nature and from human assistance (artificial earthquakes) [11].

2.2 Tsunami

Tsunami waves cannot be seen when they are far in the middle of the ocean, but once they reach shallow areas the waves will move quickly. The energy generated by a tsunami is constant with respect to the function of altitude and its continuation [10].

2.3 Earthquake resistant building structure

Earthquake resistant buildings are defined as buildings that are able to withstand the design earthquake loads without experiencing excessive damage or not collapsing due to the earthquake [12].

2.4 Structure behaviour

2.4.1 Displacement

The degrees of freedom are the degrees of independence needed to express the position of a degree's system at any time. If a reviewed point experiences a displacement of X, Y, and Z direction, then the system has three degrees of freedom. This happens because the point in question can move freely in three directions [1].

2.4.2 Inter-story drift

The inter-story drift is the horizontal displacement at the top of the reviewed level relative to the bottom [1]. The determination of inter-story drift can be seen in Fig. 1.

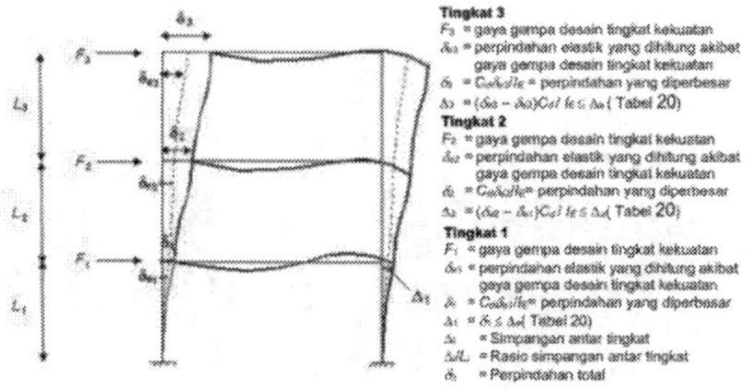

Fig. 1. Determination of Inter-story drift

2.4.3 Base shear force

Base shear force is the total design shear or lateral force that occurs at the base level [1]. The basic seismic shear force (V) in the specified direction is determined by Eq. (1-2):

$$V = C_s W \quad (1)$$

$$C_s = \frac{S_{DS}}{\frac{R}{I_e}} \quad (2)$$

2.5 Pushover analysis

Pushover analysis is a linear and nonlinear two or three dimensional static analysis method, where the influence of the design earthquake on the building structure is considered as static loads that capture the center of mass of each floor, whose value is gradually increased until it exceeds the loading, which causes the plastic hinge in the building structure, then with a further increase in load undergoes a large elastoplastic deformation until it reaches a condition on the verge of collapse [9].

2.6 Seismic fragility curve

Seismic fragility curve is a curve that shows the possibility of damage to a structure when it receives an earthquake load with a certain intensity. This curve becomes very important in assessing or evaluating the seismic performance of a structure because it displays the level of seismic fragility as a possible damage based on earthquake loads that exceed the design load on the performance or limit state of a particular structure [15]. Due to a statistical program, the probability values that form the fragility curve will be obtained as shown in Fig. 2.

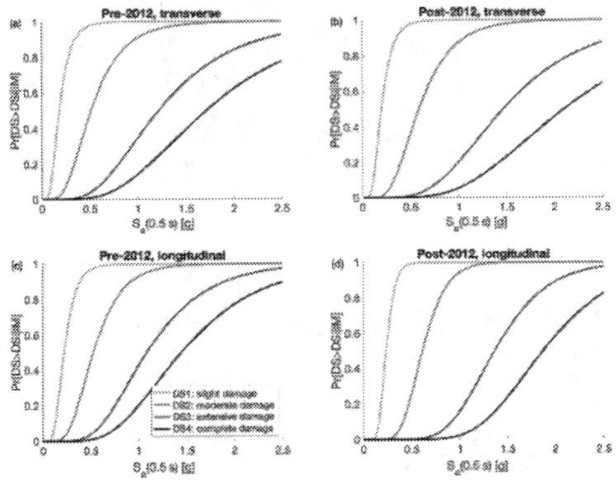

Fig. 2. Seismic fragility curves for typical buildings

The damage state of building structures according to Hazus [8] can be seen in Table 1.

Table 1. The Performance Limitations of Structure.

Damage State	Description
Slight/Minor Damage 	Small plaster cracks at corners of door and window openings and wall ceiling intersections; small cracks in masonry chimneys and masonry veneers. Small cracks are assumed to be visible with a maximum width of less than 1/8 inch (cracks wider than 1/8 inch are referred to as "large" cracks).
Moderate Damage 	Large plaster or gypsum-board cracks at corners of door and window openings; small diagonal cracks across shear wall panels exhibited by small cracks in stucco and gypsum wall panels; large cracks in brick chimneys; toppling of tall masonry chimneys.
Extensive Damage 	Large diagonal cracks across shear wall panels or large cracks at plywood joints; permanent lateral movement of floors and roof; toppling of most brick chimneys; cracks in foundations; splitting of wood sill plates and/or slippage of structure over foundations.
Complete Damage 	Structure may have large permanent lateral displacement or be in imminent danger of collapse due to cripple wall failure or failure of the lateral load resisting system; some structures may slip and fall off the foundation; large foundation cracks. Three percent of the total area of buildings with complete damage is expected to be collapsed, on average.

2.7 Tsunami fragility curve

Tsunami fragility is defined as the structural damage probability or fatality ratio with particular regard to the hydrodynamic features of tsunami inundation flow, such as inundation depth, current velocity and hydrodynamic force. In principle, the development of tsunami fragility requires that tsunami hazard information and damage data should be used synergistically [7]. The tsunami fragility curve can be seen in Fig. 3 and the level of damage is in Table 2 [13].

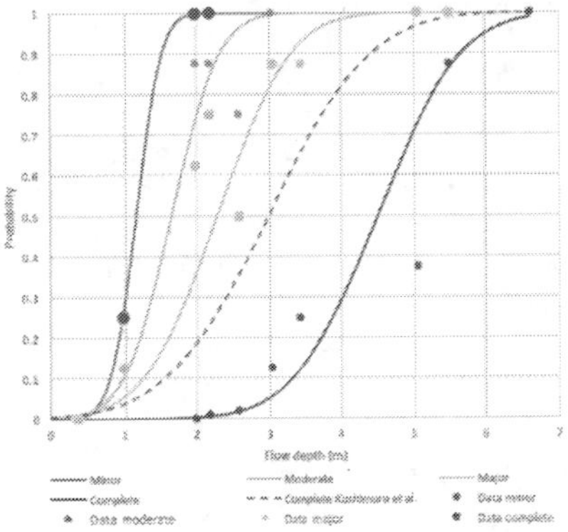

Fig. 3. Tsunami fragility curves for typical buildings

Table 2. Damage limit state of buildings due to tsunami

Classification	Damage Condition	Description
DC0	No damages	Flooded but no damages found.
DC1	Minor	Damages found windows and doors, no damage on wall and on structural component
DC2	Moderate	One side wall damages, no damage on column and beam.
DC3	Major	All walls were damaged or roofs felt down, structural components bent/deflected or broken.
DC4	Complete/ Washed away	Only floor left.

3 Research methods

3.1 Object of research

The object of research in this study is a typical two-story school building in Banda Aceh City with reinforced concrete construction. The determination of the building type is carried out by recapitulating survey data from 100 school buildings and taking two typical buildings that dominate the structural design of the building.

3.2 Research data

The types of data needed in this study are:

1. Primary data is data obtained directly from the results of filling out data forms through field surveys on school buildings in Banda Aceh City in the form of data on the dimensions of building structural elements.
2. Secondary data is supporting data and a reference for the implementation of the primary survey. This data is obtained from relevant agencies in a ready-to-use form.
3. The literature used is SNI-1726-2019, SNI-03-1727-2020, Map of Sources and Earthquake Hazards in Indonesia, as well as books related to research.

3.3 Data collection technique

Research data collection was carried out by means of a field survey. Surveys are used to collect building structure data and building vulnerability factors by looking directly at the condition of the building. Structural data collection was carried out randomly on 100 school buildings in Banda Aceh using a rapid assessment survey, so that the distribution of surveyed school buildings was achieved for each sub-district in Banda Aceh.

3.3.1 Building technical data

Typical details of building structures for various building forms in Banda Aceh are as shown in Table 3 [5].

Table 3. Typical structural detail for typical school buildings in Banda Aceh

	Materials	Typical Beam	Typical Column	Typical Joint
Pre-2012	Concrete, f_c = 21.5 Mpa	3Φ16 top	3Φ16 top	No stirrups
	Long. steel, f_y = 400 Mpa	3Φ16 bottom	3Φ16 bottom	
	Long. steel, f_y = 240 Mpa	Φ10@150 mm stirrups	Φ10@200 mm stirrups	
Post-2012	Concrete, f_c = 24 Mpa	3Φ16 top	3Φ16 top	No stirrups
	Long. steel, f_y = 400 Mpa	3Φ16 bottom	3Φ16 bottom	
	Long. steel, f_y = 240 Mpa	Φ10@150 mm stirrups	Φ10@100 mm stirrups	

3.3.2 Earthquake data

The Earthquake data used is recorded earthquake data that occurred in Sumatra – Andaman fault on December 26th, 2004 obtained from United States Geological Survey (USGS).

3.3.3 Tsunami data

The tsunami data needed to calculate the tsunami load [4] is the depth of the tsunami flow in 2004 [6] and the difference in the elevation of the building subgrade to the water level [14].

3.4 Building structure modeling

The reviewed structure in this study is the entire structure of the school building. The structural modeled using SAP2000 software as a space frame.

The data entered are detail building data, the quality of the materials used, the dimensions of the building, and the loading. The modeling structure of each school building is done with two variations of the column in the same way.

Fig. 4. 3D structure model

3.5 Loading

Loads that are inputted in the SAP2000 program are dead loads, live loads, earthquake loads, tsunami loads, and combined loads.

1. Combined earthquake loading

U_1 = 1,4 D
U_2 = 1,2 D + 1,6 L
U_3 = 1,38 D + 1 L + 0,39 Q_{Ex} + 1,3 Q_{Ey}
U_4 = 1,38 D + 1 L + 0,39 Q_{Ex} - 1,3 Q_{Ey}
U_5 = 1,38 D + 1 L - 0,39 Q_{Ex} - 1,3 Q_{Ey}
U_6 = 1,38 D + 1 L - 0,39 Q_{Ex} + 1,3 Q_{Ey}
U_7 = 1,38 D + 1 L + 1,3 Q_{Ex} + 0,39 Q_{Ey}
U_8 = 1,38 D + 1 L + 1,3 Q_{Ex} - 0,39 Q_{Ey}
U_9 = 1,38 D + 1 L - 1,3 Q_{Ex} - 0,39 Q_{Ey}
U_{10} = 1,38 D + 1 L - 1,3 Q_{Ex} + 0,39 Q_{Ey}
U_{11} = 1,08 D + 0,39 Q_{Ex} + 1,3 Q_{Ey}
U_{12} = 1,08 D + 0,39 Q_{Ex} - 1,3 Q_{Ey}
U_{13} = 1,08 D - 0,39 Q_{Ex} - 1,3 Q_{Ey}
U_{14} = 1,08 D - 0,39 Q_{Ex} + 1,3 Q_{Ey}
U_{15} = 1,08 D + 1,3 Q_{Ex} - 0,39 Q_{Ey}
U_{16} = 1,08 D + 1,3 Q_{Ex} + 0,39 Q_{Ey}
U_{17} = 1,08 D - 1,3 Q_{Ex} - 0,39 Q_{Ey}
U_{18} = 1,08 D - 1,3 Q_{Ex} + 0,39 Q_{Ey}

2. Combined tsunami loading

U_{19} = 1,2 D + 1,0 T_S + 1,0 L_{REF} + 0,25 L
U_{20} = 0,9 D + 1,0 TS

3.6 Pushover analysis

Pushover analysis is done by loading in stages, the lateral loading carried out represents the earthquake force on each floor. The lateral loading direction is carried out in direction of the main axis building. Lateral loading uses the displacement-controlled concept, which the loading process is increased gradually until the displacement target is achieved.

3.7 Tsunami loading

The tsunami load used is hydrostatic force, hydrodynamic force, and debris impact force. These forces are calculated with a flow depth of 7 meters. The hydrostatic force is input as distributed load, pyramidal shaped along the flooded column. The hydrodynamic force is input as distributed load along the flooded

https://doi.org/10.1051/e3sconf/202234002005

column. While the debris impact force is inputted as a point load. These forces are input in the y-direction. Then analysis was carried out on SAP2000 software using a combination of 19 (U_{19}) and a combination of 20 (U_{20}).

3.8 Data analysis

3.8.1 The dynamic characteristics structure analysis

Analysis of the dynamic characteristics structure is produced by performing a free vibration analysis. The dynamic characteristics of this structure consist of a natural mode shape and a natural period of vibration. In this study, free vibration analysis was performed using eigen value analysis and Ritz vector which was built in in the SAP2000 program.

3.8.2 The capacity curve and capacity spectrum curve analysis

The capacity curve is the result of pushover analysis in SAP2000. The capacity curve shows the behavior of the building structure when given an earthquake load at a certain level. The median spectral acceleration (S_a) value is obtained by converting the capacity curve into a capacity spectrum curve which is the spectral acceleration (S_a) curve to spectral displacement (S_d). The capacity spectrum curve was analyzed to obtain the median spectra displacement and median spectral acceleration values at various levels of damage [3].

3.9 Developed seismic fragility curve analysis

Seismic fragility curve shows the probability of damage to the structure under consideration when receiving earthquake loads with a certain intensity. Once the probability value obtained is used to determine the coordinates of the fragility curve as the ordinate axis (y) with the value S_a on the seismic fragility curve. The fragility curve is converted using an online graph-reader application to extract values from graphic images as shown in Fig. 5.

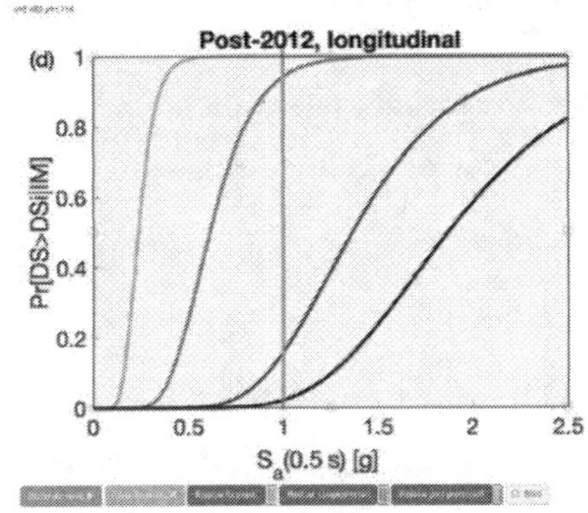

Fig. 5. Online graph-reader application display

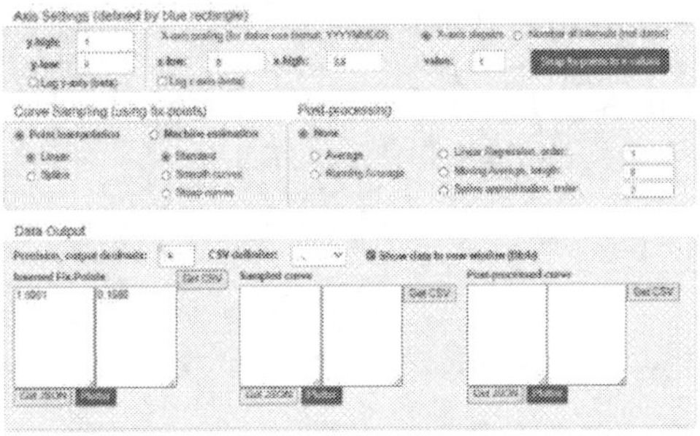

Fig. 6. Graph-reader output data results

3.10 Reinforcement analysis

After earthquake loading, structural capacity analysis is carried out on the results of displacement, shear force, and inter-story drift that occurs for the two typical buildings. The same method is also carried out on the column variation building for the two typical buildings. While for the tsunami load, the results from SAP2000 are seen for the inter-story drift that occurred. The result of this research is the comparison of the capacity structure that occurs between the existing and column variation building.

4 Results and discussion

4.1 Survey data recapitulation results

The recapitulation data was conducted to determine the object of research which is a typical school building in Banda Aceh. From the recapitulation results, an infographic is made to show the criteria for school buildings that dominate in Banda Aceh which can be seen in Fig. 6.

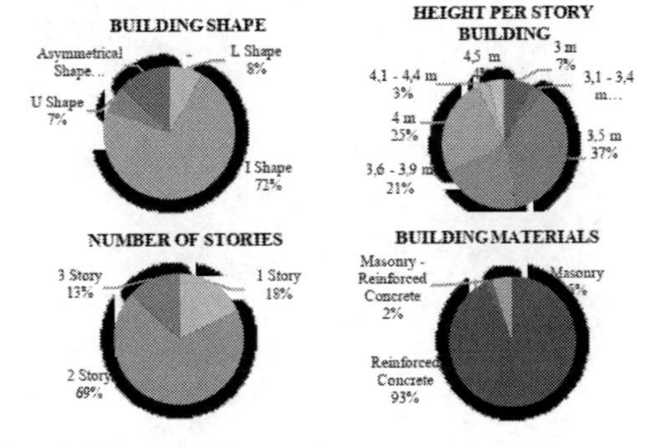

Fig. 6. Infographic data results

Based on the infographic, the building criteria obtained are in the building shape with I shape of 72%, the height of story building is 3.5 m with a percentage of 37%, a two-story building is 69%, and reinforced concrete building materials are 93%. In addition to these criteria, the selection of research objects is also influenced by the condition of the building and the vulnerabilities found during the field survey. School buildings that meet the criteria from the conclusion above are SD Negeri 48 Banda Aceh located in Kutaraja District and SD Negeri

8 Banda Aceh located in Meuraxa District. The technical data for the school buildings of SD Negeri 48 and SD Negeri 8 Banda Aceh can be seen in Table 4.

Table 4. Characteristics and building position

	Criteria	SD Negeri 48	SD Negeri 8
1	Place Coordinates	Longitude 95.2898; Latitude 5.5586	Longitude 95.31060; Latitude 5.56041
2	Construction Year	2015	2016
3	Number of stories	Two-story	Two-story
4	Height per story	3,5 meters	3,5 meters
5	Materials	Reinforced Concrete (RC)	Reinforced Concrete (RC)
6	Frame system	Space Frame	Space Frame

4.2 Number of variety analysis results

Based on the analysis results on SAP2000, it shows that the modal load participation value of the existing and column variation building in the SDN 48 and SDN 8 has met the provisions of the SNI 03 – 1726 – 2019 limit (in Table 5), which the effective mass participation factor for UX and UY is a minimum of 90%.

Table 5. Results of analysis of fundamental vibration time of structure

Building	Modal Type	Item	Static (%)	Dynamic (%)
SDN 48 (Existing Building)	Acceleration	UX	100	99,5253
	Acceleration	UY	100	98,7962
	Acceleration	UZ	1,1327	0,1718
SDN 48 (Column Variation Building)	Acceleration	UX	100	99,5654
	Acceleration	UY	100	98,2429
	Acceleration	UZ	0,7664	0,1377
SDN 8 (Existing Building)	Acceleration	UX	100	99,8135
	Acceleration	UY	100	99,7223
	Acceleration	UZ	1,0103	0,6771
SDN 8 (Column Variation Building)	Acceleration	UX	100	99,7351
	Acceleration	UY	100	99,6714
	Acceleration	UZ	0,8748	0,6763

4.3 Result of analysis of fundamental period of structure

Based on the analysis results on SAP2000, it shows that the natural period of the existing and column variation building in the SDN 48 and SDN 8 has met the provisions of the SNI 03 – 1726 – 2019 (in Table 6.) with natural period limit of 0.3759 seconds.

Table 6. Results of analysis of fundamental vibration time of structure

Building	Period	$T_{a\,izin}$	Description
	Sec	Sec	
SDN 48 (Existing Building)	0,2274	0,3759	✓
SDN 48 (Column Variation Building)	0,1728	0,3759	✓
SDN 8 (Existing Building)	0,2209	0,3759	✓
SDN 8 (Column Variation Building)	0,16054	0,3759	✓

4.4 Displacement analysis results

The maximum displacement due to the combination of x-y direction earthquake loads for existing and column variations buildings and can be seen in Table 7.

Table 7. Maximum displacement result of SDN 48

	SDN 48 (Existing Building)			SDN 48 (Column Variation Building)		
	Comb.	Joint	Displacement Max. (m)	Comb.	Joint	Displacement Max. (m)
X-direction	U_4	49	0,1198	U_4	50	0,0850
Y-direction	U_6	56	0,0839	U_6	56	0,0622

Table 7. Maximum Displacement Result of SDN 8

	SDN 8 (Existing Building)			SDN 8 (Column Variation Building)		
	Comb.	Joint	Displacement Max. (m)	Comb.	Joint	Displacement Max. (m)
X-direction	U_{16}	101	0,0763	U_7	93	0,0686
Y-direction	U_{14}	101	0,0371	U_{14}	102	0,0268

4.5 Shear force analysis results

The basic shear force (V_t) ≥ V, where the calculation of the basic shear force value for each building is as follows (eq. 1):

S_{DS} = 0,9 g
S_1 = 0,75 g
I_e = 1,5
R = 8

The coefficient of C_s must be determined by:

$$C_s = \frac{S_{DS}}{\frac{R}{I_e}} = \frac{0,9}{\frac{8}{1,5}} = 0,075$$

For structures located in areas where $S_1 \geq 0,6$ g, the C_s must be not less than:

$$C_s = \frac{0,5 \times S_1}{\frac{R}{I_e}} = \frac{0,5 \times 0,75}{\frac{8}{1,5}} = 0,0703$$

So, the value of C_s used for this analysis is 0,0703.

Table 8. Basic shear force calculation

Building	V_{syarat} / $C_s.W$	$0,85\ V_{syarat}$
SDN 48 (Existing Building)	= 0,0703 (75780,93) = 5327,399 kgf	= 0,85 (5327,399) = 4528,289 kgf
SDN 48 (Column Variation Building)	= 0,0703 (93420,93) = 6567,491 kgf	= 0,85 (6567,491) = 5582,367 kgf
SDN 8 (Existing Building)	= 0,0703 (210328,67) = 14786,106 kgf	= 0,85 (14786,106) = 12568,19 kgf
SDN 48 (Column Variation Building)	= 0,0703 (247960,67) = 17431,635 kgf	= 0,85 (17431,635) = 14816,89 kgf

Table 9. Base shear force results

Building	V	Base Shear (V_t)		$V_t > V$	
	Kgf	V_{tx}	V_{ty}		
SDN 48 (Existing Building)	4528,29	6031,03	255,52	✓	×
SDN 48 (Column Variation Building)	5,582,37	28689,31	8601,34	✓	✓
SDN 8 (Existing Building)	12568,19	316131,16	235916,26	✓	✓
SDN 8 (Column Variation Building)	14816,89	377150,87	292761,01	✓	✓

Based on the Table 9, the value of the basic y-direction shear force for the existing SDN 48 building is not greater than the value of the allowable shear force. So that this value does not meet the provisions of SNI 1726-2019. After the column variation, the result of the basic shear force obtained is greater than the value of the allowable shear force. While the basic shear force for the existing building and the column variation building at SDN 8 has met the requirements because the resulting value is greater than the allowable shear force value.

4.6 The result of the analysis inter-story drift with earthquake load

Berdasarkan SNI 1726:2019, The value of the inter-story drift must be less than 0,01 or 1% times the height of the building before the floor level under review for risk category IV buildings. The results of the calculation of the deviation between levels in the x and y directions due to the earthquake can be seen in Table 4.8 and Table 4.9.

Table 10. Inter-story drift with earthquake load x-direction

Building	Story	Displacement x-direction	Inter-story Drift	Inter-story Drift Ratio 0,010 h_{sx}	Description
		m	m	m	
SDN 48 (Existing Building)	2	0,1149	0,0572	0,035	×
	1	0,0577	0,0577	0,035	×
SDN 48 (Column Variation Building)	2	0,0691	0,0344	0,035	✓
	1	0,0347	0,0347	0,035	✓
SDN 8 (Existing Building)	2	0,0678	0,0252	0,035	✓
	1	0,0427	0,0427	0,035	×
SDN 8 (Column Variation Building)	2	0,0657	0,0337	0,035	✓
	1	0,0320	0,0320	0,035	✓

Table 11. Inter-story drift with earthquake load y-direction

Building	Story	Displacement y-direction	Inter-story Drift	Inter-story Drift Ratio 0,010 h_{sx}	Description
		(m)	(m)	(m)	
SDN 48 (Existing Building)	2	0,0838	0,0163	0,035	✓
	1	0,0675	0,0675	0,035	×
SDN 48 (Column Variation Building)	2	0,0522	0,0244	0,035	✓
	1	0,0278	0,0278	0,035	✓
SDN 8 (Existing Building)	2	0,0871	0,0352	0,035	×
	1	0,0519	0,0519	0,035	×
SDN 8 (Column Variation Building)	2	0,0265	0,0126	0,035	✓
	1	0,0139	0,0139	0,035	✓

4.7 Capacity curve evaluation results

The capacity curve is nonlinear due to an increase in load which makes the structural elements of the building change from an elastic condition to a plastic condition.

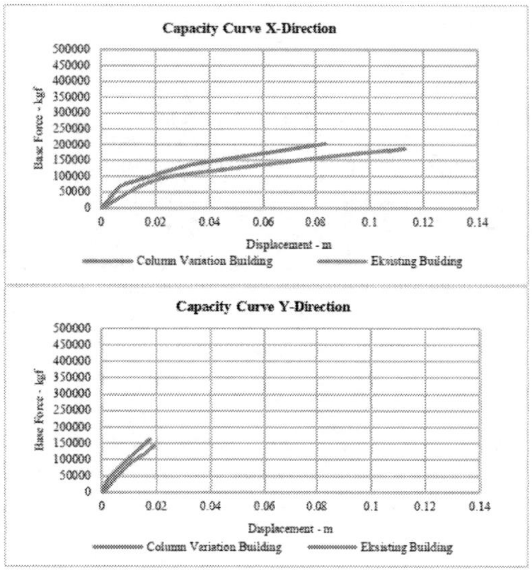

Fig. 7. Capacity curve x-y direction in SDN 48

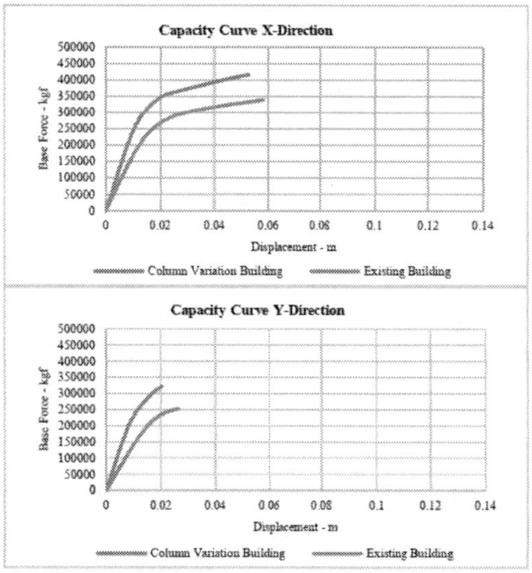

Fig. 8. Capacity curve x-y direction in SDN 8

Based on the fig. 8, the value of displacement in the x and y directions in the existing building is greater than the column variation building at SDN 48 and SDN 8. While the shear force value in the existing building is smaller than the column variation building. This is influenced by the addition of column dimensions which results in a larger cross-sectional area. Overall, these results indicate that the existing buildings and column variations in both schools are more flexible and prone to shifts in the x-direction.

4.8 Capacity spectrum results

The capacity spectrum curve aims to obtain the value of the median spectral acceleration parameter. Furthermore, on the capacity spectrum curve, the position of each damage limit is determined. Determination of the position of each damage limit on the capacity spectrum curve based on research [3].

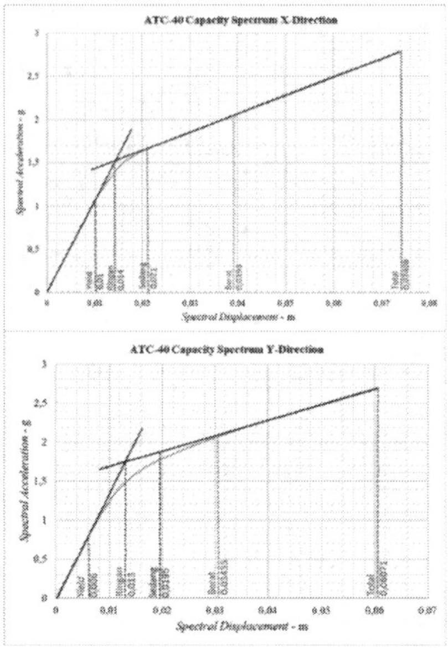

Fig. 9. Capacity spectrum x-y directions of the existing building (SDN 48)

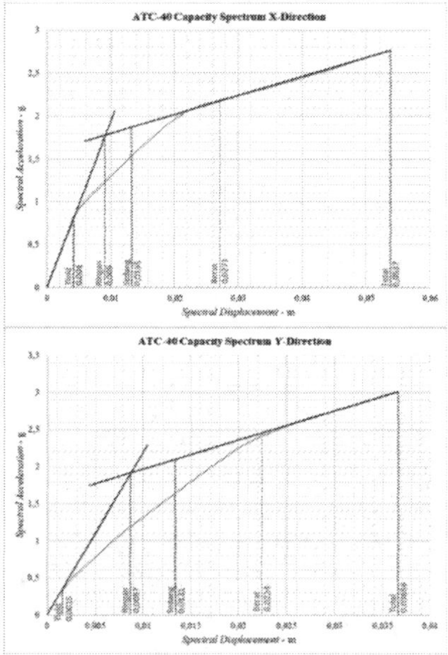

Fig. 10. Capacity spectrum x-y directions of the column variation building (SDN 48)

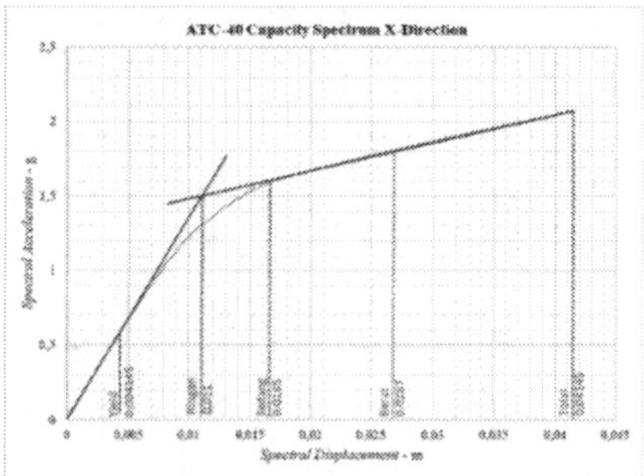

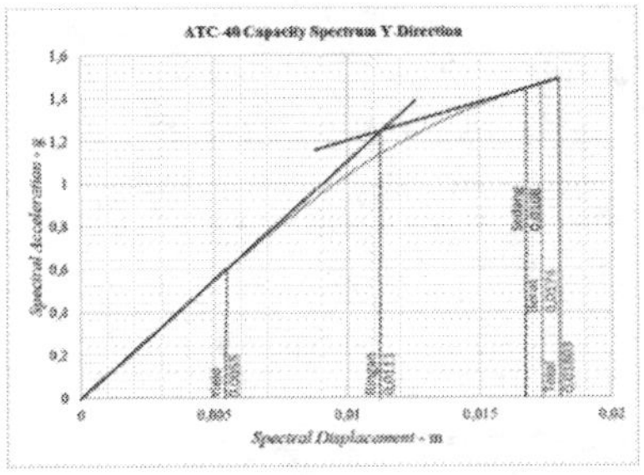

Fig. 11. Capacity spectrum x-y directions of the existing building (SDN 8)

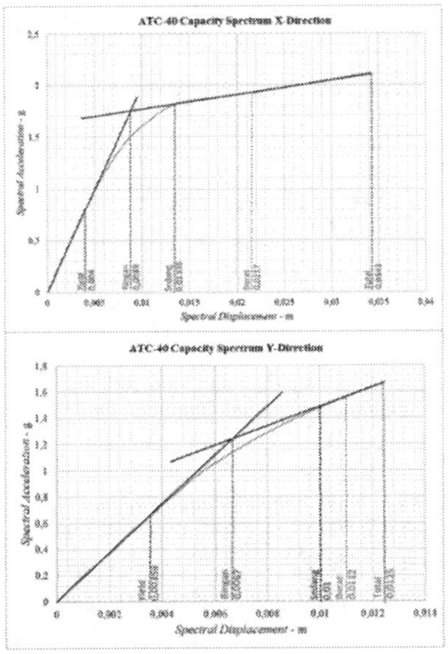

Fig. 12. Capacity spectrum x-y directions of the column variation building (SDN 8)

The seismic fragility curve in buildings is the relationship between the spectral acceleration as the horizontal axis and the probability of structural failure as the vertical axis (Fig. 9-12). Based on the identification of the capacity spectrum curve, the resulting spectral displacement (S_d) and spectral acceleration (S_a) points for each structural failure condition can be seen in Table 12 and Table 13.

Table 12. S_d and S_a values for each type of structural damage at SDN 48 building

Building	Damage state	Sd, ds (m)		Sa,ds (g)	
		x-direction	y-direction	x-direction	y-direction
SDN 48 (Existing Building)	Slight	0,0140	0,0130	1,3808	1,4681
	Moderate	0,0210	0,0195	1,6613	1,7666
	Extensive	0,0393	0,0346	2,0545	2,1640
	Complete	0,0741	0,0607	2,7928	2,6956
SDN 48 (Column Variation Building)	Slight	0,0091	0,0087	1,2440	1,1834
	Moderate	0,0135	0,0131	1,5511	1,6225
	Extensive	0,0271	0,0224	2,1898	2,4099
	Complete	0,0537	0,0366	2,7677	3,0089

Table 13. S_d and S_a values for each type of structural damage at SDN 8 building

Building	Damage state	Sd, ds (m)		Sa,ds (g)	
		x-direction	y-direction	x-direction	y-direction
SDN 8 (Existing Building)	Slight	0,0110	0,0111	1,3015	1,1303
	Moderate	0,0165	0,0168	1,5869	1,4464
	Extensive	0,0267	0,0174	1,8040	1,4694
	Complete	0,0414	0,0180	2,0706	1,4914
SDN 8 (Column Variation Building)	Slight	0,0089	0,0067	1,5053	1,1404
	Moderate	0,0134	0,0100	1,8053	1,4762
	Extensive	0,0217	0,0112	1,9262	1,5734
	Complete	0,0343	0,0125	2,1113	1,6638

4.8.1 The results of the evaluation of the developed seismic fragility curve

The required data is taken from the results of spectra response of PUSKIM (Pusat Penelitian dan Pengembangan Pemukiman). The seismic fragility curve used for this evaluation is the curve in the Gentile, et al (2019).

1. Calculation of S_a SDN 48 building
   T = 0,3759 sec
   T_0 = 0.135983 sec
   T_s = 0.679916 sec
   T_L = 14.000000 sec
   S_{ds} = 1.000124 g

Since $T \geq T_0$ or $T \leq T_s$, then $S_a = S_{ds}$, S_a = 1,000124 g. Data S_a is inputted into the online graph-reader application and generates probability values as shown in Fig. 13.

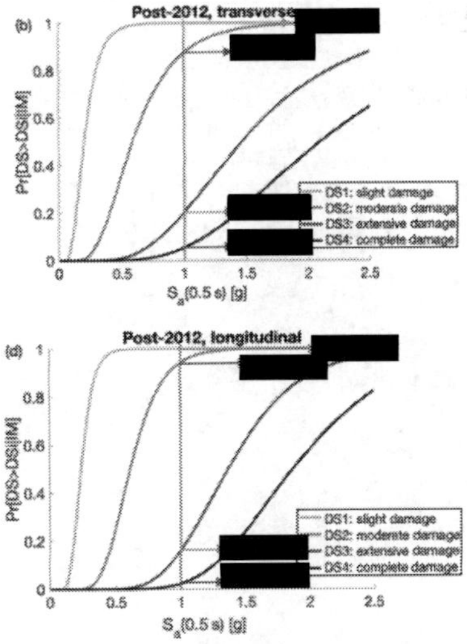

Fig. 13. Identification of the probability of the seismic fragility curve in the x and y directions of SDN 48 building

2. Calculation of S_a SDN 8 building
   T = 0,3759 sec
   T_0 = 0,135831 sec
   T_s = 0,679154 sec
   T_L = 15,000000 sec
   S_{ds} = 1,001245 g

 https://doi.org/10.1051/e3sconf/202234002005

Since $T \geq T_0$ or $T \leq T_s$, then $S_a = S_{ds}$, $S_a = 1{,}001245$ g. Data S_a is inputted into the online graph-reader application and generates probability values as shown in Fig. 14.

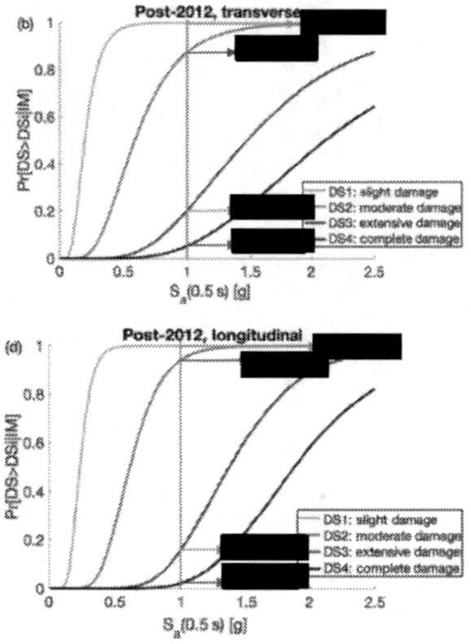

Fig. 14. Identification of the probability of the seismic fragility curve in the x and y directions of SDN 8 building.

4.9 Structural reinforcement evaluation results

Based on the results of the evaluation of the existing fragility curve, it shows that the buildings of SDN 48 and SDN 8 will experience slight and moderate conditions so that it can be recommended to carry out structural reinforcement. After adding 15% column dimensions, table 4.9 shows the results of the structural capacity by connecting the values of displacement and base reaction and in Table 14 shows the largest displacement results for the x and y directions.

Table 14. Result of displacement direction X and direction Y

Building	Column dimension	Column dimension enlargement	Displacement	
			x-direction	y-direction
	cm	%	m	m
SDN 48	30 x 30	0%	0,1198	0,0839
	40 x 40	15%	0,0850	0,0622
SDN 8	30 x 30; 30 x 40	0%	0,0763	0,0371
	40 x 40; 40 x 50	15%	0,0686	0,0268

Based on the Table 14, in the SDN 48 building, the displacement value that occurred after adding the column dimensions were reduced by 29% for the x direction and decreased by 26% for the y direction. Meanwhile, in SDN 8 building, the displacement value that occurs after adding the column dimensions is reduced by 10% for the x direction and 28% for the y direction. This shows that with the addition of column dimensions, the building structure becomes more rigid because the displacement results are getting smaller.

4.10 Tsunami load analysis results

FEMA P646 only explains how to calculate the tsunami load on the building, so that for checking, SNI 1726:2019 is used as a control for inter-story drift.

Table 15 Tsunami force calculation results

Force	Results	
	SDN 48	SDN 8
Hydrostatic force	793138,5 N	1057518 N
Hydrodynamic force	20848,2 N	20508,3 N
Debris impact force	-	62584,1 N

The results of the inter-story drift of the combined tsunami load (Table 15) that occurred can be seen in Table 16 and Table 17.

Table 16 Inter-story drift with tsunami load X-direction

Building	Story	Displacement x-direction	Inter-story Drift	Inter-story Drift Ratio 0,010 h_{sx}	Description
		m	m	m	
SDN 48 (Existing Building)	2	0,0145	0,0013	0,035	✓
	1	0,0132	0,0132	0,035	✓
SDN 48 (Column Variation Building)	2	0,0044	0,0009	0,035	✓
	1	0,0035	0,0035	0,035	✓
SDN 8 (Existing Building)	2	0,0020	0,0004	0,035	✓
	1	0,0016	0,0016	0,035	✓
SDN 8 (Column Variation Building)	2	0,0020	0,0004	0,035	✓
	1	0,0015	0,0015	0,035	✓

Based on Table 16, the result of the inter-story drift in the x-direction at SDN 48 and SDN 8 is smaller than the inter-story drift permit, so that it meets the provisions of SNI 1726-2019.

Table 17. Inter-story drift with tsunami load Y-direction

Building	Story	Displacement y-direction	Inter-story Drift	Inter-story Drift Ratio 0,010 h_{sx}	Description
		(m)	(m)	(m)	
SDN 48 (Existing Building)	2	0,0671	0,0172	0,035	✓
	1	0,0499	0,0499	0,035	×
SDN 48	2	0,0201	0,0077	0,035	✓
	1	0,0124	0,0124	0,035	✓

 https://doi.org/10.1051/e3sconf/202234002005

(Column Variation Building)					
SDN 8 (Existing Building)	2	0,0542	0,0144	0,035	✓
	1	0,0399	0,0399	0,035	×
SDN 8 (Column Variation Building)	2	0,0349	0,0137	0,035	✓
	1	0,0211	0,0211	0,035	✓

Based on Table 17, the variation of the column buildings in the two schools, the result of the inter-story drift in the y direction is smaller than the inter-story drift permit, so that it meets the provisions of SNI 1726-2019. While the results for the existing 1st floor buildings of SDN 48 and SDN 8 are greater than the inter-story drift permit. This is because the tsunami inundation reached 7 meters on the 2nd floor.

4.11 Comparison of inter-story drift between earthquake and tsunami load

Based on the results of the inter-story drift in Sections 4.6 and 4.10, a graph of the ratio of the inter-story drift to earthquake and tsunami loads can be seen in Fig. 15 and Fig. 16.

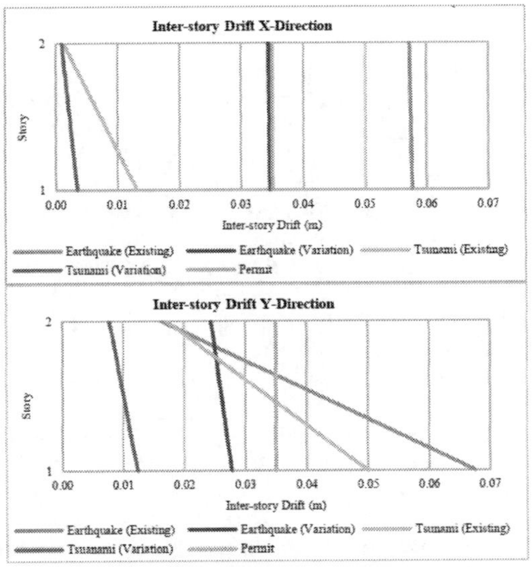

Fig. 15. The inter-story drift in the x-y directions of SDN 48

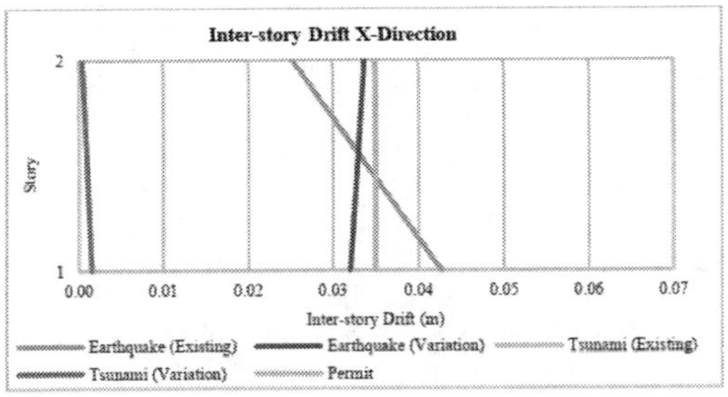

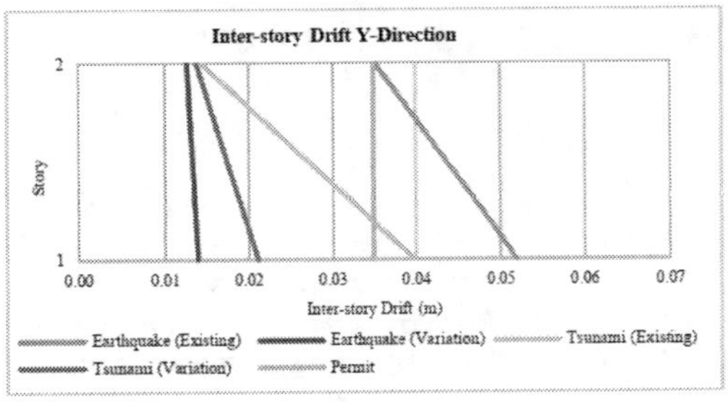

Fig. 16. The inter-story drift in the x-y directions of SDN 8

The results of the inter-story drift due to earthquake and tsunami loads in the column variations of SDN 48 and SDN 8 have met the provisions of SNI 1726-2019 because the inter-story drift is not greater than the inter-story drift permit. This shows that the addition of column dimensions affects the stiffness and behavior of the building structure, so that the building can withstand earthquake and tsunami loads.

4.12 Evaluation results of the developed tsunami fragility curve

The tsunami fragility curve used for this evaluation is the curve in the study of Syamsidik, et al (2020). The data required is the height of the tsunami inundation, which is 7 m for the buildings of SDN 48 and SDN 8. The height data is inputted into the online graph-reader application and produces a probability value as shown in Fig. 17.

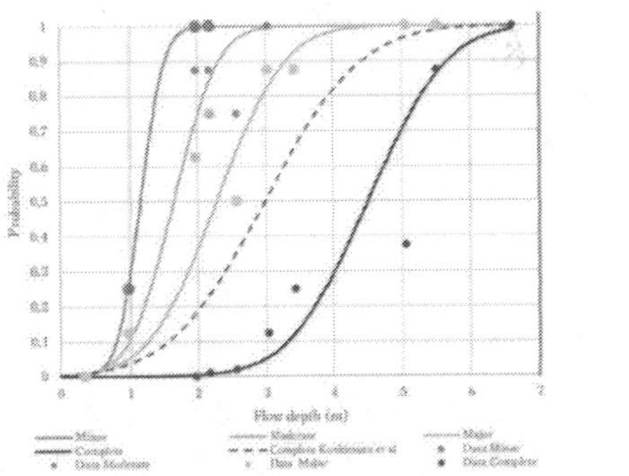

Fig. 17. Identify the probability of the developed tsunami fragility curve

The Fig.17 explains the relationship between tsunami inundation height and the probability value of the damage level to buildings. The inputted tsunami inundation height modeling that it was identified that there was no intersection between the inundation height line and the damage level. These results indicate that the probability value of the complete damage building that occurs for the SDN 48 and SDN 8 buildings is 100%. However, if viewed from the structural behavior, namely the inter-story drift of the tsunami load, the column variation buildings for the two schools were able to withstand the tsunami load. Therefore, for tsunami loading, it is necessary to re-analyze the fragility curve that adjusts to the building structure.

5 Conclusions and suggestions

5.1 Conclusions

1. Based on the earthquake and tsunami scenario in 2004, the existing buildings of SDN 48 and SDN 8 do not meet the structural boundary criteria of SNI 1726-2019. After being simulated with the addition of column dimensions, the results obtained for both buildings have met these criteria. This shows that the addition of column dimensions affects the behavior of the building structure.
2. The shear force value for the existing SDN 48 building do not meet the basic shear force requirements in SNI 1726-2019. The value obtained is very small compared to the shear force acting due to the earthquake. This is due to the small cross-sectional area of the column.
3. The inter-story drift value due to earthquake loads for the existing buildings of SDN 48 and SDN 8 have not met the provisions of SNI 1726:2019. So that the two existing buildings are recommended for structural strengthening efforts.
4. The inter-story drift value due to the tsunami load for the existing buildings of SDN 48 and SDN 8 have not met the provisions of SNI 1726:2019. This is because the tsunami inundation reached 7 meters on the 2nd floor.
5. Based on the displacement value, after the addition of the column dimensions by 15%, for the SDN 48 building, the displacement that occurs is reduced by 29% for the x direction and reduced by 26% for the y direction. While the SDN 8 building, the displacement that occurs after adding the column dimensions is reduced by 10% for the x direction and 28% for the y direction.

5.2 Suggestions

1. Using different structural modeling with different types of school buildings, such as the building shape, the height and width of the building, the number of story, or building materials.
2. Perform earthquake modeling with variations of other earthquake sources for earthquake load calculations in order to obtain different spectral response values.
3. Evaluating seismic performance by using a variety of structural reinforcement methods by reviewing other structural parts.
4. Analyze and evaluation of seismic and tsunami fragility curves according to data on school building structures.

Reference

1. BSN (Badan Standarisasi Nasional): SNI 1726:2019, Tata Cara Perencanaan Ketahanan Gempa untuk Struktur Bangunan Gedung dan non Gedung. Badan Standarisasi Nasional: Jakarta. Indonesia (2019)
2. BAPPENAS (Ministry of National Development Planning), Preliminary Damage and Loss Assesment; The December 26, 2004 Natural Disaster, Indonesia (2006).
3. X. Duan, J.W. Pappin, A Procedure for Establishing Fragility Functions for Seismic Loss Estimate of Existing Buildings Based on Nonlinear Pushover Analysis, presented on pada The 14th World Conference on Earthquake Engineering, 12-17 October 2008, Beijing (2008)
4. FEMA (Federal Emergency Management Agency), Guidelines for Design of Structures for Vertical Evacuation from Tsunami Third Edition. Department of Homeland Security FEMA Mitigation Division: Washington D.C. (2019)
5. R. Gentile, C. Galasso, Y. Idris, I. Rusydy, E. Meilianda, From Rapid Visual Survey to Multi-Hazard Risk Prioritisation and Numerical Fragility of School Buildings, Nat. Hazards Earth Syst. Sci., 19, 1365–1386 (2019)
6. H. Iemura, M. H. Pradono, M. Sugimoto, Y. Takahashi, A. Husen, Tsunami Height Memorial Poles in Banda Aceh and Recommendations for Disaster Prevention, Proceedings of the International Symposium on Engineering Lessons Learned from the 2011 Great East Japan Earthquake, March 1-4, Tokyo, Japan (2012).
7. S. Koshimura, Y. Namegaya, H. Yanagisawa, Tsunami Fragility: A New Measure to Identify Tsunami Damage. Journal of Disaster Research. Vol. 4, No. 6, 479-488 (2009)
8. National Institute of Building Science (NIBS). HAZUS (Hazard US), (1999): Earthquake Loss Estimation, National Institute of Building Sciences. NIBS:Washington D.C (2002)
9. Y. A. Pranata, Evaluasi Kinerja Gedung Beton Bertulang Tahan Gempa dengan Pushover Analysis. Jurnal. Universitas Kristen Maranatha, Bandung. (2006)
10. N. T. Sugito, Tsunami, Universitas Pendidikan Indonesia, Bandung (2008)
11. Suharjanto. Rekayasa Gempa. Kepel Press: Yogyakarta (2013)
12. I. K. Sulendra, Evaluasi Dan Tindakan Pengurangan Kerusakan Bangunan. Jurnal. Universitas Tadulako, Palu (2011)
13. Syamsidik, Benazir, M. Lutfi, A. Suppasri, dan L. K. Comfort, The 22 December 2018 Mount Anak Krakatau volcanogenic tsunami on Sunda Strait coasts, Indonesia: tsunami and damage characteristics, NHESS, 20(2), 549–565 (2020).
14. Tursina, Syamsidik, Reconstruction of the 2004 Tsunami Inundation Map in Banda Aceh Through Numerical Model and Its Validation with Post-Tsunami Survey Data, IOP Conf. Series: Earth and Environmental Science 273 (2019)
15. N. A. Wibowo, Pengembangan Kurva Kerapuhan Berbasis Incremental Dynamic Analysis Untuk Evaluasi Kinerja Seismik Jembatan Beton. Jurusan Teknik Sipil Fakultas Teknik Universitas Sebelas Maret : Surakarta (2016)

Study of vulnerability three-story school buildings with reinforced concrete construction in Banda Aceh to earthquake and tsunami hazards

Natasya Amalia[1*], *Yushar* [2], and *Winda* Yursilla[3]

[1]Student of Civil Engineering Department, Postgraduate Program, Syiah Kuala University, Jl. Syeh Abdurrauf, No. 7, Banda Aceh, 23111, Indonesia
[2]Bachelor Degree of Economic Development Department, Syiah Kuala University, Jl. Teuku Nyak Arief, Kopelma Darussalam, Banda Aceh, 23111, Indonesia
[3]Graduate of Biology Teacher Education Program, Syiah Kuala University, Jl. Tgk. Hasan Krueng Kalee, Kopelma Darussalam, Banda Aceh, Aceh 24415, Indonesia

Abstract. The city of Banda Aceh is an area located between two active faults, namely, the Aceh segment and the Seulimuem fault. In 2004, an earthquake measuring 9.3 on the Richter Scale rocked the city of Banda Aceh, followed by a tsunami on the Burma Plate. As a result, almost all school buildings were damaged. This study aims to determine the vulnerability of school buildings in Banda Aceh City to earthquake and tsunami hazards from the capacity of the building structure, so that the level of damage is obtained from the developed fragility curve. The object of this research is a typical three-story reinforced concrete school building, namely the private school of Budi Dharma and the national school of SD Negeri 50 in Banda Aceh City. The research was conducted by means of a field survey to collect data on the structure of the school building. The data consists of building data, material quality and building dimensions. Then the loadings are inputted into the SAP2000 program in the form of dead loads, live loads, earthquake loads consisting of the USGS earthquake recording and an estimated map of the Indonesian earthquake, and tsunami loads in 2017. Then pushover analysis was used to get the level of damage from the capacity curve to see the probability percentage comparison. In addition, the large value of S_a is equal to the percentage of probability as the level of damage to the building which was influenced by the building of the private school of Budi Dharma that was established before the tsunami on December 26, 2004. We concluded that the greater the value of the height of the tsunami inundation is equal to the level of light, medium, heavy and total damage, at the building of the national school of SD Negeri 50, Banda Aceh as the value of the deviation between the floors of the building from the tsunami load.

1 Introduction

Indonesian, especially Aceh Province, has a history of earthquakes due to the presence of two active faults, namely Aceh and Seulimuem segments. In 2004, an extreme earthquake measuring 9.3 on the Richter Scale rocked the city of Banda Aceh which occurred under the sea, when the Indian Plate was pushed down by the Burma Plate and was followed by a series of tsunamis. The impact of this disaster was quite significant, almost all structural and infrastructure buildings, public facilities, and other buildings were damaged.

One of the public buildings is a school building which is used for educational infrastructure. It is very important to estimate the vulnerability of school buildings against earthquakes and tsunamis because school buildings carry a high average living load. Therefore, the design of buildings in earthquake-prone areas must adhere to earthquake and tsunami standards.

Generally, a typical school building in the city of Banda Aceh consists of two types of buildings based on load-bearing systems, namely reinforced concrete and masonry. Reinforced concrete is a typical load-bearing building using reinforced concrete structures, while masonry walls use a typical load-bearing building with walls. These two typical load-bearing structures have different susceptibility to earthquake loads. This study takes a typical reinforced concrete load-bearing building.

As a result, damage to buildings due to earthquakes can be seen and the level of vulnerability predicted by referring to the fragility curve. The fragility curve is a curve that shows how big the probability of a building's vulnerability when exposed to an earthquake (NIBS, 2002). Fragile curves are needed to assess the vulnerability of existing school buildings in Banda Aceh City after the 2004 earthquake and tsunami. The fragility curve will show the level of damage to the building structure with various discrete damage conditions, such as slight damage, moderate damage, extensive damage, and complete damage. modeled using pushover analysis in SAP2000 v.20 software.

* Natasya Amalia: tasya110@gmail.com

Furthermore, the results of this study determine the vulnerability of a building obtained from the fragility curve developed and the capacity curve to see the level of damage and the probability percentage due to earthquake and tsunami loads. The results of the analysis that are reviewed are the value of building displacement and the deviation between floors from the tsunami load. So that it can be recommended improvements and various efforts to improve the structure of the building can be done to reduce these risks.

2 Literature review

2.1 Earthquake

Earthquake is a movement or vibration that occurs in the earth's crust caused by endogenous forces. Endogenous energy itself is energy from within the earth caused by changes in the earth's crust [1].

2.2 Tsunami

The tsunami, according to the Center for Volcanology and Geological Hazard Mitigation or PVMBG (2006) explains that a tsunami is a natural disaster in the form of ocean waves caused by an earthquake on the seabed and has the ability to spread at high speed, even the speed can exceed 900 km/hour. Puspito (2010) added that tsunamis can reach a height of one meter or even tens of meters from the coastline.

2.3 Structure loading

2.3.1 Dead load

Based on SNI 1727:2020 [2], dead load is the weight of all installed building construction materials, including architectural components, structural components and non-structural components as well as other installed service equipment.

2.3.2 Live load

Live loads are loads caused by users and occupants of buildings or other structures that are not included in construction loads (SNI 1727:2020) [2]. The minimum evenly distributed live load is in SNI 1727:2020 [2].

2.3.3 Earthquake loads

Earthquake loads are all equivalent static loads acting on buildings or parts of buildings that mimic the effects of ground motion due to the earthquake. Building planning against earthquake loads can be seen in SNI 1726:2019 [3].

2.3.4 Tsunami load

Tsunami waves cannot be seen when they are far in the middle of the ocean, but once they reach shallow areas the waves will move quickly. The energy generated by the tsunami is fixed to the function of altitude and its continuation (Sugito, 2008). The energy forms the forces that will become a load on the building structure.

2.3.5 Loading combination

According to SNI 1726-2019 [3], the factors and load combinations for nominal dead load (D), nominal live load (L) and nominal earthquake load (Q_E) that must be reviewed are as follows:

$$U_1 = 1{,}4\,D \quad (1)$$
$$U_2 = 1{,}2\,D + 1{,}6\,L \quad (2)$$
$$U_3 = 1{,}38\,D + 1\,L + 0{,}39\,Q_{Ex} + 1{,}3\,Q_{Ey} \quad (3)$$
$$U_4 = 1{,}38\,D + 1\,L + 0{,}39\,Q_{Ex} - 1{,}3\,Q_{Ey} \quad (4)$$
$$U_5 = 1{,}38\,D + 1\,L - 0{,}39\,Q_{Ex} - 1{,}3\,Q_{Ey} \quad (5)$$
$$U_6 = 1{,}38\,D + 1\,L - 0{,}39\,Q_{Ex} + 1{,}3\,Q_{Ey} \quad (6)$$
$$U_7 = 1{,}38\,D + 1\,L + 1{,}3\,Q_{Ex} + 0{,}39\,Q_{Ey} \quad (7)$$
$$U_8 = 1{,}38\,D + 1\,L + 1{,}3\,Q_{Ex} - 0{,}39\,Q_{Ey} \quad (8)$$
$$U_9 = 1{,}38\,D + 1\,L - 1{,}3\,Q_{Ex} - 0{,}39\,Q_{Ey} \quad (9)$$
$$U_{10} = 1{,}38\,D + 1\,L - 1{,}3\,Q_{Ex} + 0{,}39\,Q_{Ey} \quad (10)$$
$$U_{11} = 1{,}08\,D + 0{,}39\,Q_{Ex} + 1{,}3\,Q_{Ey} \quad (11)$$
$$U_{12} = 1{,}08\,D + 0{,}39\,Q_{Ex} - 1{,}3\,Q_{Ey} \quad (12)$$
$$U_{13} = 1{,}08\,D - 0{,}39\,Q_{Ex} - 1{,}3\,Q_{Ey} \quad (13)$$
$$U_{14} = 1{,}08\,D - 0{,}39\,Q_{Ex} + 1{,}3\,Q_{Ey} \quad (14)$$
$$U_{15} = 1{,}08\,D + 1{,}3\,Q_{Ex} - 0{,}39\,Q_{Ey} \quad (15)$$
$$U_{16} = 1{,}08\,D + 1{,}3\,Q_{Ex} + 0{,}39\,Q_{Ey} \quad (16)$$
$$U_{17} = 1{,}08\,D - 1{,}3\,Q_{Ex} - 0{,}39\,Q_{Ey} \quad (17)$$
$$U_{18} = 1{,}08\,D - 1{,}3\,Q_{Ex} + 0{,}39\,Q_{Ey} \quad (18)$$

ρ is the redundancy factor, which must be applied to the seismic force resisting system in each of the two orthogonal directions for all structures. The provisions for the value of can be seen in SNI 1726-2019 [3].

The load combinations for the tsunami based on FEMA P646 are as follows [4]:

$$U19 = 1{,}2\,D + 1{,}0\,T_S + 1{,}0\,L_{REF} + 0{,}25\,L \quad (19)$$
$$U20 = 0{,}9\,D + 1{,}0\,T_S \quad (20)$$

2.4 Design response spectrum

According to SNI 1726:2019 [3], the response spectrum is a spectrum presented in the form of a graph/plot that explains the relationship between the maximum response acceleration of a system of one degree of freedom due to a particular earthquake, as a function of the damping factor and natural vibration time. For periods greater than or equal to T_0 and less than or equal to T_s, the design acceleration response spectrum (S_a) is the same as S_{DS}.

2.5 Capacity curve

According to Dewobroto [5], pushover analysis is carried out by applying a lateral load in a certain pattern with a simulated earthquake load and must be given together with the combined effect of dead loads and not less than 25% of the required live load. Lateral loads must be applied to the center of mass for each story. From the pushover analysis will produce a capacity curve, which describes the relationship between the base shear and the displacement of the reference point on the roof structure. The capacity curve will show a linear condition before reaching the melting state and then behave non-linearly. This curve is influenced by the distribution pattern of the lateral force used as the thrust load. The main purpose of this earthquake susceptibility curve is to estimate the value of peak ground motion acceleration where the earthquake response of a structure exceeds the capacity resulting in structural failure.

2.6 Fragility curve

2.6.1 Seismic fragility curve

Duan & Pappin [6] in their research proposed a procedure to determine the median spectra displacement point at various damage conditions based on the results of nonlinear static pushover analysis. The steps of the procedure are described as follows.

a. Perform nonlinear static pushover analysis to get curve (spectrum)
b. Identifying the spectral displacement value on the yielding capacity (spectrum) curve. This is the median point of the yielding capacity.
c. Identify the spectral dispacement value when the first component reaches a complete breakdown condition. According to HAZUS [7] recommendation, this point is the median spectral displacement for slight damage conditions.
d. The median spectral displacement value for moderate damage conditions can be obtained by shifting the spectral displacement medium from slight damage conditions by a factor of 1.5.
e. Identify the spectral displacement with the collapse point on the capacity curve as the median spectral displacement point for complete damage conditions.
f. The median spectral displacement value for extensive damage conditions can be determined by placing it in the middle between the median points for moderate and complete damage conditions on a log scale.

Several damage limit criteria can be used in determining the performance of the structure based on the capacity curve. The limits for damage to building structures can be seen in Table 1.

Table 1. Description of structure performance limitations based on Hazus

Damage State	Description of the damage that occurred
Slight/Minor Damage	The plaster of the walls had small cracks in the corners of the door and window openings. Small cracks are considered visible with a maximum width of less than 1/8 inch (cracks wider than 1/8 inch are referred to as "large" cracks).
Moderate Damage	The plaster of the walls has major cracks, large cracks in the door and window openings, on some of the columns there are shards and cracks that are rather large but the columns can still safely stand upright.
Extensive Damage	Large cracks in the shear wall panels, large cracks in the joints between the columns and beams, several columns were damaged badly enough to no longer be safe as the main support for the structure, the floor slabs experienced large cracks, and the foundation structure suffered large cracks.
Complete Damage	The structure of the building may be in danger of collapsing due to failure of fragile walls, cracks in large foundations, some of the columns have collapsed. Three percent of the total building area with total damage is expected to collapse, on average.

The seismic fragility curve of the building will describe the probability of damage to the building based on the acceleration spectra as shown in Fig. 1.

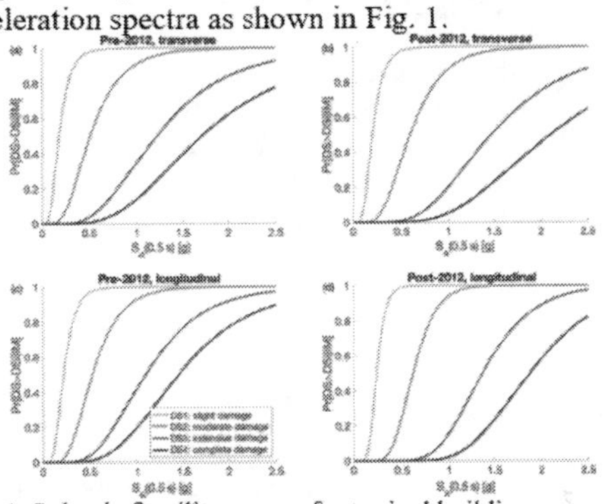

Fig. 1. Seismic fragility curves for typical buildings

2.6.2 Tsunami fragility curve

The assessment of building damage due to the tsunami can be described by means of a fragility curve. The tsunami fragility curve function was developed to estimate the impact of future tsunamis. The level of damage is classified into five kerus status from the curve, as shown in Figure 2.

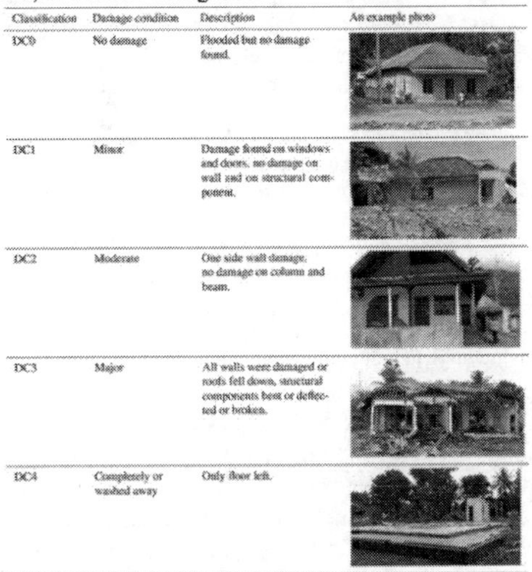

Classification	Damage condition	Description	An example photo
DC0	No damage	Flooded but no damage found.	
DC1	Minor	Damage found on windows and doors, no damage on wall and on structural component.	
DC2	Moderate	One side wall damage, no damage on column and beam.	
DC3	Major	All walls were damaged or roofs fell down, structural components bent or deflected or broken.	
DC4	Completely or washed away	Only floor left.	

Fig. 2. The performance limitations of structure

The tsunami fragility curve as shown in Figure 3.

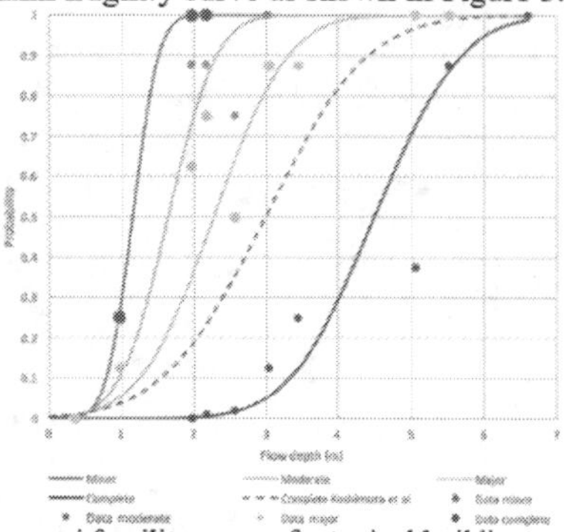

Fig. 3. Tsunami fragility curves for typical buildings

3 Research methods

3.1 Object of research

The object of research in this study is a typical three-story reinforced concrete school building located in the city of Banda Aceh using a rapid assessment survey. The study took 2 typical buildings from the survey results of 100 three-story school buildings and took the typical structure that dominates the structural design of the building, then looked at the location for building selection.

3.2 Data types and sources

The types of data needed in this research are primary data obtained directly through field surveys on school buildings in the city of Banda Aceh as well as, secondary data for supporting data together with a reference for the implementation of the primary survey. Meanwhile, the literature used in this study comes from journals, websites, books related to earthquake analysis which are included in the pushover analysis. The regulations used include SNI-1726-2019 [3] concerning Procedures for Planning Earthquake Resistance for Building and Non-Building Structures, SNI-03-1727-2013 [8] concerning Minimum Loads for Designing Buildings and Other Structures, Map Sources and Hazards of the 2017 Indonesian Earthquake, as well as books related to this research.

3.3 Data collection technique

Research data collection was carried out by means of a field survey. The survey was used to collect building structure data in the form of structural dimensions of the building and building vulnerability factors from the two school buildings surveyed. The survey was ealuated firsthand how the condition of the building. To complete the data on the survey form, a map of the fault location is needed together with tsunami data analysis by SAP2000 v.20, while the data is inputted directly from the developed earthquake and tsunami fragility curve, converted using an online graphreader application as shown in Figure 4.

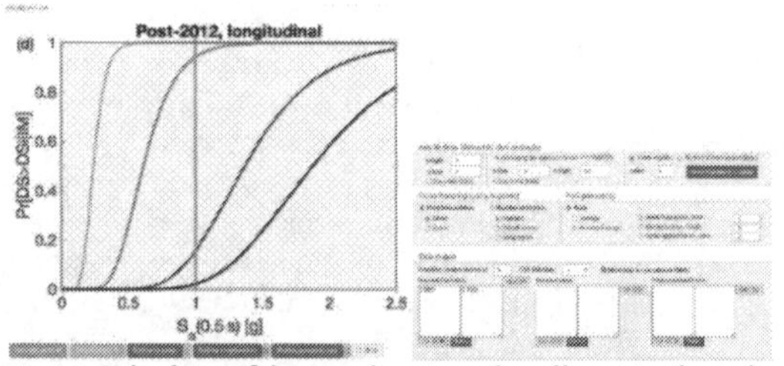

Fig. 4. Display of image input and online graphreader application data results

3.3.1 Building technical data

Based on field research, there is excess reinforcement visible in the beam so that it can be used as a source for modeling school building structures. As taken from Gentile's research [9], from the results of his research on From Rapid Visual Survey to Multi-Hazard Risk Prioritization and Numerical Fragility of School Buildings in Banda Aceh, Indonesia, typical details of building structures for various forms of buildings in Banda Aceh as shown in Table 2.

Table 2. Typical quality of concrete and steel reinforcement for school buildings in Banda Aceh.

Year	Material	Typical Beam	Typical Column	Typical Join
Before 2012	Concrete fc = 21.5 MPa	3Φ16 up	3Φ16 up	without Sengkang
	Bones Lon. fy = 400 MPa	3Φ16 down	3Φ16 down	
	Tran Bone. fy = 240 MPa	Sengkang Φ 10-150 mm	Sengkang Φ 10-200 mm	
After	Concrete fc = 24 Mpa	3Φ16 up	3Φ16 up	without Sengkang
	Bones Lon. fy =400 MPa	3Φ16 down	3Φ16 down	
	Tran Bone. fy = 240 MPa	Sengkang Φ 10-150 mm	Sengkang Φ 10-100 mm	

3.3.2 Earthquake data

Earthquake data used in this study is a scenario recording of the earthquake in the Sumatra – Andaman fault on December 26, 2004 obtained from the United States Geological Survey or USGS and earthquake forecast data obtained from the portal of the Center for Research and Development of Housing and Settlements (PUSKIM) based on maps. Indonesia earthquake 2017. Furthermore, in the discussion results chapter it is called "USGS earthquake recording" and the estimated load for the 2017 Indonesian earthquake map is referred to in the discussion results chapter as "estimated Indonesia earthquake map 2017".

3.3.3 Tsunami data

Banda Aceh is located between the 200 km long Aceh segment and the 120 km Seulimum segment. The existence of this segment results in vulnerability to earthquakes and tsunamis. The estimated height of the tsunami inundation inBanda Aceh City was obtained from the research of Iemura, et al [10].

1. Combined earthquake loading

The seismic design category for the second typical building is category D and the redundancy factor used (ρ) is 1.3. This study did not review the soil conditions, so that the lateral soil load (H) was considered 0. To obtain the ultimate load from the loads that might occur on the structure, a combination of factored loads was carried out. Combination of loading as the equation 1-18.

The results of the run analysis of each combination will shown as a combination that results in the largest structural deformation. The result will be used for a displacement control in the pushover analysis.

2. Combination of tsunami loading

The combination of tsunami loading is similar to the equation 19-20.

3.4 Pushover analysis

Analysis of pushover loading in stages, the direction of lateral loading was carried out in the direction of the main axis of the building. Lateral loading uses the displacement-controlled concept, namely, the loading process is increased gradually until the displacement target is achieved.

3.4.1 Gravity loading

In the static thrust load analysis, the percentage of the load used in the gravity load is 100% dead load (including additional dead load) and 50% live load. Gravity loading is carried out on the Load Case Name named "Grav" where the loading is carried out at the initial condition before receiving the load. In Load Applied, 2 types of working loads are used, namely dead loads and live loads with each scale factor given 1. Then the Analysis type was selected Nonlinear analysis because what will be done is nonlinear analysis.

3.4.2 Lateral loading

In this analysis, two directions of lateral loading were made as pushover loads, namely the X direction and the Y direction. Lateral loads were added to the Load Case Data table.

3.5 Data analysis

3.5.1 Analysis of the seismic performance limits of the developed capacity spectrum curve

The capacity spectrum curve that gets the value of the median spectral acceleration (S_a) parameter is one of the important parameters to get the probability of structural failure (P) which is used to analyze the seismic fragility curve. To get the median spectral acceleration (S_a) value, it is necessary to convert the capacity curve which is the basic shear force curve (V) to roof/floor displacement (Δroof) into a capacity spectrum curve which is the spectral acceleration curve (S_a) to spectral displacement (S_d) and can be seen in section 2.4 page 15. In section 2.8.1 page 19 it has been explained that the table was identified according to the analysis proposed by Duan and Pappin (2008) to obtain the values of the median spectral displacement and median spectral acceleration at various levels of damage.

The final result of this study is to analyze the level of damage obtained from the fragility curve developed by using the fragility curve from the journal Natural Hazards and Earth System Sciences by Gentile, et al. [9], a building that shows the probability of damage to the building when it receives an earthquake load with a certain intensity. Then, it is also shown when each of these buildings was built, so that the curve after or after the tsunami 2004 can be used. The fragility curve will show the percentage of the probability of damage to the building structure with various damage conditions, namely light damage, moderate damage, severe damage, and total damage. The median spectral acceleration at various levels of damage is obtained from the capacity spectrum curve. As well as reviewing the value of the acceleration spectra on the response curve of the capacity spectrum in order to see the condition of the location of the level of damage to the building.

3.5.2 Tsunami analysis

This study also was analyzed the calculation of the tsunami load for the risk of reducing damage to building structures in the event of a tsunami disaster. The displacement is obtained by entering the load on the column structure with a height as high as the depth of the tsunami flow, for each building, namely the private junior high school of Budi Dharma (Sekolah Budi Dharma) with a flow depth of 1.8 meters, the elevation of the base of the building to the water level elevation as high as 2 meters and for the national elementary school 50 (SD Negeri 50) with a flow depth of 2.6 meters, the elevation of the base of the building to the water level as high as 2 meters, the height is taken according to the sub-district because the school area was not hit by the tsunami, only entered the height to show the condition of the school building if it was hit by a tsunami disaster. After that, an analysis was carried out on SAP2000 v.20 to compare the height of the tsunami with the probability of the fragility curve that was developed (the vulnerability curve of the Volcanogenic tsunami of Mount Anak Krakatau, research by Syamsidik, et al [11] on the two typical school buildings.

4 Results and discussions

4.1 Survey data recapitulation results

The determination of school buildings was carried out from the results of data recapitulation based on the selection of buildings that are typical school buildings in the city of Banda Aceh. The results of data recapitulation in the form of infographics is shown in Figure 6 dan Figure 7.

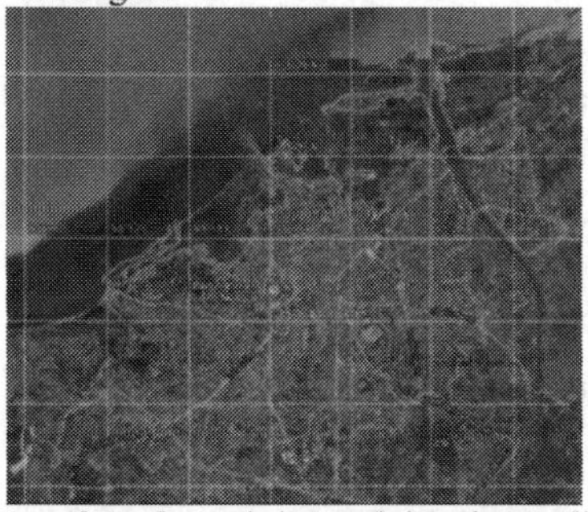

Fig. 6. Map of Banda Aceh (research location Budi Dharma School)

Fig. 7. Map of Banda Aceh (research location SD Negeri 50)

Based on the results of the data recapitulation of 100 school buildings, the selection of school buildings by looking at the relatively close distance to the shore and faults, while the selection of these buildings can be concluded that the criteria for the two forms of school buildings that dominate in the city of Banda Aceh as follow:

- Building shape : Type I (top view of I-shaped building)
- Number of floors : 3
- Height per floor : 3.5 m
- Long : 24 m and 40 m
- Wide : 9 m
- Composition material : Reinforced Concrete (RC)

School buildings that meet the criteria from the conclusions above are Sekolah Budi Dharma and SD Negeri 50.

The data obtained in the field for the Sekolah Budi Dharma and SD Negeri 50 Banda Aceh in the form of building characteristics, position, structural dimensions, and reinforcement details is shown in Tables 3 - 8.

Table 3. Characteristics and position of the Sekolah Budi Dharma and SD Negeri 50 Banda Aceh

No	Criteria	Sekolah Budi Dharma	SD Negeri 50
1	Place Coordinates	Longitudinal 95.3145, Lateral 5.5478	Longitunal 95.31128, Lateral 5.53396
2	Construction year	2001-2005	2016-2017
3	Distance to fault	3 Km	6 Km
4	Total students	- person	- person
5	Number of floors	3 floors	3 floors
6	Height/floor	3 meters	3 meters
7	Material	*Reinforced Concrete (RC)*	*Reinforced Concrete (RC)*
8	Space frame system	*Frames*	*Frames*
9	Roof shape	*Multi Pitch*	*Multi Pitch*
10	Roof type	Wood frame	Wood frame

Tabel 4. Plate reinforcement dimensions and details

Type	Plate Thickness(mm)	Position	reinforcement
Floor	120	focus	D 10-200
		Field	D 10-200
Ladder	120	focus	D 13-150
		Field	D 13-150

Tabel 5. Dimensions and details of the beam reinforcement of the Sekolah Budi Dharma of Banda Aceh

Code	Dimensions (mm)	Position	reinforcement		
			Top	Bottom	Stirrup
B	300 x 800	Tumpuan	4Φ16	4Φ16	Φ10-100 mm
		Lapangan	4Φ16	4Φ16	Φ10-100 mm
B2	300 x 500	Tumpuan	4Φ16	4Φ16	Φ10-100 mm
		Lapangan	4Φ16	4Φ16	Φ10-100 mm

Tabel 6. Dimensions and details of beam reinforcement for SD Negeri 50 Banda Aceh

Code	Dimensions (mm)	Position	reinforcement		
			Top	Bottom	Stirrup
B1	300 x 600	Tumpuan	4Φ16	4Φ16	Φ10-100 mm
		Lapangan	4Φ16	4Φ16	Φ10-100 mm

Tabel 7. Dimensions and details of the reinforcement of the Sekolah Budi Dharma Banda Aceh column

Code	Dimensions (mm)	Position	reinforcement		
			Top	Bottom	Stirrup
K1	300 x 600	Tumpuan	4Φ16	4Φ16	Φ10-150 mm
		Lapangan	4Φ16	4Φ16	Φ10-150 mm
K2	350 x 400	Tumpuan	4Φ16	4Φ16	Φ10-150 mm
		Lapangan	4Φ16	4Φ16	Φ10-150 mm

Tabel 8. Dimensions and details of the reinforcement of the SD Negeri 50 Banda Aceh column

Code	Dimensions (mm)	Position	reinforcement		
			On	Bawah	Sengkang
K1	400 x 500	Tumpuan	4Φ16	4Φ16	Φ10-150 mm
		Lapangan	4Φ16	4Φ16	Φ10-150 mm

4.2 Number of variety analysis results

Modal analysis was used to evaluate the effective mass participation of variance whose magnitude must be greater than 90%, this value is obtained directly from SAP2000, the mass variance value for the USGS earthquake record and the estimated Indonesia earthquake map 2017 produces the same value, these values is shown in Table 9 below.

Tabel 9. Effective Mass Diversity Participation the USGS earthquake record and the estimated Indonesia earthquake map 2017

Bangunan	Tipe Modal	Item	Static (%)	Dynamic (%)
SMP BUDI DARMA	Acceleration	UX	100	99,5995
	Acceleration	UY	100	99,2121
	Acceleration	UZ	0,2475	0,0467
SDN 50 Banda Aceh	Acceleration	UX	99,1824	92,3549
	Acceleration	UY	99,1355	92,7287
	Acceleration	UZ	0,1183	0,0022

Based on the table above for the USGS earthquake records and the estimated Indonesia earthquake map 2017 was reviewed in the two buildings, it shows that the mass participation value of variance has met the provisions of the SNI 03-1726 - 2019 limit, namely the effective variance mass participation factor for the x-direction and y-direction of a minimum of 90%.

4.3 Structural fundamental period analysis

In the fundamental period of the structure (the period of structural vibration) the symbol T is a very important property in the design process of earthquake-resistant building structures. The period of vibration of the structure will determine the magnitude of the earthquake load applied in the design of the structure.

From the results of SAP2000, the natural fundamental period time T = 0,518 for Sekolah Budi Dharma and T = 0,611 for SD Negeri 50 with Ta max obtained from the calculation of SNI 1726-2019. So, T analysis < Ta max, then the structure is declared to have met the conditions for the permit fundamental period obtained from the calculation of SNI 1726-2019.

Tabel 10. The results of the analysis of the fundamental vibration time of the building structure

Building	Step Num	Periode	$T_{aSyarat}$	Information
	Unitless	Sec	Sec	
SMP BUDI DARMA	1	0,31614	0,518	Fulfil
	2	0,23386		
	3	0,19872		
	4	0,18618		
	5	0,13514		
	6	0,12224		
	7	0,11696		
	8	0,10126		
	9	0,09100		
	10	0,06678		
	11	0,05571		
	12	0,03758		
SDN 50 Banda Aceh	1	0,3279	0,611	Fulfil
	2	0,2892		
	3	0,2718		
	4	0,2138		
	5	0,1923		
	6	0,1889		
	7	0,1768		
	8	0,1744		
	9	0,1643		
	10	0,1604		
	11	0,1493		
	12	0,1372		

Based on Table 10, the natural period of the building resulting from SAP2000 v.20 has also met SNI 03 – 1726 – 2019 with the limit period for the vibration of the sekolah Budi Dharma building for the USGS earthquake recording and the estimated 2017 Indonesian earthquake map of 0.518 seconds, while the vibration period limit for the building is SDN 50 Records. USGS earthquake and the estimated Indonesia earthquake map 2017 is 0.611 seconds. The limit value of the vibration period in the two buildings is different due to the different building heights.

4.4 Results of Comparative Evaluation of the Capacity Curves for the USGS earthquake and the estimated Indonesia earthquake map 2017

The pushover analysis method was aimed at building capacity curves. The capacity curve shows the relationship between the recorded displacement and the resultant shear force caused by the specified displacement target or static load to the ultimate building condition. The largest recorded displacement in the x direction in the Sekolah Budi Dharma building for the USGS earthquake record is found in combination 14 at joint 96 of 0.060599 m and the y direction is found in combination 6 at joint 119 of 0.272601 m, for the estimated Indonesia earthquake map 2017 is found in combination 4 at joint 108 of 0.22029 m and the y direction is found in combination 6 at joint 119 of 0.306956 m. While the SD Negeri 50 building for USGS earthquake records is found in combination 12 at joint 151 of 0.075015 m and the y direction is found in combination 6 at joint 137 is 0.064917 m, for the estimated Indonesia earthquake map 2017 is found in combination 12 at joint 151 of 0.039657 m and the y direction is found in combination 6 at joint 160 of 0.162109 m. Monitored displacement is the point that experiences the largest structural deformation of a combination of factored loads that follow equations (1) to (18). The result of the combined factored load alone.

4.4.1 Capacity curve for sekolah Budi Dharma of Banda Aceh

The capacity curve was showed nonlinear due to an increase in load which makes the structural elements of the building change from an elastic condition to a plastic condition, for more details the capacity curve in the x direction and y direction is shown in the figure below.

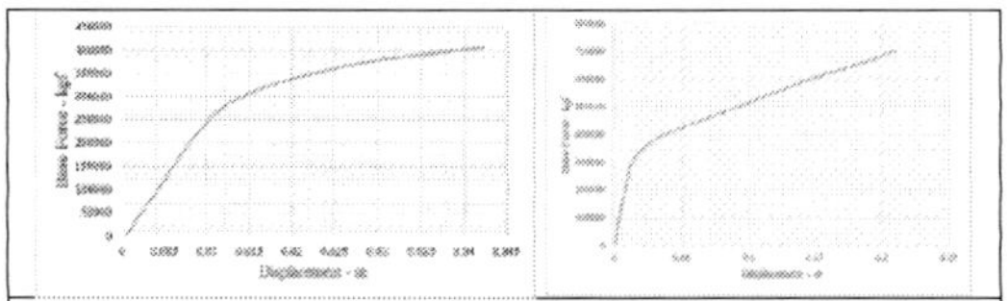

Fig. 8. Capacity curve x-direction (USGS Earthquake Record- Estimated Indonesia Earthquake Map 2017)

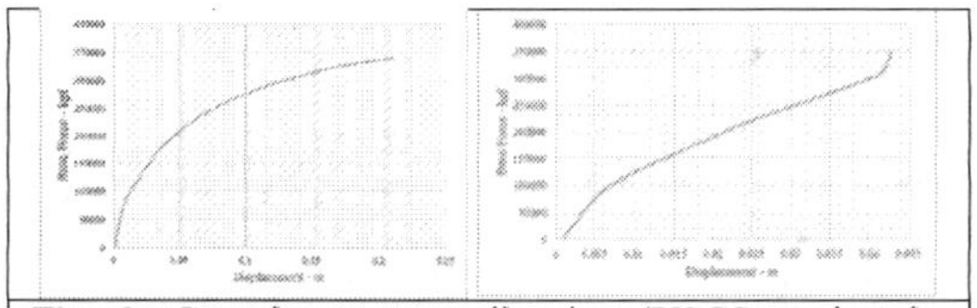

Fig. 9. Capacity curve y-direction (USGS earthquake record- Estimated Indonesia earthquake map 2017)

4.4.2 Capacity Curve for SD Negeri 50 Banda Aceh

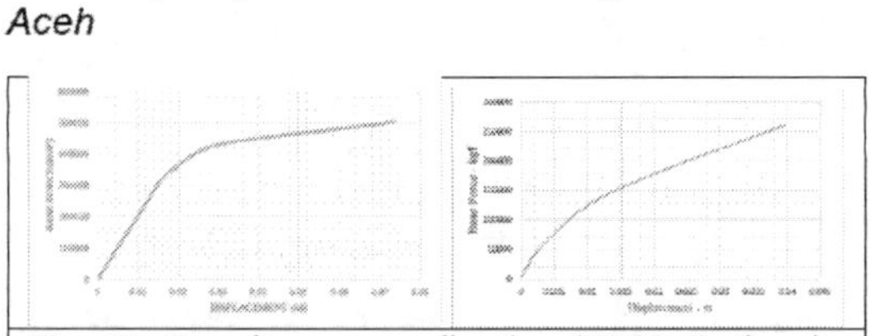

Fig. 10. Capacity curve x-direction (USGS Earthquake Record- Estimated Indonesia Earthquake Map 2017)

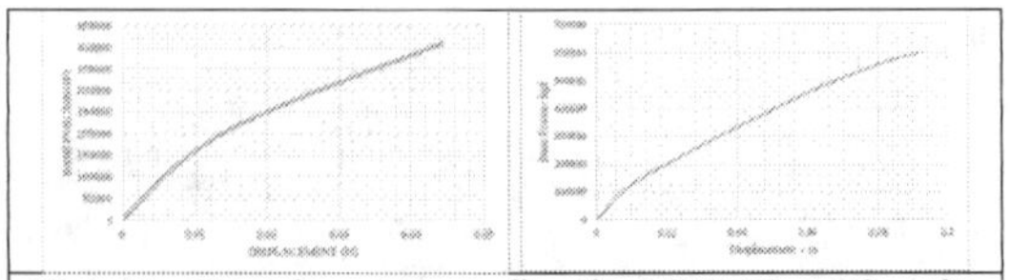

Fig. 11. Capacity curve y-direction (USGS earthquake record- Estimated Indonesia earthquake map 2017)

Based on the capacity curve as shown in Figures 8 - 11, the result indicated the pushover analysis in the x and y directions shows the maximum displacement value and the base shear value as shown in Table 11 below.

Tabel 11. Maximum displacement value and base reaction value

School building	Earthquake type reviewed	Displacement (m)		Base Shear (Kgf)	
		x direction	y direction	x direction	y direction
Sekolah Budi Dharma	USGS Earthquake Record	0,04	0,21	408276,62	339411,38
	Estimated Indonesia Earthquake Map 2017	0,21	0,04	699841,47	349533,69
SD Negeri 50	USGS Earthquake Record	0,07	0,04	502770,9	410214,3
	Estimated Indonesia Earthquake Map 2017	0,039	0,092	259991,19	598606,47

The table above shows that the Sekolah Budi Dharma building against the USGS earthquake recording is more flexible and prone to shifting in the y direction, on the contrary for the 2017 Indonesia earthquake map forecast it is more flexible and susceptible to shifting in the x direction. Meanwhile, the SD Negeri 50 building for the USGS earthquake record is more flexible and susceptible to shifts in the x direction, the estimated Indonesia earthquake map 2017 looks more flexible and vulnerable to shifts in the y direction. Based on the results of the largest value of the displacement monitor, it can be seen that the displacement in the Sekolah Budi Dharma building due to the burden of the USGS earthquake recording is smaller than the displacement due to the Puskim load, as well as the SD Negeri 50 building.

4.4.3 The median value of spectral displacement graph point and damage the state's Sekolah Budi Dharma Banda Aceh

The capacity spectrum curve is useful for getting the value of the median spectral acceleration parameter. This has been supported in the SAP2000 program.

Furthermore, to form the seismic fragility curve of the building, it is necessary to determine the position of each damage state on the capacity spectrum curve. Determination of the position of each damage limit on the capacity spectrum curve based on research [6].

Based on Duan & Pappin [6], the stages of identifying the capacity spectrum curve to obtain the median spectral displacement point value are as follows as shown in the graph below as well as determination of each damage state on the capacity spectrum curve is shown in the following figure.

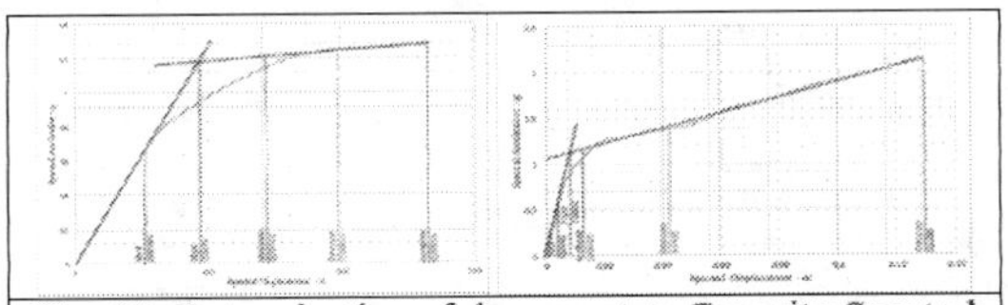

Fig. 12. Determination of damage state Capacity Spectral x-direction (USGS earthquake recording - Estimated 2017 earthquake map in Indonesia)

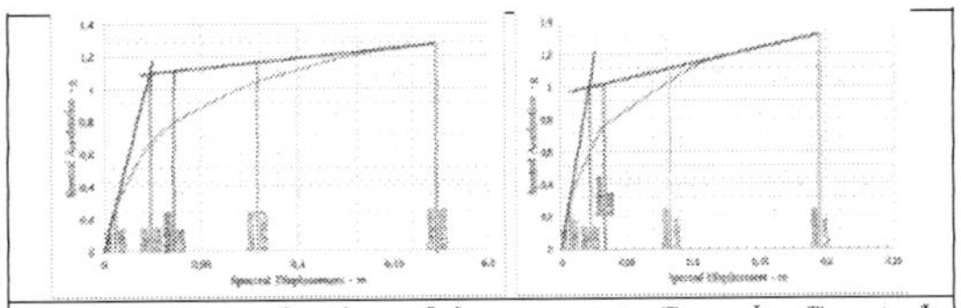

Fig. 13. Determination of damage state Capacity Spectral y-direction (USGS earthquake recording - Estimated map of Indonesia earthquake 2017)

The graph above was obtained from by the calculation of the ATC 40 Capacity Spectral curve, so we are able to determine the damage state curve for each building.

4.4.4 Chart median spectral displacement point and damage state for SD Negeri 50

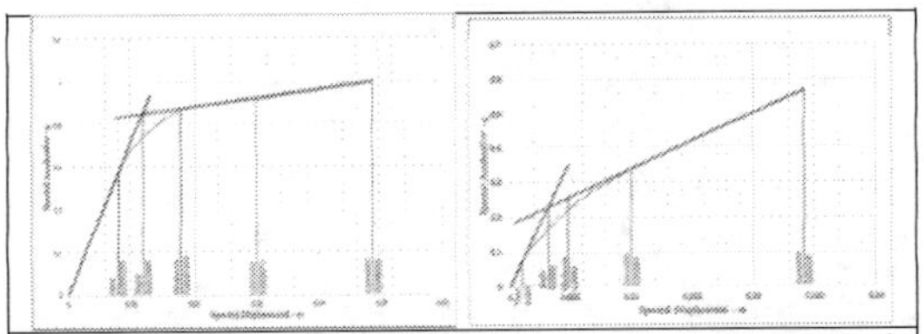

Fig. 14. Determination of damage state Capacity Spectral x-direction (USGS earthquake recording - Estimated 2017 earthquake map in Indonesia)

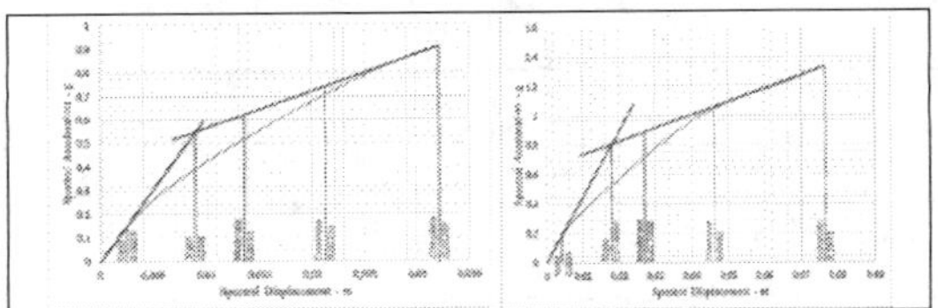

Fig. 15. Determination of damage state Capacity Spectral y-direction (USGS earthquake recording - Estimated map of Indonesia earthquake 2017)

As discussed above, the graph was obtained from the calculation of the ATC 40 Capacity Spectral curve, so that we can determine the damage state curve for each building.

4.5 The results of the comparison analysis of the damage limit of the spectrum capacity curve

The seismic capacity curve of the building is the relationship between the spectral acceleration as the horizontal axis and the probability of structural failure as the vertical axis. What is really needed in describing the fragility curve is the value of the acceleration spectra at each damage state. Based on the identification of the capacity spectrum curve in section 4.4.3 and section 4.4.4, the spectral acceleration and spectral displacement points are generated for each structural failure condition as recapitulated in the table below.

1. The results of the spectral acceleration and spectral displacement points for the Sekolah Budi Dharma Banda Aceh

Tabel 12. S_d and S_a values for each type of structural damage to the Sekolah Budi Dharma Banda Aceh .

Earthquake type reviewed	*Demage state*	S_d, ds (m)		Sa,ds (g)	
		x direction	y direction	x direction	y direction
USGS Earthquake Record	*Slight*	0,0095	0,0240	0,9545	0,6664
	Moderate	0,01425	0,0360	1,1435	0,7944
	Extensive	0,0197	0,0798	1,2480	1,0496
	Complete	0,0265	0,1724	1,2847	1,2730
Estimated Indonesia Earthquake Map 2017	*Slight*	0,0095	0,0220	0,9528	0,5960
	Moderate	0,01425	0,0330	1,1125	0,7659
	Extensive	0,0426	0,0824	1,3922	1,0315
	Complete	0,1288	0,1945	2,1558	1,3212

2. The results of the spectral acceleration and spectral displacement points for SD Negeri 50 Banda Aceh

Tabel 13. S_d and S_a values for each type of structural damage to the SD Negeri 50 Banda Aceh.

Earthquake type reviewed	*Demage state*	S_d, ds (m)		Sa,ds (g)	
		x direction	y direction	x direction	y direction
USGS Earthquake Record	*Slight*	0,0125	0,0090	0,7519	0,3916
	Moderate	0,01873	0,0135	0,8720	0,5910
	Extensive	0,0304	0,0213	0,9201	0,7066
	Complete	0,0487	0,0321	0,9953	0.9080
Estimated Indonesia Earthquak e Map 2017	*Slight*	0,0032	0,0180	0,1581	0,5060
	Moderate	0,00480	0,0270	0,2048	0,6891
	Extensive	0,0108	0,0462	0,3466	1,0278

Earthquake type reviewed	*Demage state*	S_d, ds (m)		Sa,ds (g)	
		x direction	y direction	x direction	y direction
	Complete	0,0242	0,7671	0,5678	1,3376

4.5.1 The result of the comparison of the percentage value of the probability value of the developed earthquake fragility curve

Based on subchapter 1, S_a can be searched for each building, so it can be analyze the percentage of the probability of the developed curve. Data were obtained from the results of the Spectra response program on the Puskim portal.

1. Calculation of acceleration spectra for Sekolah Budi Dharma building

Where, T = 0.518 seconds
T_0 = 0.135799 seconds
T_s = 0.678993 seconds
T_L = 14,000000 seconds
S_D = 1.001484 g

Due to T T_0 or T T_S , then the same as S_{DS}. S_a
S_a= 1.001484 g

Furthermore, to find out what percentage of the probability of a building experiencing damage limits is shown in the fragility curve below.

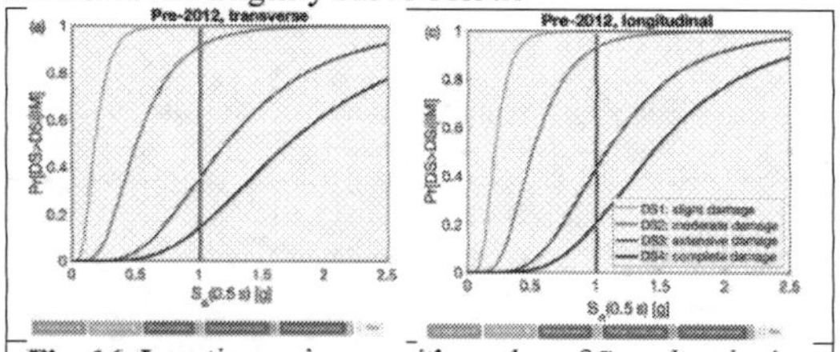

Fig. 16. Inserting an image with a value of S_a and assigning a point to follow the graph in the x-direction and y-direction

2. Calculation of acceleration spectra for SD Negeri 50 . buildings

Where, T = 0.611 second
T_0 = 0.135816 seconds
Ts = 0.679080 seconds
T_L = 12000000 seconds
SD = 1.001354 g

Due to T T0 or T≤ TS , then the same as SDS. S_a
S_a= 1.001354 g

From the calculation above, it can be seen the percentage probability of the developed curve. The curve taken by the study on the fragility curve from the journal Natural Hazards and Earth System Sciences by Gentile, et al (2019).

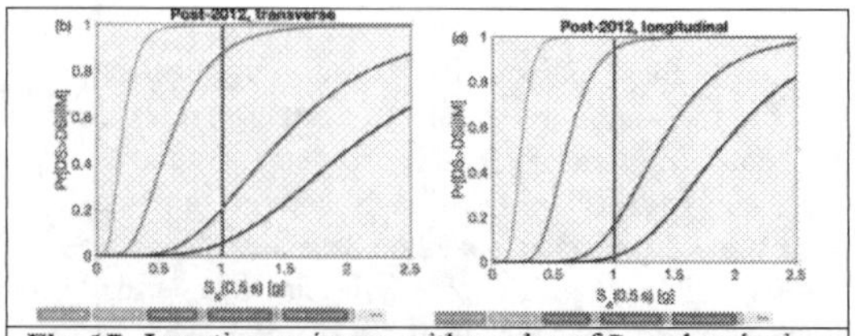

Fig. 17. Inserting an image with a value of S_a and assigning a point to follow the graph in the x-direction and y-direction

Tabel 14. Percentage probability for each type of structural damage to the building under review.

Building reviewed	*Demage state*	S_a (g)	Probabilitas (%)	
			x direction	y direction
Sekolah Budi Dharma	***Slight***	1,001484	99.7	99.5
	Moderate		91	93.6
	Extensive		35.9	43.1
	Complete		14.4	20.1
SD Negeri 50	***Slight***	1.001354	100	99.8
	Moderate		87.8	94
	Extensive		20.9	16.6
	Complete		5.7	2.4

Comparison of the value of S_a and the probability of the building as shown in Table 14, it can be concluded that the greater the value of S_a of the building (Sekolah Budi Dharma), the greater the percentage of the probability of moderate, severe and total damage. This is also influenced by the Sekolah Budi Dharma building which was established before the earthquake and tsunami on December 26, 2004. Meanwhile, SD Negeri 50 has a small value of S_a, so that the probability of total damage is getting smaller.

4.6 Tsunami load calculation results

4.6.1 The results of the analysis of the inter-story drift the tsunami load floors

The results of the deviation between floors to the combination of tsunami loads which were occur in each school is shown in Table 15 and Table 16.

Tabel 15. Deviation between floors due to tsunami direction x

School building	Floor	Tall	Displace ment x	The drift between floors	Deviation between floors permits
		(m)	(m)	(m)	(m)
Sekolah Budi Dharma	3	3,5	0,000087	-0,000067	0.035
	2	3,5	0,000154	0,000081	0.035
	1	3,5	0,000073	0,000073	0.035
SD Negeri 50	3	4	0,000078	0,000050	0,08
	2	4	0,000028	0,000022	0,08
	1	4	0,000006	0,000006	0,08

Tabel 16. Deviation between floors due to tsunami direction y

School building	Floor	Tall	Displacement x	The drift between floors	Deviation between floors permits
		(m)	(m)	(m)	(m)
Sekolah Budi Dharma	3	3,5	0,002143	0,000105	0.035
	2	3,5	0,002038	0,000615	0.035
	1	3,5	0,001423	0,001423	0.035
SD Negeri 50	3	4	0,001222	0,000077	0,08
	2	4	0,001145	0,000145	0,08
	1	4	0,001	0,001000	0,08

Based on the table above, it can be seen that the deviation between floors in the x and y directions based on SNI 1726: 2019 [3] has met the criteria because the deviation between floors does not exceed the allowed deviation between floors.

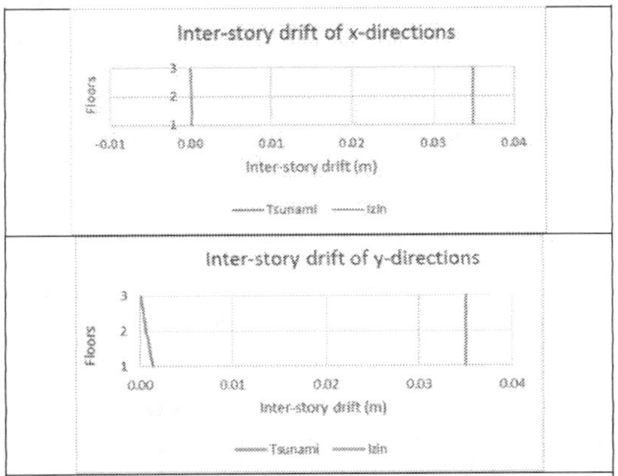

Fig. 18. The inter-story drift in the x and y directions of the Budi Dharma school

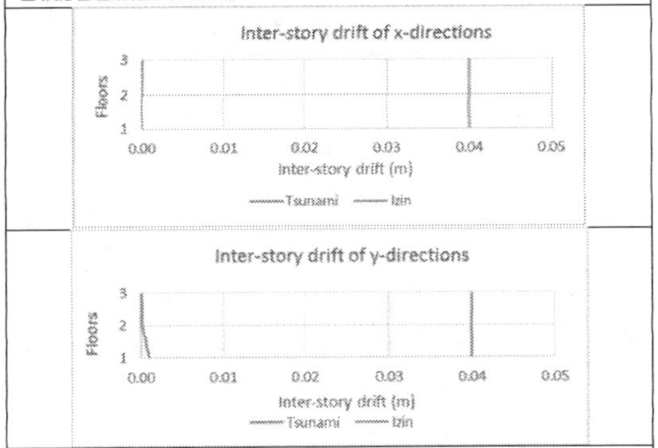

Fig. 19. The inter-story drift in the x and y directions of the SD Negeri 50

In Figures 18-19 it suggested that the deviation between floors due to earthquake loads has exceeded the permit limit based on SNI 1726:2019 [3] so that the treatment of buildings due to earthquake loads must be strengthened so that the building can withstand tsunami loads.

4.6.2 The results of the comparison of the percentage probability values of the developed tsunami fragility curve

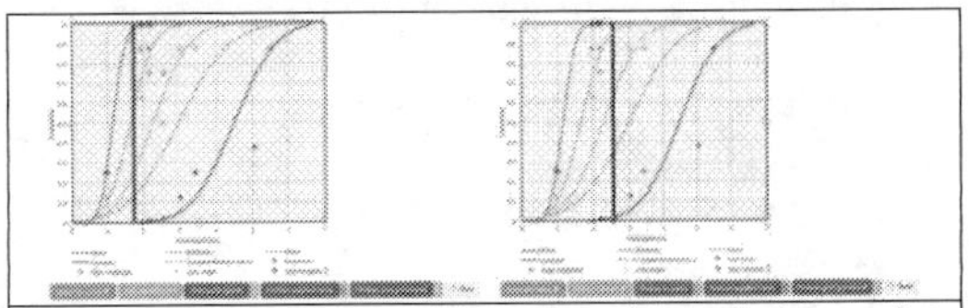

Fig. 20. Insert a picture with a value of Sa and give a point following the graph (Sekolah Budi Dharma - Sekolah SDN 50)

Tabel 17. Comparison of damage state results to probability and flow depth

Building reviewed	*Demage state*	*Flow depth* (m)	Probabilitas (%)
Sekolah Budi Dharma	***Minor***	1,8	97
	Moderate		57
	Major		25
	Complete		0.5
SDN 50	***Minor***	2,6	100
	Moderate		94
	Major		62
	Complete		1,4

Based on the results of the graphs and tables above, it can be drawn a comparison of the two school buildings if the greater the value of the tsunami inundation height on the SD Negeri 50 building, the greater the level of light, medium, heavy and total damage when given a tsunami load. At the Sekolah Budi Dharma, because it has a smaller height than SD Negeri 50, it produces a smaller percentage probability value for the level of damage.

5 Conclusions and recommendation

5.1 Conclusion

1. Based on nonlinear statistical analysis the x direction shows the maximum displacement value of the two buildings was 0.21 m and the base shear value was 699841.47 Kgf, while in the y direction the maximum displacement value of the two buildings was 0.21 m and the base shear value was 339411, 38 kgf.
2. The spectral acceleration (S_a) value of the capacity spectrum of Sekolah Budi Dharma was 1.001484 g and SD Negeri 50 was 1.001354 g.
3. The results of the comparison of the value of S_a and the probability of the two buildings when viewed against the earthquake is obtained if the value of S_a is large, the greater the percentage of the probability of moderate, severe and total damage is also influenced by the building being reviewed before the earthquake and tsunami on 26 December 2004 (Sekolah Budi Dharma).

4. The results of the comparison of the two school buildings, if the greater the value of the tsunami inundation height on the SD Negeri 50 building, the greater the level of light, medium, heavy and total damage when viewed from the tsunami hazard. At the Sekolah Budi Dharma, because it has a smaller height than SD Negeri 50, it produces a smaller percentage probability value for the level of damage.

5.2 Recommendation

1. Further research is expected to be able to make variations of earthquake modeling based on other earthquake sources in order to get different spectral response values.
2. In this study, what is analyzed is the existing building with structural data through thorough research results, it is hoped that further research will use building structural data such as draft drawings, concrete compressive strength test results, and steel tensile test results applied to which buildings will be used. make seismic fragility curves and tsunami fragility curves more accurate.

References

1. Bayong, T. HK. (2006). *Ilmu Kebumian dan Antariksa*, PT Remaja Rosdakarya, Bandung.
2. Badan Standarisasi Nasional. 2020. Standar Nasional Indonesia 1727-2020: Beban Desain Minimum dan Kriteria Terkait untuk Perancangan Bangunan Gedung dan Struktur Lain, Badan Standarisasi Nasional: Jakarta.
3. Badan Standarisasi Nasional. 2019. Standar Nasional Indonesia 1726-2019: Tata Cara Perencanaan Ketahanan Gempa untuk Struktur Bangunan Gedung dan Non Gedung. Badan Standarisasi Nasional: Jakarta.
4. Federal Emergency Management Agency. (2019). *Guidelines for Design of Structures for Vertical Evacuation from Tsunami Second Edition.* Department of Homeland Security FEMA Mitigation Division: Washington D.C.
5. Dewobroto, W. (2005). *Evaluasi Kinerja Struktur Baja Tahan Gempa dengan Analisis Pushover*, Prosiding Civil Engineering National Conference: Sustainability Construction & Structural Engineering Based on Professionalism, 27 hal, Universitas Katolik Soegijapranata, Semarang.
6. Duan, X., dan Pappin, JW. (2008). A Procedure for Establishing Fragility Functions for Seismic Loss Estimate of Existing Buildings Based on Nonliner Pushover Analysis, dipresentasikan pada *The 14th World Conference on Earthquake Engineering, Beijing,* 12-17 Oktober.
7. HAZUS (Hazard US), (1999): Earthquake Loss Estimation, National Institute of Building Sciences. NIBS: Washington D.C.
8. Badan Standarisasi Nasional. 2013. Standar Nasional Indonesia 2847 :2013: Persyaratan Beton Struktural untuk Bangunan Gedung. Badan Standarisasi Nasional: Jakarta.
9. Gentile, R., Galasso, C., Idris, Y., Rusydy, I., dan Meilianda, E. (2019). From Rapid Visual Survey to Multi-Hazard Risk Prioritisation and Numerical Fragility of School Buildings, Nat. Hazards Earth Syst. Sci., 19, 1365–1386.
10. Iemura, H., Pradono, M. H., Sugimoto, M., Takahashi, Y., dan Husen, A. (2012). Tsunami Height Memorial Poles In Banda Aceh And Recommendations For Disaster Prevention, *Proceedings of the International Symposium on Engineering Lessons Learned from the 2011 Great East Japan Earthquake*, Tokyo, Japan, March 1-4.
11. Syamsidik, Benazir, Lutfi, M., Suppasri, A., dan Comfort, L. K. (2020). The 22 December 2018 Mount Anak Krakatau volcanogenic tsunami on Sunda Strait coasts, Indonesia: tsunami and damage characteristics, NHESS, 20(2), 549–565.
12. Kostov, M. (2000). *Seismic Fragility Analysis*, dipresentasikan pada *Regional Workshop on External Events PSA*, Sofia, Bulgaria, 6-10 November.

Vulnerability study of a two-story reinforced concrete school building structure in Banda Aceh toward earthquakes and tsunami hazards

Sausan Zahrah[1*] and *Muhammad* Fawzul Aziim Afif [3]

[1]Student of Civil Engineering Department, Postgraduate Program, Syiah Kuala University, Jl. Syeh Abdurrauf No. 7, Banda Aceh, 23111, Indonesia
[2]Former Student of Civil Engineering Department, Faculty of Engineering, Syiah Kuala University, Jl. Syeh Abdurrauf No. 7, Banda Aceh, 23111, Indonesia

Abstract. Banda Aceh city is an earthquake-prone area. One is the phenomenon that a magnitude 9,3 M_w and subsequent tsunami wave occurred in 2004. The incident caused casualties and damage to facilities including educational facilities such as school buildings. After the 2004 Tsunami, affected school was rebuilt insame location as before, therefore necessary to conduct an assessment of school building to ensure safety and comfort of learning process in affected area. The purpose of this study was to determine structure capacity of reinforced concrete school building structure toward earthquake and Tsunami in Banda Aceh, so it can see contruction vulnerability based on table of damage state. Objects in this study was a typical two-story reinforced concrete school buildings in Banda Aceh. The chosen building were SDN 48 and SDN 8. Structures were modelled by using SAP2000 software. Earthquake load data using the2004 Sumatera-Andaman earthquake and Puskim earthquake-predictions which were analyzed using pushover method, and tsunami loads based on the height of inundation at location of school building. As a results, it has been developed capacity curve and spectrum capacity curve to determine damage limit and probability due to earthquake and tsunami hazards. It was found that based on spectrum capacity curve bothschool buildings in slight damage, while based on fragility curve it was found that both school building havea risk of light damage and moderate damage. Based on tsunami load, it has been found that both school building has inter-story drift that exceeds allowable limit, and based on tsunami fragility curve both schoolbuilding were predicted complete damage.

1 Introduction

Banda Aceh city is an earthquake-prone area. This is because Banda Aceh City is located between two fault segments, namely the Aceh fault segment and the Seulimuem fault segment that still active [1]. A large Earthquake has occurred in Banda Aceh with 9,3 M_w and followed by tsunami wave that devastated the city and its surroundings in 2004. The hazards caused significant loss of life and damage to buildings. The loss of life and damage to facilities that occurred in this disaster was quite significant, one of them was in education sector. After the 2004 Tsunami, schools were rebuilt in their former locations, so to ensure the safety and comfort of school buildings in the affected areas, it was necessary to conduct a vulnerability assessment of the building. The purpose of this study was to determine structure capacity of two-story reinforced concrete school building structure toward earthquake and Tsunamis in Banda Aceh. As an object in this study was a typical two-story reinforced concrete school building structure that are mostly found in the Banda Aceh city. In general, the typical structure of a school building in Banda Aceh city is a two-story reinforced concrete construction. Earthquake calculation method was used the pushover analysis method; it was used to see the fragility curve. Fragility curve is a curve that shows how much the possibility of building vulnerability when the earthquake occurred [2]. The fragility curve will show the probability of damage to the building structure with various damage conditions, namely slight damage, moderate damage, extensive damage and complete damage. In addition to calculating the earthquake, tsunami load calculations are also needed in the structural design of the building so as to reduce the risk of damage to the building structure during the earthquake and tsunami. Therefore, this study was conducted to determine whether wo-story reinforced concrete school building structure have meet the existing regulatory standards or not. So from this study can see the vulnerability that occurs for typical reinforced concrete school building to be reviewed in the region of Banda Aceh as well as immediate recommendations for improvements and efforts to improve the structure in order to reduce the risk of. The result of this research is the determination of damage states in SDN 48 and SDN 8 buildings obtained from the spectrum capacity curve is in the position of slight

* Sausan Zahrah: szahrah02@gmail.com

damage or light damage, while based on the fragility curve that has been developed the school buildings have a risk of light damage and moderate damage. In the calculation due to the tsunami load, the school building has an inter-story drift that exceeds allowable limit and based on tsunami fragility curve that has been developed, SDN 48 and SDN 8 buildings are predicted to be complete damage.

2 Literatur review

1.1 Parameters design of response spectra

According to SNI 1726:2019[3], response spectrum is a spectrum that is presented in the form of a graph or plot that describes the relationship between the maximum response acceleration of a system of one degree of freedom due to a particular earthquake, as a function of the damping factor and natural vibration time.

1.2 Limitation of fundamental periodic structures

Based on SNI 1726:2019[3], the fundamental period should not exceed the coefficient result for the calculated period limitation (Cu) according to Appendix Table B.7 and the approximate fundamental period (Ta). As an alternative to carrying out the analysis to determine the fundamental period of the structure (T) it is permitted to directly use the approximate building period (Ta) and is determined from the following equation:

$$T_a = C_t . h_n^x \quad (2)$$

1.3 Tsunami

Tsunami waves cannot be seen when they are far in the middle of the ocean, but once they reach shallow areas the waves will move quickly. Energy from the tsunami is still a function of the height and so on [4]. The energy forms the forces that will become a burden on the building structure.

1.4 Structure behavior

Structural behavior is a behavior caused by the building due to the force. This behavior includes moving and moving between floors. displacement is the deviation of a floor measured from the ground floor, while inter-story drift is the horizontal displacement at the top of the level under consideration relative to the bottom. Dimana the inter-story drift permit level (Δ_a) can be seen in the following table:

1.1 Capacity curve and spectrum capacity curve

In the study by Utomo, et al [5], in addition to showing structure behaviour on the condition of elastic, plastic and up to the collapse, pushover analysis can also produce an overview of the relationship between the base shear and the roof structure or roof displacement, this relationship is then mapped as a capacity curve. In the pushover process, the structure is pushed until it yields at one or more locations in the structure. The capacity curve will show a linear condition before reaching the melting state and after that it behaves non-linearly. While the spectrum capacity curve is the relationship between spectral acceleration (Sa) to spectral displacement (Sd), to get the spectrum capacity curve, the capacity curve needs to be changed to a spectrum capacity curve.

Table 1. Inter-story drift permit level.

Structure	Risk category		
	I / II	III	IV
Structures, other than brick shear wall structures, 4 stories or less with interior walls, partitions, ceilings and exterior wall systems designed to accommodate deviations between stories.	0,025 h_{sx}^c	0,02 $0h_{sx}$	0,015 h_{sx}
Brick cantilever shear wall structure	0,010 h_{sx}	0,01 $0h_{sx}$	0,010 h_{sx}
Other brick shear wall structures	0,007 h_{sx}	0,00 $7h_{sx}$	0,007 h_{sx}
All other structures	0,020 h_{sx}	0,01 $5h_{sx}$	0,010 h_{sx}

Source: SNI 1726:2019 [3]

1.2 Seismic fragility curve and tsunami fragility curve

According to Wibowo [6], Seismic fragility curve is a curve that shows the possibility of damage that occurs to a structure when it receives an earthquake load with a certain intensity. From the research of Gentile, et al [7]. The seismic fragility curve can be seen in Fig.1 below.

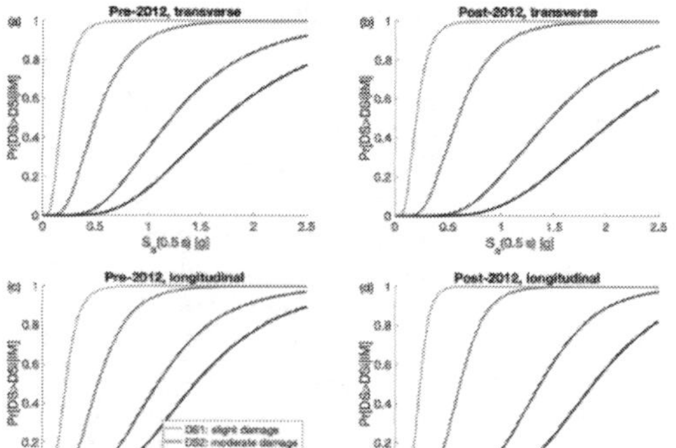

Fig. 1. Typical seismic fragility curve of school buildings. Gentile [7]

The limits of structural damage according to Hazus can be seen in Table 2.

Table 2. Description of damage states based on Hazus.

Damage states	Description of the damage
Slight/Minor Damage	Small plaster cracks at corners of door and window openings and wall-ceiling intersections; small cracks in masonry chimneys and masonry veneers. Small cracks are assumed to be visible with a maximum width of less than 1/8 inch (cracks wider than 1/8 inch are referred to as "large" cracks)
Moderate Damage	Large plaster or gypsum-board cracks at corners of door and window openings; small diagonal cracks across shear wall panels exhibited by small cracks in stucco and gypsum wall panels; large cracks in brick chimneys; toppling of tall masonry chimneys.
Extensive Damage	Large diagonal cracks across shear wall panels or large cracks at plywood joints; permanent lateral movement of floors and roof; toppling of most brick chimneys; cracks in foundations; splitting of wood sill plates and/or slippage of structure over foundations.
Complete Damage	Structure may have large permanent lateral displacement or be in imminent danger of collapse due to cripple wall failure or failure of the lateral load resisting system; some structures may slip and fall off the foundation; large foundation cracks. Three percent of the total area of buildings with Complete damage is expected to be collapsed, on average.

Source: Hazus [2]

According to Koshimura, et al [8], the tsunami fragility curve is a way to estimate structural damage due to the tsunami hazard, by integrating field surveys and analyzing data from geographic information systems. The tsunami fragility curve represents the probability of structural damage related to the inundation depth as shown in Fig. 2. Tsunami fragility curve based on Syamsidik, et al [9].

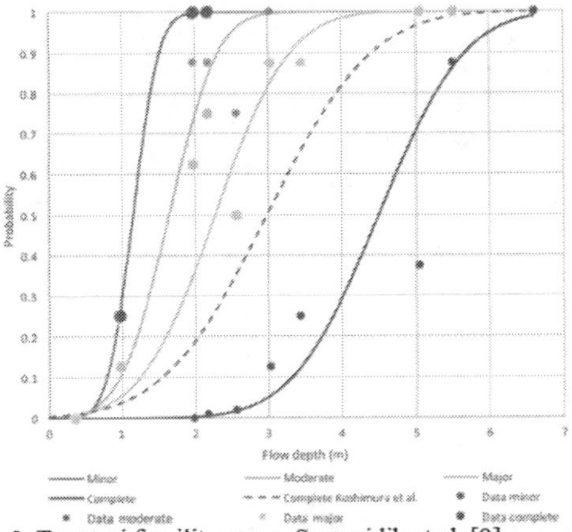

Fig. 2. Tsunami fragility curve, Syamsidik et al. [9]

Table 3. Damage states to building due to tsunami, based Syamsidik et al. [9]

Classification	Damage Condition	Description
DC0	No damages	Flooded but no damages found.
DC1	*Minor*	Damages found windows and doors, no damage on wall and on structural component
DC2	*Moderate*	One side wall damages, no damage on column and beam.
DC3	*Major*	All walls were damaged or roofs felt down, structural components bent/deflected or broken.
DC4	*Complate/* Washed away	Only floor left.

Source: Syamsidik et al. [9]

2 Research methods

2.1 Data collection techniques

Research data collection was done by field survey. Field surveys are used to collect data on building structures and building vulnerability factors by looking directly at the condition of the building. The survey was conducted on 100 school buildings in Banda Aceh City using a rapid visual survey so that the distribution of buildings was achieved for each sub-district location in Banda Aceh City.

2.1.1 Building structure data

The data of the structure building in this study uses and the data based on direct observation of the buildings as shown in Table 4 and typical details of building structures based on the research of Gentile [7], which are typical details of building structures for various forms of buildings in Banda Aceh as shown in Table 5.

Table 4. The data based on direct observation of the buildings

School	Height (m)	Columns (m)	Beams (m)	Material
SDN 48	3.5	0.3 x 0.3	0.3 x 0.4	RC
SDN 08	3.5	0.3 x 0.3	0.3 x 0.4	RC

Table 5. Typical quality of concrete and steel reinforcement for school buildings in Banda Aceh City.

	Material	Typical Beam	Typical Column	Typical Joint
Pre-2012	Concrete f_c = 21,5 MPa	3Φ16 top	3Φ16 top	No stirrups
	Long. Steel f_y = 400 MPa	3Φ16 bottom	3Φ16 bottom	
	Tran. Steel f_y = 240 MPa	Φ10@150 mm stirrups	Φ10@200 mm stirrups	
Post-2012	Concrete f_c = 24 MPa	3Φ16 top	3Φ16 top	No stirrups
	Long. Steel f_y = 400 MPa	3Φ16 bottom	3Φ16 bottom	
	Tran. Steel f_y = 240 MPa	Φ10@150 mm stirrups	Φ10@100 mm stirrups	

Source: Gentile [7]

2.1.2 Earthquake data

Earthquake data used is the Sumatra – Andaman earthquake on December 26, 2004 obtained from the United States Geological Survey (USGS) and earthquake forecast data from the 2017 Indonesia seismic hazard map obtained from PUSKIM were in the form of a response spectrum.

2.1.3 Tsunami data

The data needed to calculate the tsunami load is the depth of the 2004 tsunami flow and the difference between the subgrade of the building and the water level. The estimated height of the tsunami inundation in Banda Aceh City was obtained from the research of Iemura, et al [10].

2.2 Structure modeling

The structural system that will be modeled by using SAP2000 is space frame structure. the data needed for the structural model is building data which is the dimensions of the building, concrete materials used and also peak ground acceleration data. Structural modeling was carried out for two typical buildings.

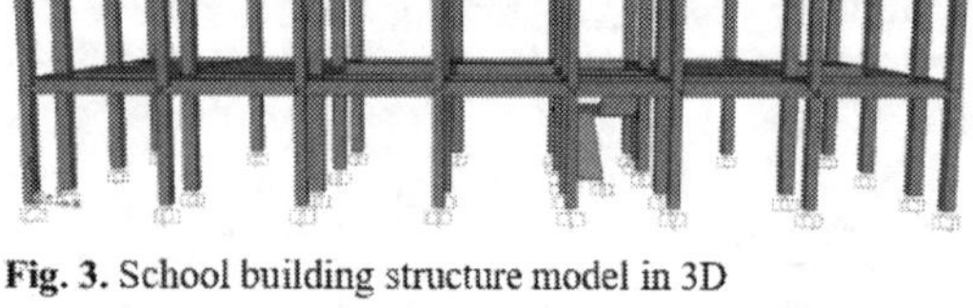

Fig. 3. School building structure model in 3D

2.3 Structure loads

The loads that are input into the SAP2000 program are loads that are reviewed in the structural design which is dead loads, live loads, earthquake loads, tsunami loads and load combination.

1. Earthquake combination loads
   U1 = 1,4 D
   U2 = 1,2 D + 1,6 L
   U3 = 1,38 D + 1 L + 0,39 QEx + 1,3 QEy
   U4 = 1,38 D + 1 L + 0,39 QEx - 1,3 QEy
   U5 = 1,38 D + 1 L - 0,39 QEx - 1,3 QEy
   U6 = 1,38 D + 1 L - 0,39 QEx + 1,3 QEy
   U7 = 1,38 D + 1 L + 1,3 QEx + 0,39 QEy
   U8 = 1,38 D + 1 L + 1,3 QEx - 0,39 QEy
   U9 = 1,38 D + 1 L - 1,3 QEx - 0,39 QEy
   U10 = 1,38 D + 1 L - 1,3 QEx + 0,39 QEy
   U11 = 1,08 D + 0,39 QEx + 1,3 QEy
   U12 = 1,08 D + 0,39 QEx - 1,3 QEy
   U13 = 1,08 D - 0,39 QEx - 1,3 QEy
   U14 = 1,08 D - 0,39 QEx + 1,3 QEy
   U15 = 1,08 D + 1,3 QEx - 0,39 QEy
   U16 = 1,08 D + 1,3 QEx + 0,39 QEy
   U17 = 1,08 D - 1,3 QEx - 0,39 QEy
   U18 = 1,08 D - 1,3 QEx + 0,39 QEy
2. Tsunami combination loads
   U19 = 1,2 D + 1,0 TS + 1,0 LREF + 0,25 L
   U20 = 0,9 D + 1,0 TS

2.4 Pushover analysis

pushover analysis was carried out by gradual increased load, lateral load was carried out representing the earthquake force on each floor. The direction of lateral loading is carried out in the direction of the main axis of the building. Lateral loading uses the displacement-controlled concept, namely the loading process is gradually increased until the displacement target is achieved.

2.5 Tsunami analysis

In analyzing the tsunami load, the used loads are hydrostatic load, hydrodynamic load and debris impact load. Hydrostatic load occurs when water moves closer to structure as a distributed load, pyramid-shaped along the flooded column. Hydrodynamic load as a distributed load along a flooded column. debris impact load is entered as point load. These loads are entered in the y direction. Next, an analysis is carried out on the SAP2000 program using tsunami combination load.

2.6 Data analysis

2.6.1 Capacity curve analysis and spectrum capacity curve analysis

In the SAP2000 program, the results of the pushover analysis obtained a capacity curve that shows the behavior of the structure when the shearing force is applied at a certain level, response spectrum curves that suitable for the existing earthquake region, yield

https://doi.org/10.1051/e3sconf/202234002007

diagrams of plastic hinges in beams and columns. An analysis of the spectrum capacity curve obtained from the pushover analysis was carried out to obtain the median spectral displacement value for each damage condition. The median spectral displacement value for each damage condition will be described in a log scale on the seismic fragility curve with a probability value of 50% and is used to determine limit state for each damage [11].

2.6.2 Seismic fragility curve analysis and tsunami fragility curve analysis

Seismic fragility curve was used to see the probability of structural failure when receiving earthquake loads with a certain intensity based on the fragility curve that has been developed. A probability value was obtained by looking at the coordinates of the fragility curve axis (y) with a value of Sa. In using the graphreader application, take one of the fragility curve was taken from previous studies. The step in reading curve values was to select a png, jpg or gif image and push 'Go', resize the blue rectangle to set the ruler for axis scaling as in Fig.4. Set values for the appropriate x and y-axis scaling. Then double click to enter a curve fix point. Click 'Generate curve' to see the output data as shown in Fig.5.

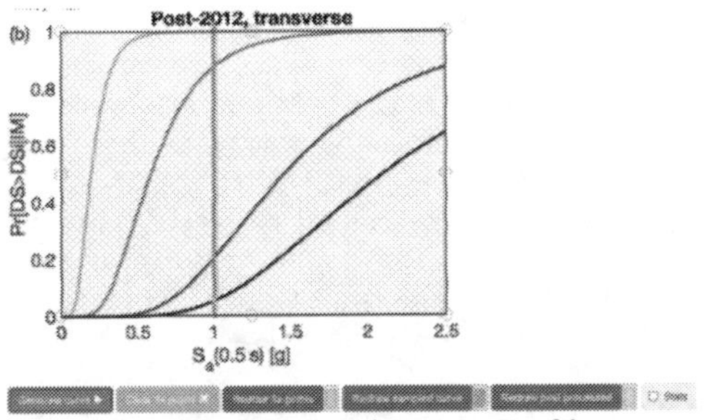

Fig. 4. Imputting Images and Dots on Graphics

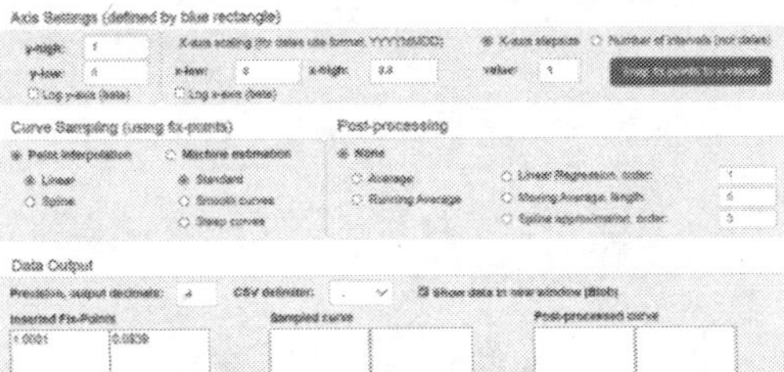

Fig. 5. Graphics output data results

From the results of this analysis, it is found that the structural performance limits indicate the probability of damage to the building structure for various damage conditions, namely slight damage, moderate damage, extensive damage, and complete damage.

3 Results and discussion

3.1 Results of school building recapitulation

Based on the results of the rapid visual survey, the criteria for school buildings in Banda Aceh are as follows.

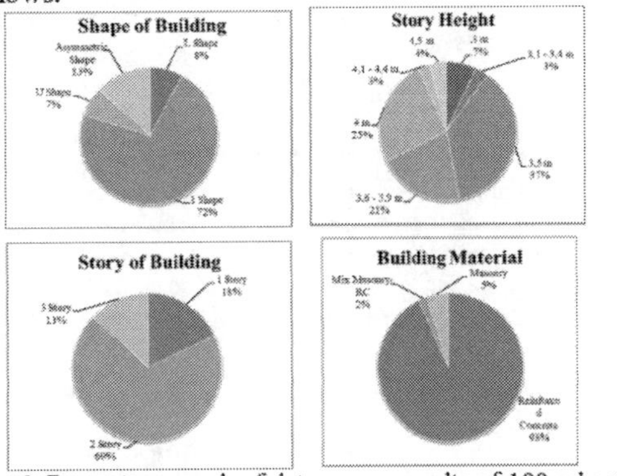

Fig. 6. Percentage graph of data survey results of 100 school buildings

It can be concluded that the typical dominant schools in Banda Aceh City are as follows:

- Building form : Tipe I (Top view)
- Number of floor : 2
- Height each floor : 3,5 m
- Constituent material : Reinforce concrete

From the survey results, the two buildings selected were SDN 48 and SDN 8. Another reason for choosing this school building is because it is relatively close to the beach and the 2004 tsunami inundation height was recorded at that location.

3.2 Fundamental period structure

From the results of SAP2000, the natural fundamental period time T = 0.2274 for SDN 48 and T = 0.2209 for SDN 8 with Ta max obtained from the calculation of SNI 1726-2019 of 0.3759. So, T analysis < Ta max, then the structure is declared to have met the conditions for the permit fundamental period obtained from the calculation of SNI 1726-2019.

3.3 The results of inter story drift analysis earthquake load

The results of the calculation of the inter-story drift for school buildings in the x-direction and y-direction can be seen in Fig. 7 and Fig.8.

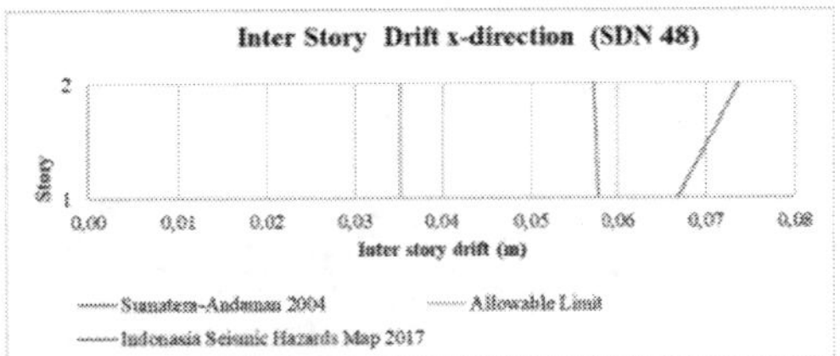

Fig. 7. The inter story drift due to earthquake load x-direction (*continued*)

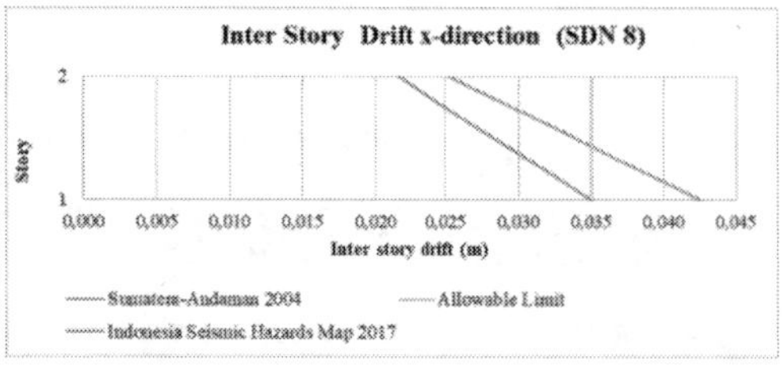

Fig. 7. The inter story drift due to earthquake load x-direction

Based on Fig.7 the results of inter story drift in the x direction at SDN 48 and SDN 8 with the Sumatera - Andaman 2004 earthquake load and from the 2017 Indonesia seismic hazard map, it can be seen that these results do not meet the inter story drift permit levels from the provisions of SNI 1726-2019.

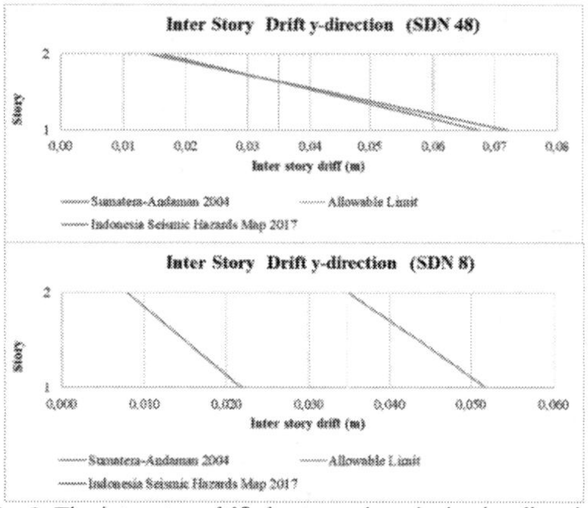

Fig. 8. The inter story drift due to earthquake load y-direction

Based on Fig. 8 the results of the calculation of inter story drift in the y direction in the SDN 48 and SDN 8 buildings with the 2004 Sumatra - Andaman earthquake load and the 2017 Indonesia seismic hazard map can be seen that these results also do not meet the inter story drift permit's levels from the provisions of SNI 1726-2019.

3.4 Capacity curve

The capacity curve is nonlinear due to an increase in load that makes the structural elements of the building change from an elastic condition to a plastic condition, to make the capacity curve clearer for the Sumatera – Andaman earthquake 2004 and earthquake load from the Indonesian seismic hazard map in the x and y directions

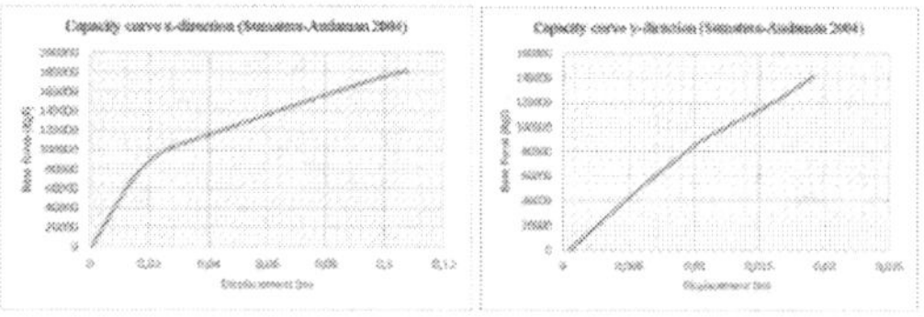

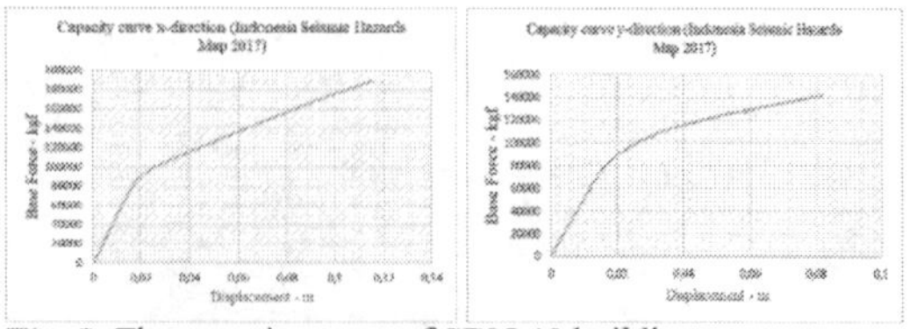

Fig. 9. The capacity curve of SDN 48 building

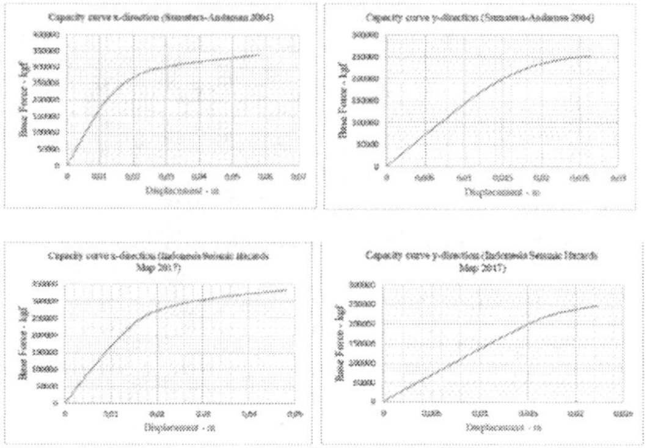

Fig. 10. The capacity curve of SDN 08 building

3.5 Spectrum capacity curve

The capacity spectrum curve is used to obtain the median spectral displacement value. The conversion of the capacity spectrum curve is carried out with the SAP2000 program from the results of the capacity curve. To determine the damage states from the capacity spectrum curve, it was carried out based on research Duan and Papin [11]. The position of each damage state on the capacity spectrum curve for SDN 48 and SDN 08 with the 2004 Andaman earthquake.

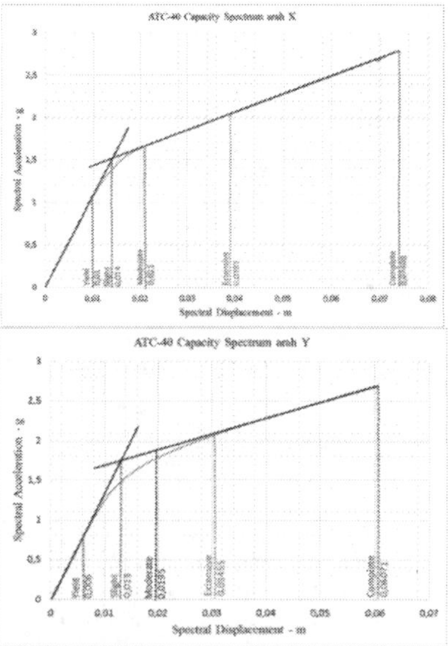

Fig. 11. Damage states determine SDN 48 (Sumatera-Andaman earthquake 2004)

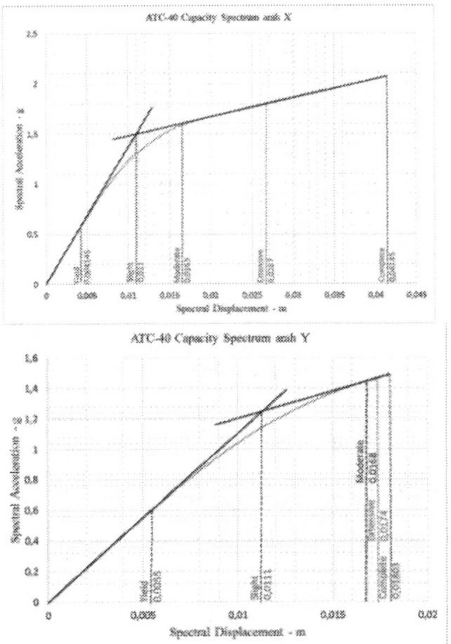

Fig. 12. Damage states determine SDN 08 (Sumatera-Andaman earthquake 2004)

Meanwhile, with the indonesia sesmic hazards map 2017, the damage states on the capacity spectrum curve for SDN 48 and SDN 8 are as follows.

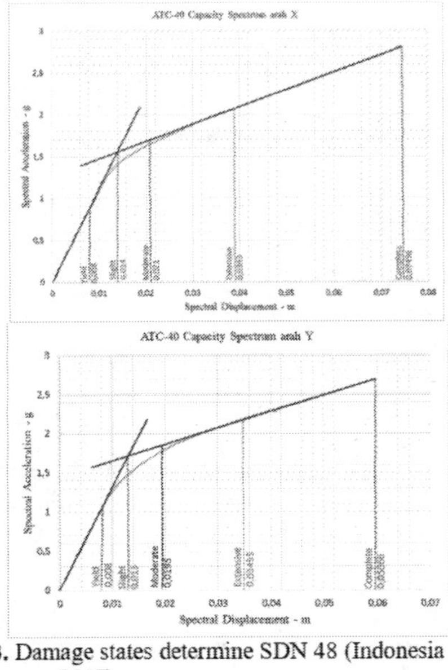

Fig. 13. Damage states determine SDN 48 (Indonesia seismic hazards map 2017)

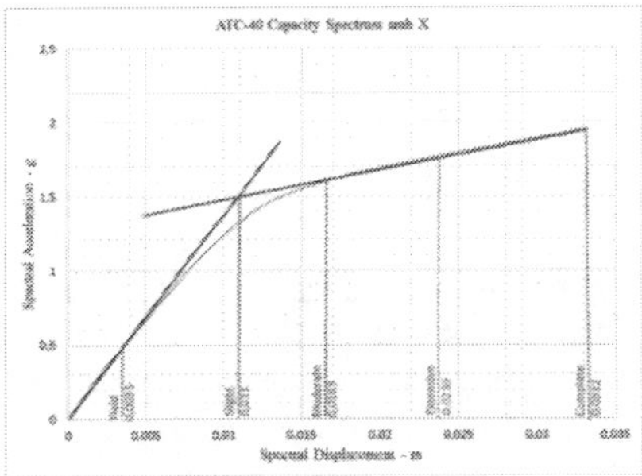

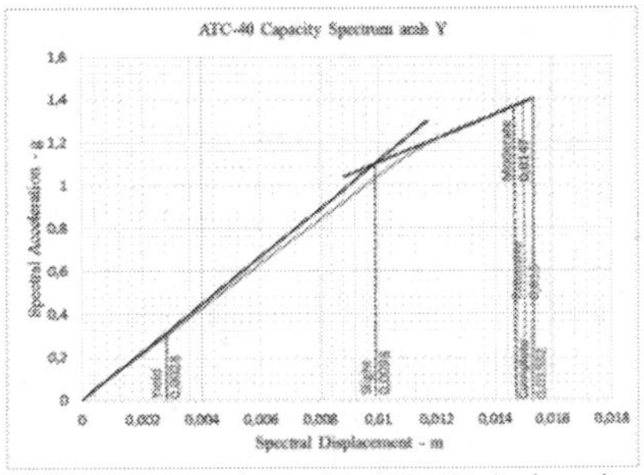

Fig. 14. Damage states determine SDN 08 (Indonesia seismic hazards map 2017)

Based on the identification of the capacity spectrum curve, the spectral displacement and spectral acceleration points for each structural failure condition are generated as shown in the following table:

Tabel 6. Sd and Sa values for various conditions of damage to SDN 48

Seismic Type	Demage state	S_{d}, ds (m)		Sa,ds (g)	
		x	y	x	y
Sumatera-Andaman 2004	Slight	0,0140	0,0130	1,3808	1,4681
	Moderate	0,0210	0,0195	1,6613	1,7666
	Extensive	0,0393	0,0346	2,0545	2,1640
	Complete	0,0741	0,0607	2,7928	2,6956
Indonesia Seismic Hazards Map 2017	Slight	0,0140	0,0130	1,3638	1,4387
	Moderate	0,02100	0,0195	1,6178	1,7674
	Extensive	0,0393	0,0346	2,0746	2,1683
	Complete	0,0741	0,0597	2,8099	2,6899

Tabel 7. Sd and Sa values for various conditions of damage to SDN 08

Seismic Type	Demage state	S_{d}, ds (m)		Sa,ds (g)	
		x	y	x	y
Sumatera-Andaman 2004	Slight	0,0110	0,0111	1,3015	1,1303
	Moderate	0,01650	0,0168	1,5869	1,4464
	Extensive	0,0267	0,0174	1,8040	1,4694
	Complete	0,0414	0,0180	2,0706	1,4914
Indonesia Seismic Hazards Map 2017	Slight	0,0110	0,0098	1,3016	1,0153
	Moderate	0,01650	0,0147	1,5914	1,3684
	Extensive	0,0239	0,0150	1,7715	1,3841
	Complete	0,0332	0,0153	1,9549	1,4006

3.6 Results of tsunami load analysis

The results of the tsunami load analysis obtained the inter-level deviation where the control for the deviation between the permit levels was used SNI 1726-2019.

 https://doi.org/10.1051/e3sconf/202234002007

Table 8. Inter story drift in x direction of SDN 48.

Floor	Height	Displacement y	Inter-floor drift	Inter-floor drift permit (0,010 h_{sx})
	(m)	(m)	(m)	(m)
2	3.5	0.014475	0.001299	0.035
1	3.5	0.013176	0.013176	0.035

Table 9. Inter story drift in y direction of SDN 48.

Floor	Height	Displacement y	Inter-floor drift	Inter-floor drift permit (0,010 h_{sx})
	(m)	(m)	(m)	(m)
2	3.5	0.067058	0.017171	0.035
1	3.5	0.049887	0.049887	0.035

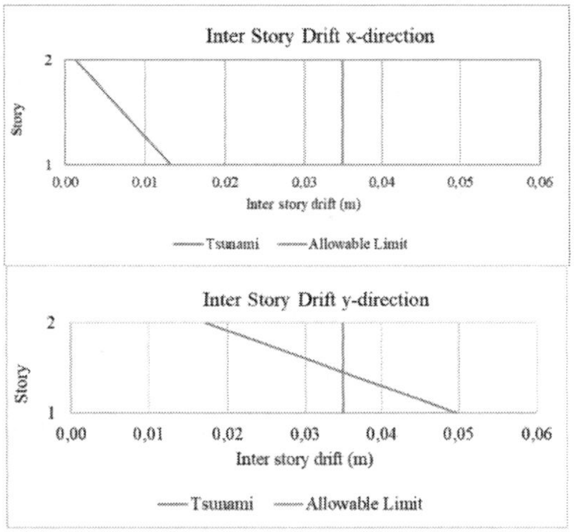

Fig.15. Inter story drift due to tsunami load at SDN 48

As for the deviation between floors for SDN 8 buildings in the x-direction and the y-direction

Table 10. Inter story drift in y direction of SDN 08.

Floor	Height	Displacement y	Inter-floor drift	Inter-floor drift permit (0,010 h_{sx})
	(m)	(m)	(m)	(m)
2	3.5	0.001959	0.000366	0.035
1	3.5	0.001593	0.001593	0.035

Table 11. Inter story drift in y direction of SDN 08.

Floor	Height	Displacement y	Inter-floor drift	Inter-floor drift permit (0,010 h_{sx})
	(m)	(m)	(m)	(m)
2	3.5	0.054206	0.014351	0.035
1	3.5	0.039855	0.039855	0.035

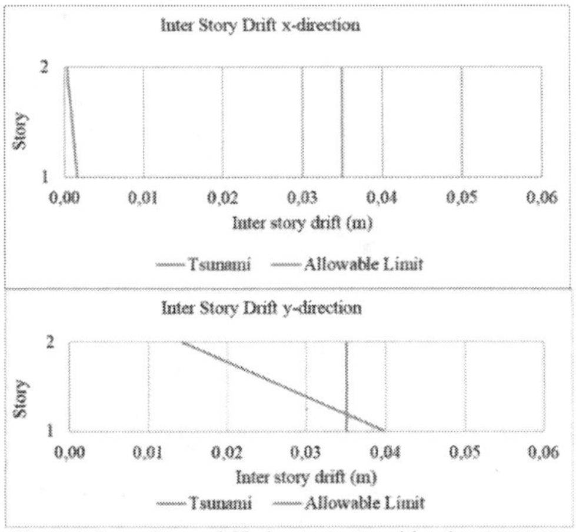

Fig. 16. Inter-Story drift due to tsunami load at SDN 08

Based on the results of the deviation between floors due to the tsunami load at SDN 48 and SDN 8, the deviation between floors in the y direction has an inter-floor deviation value that exceeds the allowable inter-floor deviation based on SNI 1726-2019. Due to the depth of the tsunami inundation at that location has reached the second floor.

3.7 Determination of damage states

The data needed is obtained from the results of the Puskim Spectra response program. In Puskim earthquake modeling, the results of the Spectra program for SDN 48 are as follows:

T = 0,3759 s
T_0 = 0,135983 s
Ts = 0,679916 s
TL = 14,00000 s
Sds = 1,000124 g

For SDN 48 buildings, the fundamental period of the building (T) is greater than or equal to T0 and less than or equal to Ts, the design acceleration response spectrum (Sa) is the same as SDS. Then the value of the acceleration spectrum (Sa) = 1.000124 g. Meanwhile, the results of the Spectra program for SDN 8 are as follows:

T = 0,3759 s
T0 = 0,135831 s
Ts = 0,679154 s
TL = 15,00000 s
Sds = 1,001245 g

For an SDN 8 building, the fundamental period of the building (T) is greater than or equal to T0 and less than or equal to Ts, the design acceleration response spectrum (Sa) is the same as SDS, so the acceleration spectrum (Sa) = 1.001245 g. Then the position of the damage state for the value of Sa is as follows:

 https://doi.org/10.1051/e3sconf/202234002007

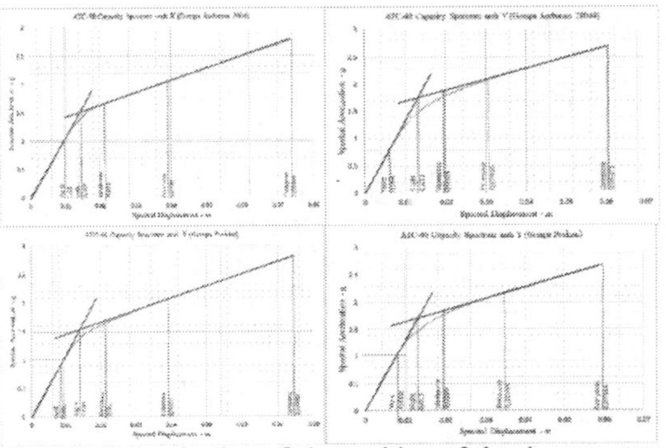

Fig. 17. Determination of the position of the damage state based on the value of Sa on the SDN 48

Based on Fig. 17 it can be seen that SDN 48 with a load of the 2004 Andaman earthquake and the Puskim earthquake at a Sa value of 1,000124 is in the x direction and the y direction is between yield and slight conditions.

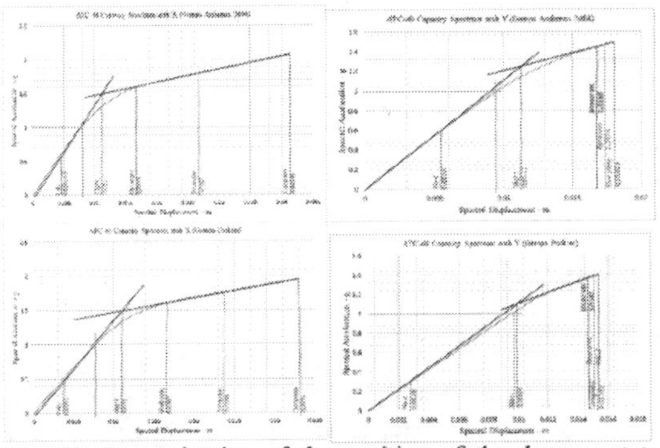

Fig. 18. Determination of the position of the damage state based on the value of Sa on the SDN 08

Based on Fig. 18 it can be seen that SDN 08 with the load of the 2004 Andaman earthquake and the Puskim earthquake at a Sa value of 1.001245 is in the x direction and the y direction is between yield conditions and close to slight damage.

3.8 Probability value based on the developed fragility curve

Based on research by Gentile [7], the probability value of building failure for each damage condition can be known by looking at the previous curve based on the obtained Sa value. Where the data Sa for SDN 48 1.000124 g is inputted in the graphreader application so that the probability value of the output data results is a curve for SDN 48 as follows:

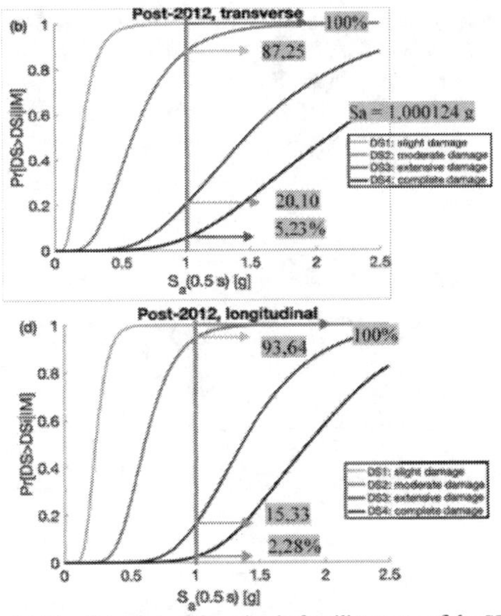

Fig. 19. Identification of the seismic fragility curve of the SDN 48 building in the x and y directions

As for the SDN 8 Sa data obtained is 1.001245 g so that the probability value obtained from the results of the curve output data for SDN 8 is as follows:

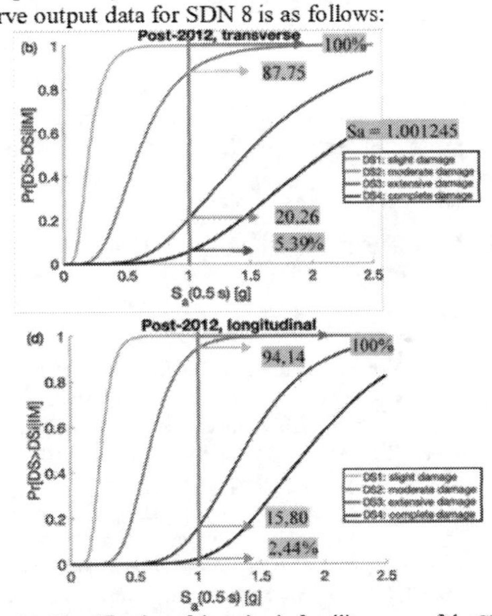

Fig. 20. Identification of the seismic fragility curve of the SDN 08 building in the x and y directions

From the results of the evaluation of the comparison of damage limits based on the spectrum capacity curve to the Sa values of 1,000124 and 1.001245 for the SDN 48 and SDN 8 buildings, it was found that the damage conditions were in the boundary between yield and slight damage, while the evaluation was carried out on the fragility curve that had been developed. against the Sa value of 1.000124 and 1.00124 for the school, it was found that large damage occurred in conditions of slight damage and moderate damage.

Evaluation of the probability of damage due to tsunami loads is carried out based on the tsunami fragility curve that has been developed by Syamsidik, et al [9].

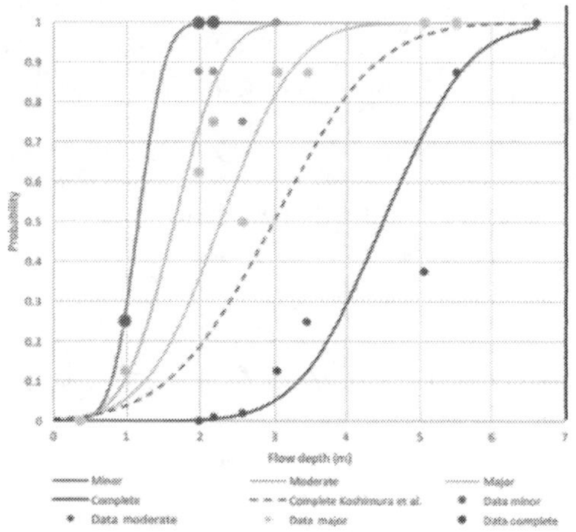

Fig. 21. Identification of the possibility of finding a tsunami for SDN 48 and SDN 08

As can be seen in Fig. 21 above that 2004 tsunami records in SDN 48' and SDN 8' buildings, which are located in Meuraxa and Kuta Raja Districts, reached 7 meters. According to vulnerability curve [9], the damage states probability for those buildings are 100%. Hence, the tsunami hit those buildings and caused total damage. As a result, the tsunami's hit left solely basement.

4 Conclusions and suggestions

4.1 Conclusions

Based on the results of the discussion and data analysis carried out, a conclusion is drawn regarding the study of the vulnerability of school buildings with reinforced concrete construction in Banda Aceh City to earthquake and tsunami hazards as follows:

1. Based on the results of structural behavior due to Sumatera-Andaman earthquake 2004 and indonesia seismic hazards map 2017 earthquake source map at SDN 48 and SDN 8, the value of inter story drift is greater than the inter story drift permit levels, then the building does not meet the provisions of SNI 1726-2019
2. Based on the results of structural behavior due to the 2004 tsunami load at SDN 48 and SDN 8, inter story drift in the y direction is greater than permits because the height of inundation at that location reaches the second floor.
3. From the results of the spectrum response Puskim program, the spectrum acceleration (Sa) for SDN 48 buildings is 1,000124 g and SDN 8 is 1.001245 g.
4. From the results of determining the level of damage states on the capacity spectrum curve with Sa values of 1,000124 and 1.001245 for SDN 48 and SDN 8 buildings, it was found that the school buildings are in a yielding position towards the level of light damage, while based on the seismic fragility curve that has been developed probability for slight damage is 100% and moderate damage reaches 90%.
5. Based on the tsunami fragility curve that has been developed with a tsunami height of 7 meters in SDN 48 and SDN 8, the probability value for each level of damage is 100%, in other words, the probability will be complate damage.

4.2 Sugesstions

From this research, several suggestions and inputs related to research can be obtained so that further research or development can be carried out, namely:

1. Using the structural data of the building itself such as detailed drawings, column compressive strength test results and steel tensile test results in order to create a more accurate seismic fragility curve.
2. Calculate earthquake modeling according to the location of the school building and the source of the earthquake.
3. Create a tsunami fragility curve based on the capacity of school buildings.
4. Make a building model without walls to see the behavior of the building due to the tsunami load.

References

1. Sugiyanto, D., Zulfakriza, Ismail, N., Adriansyah, F., Meilano, I., dan Hasanuddin, Z.A. (2011). Analisa Deformasi Permukaan Patahan Aktif Segmen Seulimum dan Segmen Aceh. Dipresentasikan pada Prosiding Seminar Hasil Penelitian Kebencanaan TDMRC-Unsyiah, Banda Aceh, 13 – 19 April 2011.
2. HAZUS (Hazard US), (1999): Earthquake Loss Estimation, National Institute of Building Sciences. NIBS: Washington D.C.
3. Badan Standarisasi Nasional. (2019). Standar Nasional Indonesia 1726-2019: Tata Cara Perencanaan Ketahanan Gempa untuk Struktur Bangunan Gedung dan non Gedung. Badan Standarisasi Nasional: Jakarta.
4. Sugito, N. T. (2008). Tsunami, Universitas Pendidikan Indonesia, Bandung.
5. Utomo, C., Susanto, R., Tudjono, S., & Wibowo, H. (2012). Evaluasi Struktur Dengan Pushover Analysis pada Gedung Kalibata Residences Jakarta. Jurnal Karya Teknik Sipil, 1(1), 1-10
6. Wibowo, N. A. (2016). Pengembangan Kurva Kerapuhan Berbasis Incremental Dynamic Analysis untuk Evaluasi Kinerja Seismik Jembatan Beton. Tugas Sarjana, Teknik Sipil Universitas Sebelas Maret: Surakarta.
7. Gentile, R., Galasso, C., Idris, Y., Rusydy, I., dan Meilianda, E. (2019). From Rapid Visual Survey to Multi-Hazard Risk Prioritisation and Numerical Fragility of School Buildings, Nat. Hazards Earth Syst. Sci., 19, 1365–1386.

8. Koshimura, S., Namegaya, Y., Yanagisawa, H. (2009). Tsunami Fragility A New Measure to Identify Tsunami Damage, Jurnal of Disaster Research Vol.4 No.6, Japan.
9. Syamsidik, Benazir, Lutfi, M., Suppasri, A., dan Comfort, L. K. (2020). The 22 December 2018 Mount Anak Krakatau volcanogenic tsunami on Sunda Strait coasts, Indonesia: tsunami and damage characteristics, NHESS, 20(2), 549–565.
10. Iemura, H., Pradono, M. H., Sugimoto, M., Takahashi, Y., dan Husen, A. (2012). Tsunami Height Memorial Poles In Banda Aceh And Recommendations For Disaster Prevention, Proceedings of the International Symposium on Engineering Lessons Learned from the 2011 Great East Japan Earthquake, Tokyo, Japan, March 1-4.
11. Duan, X., Pappin, JW. (2008). A Procedure for Establishing Fragility Functions for Seismic Loss Estimate of Existing Buildings Based on Nonliner Pushover Analysis, dipresentasikan pada The 14th World Conference on Earthquake Engineering, Beijing, 12-17 Oktober.
12. Badan Standarisasi Nasional. (2020). Standar Nasional Indonesia 1727-2020: Beban Desain Minimum dan Kriteria Terkait untuk Bangunan Gedung dan Struktur Lain. Badan Standarisasi Nasional: Jakarta.
13. Federal Emergency Management Agency. (2019). Guidelines for Design of Structures for Vertical Evacuation from Tsunami Third Edition. Department of Homeland Security FEMA Mitigation Division: Washington D.C.

Earthquake vulnerability assessment of the 6.5 Mw Pidie Jaya earthquake: Analytical-based fragility curves

Adrian Ulza[1], and *Yunita* Idris[1,2*]

[1]Tsunami and Disaster Mitigation Research Center (TDMRC), Universitas Syiah Kuala, Banda Aceh, Indonesia
[2]Department of Civil Engineering, Universitas Syiah Kuala, Banda Aceh, Indonesia

Abstract. Nearly all residential houses were damaged due to 6.5 Mw earthquake in Pidie Jaya, 2016. The state of damage can be slight, moderate, and even can be extensive which lead to the collapsing. The confined masonry structure, which commonly found in Aceh, especially for housing construction, were seemingly prone to the extensive earthquake excitation. In this paper, analytical-based fragility curves are employed to the typical house structure. To account various uncertainty, 32 ground motion records are considered in the analysis. Based on the results, the fragility curve could render different interpretation if different definition of damage intensities is used.

1 Introduction

In the aftermath of the 6.5 Mw earthquake in Pidie Jaya Regency, Indonesia, severe damages are apparent in most of the buildings, and not to mention that several buildings are completely collapsed to the ground. Unexpectedly, the damages of structure were occurred in low-rise residential houses. Based on the field investigation, the typical structural system of the housing can be classified as Confined Masonry. Moreover, despite that the magnitude of the earthquake was considerably high, poor construction practices to meet the standard and minimum design of seismic details are the suspected glitches. Since most of the residential houses in Aceh province are quite typical to the damaged confined masonry that found in Pidie Jaya, its liability is certainly compromised, and in due course, the vulnerability of such structural system is pertinent to be investigated.

Confined masonry is a non-engineered structural system, which favourably chosen as typical house construction in Indonesia and mostly in developing countries, by virtue of its material availability, construction practicality, and relatively low-cost compared to the engineered structural system, such as Reinforced Concrete Frame with Masonry Infills (RCFMI).

Although for common people, the confined masonry appears to be quite similar with RCFMI, both of structural systems have different means in engineering perspectives. Unlike the confined masonry structure, a regular building with RCFMI system will be designed and analysed by the authorized engineer to meet the local building code requirements with special structural detailing, so it can perform satisfactorily under earthquake event.

In most practices, the building is analysed as a bare frame, where the structural frame is designed to carry the gravity load and to resist the lateral earthquake forces, while the stiffness contribution from the infill wall is neglected, despite several studies has investigated that the infill wall act as a diagonal strut which significantly increase the lateral strength and deformation. Contrastingly, the brick masonry in confined masonry structure is confined with light amount of concrete and its reinforcement, having a thickness mostly as thick as brick masonry. During the earthquake excitation, the confined masonry act as whole to resist the lateral force.

Based on the experimental study conducted by Sari et al [1], it can be observed that the diagonal crack was seemed on the specimen which has similar typology and structural configuration with typical housing in Pidie Jaya, Aceh, meaning that the confined masonry also has the in-plane diagonal strut action, similar with RCFMI. In addition, there were evidences where several damaged houses in the post-event of 6.5 Mw Pidie Jaya Earthquake also shown similar failure mode as can be seen on Figure 1.

Fig. 1. Damaged House of Confined Masonry System due to 2016's Pidie Jaya Earthquake

* Corresponding Author: yunita.idris@unsyiah.ac.id

Although the presence of the out-of-plane failure of masonry wall cannot be negligible, this study is limited to observe the in-plane behaviour of confined masonry. The structural analysis will be utilized by using SAP2000. The structural model will be developed, based on the typical housing in Pidie Jaya. In the absence of local building code in assessing the existing building structure, the modelling procedures will be in accordance with ASCE 41-17 [2].

There are several procedures to assess the proneness of structure due to earthquake. In the conventional procedures, the target performance of structure can be evaluated by performing the available variants of pushover analysis techniques [3], [4], and [5]. A more robust and tedious method by incorporating the probabilistic of earthquake records can be performed by using incremental dynamic analysis (IDA) [6]. In the later study, the demanding dynamic analysis that is required in the IDA, can be evaded via SPO2FRAG [7] and taking into account the seismic fragility assessment. Taking the idea of using an equivalent single-degree-of-freedom structural model to generate the damage fractions, in order to develop the fragility function, this study will combining the capacity spectrum method and then presenting it in the form of record-to-record variability via the analytical fragility function with the fitting approach by using Maximum Likelihood Estimation (MLE), which is based on the previous study from J. W. Baker [8].

2 Methodology

a. Structural model and material properties

The structural plan of house is defined by the typology study from the typical house in Pidie Jaya, whereabouts the map that showing the study location can be seen on Figure 2. The compressive strength of concrete is based on the average results of in-situ non-destructive test that is taken on field survey on several houses. Other material properties, such as brick masonry and reinforcement bar can be determined by the typical material characteristics in Aceh. The material properties are summarized in Table 1.

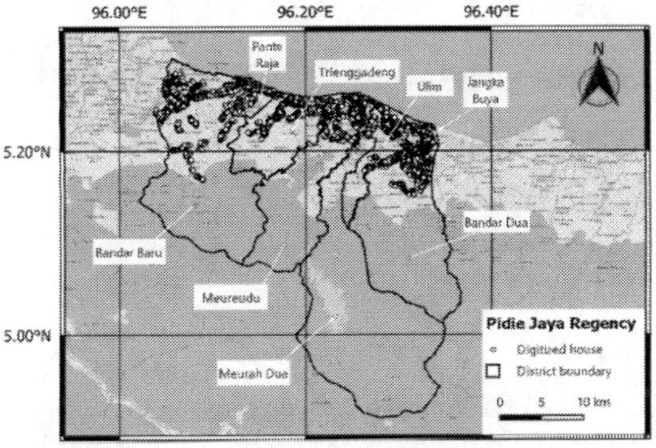

Fig. 2. Map of Pidie Jaya Regency, Aceh. Background map by © OpenStreetMap Contributors 2019. Distributed under a CC BY-SA License.

Table 1. Material properties

Parameters	Nominal
Compressive Strength of Concrete, f_c'	15 MPa
Yield Strength of Reinforcement, f_y	240 MPa
Tensile Strength of Reinforcement, f_u	300 MPa
Compressive Strength of Brick Masonry	5.3 MPa
Shear Strength of Brick Masonry	0.35 MPa
Elastic Modulus of Brick Masonry	2957 MPa

In order to construct the analytical model, simplified structural model is assumed. The confined structure of masonry is assumed as 1-dimensional frame elements of reinforced concrete column and beam with dimensions of 11x11 cm and 11x13 cm, respectively. The longitudinal reinforcement bars for column and beam are 4φ10, while the transverse reinforcement is φ6−150. To put some note, restrained fixity is assumed at the base of the column. Furthermore, to represent the diagonal strut action from the masonry, equivalent diagonal compression struts are defined and located concentrically to the reinforced concrete columns. In summary, the structural model is shown in Figure 3. The nonlinear behaviour of structural model that is developed in this study is referring to ASCE 41-17 [2] procedures, as can be illustrated in Figure 4.

In the analytical model, the infill strut is modelled as 1-dimensional elements to carry the axial compressive strength. The tension limit is predefined, to neglect its tension stiffness and avoiding it to carry the tension load. In addition, the stiffness matrix of structure is modified, so thus the equivalent strut element will only transmit the axial force rather than bending moment, somewhat represented as a pinned condition.

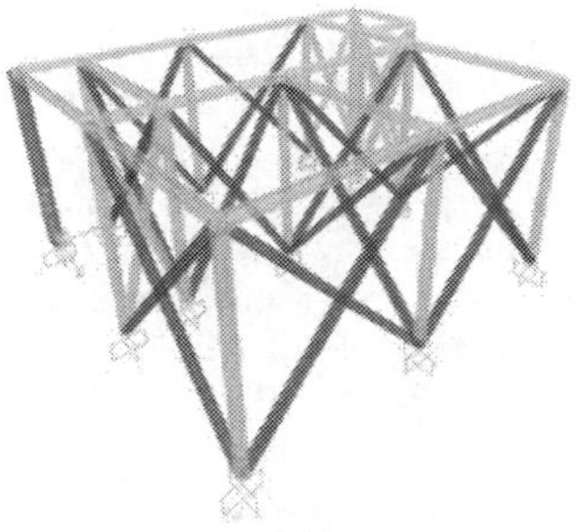

(a)

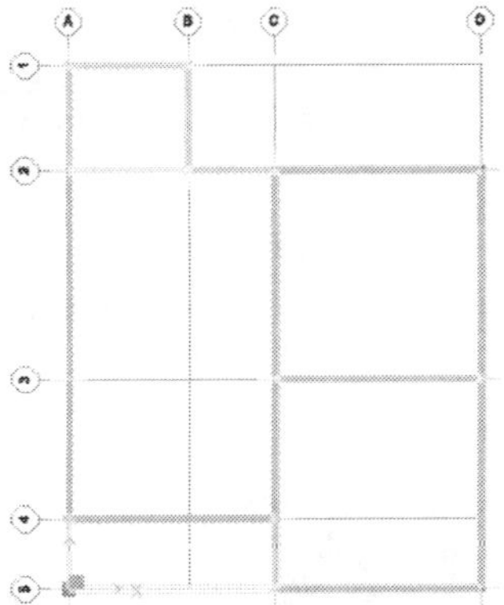

(b)

Fig. 3. Structural model (a) Perspective view; (b) Plan view

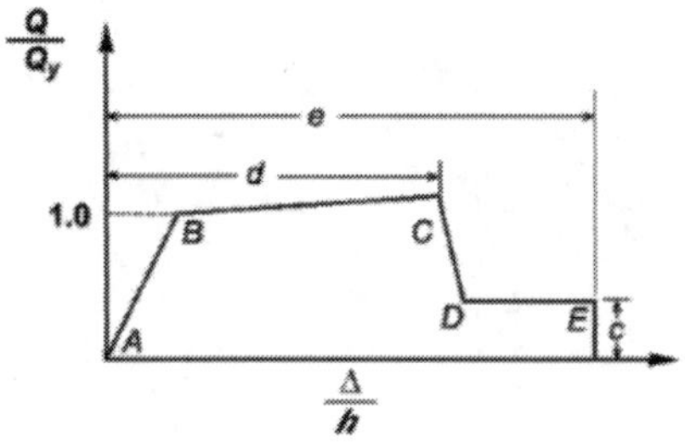

(a)

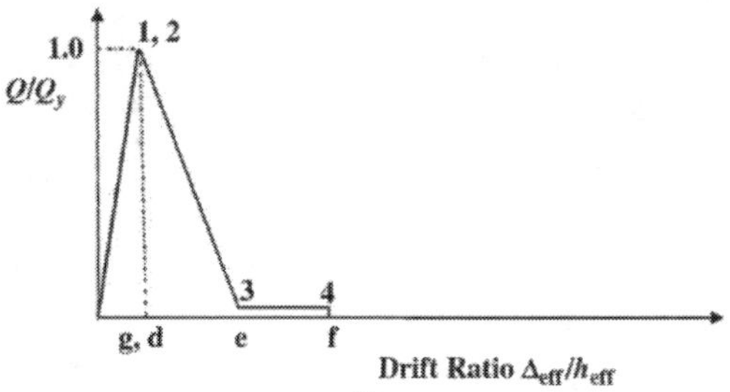

(b)

Fig. 4. Generalized force-deformation relationship for deformation-controlled, (a) Column and beam; (b) masonry wall, Source: ASCE 41-17 [2]

Based on Figure 4, the terminology of column and beam forces are determined in terms of flexural moment strength, while for the masonry wall, the force terminology is in terms of axial compressive strength. The axial tension strength for the masonry wall is neglected, so thus its out-of-plane capacity. The nominal moment strength of column and beam can be computed accordingly, with the section analysis of reinforced concrete. The compression capacity of masonry wall is defined based on the previous study of Zarnic and Gostic [9]. In addition, the deformation capacity for column, beam, and unreinforced masonry wall elements are referring to the ASCE 41-17 [2].

b. Earthquake ground motions

Due to some circumstances to obtain the actual ground motion of the 2016's Pidie Jaya Earthquake, plausible approach is engaged, instead. To investigate the vulnerability of structural model due to the variability of earthquake excitations, 32 strong ground motion records are considered, in which based on PEER NGA-West2 Database [10], and enlisted on Table 2.

Table 2. Summary of selected ground motion from PEER NGA West 2 database

No	Earthquake Name	Year	Magnitude
1	Parkfield	1966	6.19
2	Coyote Lake	1979	5.74
3	Imperial Valley-06	1979	6.53
4	Livermore-01	1980	5.8
5	Livermore-02	1980	5.42
6	Anza (Horse Canyon)	1980	5.19
7	Mammoth Lakes-02	1980	5.69
8	Mammoth Lakes-03	1980	5.91
9	Mammoth Lakes-04	1980	5.7
10	Victoria_ Mexico"	1980	6.33
11	"Westmorland"	1981	5.9
12	"Morgan Hill"	1984	6.19
13	"Hollister-04"	1986	5.45
14	"Chalfant Valley-01"	1986	5.77
15	"Chalfant Valley-02"	1986	6.19
16	"Chalfant Valley-03"	1986	5.65
17	"Chalfant Valley-04"	1986	5.44
18	"Superstition Hills-02"	1987	6.54
19	"Landers"	1992	7.28
20	"Double Springs"	1994	5.9
21	"Kobe_ Japan"	1995	6.9
22	"Duzce_ Turkey"	1999	7.14
23	"Manjil_ Iran"	1990	7.37
24	"Hector Mine"	1999	7.13
25	"Chi-Chi_ Taiwan-04"	1999	6.2
26	"Landers"	1992	7.28
27	"Tottori_ Japan"	2000	6.61
28	Parkfield-02_ CA	2004	6
29	Basso Tirreno Italy	1978	6
30	Potenza_ Italy	1990	5.8
31	Joshua Tree CA	1992	6.1
32	Darfield New Zealand	2010	7

The selected ground motion is then scaled in the spectral acceleration-period domain, to be matched to the target response spectrum. The target response spectrum is referring to the Indonesia's Earthquake Map and Response Spectra [11], in the location of Pidie Jaya. The parameters of the corresponding design response spectrum are summarized in Table 3.

Table 3. Summary of design response spectrum

Parameters	Nominal
Location	Pidie Jaya, Aceh Regency
Longitude	96.201 degrees
Latitude	5.241 degrees
PGA	0.54 g
Ss	1.275836 g
S1	0.557154 g
SD_S	0.756712 g
SD_1	0.774701 g
Site Class	SE

As noted in SNI 1726 2019 [12] and also in accordance with ASCE 7-16 [13], the design response spectrum shall be scaled by 1.5, to achieve the Risk-Targeted Maximum Considered Earthquake (MCER) Response Spectrum. The ground motion records were spectrum matched, at least over a period range between 0.2T and 1.5T for a damping ratio of 5%, as suggested in NEHRP/NIST GCR 11-917-15 [14]. Based on the modal analysis, the building fundamental period resulting in T=0.343 sec. Henceforth, rather than by using a single period scaling, the 32 ground motion records are averagely scaled in arithmetic mean, by the scaling method of the minimize Mean Squared Error (MSE) within the period range of 0.05 sec and 0.08 sec (more conservative). The generated scaling of ground motion records in this study is depicted on Figure 5.

In this study, the pushover analysis is conducted by using SAP2000 to obtain the capacity curve. As the usual capacity spectrum method, the capacity curve is translated into Acceleration-Displacement Response

 https://doi.org/10.1051/e3sconf/202234002008

Spectrum (ADRS) format to enable direct interpretation between capacity spectrum curve with response spectrum curve, where the intersection between those two curves is defined as the performance point (PP).

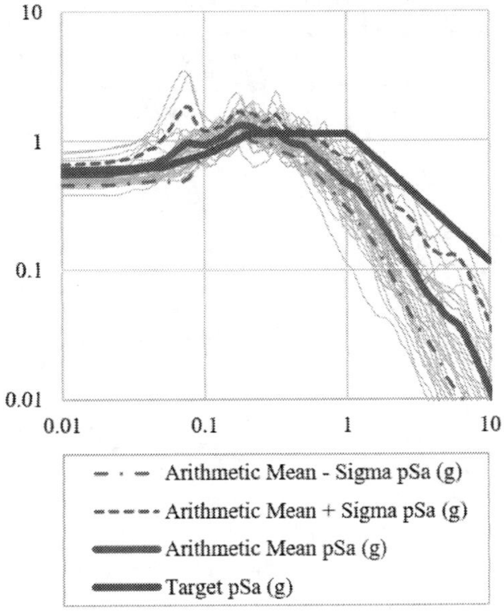

Fig. 5. Ground motion scaled to the target response spectrum within period of 0.05 sec and 0.08 sec (Re-Illustrated from the scaling results of PEER NGA-West2 Website)

Previously, equivalent linearization method, based on FEMA 440 [15] is used. However, the analysis results cannot render good results, specifically in the elastic range of the capacity spectrum. As such, the analysis is reiterated by using conventional capacity spectrum method, based on ATC-40 [16]. Due to practicality consideration, only the Arithmetic Mean pSa, Arithmetic Mean + sigma pSa, and Arithmetic Mean - sigma pSa is used as response spectrum set to be analysed with the corresponding capacity spectrum curve. The PP is estimated by the intersection between the capacity spectrum curve with the construction of inelastic response spectrum is in accordance with ATC-40 [11] method.

c. Formulation of analytical-based fragility curve function

In this study, the acceptance criteria from ASCE 41-17 are assumed as the Damage states (DS), corresponded to the later creation of fragility function's family curve, which can be classified into three categories, e.g., Life Safety (LS) as DS-1, Immediate Occupancy (IO) as DS-2, and Collapsed Prevention (CP) as DS-3. In order to render the probability earthquake intensity levels, each response spectrum set is scaled up and scaled down by using scaled factor with the increments of 0.1. The number of PPs generated is then observed, to determine the Intensity Measures (IM) of spectral acceleration and the level of DS. The construction of fragility curve is derived from the set of IM and DS pairs by adopting the fitting approach by using MLE [3]. The general cumulative lognormal distribution can be computed as follows:

$$P(DS \geq DS_i|pSa) = \Phi\left(\frac{ln(pSa/\theta_{DS_i})}{\beta_{DS_i}}\right) \tag{1}$$

Where:

Φ = standard norm. cumulative dist. func.

θ_{DS_i} = expected median value of pSa at a certain damage state; and

β_{DS_i} = natural logarithm standard dev. of pSa at a certain damage state

Meanwhile, to estimate the fragility function parameters by using the maximum likelihood function, it can be obtained as follows:

$$\{\hat{\theta},\hat{\beta}\} = arg\ arg\ max \sum_{i=1}^{m} \Phi\left(\frac{ln\left(\frac{pSa}{\theta_{DS_i}}\right)}{\beta_{DS_i}}\right) + (n-m)ln\left(1-\Phi\left(\frac{ln(pSa/\theta_{DS_i})}{\beta_{DS_i}}\right)\right) \tag{2}$$

Where:

m = ground motion at specified damage state

n = ground motion used in the analysis

3 Results and discussion

a. Results of fragility curve

The result of analytical based fragility curve is shown in Figure 6. Based on the following figure, it can be said that the probability to reach the DS-1 is relatively yield on lower pseudo spectral acceleration compared to the other DSs. Contrastingly, the DS-2 and DS-3 is likely can be triggered in relatively close range of pseudo spectral acceleration. It can be means as well, there is a high probability that the structure is having low ductility, since the extensive damage state (DS-3) can be reached sooner, upon reaching the moderate damage state (DS-2).

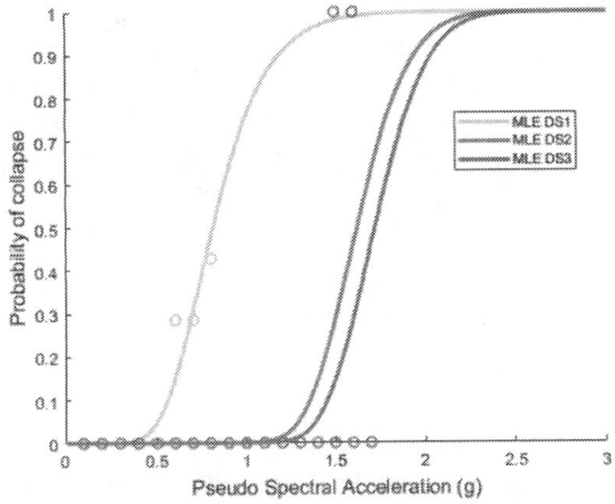

Fig. 6. Analytical-Based Fragility Curve with Acceptance Criteria of ASCE 41-17 as DS

In comparison to Figure 6, fragility curve with introducing the Inter-story Drift (IDR) as DS indices is represented in Figure 7. It can be seen that from the following curve, defining the IDR as DS can differ the intensity of damage probabilities. In fact, the DS-1 curve in Figure 7 highlight more disperse probability of failure, compared to Figure 6. Contrastingly, perhaps, the distinguish features by adopting the IDR as DS are how the DS-2 and DS-3 curves can intersect each other, which can lead to a bias to determine a firm judgement, associated to the DS probabilities. Apparently, unlike in Figure 6, where the observed fractions of DS-2 and DS-3 are solid, in Figure 7 the observed fractions of DS-2 and DS-3 are assorted within the pSa range of 1.5-1.6 g, with an intermediate probability.

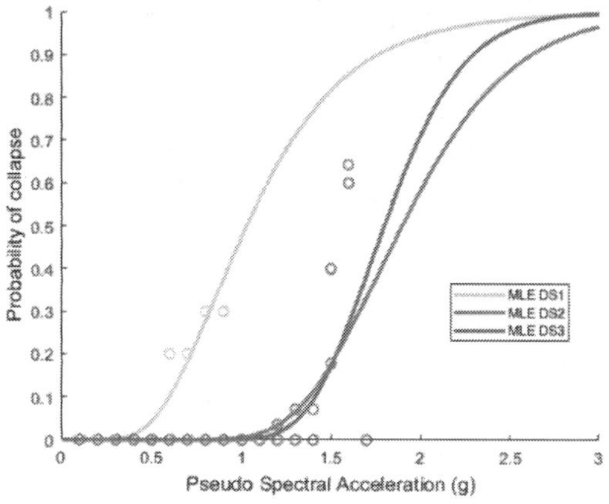

Fig. 7. Analytical-based fragility curve with inter-story drift (IDR) as DS

b. Extended discussion for the structural response and behaviour

To observe the structural response and behaviour, the capacity spectrum curve in ADRS format is shown in Figure 8. In the following figure, the structural model discussed in this study (Frame + Infill Strut) is depicted in comparison with the bare frame model without infill strut. It can be concluded that the structural model incorporating infill strut is having a higher stiffness and lateral strength capacity compared with the bare frame model. Whereas, the bare frame model is only having 10% lateral strength of the Frame + Infill strut model.

However, it should be highlighted as well that the bare frame model shows higher deformation capacity by roughly twice as much, compared the structure with infill strut model. This signifies the model with infill strut resolute brittle behaviour, compared to the bare frame model. The capacity spectrum also indicated strength deterioration type which is not preferable for earthquake-resistant structure system, where the sudden drop of strength occurred rapidly after the failure point.

To sum up, although the masonry infill contributes to the general in-plane stiffness of the structure frame, due to the lacks of reinforcement details, the column drift capacity is considerably minimum, and resulting in a brittle response manner, which is indicated from the insignificant dissipation of hysteretic energy.

In addition, the analytical results show that the failure mechanism of the structure is anticipated by the compressive strut action of infill wall. Nonetheless, the column member begins to yield, followed by the infill wall dropping its load, upon it reached the peak strength capacity, and then tailed by the sudden drop of lateral strength. The observed response from the analytical works is also compared to the field investigation, and signify seemingly similar failure mechanism.

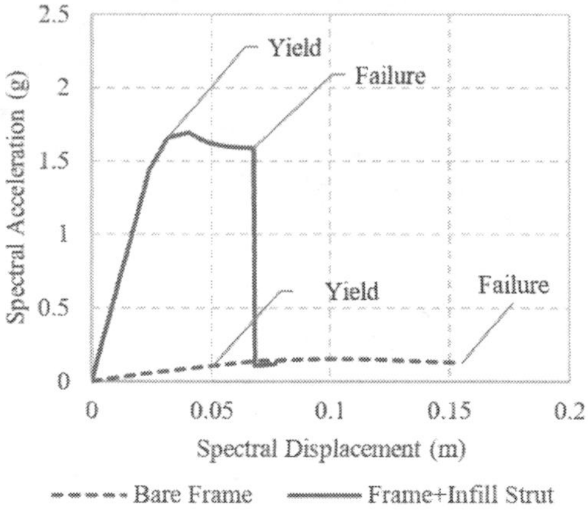

Fig. 8 Analytical-based fragility curve

4 Conclusion

This study is limited to a particular typology of the "type 36" regular house, which is commonly found in Aceh. The writers are motivated to study this typical house due to its proneness to the notably 2016's Pidie Jaya Earthquake. The structural model is developed under presumptive modelling parameters, that is also be based on plausible evidence to some extent, taking into account the common practices and also based on field observation.

The reliability of structural analysis can be somewhat influenced by the basis of analysis assumption, from the sophistication of modelling approaches to the uncertainty of earthquake characteristics. In order to eliminate the biases, in this study, earthquake vulnerability assessment is conducted by involving 32 ground motion records that is scaled to the specified target response spectrum, to include the randomness and also probabilistic consideration.

However, the writers perceive that there is always a room for improvement to enhance the results of study. Perhaps, the flaws of this study can be asserted in the limitation of selecting pertinent ground motion record. Under some circumstances, the corresponding ground motion of Pidie Jaya Earthquake is not available to be used in this study. Further study by utilizing empirical fragility curve and including analytical-based fragility curve with Peak Ground Acceleration (PGA), for more intuitive and direct interpretation.

References

1. M. I. Sari, Abdullah, M. Afifuddin, Perilaku Lateral Siklik Portal Beton Bertulang Berisi Dinding Bata Merah, Jurnal Teknik Sipil, 845-856, Universitas Syiah Kuala (2018)
2. ASCE 41-17, ASCE, Seismic Evaluation and Retrofit of Existing Buildings (2017)
3. A. K. Chopra, R. K. Goel, A Modal Pushover Analysis Procedure for Estimating Seismic Demands for Buildings, Earthquake Engineering and Structural Dynamics, 31, 461-582 (2002)
4. P. Fajfar, A Nonlinear Analysis Method for Performance Based Seismic Design, Earthquake Spectra, 16, 573-592 (2000)
5. A. C. Guyader, D. Iwan, Determining Equivalent Linear Parameters for Use in a Capacity Spectrum Method of Analysis, 132, 59-67 (2006)
6. D. Vamvatsikos, C. A. Cornell, Incremental Dynamic Analysis, 31, 491-514 (2002)
7. Baltzopoulos et al., SP2FRAG: Software for Seismic Fragility Assessment Based on Static Pushover, Bull Earthquake Eng, 15, 4399-4425 (2017)
8. J. W. Baker, Efficient Analytical Fragility Function Fitting using Dynamic Structural Analysis, Earthquake Spectra, 31, 579-599, (2015)
9. Zarnic R, Gostic S., Masonry Infilled Frames as an Effective Structural Sub-Assemblage, Proceedings of the International Workshop., Bled, Slovenia, 335-46 (1997)
10. PEER NGA-West2 Database, https://ngawest2.berkeley.edu/
11. Indonesia's Earthquake Map and Response Spectra, http://rsa.ciptakarya.pu.go.id/2021/
12. SNI 1726:2019, Tata Cara Perencanaan Ketahanan Gempa untuk Struktur Gedung dan Non-gedung (2019)
13. ASCE/SEI 7-16, Minimum Design Load and Associated Criteria for Buildings and Other Structures (2016)
14. NEHRP/NIST GCR 11-917-15, Selecting and Scaling Earthquake Ground Motions for Performing Response-History Analyses (2015)
15. FEMA 440, Improvement of Nonlinear Static Seismic Analysis Procedures (2005)
16. ATC-40, Seismic Evaluation and Retrofit of Concrete Buildings (1996)

Validating a low-cost seismometer using a shaking table

Bambang Setiawan[1,*], *Mumammad* Rizal[2], *Halida* Yunita[3], *Taufiq* Saidi[3], *Muttaqin* Hasan[3] and *Zulkifli* Zulkifli[4]

[1]Program Study of Geological Engineering, Universitas Syiah Kuala, 23111 Banda Aceh, Indonesia
[2]Department of Mechanical Engineering, Universitas Syiah Kuala, 23111 Banda Aceh, Indonesia
[3]Department of Civil Engineering, Universitas Syiah Kuala, 23111 Banda Aceh, Indonesia
[4]Department of Economic Development, Universitas Syiah Kuala, 23111 Banda Aceh, Indonesia

Abstract. Ambient vibration measurement using a seismometer represent a non-invasive and non-destructive method for seismic-related studies. This method inhibits widespread applications, i.e., assessing site fundamental frequency, estimating bedrock level, calculating building natural frequency, deducing shear wave velocity profile. However, the instrument required to carry out such ambient vibration measurement is generally expensive. Recently, there is a low-cost Raspberry Shake 3D seismometer (RS-3D) available on the market. This seismometer is based on Raspberry Pi hardware. It contains three 4.5 Hz orthogonally oriented geophones, which can be set at a sampling rate of 100 Hz and a bandwidth of -3dB points at 0.6 to 34 Hz. Thus, laboratory testing was carried out to assess this device's performance in this demanding application. Generally, RS-3D demonstrates a great performance. Further detailed results are presented in this paper.

1 Introduction

Ambient vibration measurement using a seismometer represent a non-invasive and non-destructive method for seismic-related studies, i.e., [1], [2], [3], and [4]. This method may inhibit widespread applications, i.e., assessing site fundamental frequency ([5] - [8]), estimating bedrock level ([9] - [14]), calculating building natural frequency [15], deducing shear wave velocity profile ([16] - [20]). However, the instrument required to carry out such ambient vibration measurement is generally expensive. Recently, there is a low-cost Raspberry Shake 3D (RS-3D) seismometer available on the market. This seismometer is based on Raspberry Pi hardware. It contains three 4.5 Hz orthogonally oriented geophones, which can be set at a sampling rate of 100 Hz and a bandwidth of -3dB points at 0.6 to 34 Hz. Thus, laboratory testing must be conducted to assess this device's performance in this demanding application.

The purpose of the study is to investigate the performance of a low-cost RS-3D seismometer from which further suggestions will be proposed.

2 Methodology

The methodology in this study that is begun with sensor selection and experimental setting up. Then, it is followed by data acquisition and analysis.

2.1 Sensor

A low-cost Raspberry Shake 3D seismometer (RS-3D) was used in this study. The used sensor in this seismometer is a 4.5 Hz (electronically extended down to 2 seconds) Sunfull PS-4.5B geophones, 375 Ohm. It can be set at a sampling rate of 100 Hz and a bandwidth of -3dB points at 0.6 to 34 Hz [21].

2.2 Shaking table

Multi-Axial Simulation Table (MAST) System was used in the present study. A schematic MAST System diagram, which consist of (1) Table, (2) Integrated hydraulic control manifold with optional isolation, (3) Fixed actuator assembly, (4) Close-coupled accumulation, (5) Mounting frame, (6) Optional chamber floor interface, and (7) Swivel and strut assembly, is presented in Fig. 1.

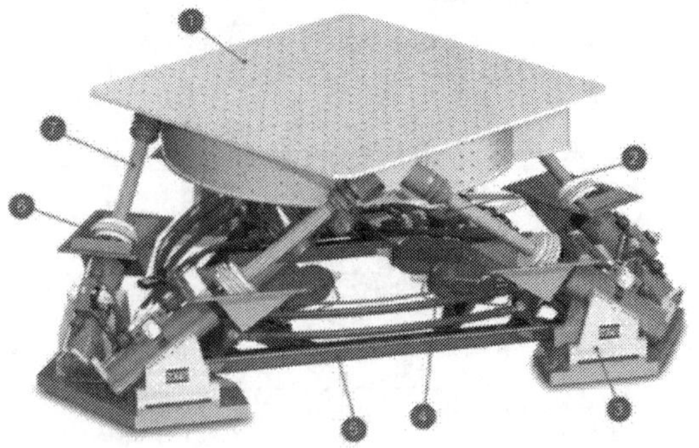

Fig. 1. Schematic system diagram of MAST System [22].

This study employed MAST System 354.20, which can provide simulation frequency of 0.8 - 100 Hz, linear displacement up to 140 mm, angular displacement up to 9 deg., linear velocity up to 1700 mm/s, and linear acceleration up to 15.9 g. Detail specification of the MAST System 354.20 can be seen in Table 1 [23].

* Corresponding author: bambang.setiawan@unsyiah.ac.id

https://doi.org/10.1051/e3sconf/202234002009

Table 1. Detail specifications of MAST System 354.20 [3].

Description	Units	Amount
Table Size (Square table)	mm	2.2 x 2.2
Simulation Frequency (bare table)	Hz	0.8-100
Simulation Frequency (max payload)	Hz	0.8-100
Maximum Payload	kg	2000
Table Mounting Pattern	mm	100 x 100
Actuator Peak Force (210 bar)	kN	67
Linear Displacement		
- Vertical (Z)	± mm	140
- Lateral (Y)	± mm	110
- Longitudinal (X)	± mm	125
Angular Displacement		
- Roll	± deg	9
- Pitch	± deg	8
- Yaw	± deg	6
Linear Velocity		
- Vertical (Z)	mm/s	1700
- Lateral (Y)	mm/s	1300
- Longitudinal (X)	mm/s	1350
Angular Velocity		
- Roll	deg/s	78
- Pitch	deg/s	70
- Yaw	deg/s	50
Linear Acceleration (bare table)		
- Vertical (Z)	g	15.9
- Lateral (Y)	g	11.6
- Longitudinal (X)	g	12.8
Linear Acceleration (max payload)		
- Vertical (Z)	g	6.0
- Lateral (Y)	g	4.2
- Longitudinal (X)	g	4.8

2.2 Experimental setting up

The overall experimental setting up is shown in Fig. 2. The RS-3D seismometer was clamped on a rock slap. The slap was tighten using a 12 mm screw bolt. A power bank of 12 AH lithium-ion battery was used to operate the RS-3D. A portable ethernet wifi-router was connected to the RS-3D to initiate the start-up of the instrument, collect the data, and shut down the instrument. The RS-3D seismometer on top of the table of the MAST System setting is shown in Fig. 3.

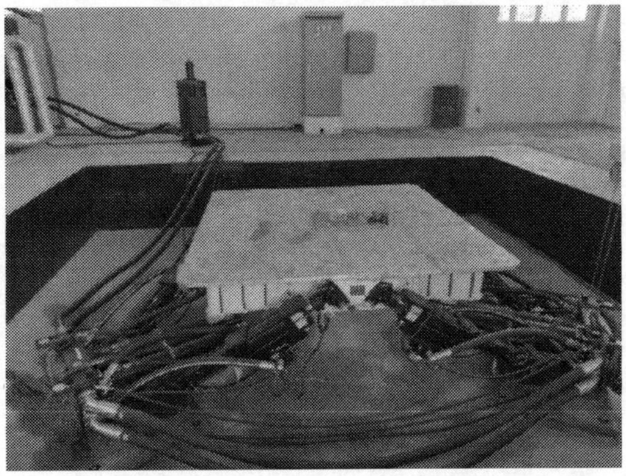

Fig. 2. The raspberry seismometer was set in the top-middle of the shaking table.

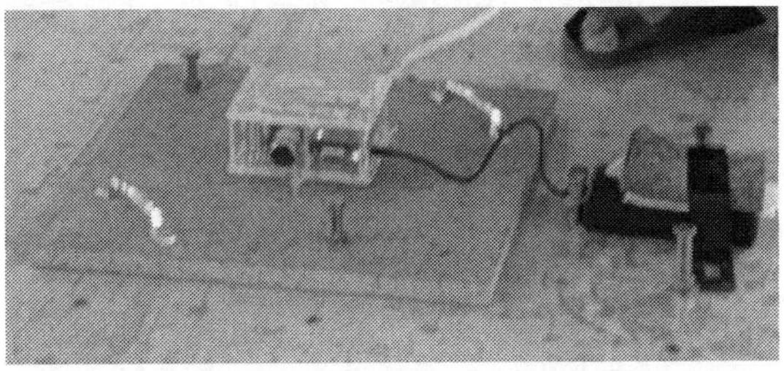

Fig. 3. Rock slap was used to tie the raspberry seismometer.

2.2 Data acquisition and analysis

The time histories data collection was recorded using RS-3D Raspberry seismometer. Four different trials were carried out, as shown in Table 2. The time histories data of these four trials are presented in Figs. 4 to 7.

Table 2. Experimental trials of the present study.

Trial No.	Frequency	Recording duration	Remarks
Trial#1	0.4 Hz	9 minutes	All axes of E-W, N-S, and U-D
Trial#2	0.4 Hz	9 minutes 18 seconds	E-W axis
Trial#3	1.0 Hz	10 minutes	E-W axis
Trial#4	1.0 Hz	10 minutes	All axes of E-W, N-S, and U-D

In this study, the experimental data were analyzed using the Geopsy software package (GSP) [24]. Two analyses were carried out. The first analysis is to discover the origin of the waves. Damping tool in the GSP was employed in the first analysis, from which the origin of the waves is concluded. The second analysis is to detect the frequency and source of the recorded waves. The Geopsy spectrum and 2D spectrum toolboxes were used to detect the frequency and source azimuth of the recorded waves. Spectral analysis using spectrum toolbox provides

https://doi.org/10.1051/e3sconf/202234002009

a means of measuring the strength of periodic components of a wave at different frequencies. Spectral analysis includes the calculation of waves in a set of sequenced data. The 2D spectrum toolbox provides the spatial distribution of the spectral analysis by means of frequency and azimuth. The used parameters in the second analysis are window length (25s), smoothing using the Konno & Ohmachi [25] smoothing approach with a smoothing constant of 40, and window type of Tukey with width 5%.

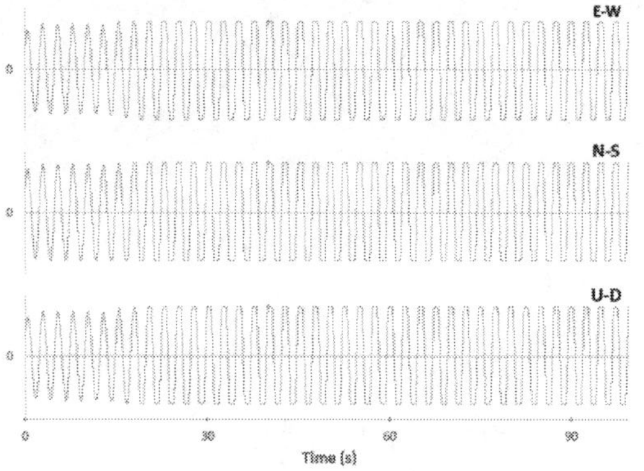

Fig. 4. Time histories data for Trial#1.

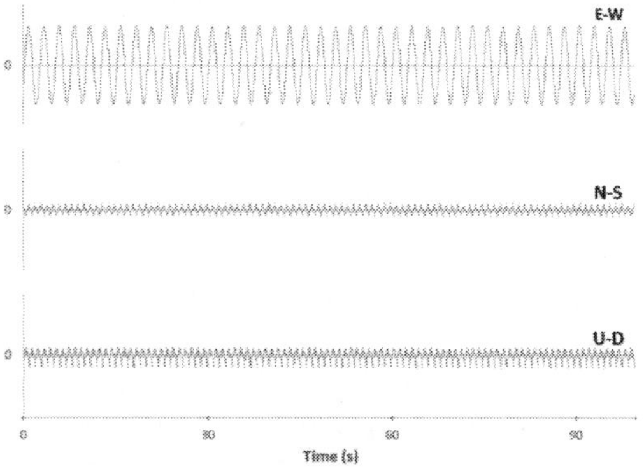

Fig. 5. Time histories data for Trial#2.

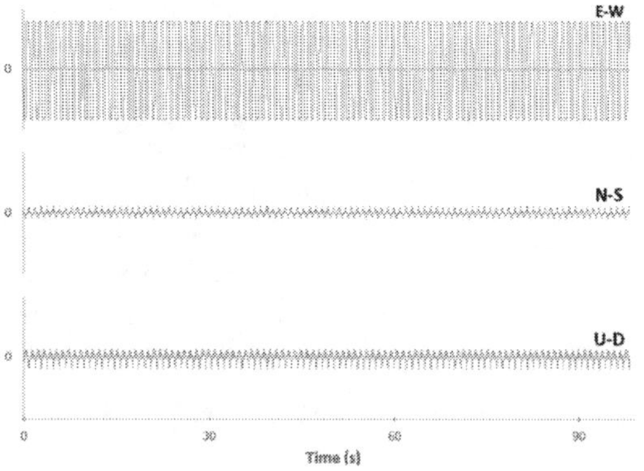

Fig. 6. Time histories data for Trial#3.

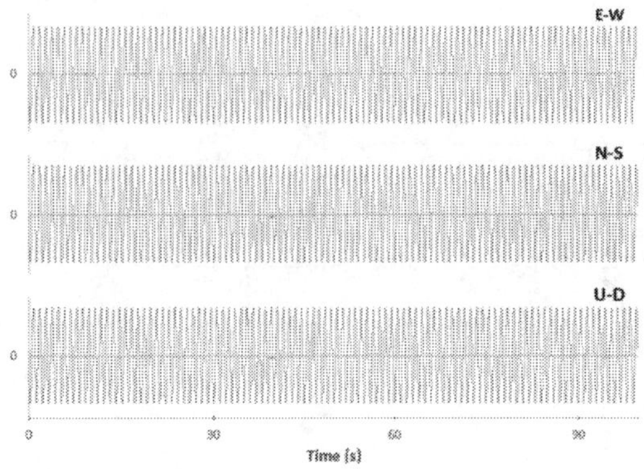

Fig. 7. Time histories data for Trial#4.

3 Results and discussion

Shaking table tests were conducted at various trials, as aforementioned above. The collected data were analysed in two steps for detecting the origin of the waves and estimating the frequency of the shaking. The results are presented in Figs. 8 to 19.

Prior to the frequency analysis, a detection of the origin of the waves was carried, which suggests harmonic waves generated from the shaking table, as damping <1% is observed except for the vertical component of Trial#2, as shown in Figs, 8, 11, 14, and 17. The spectrum analyses results are shown in Figs. 9, 12, 15, and 18. The frequencies obtained from the spectrum analyses are tabulated Table 3.

Results of 2D spectrum analysis are shown in Figs. 10, 13, 16, and 19. This 2D spectrum analysis suggests that Trial#1 and Trial#4 receive wider spatial distribution wave sources than Trial#2 and Trial#3.

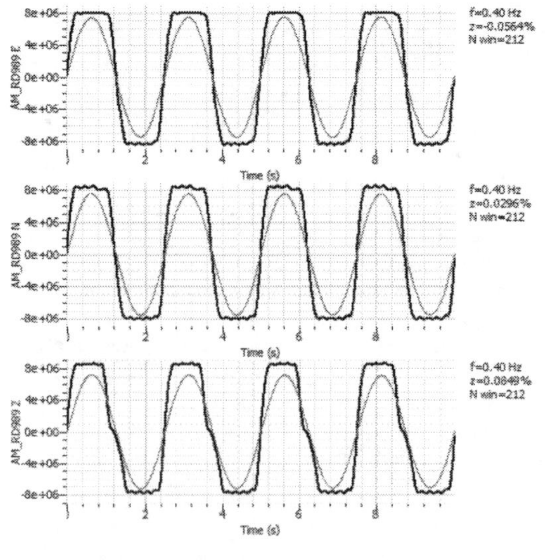

Fig. 8. Results of damping analysis of Trial #1 data.

https://doi.org/10.1051/e3sconf/202234002009

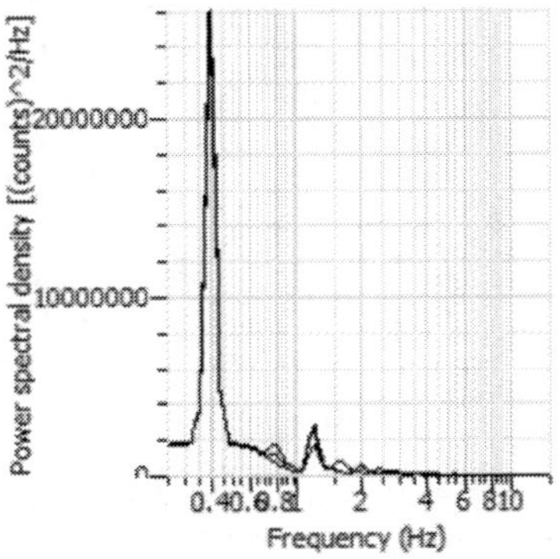

Fig. 9. Summary spectrum analysis of Trial #1 data.

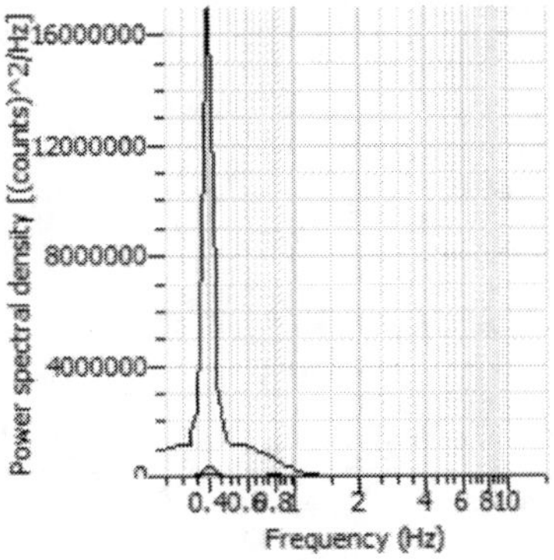

Fig. 12. Summary spectrum analysis of Trial #2 data.

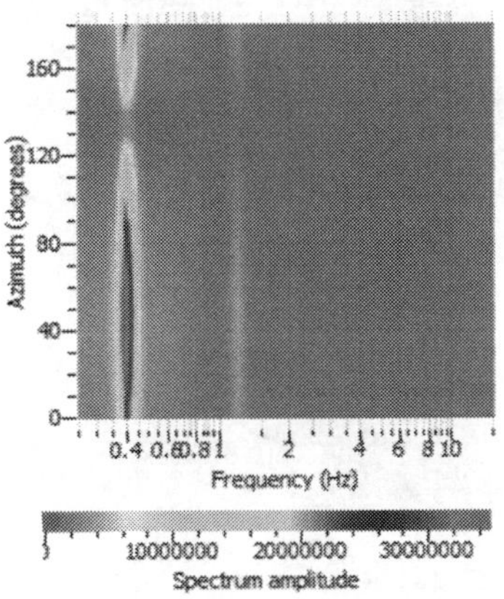

Fig. 10. Azimuth wave-source at frequency between 0.25 Hz and 15 Hz for Trial #1.

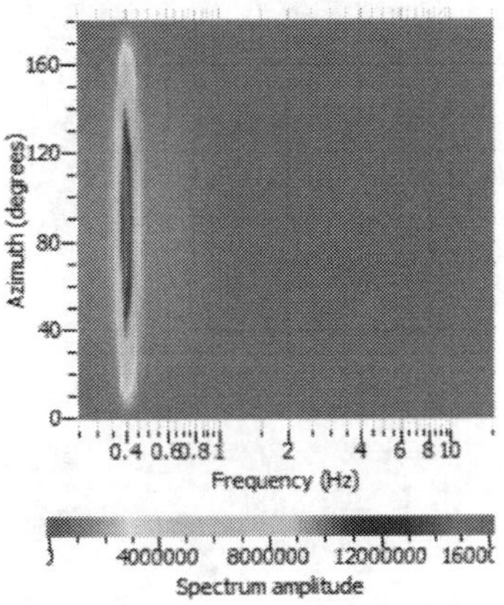

Fig. 13. Azimuth wave-source at frequency between 0.25 Hz and 15 Hz for Trial #2.

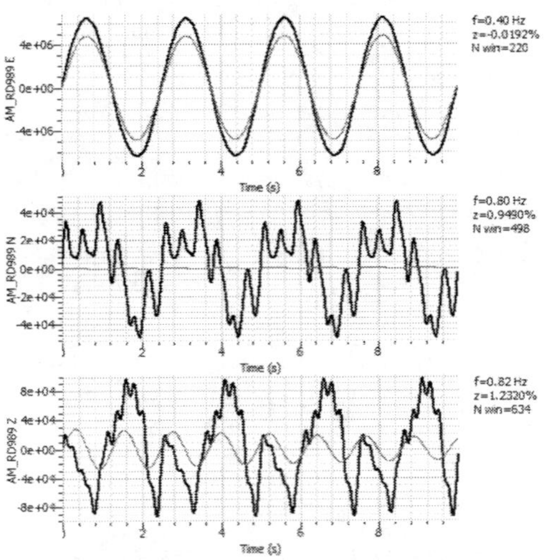

Fig. 11. Results of damping analysis of Trial #2 data.

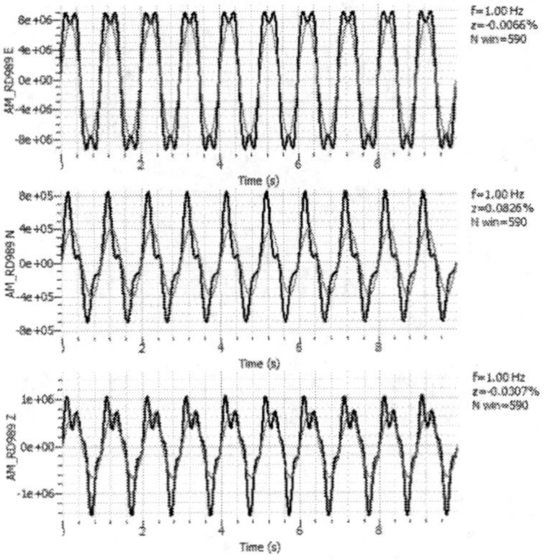

Fig. 14. Results of damping analysis of Trial #3 data.

https://doi.org/10.1051/e3sconf/202234002009

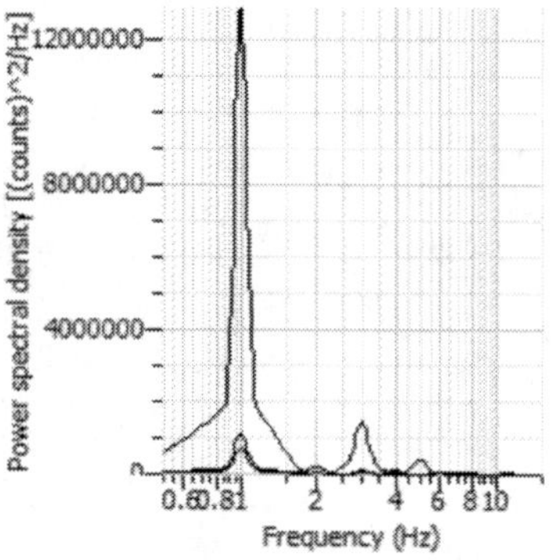

Fig. 15. Summary spectrum analysis of Trial #3 data.

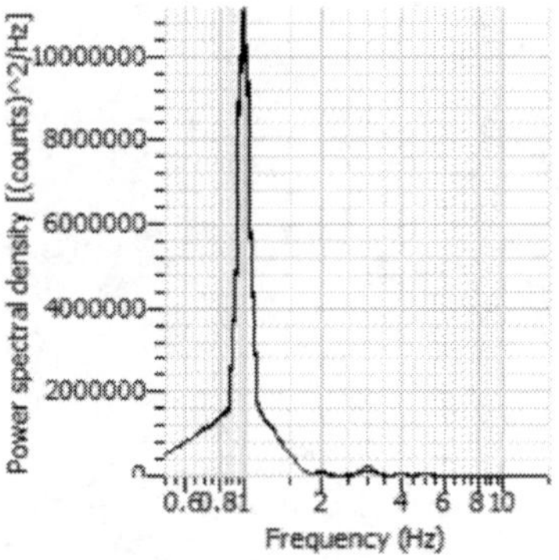

Fig. 18. Summary spectrum analysis of Trial #4 data.

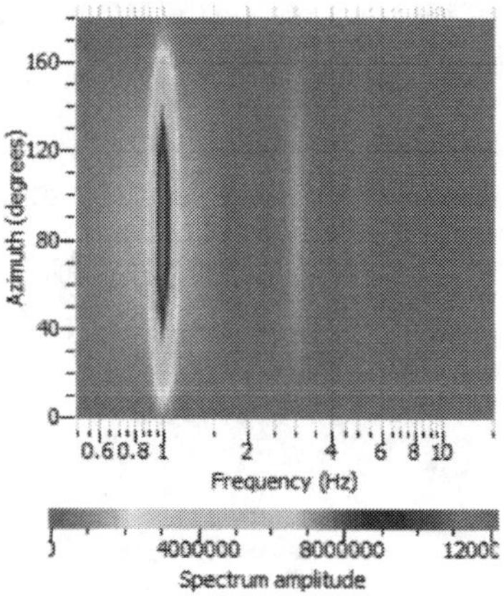

Fig. 16. Azimuth wave-source at frequency between 0.5 Hz and 15 Hz for Trial #3.

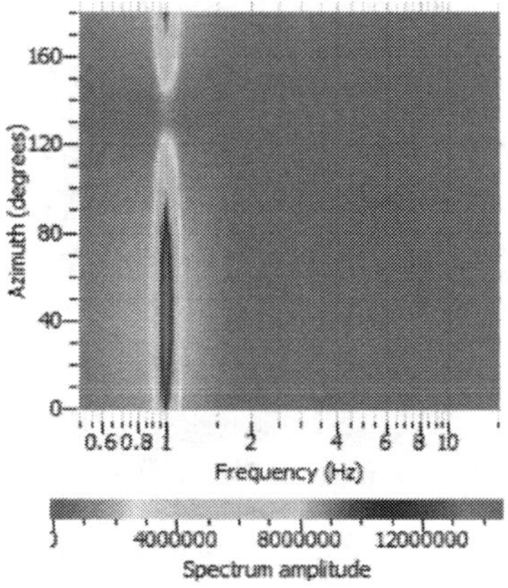

Fig. 19. Azimuth wave-source at frequency between 0.5 Hz and 15 Hz for Trial #4.

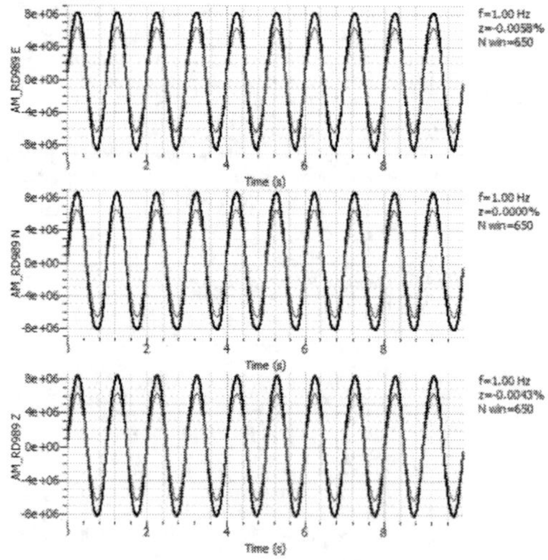

Fig. 17. Results of damping analysis of Trial #4 data.

Table 3. Frequencies obtained from the spectrum analyses.

Trial No.	**Frequency (Hz)**	
	MAST System	**Rs-3D Seismometer**
Trial#1	0.4 Hz	0.394015 Hz
		0.394015 Hz
		0.394015 Hz
Trial#2	0.4 Hz	0.394015 Hz
		0.394015 Hz
		0.394015 Hz
Trial#3	1.0 Hz	0.978717 Hz
		0.978717 Hz
		0.978717 Hz
Trial#4	1.0 Hz	0.978717 Hz
		0.978717 Hz
		0.978717 Hz

To assess the performance of the RS-3D seismometer, the Raspberry Shake vibration data results of spectrum analysis, i.e. frequency, were compared against the frequency input from the shaking table. The comparison is presented in Table 3 and Fig. 20, which suggests in a great agreement.

Additionally, root mean square error (RMSE) was calculated to measures how much error there is between two data sets of MAST system input frequency and RS-3D seismometer response frequency. In other words, it compares an input (predicted) value of MAST system frequency and an output (observed) value of RS-3D seismometer frequency. RMSE can be approximated using an Equation 1 [26], as follow:

$$RMSE = \sqrt{\frac{\sum_{i=1}^{n}(P_i - O_i)^2}{n}} \quad (1)$$

Predicted and observed values are represented by P_i and O_i, respectively. The number of observations is n.

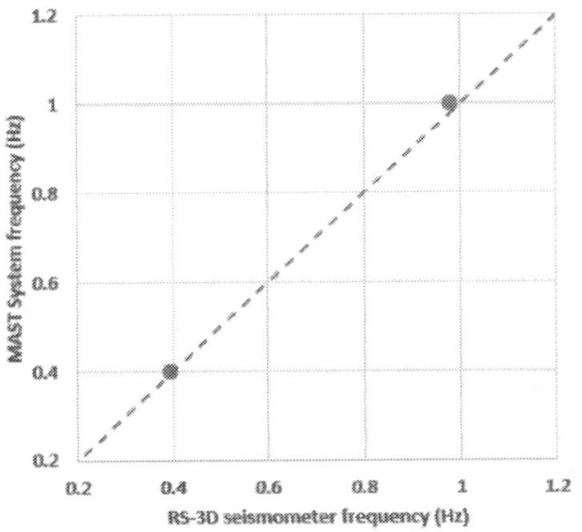

Fig. 20. The performance of the RS-3D seismometer.

The RMSE is the square root of the variance of the residuals. RMSE in this study indicates the absolute fit of the RS-3D seismometer for measuring the input frequency of the shaking table. Generally, lower value of RMSE indicates better fit. The RMSE in this study is 0.016 g which suggests a good performance of the RS-3D seismometer. Based on a rule of thumb, RMSE values less than 0.1 shows that the model can relatively predict the data accurately. Furthermore, a RMSE value of the present study (0.016) that is smaller than both standard deviation values of observed data (0.305) and predicted data (0.313) is considered good [27].

In addition, scatter index (SI), as shown in Equation 2, is also calculated to assess the performance of the RS-3D seismometer. SI is simply the RMSE divided by the mean value of the observed data [28], as follow:

$$SI = \left(\frac{RMSE}{Mean\ of\ observed\ value}\right) \times 100\% \quad (2)$$

Generally, if SI < 10% is a good instrument, SI<5% is a very good instrument. The present study estimates the SI of 2%, which suggests very good instrument.

4 Conclusion

Ambient vibration measurement using a seismometer may inhibit widespread applications. Recently, there is a low-cost seismometer (RS-3D seismometer) available on the market. RS-3D seems to be a feasible option in ambient low-noise measurement methods. However, the performance of this seismometer requires further study. This study presents the performance of this RS-3D by validating this seismometer using a shaking table. Four trials were carried out, from which twelve datasets were acquired using RS-3D seismometer. These datasets were analysed to obtain the observed predominance frequency. RMSE and SI calculations were used to deduce the comparison between input frequency of MAST system and an output frequency of RS-3D seismometer frequency. This study concludes that RS-3D seismometer provides a great performance. However, further testing using many different input frequencies is required.

Author Contributions. Authors of this paper are, as follow: Bambang Setiawan (B.S.), Muhammad Rizal (M.R.), Halida Yunita (H.Y.), Taufiq Saidi (T.S.), Muttaqin Hasan (M.H.), and Zukifli Zulkifli (Z.Z.). The authors contribution can be described, as follow: conceptualization, B.S., M.Z., H.Y., T.S., M.H., and Z.Z.; methodology, B.S., M.R., H.Y., and M.H.; software, B.S. and M.R.; validation, M.H., H.Y., and M.R.; formal analysis, M.R. and B.S.; investigation, T.S. and M.H.; resources, T.S., B.S. and Z.Z.; data curation, B.S. and M.R.; writing—original draft preparation, B.S.; writing—review and editing, B.S., M.R., H.Y., and T.S.; visualization, B.S. and M.R.; supervision, Z.Z.; project administration, Z.Z.; funding acquisition, T.S., H.Y., and M.H.

Acknowledgments. The authors wish to acknowledge the Laboratorium Geologi Rekayasa Program Studi Teknik Geologi Universitas Syiah Kuala for providing a seismometer instrument and technical staff. In addition, the first author is grateful to the Faculty of Engineering of Universitas Syiah Kuala for their support. The shaking table of MAST 354.20 was provided by Tsunami and Disaster Mitigation Research Center (TDMRC) through Proyek SBSN Kementerian Pendidikan, Kebudayaan dan RISTEK tahun 2019.

References

1. S. Bonnefoy-Claudet, C. Cornou, P-Y. Bard, F. Cotton, P. Moczo, J. Kristek, D. Fah, Geophys. J. Int. **167** (2006) 827–837.
2. V. Del Gaudio, S. Muscillo, J. Wasowski, Eng. Geol. **182** (2014) 182-200. DOI: 10.1016/j.enggeo.2014.05.010.
3. P. Di Stefano, D. Luzio, P. Renda, R. Martorana, P. Capizzi, A. D'Alessandro, N. Messina, G. Napoli, S. Todaro, G. Zarcone, Nat. Hazards Earth. Syst. Sciences, **2** (2014) 2597-2637, DOI:10.5194/nhessd-2-2597-2014.

4. A. Yulianur, T. Saidi, B. Setiawan, S. Sugianto, M. Rusdi, M. Affan, Data Brief, **30** (2020) 105554, DOI: 10.1016/j.dib.2020.105554
5. B. Setiawan, M. Jaksa, M. Griffith, D. Love, Soil Dyn. Earthq. Eng. **110** (2017) 244- 261. DOI: 10.1016/j.soildyn.2017.08.006
6. P. Gueguen, J-L. Chatelain, B. Guilier, H. Yepes, J. Egred, Soil Dyn. Earthq. Eng. 17 (1998) 329-334.
7. M. Herak, I. Allegretti, D. Herak, K. Kuk, V. Kuk, K. Maric, S. Markusic, J. Stipcevic, Bull. Earthq. Eng. **8** (2010) 483-499.
8. N. Tumur-Uyal, B. Setiawan, M. S. Oucherif, M. S., IOP Conf. Series: Materials Sci. Eng. **523** (2019) 1, 012043. DOI: 10.1088/1757-899X/523/1/012043
9. M. Ibs-von Seht, J. Wohlenberg, Bull. Seismol. Soc. Am. 89 (1999) 250–259.
10. S. Parolai, P. Bormann, C. Milkereit, Bull. Seismol. Soc. Am. **92** (2002) 6, 2521–2527
11. K. Hinzen, B. Weber, F. Scherbaum, J. Earthq. Eng. **8** (2004) 6, 909–926.
12. P. Bodin, K. Smith, S. Horton, H. Hwang, Eng. Geol. **62** (2001) 159–168.
13. M. Hellel, E. H. Oubaiche, J-L. Chatelain, D. Machane, R. Bensalem, B. Guillier, G. Cheikhlounis, Seismol. Res. Lett. **83** (2012) 5, 798 – 805.
14. B. Setiawan, M. Jaksa, M. Griffith, D. Love, Eng. Geol. **243** (2018) 145-159. DOI:10.1016/j.enggeo.2018.06.022
15. C. Salameh, B. Guillier, J. Harb, C. Cornou, P-Y. Bard, C. Voisin, A. Mariscal, Bull. Earthq. Eng. **14** (2016) 10, 2705–2730, DOI:10.1007/s10518-016-9920-9.
16. B. Bettig. P-Y. Bard, F. Scherbaum, J. Riepl, F. Cotton, C. Cornou, D. Hatzfeld, Boll. Geof. Teor. Appl. **42** (2001) 3-4, 281–304.
17. F. Scherbaum, K. Hinzen, M. Ohrnberger, Geophys. J. Int. **152** (2003) 3, 597 – 612.
18. G. Di Giulio, I. Gaudiosi, F. Cara, G. Milana, M. Tallini, Geophys. J. Int. **198** (2014) 848-866, DOI: 10.1093/gji/ggu162.
19. B. Setiawan, M. Jaksa, M. Griffith, D. Love, Soil Dyn. Earthq. Eng. **122** (2019) 16-38. DOI:10.1016/j.soildyn.2019.03.036
20. A. Yulianur, T. Saidi, B. Setiawan, S. Sugianto, M. Rusdi, J. Eng. Science Technol. **15** (2020) 5, 2871-2889
21. Raspberry Shake, (2021) Raspberry Shake 3D, https://raspberryshake.org/products/raspberry-shake-3d/
22. MTS (2021). Technical Overview. https://www.mts.com/en/products/automotive/subsystem-component-test-systems/multi-axial-simulation-table-systems#models
23. MAST (Multi-axial Simulation Table) System, (2021) Turnkey solution for replicating real-world vibration phenomena in controlled laboratory settings, 20 pp.
24. Geopsy., 2018. Geopsy home: software applications for ambient vibration. http://www.geopsy.org, accessed 1/06/2018.
25. K. Konno, T. Omachi, T. Bull. Seismol. Soc. Am. **88** (1998) 1, 228-241.
26. J. Wang, L. Aouf, Y. Jia, Y. Zhang, Remote Sensing, **12** (2020) 2858, DOI: 10.3390/rs12172858
27. M. W. Liemohn, A.D. Shane, A. R. Azari, A. K. Petersen, B. M. Swiger, A. Mukhopadhyay, J. Atmos. Sol. Terr. Phys, **218** (2021) 105624
28. L. Zambresky, GKSS Forschungszentrum Federal Republic of Germany, Research Department Technical Report No. 63 (1989) 90 pp.

Design and testing booking application for private doctors' practices to support disaster management

Dalila Husna Yunardi*, *Rahmad* Dawood[2], *Maya* Fitria[1], *Rini* Deviani[2]

[1] Department of Engineering, Syiah Kuala University, 23111 Darussalam, Banda Aceh, Indonesia
[2] Department of Informatics, Syiah Kuala University, 23111 Darussalam, Banda Aceh, Indonesia

Abstract. During the pandemic or natural disaster, people will be dispersed and locked down in their own homes or crisis centres. During these times, people are still in need of access to healthcare. While general healthcare would be provided by the government during these times, certain people within the population will need access to specialised doctors, especially children, the elderly and people who have prior or long-term diseases. During a disaster, people may not have access to these services. Therefore, a simple technology that lets them book an appointment with these specialised doctors will be designed in this research. This booking application, will be able to back up the general healthcare when it is overwhelmed, especially during a disaster. This research was carried out using qualitative research. The data collection was conducted using interview and observation. These primary data are then coded according to the general themes. The data is then used to create a usable design for the booking application prototype, using User-Centred Design as its method. After the prototype was created successfully, it was then tested for its usability with its potential users. 6 people from 3 user groups agreed to be respondents for the usability testing. Using think aloud protocols, it was found that on average 4 out of 6 people found the prototype usable and suitable for their jobs.

1 Introduction

Indonesia has different healthcare system compared to other countries in the world. The current healthcare system, dictates that patient must first go to a community health centre to be diagnosed. Then get a recommendation to see specialised doctor according to the first diagnosis made by a general practitioner in the health community centre. Patients can use the national healthcare insurance known as BPJS, that will provide them with free services.

However, sometimes patients feel as if they do not get the best service provided with BPJS, or they want more time with the specialised doctors; therefore, they have the option to go to practices in private clinics in order to see the doctors that they want. This can be achieved by calling the phone number of the desired practice and register to see the doctor. The patients are usually are given a number and they must wait before the doctor can attend to them.

This way of seeing the doctor is quite common in Banda Aceh. Some people do not want to go to the health community centres first, due to the longer waiting time and the lack of specialised doctors. In most cases, older patients are seen to prefer to go to private practices as they would like to see the same doctors for their ailments.

However, during disaster time or times like this; pandemic, it would be a lot harder to go to the private practices and wait in line. Most people are afraid to go to the private practices, as some doctors still share the space with other doctors and there are also other people waiting in line. This will increase the likelihood of being in contact with diseases that normally spread with disasters or to COVID-19. This can be managed using technology.

Thus, the research aims to design and test a booking application that caters for registering patients who would like to make an appointment with their doctors. This application will support private clinics in managing their time as well as the crowds, as people still get ill during disaster and pandemic, which means they still need to go and see their doctors.

1.1 Research objectives

The objectives of this research are as follows:

- To collect relevant data from prospective users of specialised doctor's private practices in Banda Aceh
- To observe the activities that are carried out in a day within the private practice
- To interview and confirm the activities that took place in the private practice

* Corresponding author: dalilayunardi@unsyiah.ac.id

- To process and code the data
- To apply user-centred design in building a prototype design for booking application, and
- To conduct usability testing for the prototype to establish user satisfaction.

2 Research methodology

The development of this application will focus on user experiences, which means having a deep understanding of related users, what they need, what they value, their abilities, and also their limitations on utilizing the application. Hence, this research will be conducted using qualitative methodology for the data collection and analysis. This methodology will be used alongside User-Centred Design; a methodology to design software.

2.1 Qualitative research

Shank [1] defined qualitative research as a method in which experiences of the subjects are taken systematically where those experiences will then create meaning. Shank [1] also emphasised that by choosing qualitative research methodology, one must abide by the rules set and agreed upon the people within the qualitative research community. Quantitative research focuses on meanings and the context of a phenomenon [2], which in turn supports the methodology that will be used to design the software; namely User-Centred Design (UCD). UCD requires its developer to understand what the user needs' are in order to design and build a usable software [3], therefore this type of research fits the profile of the identity of this research: interpretative and naturalistic [4].

Qualitative research provides an in depth understanding as well as interpreting meanings from phenomenon within their natural settings. These factors will contribute heavily in this research, since this research looks well into how the current situation of specialised doctor's practise are run in a day-to-day basis [4].

The sections below will explain the techniques that will be used in this research, they are: interview and observation. The later sections will also explain the software development methodology as well as the usability testing techniques.

2.1.1 Observation

Observation is one of the oldest and most fundamental of qualitative research techniques. Observation is commonly accompanied by either interview, which is also conducted in this research and document analysis [5].

Observation is when the researcher collects data by looking and listening as well as using other senses to make meaning out of a phenomenon. The process is carried out systematically and purposefully [5].

During observation, the researcher takes particular note on the activities that are carried out in a phenomenon, the actors or people involved, the sociodemographic of the people, the setting or the environment in which the phenomenon takes place and the relationship between people within the phenomenon. These things are some of the items that the researcher will take note during the course of the observation [6].

The researcher records any interactions, process or knowledge that took place whilst observation. These items are known as fieldnotes; which will act as primary data for the researcher to work on [5].

Observation ties in with UCD, as data from observation provides how the world is perceived in its actual settings. Observation will give sense towards the application that is about to be designed.

The observation took place in 3 specialised doctor's practices in the city of Banda Aceh while adhering to strict health protocols suggested by the local authorities.

2.1.2 Interview

Interview is a conversation that takes place between the interviewer and the interviewee. During this conversation, knowledge is interchangeable between the two or more people who are interacting in the interview. In qualitative research, interview has become commonly used to make sense and meaning about a phenomenon. Researchers use interview to gain understanding about a process as well as to confirm any previous knowledge gathered using other techniques. Interview can be conducted in various ways, such as interviewing people by phone, written or online surveys, meeting people face to face and record the interview or direct in-person interviews, where the subjects are interviewed directly during an observation or when an activity takes place [7].

The recorded interviews are then transcribed and coded. Coding in qualitative research is the process of generating ideas and themes that occur during the interview. During coding, the researcher takes the unprocessed data and identify the underlying concepts and categories [8]. In this research, these categories and concepts are arranged in a pyramid-like structure to find the repeated events or themes.

The data obtained from the interview will confirm the findings from the observation session. The interview will provide meaning to the activities that the people are doing. This will in turn feed the data to the UCD methodology as design requirements.

The interview will be conducted with 6 people from 3 user groups. The user groups are, receptionists, patients and healthcare providers.

Due to the reduced activities for citizens of Indonesia during the COVID-19 pandemic, the interview for this research was conducted via video conference and it was recorded.

2.2 User-Centred Design (UCD)

User-centred design is a software design methodology where the interface of the system or application is created and planned according to the needs and capabilities of the people who are going to use it [9]. Figure 1 shows the activities that are conducted in UCD.

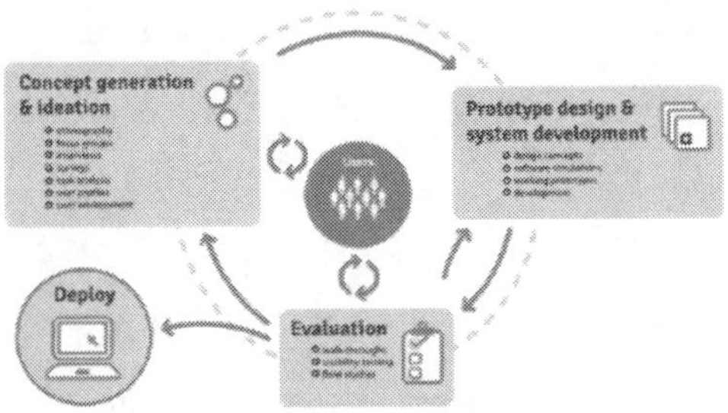

Fig. 1. User-centred design process, extracted from [10]

Figure 1 shows the various activities and stages that are conducted in UCD. The first stage is concept generation and ideation. This is where the researcher would collect qualitative data as necessary. In this research, this is done by conducting observation and interviews. The data collected from those two techniques then provide the vision of what the application would look like. Then based on the vision, the next stage of this method is to build a prototype design and system environment. In this stage, initial design concepts are introduced and a working prototype is built. Then the prototype is evaluated using usability testing with the prospective users. This stage is where the users are operating the prototype based on scenarios given. The process continues in a loop until the desired product is reached, then finally the end-product is released or deployed to the public [10].

The objective of UCD is that the design of an application is user data-driven, meaning the person in charge of the direction of the application should understand fully how the data interacted with the used before proposing a design [11].

By doing so, the expected outcome is that a design that will provide a rich and in context user experience for the users.

2.3 Usability testing

Usability testing is method that assesses the experience of users while operating a digital application [3]. In usability testing the user is given a set of tasks that s/he must complete. This method is designed to look for any part of the application that give the most problems for the users, hence giving them an inadequate user experience.

While the researcher notes the findings, the design and the experience of the application can be re-evaluated and built to suit the user better.

In this research the usability testing will be conducted with think aloud protocol via online video conference application due to the COVID19 pandemic. The possible users of this application have agreed to be recorded and have given consent for their profiles to be analysed, as long as their identities remain anonymous.

2.3.1 Think aloud protocol

Think aloud protocol is a technique within human computer interaction where the users articulate what they are thinking while carrying out tasks. Think aloud protocols are often used in training user experience (UX) professionals in their line of work and are considered as the golden standard of usability testing [12].

In this research think aloud protocol will have its own scenario devised by the researcher and will be conducted via online video conference application.

2.4 Research Environment and Sampling

The research is conducted during the COVID19 pandemic. There were 3 private practices being observed directly. These private practices are within the city of Banda Aceh.

Purposeful sampling is used to collect the samples in this research. Purposeful sampling is a way of selecting samples in qualitative research where the sample size is strategically determined to fit the needs of the research.

For the interview, there were 6 interviews conducted, where 2 of them were face to face; adhering to strict health protocols. The rest were done using online video conference software. All respondents agreed for their participation to be recorded. The 6 respondents include 2 patients, 2 security officers and 2 receptionists.

3 Results and discussion

The results of this research will be explained in further sections.

3.1 Observation results

There were 3 specialised doctor's practices that were observed in this research. They are: Cempaka Lima Clinic, NCB Clinic and Bunda Thamrin Clinic. These practices agreed to be observed in the afternoon where it is normally the peak time for patients to come for their appointments. These practices have specialised doctors from many fields, such as dermatology, oncology, obstetrics and gynaecology, dentistry, internist, ophthalmology and also a pharmacist is in place. These practices were located around Banda Aceh.

These 3 practices showed similarities in the registration process, waiting period and regulations during COVID-19 pandemic.

All 3 practices have receptionist desks up front after the front door. This is where the patients register, asked for information and check the doctor's schedule. 1 practice has a landline and the rest use a mobile phone to contact doctors and patients. 2 out of 3 practices have security officers as their first point of information and other things that are deemed important.

During observation, it was found that almost all patients have come earlier during the day to the practices to register themselves and book for an appointment for their preferred doctors. The receptionists take their

https://doi.org/10.1051/e3sconf/202234002010

personal details and write them in a book. Copies of the appointment list are distributed through the personal assistants or nurses in attending od each doctor. They will refer to this list when calling out patients' names.

Patients generally arrived about half an hour to an hour before the doctor arrives. Patients wait for their turn in the waiting room for their names. Despite the COVID-19 regulations, some patients still sit inside the waiting room. This is because they want to be there when the doctor arrives and have a greater chance of being called first even though they may be on the last of the list.

However, one of the practices' receptionists made remarks to the patients that they can wait outside or in their vehicles for the appointment times. When the doctor has arrived, they will be called through their mobile phones to attend to their appointments.

In all 3 private practices, they have printed out rules and regulations regarding the social distancing and health protocols that must be applied by all inbound patients coming into the practice. These rules are also stated by the receptionist during the registration process.

To conclude, the observation on the field revealed that the current registration process and waiting period are very much similar in every setting. Therefore, it is necessary to support this process through a technology.

3.2 Interview results

During the interview with the respondents were asked to describe these factors:

1. Short biography about themselves; this is to provide a demographic for the research.
2. Reasons for going to a specialised doctors' practices (patients).
3. The current registration procedure during COVID-19 pandemic (patients).
4. The situation during registration procedure (patients).
5. The working hour (receptionist and security officers).
6. Activities during working hour (receptionists and security officers).
7. Any other problems that they encounter during the registration and booking process.

Figure 2 shows the themes that were found during the interview.

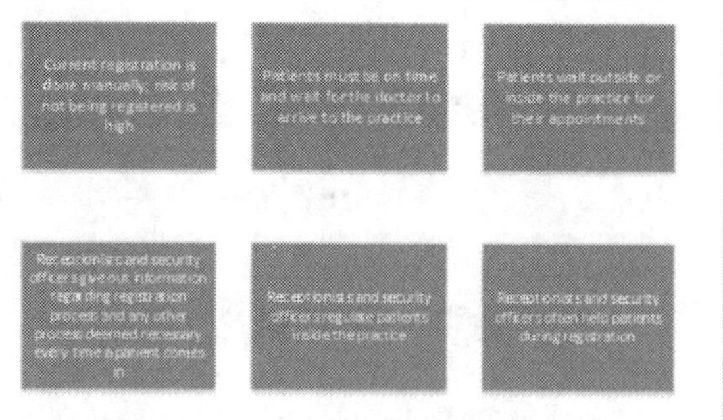

Fig. 2. themes discovered from interview process.

These themes represented the key areas that are considered to be problems within the private practices. One of the recurring themes is that the long waiting time after registration and seeing the doctor for the appointment. According to one of the patients interviewed (Male, 25), the earlier a patient registers for an appointment during that day, the quicker he gets to see the doctor.

This corroborates the waiting time between after registration time and meeting the doctor, which is about 10 minutes if the patient is an early bird. However, if the patient registers late, it is about 50 minutes up to an two hours before he can see the doctor. This also depends on the time spent between the doctor and the patients.

Figure 2 also suggests that the receptionists of the private practices had to regulate the amount of people coming into the practice. This also showed up during observation. People were waiting around the practice were waiting inside the practice for the doctor. One of the receptionists interviewed (Female, 22) mentioned that the patients stay around in the practice with the hope that they will be able to take someone else's time if another patient is late to his appointment.

Both of the security officers interviewed mentioned that they had to put extra work in regulating the patients who are coming into the practice. Due to COVID19 pandemic, there was a rule set up by the local council that people should be distanced about 2 metres from each other within public spaces. Therefore, there should a certain number of people within the private practices. However due to other illnesses that patients have and must be treated, sometimes this regulation is turned a blind eye upon.

All of the interviewees mentioned that it would be helpful if they had some sort of simple technology that can help them in their daily activities of registering and helping patients in the private practices.

3.3 Prototype

Based on the observation and interview results, a simple application can be developed. The results and themes from Figure 2 suggested that the most suitable technology for this research is a mobile application. The next section explains the visioning of the future booking application.

3.3.1 Visioning

Visioning in this case is when the functionalities of this application is proposed. According to tree diagrams in Figures 2, 3 and 4, the booking application should have 3 sides. The first side is the patient's side, then the receptionist's side and finally the doctor's or nurse's side. These 3 sides will support a more cohesive registration system that will regulate the amount of people coming in to see the specialised doctors.

This will also be a plus point during a disaster and a pandemic such as today. During a disaster, people may be displaced and confused. People with pre-existing disease may have difficulties in reaching their usual doctors. With the support of this proposed application,

https://doi.org/10.1051/e3sconf/202234002010

patients can reach and make appointments during a disaster or pandemic to see their doctors.

Figure 3 shows the proposed functionalities for the patient's side of the application. These functionalities are derived from the results of interview and observation.

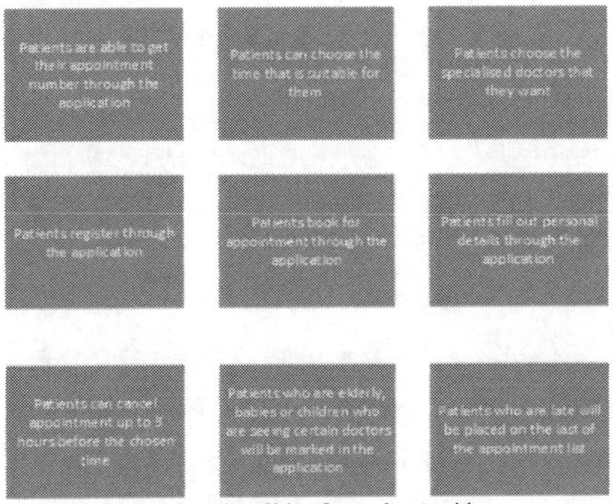

Fig. 3. Proposed functionalities for patient's side

Figure 4 shows the proposed functionalities for the receptionist's side of the application. These functionalities are derived from the results of interview and observation.

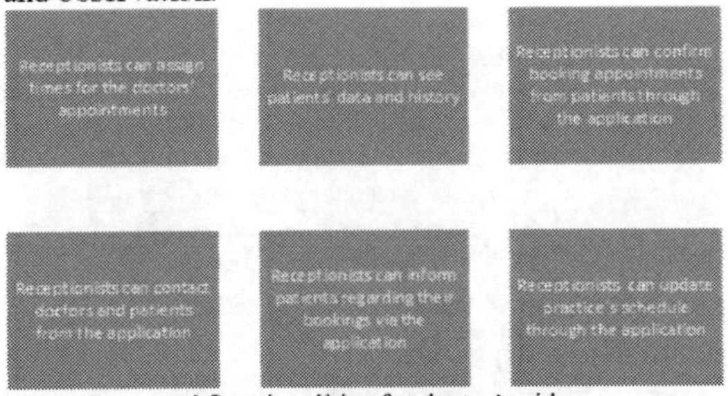

Fig. 4. Proposed functionalities for doctor's side

Figure 5 shows the proposed functionalities for the doctor's or nurse's side of the application. These functionalities are derived from the results of interview and observation.

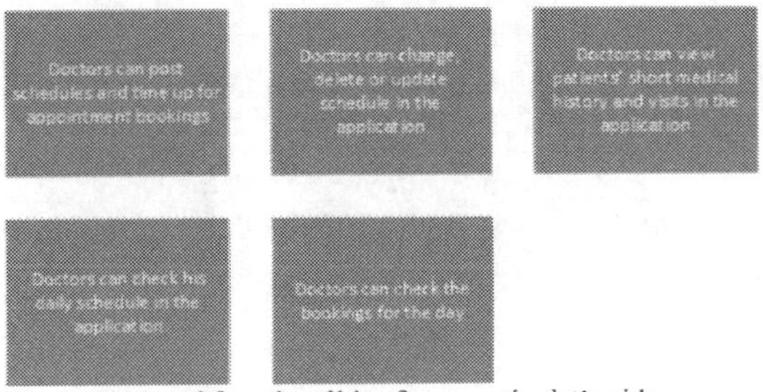

Fig. 5. Proposed functionalities for receptionist's side

With these proposed functionalities, this booking application will be able to support disaster management in various ways, especially in times when healthcare providers or specialised doctors can only be seen during certain times.

3.4 Usability testing results

The usability testing was conducted with 6 people via online video conference. More details on the users are shown in table 3.

Table 3 explains the demographic of the users. These users have agreed to be recorded and have asked to remain anonymous for this research.

Before the usability testing is carried out, a test plan must be devised. A test plan is a layout of how the usability testing will take place. The test plan will serve as a guide to the researcher to what activities and tasks must be carried out by the users. This test plan must be followed thoroughly. The goal is to discover any positive or negative findings within the application.

3.4.1 Participants

The participants of this usability testing are 6 people, where there were 3 patients user group, 1 doctor, 1 nurse/healthcare provider and 1 receptionist. Their demographics are shown in table 4.

Table 4. Participant demographics

Item	**Description**
Initials	RC
Sex	Male
Age	21
User group	Patient
Initials	AU
Sex	Female
Age	21
User group	Patient
Initials	SR
Sex	Female
Age	21
User group	Patient
Initials	EW
Sex	Female
Age	23
User group	Nurse/Healthcare provider
Initials	OR
Sex	Male
Age	29
User group	Doctor/Healthcare provider
Initials	FM
Sex	Female
Age	26
User Group	Receptionist

All participants have agreed to be recorded during the usability testing process and they also agree that what they say during this process to be written or quoted within the research.

 https://doi.org/10.1051/e3sconf/202234002010

3.4.2 Working prototype

Figure 6, 7 and 8 are the working prototype implemented based on the vision of the application.

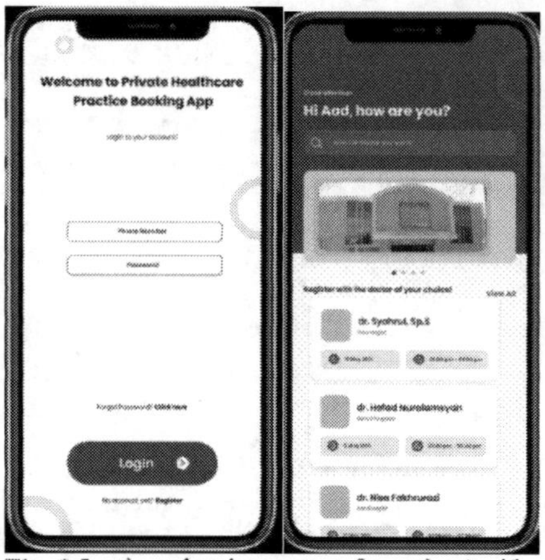

Fig. 6. Login and welcome page for patient's side of the application

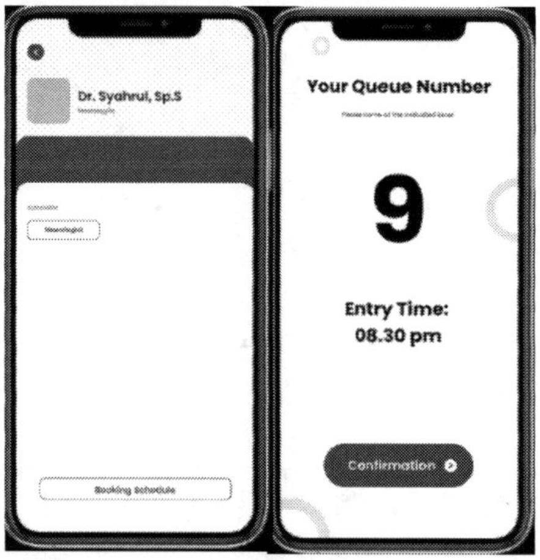

Fig. 7. Booking an appointment and getting the time slot for the appointment for patient's side of the application

Fig. 8. Login and changing schedule for the doctor's side of the application

3.4.3 Test plan

The test plan for this application is laid out in table 5.

Table 5. Test Plan for Booking Application

Item	Description
Q1	does the user understand how to use the application without help from anyone?
Goal	user understands and able to use the application without prior help
Metric	understand how to use the application without prior help
Procedure	Ask users about their perception about the application especially on the login, registration and appointment number functionalities
	Ask users to conduct think aloud protocol
Data collected	Number of users able to understand and use the application without help and comments about the application
Q2	does the user understand the goal of the application?
Goal	User understands thoroughly on the goal and objective of the application
Metric	Whether the user recognise the goal of the application or not
Procedure	Ask users about the how-to in using the application, for example using the appointment function
	Ask users to conduct think aloud protocol
Data collected	Observation on user's first impression about the application and the number of users recognise the goal of the application.
Q3	Can the user group doctor/nurses/healthcare provider clear his/her schedule without any problem?
Goal	This user group can clear its schedule without any problem
Metric	This user group able to clear the schedule
Procedure	Ask user to clear his/her schedule
	Login to profile
	Clear available schedule
	Ask user to conduct think aloud protocol
Data collected	Time it takes to carry out the clear schedule activity
Q4	Can the user group administrator delete and edit all data on patients and doctors without any problem?
Goal	This user group can delete and edit data on patients and doctors without any problem
Metric	This user group is able to delete and edit data
Procedure	Ask user to delete and edit data
	Login to profile
	Access patient's and doctor's data
	Edit patient's data
	Edit doctor's data
	Delete patient's data
	Delete doctor's data
	Ask user to conduct think aloud protocol
Data collected	Time it takes to carry out the entire process from start to finish

Table 4 explains a simplified version of the test plan for the booking application. The test was carried out thoroughly. The results as expressed in the next sections.

3.4.4 Test plan results

Q1 was asked to all participants. 4 out of 6 participants understood and able to use the application on the first glance. 2 participants did not understand how to use the application without help. The two participants who were not able to use the application; FM and SR on the first glance managed to use it again later after given some guidance.

Q2 was also carried out to all participants. 5 out of 6 participants understood the goal of the application. Only 1 participant did not get the main objective of the application. SR was the participant. During the application testing, SR was also the only one who had trouble in carrying out the booking functionality. She managed later on with some guidance.

Q3 was carried out to 2 of the participants in the doctor/nurse/healthcare provider user group, since the scheduling functionality is only available to this user group. Both participants, OR and EW were able to carry out the edit and delete process for scheduling without any hesitation. The time it took to conduct these functionalities are shown in table 6.

Q4 was tested out 2 of the participants, FM and AU. Both are from different user groups. This is done in order to observe how users from different demographic can adapt with the administrative role. Both users were able to use edit, delete and conduct administrative functionalities with minimal intervention during the testing process. The time it took to conduct these functionalities are shown in table 7.

3.4.5 Think aloud findings

During think aloud several negative and positive feedbacks were identified. From all 6 participants it was found that the simplicity of the application design made it easy to understand the functionalities provided by the application. Both receptionist and healthcare provider user group also agreed that the design of the application made them easier to recognise the goal of the application, without any previous experience.

The negative point of this application that was identified during the think aloud protocol was that the navigation of the application was not clear. This was emphasised the most in the patient user group. Even when 4 participants from this user group operated the application with general ease, there were repeated comments about the navigation of the application. One participant said "I am not sure where to begin booking for a doctor after the login page" (SR, 21).

This suggests that the user experience needs to be redesign in the future with the consideration of a wider demographic area. This result suggests that SR may need a different approach in user experience.

During the entire procedure, every participant agrees that this application will be helpful during a disaster and a pandemic. Even though both situations are dire, people with certain diseases are still present. The people are still needed to be treated. OR, 29 is a doctor. He thought that this application is extremely important to be installed in private doctors' practices. In his practice, he has many patients who are waiting in line to meet the doctors during appointment. During the pandemic, the regulation is that there should be minimal crowd and there should be a distance about 1.5 to 2 metres between people. This application can also help uphold this regulation and reduce the number of people in the private practices.

This application is also resilient. This was mentioned by FM. She had two perspectives for this application. One as a patient. During a disaster, as it happened previously in Banda Aceh. It was a lot harder to book an appointment to meet specialised doctors. Based on her own experience as a receptionist, this application is useful during emergency situation, such as during a disaster, as it can help patients book a time that is convenient for them to meet a doctor.

4 Conclusion

To conclude the design and testing of this booking application has been prolific. Necessary data has been gathered and these data can also be a starting point for future development of this application. Using UCD as the design methodology has proven to be effective, as proven in the usability testing, where all 6 participants have agreed that simple technology such as this booking application can help them in the day-to-day tasks in the specialised doctor's practice.

References

1. G. Shank, *Qualitative Research: A Personal Skills Approach*, 2nd ed. (Pearson, USA, 2005)
2. S. D. Lapan and F. J. Riener, *Qualitative Research an Introduction to Methods and Designs*, 1st ed. (Jossey-Bass, San Francisco, 2012)
3. B. Martin and B. Hanington, *Universal Methods of Design*, 2012th ed. (Rockport Publishers, Berverly, Massachusets, n.d.)
4. N. K. Denzin, K. Norman, and Y. S. Lincoln, *Handbook of Qualitative Research*, 1st ed. (Sage Publications, California, 2009)
5. L. Mckechnie, L. Baker, M. Greenwood, and H. Julien, New Review of Information Behaviour Research **3**, 113 (2002)
6. L. H. Rosen and M. K. Underwood, in *Encyclopedia of Research Design*, edited by N. Salkind J., 1st ed. (Sage Publications, California, 2010), pp. 952–954
7. L. Brinkmann, in *Interviewing: Understanding Qualitative Research*, 1st ed. (Oxford University Press, New York, 2013), pp. 26–32
8. L. Benaquisto, in *The SAGE Encyclopedia of Qualitative Research Methods*, edited by L. Given M. (Sage Publications, USA, 2008), pp. 85–88
9. M. Endsley R., B. Bolte, and D. Jones G., *Designing for Situation Awareness: An Approach*

to User-Centered Design (Taylor & Francis, Georgia, 2003)
10. M. Tara, S. Taneva, M. Casselman, M. Yeung, C. McDaniel, W. Ho, and J. Cafazzo, BIT **46**, 49 (2021)
11. K. Holtzblatt, J. B. Wendell, and S. Wood, *Rapid Contextual Design: A How-To Guide to Key Techniques for User-Centred Design*, 1st ed. (Elsevier, USA, 2005)
12. H. Sharp, Y. Rogers, and Jennifer Preece, *Interaction Design: Beyond Human-Computer Interaction*, 5th ed. (John Wiley & Sons, Indianapolis, 2019)

Post-disaster reconstruction in the rural-urban fringe following the Great East Japan Earthquake

Kenji Muroi*

Department of Social and Human Environment, Nagoya University Graduate School of Environmental Studies, Furo-Cho Chikusa Nagoya 464-8601, Japan

Abstract. Although many studies have been conducted since the Great East Japan earthquake, most have focused only on recovery processes in particular affected areas. However, southern Miyagi Prefecture belongs to the rural-urban fringe of the Sendai Metropolitan Area; hence, disaster reconstruction in this area is closely related to the urban system and long-term land development. Based on these ideas, this paper explores the interaction between wider social trends and the disaster reconstruction process. The results indicate that disaster reconstruction in this area was an extension and acceleration of existing social trends such as "selection and concentration" and "intensification of agricultural competitiveness" as a whole. However, there is a unique agricultural land-use system and related community organization rooted in the peculiar history and environmental conditions of this area. Existing disaster reconstruction policies do not pay sufficient attention to such historical contexts, which has caused several conflicts and problems. Disaster studies should pay careful attention to such tensions between wider political trends and local historical contexts while focusing on land-use changes.

1 Framework and Focus

The purpose of this paper is to elucidate social changes following the Great East Japan earthquake of March 11, 2011. I would like to begin by outlining this study's analytical framework.

Disaster reconstruction studies involve two types of frameworks. The first type focuses on policies and activities during the reconstruction process in the affected areas. In this type of study, theoretical attention has been paid to dilemmas in the context of recovery [1]. In the Japanese context, this type of study often utilizes the dichotomy between the national government and local communities. The disaster reconstruction approach in Japan is characterized as central-government-initiated and technology-dependent. That is, the Japanese approach to disaster is "top-down" wherein many decisions are deferred to higher levels in the hierarchy before municipal governments can act [2,3]. Accordingly, in the case of the Great East Japan earthquake, the central government created a basic framework for reconstruction projects and provided enormous financial support for each. Although this "supply-side" reconstruction approach led to a remarkable upsizing of projects, it did not necessarily match the local needs in affected areas. Such discrepancies between supply and demand have been critically examined through case studies focusing on particular affected areas, yielding many research findings (although most are in Japanese) [4,-8].

The second type of analytical framework focuses on the relationship between the disaster reconstruction process and long-term social trends. As Mitchell pointed out, every hazard is situated in a wider social context and thus hazard management is constrained by exogenous socio-economic factors [9]. This wider spatio-temporal analytical framework is useful not only for disaster studies but also for community and regional studies in general in that disasters often reveal structural principles and their contradictions in current society, which are overlooked during normal periods of time [10-12]. Although *3.11: Disaster and Change in Japan* by R. J. Samuels was a remarkable achievement in this respect [13], this type of study is rare. Ten years after the earthquake, the lessons from its aftermath should be reconsidered from a broad perspective.

Accordingly, this study aims to explore the characteristics of and challenges to disaster recovery from a wider viewpoint, focusing on general social trends. Specifically, I emphasize the following perspectives in the analysis.

First, I focus on the impact of regional urban systems on disaster recovery processes in particular municipalities. Here, I am using the concept of urban system in Gottmann's sense to refer to "megalopolis," the growing interconnection of cities [14]. Our research area is in the southern part of Miyagi Prefecture, in the Sendai Metropolitan Area. Compared to the northern coastal area (Sanriku region), a fishery area that adjoins the Ria coastline, industrial development and

* Corresponding author: muroi.kenji@j.mbox.nagoya-u.ac.jp

urbanization are more advanced in the south. Therefore, a viewpoint focusing on urban systems has particular significance in our study.

Second, a historical perspective is also important. The southern coastal area is located on the Sendai Plain and has long been an agricultural production site in the suburbs of Sendai city. Urbanization in this area followed agricultural land development. Although there is a distinction between urban studies and rural studies in current academic disciplines, such a distinction is not clear in the rural-urban fringe, as in our research areas. In fact, as described below, focusing on the rural roots of urbanization is of great importance in understanding the reconstruction process in this area.

Third, I focus on the reconstruction policies of municipal governments. Although centralization is a basic characteristic of the Japanese risk management system, as mentioned above, local governments also have political autonomy to a certain extent. Next, I explore the reconstruction policies of municipalities as one of the important independent variables and identify regional differences in disaster reconstruction due to such factors.

The discussion in this paper is based on a case study of Watari and Yamamoto, which are towns in Miyagi. Both municipalities are on the periphery of the Sendai Metropolitan Area and hence are suitable examples to illustrate the contradictory interaction between urban land use and agricultural land use in the process of disaster reconstruction. Our research is based on detailed interviews with community leaders of both towns and analysis of various statistics and official documents, such as the minutes of local council meetings. We also conducted a survey in 2018 on community reconstruction in Miyagi Prefecture[1)]. Our findings were derived from a comprehensive analysis of these multiple data sources.

The remainder of this paper proceeds as follows. First, I briefly examine the history of development in this area since the period of high economic growth as a social context for the 3.11 earthquake. Next, I examine the relationship between disaster reconstruction processes and regional urban systems, focusing on the characteristics of the population change after the disaster and the reconstruction policies in Yamamoto. Then, I examine changes in agricultural land use and related challenges. Finally, I reconsider the relationship between disaster reconstruction and long-term social trends in the context of the rural-urban fringe.

2 Development in the pre-disaster period

The southern coastal area in Miyagi is a low-lying flatland surrounded by the Pacific Ocean and the Abukuma River. Although the soil in the floodplain is suitable for agriculture, the low-lying coastal land experiences frequent flooding and used to be a vast wetland. Thus, improving agricultural drainage has been a long-standing challenge.

In the wake of post-war food shortages, the central government conducted large-scale irrigation and drainage projects from 1952 to 1971. More than 4,000 ha of wet rice paddies in low-lying areas in Watari and Yamamoto were converted to dryland fields during this project [16]. Further, the mechanization of agriculture and farmland consolidation were promoted also from the 1960s, and accordingly, agricultural productivity and farm management improved drastically [17-20].

The accomplishment of the goals of these agricultural development projects was followed closely by the commencement of new industrial development projects. As the Sendai Bay coastal area was designated one of the "new industrial cities" by the National Comprehensive Development Plan in 1964, industrial and transportation development projects were heavily promoted. During the early stages of development, the populations of Watari and Yamamoto decreased because both towns were separated from the Sendai urban area by the Abukuma River, which made commuting to urban areas quite difficult. However, the frequency of service and number of cars per train increased in the late 1960s, and several new stations were constructed in the 1980s [19]. With the improved commuting facilities, the populations of Watari and Yamamoto began to increase in the 1970s. Expressway and bridge-building projects were also promoted by the government from the 1980s. The Sendai Metropolitan Area was directly connected to the Tokyo Metropolitan Area by a highway in 2009. The Sendai Airport Line was developed in 2007, improving air connectivity. Such drastic traffic development meant that each municipality in the southern coastal area became a metropolitan hub, with all of them well-connected to each other across multiple transport networks.

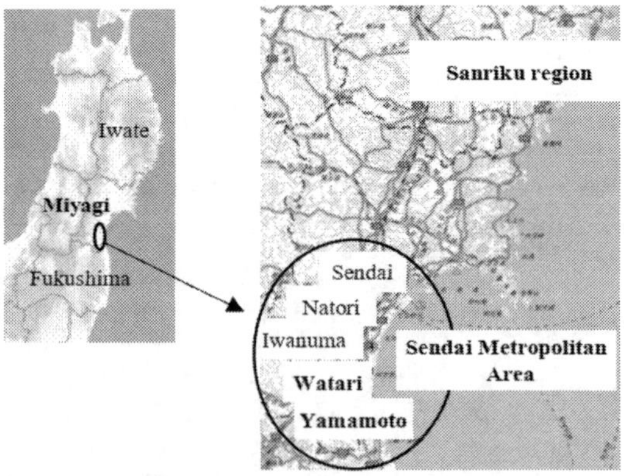

Fig. 1. Research area.
Source: Geospatial information authority of Japan.

The mechanization of agriculture contributed to a decline in the farming population as well as in agricultural development. For example, the ratio of workers in the primary industry in Watari declined from 68.5% in 1955 to 9.3% in 2010. Similarly, industrial development in the coastal areas had an adverse impact on the fishery industry, leading to its decline. By contrast, the commuter population increased drastically. The ratio of workers in Watari and Yamamoto who commuted to other municipalities reached 65.1% and

58.9%, respectively, in 2010; and their workplaces were mostly in Sendai and neighboring cities.

Against the backdrop of the expansion of the Sendai Metropolitan Area from the 1990s, housing land development advanced sporadically in the form of commuter towns. Thus, growing land-use change and social differentiation in the rural communities in the southern coastal area formed the social context in which the Great East Japan earthquake occurred.

3 Damage and population change

In this section, I will examine the characteristics of the damage and population change caused by the earthquake in the southern coastal area.

There are differences in the damage caused by the earthquake and tsunami in the Sanriku region from that in the southern coastal area. In particular, the damage caused by the tsunami was much greater in the Sanriku region than in the southern area. The impacts of a tsunami may vary depending on the scale of the seismic movement and the geographical features of the coastline. Thus, the Sanriku's complicated rocky coastline created a higher tsunami than in the southern coastal area, which led to greater damage and loss of life. Business centers and public facilities were also heavily damaged in the Sanriku region because urban functions were concentrated in coastal areas.

Meanwhile, the characteristics of damages in the south were as follows: First, the tsunami caused serious agricultural losses. While the tsunami inundation rate was less than 10% in Sanriku municipalities, in the low-lying southern region, almost half of the municipal area was inundated. The coastal plain was mainly utilized for agriculture; hence, it suffered serious agricultural damage.

Second, the social impact of the damage to the railway line was far more serious in the south. Railway lines in the southern coastal area were partially swept away by the tsunami, obstructing commuting for an extended period of time. Although damage to the railway line in central cities such as Sendai, Natori, and Iwanuma was not significant, damage in peripheral areas like Watari and Yamamoto was much heavier, such that restoration of the railway required two years in the southern part of Watari and almost five years in Yamamoto.

The long-term suspension of commuting caused massive population outflows in both towns. In fact, population decreases following the earthquake in Watari and Yamamoto were caused not only by deaths and persons gone missing due to the tsunami, and enforced relocation from high-risk areas, but also by voluntary migration due to the obstructions to commuting. As census data show, the areas they moved into were mostly Sendai and neighboring cities such as Natori and Iwanuma (Table 1). As a result, the population gap between the center and periphery of the Sendai Metropolitan Area expanded following the earthquake (Figure 2). Railway tracks suffered similar damage in the Sanriku region. However, the commuting population in Sanriku was much smaller; thus, the impact of railway service suspension was quite limited. Population changes caused by damage to commuting facilities are unique to disasters in urban areas.

Table 1. Changes in settlement sites (2010–2015).

Residence in 2010	Population outflow	Residence in 2015 (ratio to population outflow)
Watari	3,848	Sendai (24.8) Iwanuma (17.3) Natori (9.8)
Yamamoto	3,405	Watari (23.1) Sendai (21.5) Natori (10.1)

Source: National population census.

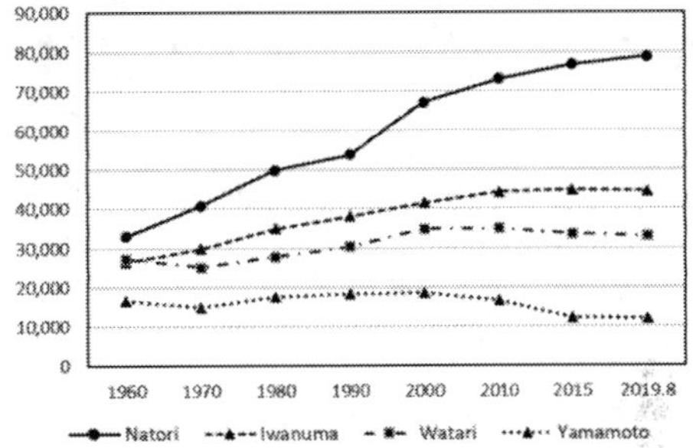

Fig. 2. Population trends.

Who remained and who moved out? Table 2 shows the population trends for coastal agricultural villages in Watari. Villages D and E were designated as risk areas; therefore, the number of households in these villages declined drastically after the earthquake. However, the population of villages C and G also decreased drastically after the earthquake. In these villages, the population increased rapidly in the 1990s. This was caused by small-scale housing development at that time, as mentioned in Section 2. This means that people who moved out after the earthquake were mostly newcomers from the 1990's.

Table 2. Change in the number of farming households by agricultural village.

		1990	2000	2010	2015
A	Farmers	39	36	29	31
	Non-farmers	36	47	53	66
B	Farmers	36	31	27	23
	Non-farmers	11	33	37	39
C	Farmers	79	68	49	20
	Non-farmers	**82**	**146**	155	88
D	Farmers	51	45	39	8
	Non-farmers	5	13	13	5
E	Farmers	44	41	36	7
	Non-farmers	11	67	69	33
F	Farmers	98	108	105	65
	Non-farmers	48	60	69	43
G	Farmers	52	48	32	24
	Non-farmers	**49**	**104**	110	53
H	Farmers	85	74	49	30
	Non-farmers	**334**	**598**	724	697

Source: Census data for agricultural villages.

This was supported by interview research in these villages. According to community leaders, while farmers remained in this area, most newcomers and commuters moved out. Although the population change in village H seems to be relatively small, it was because a new housing complex for group relocation was developed in this district. In actuality, a large number of newcomers in this village also moved away after the earthquake.

These findings show how the disaster revealed the differences between native farmers and middle-class newcomers with respect to ties to the land. As discussed later, this population decline mechanism was also confirmed in Yamamoto.

Such voluntary migration and group relocation projects due to the designation of disaster risk areas also caused an expansion in the disparity between inland and coastal areas in municipalities. Although the cause of the group relocation project was disaster risk reduction, it resulted in the concentration of urban functions in inland areas. Particularly in Yamamoto, as will be described in the next section, such consolidation of urban functions in central districts was intentionally promoted with the slogan of "compact city." In contrast, as the population declined, convenient facilities such as shops and hospitals withdrew and living conditions worsened remarkably in the coastal areas of Watari and Yamamoto.

From a broader perspective, living conditions after the earthquake worsened in the Sanriku region as a whole because of the promotion of relocation from urban coastal areas to inconvenient highlands. In contrast, living conditions in the south were significantly improved because of relocation (Table 3). Therefore, the disparity between the inland and coastal areas in the municipalities of the south tends to be underestimated in reconstruction policies. However, in Watari, where relatively many people remained in coastal areas, the living conditions worsened remarkably. In Yamamoto, although actual situations are not reflected in the quantitative data, the disparity between inland and coastal areas is higher, as will be discussed below.

Table 3. Changes in Living Conditions after the EQ.

Access to Shopping.

		Became better	No change	Became worse
North	Kesennuma	9.4	61.3	25.5
	Minami sanriku	9.1	30.3	60.6
	Onagawa	0	52.0	48.0
South	Natori	41.7	41.7	16.7
	Iwanuma	33.3	58.3	0.0
	Watari	**0**	**36.4**	**54.5**
	Yamamoto	**25.0**	**62.5**	**12.5**

Access to Hospital.

		Became better	No change	Became worse
North	Kesennuma	11.3	60.4	24.5
	Minami sanriku	15.2	30.3	54.5
	Onagawa	0	60.0	40.0
South	Natori	33.3	58.3	8.3
	Iwanuma	0	91.7	8.3
	Watari	**0**	**27.3**	**72.7**
	Yamamoto	**6.3**	**87.5**	**6.3**

Source: See note 1.

4 Disaster reconstruction as an urban renewal project

Regarding disaster reconstruction policies following the earthquake, the following basic characteristics can be observed: First, safety against tsunamis was strongly emphasized, and large-scale disaster prevention projects were aggressively promoted [15]. Specifically, while huge seawall construction and land elevation works were conducted in coastal areas, residents were prohibited from living in their original homes in areas designated as risk-prone. In addition, large-scale relocation projects were promoted. Second, a "build back better" policy was pursued. "Build back better" meant mainly economic development in the Japanese context, captured in the declaration "Towards Reconstruction: Hope Beyond Disaster" presented by the Reconstruction Design Council for the Great East Japan earthquake. This declaration emphasized both the reconstruction of afflicted areas and the revitalization of the national economy through disaster reconstruction. Based on these ideas, the government provided huge financial support; hence, reconstruction projects tended to become bloated.

In terms of the response to disaster prevention policies, such as huge construction and relocation projects, there were obvious differences between the Sanriku region and the southern coastal areas. While active protests were held in several fishing villages in Sanriku because these projects were not compatible with the needs of fishing communities, there were few protests in the south because the fishery industry had already declined following industrial development after the 1960s. Meanwhile, "build back better" policies had a greater impact in the southern areas. In this respect, in this section, I will discuss the reconstruction projects in Yamamoto that pursued the "compact city" ideal as well as the changes in agricultural land use in the next section.

Unlike other affected municipalities, Yamamoto pursued not only the reconstruction of affected areas but also drastic urban restructuring with the slogan "compact city." Compact city policies in the Japanese context denote urban planning that promotes the concentration of urban functions in a few areas for the purpose of fiscal efficiency. Given the backdrop of depopulation and the fiscal crisis, "compact city" is now a key political strategy employed by many peripheral municipalities. The compact city project in Yamamoto was as follows: Damaged existing railway lines would be relocated and restored inland, and new stations would be created. A new city center would be built around these stations by promoting large-scale housing reconstruction projects and urban development.

According to the mayor, the city administration of Yamamoto had difficulty in efficiently managing autonomous bodies in the pre-disaster period because 22 villages were scattered throughout the town. As such, the city administration saw disaster reconstruction as an opportunity for drastic urban redevelopment for more efficient town management. The location of the new Yamashita train station was particularly convenient for its connection with National Route 6 and the Yamamoto interchange. The new Yamashita station area was thus regarded as an ideal base for a "compact city," and a large-scale shopping mall and several welfare facilities were developed there. As the scale of the disaster reconstruction project in Yamamoto was larger than that of neighboring municipalities, the city administration depended heavily on nationally subsidized projects like the Tsunami Reconstruction Base Development Project.

The main aim of this large-scale project was to cope with the estimated long-term trends of depopulation and aging in Yamamoto. Yamamoto is located at the southern end of the Sendai Metropolitan Area, furthest from Sendai City, and was placed in a disadvantageous position in the urban system. Presumably, it is assumed that the disaster reconstruction plan for Yamamoto was driven not only by disaster-related factors but also by the city administration's sense of crisis caused by the peripheral location of Yamamoto in the Sendai Metropolitan Area. In Yamamoto, therefore, the disaster reconstruction plan was developed as a new version of long-term comprehensive city planning. However, the disaster reconstruction plan gave rise to various disputes and conflicts.

The first issue was related to consensus-building regarding the adaptation to the existing system. As Yamamoto's mayor had previously been a crisis manager for Miyagi Prefecture, he was familiar with national disaster reconstruction institutions and initiatives. Under the existing legal system, most disaster recovery projects depend heavily on national subsidies. Thus, the mayor prioritized rapid adaptation to the existing system and made the following statement at a plenary meeting in December 2011:

If we cannot utilize national institutions properly, it is impossible to recover from this big disaster. We must work together on these tasks to meet the government's deadline. I think there will be no going back. I will make all possible efforts.

However, such a political stance meant that the mayor tended to neglect the opinions of the affected people. Yamamoto's reconstruction plan was made in December 2011, the fastest in Miyagi Prefecture. Before making the plan, the town administration conducted a survey among citizens regarding the basic vision for disaster reconstruction. However, people's opinions can change over time. The town administration's hurried decisions meant that it was indifferent to follow-up research. Such a political stance was repeatedly criticized in the town council, and a motion censuring the mayor's political stance was passed without objection in December 2013.

The second issue was population decline. The relocation project for the railway line was conducted not only for tsunami-related disaster prevention, but also with sustainable development in mind in an era of population decline and aging. However, this large-scale reconstruction project took much longer than the railway restoration projects in other municipalities. Ironically, therefore, the population decline accelerated during the long period of suspended railway operations. Specifically, it took 5 years and 9 months for the railway to restart, and the population decreased by 30% during this period.

Population decline was also caused by another political factor, that is, the scale of the designated risk area in Yamamoto. The standard risk areas in Iwanuma and Watari were estimated to have a tsunami inundation depth of two meters, whereas the standard in Yamamoto town was just over one meter. As a result, the risk area in Yamamoto Town encompassed 1945 ha, much greater than the 842 ha in Iwanuma and 545 ha in Watari. This large area designation was presumably based not only on the scientific simulation of tsunami inundation but also on the political judgment of the town administration. That is, the political incentive to accomplish successful group relocation to the new central district had an impact on the scale of the designated risk area. The fact that almost one-third of the town's territory was designated as a risk area, in which housing reconstruction was restricted, had a significant impact on the population outflow. Therefore, some people in the affected area raised objections to the town administration and submitted a petition for a reduction of the size of the risk area.

The third issue concerning disaster recovery was the increase in the disparity between districts in Yamamoto following the earthquake. While the administration of Yamamoto designated even areas in which the estimated tsunami inundation depth was less than two meters as risk areas, it still permitted residents to live in such areas under the requirement that they raise the ground level of their homes. However, the town administration prioritized the relocation of residents to the new central districts by providing additional financial support to people who chose to relocate. Residents who chose to remain in their original locations and undertook housing reconstruction objected to this selective support. Furthermore, while the administration built raised roads as part of the "multiple defense" policy for tsunami disaster mitigation, people who rebuilt their houses in risk areas were excluded from such protection and remained exposed to tsunami risks. After the construction of the new central districts was completed, remarkable disparities between existing districts and newly developed districts in services such as shopping and kindergarten became apparent.

The compact city project mainly targeted young households whose members commuted to the Sendai Metropolitan Area. The town administration actively promoted the relocation project with the slogan "Yamamoto town is the best place for child rearing." Meanwhile, the compact city project was not necessarily attractive to the older residents who had lived in rural areas for a long time, given what the changes in lifestyle, landscape, and neighborhood communities in newer urban developments would mean for them.

Why should people engaged in agriculture and fishery have to relocate to urban centers? It is pretty good for the people whose children commute to high school in Sendai City, to be sure, but.... We should have more diverse images of a compact city, which are in sync with Yamamoto Town, for example, a compact city with an ocean view and a compact city coexisting with agriculture....

--Statement by a council member at a plenary meeting in December 2013

Therefore, some people decided to remain in risk areas even though the population declined drastically and daily conveniences significantly worsened. Other people petitioned to relocate to places other than the designated new urban centers. In this sense, the compact city project in Yamamoto revealed conflicts of interest between social groups.

5 Changes in agricultural land use

As previously mentioned, tsunamis caused large-scale agricultural damage in the southern coastal areas, and tsunami-inundated farmland areas amounted to 1826 ha in Watari and 1416 ha in Yamamoto. Almost all agricultural facilities and machines were swept away.

The town administrations of Watari and Yamamoto promoted large-scale agricultural restoration projects. These projects pursued not only agricultural restoration, but also an increase in agricultural productivity. When the Great East Japan earthquake occurred, promoting large-scale farming and boosting Japan's agricultural competitiveness in the global market were priorities for national politics against the backdrop of the Trans-Pacific Partnership (TPP) Agreement negotiations. In this political context, the Reconstruction Design Council proposed that agriculture in the southern coastal plain areas should be rebuilt as a "front runner of land-extensive farming in Japan" and large-scale grant projects for agricultural reconstruction were introduced.

Based on this scheme, both the restoration and consolidation of farmland as well as the introduction of large-scale agricultural machines and facilities, were actively promoted in Watari and Yamamoto. The total cost of these projects added up to as much as 400 million dollars in each municipality.

As a result, the scaling up of agriculture progressed rapidly. Each unit of paddy field was successfully enlarged from 30 ares to 1 ha, and fragmented land holdings were agglomerated via farmland consolidation projects. Many farmers gave up farming after the damage caused by the earthquake and delegated their farmland cultivation to a handful of full-time farmers. Although contract farming had already progressed in the pre-disaster period, the trend accelerated after the earthquake. The area under cultivation per farmer was remarkably enlarged, and new large-scale full-time farmers appeared after the earthquake. Hence, the efficiency of agricultural production improved remarkably.

Table 4. Changes in agriculture.

		2005	2010	2015
Watari (coastal area)	Operating cultivated land (ha)	641.5	705.1	628.1
	Cultivated land rented (ha)	70.0	117.2	262.3
	Farm households	316	294	164
Yamamoto (coastal area)	Operating cultivated land (ha)	785.4	760.9	357.2
	Cultivated land rented (ha)	136.9	158.2	200.8
	Farm households	400	325	83

Source: Census data for agricultural villages.

The southern coastal areas were a farming-residential mixed zone before the earthquake. Following the disaster, however, reconstruction policies led to the concentration of urban functions in inland areas. Meanwhile, vast coastal land tracts left vacant due to relocation were utilized for large-scale agricultural development projects. Although relocation projects were promoted with the aim of disaster risk reduction, this functional differentiation of land use was closely interrelated with general political trends, such as "selection and concentration," and intensification of agricultural competitiveness had been pursued since the late 1990s because of financial deterioration and TPP negotiations. These political purposes were accomplished quite successfully in disaster reconstruction projects in the southern coastal areas.

However, rapid land-use changes gave rise to certain conflicts and contradictions as well. First, the community-based farmland management system disintegrated. To examine this issue in detail, let us take the case of Nakahama District in Yamamoto as an example. Nakahama was one of the most severely damaged districts in Yamamoto, where most of the land was designated as a risk area. Although the operating cultivated land in Nakahama was reduced from 144.7 ha in 2010 to 42.8 ha in 2015, the number of farm households decreased from 61 to 5, and the rate of cultivated land rentals increased from 38.9% to 89.5%. This means that agricultural productivity in Nakahama improved drastically.

According to one of the large-scale full-time farmers, he rented various large-scale agricultural machines and facilities, such as tractors, combine harvesters, rice dryers, and so on, for free on the condition that he would cultivate large holdings of at least 35 ha for 10 years. As a result, the efficiency of agricultural production improved remarkably. However, he faced difficulties in mowing the grass on the footpaths between rice fields and in cleaning ditches. Before the earthquake such agricultural land maintenance was conducted collectively by the village. However, such village customs collapsed because of the decrease in the number of farmers. As large-scale agricultural machines were useless for mowing grass and cleaning ditches, he was forced to employ several people for such work. However, as the scale of farmland increased, farmland maintenance became much more difficult as well. In the face of this difficulty, the town administration prepared subsidies to call for participation in mowing grass; nevertheless, there were few participants, and the subsidies were refunded. As a result, the farmer was eventually forced to depend on herbicides.

The breakdown of collective farmland management not only constrained the further expansion of agricultural production, but also led to the disintegration of community cohesion. It has been pointed out that the collective activities of mowing grass and cleaning ditches are sources of communalities [21,22]. Customary practices of social gatherings after collective agricultural activities in Nakahama and other villages, such as "Ocha Nomi (tea parties)," disappeared after the earthquake. Under these circumstances, social welfare for isolated elderly people who are no longer practicing agriculture emerged as a new challenge (see Mochizuki 2020 for details) [23].

Agricultural reconstruction projects were rated highly by the town administrations of Watari and Yamamoto. The mayor of Iwanuma, for example, stated the following at a plenary meeting in March 2015:

Reconstruction of agriculture, the main industry of Watari, and of fisheries had good results all in all. ...Land consolidation of paddy fields reached 1123 ha. Infrastructural development for efficient agricultural management in international markets for agricultural commodities is making steady progress. I think agriculture in Watari will be a major source of employment and hence a growth industry.

However, there are discrepancies between this positive evaluation by the mayor and the actual situation of the farmers mentioned above.

Second, the management of coastal forests was undermined. Before the earthquake, coastal villages managed the seaside forest because it was indispensable for community life and agricultural production to mitigate the damage caused by high tides and storms. The tsunami caused by the 3.11 earthquake swept away the seaside forest, and most of the local residents left coastal areas due to relocation projects. Subsequently, Miyagi Prefecture proposed reforestation in collaboration with the private sector, and accordingly, an environmental non-profit organization (NPO) initiated the reforestation project backed by financial support from the government. This project was promoted based on the idea of environmental regeneration, and it attracted public attention. Over 5,000 people and various companies participated as environmental volunteers, and more than 40,000 trees were planted along 4 km of coastline over a span of three years. This reforestation project was rated highly and designated as a model project of "green decentralization reform" by the national government.

After the tree planting, however, the project stagnated because of a shortage of funds and manpower. In this project, the government paid attention only to tree planting, and subsidies were discontinued after the tree planting was finished. However, growing trees requires long-term maintenance. According to the members of the NPO, because tree planting is an attractive event related to environmental studies for school children and for corporate social responsibility or CSR activities, there were many such participants. However, regular maintenance of trees, such as undergrowth cutting and pruning, is not as attractive for such people; hence, the number of participants also decreased drastically. A few farmers remained in the coastal area, but they could not afford to cope with forest management due to the expansion of agricultural production. Although reforestation-related difficulties were brought up in the town council meetings, the town administration did not have a positive attitude because of financial concerns and unclear role divisions between the Town and the Prefecture. Against the backdrop of such difficulties, members of the NPO started to grow peanuts in abandoned farmland to gather funds for management activities; however, they had difficulties producing a good yield because they were not farmers.

Coastal forests have played a significant role in mitigating damage caused by blowing sand and storm surges in this area. Thus, customary management of the coastal forest, which is indispensable for agricultural production, became a "disaster subculture" [24] in this area. The case study mentioned above depicts the difficulties involved in replacing such a community-based disaster subculture with environmental volunteer activities. Furthermore, while the government emphasized disaster prevention projects focusing only on tsunamis and efficient agricultural development, such reconstruction projects had negative outcomes with respect to disaster risk reduction in the coastal area due to their inattention to multifunctional rural villages.

Third, land use changes at relocation sites caused problems as well. The place where the reconstruction housing complex was built used to be paddy fields before the earthquake. As mentioned in Section 2, the affected areas in the south were low-lying areas; hence, agricultural drainage systems had been developed through long-term efforts and ingenuity. However, group relocation projects were promoted without sufficient attention to such historical contexts.

As a result, urban flooding occurred repeatedly at the relocation site in Watari. The place where the reconstruction housing complex was constructed used to be paddy fields before the earthquake. However, building lots were developed without proper connections to existing agricultural drainage channels, and thus floodwater overflowed from the reservoir around the housing complex. When one councilor mentioned the installation of drainage pumps after the flooding, the town administration rejected the argument because the reservoir was designed in accordance with the guidelines of the Prefecture.

In Yamamoto, as mentioned, group relocation projects were promoted as urban restructuring projects around newly constructed stations. During the reconstruction projects, however, it turned out that one of the relocation sites was in a potential tsunami inundation zone. To deal with this problem, the town administration planned to raise the foundation of the housing complex. However, although this land elevation work was beneficial for people who settled in the new housing complex, it might have caused new flooding risks for people who lived in nearby low-lying areas.

Therefore, neighborhood community associations opposed the construction plan and submitted a petition to the town council. As a result, the project was delayed by almost a year.

In rural-urban fringes such as Watari and Yamamoto, urbanization has historically progressed in concert with agricultural land development. However, after the earthquake, housing complex construction projects followed by group relocation schemes were conducted by the Department of City Planning; hence, the reconstruction plan was made without sufficient attention to previous agricultural land use. As illustrated, the problems mentioned above were caused by contradictions between administrative rules and the historical contexts of land use in this area.

6 Conclusion

Although most studies on the Great East Japan earthquake have been case studies focusing on recovery processes in specific affected areas, this paper aimed to explore the interrelationships between long-term social trends and the disaster reconstruction process. Based on the analysis, I would like to highlight several points below.

First, there is an association between the urban system and disaster reconstruction. Previous studies have mainly examined housing reconstruction following group relocation projects after the EQ. However, our research revealed that a considerable number of people who had lived outside of the disaster risk area also voluntarily migrated to other areas. This population outflow was mainly caused by a disturbance in commuting due to railway line damage.

However, the root causes of the population decline after the EQ can be traced to regional urban development since the 1960s. In Watari, the majority of the newcomers from the 1990s left; hence, housing development that turned towns into bedroom communities in the pre-disaster period was closely connected with population changes after the EQ. Accordingly, reconstruction revealed differences between social strata in terms of ties with the land.

Disaster reconstruction in Yamamoto was a more extreme case. Although the urban renewal project in Yamamoto has been promoted with the aim of adapting to competitive urban systems, prolonged construction periods due to large-scale projects ironically caused a drastic population outflow. As a result, the population gap between the center and periphery in the Sendai Metropolitan Area and the gap in living conditions between coastal rural areas and inland city centers inside the municipalities have expanded interactively following the EQ.

Second, there is also a relationship between agricultural land use and disaster reconstruction. Disaster reconstruction policies following the EQ tended to prioritize economic development, with the slogan of "build back better," as well as technological disaster mitigation. In the context of agricultural reconstruction, such a tendency was prominent, and large-scale projects for the improvement of agricultural efficiency were strongly promoted. In contrast, the community-based farmland management system was disintegrated due to the advancement of contract farming and large-scale relocation projects. This led not only to a disturbance with respect to further development of agricultural production, but also to the disintegration of community cohesion. Furthermore, land use specialized only for efficient agricultural production led to a decline in disaster subculture, such as community-based management of disaster prevention coastal forests. The rapid conversion of paddy fields to residential lands in relocation sites has also caused urban floods.

In summary, disaster reconstruction in the rural-urban fringe was characterized by the concentration of urban functions in inland areas through large-scale relocation projects and large-scale agricultural development projects in the vast coastal land left vacant due to relocation. This functional differentiation of land use was closely interrelated with general political trends such as "selection and concentration" and intensification of agricultural competitiveness against the background of financial deterioration and Trans-Pacific Partnership negotiations since the late 1990s. Put shortly, the reconstruction process in this area was an extension and acceleration of existing trends as a whole.

However, there is a unique agricultural land-use system and related community organization rooted in the peculiar history and natural environment of the coastal area. Existing disaster reconstruction policies have not paid sufficient attention to such historical contexts. Although the "multifunctionality of agriculture" is emphasized as a new ideal in current agricultural policies, it has been overlooked in disaster reconstruction policies. Hence, vulnerabilities such as urban flooding and the decline of disaster subcultures were newly created. In other words, disaster studies should pay careful attention to the tensions between wider political trends and local-historical contexts. Such a viewpoint is significant not only for disaster studies, but also for environmental studies in general.

Notes

1) This survey was conducted from May to August 2018, and targeted 573 communities (neighborhood association districts) in 10 affected municipalities in Miyagi prefecture, as described below.

Municipality	Number of surveys distributed	Valid responses	Collection rate
Kesennuma city	204	106	52.0%
Minamisanriku town	75	34	45.3%
Onagawa town	33	25	75.8%
Ishinomaki city	87	62	71.3%
Higashimatsushima city	69	38	55.1%
Shichigahama town	16	8	50.0%
Natori city	21	12	57.1%
Iwanuma city	28	12	42.9%
Watari town	15	11	73.3%
Yamamoto town	25	16	64.0%
Total	573	324	56.5%

References

1 I. Davis and D. Alexander, *Recovery from Disaster.* Routledge (2016)
2 N. R. Britton, "National Planning and Response," H. Rodriguez, E. L. Quarantelli, R. R. Dynes eds., *Handbook of Disaster Research.* Springer, pp. 347-367 (2007)
3 S. Tanaka, Y. Kuroda, N. Yokota, A. Oyane, *The Great East Japan Earthquake: Disaster Risk Reduction and Disaster Relief.* Yuhikaku (2019)
4 K. Ueda and H. Torigoe, "Why do Victims of the Tsunami Return to the Coast?," *International Journal of Japanese Sociology*, **9**, pp. 21-29 (2012)
5 T. Kawamura, T. Okamoto, K. Yoshino eds., *Regeneration from "3.11".* Ochanomizu Shobo (2013)
6 K. Hasegawa, T. Hobo, H. Ozaki, *Recovery from the Great East Japan Earthquake at the Crossroad.* University of Tokyo Press (2016)
7 Kansai University Faculty of Societal Safety Sciences, *The Great East Japan Earthquake: Evaluation after 5 Years.* Minerva Shobo (2016)
8 H. Yoshino and M. Kato eds., *The Great East Japan Earthquake: Reconstruction and Prospect.* Yuhikaku (2019)
9 J. K. Mitchell, N. Devine, K. Jagger, "A Contextual Model of Natural Hazard," *Geographical Review*, **79(4)**, pp. 391-409 (1989)
10 E. R. Wolf, "Distinguished Lecture: Facing Power – Old Insights, New Questions," *American Anthropologist*, **92(3)**, pp. 586-596 (1990)
11 A. Oliver-Smith, "Global Changes and the Definition of Disaster," E. L. Quarantelli ed., *What Is a Disaster?.* Routledge, pp. 177-194 (1998)
12 P. Steinberg and R. Shields eds., *What is a City? Rethinking the Urban after Hurricane Katrina*. The University of Georgia Press (2008)
13 R. J. Samuels, *3.11: Disaster and Change in Japan.* Cornell University Press (2013)
14 J. Gottmann, *Megalopolis.* The Twentieth Century Fund Inc. (1961)
15 J. Coaffee and P. Lee, *Urban Resilience.* Palgrave (2016)
16 Miyagi Prefecture, *History of Agricultural Land Development in Miyagi Prefecture.* (1994) (in Japanese)
17 Watari Town, *History of Watari Town: Volume 2.* (1977) (in Japanese)
18 Yamamoto Town, *History of Yamamoto Town: Volume 2.* (1986) (in Japanese)
19 Watari Town, *History of Watari Town: Contemporary History.* (2008) (in Japanese)
20 Yamamoto Town, *History of Yamamoto Town: Volume 3.* (2005) (in Japanese)
21 S. Shogenji, *Agriculture and Humans.* Iwanami Shoten (2013) (in Japanese)
22 N. Fukuyo, *Building Resilient Community.* Nihon Keizai Hyoronsha (2020) (in Japanese)
23 M. Mochizuki, *Sociology of Disaster Reconstruction and Life.* Ochanomizu Shobo (2020) (in Japanese)
24 D. E. Wenger and J. M. Weller, "Disaster Subculture: The Cultural Residues of Community Disasters," *Preliminary Papers*; **9**, Disaster Research Center, pp. 1-18 (1973)

E3S Web of Conferences **340**, 03002 (2022) https://doi.org/10.1051/e3sconf/202234003002
The 13th AIWEST-DR 2021

Disaster mitigation efforts based on community in coastal area of desa Berdaya Mirring

Nurmansyah[1,*], *Neneng* Alghina Micha Grandisa[1]

[1]Rumah Zakat, Batu Kencana Street No.6, Bandung, Indonesia

Abstract. Indonesia is one of the countries that is categorized as a disaster-prone country. BNPB noted an increase in the number of disasters in Indonesia in the last 10 years, including earthquakes, volcanic eruptions to tidal waves and abrasion. West Sulawesi is one of the provinces prone to disasters, one of which is in Mirring Village, Polewali Mandar. Mirring Village is a coastal area that is prone to coastal abrasion and tidal waves, therefore the area is planted with mangroves to withstand tidal waves from the sea. Unfortunately, the people of Mirring village did not understand the function of the mangrove until finally Rumah Zakat through Relawan Inspirasi came to provide education regarding the use of mangroves, disaster mitigation to become a disaster-resilient family and an economically resilient family. This research is a descriptive study using a qualitative approach. The data collected with interview with 17 respondents. The results show that the community welcomes the community empowerment program towards a disaster-resilient village by Rumah Zakat. The community has understood the function of mangroves and participated in disaster training to become disaster-resilient families. The community also initiated the existence of crab ponds and mangrove cultivation which later could become their source of income and form an economically resilient family.

1 Introduction

Every year, Indonesia experiences various natural disasters that occur one after another. This happens because geologically, Indonesia is located in the ring of fire, the path of volcanoes that surrounds the scope of the Pacific. Moreover, Indonesia is an archipelagic country that has a large sea and is located at the confluence of three plates; the Pacific, Indo-Australian and Eurasian plates.

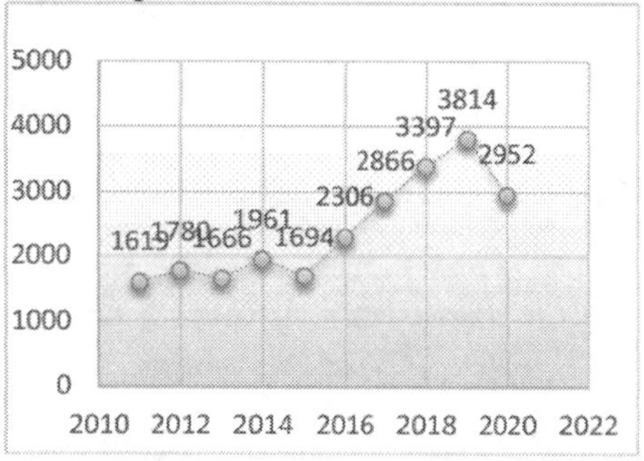

Fig. 1. Number of disasters in Indonesia 2010-2020
Source: https://gis.bnpb.go.id/, 2021

Figure 1 shows the number of disasters in Indonesia in the last 10 years. The number of disasters has increased quite high every year such as earthquakes, volcanic eruptions, forest fires, droughts, floods, landslides, cyclones and tidal waves and abrasion.

One of the natural disasters that was quite shocking was the tsunami in the Sunda Strait. The tsunami on December 22, 2018 occurred without earthquake, but was caused by volcanic activity, an underwater landslide on Mount Anak Krakatau. This is considered logical because volcanic eruptions can cause the mountain body to slide and the sea column to become unstable which results in rising the sea waves. Julius said that the Sunda Strait tsunami event reminded us that Indonesia is a disaster-prone country which is quite complex. Therefore, community preparedness training is needed.

Rumah Zakat as an NGO realizes the importance of community empowerment, both economically and socially. Indonesia as one of the disaster-prone countries needs to prepare its community for disaster preparedness training, especially those in coastal areas.

Desa Berdaya Rumah Zakat is one of the community empowerments programs, specifically for disaster mitigation, Rumah Zakat has a Disaster Resilient Village (Desa Tangguh Bencana), to form empowered and resilient communities in coastal areas

Desa Berdaya Mirring is one of the villages empowered by Rumah Zakat which is located in Tappina Hamlet, Polewali Mandar, West Sulawesi. This village is one of the areas prone to disasters, such as abrasion and tsunamis that occurred in other coastal

*Corresponding author: nurmansyah@desaberdaya.org

areas. The following is a history of disasters in West Sulawesi.

Table 1. Disaster history in West Sulawesi.

Year	Type of Disaster	Impact
1967	6.3 Magnitude Earthquake and Tsunami	13 people died
1969	6,9 Magnitude Earthquake and Tsunami	64 people died, 97 were injured, 1287 were damaged in 4 villages. 4 m Tsunami in Pelattoang and 1.5 m in Parasanga and Palili
1983	6,7 Magnitude Earthquake	No tsunami occurred but many houses were damaged
2018	Coastal Abrasion	In the last 10 years, nearly 10 meters of coastal area have been eroded
2020	4 meters wave heigh	Residential settlements along the Polewali Mandar coast are damaged
2021	6, 2 Magnitude Earthquake	107 people died, 3,369 injured, 18,215 houses damaged, 70,496 refugees

Source: Various News Sources, 2021

Desa Berdaya Mirring is surrounded by the ocean and has extensive mangroves. Even so, the surrounding community has not maximized the potential of the sea nor planting mangroves for people's lives, including for the community's economy. Not only that, public education about disasters is still very low, so disaster mitigation in the village is still lacking.

Various natural disasters that have occurred recently have made Indonesia the country with the second largest number of disaster victims in the world, namely 160 thousand people. Changes occurred in 2018, making Indonesia the first country with the most casualties [5]. Nugroho et al (2018), stated that the level of Indonesia's preparedness in dealing with disasters is still not optimal, this can be seen from the behavior of people who are still nervous during disaster evacuation.

This indicates that disaster communication through disaster mitigation still needs improvement. Mitigation is a continuous preventive action with the aim of reducing the impact of disaster risks that threaten life and property [6]. Mitigation efforts are not only the task of the government. Therefore, mitigation communication has a role to synergize policies between the government and non-government institutions involved in disaster management [3]

The central government, through BNPB (National Disaster Management Authority), encourages the formation of a strategy for the formation of a Disaster Resilient Village, which is contained in the Regulation of the Head of BNPB Number 1 of 2012 [1]. Furthermore, BNPB emphasized that efforts to reduce disaster risk are not only the task of the government, but also the direct involvement of individuals, families, and communitiesin dealing with disasters. Social values in society are the basic capital in the development of Disaster Resilient Village. An important step that must be taken is mapping the nature and characteristics of the area to understand the level of community capacity and vulnerability [4].

Therefore, Rumah Zakat present to support the government's efforts to improve community-based disaster mitigation communications in Desa Berdaya Mirring. Furthermore, the natural potential of Desa Berdaya Mirring, such as mangroves, can be used as one of the economic resources of the community so that they can create disaster-resilient families and economically resilient families.

The disaster that occurred caused a lot of losses, especially for the affected around the disaster location. Losses caused include loss of residence, property, life and affect the economy of the community. Moreover, people are still confused during the evacuation process and their understanding of disaster is still low. This requires serious efforts from the government and non-government institutions to anticipate a crisis. Therefore, the government and non-government institutions jointly carry out disaster mitigation programs, one of which is through the Disaster Resilient Village which is expected to form a disaster-resilient family and an economically resilient family.

2 Purpose

This study aims to determine community-based disaster mitigation process in Desa Berdaya Mirring assited by Rumah Zakat. In addition, the study will look at the process of Desa Berdaya Mirring assited by Rumah Zakat towards disaster-resilient families and economically resilient families.

3 Method

This study uses a qualitative approach. The type of research used is descriptive research using survey methods. The survey was conducted through interviews to obtain the necessary information. Interviews were conducted on 17 respondents consisting of 14 beneficiaries, 1 Relawan Inspirasi in Desa Berdaya Mirring, 1 Hamlet Head and Village Head. In addition, this study uses a study of documentation obtained from journals, books, news and other supporting documents. The study was conducted for approximately 3 months.

4 Results

Desa Berdaya Mirring is one of the villages located in Tappina Hamlet, Mirring Village, Binuang District, Polewali Mandar Regency and is located on the coast. As a village located on the coast, Desa Berdaya Mirring has abundant natural potential. In addition to natural potential, coastal areas are vulnerable to natural disasters such as abrasion and tsunami, this can be anticipated by planting mangroves around the coast.

Initially, Desa Berdaya Mirring's community did not understand the function of mangrove planting as a disaster mitigation effort. In fact, 80% of the people make mountains as a source of life, they are quite afraid to get close to the sea. Whereas mangroves have many benefits if used properly, including for disaster mitigation and the economic life of the community.

This research uses Risk and Crisis Communication Management Theory from Matthew Seeger and the concept of disaster mitigation from George D. Haddow and Kim S. Haddow. The combination of these two theories can create disaster mitigation communication [7]. Two main things in disaster mitigation are program activities and communication strategies.

4.1 Program activity

One of the issues discussed in this research is the program activities carried out in Desa Berdaya Mirring as a Disaster Resilient Village. Desa Berdaya Mirring for the first time carried out a disaster mitigation communication program, in August 2020. Desa Berdaya Mirring was chosen as a Disaster Resilient Village because of its location around the Makassar Strait and as a coastal village that has mangroves but its sustainability is threatened due to damage caused by garbage and the people there who do not understand the function of mangrove planting.

Coastal communities do not understand that they have to maintain and cultivate mangroves, whereas this will protect them from disasters originating from the sea such as abrasion and tsunami. Mangroves can also have economic value for the community if they are processed into food. Several natural disasters that have occurred in Desa Berdaya Mirring include hurricanes, tidal winds and coastal abrasion. This disaster caused some damage which eventually moved people to start caring about the environment and disasters.

At first, the people of were afraid to approach the sea. They don't even understand the function of planting mangroves. Some people regard mangroves as shrubs and use them as firewood for sale. For their livelihoods and daily needs, 80-90 percent of the people choose to go to the mountains which are quite far from their homes or go to the sea to look for fish. In addition, the community does not understand disaster mitigation at all. What should they do when a disaster strikes? Or signs when disaster will come.

Regarding to that, Rumah Zakat, which is represented by Relawan Inspirasi as a community mobilizer, has begun to carry out disaster mitigation activities. This disaster mitigation activity is not only the duty of the government or non-government institutions, but also need active participation of the community which is vital for the success of the disaster mitigation communication program.

The first thing that was done by Relawan Inspirasi was conducting socialization to several figures and the local government. This is carried to invite all parties to cooperate and actively participate in this disaster mitigation communication program. This initiation was welcomed by village leaders and local government.

The next step is socialization to the community with the assistance of several parties such as the Environment Service, Forest Farmers Service and PMI. After the socialization, the First Emergency Training was conducted. This training is given to farmer groups because these groups are involved in mangrove cultivation. After the training, the farmer groups will provide training to their respective families. This farmer group consists of 35 people, if the assumption is that one family consists of two people, then a Disaster Resilient Family of 70 people will be formed.

The Disaster Resilient Family consist of 70 members will later become Agents of Change and provide public education about disaster resilience. There are three things that are included in the education and training component for families, namely the introduction of standby bags, directions for disaster evacuation and early warning signs.

In addition to forming a Disaster Resilient Family, Rumah Zakat has the initiative to form an Economic Resilient Family. It aims to form an independent and economically empowered community or family. Since the empowerment activity carried out in Desa Berdaya Mirring, the community welcomed it well, so that it is created smooth communication between Relawan Inspirasi and community. Relawan Inspirasi provides insight related to the economic potential in the growth and development of mangroves. This makes people do not need to go to the mountains or look for fish in the sea to meet their daily economic needs because later there will be a lot of marine life around the mangroves. To make this training a success, Relawan Inspirasi was assisted by several parties, one of which was a professor from Hasanuddin University, who assisted group members in crab cultivation. This is in line with the idea that emerged from the community to accelerate the process of crab cultivation around the mangrove area. Rumah Zakat hopes that the Empowered Village program in Mirring Village can increase public awareness about the importance of disaster prevention efforts in coastal areas and can form disaster-resilient families and economically resilient families.

4.2 Communication strategy

Communication strategy is important in disaster mitigation, in Perka BNPB No.1 of 2012 states that community-based disaster risk reduction places people living in disaster-prone areas as the main actors and actively participates in a sustainable and efficient manner. This means that local people know more about the nature around where they live, they also know the forms of activity of the community, therefore in this case we are bringing together local wisdom which we link with disaster mitigation activities. For example, in Mirring Village there is a season map where the community is busy with harvesting from their gardens in March, so we cannot carry out emergency first aid training activities in the water, but we can synchronize it with where they have a lot of activity, namely in the mountains. We also can't immediately change the habits of the people who mostly make a living in the mountains.

Besides that, the community is given the opportunity to monitor and evaluate the ongoing program. So, they welcome to this program. They are very cooperative and actively participate in this disaster mitigation activity. Relawan Inspirasi and the community regularly hold meetings every week to evaluate and discuss the next program to be carried out. In addition, several parties who were deemed to have an important role such as the Hamlet Head and the Village Head also supported the success of this activity.

Rumah Zakat is the only non-government institution involved in empowering and educating the community about disaster mitigation in Desa Berdaya Mirring. Relawan Inspirasi as the party appointed by Rumah Zakat assists the community in these various mitigation activities. The mitigation carried out involves the Department of the Environment for mangrove cultivation and PMI (Indonesian Red Cross) for disaster training. Desa Berdaya Mirring does not yet have social media or a Disaster Report application so all information is still conveyed directly through regular meetings or during training. The training is carried out on the sidelines of the community's free time so that it does not interfere with the community's economic activities.

Disasters can happen to anyone and anywhere. What we can do is reduce the risk of the impact by increasing the competence of the community through disaster training. There needs to be togetherness and active participation from various parties, especially the community.

5 Conclusion

This study discusses efforts to improve community-based disaster mitigation carried out in Desa Berdaya Mirring with the ultimate goal of forming a disaster-resilient family and an economically resilient family. Rumah Zakat through Relawan Inspirasi makes several efforts in accordance with the combination of two theories: Risk and Crisis Communication Management and disaster mitigation. Steps taken by Relawam Inspirasi include socializing the importance of mangroves for disaster mitigation. Furthermore, the community is given training on mangrove cultivation and initiates the construction of crab ponds that can become an economic source for the community. The existence of crab pond innovation as a form of mangrove cultivation, is able to move the community to participate actively. In addition to socialization, Desa Berdaya Mirring has conducted emergency first aid exercises, especially for conditions around the sea as one of the efforts to increase public awareness and concern for disaster hazards.

References

1. Buchari, A., Santoso, M. B., & Marlina, N. (2017). Pengembangan Kapasitas Kelembagaan Desa Tangguh Bencana di Kabupaten Garut (Studi Kasus di Desa Pasawahan Kecamatan Tarogong Kaler). *Jurnal Analisis Kebijakan dan Pelayanan Publik*, 49-62.
2. Kumparan. (2018, Oktober 2). *BNPB: Budaya Mitigasi Bencana Masyarakat Indonesia Lemah*. Retrieved from https://kumparan.com/: https://kumparan.com/kumparannews/bnpb-budaya-mitigasi-bencanamasyarakat-indonesia-lemah-1538494508088227510
3. Lestari, P., Prabowo, A., & Wibawa, A. (2012). Manajemen Komunikasi Bencana Merapi 2010 pada saat Tanggap Darurat. *Jurnal Ilmu Komunikasi*, 173-197.
4. Maarif, S., Damayanti, F., E, S. D., & A, W. P. (2012). Initiation of the Desa Tangguh Bencana Through Stimulus-Response Method. *Indonesian Journal of Geography*, 173-182.
5. Setyawan, F. A. (2019, Agustus 01). *BNN Sebut Indonesia Nomor Wahid Jumlah Korban Bencana 2018*. Retrieved from https://www.cnnindonesia.com/: https://www.cnnindonesia.com/nasional/20190801143833-20-417449/bnpb-sebut-indonesia-nomor-wahid-jumlah-korban-bencana-2018
6. Wardyaningrum, D. (2014). Perubahan Komunikasi Masyarakat dalam Inovasi Mitigasi Bencana di Wilayah Rawan Bencana Gunung Merapi . *ASPIKOM*, 179-197.
7. Yunia, A., Pinariya, J. M., Forceila, D., & Ivana, L. (2020). Program Berbasis Masyarakat dalam Upaya Pengurangan Risiko Bencana di Kabupaten Pandenglang. *Journal of Communication Studies*, 172-189.

The implementation of disaster education after the sixteen years of the 2004 Indian Ocean Tsunami in Aceh-Indonesia: Progress or regress?

Syahrul Ridha[1,2,3*], *Alfi* Rahman[2,3], *Abdul* Wahab Abdi[1], *Puspita* Annaba Kamil[4]

[1]Department of Geography Education, Universitas Syiah Kuala, Banda Aceh-Indonesia
[2] Pusat Riset Ilmu Sosial dan Budaya (PRISB), Universitas Syiah Kuala, Banda Aceh-Indonesia
[3]Tsunami and Disaster Mitigation Research Center (TDMRC), Universitas Syiah Kuala, Banda Aceh-Indonesia
[4]Department of Geography Education, STKIP Al-Washliyah, Banda Aceh-Indonesia

Abstract. Strengthening disaster education for the community is an effort to build awareness, knowledge, and actions that need to be taken before, during, and after a disaster. Disaster risk communication plays an important role in disaster education as an effort to provide views about disasters to the community. This study aims to look at the progress of the implementation of disaster education school-based after the 2004 Indian Ocean tsunami. Progress is assessed through two aspects, namely: (1) the national curriculum in education unit level, namely Elementary Schools; Junior High School; and Senior High Schools, and (2) the implementation of disaster education school-based in the learning process. The national curriculum will synthesize by using a comprehensive literature review in national curriculum documents. This study applied a content analysis method in analyzing the qualitative data. The questionnaire survey was conducted at (*Sekolah Siaga Bencana*, SSB) in two districts affected by the 2004 tsunami. A literature study on the national curriculum documents in schools shows that disaster education has been integrated into the curriculum; for example, learning in elementary schools uses a thematic learning approach following factual conditions in the field, such as learning related to disasters. In junior high schools, the curriculum is integrated in an integrated manner in social science learning to provide awareness and knowledge about disasters. Meanwhile, the high school curriculum is integrated into geography subjects to increase knowledge and skills about disasters and make decisions before, during, and after a disaster occurs. This progress has occurred since the change in the curriculum from the KTSP curriculum (2006-2012) to the National Curriculum (2013). However, there are still several obstacles in implementing disaster education in Aceh, such as a lack of facilities for disaster learning. Therefore, a synergy between the government and schools is needed to improve disaster education school-based implementation within the national curriculum.

1 Introduction

Increased attention to disasters is caused by awareness of disaster hazards. The danger of disasters is a part of human life that cannot be predicted when, where, and how big the impact will be. The character of a disaster is described as an event that produces an impact on the community to produce a response to the incident. The development of disaster knowledge has shaped people's perceptions of disasters; even until now, disasters are understood as unpredictable events.

Strengthening disaster education for the community is an effort to provide awareness and knowledge of disasters and the actions that need to be taken before, during, and after disasters occur [1,2]. Communicating disaster risk plays an important role in disaster education as an effort to provide views about disasters to the community [3]. Increased attention to disasters is caused by the awareness of the dangers of disasters that can be communicated in people's lives, which often cannot be predicted when, where, and how big the impact will be [4,5].

Currently, disasters are one of the issues that are of concern to the world. It is proven by several international policies that focus on disaster management, which is accompanied by the emergence of institutions that specifically examine disaster issues at the national and international levels. For example, the Tsunami and Disaster Mitigation Research Center (TDMRC) was established by Universitas Syiah Kuala after the 2004 Indian Ocean tsunami disaster. In addition, the International Research Institute of Disaster Science (IRIDeS) was established by Tohoku University after the great earthquake and tsunami disaster. East Japan in 2011. After the 2004 Indian Ocean tsunami, Aceh experienced a great downturn in the fields of economy, infrastructure, health, and even education [6]. With national and international assistance, the Acehnese people continue to strive to rise from adversity.

After sixteen years of the tsunami disaster, there have been many developments felt by the people of Aceh, one of which is disaster education. In the national curriculum, disaster education is one of the materials studied in high school [7,8]. Disaster material is studied in class XI in ge-

*Corresponding author: syahrul.ridha@unsyiah.ac.id

ography. The competencies expected in this disaster material consist of two, namely, knowledge competencies and skills competencies. Knowledge competence expects students to be able to analyze the types and management of natural disasters through education, local wisdom, and the use of modern technology, while the competency skills are to make sketches, plans, and/or maps of potential disasters in the local area as well as disaster mitigation strategies based on these maps [9]. Thus, disaster education in Indonesia is developing well, and this can be indicated by the integration of disaster education into the national curriculum.

Disaster education in schools aimed at young children is important in disaster management, and they have a future that is important to save and for better disaster management [5,1]. Along with the advancement of science and technology, there is an increasing understanding of the community that disasters can be reduced so that they need to be managed properly. In addition, disaster education has also been studied in various disciplines, such as disaster risk management and environmental studies, but disaster education is a new question in the world of education, and the literature on the concept of disaster education in education is still minimal. [10]. Therefore, this study intends to look at the progress or setbacks of the implementation of disaster education after the 2004 Indian Ocean tsunami. Progress and setbacks are assessed from two aspects of education, namely: (1) the national curriculum in elementary schools; Junior high school; and Senior High Schools, and (2) Implementation in learning in schools.

2 Method

The national curriculum is synthesized using a comprehensive literature review in the form of national curriculum documents, namely: elementary school, junior high school, and high school curriculum. During the literature review, this study applied the content analysis method to analyze the qualitative data collected. Meanwhile, the implementation of strengthening disaster education in the teaching and learning process in schools was carried out between July 2019 to March 2020 through a questionnaire service involving 40 Disaster Preparedness Schools (SSB) that had been established and developed by the Tsunami and Disaster Mitigation Research Center (TDMRC) of Universitas Syiah Kuala consisting of teachers and students. The questionnaire survey was conducted in two areas where the Disaster Preparedness School was badly affected by the 2004 tsunami. Descriptive statistics were used to further analyze responses and interviews from teachers and students.

3 Result and discussion

3.1 National curriculum

3.1.1 Elementary school national curriculum

The literature study of the national curriculum document in schools shows that disaster education is integrated into the curriculum; for example, learning in elementary schools uses a thematic learning approach according to factual conditions in the field, such as learning related to disasters. Students are expected to have competencies that can preserve the environment wisely and responsibly. These competencies are applied through the scope of material for connectivity between spaces and tackling environmental problems wisely in life [11]. Thus disaster education can be applied in basic education by providing factual, conceptual, procedural, and metacognitive knowledge at the basic level regarding disaster science, disaster technology, art, and culture for disaster preparedness, and being able to relate the above knowledge in one's context, family, school, community, and natural environment [12].

3.1.2 Junior high school national curriculum

In junior high school, the curriculum is integrated in an integrated manner in social science learning to provide awareness and knowledge about the disaster. Students are expected to have competence about the concept of space (location, distribution, potential, climate, morphology, geology, environment) and interactions between spaces and their influence on human life in the aspect of disaster risk reduction [9]. In addition, students are expected to understand spatial aspects and inter-space connectivity in regional and national scopes on the changes and sustainability of post-disaster community life that can affect geographical, economic, and cultural conditions [11]. Various initiatives can be carried out by the school community, either directly or indirectly, and contribute to strengthening disaster preparedness capacity. As a social subsystem that is inseparable from society, schools have the potential to play an active role as the main actor in building the response and preparedness of young people to disasters through disaster education in junior high schools [13].

3.1.3 High school national curriculum

Meanwhile, the high school curriculum is integrated into geography subjects to improve knowledge and skills about disasters and the ability to make decisions before, during, and after disasters occur [2,14]. While in other subjects, it does not contain disaster material. In addition, there are no specific subjects related to increasing disaster knowledge and skills. This progress has occurred since the curriculum change from the Education Unit Level Curriculum (KTSP) in 2006-2012 to the 2013 National Curriculum. The standard of learning content related to disaster education in geography learning is that students understand and can implement disaster mitigation and adaptation [11]. In addition, the basic competence of disaster knowledge is that students can analyze the types and management of natural disasters through education, local wisdom, and the use of modern technology. In addition to knowledge, students are also expected to have skills in making sketches, plans, and/or maps of potential local disasters and disaster mitigation strategies based on these maps [9].

If studied geologically, climatologically, and socio-culturally, Aceh Province is an area prone to disasters, both natural and social disasters. It is necessary to plan, integrate, direct, coordinate a sustainable disaster education to provide knowledge to the public about hazards and disaster risk reduction through disaster education. The principles of implementing disaster education are:

- Disaster education for all students regardless of gender, ethnicity, religion, race, and descent.
- Disaster education is held at all types and levels of education.
- Disaster education is carried out in a directed, planned, programmed, integrated, coordinated, professional, and comprehensive manner.
- Disaster education is based on Islamic values and other religious values adopted by the Indonesian people by upholding human rights, customs, and local wisdom.
- The implementation of disaster education is carried out in an integrated manner between lines, types, and levels of education.
- The implementation of disaster education in the formal, non-formal, and informal education channels is carried out by adjusting the time, place, learning facilities, and infrastructure, educators and education staff, forms, other learning resource programs with the conditions of student difficulties.

3.2 Implementation in learning at school

The implementation of disaster education in schools is based on lesson hours in each education unit, namely elementary schools, junior high schools, and high schools. In addition, the implementation is also carried out separately as extracurricular learning in disaster-prepared schools (*SBB*) [15,16]. However, there are still some obstacles in implementing disaster education in Aceh, such as the lack of facilities for disaster learning. This is indicated by the absence of a disaster laboratory, fewer study hours. In addition, teachers in schools are less able to use effective learning time due to a lack of training in disaster education learning in schools.

The role of knowledge in disaster education is the basis or beginning of the actions and awareness of each individual [17,18]. The existence of knowledge capacity can be used as the basis for each individual in taking action before, during, and after a disaster occurs [19,20]. In disaster management, individual or community knowledge is crucial to disaster preparedness [21,22]. In disaster preparedness planning, it is necessary to know in advance the types of hazards that occur in an area, for example, geological hazards such as volcanic eruptions; tsunamis; earthquakes, environmental hazards such as forest fires; floods; landslides; and storms [23,2].

When disaster is discussed in the context of geography, it cannot be separated from the concepts of physical and human geography, especially when analyzing disasters where nature and culture are strong factors (Fig.1). Such an integrative approach that considers the reciprocal interaction between nature and culture can solve problems, especially in environmental problems that can be implied in education [23]. In addition to knowing the disaster hazard, it is also necessary to know the causes, vulnerabilities, community resilience, duration, and area of the disaster [24,25]. Therefore, government policies, emergency planning, warning systems, and resource mobilization are the main things to deal with these disasters [15,5].

Disaster preparedness education can be implemented through various types of education, both formal, non-formal, and informal [10]. Disaster preparedness education can be implemented by integrating it into the existing learning curriculum in schools. The objectives of disaster education in Aceh-Indonesia are as follows [5].

- Cultivate human values and attitudes as well as concern for disaster risk.
- Develop an understanding of disaster risk, social vulnerability, physical vulnerability, behavioral vulnerability, and motivation.
- Develop attitudes and improve knowledge, skills for disaster risk reduction, responsible natural resource and environmental management, and adaptation to disaster risk.
- Develop efforts for individual and collective disaster risk reduction.
- Improve disaster preparedness knowledge and skills.
- Improve disaster emergency response capacity.
- Develop readiness for rehabilitation and reconstruction after a disaster occurs.
- Improving individual and family disaster preparedness when a disaster occurs.

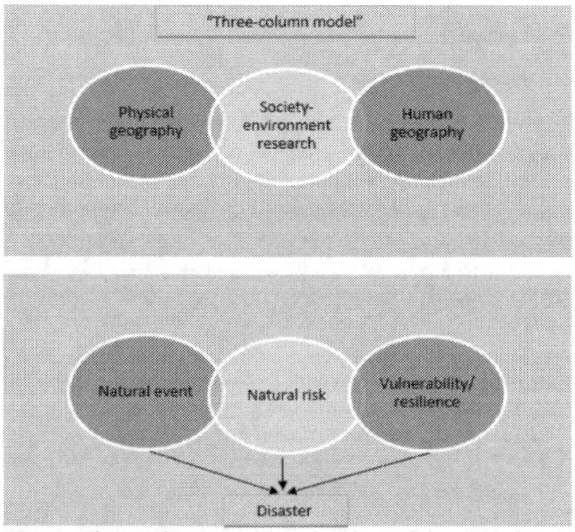

Fig. 1. Three-column model and object of investigation in disaster studies, adapted of [23].

The implementation of education must be adjusted to the characteristics and needs of schools and regions [20]. In practice, schools can collaborate with the government and research institutions at universities nationally and internationally. For example, the establishment of the TDMRC at Universitas Syiah Kuala, the IRIDeS at Tohoku University. So that the goals of disaster education can be achieved optimally to prepare a young generation that is strong, intelligent, creative, and innovative in disaster risk reduction, basic education related to disasters can be provided in the context of the implementation of disaster management, including the following [5].

- Knowledge of the type, source, magnitude, impact, and signs of disaster hazard.
- Access for all school components to increase knowledge capacity, understanding, and preparedness skills that can be obtained through the provision of materials, assignments, training, and extracurricular activities.
- Knowledge of the history of disasters that have occurred in their environment or area.
- Knowledge of the vulnerabilities and capacities of the surrounding environment.
- Knowledge of what can be done to minimize the impact of disasters.
- Skills provision for all components in carrying out disaster risk reduction efforts.
- Knowledge of local wisdom for disaster risk reduction.

4 Conclusions

The explanation above shows that there has been progress in disaster education in Aceh after the 2004 Indian Ocean tsunami. This is indicated by national curriculum documents, both at the primary and secondary education levels. Students are expected to have attitudes, knowledge, and skills in dealing with disasters before, during, and after disasters occur in the curriculum. Disaster learning can be done intracurricular, extracurricular, and co-curricular. In addition, disaster education in Aceh with the aim of (1) having faith and piety to *Allah Subhanahu Wa Ta'ala*, (2) having noble character, (3) having knowledge of disaster, (4) having attitude and concern for disaster risk, (5) disaster response and preparedness (6) can adapt to major and sudden changes. However, the implementation of disaster education in both intracurricular, extracurricular, and co-curricular learning has not been optimally implemented, especially in extracurricular and co-curricular learning experiences. Therefore, a synergy between the government and schools is needed to improve the disaster learning experience in schools following the objectives of the national curriculum, both intracurricular, extracurricular, and co-curricular. For example, making a disaster science project competition among students at school is expected to be held in the Aceh tsunami commemoration.

References

1. P. A. Kamil, S. Utaya, Sumarmi, and D. H. Utomo, Int. J. Disaster Risk Reduct. **43**, 101411 (2020)
2. P. A. Kamil, S. Utaya, Sumarmi, D. H. Utomo, A. W. Abdi, and S. Ridha, IOP Conf. Ser. Earth Environ. Sci. **630**, 012018 (2021)
3. A. Rahman and K. Munadi, IOP Conf. Ser. Earth Environ. Sci. **273**, 012040 (2019)
4. W. Adiyoso and H. Kanegae, in *Disaster Risk Reduct. Indones. Prog. Chall. Issues*, edited by R. Djalante, M. Garschagen, F. Thomalla, and R. Shaw (Springer International Publishing, Cham, 2017), pp. 561–587
5. W. Adiyoso, *Manajemen Bencana Pengantar Dan Isu-Isu Strategis* (Bumi Aksara, Jakarta, 2018)
6. N. Ismail, K. Okazaki, C. Ochiai, and G. Fernandez, Int. J. Disaster Risk Reduct. **28**, 439 (2018)
7. S. Ridha, S. Utaya, S. Bachri, and B. Handoyo, IOP Conf. Ser. Earth Environ. Sci. **273**, 012035 (2019)
8. S. Ridha, S. Utaya, S. Bachri, B. Handoyo, P. A. Kamil, and A. W. Abdi, IOP Conf. Ser. Earth Environ. Sci. **630**, 012017 (2021)
9. Permendikbud RI No. 24, (2016)
10. K. Kitagawa, Educ. Sci. **11**, 1 (2021)
11. Permendikbud RI No. 21, (2016)
12. Permendikbud RI No. 20, (2016)
13. P. A. Kamil, S. Utaya, Sumarmi, and D. H. Utomo, IOP Conf. Ser. Earth Environ. Sci. **412**, 012016 (2020)
14. S. Ridha and P. A. Kamil, J. Geogr. **120**, 72 (2021)
15. R. S. Oktari, K. Shiwaku, K. Munadi, Syamsidik, and R. Shaw, Int. J. Disaster Risk Reduct. **12**, 300 (2015)
16. R. S. Oktari, K. Shiwaku, K. Munadi, Syamsidik, and R. Shaw, Int. J. Disaster Risk Reduct. **29**, 3 (2018)
17. S. Fuhrmann, L. D. Stone, M. C. Casey, M. D. Curtis, A. L. Doyle, B. D. Earle, D. D. Jones, P. Rodriguez, and S. M. Schermerhorn, J. Geogr. **107**, 112 (2008)
18. A. Gampell, J. C. Gaillard, M. Parsons, and L. Le Dé, J. Geogr. 1 (2020)
19. S. Adella, M. Fauzan Nur, N. Nisa, A. Tirtana Putra, F. Rizki Ramadhana, F. Maiyani, K. Munadi, and A. Rahman, IOP Conf. Ser. Earth Environ. Sci. **273**, 1 (2019)
20. T. H. Setiawan, G. H. Salim, M. Wimala, A. F. Van Roy, and Y. L. D. Adianto, Int. J. Disaster Manag. **3**, 53 (2020)
21. K. Brundiers, Int. J. Disaster Risk Reduct. **27**, 406 (2018)
22. A. Gampell, J. Gaillard, M. Parsons, and L. L. Dé, Aust. J. Emerg. Manag. **35**, 44 (2020)
23. L. Mönter and K.-H. Otto, J. Geogr. High. Educ. **42**, 205 (2018)
24. F. I. Rivera and N. Kapucu, *Disaster Vulnerability, Hazards and Resilience: Perspectives from Florida* (Springer International Publishing, Switzerland, 2015)
25. K. Kitagawa, Nat. Hazards **105**, 3099 (2021)

Analysis of village owned enterprises (*BUMDes*) financial performance before and during COVID-19 pandemic

Raida Fuadi[1*], *Linda*[1], *Gamal* Batara[1], *Nurma* Sari[2]

[1]Accounting Department, Universitas Syiah Kuala, Banda Aceh, Indonesia
[2]Islamic Economic Department, Universitas Syiah Kuala, Banda Aceh, Indonesia

Abstract. COVID-19 outbreak had brought a significant impact on economic dynamics in Indonesia, including rural economy. Rural economy continues to decline which is marked by an increase in the poverty rate and a decrease in the purchasing power of rural communities. This study aims to analyse *BUMDes* financial performance before and during COVID-19 pandemic in Banda Aceh city. *BUMDes* or Viilage-owned Enterprises are village business institutions managed by village's community and village's government in order to strengthen the village economy based on needs and potential of the village. The indicators of financial performance used in this study are sales and the ability of *BUMDes* to pay their debts. This study uses primary data. The population are managers of *BUMDes* located in Banda Aceh city. The sampling method used in this study is random sampling. Analysis of differential test between *BUMDes* financial performance before and during COVID-19 pandemic is performed using paired sample t-test. The financial performance variables are sales turnover and liquidity. *BUMDes* sales turnover is measured by the number of sales, while *BUMDes* liquidity is measured by *BUMDes* ability to pay salaries and also *BUMDes* ability to pay operational debts. The results show that there is a significant difference between the number of sales before COVID-19 pandemic and the number of sales during COVID-19 pandemic. The descriptive statistics shows that the average sales before COVID-19 was higher than the average sales during COVID-19 pandemic. This significant difference is indicated by a p-value < 0.05, which is 0.05. Meanwhile, *BUMDes* ability to pay its employees' salaries and *BUMDes* ability to pay debts are not significant, because the p-values > 0.05, which are 0.83 and 0.25 respectively. The descriptive statistics shows that the average *BUMDes* ability to pay debts before COVID-19 pandemic was higher than the average ability to pay debts during COVID-19 pandemic.

1 Introduction

Indonesia government since 2015 has distributed extra amount of money and authority to villages around Indonesia. The goal is to support villages so that they are able to manage their potentials to improve rural economy as well as the welfare of rural communities. Villages economics are the foundation of country economics development, therefore Indonesia government delegates fund management authority to villages, so that villages are able to manage their own economic potential through *BUMDes* (Village-owned enterprises). *BUMDes* accommodates the economic activities of rural community in a form of business entity that is managed professionally. The main business activities of *BUMDes* however still rely on the original potential of the village. This hopefully will bring benefits for rural communities by making community businesses more productive and effective.

Aceh is the second province with the most *BUMDes* in Indonesia, with 7,831 *BUMDes*. The first region is East Java with 8,035 *BUMDes* [1].

According to the Head of Village and Community Office (DPMD) (2018), the problem in Aceh is that the number of poor people in Aceh continues to increase while village funds transferred by central government are also increasing every year. Data obtained from the Village and Community Empowerment Office (DPMD) of Banda Aceh, currently there are 28 active BUMDes and 62 inactive BUMDes in Banda Aceh. DPMD Office needs to conduct an evaluation of 2018 programs and socialize their next programs in for 2019, to prevent other BUMDes bankruptcy. BUMDes should be able to become a platform for economic empowerment of the poor community. Thus, BUMDes should be assisted and monitored so that they get a clear direction. Their performances should also be assessed, to identify how far BUMDes could bring benefit to the community by increasing sales, so that they can bring welfare to their workers and also to the local community [2].

While Aceh government had not been able to maximize the performance of BUMDes, COVID-19 outbreak started in mid-2020. As a result of the COVID-19 pandemic, some fearful predictions have emerged from business actors. This is because the increasing amount of budget for health care sector. This pandemic has brought social and political uncertainty to countries around the world and has an impact on the global economic crisis. According to [3], policy makers have

* Corresponding author: lindarisyad@unsyiah.ac.id

to make an analysis of economic predictions due to the effects of COVID-19 because the economic crisis has emerged and progressed very quickly. Business actors need to conduct surveys that can assess future steps in the midst of this uncertainty. The founder of the great e-commerce, Alibaba, Jack Ma, said that 2020 is the year of survival, while 2021 will be the recovery year for business to start gaining profit again.

Village fund is allocated from Indonesia State Budget (APBN). This village fund is directly transferred by the central government to the village account. One of the allocations is for village's *BUMDes* activities. However, since the COVID-19 outbreak, the allocation has changed significantly, due to the urgency for healthcare allocation. Based on the [4], the minimum allocation of village funds for COVID-19 is 8%, while for the maximum amount is not limited.

With the increasing allocation for COVID-19 healthcare, the allocations for other posts are decreased, including allocation for *BUMDes* activities. On the other hand, *BUMDes* are expected to trigger and drive the village economy in the midst of economic downturn during COVID-19 pandemic. The results of research by [5] shows that most of the micro, small and medium-sized enterprises in Pakistan are facing financial problems, supply chain cycle disruptions, demand declines, sales declines and profit declines during this COVID-19 pandemic. 83% of companies reported that they were not prepared to handle the situation and that they had not planned their next steps to face the pandemic.

[6] conducted a qualitative research on *BUMDes* in Wiau Village, Southeast Minahasa Regency, Indonesia. The research uses an approach proposed by George Terry regarding management functions, such as planning, implementation, and supervision. From the planning function, there have been changes in *BUMDes* revenue target. Previously, the village's *BUMDes* income streams are from savings and loans business and also from equipment rental used for traditional community events. The revenue target was extended from the previous target and even removed from the target. In terms of program implementation during the COVID-19 pandemic, they still rely on the type of business that is still in accordance with the regulations. The financial report shows a decrease in *BUMDes* income due to small scale of activities during COVID-19 pandemic.

On the other hand, there is a village that could adapt and survive during COVID-19 pandemic. A qualitative research by [7] on *BUMDes* in Sukorejo village concluded that during the COVID-19 pandemic, the economic performance in the village did not experience an economic downturn. During the pandemic, Sukorejo village optimizes the potential of its local strengths, both from social capital and economic capital. One of the village's capitals is agriculture. In addition, the Sukorejo village entrepreneurship strategy is able to stimulate *BUMDes* to develop a business strategy in the midst of COVID-19 pandemic. Economic activities in a situation of "physical distancing and lock down" only change the way or method of approaching *BUMDes* business.

From the discussion above, this study aims to analyse the financial performance of *BUMDes* before and during COVID-19 pandemic in Banda Aceh. Banda Aceh is the capital city of Aceh Province and Aceh Province is one of the provinces in Indonesia that receives the largest village funds in Sumatera due to its large number of *BUMDes* [8].

2 Literature review

2.1 Economics uring COVID-19 pandemic

The current COVID-19 pandemic is having a bad impact on the Indonesian economy. The Organization for Economic Co-operation and Development (OECD) reports that the COVID-19 pandemic has resulted in the threat of a major economic crisis marked by the cessation of production activities in many countries, falling levels of public consumption, loss of consumer confidence, and falling stock exchanges that ultimately lead to uncertainty [9]. In the release of Indonesia's economic growth in the first quarter of 2020 issued by the Central Statistics Agency (BPS), it is clear that the COVID-19 Pandemic has caused damage to the Indonesian economy. The growth rate for Indonesia's Gross Domestic Product (GDP) in Q1 2020 was recorded at only 2.97% (year-on-year). This figure is the lowest growth rate since 2001. The Ministry of Finance and Bank Indonesia previously had predicted growth in the range of 4% -5% in Q1 2020. The GDP growth during COVID-19 was much lower than predicted. On the expenditure side, the biggest contraction in GDP was recorded in household consumption expenditure, which worsened by 2.84%, representing the largest contraction in consumption since 1999.

The sharp decline in consumer household spending can be attributed to at least two things. First, it is caused by an increase in the number of unemployed, which directly impacts decreasing income and household consumption expenditure. Second, it is caused by an increase in uncertainty due to the COVID-19 pandemic. This increased uncertainty has led to a shift in consumption to precautionary savings by households whose income has not been greatly affected by this pandemic. However, not all sectors in the economy have heterogeneous impacts. According to [10], some of the sectors severely affected were the transportation sector (1.27% from the previous 7.55%), the construction sector (-2.41%), and the manufacturing industry (-1.47%). Various innovations that are more directed at the defence of people's welfare are needed during the COVID-19 pandemic because, in the end, the people's purchasing power is better able to help the economy move and save business growth, business profitability, and national economic growth [11].

2.2 *BUMDes* sales during COVID-19 pandemic

The economic impact of COVID-19 can emerge a major economic crisis. This is indicated by a decreasing production activities, decreasing people's purchasing power, falling stock market, which leads to uncertainty

[12]. During the COVID-19 pandemic, MSMEs are also affected by this pandemic. The Minister of Cooperatives and MSMEs, Teten Masduki, stated that based on the results of a survey, COVID-19 pandemic has a major impact on the sustainability of MSMEs [13]. 47 percent of MSMEs had to go out of business due to the impact of the COVID-19 pandemic. The sustainability of MSMEs is also affected by this pandemic. MSMEs are experiencing cash flow health problems and weakening on the demand side which suppresses the sustainability of MSMEs.

During the monetary crisis, net income decreased significantly due to weakening people's purchasing power and increased interest costs, so that a firm's profitability decreased significantly. When people's purchasing power decreases, it will certainly impact the company's total sales. When the company's sales decline, its profit will also decrease if the company is unable to minimize operating costs incurred. The results of [14] show that the profitability ratio, measured using returns on assets (ROA), of service companies before the monetary crisis and during the monetary crisis differ significantly, where there is a decrease in the average ratio. [15] also states that there was a decline in public companies' financial performance on the Jakarta Stock Exchange for all sectors during the crisis, especially at the level of ability to generate profits, as indicated by a significant decrease in ROA. In addition, [16] show that the COVID-19 pandemic has a significant negative impact on listed Chinese companies' performance due to a decrease in the value of total revenue, which has impact on reducing ROA. Thus, the first hypothesis can be formulated as follows:

H1: There is a difference between *BUMDes* sales before and during the COVID-19 pandemic

2.3 *BUMDes* ability to pay debts during COVID-19 pandemic

World Bank Managing Director, Mari Elka Pangestu, estimated that Indonesia's economic growth could weaken below 5% in the first quarter of 2020 [17]. The financial performance of the industrial sector will also certainly be affected by the economic crisis [18]. This economic crisis will certainly lead to a decline in industrial sector sales due to decreased purchasing power [19]. If sales in the industrial sector experienced a decrease during the economic crisis due to the COVID-19 pandemic, this would certainly affect the company's financial condition in general, including the value of the company's current assets. The components of current assets that are significantly affected by sales changes are the cash component resulting from cash sales and the value of trade receivables as a result of credit sales. When current assets experience significant changes, this will ultimately affect the value of the current ratio. [20] show differences in financial performance, including the current ratio in the period before and after the monetary crisis. The result of [14] also shows that there was a significant decrease in a firm's liquidity ratios, measured using the current ratio, during the monetary crisis. Therefore, the second hypothesis can be formulated as follows.

H2: There is a difference between *BUMDes* ability to pay debts before and during COVID-19

2.4 *BUMDes* ability to pay salaries during COVID-19 pandemic

In addition to the ability of *BUMDes* to pay operational debt, this study will also look at *BUMDes* ability to pay salaries as an indicator of *BUMDes* liquidity. [21] in his qualitative research concluded that the MSMEs had experienced problems in paying the salaries of their employees during the COVID-19 pandemic. This happened because at that time there was no income due to the absence of demand from consumers. In order to reduce their operational expenses, the MSMEs postponed the payment of employee salaries temporarily for one month.

Similar to the indicator of the ability to pay operational debt, the ability to pay salary expenses also experienced differences between before and during COVID-19. Therefore, the third hypothesis is formulated as follows.

H3: There is a difference between *BUMDes* ability to pay salaries before and during COVID-19 pandemic

3 Research methods

This study analyses the financial performance of *BUMDes* that were active in Banda Aceh before the time of COVID-19 outbreak. The financial performance indicators in this study are sales and liquidity. The data used in this study is primary data, using Likert scale, and in the form of cross section.

The sales variable refers to the research of Rofiqoh [15] and Pakpahan [12] the liquidity variable refers to the research of Istiningrum [14] and Sari [21].

Sales variable was measured using primary data by distributing questionnaires to respondents whereas measured using a Likert scale ratio. Questionnaire questions are a description of the total sales of BUMDes 19 and how consumer demand before the Covid pandemic was for goods and services sold by BUMDes. The liquidity variable also uses secondary data, using the Likert Scale. The liquidity variable in the questionnaire distributed to respondents is to measure the extent to which BUMDes can pay off trade debts and pay off workers' salary debts before the time of the covid 19 pandemics.

The population are the Head, Treasurer, and Secretary of *BUMDes* with total of 84 respondents. Sampling method used in this study is simple random sampling, using the Slovin formula:

$$n = \frac{N}{1+Ne^2} \tag{1}$$

n = sample size
N = population size
e^2 = margin of error (5%)

The population in this study is shown in Table 3

Table 3. Population of *BUMDes* in Banda Aceh.

No	Village	*BUMDes* Name
1	Ateuk Jawo	Atra Gampong
2	Ateuk Deah Tanoh	BUMG Ateuk Deah Tanoh
3	Peunayong	Hareukat Anak Naggroe
4	Laksana	Mandiri
5	Keuramat	Keuramat Jaya
6	Beurawe	Hareukat Bersama
7	Kuta Alam	Meukuta Alam
8	Bandar Baru	Beru Beurukat
9	Lampulo	Jroh Naguna
10	Lambaro Skep	Semangat Baru
11	Cot Lam Keuh	Ceko
12	Gampong Baro	Desa Baro
13	Punge Ujung	Punge Ujung
14	Tibang	Tibang Makmu
15	Deah Raya	Beu Sejahtera
16	Panteriek	Meuligo Intan
17	Blang Cut	Makmue Sejahtera
18	Lampaloh	Sejahtera Bersama
19	Geucu Inem	Cemerlang
20	Lhong Raya	BUMG Lhong Raya
21	Peunyerat	Hudep Saree
22	Lhong Cut	Sejahtera
23	Lamjamee	Lamjamee Mandiri
24	Lampoh Daya	Lampoh Daya
25	Emperom	BUMG Emperom
26	Geuce Menara	Kembang Perdana
27	Doy	Ade Beurata
28	Ie Masen Ulee Kareng	Al Hidayat
	Total Respondents	28 x 3 respondents = 84

Each BUMDes is represented by 3 respondents, thus the total respondents are 84. Based on Slovin equation, minimum number of samples in this study is 45 respondents.

The data analysis uses paired sample t-test difference test. Paired Sample T-Test is a test used to compare the difference between two means of two paired samples with the assumption that the data are normally distributed. Paired samples come from the same subject, each variable is taken in different situations and circumstances. Before carrying out the Paired sample t test, the data normality test must be first carried out, because the paired sample t-test is a parametric test.

The questionnaire in this study is also tested for validity and reliability. Validity shows the accuracy or level of accuracy of the measuring instrument. Reliability test is conducted to measure the consistency of measuring instrument in measuring respondent's consistency in answering question items in questionnaires or research instruments [22].

4 Result

4.1 Respondents

We sent questionnaires to 84 individuals. However, there are only 54 respondents that fully fill the questionnaires. The number of samples still meet the minimum number of samples based on Slovin equation which is 45 samples. The respondents. include the Heads, Secretaries, and Treasurers of *BUMDes* in Banda Aceh city. Respondents come from various backgrounds with details shown in Table 1.

Table 1. Respondents' Job.

No	Job	Frequency	Percentage
1	Civil servant	11	20%
2	Private company employee	7	13%
3	Entrepreneur	12	22%
4	Only work at *BUMDes*	10	19%
5	College student	5	9%
6	Others	9	17%
	Total	54	100%

4.2 Data analysis result

Since the data used in this study is primary data using questionnaire, validity and reliability tests need to be conducted first. The result of the validity test shows loading factor 0.606 and the reliability test shows Cronbach's Alpha 0.711, which means the questionnaire is valid and reliable. The result of descriptive statistics is as shown in Table 4.2.

Table 4.2 Descriptive Statistics.

Test	n	Descriptive Statistics (Mean; Std Deviation)	Paired T-Test		
			T	df	Sig.(2-tailed)
Sales Before COVID-19	54	3.278 ; 1.277	4.069	53	0.000
Sales During COVID-19		2.333 ; 1.043			
Liability Payment Before COVID-19	54	3.000; 0.957	1.160	53	0.254
Liability Payment During COVID-19		2.889; 0.919			
Payroll Payment Before COVID-19	54	3.222; 1.245	0.215	53	0.831
Payroll Payment During COVID-19		3.194; 1.167			

The standard deviation is a reflection of the average deviation of the data from the mean. The standard deviation can describe how much the data varies. If the standard deviation value is greater than the mean value, it means that the mean value is a poor representation of the overall data. However, if the standard deviation value is smaller than the mean value, this indicates that the mean value can be used as a representation of the entire data. Based on Table 4.2, each variable has standard deviation which is smaller than its mean value.

This indicates that the mean value can be used as a representation of the entire data.

The average sale before COVID-19 pandemic was at 3.278 while the average sale during COVID-19 pandemic was at 2.333, which means the sale has decreased during COVID-19 pandemic. The difference in sales before and during COVID-19 pandemic has a significant effect, with a p-value of 0.000. This means that Hypothesis H1 is accepted, there is a significant difference between *BUMDes* sales before and during the COVID-19 pandemic. The results of this study are in line with Wojongan's (2021) research, where due to COVID-19 pandemic, the revenue target was extended from the previous target and even removed from the target. Previously, village income came from several business activities and showed income figures that were in line with expectation, but during the COVID-19 pandemic, it was necessary to diversify products to adapt to the needs of today's society.

The large-scale economic crisis due to COVID-19 has caused many activities to be reduced and even stopped. People's purchasing power has decreased and affecting MSMEs business continuity. MSMEs are the most type of enterprises in Indonesia and have also been proven to be resistant to various economic crises. But, during this COVID-19 pandemic, MSMEs are experiencing cash flow problems and weakening on the demand side which suppresses the sustainability of MSMEs. Product sales activities produced by various types of business activities are closely related to human purchasing power. When people's purchasing power decreases, it will certainly impact the company's total sales. When the company's sales decline, its profit will also decrease if the company is unable to minimize operating costs incurred [12].

BUMDes liquidity indicators are the *BUMDes* ability to pay debts and ability to pay BUMDes employees' salaries. *BUMDes* ability to pay payables shows a different means between before COVID-19 and during COVID-19 pandemic. The average value of BUMDes' ability to pay debts before COVID-19 was 3.000 while during COVID-19 pandemic was 2.889. The difference is not significant with the p-value is > 0.05, which is 0.254.

As the second liquidity variable, *BUMDes* ability to pay salaries also does not show a significant difference between before and during COVID-19 pandemic. The average value of the ability of *BUMDes* to pay salaries before COVID-19 was 3.222, while during Covid, it was 3.194. The average difference was not significant where the p-value was 0.831. Based on the results, H2 and H3 cannot be accepted.

In theory, the financial performance of the industrial sector will certainly be affected by the economic crisis. This economic crisis will certainly cause a decline in sales in the industrial sector due to a decrease in purchasing power. If sales in the industrial sector experience a decline during the economic crisis due to the COVID-19 pandemic, this will certainly affect the company's financial condition in general, including the company's liquidity value. When current assets experience significant changes, this will ultimately affect the value of the liquidity ratio [18, 19]. In line with the results of [21] research, the MSMEs had experienced problems in paying their employees' salaries during the COVID-19 pandemic. This happened because at the time there was no income due to the absence of demand from consumers. In order to reduce their operational burden, the MSMEs postponed the payment of employee salaries temporarily for one month.

Until now, all *BUMDes* in all Indonesian regions have received assistance from the government. The initial capital of *BUMDes* comes from village's investment to BUMDes which is allocated through the Village Revenue and Expenditure Budget (APB Desa). The Village Law which is clarified through the Regulation of the Minister of Villages PDTT Number 21 of 2015 concerning Priority Determination of the Use of Village Funds, states that one of the priorities for the use of village funds is the establishment and development of *BUMDes*. The existence of capital participation or investment from the village government has increased the cash amount received by BUMDes. Therefore, despite the decline in sales, BUMDes in Banda Aceh City are still able to pay their debts and their employees' salaries. Thus, there is no difference in *BUMDes* liquidity before and during COVID-19 pandemic.

5 Conclusion

Currently *BUMDes* face many problems due to COVID-19 pandemic, especially the decreasing people's purchasing power. This decline in purchasing power has affected *BUMDes* in the city of Banda Aceh. There are several *BUMDes* in Banda Aceh that have closed their business activities because their sales have decreased drastically. In theory, when people's purchasing power decreases, it will certainly impact the company's total sales. When the company's sales decline, its profit will also decrease if the company is unable to minimize the operating costs. This analysis is based on the results of the paired sample t-test, where there is a significant difference in the number of sales between before and during COVID-19 pandemic.

The results of the different Paired sample t-test for the liquidity variable did not show a significant difference between before and during COVID-19 pandemic. *BUMDes* in the city of Banda Aceh are still able to pay their employee's salaries and are also still able to pay their business debts. This is most likely because *BUMDes* still receive some fund allocation from the village fund. The initial capital of BUMDes comes from village investment/capital participation which is allocated through the Village Revenue and Expenditure Budget (APB Desa). Hopefully, *BUMDes* will still receive fund allocation from APBDes funds, especially during crisis like COVID-19 pandemic.

References

1. H. Handi, Tujuh Provinsi dengan BUMDes Terbanyak, beritadaerah.co.id (2021)

https://www.beritadaerah.co.id/2021/06/30/tujuh-provinsi-dengan-bumdes-terbanyak/
2. Serambi Indonesia, Saatnya BUMDes Jadi Penggerak Ekonomi Desa (2019) https://aceh.tribunnews.com/2019/01/28/saatnya-bumdes-jadi-penggerak-ekonomi-desa
3. S. R. Baker, N. Bloom, S. J. Davis, S. J. Terry, Covid-Induced Economic Uncertainty, NBER Working Paper No. 26983 (2020)
4. Ministry of Finance of the Republic of Indonesia, Minister of Finance Regulation Number 222/PMK.07/2020 concerning Village Fund Management (2020)
5. M. Shafi, J. Liu, W. Ren, Impact of COVID-19 Pandemic on Micro, Small, and Medium-sized Enterprises Operating in Pakistan, (2020)
6. A. V. E. Wojongan, Pengelolaan Badan Usaha Miliki Desa (BUMDes) di Tengah Pandemi COVID-19 (Studi di Desa Wiau Kecamatan Posumaen Kabupaten Minahasa Tenggara), Jurnal Politico Vol 10 No 4 (2021)
7. V. A. Rahmi, M. Z. Fathoni, H. Ismanto, Pendekatan Bisnis BUMDES Berkemajuan di Kondisi Wabah Pandemi COVID-19, Jurnal Riset Entrepreneurship Vol 3 No 2 (2020)
8. Handi, Tujuh Provinsi dengan BUMDes Terbanyak, beritadaerah.co.id (2021) https://www.beritadaerah.co.id/2021/06/30/tujuh-provinsi-dengan-bumdes-terbanyak/
9. OECD (Organisation for Economic Co-operation and Development), Coronavirus (COVID-19): SME Policy Responses (2020)
10. M. I. Modjo, Memetakan Jalan Penguatan Ekonomi Pasca Pandemi, The Indonesian Journal of Development Planning Vol IV No 2 (2020)
11. T. Ibn-Mohammed et al., A Critical Analysis of the Impacts of COVID-19 on the Global Economy and Ecosystem and Opportunities for Circular Economy Strategies, Resources, Conversation and Recycling Volume 164 (2020)
12. A. K. Pakpahan, COVID-19 dan Bagi Usaha Mikro, Kecil, dan Menengah, Jurnal Ilmiah Hubungan Internasional Edisi Khusus (2020)
13. D. R. Cahyani, 47 Persen UMKM Bangkrut Akibat Pandemi Corona, bisnis.tempo.co (2020) https://bisnis.tempo.co/read/1344540/47-persen-umkm-bangkrut-akibat-pandemi-corona/full&view=ok
14. A. A. Istiningrum, Perbandingan Kinerja Keuangan Perusahaan Jasa yang Terdaftar di BEJ Sebelum dan Selama Krisis Moneter, Jurnal Pendidikan Akuntansi Indonesia Vol. IV No. 1 (2005)
15. I. Rofikoh, Pengaruh Krisis Moneter Terhadap Kinerja Perusahaan Publik di Bursa Efek Jakarta, Journal of Accounting and Investment Vol 2 No 2 (2001)
16. H. Shen et al., The Impact of the COVID-19 Pandemic on Firm Performance, Emerging Markets Finance and Trade Vol 56 Issue 10 (2020)
17. C. I. Burhanuddin, M. N. Abdi, Ancaman Krisis Ekonomi Global dari Dampak Penyebaran Virus Corona (COVID-19). Jurnal AkMen, 17(1) (2020)
18. S. Devi et al., The Impact of COVID-19 Pandemic on the Financial Performance of Firms on the Indonesia Stoch Exchange, Journal of Economics, Business, and Accountancy Ventura Vol. 23 No. 2 (2020)
19. C. Wijayangka, Analisis Kinerja Keuangan Perusahaan Sektor Manufaktur Akibat Krisis Keuangan Amerika. Jurnal Manajemen Indonesia, 14(2) (2014)
20. F. M. Bintang, A. Malikah, Afifudin, Pengaruh Opini Audit Tahun Sebelumnya, Debt Default, Rasio Likuiditas, Rasio Leverage Terhadap Opini Audit Going Concern. E-JRA 8(10) (2019)
21. R. N. Sari, Dampak Pandemi COVID-19 Terhadap UMKM Batik di Kabupaten Sumenep, RISTANSI: Riset Akuntansi, 1(1) (2020)
22. J. F. Hair et al., Multivariate Data Analysis, Prentice-Hall Inc, New Jersey (2010)

 https://doi.org/10.1051/e3sconf/202234003005

Disaster education through the adaptation of the werewolf game and acehnese local wisdom

Ghazi Maulana[1], *Uswatun* Hasanah[2], *Salwa* Hanum[3], *Syifa* Urrahmi[3], *Teuku* Afzal Dzikri[3], and *Rina* Suryani Oktari[4*1]

[1]Faculty of Medicine, Syiah Kuala University, Banda Aceh
[2]Faculty of Dentistry, Syiah Kuala University, Banda Aceh
[3]Faculty of Industrial Engineering, Syiah Kuala University, Banda Aceh
[4]Tsunami & Disaster Mitigation Research Centre (TDMRC) and Department of Family Medicine, Faculty of Medicine, Universitas Syiah Kuala, Banda Aceh

Abstract. Aceh is one of the provinces in Indonesia that is vulnerable to several disasters caused by natural hazards. The earthquake and tsunami in 2004 have given considerable losses to the people of Aceh and impacted people's lives and education process. Some of the impacts include the deaths of tens of thousands of students and teachers and 1,488 schools damaged, which disrupted the education process of 150,000 students. In addition, the people of Aceh are also faced with disasters caused by non-natural hazards, namely is the coronavirus disease (COVID-19) pandemic. Aceh has local wisdom, such as *Rumoh Aceh, patchouli, jamblang*, and *Nandong Smong*, which has been known from ancient times across generations and has been proved by research for their benefits in disaster mitigation. This study had developed a game to introduce disaster mitigation combined with local wisdom in Aceh called Serigala Siaga Bencana (SERGANA), adopted and modified from the werewolf game. The validity test of this game involved 15 students from Dayah Darul Hikmah, who were then fostered as a team of students for disaster preparedness with knowledge, understanding, and practice of preparedness in dealing with disasters. The results showed an increase in the score from pre-test to post-test after playing the game SERGANA.

1 Introduction

The 2004 tsunami caused considerable losses to the people of Aceh. This incident had an impact on people's lives and the implementation of education, including killing tens of thousands of students and teachers. In addition, 1,488 schools were damaged, causing around 150,000 students to be disrupted in their educational process [1].

Currently, the people of Aceh are also faced with disasters caused by non-natural hazards, namely is the coronavirus disease (COVID-19) pandemic. Based on data until September 12, 2021, the number of confirmed cases reached 35,909 people, with 5,879 people treated, 28,337 people recovered, and 1,693 people died in Aceh [2]. The occurrence of earthquakes and tsunamis in the era of the COVID-19 pandemic will undoubtedly have an increasingly lousy impact on the community.

Aceh has local wisdom which based on research by experts, has various benefits in anticipating potential disasters such as *Rumoh Aceh, patchouli, jamblang, and Nandong Smong* from Simeulue, which have been known for generations [3-5]. Local wisdom is very suitable to be adopted and used to preserve and introduce Acehnese culture to the community, especially millennials who have been distracted by globalization [6].

Dayah is one of the educational institutions that has a great influence on the lives of the Acehnese people. This is because the dayah has a role in building the Acehnese community, conveying Islamic da'wah, and educating the santri in Aceh. The uniqueness of the dayah in providing Islamic values is very relevant to the times, without destroying the local culture and language [7].

Dayah Darul Hikmah is one of the educational institutions in Aceh which is about 1,4 kilometers from the coast. In the 2004 tsunami incident, all the facilities and infrastructure owned by *dayah* were destroyed entirely. As a result, all students/*santri* and teachers and *Ustaz-ustazah* who served in this *dayah* also died. Lack of knowledge and preparedness in dealing with disasters is the main factor that causes many casualties. Based on the results of observations and interviews with the leaders of the *dayah*, it is known that disaster education or socialization has never been carried out and there are no disaster mitigation facilities and infrastructure. Therefore, systematic and integrated efforts are needed to carry out mitigation and preparedness in anticipating

* Corresponding author: okta@unsyiah.ac.id

disasters in *Dayah* Darul Hikmah, especially during the COVID-19 pandemic.

This study aims to develop a game-based method to provide education to improve preparedness and disaster risk reduction. This method was adapted from a werewolf game that was modified to contain elements of disasters. The application of this method is expected to increase the attractiveness and understanding of students about disasters. In addition, this method also adds an element of knowledge and the role of several institutions directly involved in disasters. The game can be played both offline and online. Thus, in addition to entertaining, this game can also increase the knowledge of the players about disasters.

2 Methods

This study uses a Research and Development (R&D) approach to produce a disaster education game that adopts the werewolf game and local wisdom in Acehnese. The method developed is in the form of the *Serigala Siaga Bencana* or abbreviated as SERGANA. Game testing was carried out at *Dayah* Darul Hikmah, a boarding school located very close to the beach, approximately 1.4 kilometers. The location of *Dayah Darul Hikmah* is located in the red zone, which indicates the area is vulnerable and at high risk of being hit by a tsunami. As a result, 15 students of *dayah* were fostered as a team of *santri* for disaster preparedness playing this game accompanied by providing education on knowledge, understanding, and practice of preparedness in dealing with disasters. After the students played the game and gave the education, it was continued with the measurement of knowledge in the students through post-test work.

Research and development approaches of Borg and Gall (1989) were used in the process of developing SERGANA [8]. This study involved only three (see Figure 1) of the total 10 steps proposed by Borg and Gall, namely: i) Preliminary study, ii) Model's development, and iii) Model's testing.

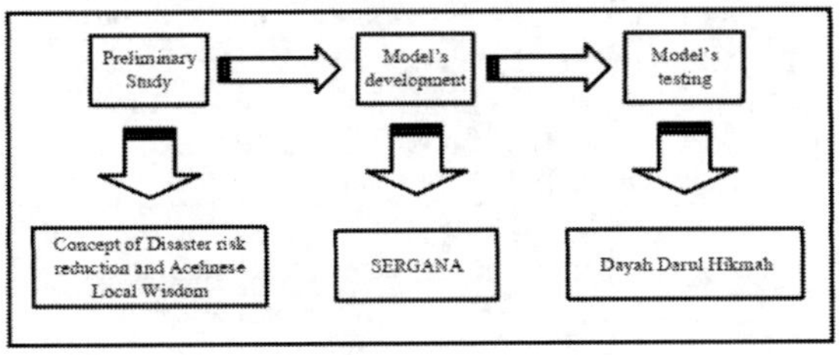

Fig. 1. Research and Development Process of the study. (Source: Modified from Borg and Gall, 1989)

3 Result and discussion

3.1 The concept of SERGANA

This study has developed a game to introduce disaster mitigation combined with local wisdom in Aceh named Wolf Prepared for Disaster or *Serigala Siaga Bencana* (SERGANA). This game was adopted and modified from the werewolf game so that it is suitable for playing both offline and online. In addition to entertaining, this game would also increase the knowledge of the players regarding disasters. Six to 12 people can play this game. Players in this game are divided into five major roles. Role I is the "Moderator" who directs the game and explains what the players are doing. Role II is "BMKG" who knows the role of one of the players when it is his turn. BMKG is Meteorology, Climatology, and Geophysics Agency, a government agency that conveys information and early warnings to related agencies and the public regarding disasters. Role III is "BNPB" that protects one of the players or itself from the threat of disaster. BNPB is the National Agency for Disaster Management, a government agency tasked with coordinating disaster management. Role IV is "Disaster", which consists of tsunamis, earthquakes and COVID-19. All disaster roles know each other but can only launch one type of disaster on other players, which disaster is determined randomly. Role V is "Citizen", who is the community, which will perform the disaster mitigation.

This game introduces several disaster mitigation efforts—*First*, *Nandong Smong*'s Song (tsunami alert). *Nandong Smong* is one of the local wisdoms from Simelue-Aceh in dealing with the tsunami disaster [9,10]. *Nandong Smong* is a traditional art that is very popular among the people of Simeulue, which contains education on preparedness in dealing with the tsunami [11-13]. *Second*, establishing *Rumoh Aceh* (Acehnese traditional house). *Rumoh Aceh* is determined to have a role in disaster mitigation. Previous research using observations on the Rumoh Aceh model mockup concluded that Rumoh Aceh is earthquake-resistant [4]. This is because the structural elements are able to respond to earthquakes in longitudinal and transverse directions, which are determined by a relationship that forms the construction in the legs, body, and head area [14,15]. Third, consuming *jamblang* (*syzygium cumini*) and processed patchouli (health supplements) Based on various studies, it is said that patchouli and jamblang have multiple health benefits. For example, patchouli is often used as a traditional medicine that functions as an antidepressant, decongestant, headache medicine, anti-inflammatory, anti-inflammatory, antiasmatic, antimicrobial and others [5,16]. *Jamblang* also contains antioxidants close to vitamin C with IC50 score of 8.85, which is categorized as very active antioxidants, which increase the body's immunity and functions as anti-viral, anti-bacterial, anti-asthmatic, anti-allergic, infectious drugs, and others [3,17].

In addition, the SERGANA game also has a disaster preparedness mode and a post-disaster mode that residents can choose directly. Preparedness mode consists of: i) preparing shelter in a place that feels strong to avoid building debris such as under tables and others when an earthquake strikes, ii) following directions when a Tsunami occurs, iii) using health protocols and iv) avoiding when the Covid-19 pandemic strikes [18-19]. While the post-disaster modes include i) staying away from buildings, ii) staying away from the coast, and iii) doing self-isolation [18,20]. Residents who choose a package that is suitable for the disaster will be declared safe. Through this game, it is hoped that it can increase students understanding of disaster mitigation science with local wisdom in Aceh. This game is also a fun and entertaining learning tool without feeling overwhelmed. This game is expected to continue to increase its existence and spread to other circles, so that knowledge about disaster mitigation can spread more quickly.

3.2 Result of model development of SERGANA

Several activities have been carried out at the SERGANA development stage, including discussing the method of implementing the SERGANA game with BPBA (Aceh Disaster Management Agency), *Dinas Dayah*, and teachers, conducting a pretest to test the level of knowledge related to disasters, small-scale group trials, post-test, large-scale game testing, and product improvement.

Table 1. Pretest and Post-test score

No	Initial Name	Grade	Pretest (%)	Post-test (%)	Post-test/Pre-test (%)
1	AA	8-1	55	70	27,27
2	AH	8-2	55	85	54,55
3	AN	8-1	45	55	22,22
4	FA	8-2	60	75	25
5	H	7-1	65	80	23,08
6	KPM	8-2	70	75	7,14
7	KK	8-1	60	90	50
8	NM	8-1	65	85	30,77
9	NAA	8.1	65	85	30,77
10	RA	8-2	70	90	28,57
11	RP	8-1	65	85	30,77
12	RT	7-1	40	75	87,5
13	SAY	8.3	50	75	50
14	SM	7-2	30	75	150
15	TH	7-3	35	-	-
Average (%)			55,33	78,57	42,00

SERGANA consists of several modes and two phases, namely the Preparation phase and also the News phase. Every time the News phase arrives, all players discuss to determine which player is expelled for being accused of causing disaster through voting as a form of anticipation. At the same time, each phase of the Preparation of the Calamities designates one by one the players who will be hit by it. The game will end when a disaster hits all citizens or all Disasters have been eliminated from the game. Through the SERGANA game, we hope to add to the public's knowledge about disasters and introduce elements of Aceh's local wisdom.

3.3 Test model of SERGANA

The SERGANA game is played by 6 to 15 students who will be fostered as a disaster preparedness student team with knowledge, understanding, and preparedness practices in dealing with disasters. The game is carried out for 15-20 minutes so that the students: 1) guide the game; 2) plays the role well; 3) know the flow of the game from beginning to end. This game can be played anywhere by students.

The development of the SERGANA game which was adapted from the werewolf game, provides benefits for the students who play it. One of the students who joined the team said:
"This game is fun and exciting because I can directly play the role and know the institutions in disaster as well as local wisdom from Aceh itself. This game can also be played by my friends and I in our spare time".

Fig. 2. Students Playing SERGANA at Dayah Darul Hikmah. (Source: Author Documentation)

In addition, the school is also grateful to the implementers of SERGANA because students have good knowledge of local Aceh wisdom such as *nandong smong, rumoh aceh, jamblang,* and *patchouli* as well as institutions related to disasters so that they can increase their preparedness against earthquake, tsunami, and Covid-19. The SERGANA game consists of several phases, as described in the Table 2.

Table 2. SERGANA game phase.

Phase	Description
Preparation Phase	In this phase, the moderator directs the players by giving each player a role card and giving instructions according to the role they get
News Phase	This phase provides information about what players have been eliminated when the preparation phase is complete along with other additional information.
Discussion Phase	In this phase the players are given time to have discussions to determine which players will be eliminated
Elimination Phase	In this phase the moderator directs players to vote for players to be eliminated
Victory phase	This phase determines the victorious player who wins between the disaster and the citizens

4 Conclusion

This study has developed an attractive, sustainable concept of disaster education and adapts Aceh local wisdom through the SERGANA game which is modified from the werewolf game. The development of the SERGANA game adopted from the werewolf game can be used as an indicator to enhance the role of local wisdom, such as *nandong smong*, *rumoh aceh*, *jamblang* and *patchouli* in the disaster education and pandemic mitigation model. This research is the initial stage of the development of SERGANA using a qualitative approach. SERGANA quantitative evaluation of development effectiveness is a study that will be carried out in the next step. The results showed an increase in the score from pre-test to post-test after playing the game SERGANA, which was 42%.

Acknowledgments. The authors would like to give the sincerest gratitude to *Dayah Darul Hikmah* for continuous support and the warm welcoming. This study is part of the program funded by the Indonesian Ministry of Research, Technology, and Higher Education 2021.

References

1. Syamsidik, R.S. Oktari, A. Nugroho, M. Fahmi, A. Suppasri, K. Munadi, R. Amra, Fifteen years of the 2004 Indian Ocean Tsunami in Aceh-Indonesia: Mitigation, preparedness and challenges for a long-term disaster recovery process. *International Journal of Disaster Risk Reduction*, **54**, 102052 (2021)
2. Pemerintahan Aceh. Info COVID-19. URL: https://covid19.acehprov.go.id. Accessed on 20th May 2021 (2021)
3. S. Nurhalisa, I. Ibrahim, I.A.P. Paerah, Formulasi Kapsul Daun dan Biji Jamblang (Syzigium cumini L.) sebagai Antioksidan Alami dari Desa Pallantikang Kabupaten Maros. Jurnal Medika Hutama, **2**(2), 711-720 (2021)
4. E. Meutia, Pemetaan Sistem Struktur Konstruksi Rumah Tradisional Aceh dalam Merespon Gempa. Jurnal Arsitektur dan Perkotaan "KORIDOR", **8**(1), 62-69 (2017)
5. M. Tahir, A. Muflihunna, Syafrianti, Penentuan Kadar Fenolik Total Ekstrak Etanol Daun Nilam (Pogostemon cablin Benth.) dengan Metode Spektrofotometri UV-VIS. Jurnal Fitofarmaka Indonesia, **4**(1), 215-218 (2017)
6. D.K. Marjanto, B. Setiawan , S.D Kusumah, B. Utama , S. Biantoro, et al. Kearifan Lokal & Lingkungan. 1st ed. Jakarta: PT GADING INTI PRIMA (2013)
7. Kantor Wilayah Kementerian Agama Provinsi Aceh. Sejarah Dayah. (2021)
8. W.R Borg and M. Gall. Education Research: An Introduction (4th edition). New York. Logman publisher. (1989)
9. T. Husin. Kapita Selekta Hukum Adat Aceh Dan Qanun Wali Nanggroe. Revisi. Banda Aceh: Bandar Publishing (2016)
10. A. Kurniasih, J. Marin, R. Setyawan. Belajar dari Simeulue: Memahami Sistem Peringatan Dini Tsunami di Indonesia. J Geosains dan Teknol. **3**(1):21 (2020)
11. I. Maulana, L. Raswita, S. Karlina, M.H.A Mukarram, R.S. Oktari. PASMINA: Arts-based tsunami education with an Islamic theme adopted from Nandong Smong. In *IOP Conference Series: Earth and Environmental Science* (Vol. 708, No. 1, p. 012094). IOP Publishing (2021)
12. R. H. Sari, T. Husin, Syamsidik, Kearifan Lokal Smong Masyarakat Simeuleu dalam Kesiapsiagaan Bencana 12 Tahun Pasca Tsunami. Jurnal Ilmu Kebencanaan (JIKA), **3**(1), 27-32 (2016)
13. A.N. Gadeng, E. Maryani, E. Ningrum. The Simulation of Smong in Geography Learning to Enhance Understanding of Disaster. IOP Conf Ser Earth Environ Sci. Jun 28;**286**(1) (2019)
14. E. Meutia. Identifikasi Model Proporsi Bangunan Arsitektur Tradisional Aceh. In: Prosiding Seminar Kearifan Lokal dan Lingkungan Binaan. Medan; p. 145–55 (2017)
15. E. Meutia. Studi Sistem Struktur Dan Kontruksi Rumah Tradisional Aceh: di Daerah Pesisir Dan Pegunungan. In: Prosiding Seminar Nasional Kota dan Permukiman Lestari. Banda Aceh; (2015)
16. C.W. Li, X.L. Wu, X.N. Zhao, Z.Q. Su, H.M. Chen, X.F. Wang, et al. Anti-inflammatory property of the ethanol extract of the root and rhizome of Pogostemon cablin (Blanco) benth. Sci World J. 2013;(2013)
17. A.N. Sari. Potensi Antioksidan Alami pada Ekstrak Daun Jamblang (Syziguym cumini (L.) Skeels). Eksakta. ;**18**(2):107–12. (2017)
18. Badan Nasional Penanggulangan Bencana. Buku Saku Tanggap Tangkas Tangguh Menghadapi Bencana. Jakarta Timur: Pusat Data Informasi dan Komunikasi Kebencanaan BNPB; (2020)
19 Kepala Subdit KIE Kesehatan, Kepala Seksi Strategi Komunikasi Kesehatan, Kepala Seksi Penyebarluasan Informasi Kesehatan, Damawanti B, Rabina T, Sinansari, et al. Tanya Jawab Seputar Virus Corona. Jakarta: Kementerian Kesehatan RI, Direktorak Jendral, Direktoran Promosi Kesehatan dan Pemberdayaan Masyarakat; (2020)
20 M. Yahya, F. Fathahillah, R. Megavitry. Penerapan Alat Cuci Tangan Guna Pencegahan Penyebaran Virus COVID-19 di Desa Lagaruda. Dedikasi. **22**(2):143–7 (2020)

Conceptual framework of computer application for promoting tsunami evacuation route

Juliana Fisaini[1,2*], *Yunita* Idris[1,2], and *Nazriatun* Nisa[2]

[1]Civil Engineering Department, Universitas Syiah Kuala, Jalan Tgk. Syech Abdurrauf No. 7 Darussalam, Banda Aceh 23111, Indonesia

[2]Tsunami and Disaster Mitigation Research Center (TDMRC), Universitas Syiah Kuala, Jl. Hamzah Fansuri No.8, Kopelma Darussalam, Kota Banda Aceh 23111 Indonesia

Abstract. Disaster resilient is one of development priority in Indonesia. Programs related disaster risk reduction have been considered as important agenda by government and community, especially in cities that hit by earthquake and tsunami. Serious games have been used in many fields, including disaster management. This paper aims to introduce the conceptual framework of a computer application for promoting tsunami evacuation route in Baitussalam, Aceh Besar, Indonesia. The computer application is built based on serious games concept. Activity Theory-based Model of Serious Games (ATMSG), which consist of the gaming activity, the learning activity and the instructional activity in designing process, employed as the methodology to build the framework. The result present ini Serious Game diagram which explain the activities in the game. The activities deliver through scenarios which categorized based on land used and activities of the local community.

1 Introduction

The devastated Indian Ocean tsunami in 2004 has brought the awareness of disaster risk reduction for the government of Indonesia and international community. Efforts on implementing disaster risk management programs have become important agenda. It can be seen from the shift of paradigm in disaster management in the recent decades. Before, disaster management focused on emergency response. Nowadays, disaster management is focusing on increasing the awareness of the communities in every possible aspect [1]. Including designing and creating tsunami evacuation route plan (network). It has significant role for people who live in coastal areas which is tsunami-prone areas and vertical evacuation building (escape building) is not available.

However, 16 years after tsunami hit the city, the local community has lack of information regarding the network of tsunami evacuation route. The new residents have come to occupy areas prone to tsunami. The growth of communities because of new residents make them more vulnerable [2] because many part of the communities that have never experienced the tsunami, the greater number of people settled in high-risk area, the higher the probability of fatalities because of the disaster. The larger population of people living in high-risk areas, the higher the probability of death.

This study conducts in Kecamatan Baitussalam, Aceh Besar, Indonesia. This area is located in the coastal area, situated directly adjacent to Andaman Sea and hit by India Ocean tsunami in 2004. No vertical evacuation building is built in this area until now. The only way to evacuation from tsunami in this area is through tsunami evacuation route to reach the green area.

Besides, this abandoned neighbourhood due to 2004 Indian Ocean tsunami has been replaced by the new residential area. Several residential areas have been developed in this area, some of them are very close to the coast. This phenomenon cause fear that the new comers to this area who are not familiar with their neighbourhood situation. Thus, it is important to promote the tsunami evacuation route to this community. In some level, the knowledge regarding disaster and tsunami evacuation route contribute in evacuating people and reach the safe area in time [3].

In early of 2020, the world facing the pandemic situation making it hard to conduct drills or simulation for tsunami disaster which needed many people to get together at the same place. At the same time, disaster keep on occurring, which makes the community twice more vulnerable. Serious game can be used as an alternative option to conduct drill or simulation to increase the awareness of the community.

This paper aims to introduce the conceptual framework of a computer application for promoting tsunami evacuation route. The computer application is built based on serious games concept, which the main purpose is to deliver implicit messages through entertainment [4]. In this case, promoting tsunami evacuation routes in a coastal area and increasing the awareness of disaster preparedness during playing a game. In addition, this media also supports the efforts to

* Corresponding author: j.fisaini@unsyiah.ac.id

sustain the campaign of disaster risk reduction during Covid-19 pandemic.

2 Literature review

2.1 Serious games on disaster management

Serious games have been known as one of the tools that can be used to introduce and learn something, including in the field of disaster mitigation [4]. Besides, it is also managed as a tool to increase the awareness and knowledge of the community [5]. Several serious games related to disaster have been developed [5] [6][7].

Serious games related to geological hazard are more likely to be found. Meester [8] developed "Disaster in my Backyard" to introduce 5 important topics regarding disaster management which consist of situational awareness, decision-making, coordination and communication, planning, and rescue. Mossoux et al [6] created a board game named Hazagora which aim to raise the awareness about geohazards and disaster risk reduction strategies. The purpose of the game is to develop a resilient community in the face of periodic geological hazards. This game is design based on reality which displays a volcanic island divided into different land cover areas. Multiplayer game 5 to 10 people. Wang et al [9] has developed Defying Disaster, a 2D side scroller serious game designed to teach people how to prepare for and handle an earthquake. Based on test play to 30 persons, the participants believe that learning about disaster topic through serious game is more effective that reading materials. Gampell et al [7] explored disaster video games in two New Zealand museums named Te Papa in Wellington and Quake City in Christchurch, named Quake Safe House (QSH).

In addition, serious games in hydrometeorological hazards has been attracted researchers to involve in the field as well. Meng-Han Tsai [5] created Battle of Flooding Protection, a serious game to educate disaster prevention and Delima [10] built DisCoord, a serious game first implemented in the landslide-and flood-prone and useful as a public pedagogy intervention as they bring different forms of knowledge together in a public space and facilitate co-learning.

Hawthorn et al [4] reviewed and collected data of serious games with main purposes related to Tsunami Risk Communication. Academic database and non-academic source i.e. YouTube and Google Play Store includer. Based on the search, several serious games such as Disaster Master (undated), Stop Disasters, Earth Girl and Earth Girl 2, Disaster Hero (Legacy Interactive) and Tanah – The Tsunami and Earthquake Fighter are listed. This work indicates that the demand of serious games in the field of disaster has amplified with the possibility of increasing people awareness of disaster damages and escalating the frequency of disaster.

2.2 Disaster resilience

Since the devastated tsunami in 2004, the paradigm of disaster management in Indonesia has shifted from focusing on emergency response to preparedness in coping the disaster [11]. This changing is followed by mainstreaming disaster management into every level. So that, disaster management is not only the government responsibility, but also NGO – Non Government Organization, private sector and the communities. This is in accordance with the goals of Sendai Framework for Disaster Risk Reduction (SFDRR), to build community that is resilient to disasters [12].

Based on the UNDRR terminology, resilience is the ability of a system, community or society exposed to hazards to resist, absorb, accommodate, adapt to, transform and recover from the effects of a hazard in a timely and efficient manner, including through the preservation and restoration of its essential basic structures and functions through risk management.

Community disaster resilience (CDR) is the capability of the community to be able to recover in coping the disaster [13]. In the recent year, the definition of community disaster resilience not only concentrating on the ability of the community to recover from the disaster but starting to include readiness, mitigation and disaster adaptability within it [13]. The resilience of community toward disaster will be able to achieve by improving their capacity and knowledge regarding disaster in their neighbourhood. Education and capacity building to enhance a better understanding of disaster hazard and disaster risk at their neighbourhood can be perform in disaster prone community [12].

2.3 Land use, activity center and route choice model

Tsunami evacuation routes plays significant roles in the prone tsunami area with no vertical evacuation building available such as in District of Baitussalam, Aceh Besar, Indonesia. It becomes the only alternative of tsunami mitigation instrument. Planning of tsunami evacuation route covers both route for vehicles and pedestrians [14], however in South East Asia countries, where most road infrastructure are narrow, the effort of tsunami mitigation mostly by running and walking (pedestrian) [15]. L¨ammel et al (2011) also agree that mass evacuation become a complex problem in transport planning [15].

When the disaster come, there is an urgent to evacuate people in large number from vulnerable area (red zone) to the safe area (green zone) to reduce the disaster risk [15]. In evacuating people, especially in the area with no vertical evacuation shelter, road assigned as tsunami evacuation routes become very important. In planning of evacuation route, road network which perpendicular to the coastal line turn itu the first consideration [16] and the shortest route to reach safe area or evacuation shelter. In transport planning, the shortest route is not only determined by the shortest distance [15] but also consider the shortest travel time. This theory is known as Nash Equilibrium Approach [15]. Besides, the closest evacuation places and the shortest routes also able to be calculated using the Dijkstra method [17].

Researches on route planning in several countries with high tsunami potential have been performed. Wargadalam et al stated that lack of information of shelter and tsunami evacuation route relatively increase the number of the victims based on the tsunami incident

https://doi.org/10.1051/e3sconf/202234003006

in Pangandaran, Indonesia. Lack of information of shelter location and evacuation routes lead to traffic congestion during the mitigation process [18]. Takabatake et al [14] simulated the mobility behaviour of local residents and tourists in tsunami evacuation. In this model, the behaviour of local residents and visitors are differentiate based on their familiarity to the city of Kamakura, Japan. It is assumed that local residents have information of shortest route to evacuate and know the location of shelters, while visitors follow the crowds in evacuation process and run to higher land. Result shows that in the bottle neck area wherepassed by visitors during evacuation are congested and the speed of evacuation decrease dramatically.

Scheer et al [19] developed generic framework for tsunami evacuation planning. Several essential items for tsunami evacuation map listed in the framework, i.e location of evacuation shelter, travel time to reach the shelter and safe place or green zone of tsunami hazard. In details, the tsunami evacuation map should exhibit the direction to escape, the availability of roads and/or major paths (evacuation routes) in this direction, information of the number of people that will use these evacuation routes simultaneously, the minimum travel speed (of the evacuees) to count with and the distance towards the next safe destination (including locations of horizontal and vertical shelters) [20].

Álvarez [3] explained about the utilize of pedestrian di Kota Iquique, Chili which potentially obstruct evacuation process if tsunami hit the city. Several reasons of this possibility are physical condition of pedestrian surface, lack of pedestrian maintenance, complex design of pedestrian i.e. stairs and the utilization of pedestrian for restaurant dan parking area. Sahal et al explores the alternative of pedestrian evacuation planning in the coastal area of Rivera River, French Mediterranean coast. Using RouteFinder and SimWalk to identify the routes to the shelter from the beach and estimate travel time by walking to the shelter [21]. Hector R. Lim Jr develop a model of route choice behaviour of evacuees in highly urbanized area based on the case study in Bagong Silangan, Quezon City, Philippines [22].

3 Methodology

3.1 Study area

This study is located in Kecamatan Baitussalam, Aceh Besar, Indonesia, with an area of 20.84 km2, which is divided into 13 (thirteen) villages with population of 19,264 people. This area is mostly located in the coastal area as shown in Fig. 1. The targeted location are in Kajhu Village and Cadek Village.

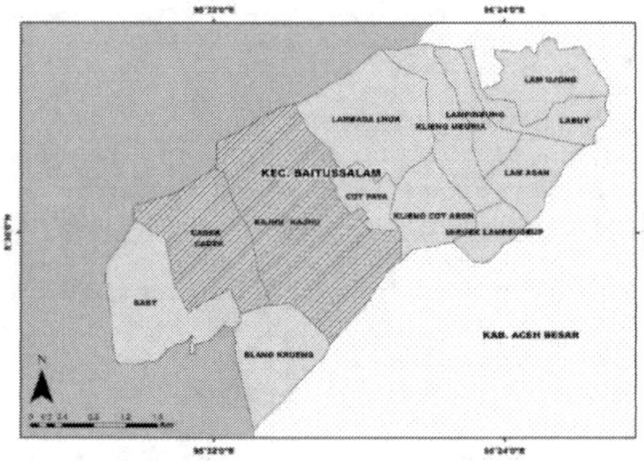

Fig. 1. Study Location

3.2 Framework development process

The framework development of this computer application is adapted from the work of Morschheuser et al [23]. This work reviewed serious game models and proposed a method to develop a serious game. The method consists of seven phases as shown in Fig. *2*.

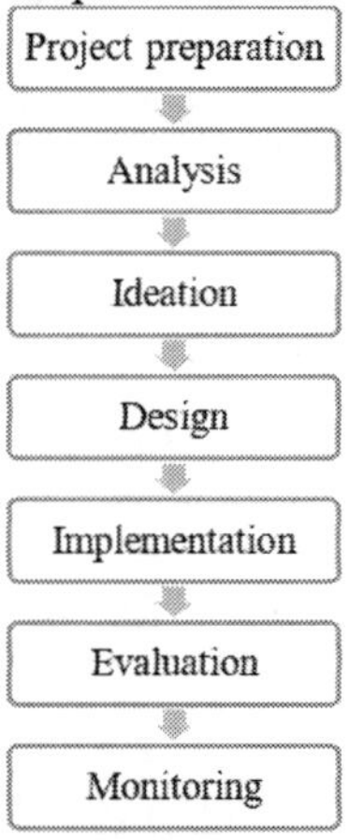

Fig. 2. Design game method (morschheuser).

The seven phases consist of Project preparation, Analysis, Ideation, Design, Implementation, Evaluation and Monitoring. We modified the phases by adding promotion phase. Promotion has significant role in disaster management game since the targeted users are from large community. All people who live and visit tsunami prone are required to know the existing of the game.

Since the purpose of the game is to introduce tsunami evacuation route, we deliver the insight of tsunami evacuation route through several scenarios. Scenarios cover areas based on land used of the area, for example education, commercial, residential and worship center.

The framework will be implemented in the form of computer application especially computer video game with single player. The user, namely targeted group, will experience the realm in the gamec as they visited the location since the setting of the serious game is develop based on the reality of Kecamatan Baitussalam, Aceh Besar, Indonesia. This consideration is chosen to fulfil the general purpose of serious game, to simplify the reality.

3.3 Serious game design

In designing a serious game, the general network of the game plays an essential process of development. The initial network of the game story considers as the groundwork in creating serious game diagram. In order to establish the development process systematically, a serious game design methods/framework is employed. Game design framework used to be a tool to analyse games. The rise of framework development to analyse and develop a game has been began in 1960s. LeBlanc, Hunicke and Zabek introduce MDA (Mechanics-Dynamics-Aesthetics), a framework focuses on the design of games for entertainment [24]. Later on in 2015, [24] presented DPE (Design - Play - Experience) framework. The framework depicts the sub components of serious game design, including the Learning, Storytelling, Gameplay, and User Experience layers.

In the domain of serious games, several methods and frameworks have been introduced to analyze and design the game. Each method has advantages and disadvantages. Recently, ATMSG was establish to assist researchers from multidisciplinary field to analyse and design the serious game [25].

The framework is developed based on Activity Theory-based Model of Serious Games (ATMSG), which consist of the gaming activity, the learning activity and the instructional activity in designing process [25]. This method considers the role of users as well as the serious game play the similar position in delivering lessons embedded in the application, which named instructional activity. Instructional activity consists of intrinsic instruction (task in the serious game) and extrinsic instruction (the instruction that informed by instructor).

The process serious game design using ATMSG employs 4 steps which classified into 2 phases. First phase named Analyze Activities which is the first steps of the process, to identify and describe the activities in the activity network. Second phase named analyze action, consist of step 2 – represent game sequence, step 3 identify action, tool and objectives dan step 4 provide description of the implementations.

In order to build the serious games, several stages are employed. These stages derived from the methods used. In general, it is necessary for serious game designers to determine the objectives of the serious game in the first place. The serious game named SG EvaNami – Promoting Tsunami Evacuation Route aims to provide virtual drills to endorse the existing tsunami evacuation network to the inhabitants around tsunami prone area. The game is developed to create the awareness of tsunami risk to the residents in the targeted area.

3.4 Coordination of stakeholders

In order to achieve the goal, coordination among stakeholders become an important aspect. To acquire the support in collecting data, we cooperate with several stakeholders such as head of district, representatives from Regional Board for Disaster Management in Aceh Besar, Indonesia (BPBD Aceh Besar), and two headmen of the targeted area. In consequence, these partnerships intensify the public involvement in developing the framework. The collaboration among stakeholders also stimulates the public participation.

In preparation of developing the framework, discussion with the headmen and head of district has been initiated. The main topic of the discussion is to measure the tsunami preparedness in Baitussalam area. To quantify the information, a questionnaire survey is implemented. The questionnaire survey will be explained in the next chapter.

Furthermore, to obtain the information regarding existing tsunami evacuation network in Baitussalam area, cooperation with Regional Board for Disaster Management in Aceh Besar, Indonesia (BPBD Aceh Besar) is established. The representatives of local government agency who responsible for disaster management hand over the existing tsunami evacuation routes in the area of Regency Aceh Besar. It is believed that the existing tsunami evacuation routes in Aceh mostly not relevant to the current condition. Since these data is employed as the foundation of the design game, to guaranty the accuracy of the data, we organized a tsunami evacuation routes survey.

3.5 Tsunami evacuation route survey

The tsunami evacuation routes survey is divided into two categories: survey of the properness of the existed tsunami evacuation network and survey of the alternative tsunami evacuation route for serious games scenario. The survey of the existence of existed evacuation route aims to ensure that the evacuation routes listed on the map are still available and usable. It is essential to assess the availability and quality of the available network to minimize the fear of existing network does not meet the requirements of tsunami evacuation routes characteristics or the existed tsunami evacuation routes remain in appropriate condition.

In addition to utilizing the existing tsunami evacuation routes, this research also maps several alternative routes that can be used in the event of tsunami disaster. The alternative tsunami evacuation route was chosen based on an imaginary line built from the tsunami wave flow boundary that occurred in 2004. So that the selected meeting point is not in tsunami prone zone.

This survey conducted to ensure that the specified meeting points are located in tsunami green zone. The alternative routes and selected meeting points embed in the serious game EvaNami through several scenarios.

The mapping of the alternative tsunami evacuation route conducted using android-based application, Avenza Maps. This application used to map the paths traversed by surveyors. This application recorded the distance and travel time to the designed meeting points. The survey is conducted by 6 surveyors within 3 days.

https://doi.org/10.1051/e3sconf/202234003006

3.6 Tsunami preparedness questionnaire survey

To comprehend the preparedness of the community toward tsunami, the questionnaires were delivered. In the first phase, the questionnaires distribute through an online survey platform named Google Form. We shared the survey through Facebook and WhatsApp. Due to the limited access of targeted community to social media, 4 surveyors were sent to the targeted area for the second phase of questionnaire survey. 30 people were interviewed regarding their readiness of facing tsunami in the future.

The questionnaires were adapted from Individual and Household Preparedness Framework in Anticipating Natural Disasters, an assessment developed by Indonesian Institute of Sciences (LIPI) in collaboration with UNESCO/ISDR in 2006 [26] [27]. This framework aims to measure the level of knowledge/capacity, vulnerability and readiness of community in dealing with earthquake and tsunami. The questions comprise of 3 groups which consist of the characteristics of participants including experience of the 2004 Indian Ocean Tsunami in the area and duration of residency in the area, knowledge related to disaster especially tsunami and household disaster preparedness plan.

3.7 Evaluation process

The process of game evaluation is conducted through a Focus Group Discussion (FGD) involved stakeholders and community representatives. The evaluation is done by conducting pre-test, test play, and post-test. Pre-test and post-test are questionnaire-based test.

4 Result and discussion

The designed computer application aims to promote tsunami evacuation route. The application is built based on the concept of serious game and route choice model for tsunami evacuation.

4.1 Tsunami evacuation route in Baitussalam Aceh Besar

Regional Board for Disaster Management in Aceh Besar, Indonesia revealed that one of our targeted area has been assigned as one of Desa Tangguh Bencana (Resilient Village). Desa Tangguh Bencana (Destana) is a national government program through Indonesian National Board for Disaster Management (BNPB). The program is built the capacity of the village to be able to comprehend the disaster hazard in their area and be able to organize the human resources in the community to decrease the vulnerability as well as increase the capacity to reduce the disaster risk. Thus, our project is believed to have synergy to the existed government program in the area since we have similar goal: to create resilient community in disaster prone area.

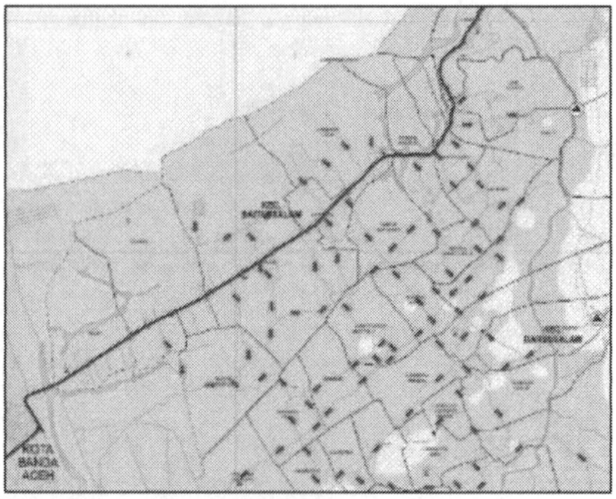

Fig. 3. Tsunami Evacuation Route.
Source: BPBD Aceh Besar, Indonesia

Furthermore, to support the SG EvaNami framework development, the Regional Board for Disaster Management in Aceh Besar, Indonesia (BPBD Aceh Besar) provided the existed tsunami evacuation map of Kecamatan Baitussalam, Aceh Besar, which shown in Figure 1. Figure 1 display the tsunami evacuation route existed in Kecamatan Baitussalam, Aceh Besar Indonesia. Tsunami evacuation route signs appears to direct people to avoid the coastal area. The evacuation should be done by running vertically to the coastal line. However, the information of tsunami evacuation shelter (meeting point) and the shortest routes are not available. There also no information regarding the green zone of tsunami. This map is not available in the public area of Baitussalam, Aceh Besar.

4.2 Game objective

Every serious game has specific purpose to be achieved. In general, serious game in the field of disaster aims to improve the awareness of disaster risk and resilience. Several developed serious games have very specific goals, for example DisCoord aims to educate the community about the landslide and flood-prone [10], Defying Disaster - a 2D side scroller serious game designed to teach people how to prepare for and handle an earthquake [24] and some others mentioned in Introduction. It is also applied to SG EvaNami, which framework is introduce in this paper, the SG EvaNami is designed with the purpose to promote tsunami evacuation routes in the coastal area. In addition, the framework of SG EvaNami also linked to website EvaNami, where the information of tsunami knowledge is available. Besides, the website also provides quiz to test the preparedness of community in facing tsunami in the future. The website can be accessed for free by the community. This digital media also benefit to forward the information regarding disaster to the next generation. In addition, in the Covid19 pandemic era, which physical distance is promoting and the crowd is prohibited [28], it is can be used as a virtual drill for tsunami evacuation.

4.3 Targeted users

Development of serious games with specific purpose mostly associate with stakeholders in the field. SG EvaNami which focus on promoting tsunami evacuation routes have targeted users whom have been determined. At the first phase, the serious game intend to be introduce to stakeholders in related sector, such as local government, BPBD Aceh Besar, researchers, academicians, and headmen of the targeted area. In the next phase, this drill is expected to be used by the community from various ages including students and elderly with computer dan smartphone literacy.

Children under 12 years old who are able to read and operate computer or and smartphone will be able to play SG EvaNami under supervision of parents. There is a scenario in the SG EvaNami which placed the school as the base area. This scenario represents the education center/land used as disaster prone area Teenagers older than 12 years old, adult both men and women, as well as elderly are assumed to have activities in the residential area, commercial area and worship center. Therefore, the proposed scenarios in the framework consider these activity center as the basis area or starting points in the scenarios.

4.4 Conceptual framework of SG EvaNami

The conceptual framework of SG EvaNami – Promoting Tsunami Evacuation Route is presented in Fig. 4. The framework displays the development process of the serious games divided into six stages. The first stage is Ideation. In this stage, we explore general plan of the serious game and identify data need for the game development.

The second step, context and user analysis. Context analysis begin by conducting preliminary survey. The preliminary survey consists of community capacity toward disaster survey which aim to identify the level of community resilience of targeted community (game user). Furthermore, in this stage we also perform validation of imaginary line of tsunami inundation boundary and data collection of existing tsunami evacuation route and survey alternative tsunami route. The third stage is the design process. We employ ATMSG method to design the serious game. Game development is implemented by using Unreal Engine.

The fifth stage is evaluation. The evaluation process is done through a focus group discussion involved stakeholders and community representatives as targeted user. The activity of evaluation cover pre-test, test play, and post-test. Questionnaires regarding the knowledge of tsunami evacuation route distributes to the FGD participants before and after playing the game to test the impacts of playing the game for players and user experience of the game. Participants are requested to present their opinion in order to develop better version of the game. The last stage is Promotion. This activity intends to socialize the media to the public. The promotion is conducted by sharing the information through various platform.

4.5 Proposed evacuation route employed in SG EvaNami

In the design process, scenarios in the game is determined by considering the center of activity of the community. In general, people have do go to school to get education. Women in the area is dominated by stay at home mother, which brought us to the consideration mostly women will go to shopping area (commercial area). Furthermore, since the targeted area is in Aceh where the Islamic sharia law is implemented, we add mouse or worship center as one of many scenarios. Consequently, five scenarios are proposed, which 5 location Residential Area in Desa Cadek, SD Mon Singet, Pasar Kajhu, Pasar Ikan Lam Ujung, Perumahan di Lambada Lhok and Mesjid Labuy as the basis area in the scenarios. These areas are chosen due to the location close to the coastal area and hit by tsunami in 2004.

Furthermore, the shelter or meeting point for evacuation in each scenario is determine. The location of meeting point is decided base on the tsunami evacuation shelter criteria or located in the green zone from tsunami. Grounded on the two locations, we proposed several tsunami evacuation routes to connect the red zone to the green zone.

Fig. 5 shows the proposed evacuation routes with several alternatives. The information of road network acquired from Google Map. Surveyors did site visit (survey) to ensure the quality of the road meet the characteristics of tsunami evacuation routes, at least for pedestrians. By using the concept of route choice model which the best route is considered by the shortest route [15]. Besides, several alternatives provided which have possibility that during the congested hours, the alternative routes serve shorter time to reach the evacuation shelter. This also derive from the concept of route choice model [22].

Besides the shortest path and shortest travel time to reach the evacuation shelter, another consideration take into account in planning the evacuation route is the route should be perpendicular to coastal lines [16] and the coordinate of the shelter area have to be in the green zone of tsunami (based one tsunami flow in this location in 2004).

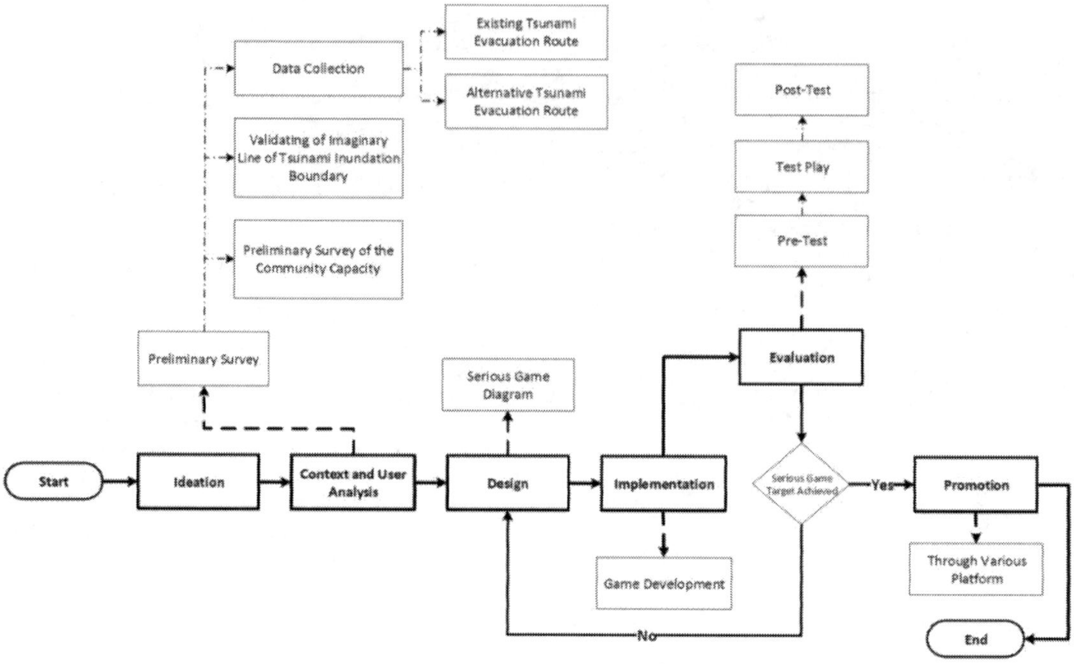

Fig. 4. Conceptual framework of SG EvaNami.

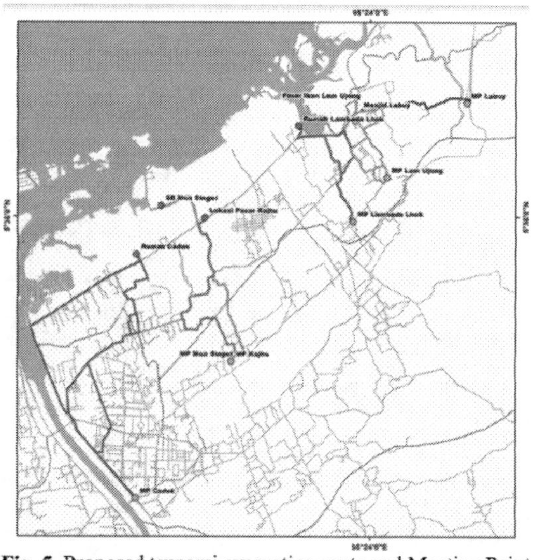

Fig. 5. Proposed tsunami evacuation route and Meeting Point in SG EvaNami.

5 Conclusion

To conclude, this paper presents the conceptual framework of a computer application. The computer application includes in a computer video game aims to promote tsunami evacuation route. The framework is develop based on existed methods of developing a serious game with some modification to achieve the main goal of game. Besides, the concept of land used, route choice model, and tsunami evacuation route requirements employed. It is expected that the serious game will be able to be a tool in promoting tsunami evacuation route to both for recent community and the next generation. This computer application preserves as a drill in disaster mitigation during Covid19 pandemic, where large scale social restriction is implemented.

The conceptual framework describes the whole process of developing the game. It consists of six stages, from ideation to socialization. In developing the serious game, stakeholders are involved. The community capacity in tsunami preparedness are analyzed through questionnaires survey. In designing scenarios, tsunami evacuation route survey and alternative road are explored, and the evaluation process involved stakeholders as well as targeted users to get the feedback for the game. It is believed that this computer application advances the benefit of serious game in disaster management field.

References

1. B. Ryan, K. A. Johnston, M. Taylor, and R. McAndrew, Int. J. Disaster Risk Reduct. **49**, 101655 (2020)
2. B. Adu-Gyamfi and R. Shaw, Int. J. Environ. Res. Public Health **18**, (2021)
3. G. Álvarez, M. Quiroz, J. León, and R. Cienfuegos, Nat. Hazards Earth Syst. Sci. **18**, 2027 (2018)
4. S. Hawthorn, R. Jesus, and M. A. Baptista, Int. J. Serious Games **8**, 21 (2021)

5. M. H. Tsai, Y. L. Chang, J. S. Shiau, and S. M. Wang, Int. J. Disaster Risk Reduct. **43**, 101393 (2020)
6. S. Mossoux, A. Delcamp, S. Poppe, C. Michellier, F. Canters, and M. Kervyn, Nat. Hazards Earth Syst. Sci. **16**, 135 (2016)
7. A. V. Gampell, J. C. Gaillard, M. Parsons, and L. Le Dé, Int. J. Disaster Risk Reduct. **49**, 1 (2020)
8. K. Meesters, L. Olthof, and B. Van de Walle, Proc. Eur. Conf. Games-Based Learn. **2**, 714 (2014)
9. Y. Chen, T. Wang, X. Wu, and Y. Xie, 109 (2015)
10. G. Delima, L. Jacobs, M. Loopmans, M. Ekyaligonza, C. Kabaseke, M. Kervyn, and K. Mertens, Int. J. Disaster Risk Reduct. **60**, 102303 (2021)
11. H. Fahlevi, M. Indriani, and R. S. Oktari, Jamba J. Disaster Risk Stud. **11**, 1 (2019)
12. A. A. Shah, R. Shaw, J. Ye, M. Abid, S. M. Amir, A. K. M. Kanak Pervez, and S. Naz, Int. J. Disaster Risk Reduct. **34**, 165 (2019)
13. R. S. Cox and M. Hamlen, (2015)
14. T. Takabatake, T. Shibayama, M. Esteban, H. Ishii, and G. Hamano, Int. J. Disaster Risk Reduct. **23**, 1 (2017)
15. G. Lämmel, 1 (2011)
16. F. Kitamura, D. Inazu, T. Ikeya, and A. Okayasu, Int. J. Disaster Risk Reduct. **45**, 101519 (2020)
17. R. D. Prayogo and S. A. Karimah, 2020 8th Int. Conf. Inf. Commun. Technol. ICoICT 2020 0 (2020)
18. R. Wargadalam, H. Nakanishi, Y. Vidyattama, J. Black, and Y. Suenaga, IOP Conf. Ser. Earth Environ. Sci. **630**, (2021)
19. S. Scheer, A. Gardi, R. Guillande, G. Eftichidis, V. Varela, B. de Vanssay, and L. Colbeau-Justin, *Handbook of Tsunami Evacuation Planning* (2011)
20. S. J. Scheer, V. Varela, and G. Eftychidis, Phys. Chem. Earth **49**, 79 (2012)
21. A. Sahal, F. Leone, and M. Péroche, Nat. Hazards Earth Syst. Sci. **13**, 1735 (2013)
22. H. R. Lim, M. B. B. Lim, and M. Piantanakulchai, Asia Pacific Manag. Rev. **24**, 98 (2019)
23. B. Morschheuser, K. Werder, J. Hamari, and J. Abe, Proc. Annu. Hawaii Int. Conf. Syst. Sci. **2017-Janua**, 1298 (2017)
24. Y. Chen, T. Wang, X. Wu, and Y. Xie, 109 (2015)
25. M. B. Carvalho, F. Bellotti, R. Berta, A. De Gloria, C. I. Sedano, J. B. Hauge, J. Hu, and M. Rauterberg, Comput. Educ. **87**, 166 (2015)
26. T. Triyono, W. Widayatun, and D. Hidayati, (2011)
27. LIPI, UNESCO, and UN-ISDR, *Kajian Kesiapsiagaan Masyarakat Dalam Mengantisipasi Bencana Gempa Bumi & Tsunami* (2006)
28. J. Brinkman and K. Mangum, (2020)

Research in disaster and social capital: Mapping out the themes and findings

Ratna Mulyany[1 1*], *Zata* Hulwani Zuhri [2], and *Heru* Fahlevi[3]

[1,2,3] Department of Accounting, Universitas Syiah Kuala, Jl. Tgk. Syech Abdul Rauf, Banda Aceh, 23111, Indonesia

Abstract. Research in the field of disaster has been growingly and widely conducted primarily in response to the local and global disaster phenomena which has also witnessed an increasing trend over the time. Meanwhile, social capital has been recognized as a key element in disasters' recovery and resilience. This study employs qualitative standpoint to identify the findings and themes of research linking the issues of disaster and social capital. In selecting and critically appraising all the relevant studies in the topic of disaster and social capital, systematic review method was adopted. Using the Scopus database to identify relevant articles, this study reviewed 39 articles published in (accounting and business) journals indexed in Scopus from 1990 to 2020. The finding identifies several research themes within the interaction of disaster and social capital. Five sub-clusters were identified consisting of research focused on bridging social capital, bonding social capital, the factors, the role, social participation, and obstacles. Findings from each sub-cluster were mapped out and it generally indicates somehow synchronous patterns of findings from each cluster indicating the various essential roles of social capital were highlighted for the disaster resilience and recovery process.

1 Background

The occurrences of disasters, such as earthquakes, hurricanes, floods and droughts, and also Tsunami have the power not only to upend the lives of local residents but also bring the direct and indirect impacts through all of the life facets, as well as economy. This catastrophe has been a wide concern and critical issue to field study, not only science, economics, but also through all of the field of studies. Moreover, the number of studies regarding the disaster keep growing, and most of that literature often link to the social capital which commonly play a role in disaster relief [1-3]. Social capital provides economic and social benefits both to the individuals involved and to other parties such as growing solidarity, easy access to information, diffusion of information, and togetherness in completing social tasks [4]. Therefore, social capital according to [5-7]. has been identified as a key factor in disaster recovery and resilience.

Research in the field of disaster has been growingly and widely conducted primarily in response to the local and global disaster phenomena which has also witnessed an increasing trend over the time [8]. On the other hand, social capital has been recognized as a key element in recovery and resilience to disasters [2]. There has also been extensive research on social capital carried out by various authors from various fields of science dominated mainly by the fields of sociology, economics, organization and politics [9]. The pandemic does not stop some parties from helping each other and this action is a form of social capital. The successful social action in delivering help for disaster relief happened for example in the Philippines during the pandemic. In November 2020, Typhoon Ulyssess flooded communities and damaged the local property in Luzon, Philippine. With the threat of Covid-19 pandemic the Filipinos succeeded in raising donation and relief initiatives through social media [10]. This is also done in Indonesia. As Indonesia has experienced more than hundreds of disasters during the pandemic, the social action was apparent among general public and also corporate. Such social action even involves large scale corporations operating in the country [11].

Intertwining these two major constructs of disaster and social capital, a number of research have looked into the different interaction between disaster and social capital and it is therefore seen interesting to map out the progress of literature discussing the interaction between the two. Motivated by this notion, this study aims to identify the themes and findings of research linking the issues of disaster and social capital. It seeks to classify the clusters of research linking disaster and social capital. It then outlines important findings for each of clusters identified. The rest of this paper is structured as follow. Next section briefly describes the research method of this study followed by result. The last section concludes.

2 Method

Systematic literature review method was used in selecting and critically appraising all the relevant studies

* Corresponding Author: ummuhaziq@unsyiah.ac.id

in the topic of disaster and social capital. This method enables the discovery of the patterns of prior findings, comprehend the depth of the existing knowledge in the filed and identify gaps for future investigations [43]. This kind of research method revolves around identifying and fundamentally assessing the relevant research which the objective is to identify all empirical evidence that fits predetermined inclusion criteria to answer a particular research question or hypothesis [12]. Essentially the foundation of all types of research lies in the literature reviews yet bias has to be minimized in reviewing articles by using the systematic methods, hence the reliable findings will be provided from which conclusions can be drawn and decisions made [13]. This study used the Scopus database to collect articles on the intertwine between disaster and social capital. Using keywords search on title and abstracts ('Disaster' and 'Social Capital) and year filter from 1990 to 2020. A further filtering process was performed by classifying and coding papers into three categories which are 0 for irrelevant articles, The systematic review procedure which was performed resulted in 724 articles being collected. From 724 articles, 561 articles were removed as the article topics are outside our research objective and 39 articles were considered as directly relevant to the research objective.

3 Result

An in-depth exploration into the research focus in this field leads to the identification of several research sub-clusters and themes within the interaction between disaster and social capital. Despite the various dimensions that may be explored from research linking the two notions of disaster and social capital, the focus of this study is specifically to look into clusters of the studies and subsequently map out the findings from the literature. As depicted in Table 1, five sub clusters were identified consisting of bridging social capital, bonding social capital, the factors, the role, social participation, and obstacles. On the first two sub clusters, while bridging social capital is concerned about the connections between individuals who are dissimilar with respect to socioeconomic and other characteristics, while bonding social capital is with regards to the connections within a group or community characterised by high levels of similarity in demographic characteristics, attitudes, and available information and resources. Bridging and bonding social capital have been noted to play role differently within disaster and they have been two of the most focused research areas in the relationship of disaster and social capital.

Interestingly quite synchronous findings can be identified within each sub-cluster. For example, studies exploring the bridging social capital found that bridging and linking social capital has resulted into many advantages to the process of disaster recovery among others it increased the aid amount [16], [19], it also significantly minimize the impact of life stress among the survivors [20], and also creates longer term survival and wider neighborhood and community revitalization [21]. Likewise with bonding social capital, studies in this sub-cluster found the range benefits of the establishment of bonding social capital in which it protected the survivors of a disaster against functional capacity decline [41], played a crucial role throughout recovery phases and as immediate support [21], [22], provided the reliance for disaster-prone and susceptible to natural hazards communities [23]. Other studies for instance examined important elements related to bonding social capital such as risk perceptions and self-efficacy [24] as well as culture which is deemed relevant in the bonding social capital discussion [23].

Other stream of research connecting the disaster and the concept of social capital were concerned about the various factors which may contribute to the successful relationship between disaster resilience and social capital. For example studies indicating the importance of place and geographical location for building up social capital and community disaster resilience[26],[27]. Other studies within this sub-cluster found livelihood recovery factors; formal network in the community, leadership and trusteeship of community-based organizations, capital endowments, social cooperation networks, transportation convenience, and skills acquired from education and rural–urban migration [28], [29].

Another major theme identified from the review is research investigating the various role of social capital in the recovery of disaster. It was found that social capital played an instrumental role in personal, household, and community affected by disaster to recover more easily and quickly [19], [35-37], strengthened the collaborations and partnerships among disaster organizations, mobilizes the public as disaster volunteers, strengthens community resilience and deepens family ties [38], built self-reliance, enhances coping and quickens recovery from disaster [38], and reduced barriers to collective action speeding up recovery and connecting to aid organizations also protecting mental health among survivors with severe disaster damage [39], [40].

The focus on the dimension of social capital such as social participation was also one of the area of research in that social participation was found to protected the survivors of a disaster against functional capacity decline [41] and to relocate functional ability [41]. Additionally there has been also emphasis on the challenges in the role played by social capital in the context of disaster for instance the difficulties in internal and external communication (language barriers and generational divides), and divided memberships of religious institutions can limit/erode the linking and bridging social capital [42].

Table 1. Themes and Finding.

Cluster Finding	Sub-Clusters	Main Findings
Social Capital	Bridging Social Capital	- Individuals, communities, government and non-government institutions are worked to build awareness, strengthen, and expand social networks [14-18] - Having a large network resulted in significantly increased the amount of aid received [16], [19]. - Bridging social capital is more important in prevention and intervention efforts [20]. - With bridging, social capital had a significant buffering effect against reducing the impact of life stress within the survivors [20]. - Bridging and linking social capital offered pathways to longer term survival and wider neighbourhood and community revitalization [21].
	Bonding Social Capital	- Bonding social capital played a crucial role throughout recovery phases and as immediate support [21], [22] - A disaster-prone and susceptible to natural hazards communities relied on their social capital (bonding social capital) [23]. - Both risk perceptions and self-efficacy as bonding social capital have positive links to social capital [24]. - The effect of the amount of social capital available through bonding networks embedded in pre-disaster networks and the resources affected [25]. - Culture is relevant in the bonding social capital discussion [23].
	The Factors	- The importance of place and geographical location are the ground upon which social capital and community disaster resilience are built [26],[27]. - Livelihood recovery factors; formal network in the community, leadership and trusteeship of community-based organizations, capital endowments, social cooperation networks, transportation convenience, and skills acquired from education and rural–urban migration [28], [29]. - Self-reliance within community factors; shared identity as a close-knit, family-oriented community of hard workers [30]. - The three networks (bonding, bridging, and linking) demonstrate different structural signatures composed of differing forms of social capital that arise following a natural disaster [31]. - Disaster significantly increased social capital investment rate [32]. - Local planning is needed to have greater recognition and anticipation of the potential social capital [33]. - Mutual trust and mutual help generate good bonding relationship community [34].
	The Role	- Social capital played an instrumental role in personal, household, and community affected by disaster to recover more easily and quickly [19], [35-37]. - Social capital builds collaborations and partnerships among disaster organizations, mobilizes the public as disaster volunteers, strengthens community resilience and deepens family ties [38]. - Social capital builds self-reliance, enhances coping and quickens recovery from disaster [38]. - High levels of social capital reduced barriers to collective action speeding up recovery and connecting to aid organizations also protecting mental health among survivors with severe disaster damage [39], [40].
	Social Participation	- Social participation protected the survivors of a natural disaster against functional capacity decline [41]. - Social participation could relocate functional ability [41].
	Obstacles	- The difficulties in internal and external communication (language barriers and generational divides), and divided memberships of religious institutions can limit/erode the linking and bridging social capital [42].

4 Conclusion

Research investigating the intertwined interaction between disaster and social capital has flourished especially taking the settings from the geographical areas which are prone to disaster. There have been several themes identified within the topic of social capital and disaster indicating the diverse emphasis being placed in the effort to connect the two dimensions of disaster and social capital. Despite the various dimensions that may be explored from research linking the two notions of disaster and social capital, the focus of this study is specifically to look into clusters of the studies and subsequently map out the findings from the literature. The finding identifies several research themes within the interaction of disaster and social capital. Five sub-clusters were identified consisting of research focused on bridging social capital, bonding social capital, the factors, the role, social participation, and obstacles. Findings from each sub-cluster were mapped out and it generally indicates somehow synchronous patterns of findings from each cluster indicating the various essential role of social capital were highlighted for the disaster resilience and recovery process.

Acknowledgments. This paper is part of a research project funded under the Hibah Penelitian Lektor Universitas Syiah Kuala 2021. The authors are grateful to the grant given by the Institute for Research and Community Service or *Lembaga Penelitian dan Pengabdian kepada Masyarakat (LPPM)* of Universitas Syiah Kuala, Aceh, Indonesia.

References

1. R. Rustinsyah, R. A. Prasetyo, M. Adib, Int. J. Dis. Risk. Red. **52**. (2021)
2. S. Panday, S. Cushton, J. Karki, Balen, A. Barnes, Int. J. Dis. Risk. Red. **55** (2021)
3. D. Hidayati, MATEC Web Conf. **229**. (2018)
4. P. S. Adler, S. Kwon, Social capital: The good, the bad, and the ugly. In E. Lesser (Ed.), Knowledge and social capital: Foundations and applications. Boston, MA: Butterworth- Heineman. 89-115. (2000)
5. F. Albrecht, Dis. **42**. (2018)
6. D. P. Aldrich, M. Meyer, Amer. Behv. Sci. **59**. (2015)
7. R.L. Hawkins, K. Maurer, Bri. J. Soc. Work. **40**. (2010)
8. H. Fahlevi, M. Indriani, R. Mulyany, Nadirsyah. *What is the Role of Accounting in Disaster Recovery and Relief? A Literature Review*, in IOP Conference Series: Earth and Environmental Science. 10-12 October 2018, Banda Aceh, Indonesia, **273** (2019).
9. J. Wallis, P. Killerby, B. Dollery, Int. J. Soc. Eco. **31** (2004)
10. RAPPLER.COM. Filipino bayanihan spirit brings relief, attention to Typhoon Ulysses victims. Retrieved on 19 August 2021 from RAPPLER. COM: https://www.rappler.com/moveph/filipinos-bayanihan-brings-disaster-relief-attention -typhoon-ulysses-victims. (2020, November 17).
11. T. J. Post,. Natural disaster survivors celebrate togetherness during Ramadan. Retrieved from The Jakarta Post: https://www.thejakartapost.com/adv/2021/04/15/natural-disaster-survivors-celebrate-togetherness-during-ramadan-thanks-to-abc.html. (2021, April 15)
12. H. Snyder. J. Bus. Res. **104** (2019)
13. D. Maher, A. Liberati, J. Tetzlaff, D.G. Altman. An. Int. Med. **151** (2009)
14. J. Mimaki, Y. Taleuchi, R. Shaw. J. Coas. Conserv. **13**, 207-2015 (2009)
15. N. Pitas, C. Ehmer, Amer. J. Heal. Prom. **34**, 942 (2020)
16. C. Zhang, Y. Fang. Eco. Indic. **116** (2020)
17. W. T. Story, H. Tura, J. Rubin, B. Engidawork, A. Ahmed, F. Jundi, T. Iddosa, T. H. Abrha. Soc. Sci. Med. **257** (2020)
18. A. Tammar, S. S. Abosuliman, K. R. Rahaman. Sust. **12** (2020)
19. J. Wei, Y. Han. Sust. **10**, 1 (2018)
20. H. Rafiey, F. Alipour, R. LeBeau, Y. Salimi, S. Ahmadi. Psy. Trauma: Theo. Res. Prac. Pol. (2019)
21. R. B. Bhandari. Dis. Prev. Manag. **23**, 314 (2014)
22. K. Nguten-Trung, H. Forbes-Mewett, D. Arunachalam. Int. J. Dis. Risk. Red. **46** (2020)
23. S. C. Bernados Jr, L. A. Ocampo, E. A. Pilapil, N. F. Zamora. Int. J. Soc. Eco. Sust. Dev. **11**, 26 (2020)
24. P. Hudson, L. Hadedoorn, P. Bubeck. Int. J. Dis. Risk. Sci. **11**, 251 (2020)
25. A. Mukherji. J. Civ. Soc. **10**, 119 (2014)
26. R. S. Cox, L. E. Perry. Ame. J. Com. Psy. **48**, 395 (2011)
27. S. Naithani, A. K. Saha. Dis. Pre. Manag. Int. J. **30**, 179 (2020)
28. Y. Minamoto. Dis. Pre. Manag. **19**. 548- (2010)
29. W. Liu, J. Li, L. Ren, J. Xu, C. Li, S. Li. Soc. Indi. Res. **150**, 977 (2020)
30. E. Chamlee-Wright, V. H. Storr. T. Socio. Rev. **59**, 266 (2011)
31. L. Cheung, J. R. McColl-Kennedy, L. Coote. J. Ser. Mar. **41**, 438 (2017)
32. E. Yamamura. Reg. Sci. **95**, S143 (2016)
33. M. B. LaLone. J. Appl. Soc. Sci. **6**, 209 (2012)
34. S. Sadeka, M. S. Mohamad, M. S. K. Sarkar, A. Q. Al-Amin. Soc. Ind. Res. **150(2)**, 479 (2020)
35. M. Masud-All-Kamal. S. M. M. Hassan. Nat. Haz. **93**, 1547 (2018)
36. M. S. Akbar, D. P. Aldrich. Disasters. **42**, 475 (2017)
37. D. P. Aldrich. Nat. Haz. **56**, 595. (2010)
38. N. W. Chan, R. Roy, C. H. Lai, M. L. Tan. Int. J. Wa. Res. Dev. **35** (2018)
39. N. Tsuchiya, N. Nakaya, T. Nakamura, A. Narita, M. Kogure, J. Aida, C. Tsuji, A. Hozawa, H. Tomita. Psy. Cli. Neu. **71**, 52 (2017)
40. D. P. Aldrich. J. Civ. Soc. **7**, 81 (2011)
41. K. Gero, H. Hikichi, J. Aida, K. Kondo, I. Kawachi, Ame. J. Epid. **189**, 1369 (2020)
42. C. Ngin, J. H. Grayman, A. Neef, and N. Sanunsilp. Nat. Haz. **103**, 299 (2020)

Assessing knowledge creation process: A case of the National Disaster Management Agency of Gambia

Mbassi Saneh[1], *Alfi* Rahman[2,3*], and *Khairul* Munadi[3]

[1]Graduate Program in Disaster Science (MIK) Universitas Syiah Kuala, Darussalam, Banda Aceh, Indonesia
[2]Pusat Riset Ilmu Sosial dan Budaya (PRISB), Universitas Syiah Kuala, Darussalam, Banda Aceh, Indonesia
[3]Tsunami and Disaster Mitigation Research Center (TDMRC), Universitas Syiah Kuala, Darussalam, Banda Aceh, Indonesia

Abstract. The Disaster Risk Management policy and strategy of Gambia's National Disaster Management Agency (NDMA) have technical challenges and require knowledge creation to enhance capacities to lessen the adverse effect of risks and probabilities of disasters in The Gambia. Also, limited or no resources and leadership support to knowledge creation. The research aimed to assess the current condition of knowledge creation process in NDMA on socialisation-sharing experiences from tacit knowledge, externalisation-voicing tacit to explicit, combination-structuring explicit, and internalisation-expressing explicit to tacit (SECI) and role of leadership in the knowledge creation process. Detailed interviews, questionnaires, and WhatsApp call employed with four questions targeted five staff informants from NDMA and eight partner institutions with one informant from each. The research adopted a qualitative case study and content analysis to analyse the data using Voyant tools looking at the subject and concept. The research found that SECI has unearthed NDMA strengths, weaknesses, opportunities, threats, and leadership bureaucracy coupled with fear factors to build capacities leading to coup d'état for positions. Therefore, SECI has been unnoticed and uncoordinatedly implemented from "Happy Corner" with experiential knowledge and the National Platform for disaster risk reduction (DRR).

1 Introduction

The National Disaster Management Agency (NDMA) of Gambia was established by an Act of parliament 2008 and mandated to coordinate disaster management issues in The Gambia with Strategy and Policy frameworks. It has the political will from the highest office of the Vice President as the overall supervisor and the chairperson of the disaster governing council, which comprises all government ministries, private sectors, academic institutions, Red Cross Society, religious leaders, NGOs, civil society, media, and UN agencies. The day-to-day administrative affairs are under the purview of the Executive Director as liaison to multifaceted institutions for the implementation of disaster activities in building a resilient country. The agency has decentralised structures at regional and municipal, district, and village levels in the country's seven administrative regions: the Governors and Mayors, Chiefs, and Alkalos (village head) as administrative chairpersons.

With established structures and political will, NDMA performance has many challenges, especially technical capacities in Disaster Risk Management and Disaster Risk Reduction (DRM/DRR). The performances are required knowledge for continued innovation, education, and understanding of emerging hazards which is weak in NDMA. Also, frequent leadership changes, limited human and financial resources, and policy and governing gap to act as mentioned in the Act 2008 [1]. The 2012 and 2017 capacity and needs assessment highlighted these challenges in NDMA and partner's roles and responsibilities for operations not fully implemented due to inadequate knowledge, skills, and experiences. Thus, due to the knowledge capacities gap in individual, institutional, environmental, and policy creations, utilisation and retention need to be upgraded by creating new opportunities and building blocks for the creation of new knowledge for the personal and institutional growth of technicians trained and aware of the past, present, and probable future of DRM/DRR. And this will improve operations and implementations at all levels [2].

Therefore, a report on capacity and needs assessment and the revised strategy and policy seriously mentioned NDMA and its partner capacity to be continuously strengthened to implement DRM/DRR. In recent years, the Gambia has been exposed to hazards and vulnerable because the country is geographically right of the River Gambia flood plain and makes the country experiencing yearly flooding. The Gambia is encountering both hazards-natural and human-induced. It affects thousands of individuals yearly, such as flooding, windstorms, disease outbreaks, deforestation, oil spillage, bush and domestic fires, droughts, pest invasion, urban settlements, industrial pollution, toxic waste, and accidents. And of the recent political impasse in the

*Corresponding author: alfi.rahman@unsyiah.ac.id

2016 presidential election brought political conflict. The NDMA couldn't coordinate due to the fear factor under the purview of the president's office during the period.

The intensity and the scale of disasters defer in the urban and rural areas and paralysed community services, loss of lives and environmental damages, infrastructures, and the incomes of the people and country. Besides, Disasters are unexpected, unpredicted, and surprising events with a devastating intensity which is stressful on communities and people. Therefore, the know-how (skills) on knowledge and creation is incumbent on every institution reducing the impact of disasters through continuous knowledge creation on unforeseen hazards. To help protect safer and build resilient communities and the country from the effects. Knowing and knowledge of skills and experiences of employees is paramount via continuous knowledge creation [3,4]. Negligence to the consideration of such can lead to inefficacy and failure. Therefore, investing in human and technological capital in achieving NDMA's vision is a worthy cause. DRM requires continuous knowledge creation for a sustainable environment, society, and development [5]. Hyogo Frameworks for Action 2005-2015, Sendai Framework 2015-2030, and ECOWAS Policy for DRR 2006 are also highlighted as above. Both concepts require a holistic approach with appropriate knowledge for the implementation of DRM/DRR. Experiences and training exercises on knowledge learned before disaster risk reduction are paramount to share and transfer to bridge the institution's knowledge gap. In a nutshell, DRM/DRR matters are multi-sectoral and multi-dimensional, which call for absolute mainstreaming of all stakeholders for sustainable development [3].

The institution challenged by "knowledge society" and knowledge (know-how) for effective performance belongs to a person equipped with knowledge [6]. Therefore, knowledge creation is about socialisation being the root of a tree that branches to the SECI process to give detailed information worthy when meeting with people at a place. Further stated knowledge is tacit and explicit interaction: inbuilt thoughts that form the understanding when communicated with a clear and meaningful explanation to audiences [7].

The knowledge creation (KC) process is unstructured and unmethodical when introduced to institutions, groups, and individuals for action. Because it requires creativity to modify the ideas to suit the needs for application [5]. To enable the KC process, there should be socialisation and connection between tacit and explicit knowledge, followed by knowledge assets experiential, conceptual, systemic, and routine to facilitate the KC process of an institution. KC is convenient by the context in which and manner the task is done by individuals and groups using knowledge as the driving force to create new knowledge. Cognisant that creating new knowledge cannot be divorced from tacit and explicit [7]. KC process theory includes Socialisation, Externalisation, Combination, and Internalisation (SECI) which spearhead activities that enable the creation, transfer, and utilisation of knowledge captured from the Ba. Ba is an open place where people meet, interact, and discuss.

Thus, new knowledge created, shared, and used either analogue (tacit) or the combination (explicit) pass through Information Technology (IT) infrastructure, in which the context and content must be clear to your partner [7].

Currently, there is growing interest in disaster KC and sharing from lessons and best practices of DRM/DRR strategies to reduce the impact of disasters, informing disaster risk management and disaster risk reduction [3,5]. For this reason, KC is relevant in the DRM/DRR strategy, whereby institutions and individuals KC have different approaches when introduced for the implementation process. In this sense, KC cannot be over sighted in the DRM/DRR strategy to elevate the current status of the NDMA. Continuous promotion of knowledge and creation for the performance of employees and the installation of IT infrastructures software monitoring the trend of disasters in the country and knowledge as the knowledge management system (KMs).

Achieving the vision of NDMA DRM/DRR strategy, knowledge management is an integral part of increasing the potential of existing knowledge and transforming new knowledge into action through Information Communication Technology (ICT). This process is easy, quick, accessible, and reliable for communicating and monitoring disaster risk management and disaster risk reduction and the knowledge asset of an institution.

The aims of this study are to assess the current condition of the KC process on SECI model and to highlight the role of leadership in the KC process in NDMA.

2 Knowledge creation

2.1 Knowledge creation theory

2.1.1 Knowledge

Leadership has little comprehension of how to create and manage knowledge amicably. Institutions challenged by "knowledge society" and knowledge (know-how) for effective performance belong to a person equipped with knowledge. Therefore, knowledge is about socialisation being the root of a tree that branches to the SECI process to give detailed information worthy for the purpose at a particular time of meeting with people anywhere. Nonaka et al. (2000) further stated there are two types of knowledge - tacit and explicit interaction are inbuilt thoughts that form the understanding when communicated with a clear and meaningful explanation to audiences [6]. Knowledge is epistemological that transforms tacit and explicit for action. Ontological is the content and context of the subject reality from socialisation from individual knowledge to group [5]. Knowledge is a process within the human that is "justifying personal belief" towards the "truth" and evidence (s) for people to believe [6]. Knowledge links to performances and outputs via learning to attain, use, and value-added.

Knowledge translates to "know-what, know-how, know-who, and know-why'' [8]. For example, the perception of the institution's knowledge is skills and attitudes, while strategic perception defines knowledge as institution resources to attain goals. Knowledge is produced from processed data to generate information [3]. The analytical knowledge block hierarchy portrayed the flow of steps from data, information, knowledge, wisdom, and decisions. Knowledge is often confused with information. Knowledge at times is defined as an IT database system that processes data and information stored for retrieval. Contrary, data is raw without context to give you a meaning for interpretation. Example numbers or reports that signify nothing, but if the context and the value are visualised, you can have the content of the information processed to give a simple, meaningful, accurate, realistic, and timely description for dissemination or dissemination application. Example who, where, when, and what, which turns to knowledge by analysing the information. Knowledge is the information you analyse and use to achieve meaning. Information is transformed into knowledge for action and can be transferable. Wisdom is the application of knowledge for decision-making in any given situation.

In the same vein, resilience is compounded with utilising existing knowledge to produce new knowledge by investing in human and technology resources with the provision of data, information, know-what, and know-how worth consideration for implementing the DRM/DRR. To achieve the goal of resilience, resources must be available including, knowledge to withstand circumstances [9–12]. This cannot be in the absence of knowledge as institutional resources in a competitive environment. Knowledge is created through physical, technological, and human resources [1]. The knowledge produced resources and ploughed to increase physical, technical, and human capital to give supplementary resources in disaster risk management and disaster risk reduction (create new knowledge) to improve decision making. The figure below shows knowledge ladder flow in DRM/DRR for resilience building.

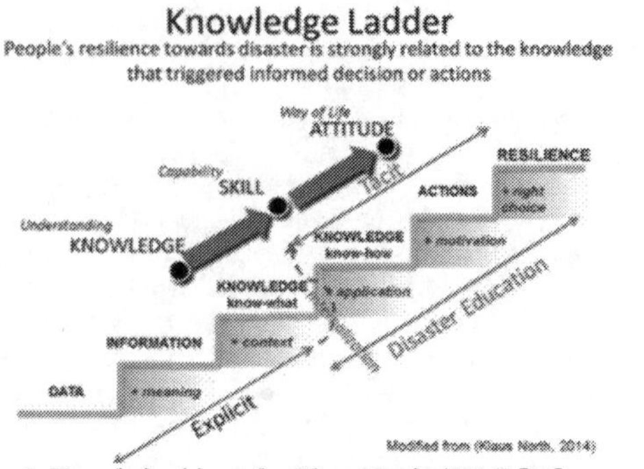

Fig. 1. Knowledge hierarchy Klaus North, (2014) [13]

Knowledge hierarchy corresponding below.

- **Know-what** defines the action to comprehend knowledge by learning, extensive training for certification, and analysing to improve the institution's quality (Cognitive knowledge).
- **Know-how** is the ability to practice and apply knowledge gained from the theory. That is capability and competence for the implementation of the knowledge for problem-solving.
- **Know-why** is the action that motivates decision-making for resilience building from the information received. The same as understanding the system of the cause and effect of underlying-problem.
- **Ba** is also defined as data that provides information, knowledge, and action to create new knowledge of wisdom and decision.

2.1.2 The SECI process

Socialisation is from tacit-to-tacit knowledge from interactions with individuals or groups of people to share experiences whereby you capture and disseminate new knowledge. During this time, communication is difficult for people to understand because the information given is chaotic due to the unstructured presentation. The process can be shared through actions, procedures, routines, practice, and ideas from the open place [7].

Externalisation tacit knowledge is interpreted in a comprehensive method to be understood. It is communicated with a group of people from tacit to explicit. The two externalisation factors are **(1)** Expressing tacit knowledge to explicit from technical ideas and images that form meaningful words. For example, dialogue "listening and contributing to the benefits of all informants." **(2)** Interpreting tacit clearly through rational thinking to capture the ideas shared [7].

Combination is explicit to explicit knowledge interaction. Explicit converted to systematic and context-specific to produce training manual, article, or report for use. This knowledge is acquired both inside and outside of the institution. It accumulates to form new knowledge from a combination of different sources to a new context. The knowledge generated from externalisation is transformed into digital or correspondent material to capture, disseminate, and edit. The materials are shareable and used for communication [7].

Internalisation is to convert explicit to tacit through institutions and individual knowledge during training, learn-on-the-job, and simulation exercise. Knowledge creation is a cycle within the SECI process. Internalisation is the technical know-how of the knowledge gained [6,7].

The SECI process spearhead activities that enable the creation, transfer, and utilisation of knowledge acquired from the Ba. The tacit and explicit are substituted and changed based on circumstance and environment. It transforms knowledge from conceptualisation (theory) to actualisation (practice) [6]. Knowledge assets emanate from KC that connect to humans and technology. It happens via socialisation of knowledge, sharing experiences from tacit. Externalisation, voicing tacit to explicit; combination-structuring explicit; and internalisation-applying new knowledge learned from explicit to tacit with conducive environment and leadership support [1]. Knowledge creation and sharing are the institution's limelight that forms the knowledge management strategy for sustainable development [14,15]. Knowledge is about knowing via socialisation.

2.1.3 Ba

Ba as a "shared space for an emerging relationship". This could be in an office, "bantaba" (community gathering in villages), market, street, or social media to share ideas and experiences [7]. Information and knowledge acquired from Ba are chaotic, unstructured, and difficult to communicate for understanding. Thus, it is a platform to share context that enables and stimulates new knowledge creation [6,7,16,17]

2.1.4 Knowledge Assets (KAs)

Knowledge assets are the inputs, outputs, and moderating factors of the knowledge-creating process. As a result, KAs encompass the content of the knowledge creation process. Assets are specific resources that are essential and worthy for disaster risk management and disaster risk reduction in disaster management institutions. Despite this, KAs are intangible, tacit and dynamic. And difficult to capture, but the result is from achievements made from performance capabilities. Knowledge assets are in two forms: human (soft) skills, experiences, norms, values, and IT (hard) technology for codification, storing, collection, and retrieval [18]. These components form the effective operations of an institution.

KC highlighted four types of knowledge assets: Experiential KAs, Conceptual KAs, Systemic KAs, and Routine KAs. The research found that the KC process consists of Ba, SECI, and KAs. In utilising the existing knowledge assets, new knowledge is created via the SECI process, which energises Ba.

2.2 Knowledge creation (KC) in DRM/DRR

Knowledge Creation is a new discipline in the DRM/DRR, and very few research materials are available. Disaster risk management and disaster risk reduction require sufficient knowledge creation and sharing for the institution's operations.

Disaster risk management, reduction, and knowledge creation are relevant in any given situation of development and growth. For the past years, the progress in knowledge and use of DRM/DRR knowledge is slowly coming up in disaster operations. This is because Priority three of the Hyogo Frameworks for Action (HFA) 2005-2015 emphasised the use of knowledge, innovation, and education in DRM/DRR [19]. It also stated in the Sendai Framework for Disaster Risk Reduction (SFDRR) 2015-2030 priority one [20], understanding disaster risk, and priority three investing in DRR for resilience (HFA 2005-2015 and SFDRR 2015-2030) are essential. Knowledge creation influences individuals and the environment from interaction and exceeding beyond the boundary to seek knowledge.

Some articles highlighted knowledge and creation as the most relevant source for competitive strength. Thus, the institution initiates and delineates bottlenecks and cultivates and uses new knowledge to solve the SECI problem. KC has not been applied and impacted to institution's growth between top management and knowledgeable workers and partner institutions in DRM/DRR [6,7,16,17]

Knowledge significantly contributes to DRM/DRR, which supports KC and its influence on the institution and is positively impacted. Disaster risk management and reduction researchers revealed that DRM/DRR requires interaction with authority, groups, workforce, resources, partnerships, and systems to avoid failure rather than boost the competitive environment [7]. KC in DRM/DRR "enables mixing and matching solutions across different disasters" for leadership to monitor and coordinate disaster events [21]. The Focus was more on knowledge management which focuses on the institution knowledge such as acquisition, accumulation, use, and reuse of present knowledge, abandoning the DRM/DRR knowledge creation process to give a chance to new knowledge from the tacit. Instead, the shift was mainly on explicit knowledge.

So far, from the few kinds of research, KC is evolving intensely in many knowledge management institutions. Notwithstanding, there are gaps in KC, such as epistemological, which is a different understanding of types of knowledge relevant to the institution. The strategic component looks at the lack of understanding of the skill and expert knowledge on DRM/DRR, whereby the institutional gap focuses on how institutions absorb and transfer knowledge and the right human resource. These four gaps are cross-cutting for the operationalisation and implementation of DRM/DRR. Despite the gaps, KC is becoming increasingly important in all DRM/DRR strategy and policy angles to do away with the old fashion of doing things differently to achieve better performance [22].

2.3 Leadership

Leadership in disaster risk management and risk reduction KC is central because they provide the environment, are proactive, creative, and learn beyond the boundary for better improvement. The leadership has a role in KC by defining, redefining, or developing a knowledge vision to establish knowledge assets or the existing knowledge assets to create new knowledge of the institution through the SECI process from the Ba. The knowledge created becomes part of the institution's knowledge assets. Leadership covers all the aspects of DRM/DRR coupled with leading knowledge vision promoting, sharing, creating, energising Ba and continuous development of knowledge spiral of KC from central and distributed (top and middle) leadership by reading the situation and leading to the three elements of KC [23].

Through leadership questioning on "What are we?", "What should be created?", "How can we do it?", "Why are we doing this?" and "Where are we going?" will determine and monitor the institution knowledge creation from the past knowledge that is limitless and to give new knowledge for development and retention with the involvement of top and middle management and the staff in the KC process fostering commitment to the KC process [7]. Central and distributed leadership increases the institution effectiveness coupled with a focus-oriented investing in skills development, structures in place, and technology that will create cultural learning system [23]. Thus, KC is up-to-date by an institution vision outlined informing leadership on strategic objectives [5,7]. The leadership should always take urgent decision making during an emergency, even sometimes when there is no information at that particular moment but applying critical thinking for action is necessary.

The SECI connects with positive leadership as the leader is the coordinator, director, facilitator, and evaluator of programmes and achievements [23]. There is no one size fit in disaster risk management and disaster risk-reduction leadership rather proactive actions to reduce the situation impact accordingly. In this case, some studies employed strategic leaderships, which constitutes the institution's strategy, policy formulation, vision, mission, fostering culture, and knowledge creation platform. Leaders at the apex of decision-making should not underrate the power of knowledge creation, which lacks the entrustment of care, trust, love, support, and concern in promoting human development at all levels in advancing growth [23].

As far as there is a new paradigm shift of KC from the traditional sense of "management" controlling the flow of information from a top-down approach rather "middle-up-down" approach to the knowledge creation process has to be recommended [7]. The knowledge creation process facilitated by middle-up management leadership to take ownership of the three fundamentals of KC as a vision to inform and know the types of knowledge assets required for operationalisation through a self-internal assessment to attain a knowledge vision.

Leadership impacted the institution's effectiveness by formulating strategy, vision, mission and fostering institutional culture. Therefore, strategic leadership in SECI relates to motivating, inspiring, and empowering staff coupled with a conducive environment [23]. The research focused on strategic leadership to help shape institutional development and growth because of the multidimensionality of the institutions and will reduce back and forward line ministry decision-making bureaucracy. Therefore, understanding the role of leadership can influence leadership perceptions [24].

The management of institutions needs to bank on knowledge to generate a plan for justifiable development in disaster risk management and disaster risk reduction. Disasters are emerging with complexity due to extreme climate events. Globalisation is increasing with natural hazards and anthropogenic activities deterring social-economic development and loss of lives in all aspects. Changes in information technology are constant and diverse. Therefore, leadership in KC's strategic leadership believes that applying the current dynamic environment to enhance institutional effectiveness in DRM/DRR is important and timely. On the contrary, leadership bureaucracy also hinders the institution's operationalisation [25]. The dimension of leadership eases the smooth running of the institution.

3 Methodology

The research adopted a qualitative case study and content analysis approach to analyse the data using Voyant tools looking at the subject and concept of the KC SECI process in NDMA. The method applied for assessment gives the researcher the ability to comprehend the informant's perspectives. During this process, the validity and reliability of the data matter. Stratified random sampling was applied for NDMA and partners for the data collection. And Five (5) informants were selected from NDMA with two senior central staff from administration, finance, and three regional coordinators Lower River Region, West Coast Region, and Upper River Region. The chosen staff have experiential knowledge of the DRM/DRR issues in The Gambia and a long working relationship with the NDMA from its inception to the research period. They are involved in DRM/DRR implementations and operations at central and regional levels with knowledge of on-the-job training. They helped to have access to information and feedback during the data collection within the institution and partners.

Open-ended questionnaires were drawn and administered through the use of Google Forms electronic questionnaire. This electronic questionnaire gives easy accessibility to informants to answer at their own suitable time. Four (4) questions targeting NDMA but during the process to have more information and balancing the data, institution partners co-opted. Eight (8) institutions with one (1) representative identified and including NDMA gives a total of nine (9) institutions taking part as in Table 1.

Table 1. Targeted institutions for the assessment of KC

No	Institution/Informants	Total
1	NDMA	5
2	Gambia Red Cross Society	1
3	National Environment Agency	1
4	Gambia National Nutrition Agency	1
5	Gambia Bureau of Statistic	1
6	Ministry of Health and Social Welfare	1
7	Ministry of Agriculture	1
8	Village Development Committee	1
9	University of Gambia	1

Partners chosen are members of the National Platform for DRR with comparative advantages in financing, expertise & technical knowledge, equipment, infrastructure, leadership style, and institutional culture for being proactive in their respective fields. They are multi-sectoral national platform committees established to help central coordination, advice, and analysis on priority areas for action to the central government towards its national development plan formulations, implementations, and the overall achievement of the Sustainable Development Goals for disaster management issues. The synopsis of these partners' institutions is below.

4 Result

The Act of parliament mandated NDMA to coordinate disaster management issues in the Gambia. The Vice President is the overall supervisor and the chairperson of the disaster governing council, and the Executive Director works in close liaison to multifaceted institutions to implement disaster activities in building a resilient country. There are decentralised structures at regional and municipal, district, and village levels under the purview of Governors and Mayors, Chiefs, and Alkalos.

Disaster risk management and disaster risk reduction require sufficient knowledge and knowledge creation for the institution's operations and implementation. For the past years, the progress in knowledge and use of DRM/DRR knowledge is slowly coming up in disaster operations. There are gaps in NDMA KC, such as epistemological, human resource, strategic, institutional, and financial commitment. Despite the gaps, KC is becoming increasingly important in all DRM/DRR strategy and policy angles to do away with the old fashion of doing things differently to achieve better performance [22]. The research has shown that SECI has been unnoticed and uncoordinatedly implemented in NDMA "Happy Corner" and the National Platform for DRR. SECI unearthed NDMA's strengths, weaknesses, opportunities, and threats. There is a lack of leadership support and commitment for KC whereby no or limited resources are budgeted for staff growth. Fear factors building capacities would lead to coup d'état of positions by juniors who acquired knowledge and new knowledge for a competitive environment.

4.1 Knowledge creation process in NDMA

4.1.1 Knowledge

According to informants, knowledge is a critical motivation for people. Knowledge has to be shared, which gives information and enhances capacity versa-visa. As a result, information is communicated, including ideas and research, to establish institutions' vision, mission, and objectives for operations. In this case, knowledge surfaced 23 times. It means that knowledge is relevant to NDMA and partners, as the quotes below by the informants.

Fig. 2. Knowledge in relation to the KC process of SECI in NDMA

4.1.2 Knowledge creation (KC)

The KC code and frequency count 11 times. They further mentioned its importance to bridge the knowledge gap of an institution through information, experience, and sharing formally or informally from the interactions. Knowledge creation in DRM/DRR promotes staff and community resilience as drivers to sustainability to development and economic health by reducing poverty.

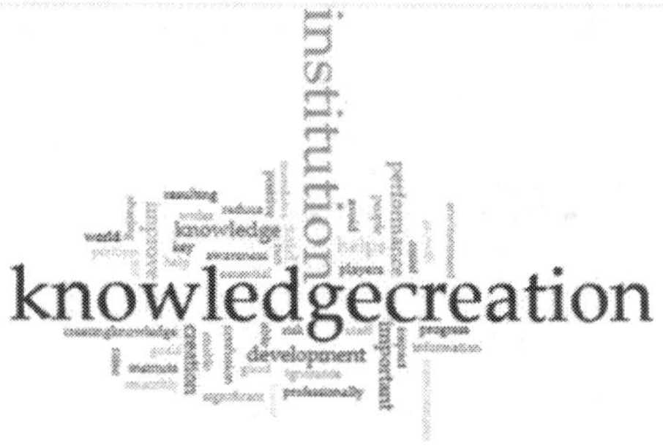

Fig. 3. Knowledge creation in NDMA

4.1.3 SECI

The explanation and understanding of SECI are in tandem with a literature review of the research, which influences individuals and the environment. It is from interaction and exceeding beyond the boundary. It can be learning, training, apprenticeship from the institution, traditional home, and community. Informants further mentioned SECI could be formal and informal consultations, commitment, and political will. The most frequent word in the corpus is SECI 34 times.

Fig. 4. SECI in NDMA

As a result, the framework developed and modified is relevant to informed decision-making in the KC process as leadership in the centre enables SECI of the institution.

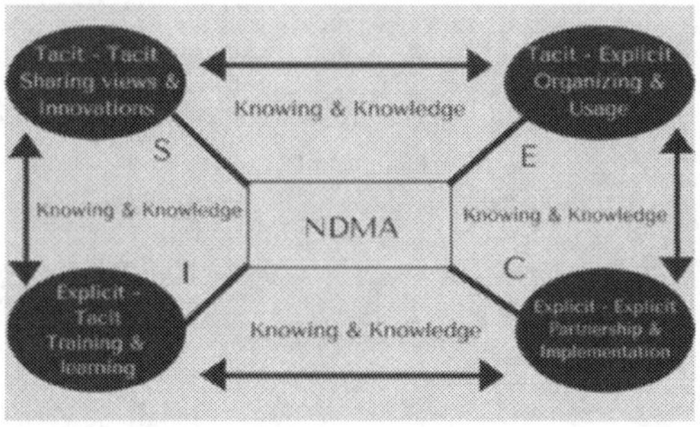

Fig. 5. SECI framework adapted and developed during the assessment Nonaka et al. (2000) modified by Mbassi Sanneh 2021

The framework developed will serve as a guide to influence and strengthen NDMA's revised strategy and policy. The relevance of knowledge creation through the use of knowledge, innovation, education, and investment.

4.1.4 Socialisation, originating Ba and experiential knowledge asset.

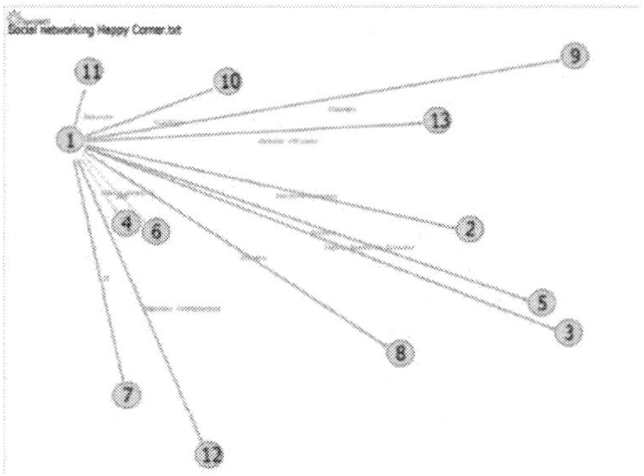

Fig. 6. "Happy Corner" networking in NDMA

The "Happy Corner" (internal socialisation) is an opportune place that enables knowledge growth where dialogue on tacit knowledge emanates from sharing experiences unsystematically at NDMA [26,27].

4.1.5 Externalisation, interacting Ba and conceptual knowledge asset

The collaboration and communication with a group of people – partners to create knowledge. NDMA works with multidisciplinary institutions on disaster risk management and disaster risk reduction, which led to the establishment of the National Platform for disaster risk reduction (DRR) for networking which includes government ministers, private sectors, academic institutions, Red Cross Society, Religious leaders, NGOs, civil society, media, and UN agencies. This platform is a multidisciplinary stakeholder national structure that helps to advocate DRR at all levels. They are responsible for coordination, advice, and analysis on priority areas for action. Success national platform depends on national ownership and leadership of the DRR process [28]. The partners have comparative advantages on financing, knowledge, expertise, equipment, infrastructure, leadership style, and institutional culture for being proactive in dealing with DRM/DRR based on the component such as resource mobilisation, advocacy, coordination, and technical committees. The National Platform for DRR was established and officially declared in March 2011 as a national committee for DRM/DRR. It was a piloted project in the Gambia for other West African countries to learn from best practices with funding and partnership support from ECOWAS. They aim to create a partnership structure that will influence political and legal commitment for DRM/DRR understanding and strengthen the scientific knowledge of DRM/DRR governance among the public and authorities and mainstream it to the national development plan for sustainable development of the country.

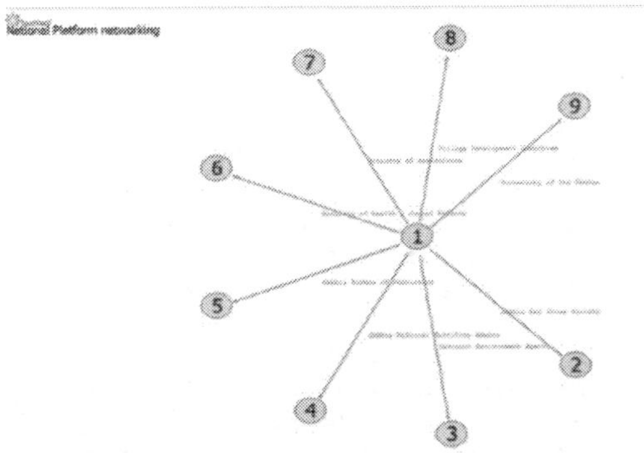

Fig. 7. The national platform for DRR networking in NDMA

4.1.6 Combination, systemic Ba and systemic knowledge asset

Combination of ideas from the National Platform for DRR minutes, reports, policy, strategy, or work plans to systematically codify for respective institutional plans aligned with international frameworks for DRM/DRR [29].

In this process, leadership must base on a shared understanding of DRM/DRR with the multi-sectoral and multidisciplinary structure to actively facilitate central and distributed leadership for all partners' involvement and participation for a better combination of concepts on ideas and suggestions the National Development Plan for sustainable development. Partners involved in planning concepts and implementing demonstrations from the SECI to transfer knowledge [25] through leadership support for KC for a better result. The system is more structured and efficient for learning, sharing plans, policy, and strategy to improve decision-making. It helps in capacity building and easy retrieval of documents and records keeping. The combination is processed through partnership support to NDMA in leveraging knowledge from existing knowledge and KAs to plan for improvement. Tacit to explicit knowledge transform to create skill and experience via combination for staff use and institutions. If not, the knowledge will gradually disappear in NDMA and partner institutions without proper documentation and record-keeping. The National Platform for DRR is the platform for the SECI, more so the externalisation and combination through interaction and collaboration with partner institutions to implement DRM/DRR activities regardless of comparative advantages.

4.1.7 Internalisation, exercising Ba and routine knowledge asset

In NDMA, learn-on-the-job, simulation exercises, training of trainers on comparative expertise skills, and internship to learn new knowledge for better operations of the institution are semi-dormant because the leadership is not investing in KC. Secondly, partner institutions are having similar challenges. The platform is where partners learn new knowledge through socialisation and practice during interactions to domesticate explicit knowledge to their tacit knowledge for internal implementation. This is not happening due to the fear factor as in the quotes.

Thereby, the NDMA's existing knowledge mapped out surfaced new knowledge availability after reviewing the SECI process that energises the "Happy Corner", National Platform, and KAs at NDMA [1,7,17]. The mapping supported NDMA leadership to know the competencies and capacities of NDMA and partners. It helped identify and define the knowledge vision relevant to the agency, and equally, communication and accessibility are genuine in KAs [30].

4.2 Role of leadership in KC process in NDMA

4.2.1 Leadership

Creating the environment to implement creativeness and ideas from lessons learned beyond the boundary is for better improvement. Therefore, leadership support creating the environment to implement lessons learned beyond for better improvement is relevant. Leaders define, redefine, and develop existing knowledge assets to create new knowledge from the SECI process. Notably, leadership covers all the aspects of DRM/DRR coupled with leading knowledge vision promoting, sharing, creating, energising Ba and continuous development of knowledge spiral of KC from central and distributed (top and middle) leadership by reading the situation and leading to the three elements of KC [7,23].

The knowledge vision defines the type of knowledge required for an institution to create. Through leadership questioning, "What are we?", "What should be created?", "How can we do it?", "Why are we doing this?" and "Where are we going?" Therefore, leadership constitutes strategy, policy formulation, vision, mission, fostering culture, and knowledge creation processes.

Leadership connects to Ba, SECI, and KAs as the three elements of the KC process. The report also mentioned that primary responsibility in DRM/DRR relies on strong leadership. In Fig. 8, the keyword is highlighted 30 times and emphasised by informants how leadership can facilitate KC for the growth of any institution, as mentioned below.

Fig. 8. Leadership role in KC process in NDMA

Table 2. Summary of leadership dimensions in NDMA KC

Condition	Leadership
Collaboration	Interactions with transparent leadership is weak.
Beliefs	"Fear factor" to be challenged by staff when capacities are built.
Process	Learning best practices and accept critics of practicing and formalising the theory to a situation.
Authority	"Information is monopolised".
Skills	Leadership powers are not utilise at central and distributed leadership.
Development	Formulate strategic issues and develop programmes. Investing in skills, development, structures in place, and technology create a cultural learning system.

5 Discussions

KC process is fragmented in NDMA and not coordinated between humanitarian workers and skilled workers. Therefore, there is a gap in information sharing and coordination in DRM/DRR. Lack of this could be an inadequate performance of current DRM/DRR operations in The Gambia. KC is essential in DRM/DRR to learn from lessons and adopt best practices for skills improvement and enhance knowledge creation by building a culture of learning from pre, during, past, and future hazards and disasters. Through KC, adequate knowledge can be accessed and utilised at all levels by empowering and protecting staff and the institution as a whole. "Ignorant is no excuse to the law". The research discovered that NDMA and partners have experiential knowledge.

As mentioned above, decision-making and time-sensitive are very dynamic in DRM/DRR operations. Therefore, a single delay in dissemination would negatively impact the decision, which will affect embracing the KC process in NDMA DRM/DRR. When the know-how is there, it makes a fast decision for action in all operations for better achievement. So, therefore, KC is a gateway for knowledge acquisition and sharing through interdisciplinary contacts by developing shared understanding towards a common purpose. KC reinforces the working practices and allows learning and applying new knowledge in the work systems. Nevertheless, despite the hidden KC process in NDMA and its partners, the researchers noticed that the SECI was implemented in an uncoordinated manner both internally and externally without leadership becoming aware of and not knowing the framework for KC and its relevance. Finally, for the implementation of KC to be effective, the role of middle management is to interpret and communicate the KC process horizontally and vertically through leadership interactions in meetings, conferences, or any place of Ba for learning.

There are numerous barriers to KC that hinder the effective knowledge creation of an institution that needs to uninstall obstacles of lack of leadership support and discouragement, trust, care, and love. There is a lack of transparency and accountability by the leadership. For example, within NDMA, programmes or information will not reach individuals concerning the matter; instead, you hear it from hearsay. Such would inspire discouragement, lack of trust, care, and love for socialisation in the working environment and hindered performance delivery for growth. Also, bottlenecks are lack of the know-how to tap the unnoticed knowledge creation process, information sharing on time, bureaucracy, capacity building, poor infrastructure, resistance to change, and implementation of the words into actions.

Furthermore, the human factor could be barriers, lack of individual self-confidence, fear of criticism, bias concerning strict orders, and obsolete systems and procedures of an institution that can deter KC of NDMA and partners. Inadequate information sharing, interaction, and communication monopolised. Research results, "trained human resource" and financial commitment are currently challenging in the country DRM expressed by staff. To avert such challenges, they recommended "staff and volunteer training" to enhance the capability to achieve NDMA goals.

The political interference in the operations of NDMA malnourished the management and leadership style as the central office mandated to coordinate all disasters in the country. Parse politics have no direct implications on DRM/DRR but rather on institutional formation, management, operations, and laws governing the institution. It is high time to revisit and unattached humanitarian operations from political affairs to effectively and transparently implement according to national, regional, and international guidelines irrespective of political affiliation and tribe. Disaster risk management and disaster risk reduction should not be influenced or dominated by politics.

Resonate from the beginning discussing the research benefit NDMA knowing the KC process available from the proposed framework SECI easy implementation. The leadership can achieve its vision to protect and build resilient communities with appropriate knowledge and generally strengthen the growth of the agency and partners for efficient and effective DRM/DRR operations and implementations. Research paper for knowledge management to be conducted by NDMA using this research for reference as the baseline assessment.

6 Conclusion

There should be continuous improvement in performance by leveraging KC and striving towards DRM/DRR to achieve the institution's objectives. KC will enhance operational excellence and leadership commitment. NDMA and partners need KC theory for successful implementation to address the four stages of knowledge conversion. It will identify tacit and explicit knowledge needs and the current condition of the KC process and provide leadership with good guidance (proposed framework) that would contribute to achieving NDMA objectives through SECI. HFA 2005-2015 and SFDRR 2015-2030 country assessment report mentioned why the framework objectives are not attainable according to the time frame due to capacity constraints of countries in the implementation of priorities. The Gambia contributes to the delay – capacity constraints of right human resources, strong leadership, and finance.

The KC process is transcending individual boundaries for new knowledge to create through the SECI process. Therefore, the assessment highlighted the underlying factors and explored issues affecting the current condition of KC in NDMA and partner institutions. The current condition of the KC process on SECI has been unnoticed and uncoordinatedly implemented in NDMA "Happy Corner" and the National Platform for DRR and the new spiral of knowledge creation.

The leadership is weak with frequent consultation to the vice president's office and leads to NDMA bureaucracy in operations, implementation, and frequent leadership changes by the previous and current government. There are leadership constraints for capacity building with no or limited resources budgeted. There is no specific budgetary allocation for DRM/DRR KC and even for the whole DRM/DRR like in other countries advocating for a certain percentage of the national budget investment to DRM/DRR programmes. The only allocation for NDMA is the emergencies response account. Another challenging issue at hand is that NDMA and many other agencies are under the purview of the vice president. And the DRM/DRR projects can be influenced by politics by other players in DRM/DRR field. There is a lack of commitment, support, lacking self-confidence, and fear factor to build staff capacity that would lead to coup d'état positions by juniors. The unwillingness to cope with new knowledge, inadequate or sometimes no information sharing, interaction, and communication monopolised. In a nutshell, NDMA leadership is a challenge with institutional management, financial, and knowledge creation.

References

1. Afri Consultan Group, *Needs Assessment for Disaster Risk Management, Recovery Planning and Checklists for Preparedness in the Face of Climate Risks* (2017)
2. CADRI, *Disaster Risk Reduction Capacity Assessment Report, the Gambia* (2012)
3. D. Amaratunga and R. Haigh, *Post-Disaster Reconstruction of the Built Environment* (Wiley-Blackwell, Oxford, UK, 2011)
4. P. Akhavan, S. Ghojavand, and R. Abdali, Journal of Information & Knowledge Management **11**, 1250012 (2012)
5. C. Bratianu, I. O.-M. & Marketing, and undefined 2010, Core.Ac.Uk (n.d.)
6. I. Nonaka, I. o Nonaka, N. Ikujiro, and H. Takeuchi, *The Knowledge-Creating Company: How Japanese Companies Create the Dynamics of Innovation* (1995)
7. I. Nonaka, R. Toyama, and N. Konno, Long Range Planning **33**, 5 (2000)
8. D. Bennet and A. Bennet, VINE **38**, 72 (2008)
9. A. Rahman and K. Munadi, in *IOP Conference Series: Earth and Environmental Science* (IOP Publishing, 2019), p. 12040
10. A. Rahman, A. Sakurai, and K. Munadi, International Journal of Disaster Risk Reduction **29**, 13 (2018)
11. A. Rahman, A. Sakurai, and K. Munadi, in *IOP Conference Series: Earth and Environmental Science* (IOP Publishing, 2017), p. 012018
12. Fachrurradhi, N. R. M. Susanti, A. Pratama, R. Fitri, S. Raudhiah, A. Rahman, and K. Munadi, in *IOP Conference Series: Earth and Environmental Science* (IOP Publishing, 2019), p. 12043

13. K. North and G. Kumta, *Knowledge Management : Value Creation through Organizational Learning* (Springer, 2014)
14. P. Akhavan, S. Ghojavand, and R. Abdali, Journal of Information and Knowledge Management (2012)
15. R. K. and F. Bulutlar, The Influence of Knowledge Sharing on Innovation (2010)
16. I. Nonaka, Organization Science **5**, 14 (1994)
17. I. Nonaka and H. Ayano, *Social Innovation Creating New Knowledge for New Social Value* (Graduate School of International Corporate Strategy, 2010)
18. K. Seneviratne, D. Baldry, and C. Pathirage, International Journal of Strategic Property Management (2010)
19. IRIDeS, *"Hyogo Framework for Action 2005-2015: Building the Resilience of Nations and Communities to Disasters" HFA IRIDeS Review Report Preliminary Focusing on 2011 Great East Japan Earthquake* (Sendai, 2013)
20. UNISDR, *The Sendai Framework for Disaster Risk Reduction 2015–2030* (2015)
21. S. H. Othman and G. Beydoun, ACIS 2010 Proceedings (2010)
22. K. Albris, K. C. Lauta, and E. Raju, International Journal of Disaster Risk Science **11**, 1 (2020)
23. G. von Krogh, I. Nonaka, and L. Rechsteiner, Journal of Management Studies **49**, 240 (2012)
24. J. Hartley and L. Rashman, in *Connecting Knowledge and Performance in Public Services*, edited by K. Walshe, G. Harvey, and P. Jas (Cambridge University Press, Cambridge, n.d.), pp. 145–172
25. E. C. K. Cheng, in (2015), pp. 11–23
26. M. L. Farnese, B. Barbieri, A. Chirumbolo, and G. Patriotta, Frontiers in Psychology (2019)
27. J. H. Song, D. Uhm, and S. W. Yoon, Leadership and Organization Development Journal (2011)
28. ISDR, *UNISDR Informs: Disaster Risk Reduction in Asia & Pacific* (2006)
29. J. Hoon Song, D. Uhm, and S. Won Yoon, Leadership & Organization Development Journal **32**, 243 (2011)
30. B. Marr, G. Schiuma, and A. Neely, International Journal of Business Performance Management **4**, 279 (2002)

Towards sustainable cities and communities: Is psychological preparedness include in the disaster risk reduction policy framework in Indonesia

Listyati Setyo Palupi[1*2345]

[1]Psychology Department, Universitas Airlangga, Jl. Airlangga 4-6, Surabaya, East Java, Indonesia
[2]Addiction Study Centre, Jl. Airlangga 4-6, Surabaya, East Java, Indonesia
[3]Health and Environmental Sustainability (HES) Research Group, Jl. Airlangga 4-6, Surabaya, East Java, Indonesia
[4]Technology and Psychology of Social Change Research Group, Jl. Airlangga 4-6, Surabaya, East Java, Indonesia
[5]Direktorat Riset dan Pengabdian Masyarakat (DRPM) RISTEKDIKTI Indonesia Penelitian Dasar (PD) research funding scheme 2021 (No. 528/UN3.15/PT/2021)

Abstract. Indonesia geographical location that surrounded by Australia plate, Pacific plate, and Eurasia plate has made the country exposed to disasters. Natural hazard that often happen in Indonesia are flood, volcanic eruption, earthquake, landslide, land fire, cyclone, extreme wave. The impact of disaster was devastating economically, socially and psychologically for affected community as well for the whole nation. Indonesia National Disaster Management Authority (NDMA) has noted that average losses caused by disaster is 30 trillion rupiah per year. Studies found that preparing individual to face disaster could help to reduce the impact of the disaster. However, the effort is still not enough and yet Indonesia is still suffering a great loss due to disasters. The disaster preparedness program that was available was mostly focused in technically prepare people and neglect the importance of psychologically prepare people to face disaster. The purpose of the study is to analyse current disaster risk reduction policy framework in Indonesia, examine whether the current policy framework has address and includes psychological preparedness for disaster and propose a solution for designing appropriate policy to reduce psychological impact of disaster.

1 Introduction

Indonesia geographical location that surrounded by Australia plate, Pacific plate, and Eurasia plate has made the country exposed to disasters. Natural hazard that often happen in Indonesia are flood, volcanic eruption, earthquake, landslide, land fire, cyclone, extreme wave. It has been noted that during 10 year period from 2010-2020, natural disaster has caused 1.183 people died in Indonesia [1]. Indonesia National Disaster Management Authority (NDMA) has noted that average losses caused by disaster is 30 trillion rupiah per year [2]. Disasters has a devastated impact for Indonesian people not only in terms of economically but also socially and psychologically. Effort has been placed and policy has been implemented to reduce the risk and impact of disaster in order to reach one of the Sustainable Development Goals in achieving sustainable cities and communities.

One of priority for action in disaster risk reduction that are applied by Government Indonesia is improving disaster preparedness to respond effectively to a disaster. However, the effort is still not enough and yet Indonesia is still suffering a great loss due to disasters. The disaster preparedness program that was available was mostly focused in technically prepare people and neglect the importance of psychologically prepare people to face disaster [3]. According to Tanesab [4], regulatory system and policy practices of disaster risk management should be strengthened and synchronized in order to increase the effectiveness of disaster risk reduction in Indonesia [4]. Thus, it is important to analyze the disaster risk reduction policy that has been implemented in Indonesia in order to reduce the risk of disaster and achieve sustainable cities and communities as state in the Sustainable Development Goals. Policy analysis has been done to examine the disaster policy implemented in Indonesia. However, most of the study is mostly focus on insurance or tax policy that concern with post disaster financing to reduce the risk of loss caused by natural disaster [5,6]. Most research is focused more on management, legislative processes and risk reduction programs. There are five main aspects of the disaster management system that require renewal, namely institutional aspects, budgetary aspects, disaster management implementation aspects, policy aspects, and disaster types aspects [7].

The purpose of the study is to analyze current disaster risk reduction policy framework in Indonesia, examine whether the current policy framework has address and includes psychological preparedness for disaster and propose a solution for designing appropriate policy to reduce psychological impact of disaster. This

*Corresponding author: listyati.palupi@psikologi.unair.ac.id

paper presents an overview covering (1) Risk of Disaster in Indonesia (2) factors that influences disaster in Indonesia, and (3) current disaster risk reduction policy framework in Indonesia. In conclusion, a series of suggestions to address issues related to mangrove policy framework in Indonesia are offered.

2 Method

This research used document analysis method which is a systematic procedure for reviewing or evaluating documents—both printed and electronic (computer-based and Internet-transmitted) material [8,9]. The analytic procedure entails finding, selecting, appraising (making sense of), and synthesizing data contained in documents. Document analysis yields data—excerpts, quotations, or entire passages—that are then organized into major themes, categories, and case examples [8]. Content analysis was used to analyze the contents of the policies and legislations related to disaster risk reduction in Indonesia. Indonesian legal system nowadays has seven layer or hierarchy which are basic constitution, general people assembly decree, Laws and Acts, government Regulations substitute law or Act, government regulations, presidential decrees, and regional government regulations. However, this research only focuses on Acts in the central government level. Source of data that used in this research is policies documents.

3 Results

3.1 Risk of disaster in Indonesia

Indonesia is recognized as natural disaster-prone country due to its location that are surrounded by three big tectonic plate which are Australia plate, pacific plate, and Eurasia plate [3]. The Indonesia National Disaster Management Agency has recorded 11,274 disasters happen in 4 years periods from 2015 until 2019 [10].

Several province that affected by disaster during this period are Aceh, North Sumatra, West Sumatra, Riau, Jambi, South Sumatera, Bengkulu, Lampung, Bangka Belitung, DKI Jakarta, West Java, Central Java, Yogyakarta, East Java, Banten, Bali, Java, Sulawesi, West Nusa Tenggara, East Nusa Tenggara, West Kalimantan, South Kalimantan, East Kalimantan, North Kalimantan, North Sulawesi, Central Sulawesi, South Sulawesi, West Sulawesi, Gorontalo, Maluku, North Maluku, West Papua, and Papua [11]. Natural disaster that happens in Indonesia the last year are earthquake, volcano eruption, landslide, flood, drought, extreme weather, tidal wave, forest and land fire and plague [11].

3.1.1 Earthquake

The National Disaster Management Agency has mapped the area that are prone to earthquake. Indonesia island that are indicated as earthquake prone areas are Sumatra, Java, Bali, Nusa Tenggara, Maluku, Sulawesi and Papua [10].

3.1.2 Tsunami

Tsunami is generated from deformation of seabed caused by an earthquake which trigger huge sea waves when it happens in the sea. Tsunami will have a tremendous effect especially for the coastline area. Indonesia is an archipelagic country with 13.700 island that have the second largest coastline in the world [12]. Thus, most of Indonesia region becomes prone to tsunami disaster.

3.1.3 Volcanic eruption

Indonesia is in the third rank of country with the most volcanoes in the world [13]. It is recorded that there are 129 active volcanoes from 500 volcanoes that exist in Indonesia. The region that are identified prone to volcanic eruption are Sumatera, Java, Lombok, Bali, Sumbawa, Flores, Banda Sea, Sulawesi, Sangir island, Halmahera [10].

3.1.4 Avalanche

Landslide could be cause by volcanic eruption or earthquake. Since Indonesia has 139 volcanoes therefore the country has regions that are exposed to landslide. The National Disaster Management Agency has identified the Indonesia regions that are avalanche-prone area West Java, Central Java, and East Java [14].

3.1.5 Flood

Flood is also one of the disasters that happen frequently in Indonesia. Indonesia has two seasons, i.e. the hot season and rainy seasons that could potentially create hydro-meteorological hazards such as flooding and drought. Aceh, North Sumatera, Riau, Jambi, South Sumatera, Bengkulu, Lampung, Bangka Belitung, West Java, Central Java, East Java, Yogyakarta, Banten, West Nusa Tenggara, West Kalimantan, Central Kalimantan, South Kalimantan, East Kalimantan, Central Sulawessi, West Sulawesi, Northwest Sulawesi, North Maluku, West Papua, Papua is considered to have a high risk of flood disaster [15].

3.1.6 Drought

Drought is another hydrometeorological hazards that often impacts Indonesia. Several districts/cities in Indonesia that are identified as drought zone are West Sumatra, Riau, Jambi, South Sumatra, Bengkulu, Lampung, Bangka Belitung Islands, Riau Islands, nearly all districts/cities in Java Island, except for those situated in the easternmost part of Java Island, and several districts/cities in East Nusa Tenggara Province, West Kalimantan, Central Kalimantan, South Kalimantan, East Kalimantan, North Sulawesi, South Sulawesi, West Sulawesi, North Maluku and Papua [10].

3.1.7 Forest and land fire

Indonesia is position at the top 8 of countries with the widest forest area [16]. Forest fire is one of natural hazard that often strike Indonesia regions several regions that are in high risk of forest and land fires include Sumatra islands, Kalimantan Islands, Sulawesi islands, East Nusa Tenggara islands, and Java Island [17].

3.2 Factors that influence disaster in Indonesia

Natural hazard is often happened in Indonesia. There are several factors that influence the magnitude impact of disaster toward Indonesian society. Society vulnerability to natural hazard could be grouped into physical, social, and economical vulnerability [18]. The distance of a community to the source of hazard, poorly planned development, poor quality infrastructure, ground factors, building physical conditions are factors that influence the society vulnerability toward disaster [10,19]. Social and economic vulnerability that influence the community susceptibility to disaster is socioeconomic status, marital status, place of residence, age head of household, built environment [20].

3.3 Current disaster risk reduction policy framework in Indonesia

There are 2 parliamentary laws, 4 Presidential Decree, 3 government regulations, 1 ministerial decree and 102 regulations that relate to disaster risk reduction activities in Indonesia. However, this study is only focus in analyzing 2 parliamentary laws that become the basis in all of disaster risk reduction effort and activities which are Act No. 24 of 2007 and Act No. 26 of 2007.

3.3.1 Law of the Republic of Indonesia number 24 of 2007

Act Number 24 of 2007 is the central law that regulate the disaster management activities which consists of 13 Chapters and 80 Articles. The main material outline the principles and aims of disaster management, inter organization role in disaster management which include responsibility and authority of the government and regional government, the role of National Disaster Management of Indonesia at the state and regional level, the right and obligation of the society, disaster management scope, disaster management stage and the description of action in each of the stage; financial source of disaster management activities, and the role of international organizations in disaster management implementation.

Law No 24 of 2007 state that the central government is authorized to declare the status and level of disaster by considering the number of people impacted, the infrastructure damage, loss of property, facilities damage, the coverage of the disaster, and social and economic impact of disaster [21]. Study found that psychological preparedness for disaster could influence the magnitude of disaster impact for the community [22–24]. Beside of that, psychological preparedness for disaster also correlated with resilience [25]. Therefore, psychological preparedness for disaster could become one of indicators that could be use to determine the status and level of disaster. However, study found that although psychological preparedness for disaster has been acknowledged as an important factor in disaster management that could influence the magnitude of disaster impact for the community nevertheless psychological preparedness for disaster is remain unnoticed [3]. It is also true that psychological preparedness for disaster is still forsaken in the Law No 24 of 2007. Therefore, disaster preparedness of the community should be included as one of measured indicator to declare the status and level of disaster. Hence, disaster preparedness should be state in Chapter 3 Article 7 Paragraph 1 and 2.

The role and authority of government and supporting agency also stated in of the Act No 24 of 2007. The National Disaster Management Agency according to Chapter 3 article 12 is assigned to manage the disaster management activities which includes providing guideline and direction on disaster prevention, emergency responses, rehabilitation, and reconstruction phase [21]. Although it has mention disaster prevention as part of the disaster management however disaster preparedness is not yet account. Study found that psychologically preparing the community to face disaster will have a significant influence in reducing the mental heath risk of being exposed to disaster. Therefore, considering the importance of psychological preparedness for disaster, the disaster preparedness especially psychological preparedness for disaster should also be assign as the National Disaster Management Agency task that state in Chapter 3 Article 12 paragraph 1 of Law No 24 of 2007. Furthermore, psychological preparedness for disaster should also inserted into disaster management planning that stated in Chapter 3 Article 36 Paragraph 4 of Law No 24 of 2007. In addition, the psychological preparedness for disaster could also be included as part of alertness that could be clearly describe in Article 45 Paragraph 2, Article 45 Paragraph 2 and Article 47 Paragraph 2 of Law No 24 of 2007.

3.3.2 Law of the Republic of Indonesia number 26 of 2007

The Law on Disaster Management Number 26 of 2007 is law that stipulate the spatial planning regulation in Indonesia. The Act consist of 80 articles divided into XIII Chapters. The main point stressed in the Act are the spatial planning principles, spatial planning classification, the role and responsibility of the authority in spatial planning, establishment of spatial planning, spatial management, supervision in spatial planning, the right and role of the community.

Spatial management regulation is an effort to promulgate the legal basis for the Government, local government and society in spatial management. The Act regulate the spatial planning in Indonesia and state that the spatial management should follow several principles which are integrity, compatibility, harmony and balance, sustainability, productivity and profitability, openness, togetherness and partnership, protection of public interest, legality and justice, and accountability.

The Law No 26 of 2007 has taken into account disasters factors in planning the spatial structure and pattern. Disasters has stated as part of Law No 26 of 2007 and mention in Article No 6 Paragraph 1 about Spatial Planning Classification, Article 20 Paragraph 5 about national spatial plan, Article 23 Paragraph 5 about provincial spatial plan, Article 26 Paragraph 6 about regency spatial plan, and Article 28 Paragraph 1 about municipal spatial planning [26]. This is shown that disasters are account in the process of designing a spatial plan for rural area, agripolitan area, urban area, metropolitan area, megapolitan area. However, disaster is not specified in the spatial management establishment, execution, and supervision.

Considering the devastating impact of disaster. Therefore, disaster should also include in Chapter 3 article 4 of Law No 26 of 2007 as one of the factors that used to classify spatial management. Furthermore, disaster should also be use to determine the content of spatial planning including the space structure and space pattern especially in province, regency, and municipal planning which is state in Chapter IV article 7 of Law No 26 of 2007. Beside including in Chapter III and IV of Law no 26 of 2007, disaster should also be include in execution of spatial management that stated in Chapter VI Section 1 Section 2, section 3, section 4, and section 5.

4 Conclusions

Based on the analysis of current Indonesia government legal framework that associated with disaster management, there are several recommendations that could be implemented to improve the effectiveness of disaster risk effort by specifically describe the focus of disaster risk reduction effort to not only focus in reducing the risk in general but also to include psychological preparedness as an integrated part in the disaster management planning.

Based on the analysis of potential disaster that might strike Indonesia as a result of geographical location and also the document analysis of Government Law that govern disaster management, this article formulated several key recommendations to strengthen the Indonesia's legal framework for Disaster Management as mean to reduce the risk of disaster for the nation, as well as addressing some of the major gaps especially in related to psychological preparedness for disaster. The following are the recommendations:

1. Amendments of Disaster Management Law which are Law No 24 of 2007 by including disaster preparedness especially psychological preparedness for disaster in the Chapter and Article specifies below:
   a. Chapter III Article 7 paragraph 1 and 2,
   b. Chapter III article 12 paragraph 1,
   c. Chapter III Article 36 Paragraph 4, No 24 of 2007,
   d. Chapter VII Article 45 Paragraph 2, Article 45 Paragraph 2 and Article 47 Paragraph 2.
2. Amendments Law No 26 of 2007 that govern spatial management planning to support the Law No 24 of 2007 by integrating disaster factor and propose the following amendments:
   a. Chapter III article 4,
   b. Chapter IV article 7,
   c. Chapter VI Section 1,
   d. Chapter VI Section 2,
   e. Chapter VI section 3,
   f. Chapter VI section 4,
   g. Chapter VI section 5

References

1. D. B. Raharjo and S. Aranditio, 1 (2021)
2. R. Dwinanda, 1 (2021)
3. L. S. Palupi, in *Sustain. Islands Dev. Initiat. – Int. Conf. 2019* (IOP Conference Series: Earth and Environmental Science, 2021), pp. 1–9
4. J. P. Tanesab, Int. J. Disaster Manag. **3**, 1 (2020)
5. Kalfin, S. S. Sukono, M. Mustofa, and A. T. Bon, 1825 (2020)
6. D. Dubelmar, M. Astrin, D. Kartini, and S. Mareli, **5**, 243 (2020)
7. T. A. Syadzily, 1 (2020)
8. G. A. Bowen, Qual. Res. J. **9**, 27 (2009)
9. S. L. Dalglish, H. Khalid, and S. A. McMahon, Health Policy Plan. **35**, 1424 (2020)
10. S. Maarif, *National Disaster Management Plan (Rencana Nasional Penanggulangan Bencana 2015-2019)* (2014)
11. BNPB, (2021)
12. Oishimaya Sen Nag, WorldAtlas 1 (2020)
13. Horsemart, 1 (2015)
14. felicia F. Facette, 1 (2018)
15. Badan Nasional Penanggulangan Bencana (BNPB), *Indeks Risiko Bencana Indonesia (IRBI) Tahun 2020* (Badan Nasional Penanggulangan Bencana (BNPB), 2020)
16. Worldometer, 285 (2019)
17. S. P. Nugroho, Badan Nas. Penanggulangan Bencana 5 (2017)
18. National Agency for Disaster Management of Indonesia, *Indonesia National Disaster Management Plan 2010-2014* (Jakarta, 2010)
19. G. Yüce, G and Arun, in *15 WCEE* (Lisboa, 2012)
20. M. Llorente-Marrón, M. Díaz-Fernández, P. Méndez-Rodríguez, and R. G. Arias, Sustain. **12**, 1 (2020)
21. Republic of Indonesia, *Law of The Republic of Indonesia Act No 24 of 2007* (Indonesia, 2007), pp. 588–595
22. S. A. Morrissey and J. P. Reser, Aust. J. Emerg. Manag. **18**, 46 (2003)
23. J. Roudini, H. R. Khankeh, and E. Witruk, Heal. Psychol. Open **4**, 1 (2017)
24. J. Roudini, H. R. Khankeh, E. Witruk, A. Ebadi, K. Reschke, and M. Stück, Heal. Emergencies Disasters Q. **2**, 165 (2017)
25. L. S. Palupi and M. Noor Rahman Himawan, in *E3S Web Conf.*, edited by B. W. Sudarno and T. . Putranto (E3S Web Conference, 2020), pp. 0–5
26. Republic of Indonesia, *Law of The Republic of Indonesia Act No 26 Year 2007* (Indonesia, 2007), pp. 1–95

Smong means more than tsunami: The understanding of tsunami in the Indonesian context

Alfi Rahman[1,5*], *Aiko* Sakurai[2], *Stephen* Sutton[3], *Mahdi* Syahbandir[4], and *Nofriadi* Nofriadi[4]

[1]Pusat Riset Ilmu Sosial dan Budaya (PRISB), Universitas Syiah Kuala, Darussalam, Banda Aceh, Indonesia
[2]Department of Social Sciences, Toyo Eiwa University, Japan
[3]College of Health and Human Sciences, Charles Darwin University, Darwin, Australia
[4]Faculty of Social and Political Sciences, Universitas Syiah Kuala, Darussalam, Banda Aceh, Indonesia
[5]Tsunami and Disaster Mitigation Research Center (TDMRC), Universitas Syiah Kuala, Banda Aceh, Indonesia

Abstract. *Smong* is an indigenous knowledge related to the tsunami. The word of *smong* comes from the Devayan language of Simeulue and refers to the complex of earthquake/sea receding/giant wave that is typical of tsunami events in Indonesia. Now their pride in the word *smong* has spread to Aceh and is being adopted in their efforts to adapt to earthquake and tsunami disasters. Simeuluean people, grateful to their nation for the reconstruction after 2004 and want to grow their pride in *smong* to a national level. They wanted to make a gift of *smong* to be owned by the Indonesian people. To do this, they feel this study initiates that the word *smong* should be part of the narrative in Indonesian as the basis for a new understanding of the tsunami disaster by Indonesian people. Initiating the word *smong* in Indonesian vocabulary does not mean removing the word tsunami that now an international word. But the acceptance of *smong* should be a source of pride in the protection of our greatest wealth of people.

1 Introduction

A tsunami is a rare but potentially devastating natural hazard that may impact a large proportion of the world's coastal population. To the extent that the public has any awareness of the tsunami phenomenon, they have traditionally been associated with the islands and rim of the Pacific Ocean, where their origin was attributed to large, distant, underwater earthquakes [1]. While the preponderance of tsunami is initiated in the Pacific [2], the tsunami on 26 December 2004 provided a tragic reminder that the Indian Ocean is also a source of tsunami hazards, with the Sunda Megathrust, in particular, generating irregular but devastating tsunami [3–5].

Over the past 2000 years, there have been 716,548 deaths attributed to tsunamis from these two oceans [1]. The largest death toll from a single event is concentrated in Aceh, Indonesia, where more than 230,000 people were killed by the 2004 Indian Ocean tsunami [1,6,7].

However, Indonesia's tsunami exposure is not limited to the oceanic tsunami; in addition to being bookended by tsunamigenic oceans, extensive faulting under the smaller water bodies between the archipelago's exposes a large proportion of the country's coastline to near- or mid-field tsunami. In this seismically intense environment, every coastal community in Indonesia with 16,056 islands [8] will have been required to confront the reality of tsunami at some stage, and it is important both locally and nationally to understand how local communities accepted and perceived tsunami events in helping them to make decisions and actions about future similar situations.

Unfortunately, the periodicity of tsunami is highly irregular, with gaps of many hundreds, even thousands of years, not being uncommon. Over these extensive timeframes, community knowledge of the concept of the tsunami may atrophy with traditional stories or songs sequestered in arcane folklore and not the subject of everyday risk-management conversations. Confoundingly, the occurrence of a large event such as the 2004 Indian Ocean tsunami does not mean Indonesian's will now be safe for hundreds of years [9] and Palu and Donggala tsunami of September 2018 show[10].

The tsunami experiences in Indonesia have impacted community understanding of the phenomenon and have also contributed to the significant change in scientific paradigms of disaster [11,12]. Provoked by the 2004 tsunami, a range of scientific disciplines have sought to refine our understanding of the cause, timing, impacts, and future potential for tsunami in Indonesia and potentially elsewhere [4]. The finding that many locations on the Indian Ocean coast areas of Indonesia have been impacted upon by several tsunamis [4,13] has focused attention on gaps in scientific understanding as well as disaster risk reduction (DRR) processes calls for dramatic improvement in disaster education [14].

* Corresponding author: alfi.rahman@unsyiah.ac.id

These sentiments have elicited strong policy support at a national and international level, with the United Nations International Strategy for Disaster Reduction (UNISDR) identifying the nexus between the occurrence of natural hazards and their impact on physical, social, economic, and environmental vulnerabilities of exposed populations [15]. The complexity of the tsunami hazard, i.e., that it can be triggered by a range of factors not limited to the submarine earthquake and that it can impact different sections of coast in different ways, contributes to its potential for causing deaths and social and economic destruction [12]. This could also be pointed from what had happened in Palu, Donggala, and the tsunami in Sunda Strait on 24 December 2018[16–19].

But, these complexity events like the 2004 Indian Ocean tsunami show that there is no inevitability of destruction and chaos. Apart from indirect positive effects such as the stimulus to science [12,14,20] and the crystallization of international policy, the more holistic context for natural hazard (and tsunami in particular) research has provided an examination of direct examples of effective disaster mitigation practices.

Because of their context, these (rare) examples tend to be categorized as traditional or indigenous knowledge [21], and the good sense of remote local populations is lauded as instructive. Within the more holistic context promoted by UNISDR, however, understandings from indigenous knowledge and scientific research can perhaps be combined to develop a more comprehensive understanding of the natural hazard such that effective strategies could be generated and sustained to reduce or eliminate the risk of impacts. In many parts of Indonesia, the local communities have practiced their indigenous knowledge in reducing the worst impact of disasters.

2 Research purpose

This paper is based upon research into one famous example of indigenous knowledge that saved lives from the 2004 tsunami: Simeulue Island. The study discusses findings derived from two parallel research programs, particularly one compelling relating to the Simeulue community's awareness of the nature of tsunami risk and how to respond to it. A word to encompass the tsunami phenomenon is central to the Simeulue story. This study argues that, by extension, a new word for this most devastating of natural hazards should be the beginning of a new Indonesian narrative about disaster risk reduction. The word that saved people on Simeulue is *smong*.

3 Research method

This study presents results deriving from two parallel research programs focusing on Simeulue Island. Both applied a qualitative research model, with one focusing on semi-structured interviews with key personnel (current and ex-government officials) while the second conducted narrative interviews with members of the general population. The data was collected in 2016 and 2017.

The data was analyzed using a variety of qualitative research approaches, including narrative analysis and ground theory. The interviews were transcribed and coded using qualitative research software (Nvivo™) to identify variously the key social/cognitive factors in the success of the people of Simeulue in 2004 and the system of information management used in the disaster risk reduction program during the 20th Century.

The analysis of more than 80 interview transcripts included the consideration of recurrent and repeated and shared concept themes, phrases, and observations from both emic and etic perspectives [22,23]. These have been examined and triangulated with a view to building middle-range theory [24,25] with cross-cultural applicability.

This paper refers to one theme that arose from the expressed concerns of individual interviewees and from the breadth of observations of the population in which the research effort was embedded.

4 Tsunami as a phenomenon and a name

The term tsunami has now gained international currency, a legacy of the frequency and length of historical documentation of the natural hazard. Japan appears to have the highest global frequency, with 25.2% of all tsunami events impacting the Japanese coast [1,7].

A hundred years ago, researchers began to use the term tsunami to describe the phenomena. According to Cartwright and Nakamura (2008), the first occurrence of the word tsunami was found in a journal kept by a retainer of the shogun Tokugawa Ieyasu [26]. In receiving news about the Sanriku earthquake of 2 December 1611, fishermen around the Sanriku region called it a tsunami, and they might have used the term because its effects were most evident when they arrived back at their harbor, only to find their villages had been washed away [26].

The term *tsunami* derives from two Japanese words and means harbor (tsu) and wave (nami) [1,7,27–29]. But the Japanese language also has other terms to describe tsunamis, such as *onami* (large wave), *shikai namisu* (waves rise in all direction), and *kaisho* (roaring and resounding sea) [26].

Many languages from countries with a strong maritime connection also has a special expression for this type of disaster coming from the sea such as tidal waves, seismic sea waves (English), *raz de mare, vagues sismiques* (French), *flutwellen* (German), *maremoto* (Spanish), *vloedgolf* (Holland), *hai-i* (Chinese), *loka* (Fijian) [7]. Subsequently, the Japanese term has been adopted worldwide, and many scholars and scientists have used tsunami to describe the phenomena despite the presence of many local alternatives.

The most commonly accepted definition by natural scientists of a tsunami is a wave or series of waves in a wave train generated by the sudden, vertical displacement of a column of water [1]. This displacement could be due to seismic activity, explosive volcanism, landslides above or below water, asteroid impact, or certain meteorological phenomena [1,30]. Tsunamis could be generated in oceans, bays, lakes, rivers, or reservoirs, each of which lends its own character to the manifestation of the natural hazard [1].

5 The 2004 Indian Ocean tsunami and the story of *smong*

On Sunday morning, 26 December 2004, an earthquake of M_w of 9.2, followed by a tsunami, devastated Aceh Province in Indonesia and other coastal countries around the Indian Ocean [31,32]. The epicenter of this earthquake was located 250 km Southwest of the Aceh and created a 1,500 km rupture along the Sunda megathrust [31]. Slippage along the fault of up to 20 meters resulted in a series of powerful waves that pounded the coastal areas along the Indian Ocean [33]. The tsunami violently swept up to 6 km inland in the provincial capital Banda Aceh and inundated over 500km of mainland Aceh's coast and nearby islands such as Simeulue [34]. More than 230,000 people were certainly killed, and an additional 93,285 people were declared missing [35]. Some 500,000 survivors lost their homes, while as many as 750,000 people lost their livelihood [35,36].

On that Sunday morning, 2004 Sahmadi[†] was a student living with friends in Banda Aceh. He and his friends were shocked by a massive earthquake. The quake went on for a long time, leaving them disoriented and afraid. As it came to an end, Sahmadi yelled *'Smong,! Smong!'* and urged them to run to high ground. But his friends didn't move. Instead, they started laughing at the word *smong* and called him a rural hillbilly. Before long, he persuaded them to evacuate to high ground with the result that they were among the survivors of the massive wave that destroyed Banda Aceh and killed more than 230,000 people.

Sahmadi comes from the island of Simeulue. It is the northernmost and lasts in the long chain of islands between Sumatra and the Sunda Megathrust. In 2004 it had a population of about 70,000 people and lay only 50 km from the tsunamigenic 9.2 M_w earthquake. In the days following the disaster, during the communications blackout caused by the destruction of infrastructure, it was generally assumed that Simeulue was 'drowned', and those massive fatalities were to be expected. In fact, Sahmadi heard media reports that said as much. Against this background, surprise exceeded relief when it was found that in fact, there were hardly any fatalities on Simeulue at all. The most commonly quoted figure is seven, while locals dispute this and say six were killed by the earthquake and only three by the actual tsunami [37]. Anecdotes tell of the shock of Indonesian military helicopter pilots seeing thousands of people smiling and waving from the hills near the coast of Simeulue. The locals also say that when SBY (Susilo Bambang Yudiyono), the former president of Indonesia, heard that so few had died, he said, "Impossible!".

But it clearly was 'possible', and before long, researchers were working beside aid workers to find out the secret of the survival of the people of Simeulue. What they found seemed simple; the locals have a traditional story about *smong* - the giant wave that follows an earthquake and recession of the seawater which runs up onto the land, destroying buildings and drowning everything in its path-. *Smong* is the word for this phenomenon in the local languages of Simeulue. Kähler describes the local language of Simeulue as 'syncretic', one that absorbs words and influences from interaction with other languages [38]. But the word *smong* does not occur in the many languages of the traders, travels, and migrants that have influenced Simeulue, and it seems to be an entirely local invention. No-one knows exactly where the word came from or how long it has existed, and some locals report that it is an onomatopoeic process [39] and represents the deep booming explosion of the wave[‡]. The origin can only be speculative, but it is clear that the word was an important part of the culture of the island from the early 20th Century.

In the 20th Century, the history of *smong* pivots on a similar massive earthquake and tsunami. The *smong* occurred on Friday, 4 January 1907 [40]. This wave is reported to have been far larger than that in 2004 and killed a great many people, almost wiping out the entire population [39,41]. The survivors, those who, by luck or good judgment, ran to the nearby hills, felt compelled to urge their descendants to be ready for *smong*: if there is a big earthquake and the sea recedes, run to the mountains. This story is the basis of the 'kearifan smong' - the wisdom of *smong*- and it is the core of the possibility of saving the lives of everyone in Simeulue in 2004.

Upon examination, the story of *smong* is not simple but quite profound. First, it is probably the only word in any language specifically for the phenomenon now widely referred to as 'tsunami'. Coming from Devayan, the main language of Simeulue, *smong* specifically refers to the complex of earthquake/sea receding/giant wave that is typical of near-field tsunami events in Indonesia (see Figure 1). This clarity of meaning is not shared by other terms. In the English-speaking world, the common term for the massive wave event used to be 'tidal wave'. This was a clear misnomer insofar as the giant wave was not generated by or even related to the tide. As mentioned, the widespread adoption of the Japanese tsunami also incorporates a misnomer. 'Tsu' is the word for harbor, and 'nami' means wave. That is, 'tsunami' means 'harbor wave'. The term was adopted by fishermen returning from the sea, where they had not noticed any unusual wave activity, only to find the port and its town devastated by a large wave [26]. They concluded that the wave affected the harbor and hence the name. Of course, as the Indian Ocean tsunami event of 2004 shows, the wave affects entire coastlines and is not restricted to harbors and ports.

† Not his real name. Interview recorded 9 December 2016

‡ Interview recorded 6 August 2017

 https://doi.org/10.1051/e3sconf/202234003010

In contrast, the word *smong* precisely refers to the phenomenon of a massive wave generated by an earthquake. This is much more than a linguistic nicety. Anthropologists have long demonstrated the important connections between language and culture' and the importance of efficiently sharing a concept or idea. In a discussion that pre-supposes aspects of Kahneman and Tversky's identification of the availability heuristic in human cognition [42], Brown and Lenneberg state that more nameable categories of phenomena are closer to the top of the human cognitive 'deck' [43]. These ideas still have currency in consideration of language and human perception, supporting the straightforward concept that the ready availability of an often-used term aids an individual's ability to access memories and process information about the defined phenomenon [44]. Simeulue appears to be one of the few places on earth where the culture specifically provides socio-cultural and individual space for the giant disastrous wave phenomenon. In interviews, people from Simeulue often express more than simply an 'awareness' or 'knowledge of *smong*; in attributing it to the reason for their survival in the 2004 tsunami [17].

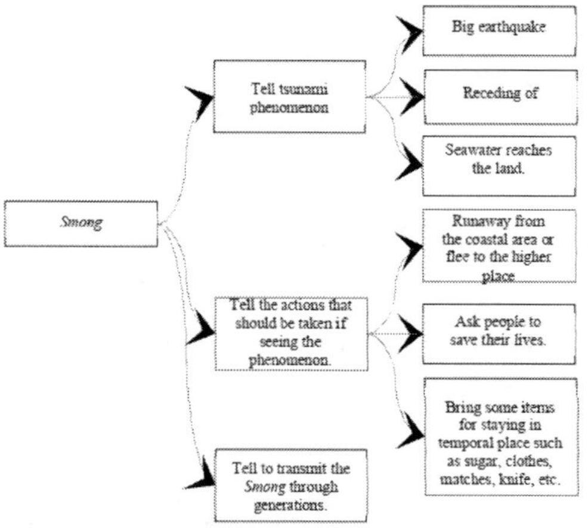

Fig. 1. The summary of the *smong* means in *nafi-nafi*. Adapted from Rahman (2018) [45]

This ownership seems to be particularly strongly captured in the traditional songs of Simeulue, the *nandong*, the classic vehicle for the transmission of the *smong* story. The *nandong* is a traditional song format in the local language. A sort of Simeulue 'Blues', it commonly incorporates sad and mournful tales of love and loss, and a 'standard' is, of course, the massive loss of family and loved ones during the *smong* event in 1907. It is hard to imagine anyone hearing the *nandong smong* failing to remember its unsettling messages.

The *nandong* is performed at special events and is a staple at weddings where extended families and community members get together to celebrate. But it is also simply part of the weekly dialogue of families in their homes where grandma or grandpa might sing *nandong* with their grandchildren. More recently, local musicians and historians have collaborated to record a more modern version of the traditional song [37]. Adding verses to the story while maintaining the original words in the style of popular music in Indonesia.

The original words of the 'nandong smong' extend beyond mere description of the phenomenon but draw the listener to ponder their personal relationship with nature, particularly the giant destructive wave. Table 1 below sets out four concluding lines from the *nandong*, first in the most widely used local language, Devayan, then Indonesian, and then English.

Table 1. The *smong* in nandong lyric which contains the metaphysical

Devayanese	Indonesian	English
Smong dumek-dumek mo	*Smong mandi-mandimu*	*Smong* is your bath
Linon uwak-uwak mo	*Gempa ayunanmu*	Earthquake is your (rocking) cradle
Kilek suluh-suluh mo	*Kilat suluh-suluhmu*	Lightning is your torch
Eklaik kedang-kedang mo	*Petir gendang-gendangmu*	Thunder is your drum

Data source: re-performed on 8 January 2017, by a former Simeulue Regency leader who served from 2001 to 2006 [37,46]

6 Discussion

The sentences of the *smong* in nandong are extraordinary. While other parts of the *nandong* provide a warning about the *smong* and the need to run to the mountain after the signs of earthquake and sea recession, these four lines seem to encourage people to embrace natural hazards, with *smong* as merely the beginning. To 'get into' the *smong* as if it is your bath! Is a number of those interviewed stressed this clear metaphysical element? Mr. Haris[§] has a strong interest in the traditional music of Simeulue and described the 'smong dumek dumekmo' passage as being critically important and something young people need to understand; "It has philosophical meaning," he said.

This indicates the degree of sophistication of the 'wisdom of *smong*'. Through the telling of stories and singing of songs, every individual on the island has a readily accessible understanding of the physical manifestation of the natural hazard and a philosophical and metaphysical connection. These deep cognitive linkages pivot on the word *smong*.

On the other hand, the Simeuluean people could also be neglected to consider the disaster prevention or mitigation countermeasure held by related stakeholders due to already having their *smong* knowledge and

[§] Not his real name. Interview record 6 August 2017

feeling that *smong* could help them face the future disaster.

There is a definite metaphysical element that has to be verified to get into "the *smong* is your bath" that could be referred to the *nandong* It does indeed go on to encourage people that the best thing to do in case of the *smong* is to run to the mountains. It is hard to imagine people hearing that *smong* lyric in *nandong* can fail to translate the *smong* DRR message [45].

The Simeuluean people seem very optimistic about the success of the *Smong* story in 2004, but there is a probability of the decreasing of the recognition of the *smong* story and DRR efforts. Even though the *smong* is much more than a word, it also had an extraordinary experience of saving lives. However, further development of the *smong* story as DRR efforts is needed because it is difficult to believe that the *smong* story can still be reliable and applicable to a future disaster.

Smong encapsulates the phenomenon, and the story of *smong* resonates with all the people of Simeulue. The word was universally known in 2004 and given a wide airing in national and international media throughout 2005.

In more recent years, the combination of ubiquitous digital media and entertainment and constant references to 'tsunami' may dilute the knowledge and interest in *smong* in Simeulue.

7 Conclusions

The *smong* story demonstrates the important connections between indigenous knowledge and DRR. This research has demonstrated that having a word that encapsulates the phenomenon and is widely known and accessible in the community is one practical example of operationalizing indigenous knowledge to save lives.

One way to address this threat is to grant the word *smong* the world recognition it deserves. As the word that encapsulates the giant destructive wave phenomenon, 'smong' should be recognized worldwide. And if not everywhere, at least in Indonesia as its ' homegrown' word for this natural phenomenon. Many people in Simeulue express their undying gratitude to the world who came to their aid to rebuild after their island was devastated. They wish to give something back, but they especially want to show their appreciation to their Indonesian countrymen. They wish to give the gift of *smong* in contributing to DRR.

Making 'smong' the Indonesia narrative for tsunami would not only be a huge gift to the people of Indonesia, but it would also be a fitting recognition of the remarkable resilience of the people of Simeulue.

Acknowledgment. The authors gratefully acknowledge the Institute for Research and Community Services (Lembaga Penelitian dan Pengabdian kepada Masyarakat, LPPM) Universitas Syiah Kuala, who fully supported and funded this research.

References

1. E. Bryant, *Tsunami: The Underrated Hazard* (Springer, 2014)
2. B. K. Rastogi and R. K. Jaiswal, Science of Tsunami Hazards (2006)
3. A. J. Meltzner, K. Sieh, H. W. Chiang, C. C. Shen, B. W. Suwargadi, D. H. Natawidjaja, B. Philibosian, and R. W. Briggs, Journal of Geophysical Research: Solid Earth (2012)
4. C. M. Rubin, B. P. Horton, K. Sieh, J. E. Pilarczyk, P. Daly, N. Ismail, and A. C. Parnell, Nature Communications **8**, 16019 (2017)
5. K. Sieh, in *From the Ground Up* (2018)
6. BRR, *Supervision Eradicating Corruption with No Tolerance*, BRR Book S (BRR NAD-NIAS, 2009)
7. V. K. Gusiakov, The Sea **15**, 23 (2009)
8. BPS Pusat, *STATISTIK INDONESIA: Statictical Year of Indonesia 2018* (BPS-Statistics Indonesia, 2018)
9. K. Sieh, P. Daly, E. Edwards McKinnon, J. E. Pilarczyk, H. W. Chiang, B. Horton, C. M. Rubin, C. C. Shen, N. Ismail, C. H. Vane, and R. M. Feener, Journal of Geophysical Research: Solid Earth (2015)
10. A. Rahman, The Conversation (2018)
11. H. Rodriguez, T. Wachtendorf, J. Kendra, and J. Trainor, Disaster Prevention and Management: An International Journal **15**, 163 (2006)
12. E. N. Bernard and A. R. Robinson, The Sea **15**, 1 (2009)
13. K. Monecke, W. Finger, D. Klarer, W. Kongko, B. G. McAdoo, A. L. Moore, and S. U. Sudrajat, Nature (2008)
14. C. E. Synolakis and E. N. Bernard, Philosophical Transactions of the Royal Society A: Mathematical, Physical and Engineering Sciences (2006)
15. UNISDR, 1 (2013)
16. A. Rahman and A. S. Sutton, The Conversation (2018)
17. A. Rahman and K. Munadi, in *IOP Conference Series: Earth and Environmental Science* (IOP Publishing, 2019), p. 12040
18. Fachrurradhi, N. R. M. Susanti, A. Pratama, R. Fitri, S. Raudhiah, A. Rahman, and K. Munadi, in *IOP Conference Series: Earth and Environmental Science* (IOP Publishing, 2019), p. 12043
19. S. Adella, M. F. Nur, N. Nisa, A. T. Putra, F. R. Ramadhana, F. Maiyani, K. Munadi, and A. Rahman, in *IOP Conference Series: Earth and Environmental Science* (IOP Publishing, 2019), p. 12041
20. Fatimahsyam, S. I. Shadiqin, and A. Rahman, Humanities & Social Sciences Reviews **6**, 16 (2019)
21. R. Shaw, N. Uy, and J. Baumwoll, United Nations International Strategy for Disaster Reduction, Bangkok (2008)
22. J. W. Berry, International Journal of Psychology (1989)

23. H. Helfrich, Culture and Psychology (1999)
24. R. Boudon, Contemporary Sociology (1991)
25. R. K. Merton, *Social Theory and Social Structure* (The Free Press, 1968)
26. J. H. E. Cartwright and H. Nakamura, Notes and Records of the Royal Society of London **62**, 151 (2008)
27. E. O. L. Dictionaries, (n.d.)
28. B. Zimmer, (n.d.)
29. S. S. Duffin, NPR (2011)
30. J. Kafle, P. R. Pokhrel, K. B. Khattri, P. Kattel, B. M. Tuladhar, and S. P. Pudasaini, Annals of Glaciology **57**, 232 (2016)
31. M. Chlieh, J.-P. Avouac, V. Hjorleifsdottir, T.-R. A. Song, C. Ji, K. Sieh, A. Sladen, H. Hebert, L. Prawirodirdjo, Y. Bock, and J. Galetzka, Bulletin of the Seismological Society of America **97**, S152 (2007)
32. A. J. Meltzner, K. Sieh, M. Abrams, D. C. Agnew, K. W. Hudnut, J.-P. Avouac, and D. H. Natawidjaja, Journal of Geophysical Research: Solid Earth **111**, n/a (2006)
33. C. Subarya, M. Chlieh, L. Prawirodirdjo, J.-P. Avouac, Y. Bock, K. Sieh, A. J. Meltzner, D. H. Natawidjaja, and R. McCaffrey, Nature **440**, 46 (2006)
34. BRR, *Aceh and Nias One Year After The Tsunami: The Recovery Effort and Way Forward* (2005)
35. BRR, *Tsunami: From Disaster to the Emergence of Light*, BRR Book S (BRR NAD-NIAS, 2009)
36. BRR, *Economy: Turning the Wheel of Life*, BRR Book S (BRR NAD-NIAS, 2009)
37. A. Rahman, A. Sakurai, and K. Munadi, International Journal of Disaster Risk Reduction **29**, 13 (2018)
38. C. Hooykaas, Bulletin of the School of Oriental and African Studies **25**, 421 (1962)
39. T. A. Sanny, *The Smong Wave from Simeulue: Awakening Strategic Development of Regency of Simeulue* (Pemerintah Kabupaten Simeulue, Simeulue, 2007)
40. H. Kanamori, L. Rivera, and W. H. K. Lee, Geophysical Journal International (2010)
41. B. G. McAdoo, L. Dengler, G. Prasetya, and V. Titov, Earthquake Spectra **22**, 661 (2006)
42. A. Tversky and D. Kahneman, Science (1974)
43. R. W. Brown and E. H. Lenneberg, Journal of Abnormal and Social Psychology (1954)
44. A. Majid, S. G. Roberts, L. Cilissen, K. Emmorey, B. Nicodemus, L. O'Grady, B. Woll, B. LeLan, H. de Sousa, B. L. Cansler, S. Shayan, C. de Vos, G. Senft, N. J. Enfield, R. A. Razak, S. Fedden, S. Tufvesson, M. Dingemanse, O. Ozturk, P. Brown, C. Hill, O. Le Guen, V. Hirtzel, R. van Gijn, M. A. Sicoli, and S. C. Levinson, Proceedings of the National Academy of Sciences (2018)
45. A. Rahman, Development Processes of Indigenous Knowledge on Tsunami Risk Reduction to Increase Community Resilience: The Case of the Smong in Simeulue Island, Aceh, Indonesia, 2018
46. A. Rahman, A. Sakurai, and K. Munadi, in *IOP Conference Series: Earth and Environmental Science* (IOP Publishing, 2017), p. 012018

The change of land use and land suitability of agriculture due to flow liquefaction in Palu and Sigi valley after the Palu earthquake in September 2018

Sukiman Nurdin[1*], *Amar* Akbar Ali[2], *F.A* Marhum[3], *Sriyati* Ramadhani[1]

[1]Civil Engineering, Tadulako University, Palu, Central Sulawesi 94118, Indonesia
[2]Architectural Engineering Tadulako University, Palu, Central Sulawesi 94118, Indonesia
[3]Geologycal Engineering Tadulako University, Palu, Central Sulawesi 94118, Indonesia

Abstract. After the Palu earthquake in 2018, there were many paddy fields has been impacted by liquefaction in three locations, namely, Sibalaya is around 30 Ha, Jono Oge is about 1784 ha, and Petobo was affected around 388 Ha. The entire paddy fields affected are about 2.202 Ha. The procedure model was used to analyse the most suitable location for the rainfed agricultural and irrigated crops after the Palu Earthquake in September 2018. Four levels of land suitability class were pre-pared based on five criteria: rainfall, temperature, slope percentage, soil types, and the spatial distribution of groundwater wells. The process in land suitability analysis with GIS after the suitability class of each parameter is an overlay process after some exploration and testing with the Land; the results show about 65% of the paddy fields land is in category S2; this limiting factor will affect its productivity, requiring additional inputs. The land users themselves can usually overcome these boundaries. And others about in class S3 is about 35% Paddy fields Land has a heavy limiting factor, which will significantly affect its productivity, requiring more additional input than Land Classified as S2. To overcome the limiting factor in S3 requires high capital, so there needs to be assistance or intervention (intervention) from the government or the private sector.

1 Introduction

1.1 Background

Indonesia lies in a tectonically active region and is exposed to a range of related natural hazards. On 28 September 2018, a 7.4 magnitudes earthquake occurred in Central Sulawesi province, causing landslides, major liquefaction events and a near-field tsunami, which struck the coast of Palu City and Donggla District. According to the National Disaster Management Agency (BNPB) the death toll is 2,256, while the number of missing persons total 1,309, and over 220 thousand people have been displaced[1][2][3].

The disaster has caused extensive damage to human settlements and buildings (houses, government offices, schools, hospitals and health centres) and infrastructure (roads, water supply, sanitation and irrigation facilities). The extent of the damage is estimated as follows[4][5][6]. Approximately 69,000 houses have been damaged, of, which 3,800 houses were damaged by liquefaction; 1,509 schools have been damaged and (temporarily) closed; 67 health centers and 17 hospitals suffered damages, limiting (or preventing entirely) their ability to provide health services; Some 42 km of raw water transmission and 100 km of water supply systems were damaged, causing people to rely on groundwater resources;Some 95 km of the canals in the Gumbasa irrigation system have been damaged, and the entire 8,180 command area is (for now) not being irrigated; and some 16 km of river works and about 20 km of coastal protection works have been damaged, exposing some 30,000 households to flooding risks.

The horizontal movement (strike-slip) had occurred along the Palu Koro geological fault and estimated (based on satellite imagery) that the Earthquake had caused 4-6m of horizontal movement at the surface of the fault line[[7][8][9].

On 28 April 2019, the Gumbasa area suffered from flooding of the Bangga River. There were no casualties, but the Bangga township was flooded by 1½-3 meters of water that left houses covered in 1-2m of sand and destroyed local tree plantations (mainly cocoa was affected). Other villages were also affected by floods on 28 April 2019, although not as severe as at Bangga. In June 2019, the construction of a large dike started to contain the Bangga River and prevent future catastrophic floods[10][11].

Land suitability evaluation for agricultural land-use planning has five physical control factors: rainfall, temperature, slope percentage, soil types, and the distribution of groundwater wells, which are used to decide the best agricultural land use for a particular area, for both rainfed agricultural and irrigated crops. Weighted Overlay tools were used to locate the most suitable site for rainfed agricultural and irrigated crops. Inland suitability, modelling all the five factors are Weighted based on their level of influence using multi-

* Corresponding author: s.nurdin@gmail.com

criteria evaluation to generate a land suitability map. Mapping agricultural land is thus vital to locate and rank which areas are highly suitable and less suitable.

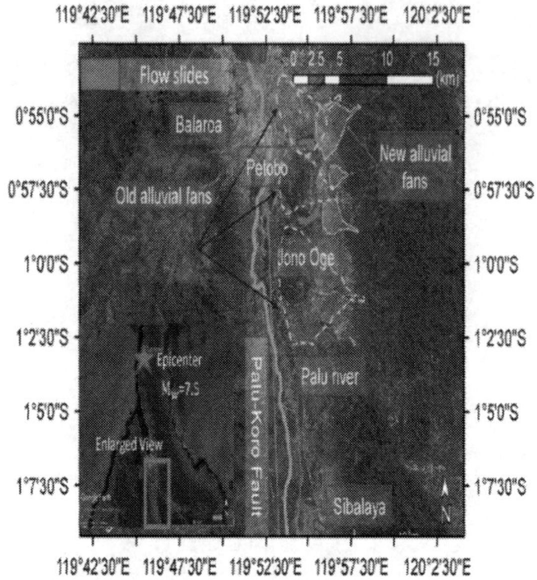

Fig. 1. Locations of liquefaction cross the paddy field along the main canal of Gumbasa Irrigation
Source: modified from [9].

1.2 Location of study

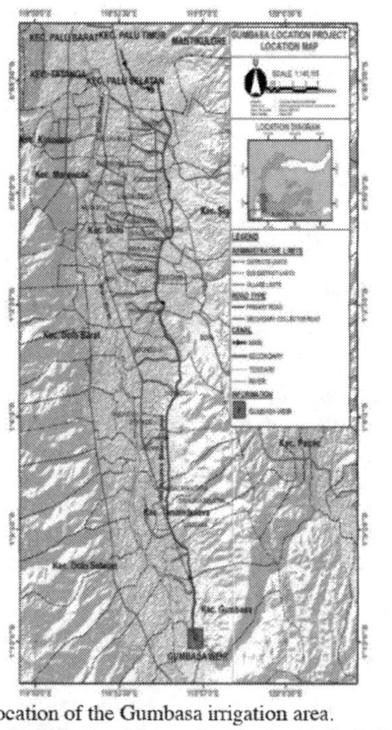

Fig. 2. Location of the Gumbasa irrigation area.

The Gumbasa irrigation system is situated on the right bank of the Palu River between 1 and 40 km South of Palu city in Palu City and Sigi District in Central Sulawesi Province (see Figure 2). The Gumbasa Irrigation System belongs to the Palu river basin managed under the BWSS III, containing the Palu Lariang, Parigi Poso and Kalukku Karamariver basins. The Gumbasa irrigation system is more extensive than 3,000 Ha and is classified as a central government scheme.

1.3 Purpose of land use and land suitability assessment

The main target of this paper is to locate the optimum Land suitable for agriculture, both rainfed and irrigated, while the second goal is to present the classification and change detection of land use after the Palu Earthquake in September 2018 Palu and Sigi Valley.

2 Methods of land suitability assessment

The procedure model was used to analyze the most suitable location for the rainfed agricultural and irrigated crops after the Palu Earthquake in September 2018. There are four levels of land suitability class as shown in. The suitability was prepared based on five criteria: rainfall, temperature, slope percentage, soil types, and the spatial distribution of groundwater wells. The process in land suitability analysis with GIS after the suitability class of each parameter is an overlay process. or overlap between parameters. This process is carried out to combine all the information from the parameters used in the suitability analysis

2.1 Preparation stage

Soil maps, legends and soil characteristics are land/soil resource data needed in assessing land suitability for strategic agricultural commodities. The data is collected in a database system and integrated with other data such as climate and slope details.

The stages of compiling land suitability maps for strategic agricultural commodities on a scale of 1:50,000 include:

1. Preparing a land map: Prepare a 1:50,000 scale soil map, especially in shapefile (.shp) format and the map legend, and a dataset of soil characteristics used for land suitability assessment of strategic agricultural commodities.
2. Checking topology, polygons, land characteristics: At this stage, a re-check is carried out on the soil map unit and the soil/land characteristics dataset, which will be used to assess land suitability for strategic agricultural commodities. Format and topology checks are carried out on each District/City level land map, as follows: First, the spatial and topological data formats need to be improved (especially polygon slivers and double lines/overlaying vertices). Second, clarify the number of *Satuan Peta Tanah* (SPT; soil mapping unit) with its legend.
3. Verifying land characteristics dataset: Verifying the completeness of land characteristics data is carried out on each SPT. The result of the activities

at this stage is a semi-detailed soil map with a scale of 1:50,000 with a complete dataset of land characteristics. This data will be used for the land suitability assessment process for strategic agricultural commodities.

4. Preparing the land suitability map format and layout (template): The design and layout of the 1:50,000 scale land suitability map using the Indonesian topographical base map (RBI) at a scale of 1:50,000 and a scale of 1:25,000 published by the Geospatial Information Agency (BIG). The RBI map is a national map standard that all ministries and institutions must refer to to support the one map policy. District/city and sub-district boundaries follow the division of administrative areas following the directions of the Sigi District spatial plan for 2020-2040

2.2 Stage of assessment and presentation of land suitability results

The stage of assessment and presentation of land suitability results is based on existing data such as data on temperature, rainfall, soil texture, and slope. These data will then be overlaid/superimposed to produce land suitability data for agriculture in strategic areas.

Land suitability assessment activities are carried out on strategic agricultural commodities: rice, corn, soy beans, shallots, red chillies, oil palm, cocoa, sugar cane, and forage fodder.

This land suitability assessment is carried out with a matching system. This means that land qualities/land characteristics are matched with land use requirements, including plant growth requirements, environment and management, using the Land Suitability Assessment System (SPKL) program package version-2[12]

Reference is made to the Land Suitability Criteria for Agricultural Commodities [12] with several modifications and improvements to the 2016 FGD results. Land suitability is evaluated on the current land use (existing land use), an assessment before land management actions are carried out. This land suitability class describes the actual land potential, utilization constraints and required improvements. The results of the SPKL assessment produce tabular data on the land suitability class of each commodity based on the land map unit. The land suitability class is symbolized by S1, S2, S3, and N to indicate the level of suitability.

To present land suitability assessment results, a table of actual land suitability classes (single crop) for strategic agricultural commodities is combined with SPT using GIS (Geographic Information System) so that land suitability maps of strategic farm commodities are arranged. The map is then ready for field verification [13].

2.3 Field verification stage

Field verification aims to check the results of the land suitability assessment with the performance of the strategic agricultural commodities being assessed. Validation is carried out mainly in strategic farm commodities' potential areas or production centers. For example, suppose the land suitability assessment results are not following the reality of plant growth in the field. In that case, it is necessary to investigate[14]: 1) land suitability parameters and criteria, 2) land characteristics, 3) land management techniques and 4) plant growth and production.

2.4 Land suitability map preparation stage

After field verification, the land suitability map is corrected using the data from the field verification, and then a suitability map for strategic agricultural commodities is drawn up. Furthermore, it is laid out in an atlas format that refers to the standard form set by the BIG by including essential map components derived from the Indonesian Topographic Map (RBI), Digital Elevation Model (DEM), and administrative boundary maps. Each map uses district/city administrative boundaries. The map legend is compiled by presenting the land suitability class. Colours and codes symbolise class distinctions. A description of each colour and symbol is displayed in the form of a map legend.

2.5 Data and information management and update stage

The land suitability map that has been prepared and the database built is continuously updated with data and information on land resources based on data and land suitability criteria by the development of science and technology (varieties, cultivation technology and land management). The existing maps and data are actual and up-to-date sources of information to support food self-sufficiency efforts and the development of agricultural areas. With the development of science and technology, land suitability maps need to be updated, including data input regularly as required.

3 Land suitability criteria and weighting

3.1 Land suitability concept

Land suitability is the degree of suitability of a plot of land for a particular use. The suitability of the land can be assessed for its current condition (actual land suitability) or after improvements have been made (potential land suitability). Actual land suitability is land suitability based on data on biophysical properties of soil or land resources before the land is given the necessary inputs to overcome obstacles. The biophysical data is in soil and climatic characteristics related to the requirements for plant growth being evaluated. Potential land suitability describes the land suitability achieved if improvement efforts are made. The land being assessed in that case can be in the form of conversion forest, abandoned or unproductive land [15].

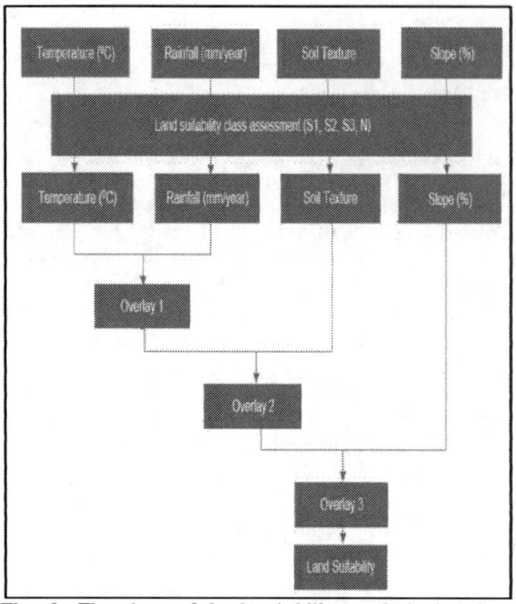

Fig. 3. Flowchart of land suitability analysis based on guidelines for the assessment of land suitability for strategic agricultural commodities, semi-detailed level 1:50,000 scale [12].

Land suitability classes can be divided into categories of land suitability based on the quality and characteristics of the land. Knowing the limiting factors will facilitate detailed interpretation in land use planning.

- Class S1, very suitable: The land has no significant limiting factor for sustainable use or only a minor limiting factor that will not reduce land productivity significantly.
- Class S2, moderately suitable: The land has a limiting factor, which will affect its productivity, requiring additional input. The farmers themselves can usually overcome these barriers.
- Class S3, marginally suitable: The land has a severely limiting factor, and this limiting factor will affect its productivity, requiring additional inputs that are more than land classified as S2. To overcome the limiting factor in S3 requires high capital, so there is a need for assistance or intervention by the Government or the private sector. Without this assistance, farmers will not be able to cope.
- Class N, not suitable: Land that is not suitable because it has a weighty limiting factor and factors that are difficult to overcome.

Table 1. Table of criteria and weighting of land suitability for irrigated rice [12].

Land use requirements/characteristics	Land suitability class			
	S1	S2	S3	N
Temperature (tc)				
Annual Average Temperature (°C)	25 – 28	>28 – 30	>30 – 33	>33
		23 – <25	21 – <23	<21
Availability of water (wa)				
Number of Wet Months (>200 mm/month)	6 – 8	4 – <6	2 – <4	<2
			>8-10	>10
Rooting media (rc)				
Drainage	hampered, a bit hampered	good	very slow, good, rather fast	fast
Texture	smooth, slightly smooth	medium	rather rough	rough
Crude material (%)	< 3	3 – 15	15 – 35	> 35
Soil depth (cm)	> 50	40 - 50	25- 40	< 25
Peat:				
Thickness(cm)	< 50	50 - 100	100 – 150	> 150
Maturity	Saprik	saprik, hemic	Hemic	Fibric
Nutrient retention (nr)				
Soil CEC (cmol/kg)	> 16	5 – 16	< 5	-
Base saturation (%)	>50	35 – 50	< 35	-
pH H2O	5,5 - 7,0	4,5 – 5,5 7,0 – 8,0	< 4,5 > 8,0	-
C-organic (%)	> 1,2	0,8 - 1,2	< 0,8	-
Nutri available (na)				
N total (%)	medium	low	very low	-
P2O5 (mg/100g)	high	medium	low-very low	-
K2O (mg/100 g)	medium	low	very low	-
Toxicity (xc)				
Salinity (dS/m)	< 2	2 – 4	4 – 6	> 6
Sodicity (xn)				
Alkalinity/ESP (%)	< 20	20 - 30	30 – 40	> 40
Sulphidic hazard (xs)				
Sulphidic depth (cm)	> 100	75 - 100	40 – 75	< 40
Erosion hazard (eh)				

3.2 Rainfall data

Rainfall data is based on Bora Station data contained in the ESP Gumbasa Hydrology Report. In the Gumbasa irrigation area, there is only one station. Based on this data, the rainfall data in the Gumbasa irrigation area is classified from 2002 to 2019. For more details, see the following table.

Tabel 2. Gumbasa Irrigation Area rainfall data table in 2002-2019.

Year	Monthly rainfall (mm)										
	Jan	Feb	Mar	April	May	Jun	Jul	Aug	Sep	Oct	Nov
2002	110.00	9.70	44.10	145.60	127.00	135.00	20.00	0.00	5.00	0.00	126.80
2003	42.70	93.00	102.00	22.00	63.00	22.50	25.00	50.50	57.50	37.00	25.00
2004	41.00	6.00	37,00	102.00	51.00	15.00	10.00	0.00	19.00	0.00	20,30
2005	43.00	10.00	27,00	44.00	90.00	31.00	18.00	24.00	10.00	46.00	0.00
2007	50.00	35,40	19.80	18.00	43.00	72.80	21.00	131.00	34.00	0.00	0.00
2008	34.00	12,20	14.00	22.80	28,80	37,00	24.00	86.00	17,00	20.00	74.00
2011	0.00	0.00	0.00	0.00	0.00	0.00	0.00	103.20	256.10	102.00	42.80
2012	83.60	46,60	48,40	51.80	17.80	50,30	152.30	36.30	10,10	75,30	23.70
2013	35,40	30,80	35,60	85,70	70,30	62.00	146.60	84.00	51.00	66.10	98,90
2014	101.60	62.50	107.70	16,10	87.50	37.60	28,40	91.90	37,00	33.60	44.70
2015	67.30	57.50	120.90	104.00	37.80	56,70	24.40	15,30	10,80	47.46	122.40
2016	41.10	49.00	33.80	63.50	0.00	57.70	171.00	30,70	166,80	75,50	43.40
2017	50.00	31.00	92.00	57.50	98.00	96.00	138.50	143.00	86.00	57.50	45,50
2018	25,50	20.00	87.50	49.50	131.00	125.00	62.00	53.00	109.00	68.00	32.50
2019	39,50	68.00	43.50	87.00	32.00	270.00	25.50	20.00	34.50	165.50	51.00
Average	50.85	35.45	54.22	57.97	58,48	71.24	57.65	57.93	60.25	52.93	50.07

Source: Bora Station Based on Analysis Results of Hydrologists, 2021.

3.1 Temperature data

The temperature conditions in the Gumbasa irrigation area are very diverse, and this is based on data from the BMKG (Meteorology, Climatology and Geophysics Agency) with the Bora Station branch issuing temperature data from 2002 to 2019 in the Gumbasa irrigation area located in 7 sub-districts located in Sigi District and Palu City. Temperature conditions are critical in the land suitability assessment process because this is a significant factor in assessing the land potential and advantages in strategic agricultural development.

 https://doi.org/10.1051/e3sconf/202234004001

Table 3. Table of temperature data for Gumbasa Irrigation Area in 2002 – 2019.

Bora Station Climatology Data Period 15 Daily Years																	
Year	2002	2003	2004	2005	2006	2007	2008	2009	2010	2011	2012	2013	2014	2015	2016	2017	20
Period	I	I	I	I	I	I	I	I	I	I	I	I	I	I	I	I	I
RH	78.88	79.12	79.40	77.88	78.30	65.83	68.91	81.45	84.82	84.74	81,12	81.80	79.75	73.35	81.38	66.47	67.
Max Temperature	36.68	36.42	37.17	36.75	38.00	34.25	37.20	37.17	37.17	36.72	37.53	37.92	38.25	31.80	33.42	33.42	23.
Min temperature	19.83	20.00	20.00	19.83	20.00	18.33	20.00	18.92	19.92	20.48	23.32	24,16	24.22	22.87	24.60	22.32	15.
Average Temperature	28,21	28,21	28.58	28.28	29.00	26.29	28.03	28.54	28.84	28,60	30.42	30.59	31.24	27.38	29.01	27.87	19.
Wind velocity	72.32	60.71	61.43	47.98	65.12	109.02	109.65	66.94	0.00	24.78	60.42	64.09	58.27	67.45	38.53	54.85	486.
Exposure Time	61.29	82.23	53.18	57,40	60.26	44.46	18.41	67.77	56.57	46.96	57.59	51.85	52.81	53.93	49.04	54.38	0.5
Year	2002	2003	2004	2005	2006	2007	2008	2009	2010	2011	2012	2013	2014	2015	2016	2017	20
Period	II	II	II	II	II	II	II	II	II	II	II	II	II	II	II	II	II
RH	76.76	78.91	78.62	78.11	78.86	67.01	70.02	83.09	84.19	88.00	80.85	83.12	79.59	79.41	81.29	66.71	67.
Max Temperature	36.76	36.00	37.17	36.92	38.17	34.93	37.56	36.67	37.33	36,60	37.83	36.91	36.86	31.58	31.83	33.48	24.
Min temperature	19.87	19.97	19.83	19.97	20.05	18.47	20.07	20.00	19.75	20.88	23.54	24.77	24.69	22.86	23.82	22.18	15.
Average Temperature	28,21	27.38	28,60	28.42	29.08	26.70	28.81	28.33	28.54	28.72	30.69	28.84	30.88	27.22	27.82	27.83	30.
Wind velocity	71.03	62.03	63.84	40.42	51.38	81.15	105.15	86.07	0.00	28.47	62.85	54.39	61.59	68.57	33.76	56.83	474.
Exposure Time	59.93	54.42	64.24	59,05	57.06	44.91	18.77	51.67	55.89	48.62	56.16	54.89	60.08	58.53	44.63	53.60	0.5

3.2 Soil type data

Concerning soil mapping units (SPT) on the format and topology found in the Gumbasa irrigation area, there are 14 types of SPT. This is based on data published by the Center for Agricultural Land Resources (BBSDLP) in 2016, specifically for Sigi District and Palu City. The soil mapping unit is very influential in developing agriculture in the strategic area, especially for developing irrigated rice fields in the irrigation area. In scoring the land map unit (SPT) following the technical guidelines issued by BBSDLP in 2016. More details can be seen in the following table.

Table 4. Table of Soil Mapping Units (SPT) in Gumbasa Irrigation Area

No.	SPT	Land unit	Soil texture	Area (ha)
1	1	Aluvial Ustik dan Kambisol Eutrik	Medium texture	1,200.92
2	10	Gleisol Fluvik dan Kambisol Ustik	Smooth texture	2,150.49
3	11	Aluvial Ustik	Medium texture	388.18
4	13	Kambisol Eutrik dan Kambisol Gleik	Smooth texture	1,150.71
5	14	Gleisol Eutrik, Gleisol Fluvik, dan Molisol Gleik	Smooth texture	374.28
6	15	Kambisol Eutrik dan Kambisol Gleik	Slightly smooth texture	130.86
7	2	Kambisol Eutrik dan Aluvial Eutrik	Medium texture	319.66
8	24	Kambisol Ustik dan Molisol Ustik	Medium texture	561.78
9	4	Gleisol Fluvik	Smooth texture	957.48
10	40	Kambisol Distrik dan Podsolik Haplik	Smooth texture	20.57
11	5	Gleisol Fluvik dan Aluvial Gleik	Smooth texture	4,515.68
12	5555	Settlement (X2)	Rough	1,084.51
13	6	Kambisol Eutrik dan Gleisol Fluvik	Slightly smooth texture	1,861.76
14	9999	Body of Water (X3)	Rough	737.12
Total				15,464

Source: GIS analysis based on [16].

4 Result And Discussion of land suitability assessment

4.1 Overlay land suitability parameters

The following process in analyzing land suitability with a geographic information system (GIS) after assessing the suitability class of each parameter is the process of overlaying or overlapping between parameters. This process is carried out to combine all the information from the parameters used in the suitability analysis. In this exercise, the overlay process used is 'union'. In simple terms, the union process can be illustrated as follows:

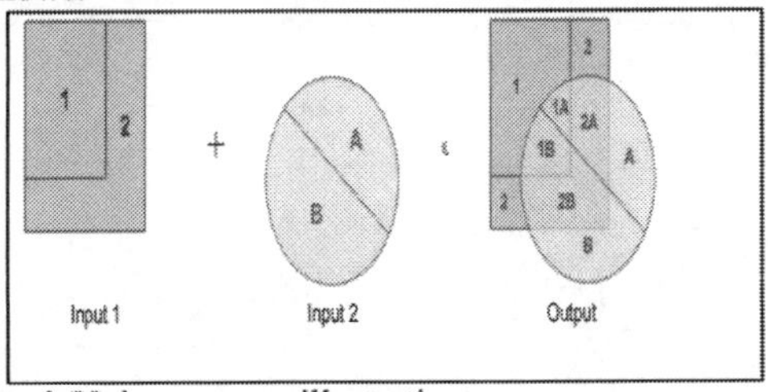

Fig. 4. Union process illustration.

In the union process, all data attributes from both inputs will be displayed on the output, both overlapping or not. As illustrated above, the following is an example of an overlayed data attribute with union:

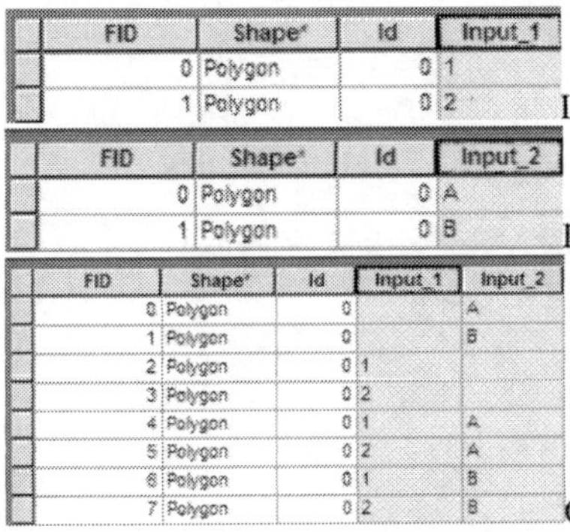

FID	Shape*	Id	Input_1
0	Polygon	0	1
1	Polygon	0	2

Input 1

FID	Shape*	Id	Input_2
0	Polygon	0	A
1	Polygon	0	B

Input 2

FID	Shape*	Id	Input_1	Input_2
0	Polygon	0		A
1	Polygon	0		B
2	Polygon	0	1	
3	Polygon	0	2	
4	Polygon	0	1	A
5	Polygon	0	2	A
6	Polygon	0	1	B
7	Polygon	0	2	B

Output

Fig. 5. Illustration of the data attribute of the union process.

3.1 Reclass land suitability data overlay results

After going through the overlay process, the next stage of land suitability analysis with a geographic information system (GIS) is reclassifying the suitability class of the overlay result. The determination of the new suitability class from the two old suitability classes is based on conformity. The S1 suitability class is only formed by the S1 and S1 classes. If there are two different land suitability classes, the new land suitability class is higher. This will affect the land suitability class because the suitability class will affect its use in the existing land use conditions. The following is a matrix of determining the new suitability class from two different suitability classes[16]:

Table 5. Table of slope data for Gumbasa Irrigation Area.

Class Suitability	S1	S2	S3	N
S1	S1*1	S2*2	S3*3	N*4
S2	S2*2	S2*2	S3*3	N*4
S3	S3*3	S3*3	S3*3	N*4
N	N*4	N*4	N*4	N*4

Source : [16]

4.2 Soil Map Analysis Gumbasa Irrigation Area

From the map, it can be seen that most of the cambisolustik and district cambisol soils are located in the eastern area of the irrigation location, where this area is adjacent to the main channel of DI Gumbasa. A higher slope than other areas also characteristic this area.

The dominance of cambisol material at this location is probably due to sand boiling during the earthquake in September 2018. Sand boiling occurs when sand and gravel material rises to the surface due to increased soil pore water pressure triggered by earthquake forces.

The type of material underneath on the east side is dominated by uitric cambisol and uitric gleisol, where this area is very dominant in the northern location of the test area, which includes the Maranata, Jono Oge and Petobo areas. In contrast, fluvic gleisol layers are primarily found in the Biromaru and Dolo areas, and a small part is in DI Gumbasa. For ustic alluvial soil texture, the dominant distribution is in the area along with the test site, located in the western region, which is close to the Palu river channel in Figure 6

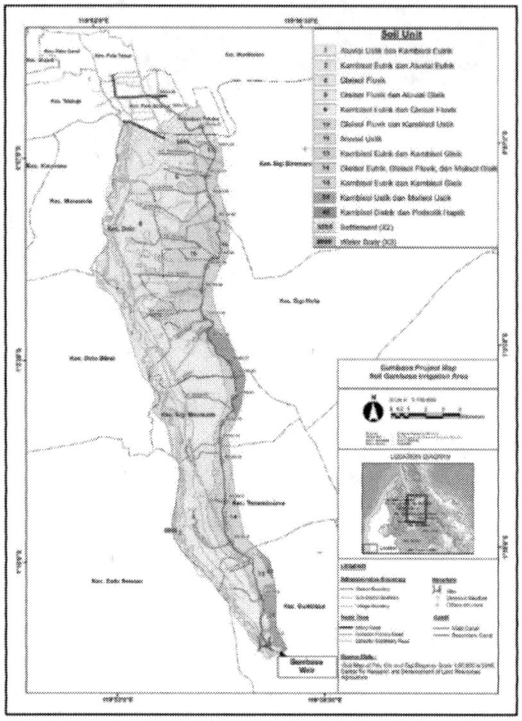

Fig. 6. Slope map of Gumbasa Irrigation Project.

3.1 Slope data

In classifying slopes for agricultural development, strategic areas are divided into four classes of slopes. It serves as a parameter in measuring the level of land suitability of the Gumbasa irrigation area as a boundary area. This is related to the development of strategic agriculture to find places suitable for use in irrigated rice fields. For more details, see the following table

Table 6. Slopes Classifying of Gumbasa Irrigation Area.

No	Slope (%)	Area (ha)	Category
1	< 3 %	12,947.82	Highly suitable
2	3 – 5 %	2,240.94	Moderately suitable
3	5 – 8 %	202.09	Marginally suitable
4	>8 %	73,15	Not suitable
Total		15,464	

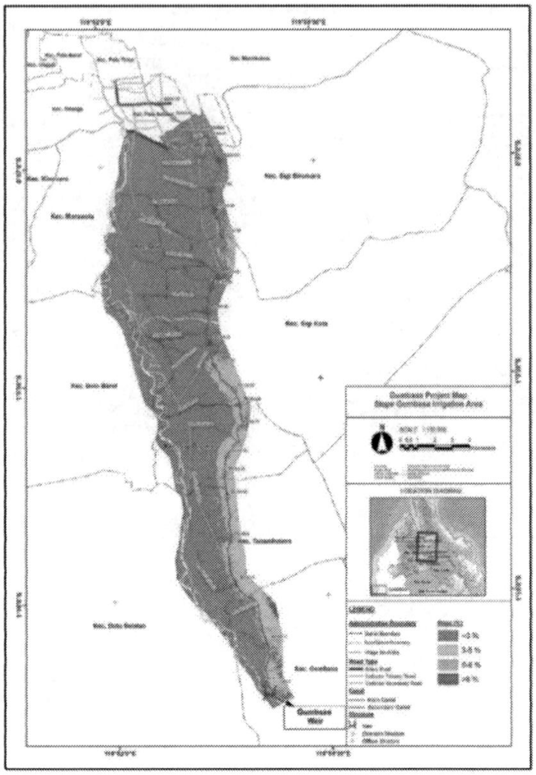

Fig.7. Slope map of Gumbasa Irrigation Project.

4.3 Land use data

The land use update of the Gumbasa Irrigation project is based on manual on-screen digitisation using Sentinel multi-temporal satellite imagery from 2016-2020 with a resolution of 10m. Digitisation activities for updating land use are based on different images using satellite data from various years. The original data on the land use of Gumbasa refers to the Sigi District Spatial Plan (RTRW) for 2020-2040 and the RTRW of Palu City in 2018-2038, as for the update of land use with high-resolution orthophoto images that BIG issued in 2018 with a resolution of 1m with ten land use classes, namely forests, transportation areas, grasslands, settlements, plantations, rice fields, shrubs, rivers, vacant land and fields/fields.

4.4 The results of land use

The change analysis is carried out by comparing the service from 2016 to 2020. By using the overlay technique, land-use change in Gumbasa is presented in the following table.

https://doi.org/10.1051/e3sconf/202234004001

Tabel 7. Land change 2016-2020 in Gumbasa irrigation.

No.	Land use	2016	2017	2018	2019	2020
1	Forest	148	148	148	148	148
2	Transportation area	11	11	11	11	11
3	Grassland	0,49	0,49	0,49	-	-
4	Settlement/activities place	1,260	1,423	1,239	1,271	1,343
5	Plantation/tree crop	2,560	2,549	2,550	2,538	2,650
6	Paddy fields	6,794	7,079	6,575	6,418	6,585
7	Shrubs/bushes	819	916	845	873	1,265
8	River	456	447	407	427	418
9	Bare land	34	149	58	294	287
10	Dry land agriculture	1,365	624	1,612	1,465	737
		15,464	15,464	15,464	15,464	15,464

The analysis results show that changes in land use indicate a decrease in land use for agricultural cultivation areas, including rice fields, plantations and dry fields/fields. In contrast, the share of abandoned land in the form of shrubs and vacant land has increased after the natural disaster that occurred on 28 September 2018. The conversion of agricultural fields into residential areas in the Gumbasa irrigation area is a significant issue for irrigated agriculture, especially from 2016-2017. This trend decreased significantly due to liquefaction in 2018. Similarly, for almost all land use trends related to agricultural and non-agricultural land use, the natural disasters in 2018 had a major influence.

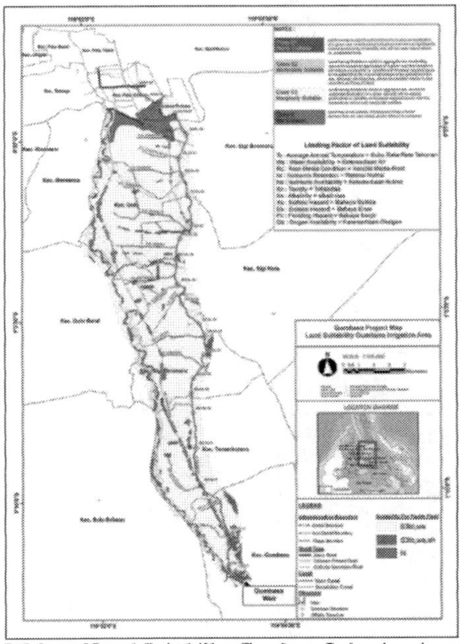

Fig. 8. Map of Land Suitability Gumbasa Irrigation Area.

The land suitability class criteria found in the Gumbasa irrigation area and the limiting factors that affect land suitability can be seen in the following table.

Table 8. Land suitability assessment Gumbasa Irrigation Area.

No	Land Suitability assessment class	Area (ha)
1	S3tc, wa	1,902.32
2	S3tc, wa, eh	13,364.47
3	N	1,902.32
Total		15,464

Source: GIS analysis 2021

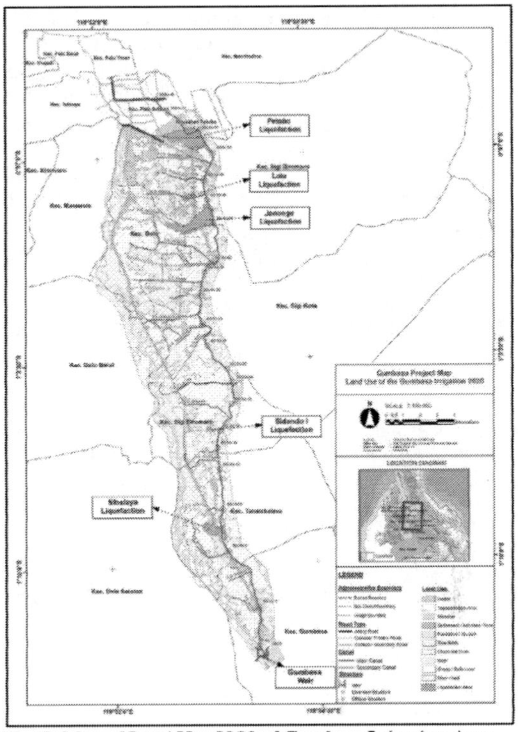

Fig. 9. Map of Land Use 2020 of Gumbasa Irrigation Area.

From the map, it can be seen that changes in agricultural land use caused by flow liquefaction in 3 main areas, namely Sibalaya, Jono Oge and Petobo are main reason that cause the changes in the area of paddy fields in the Gumbasa irrigation area. The second factor is a large number of shifts of agricultural land into residential land, especially the area Petobo in Palu City and the Biromaru and Dolo areas in Sigi Regency, especially those bordering Palu City.

5 Concluding remarks

The following are the conclusions of the Gumbasa Land Use and Land Suitability research:

1. The map projection used in the Gumbasa project follows the Indonesian National Standard (SNI) for Specifications for Presentation of topographic maps or base maps at a scale of 1:25,000 and for 1:1:50,000 (SNI 6502.2:2010 and SNI 6502.3:2010), which officially uses a map projection system UTM. Gumbasa is located in the UTM 50S zone; topographic/slope data is based on 1m resolution of LiDAR BIG (National Geospatial Information Agency) data captured in 2018. At the same time, thematic data is classified based on project clusters in relation to Project supporting data analysis.
2. The result of the land suitability assessment map is a map of the calculation results based on the technical document Guidelines for Land Suitability Assessment for Strategic Agricultural Commodities at the Semi-Detailed Level 1:50,000 Scale issued by the Center for Agricultural Land Resources in 2016 related to the development of irrigated paddy

3. Based on the results of the overlay (overlapping) assessment of land use and land suitability in the Gumbasa irrigation area, there are several inhibiting factors, namely the lack of water availability based on rainfall data for the last 15 years, it is known that the number of months >200mm/month there are only two months in a year. The lack of wet months will affect the temperature of the related data. The Land Use and Land Suitability assessment results are included in the S3 category (marginally appropriate) for class/category classification.
4. Flow liquefaction triggered by the 2018 earthquake in 3 main areas, namely Sibalaya, Jono Oge and Petobo, causes the changes in the area of paddy fields in the Gumbasa irrigation area.

References

1. H. B. Mason, "Geotechnical Reconnaissance: The 28 September 2018 M7.5 Palu-Donggala, Indonesia Earthquake," no. September 2018, pp. 1–86, 2019, doi: 10.18118/G63376.
2. D. H. Natawidjaja *et al.*, "The 2018 Mw7.5 Palu 'supershear' earthquake ruptures geological fault's multisegment separated by large bends: Results from integrating field measurements, LiDAR, swath bathymetry and seismic-reflection data," *Geophys. J. Int.*, vol. 224, no. 2, pp. 985–1002, 2021, doi: 10.1093/gji/ggaa498.
3. B. Mason, D. Hutabarat, and W. Prakoso, "Geotechnical Reconnaissance: The 28 September 2018 M7.5 Palu-Donggala, Indonesia Earthquake PacTrans-SSI Bridge Project View project Aceh-Pidie Mw 6.5 Earthquake Aftershock 2016 Monitoring View project," *Rep. Geotech. Extrem. Events Reconnaissance, Geotech. Extrem. Events Reconnaiss. Assoc.*, no. April, pp. 1–86, 2019, [Online]. Available: https://www.researchgate.net/publication/332221717.
4. "BRIEF EXPLANATION PALU-KORO FAULT Summarized from Palu Earthquake Study by PUSGEN, 2018," 2018.
5. Kementerian PPN/ Bappenas, "Rancangan Rencana Induk Pemulihan dan Pembangunan Wilayah Pasca Bencana di Provinsi Sulawesi Tengah," p. 74, 2018.
6. S. Nurdin *et al.*, "Proceeding 23rd Annual National Conference on Geotechnical Engineering," no. November 2019.
7. D. Rohit, H. Hazarika, T. Maeda, W. O. Sumartini, and T. Kokusho, "Forensic Investigation of Flow-slides Triggered by the 2018 Sulawesi Earthquake."
8. M. Okamura, K. Ono, A. Arsyad, U. S. Minaka, and S. Nurdin, "Large-scale flowslide in Sibalaya caused by the 2018 Sulawesi earthquake," *Soils Found.*, 2020, doi: 10.1016/j.sandf.2020.03.016.
9. H. Hazarika *et al.*, "Large distance flow-slide at Jono-Oge due to the 2018 Sulawesi Earthquake, Indonesia," *Soils Found.*, vol. 61, no. 1, pp. 239–255, 2021, doi: 10.1016/j.sandf.2020.10.007.
10. H. B. Mason *et al.*, "East Palu Valley flowslides induced by the 2018 MW 7.5 Palu-Donggala earthquake," *Geomorphology*, vol. 373, 2021, doi: 1016/j.geomorph.2020.107482.
11. D. Rohit, H. Hazarika, T. Kokusho, S. Nurdin, and T. Maeda, "Forensic Investigation of Flow-slides Triggered by the 2018 Sulawesi Earthquake," 2018.
12. B. B. P. dan P. S. L. Pertanian, "Bbsdlp 2015," *J. Chem. Inf. Model.*, vol. 53, no. 9, pp. 1689–1699, 2013.
13. J. Liu, "A GIS-based tool for modelling large-scale crop-water relations," *Environ. Model. Softw.*, vol. 24, no. 3, pp. 411–422, 2009, doi: 10.1016/j.envsoft.2008.08.004.
14. A. AL-Taani, Y. Al-husban, and I. Farhan, "Land suitability evaluation for agricultural use using GIS and remote sensing techniques: The case study of Ma'an Governorate, Jordan," *Egypt. J. Remote Sens. Sp. Sci.*, vol. 24, no. 1, pp. 109–117, 2021, doi: 10.1016/j.ejrs.2020.01.001.
15. R. Taghizadeh-Mehrjardi, K. Nabiollahi, L. Rasoli, R. Kerry, and T. Scholten, "Land suitability assessment and agricultural production sustainability using machine learning models," *Agronomy*, vol. 10, no. 4, pp. 1–20, 2020, doi: 10.3390/agronomy10040573.
16. L. Tahunan, "Bbsdlp 2016," 2017.

Multi-indicator building vulnerability index for assessing tsunami-induced building damages

Mohd Muhaimin Ridwan Wong[1,2*], *Nordila* Ahmad[2], *Syamsidik*[3], *Anawat* Suppasri[4], and *Maidiana* Othman[2]

[1]Humanitarian Assistance and Disaster Relief Research Centre, National Defence University of Malaysia, Kem Perdana Sungai Besi, 57000 Kuala Lumpur, Malaysia
[2]Department of Civil Engineering, Faculty of Engineering, National Defence University of Malaysia, Kem Perdana Sungai Besi, 57000 Kuala Lumpur, Malaysia
[3]Tsunami and Disaster Mitigation Research Centre (TDMRC), Jalan Professor Doktor Ibrahim Hasan, Ulee Lheue, Meuraxa, Kota Banda Aceh, 23232 Aceh, Indonesia
[4]International Research Institute of Disaster Science (IRIDeS), Tohoku University, 468-1 Aoba, Aramaki, Aoba-ku, Sendai, Miyagi, Japan 980-8572

Abstract. While relatively uncommon, tsunami typically leads to extreme and devastating impacts to human lives and development at far distances. Various international bodies and academic research have highlighted the importance of understanding risk and its determinants for effective risk management. Currently available methodologies to assess tsunami risk and the effect of vulnerability towards the resulting impact present opportunities for improvement particularly in their adaptability, accuracy and generalisability for pre- and post-disaster applications, as well as across temporal and geographical boundaries. This paper discusses the potential of a methodology to assess tsunami risk for buildings based on their vulnerability level, which in turn is determined through a set of vulnerability indicators. A vulnerability index equation consisting of four indicators namely building materials, number of storeys, coastal protective structures and coastal vegetation was formulated through the Analytical Hierarchy Process. The index is then applied for two major yet distinctly different tsunami events for the purpose of examining potential links between hazard intensity measure, building vulnerability and recorded impacts while also establishing basis for generalisability. While certain correlations can be observed, the findings are still at preliminary stage and thus approaches to further refine the index were proposed.

1 Introduction

For the last two decades, 271 tsunami events have been recorded resulting in over 300,000 casualties associated directly with these tsunami disasters [1]. Major tsunami events such as the 2004 Indian Ocean tsunami, 2010 Chile tsunami, 2011 Japan tsunami and the 2018 Sulawesi tsunami accounted for majority of these casualties. Aside from casualties, these tsunami disasters also caused major losses and setback due to properties and infrastructure destruction as damages amounted to billions of dollars in some cases. The 2011 Japan tsunami resulted in approximately US$220 billion of damages (US$37.5 billion of which were insured), while the February 2010 Chile tsunami disaster incurred damages up to US$30 billion (with US$8 billion damages insured) [1, 2].

An important approach in reducing disaster risks and impacts including for tsunami inarguably lies in understanding these risks, their determinants and how they revolve around and interact with all aspects of anthropogenic development. The United Nations Office for Disaster Risk Reduction (UNDRR) strongly advocates for effective and comprehensive disaster risk management achievable through understanding disaster risk drivers which consist of hazards, exposure and vulnerabilities [3]. Subsequent to understanding these risk determinants, impacts both direct and indirect, tangible and intangible, can be better managed via various preparedness and mitigation strategies.

In line with the above concept, tsunami studies over the past several decades have also been advancing our understanding on tsunami hazard and its impacts. Quantification of tsunami magnitude and intensity had begun since early in the 20th century as Sieberg introduced the first tsunami intensity scale, and this was followed by many other variations based on tsunami events across the globe. More recently, in addition to tsunami intensity models, this effort to improve and communicate the understanding of tsunami hazards and its impacts have been broadened as more approaches are made available. Such approaches include probabilistic risk models [4-6], deterministic impact framework [7], multi-hazard impact models [8, 9], vulnerability assessment models [10, 11], and tsunami loss estimation models [12, 13].

Despite these developments, most tsunami risk and impact assessment methodologies or models remain specific and more relevant to certain geographical scope or tsunami event. Others that are more generalisable can

* Corresponding author: waiminwong86@gmail.com

be constrained in application as well as in correlating between different tsunami risk determinants and observable impacts. The main aim of this paper is to offer a preliminary discussion of a tool to assess building vulnerabilities against tsunami hazard with varying magnitudes, as well as relate these variables to potential damages that can be observed. Thus, presented in this paper are also attempts at addressing two research questions formulated based on the primary aim. The first of these is that considering well-known and established vulnerability and/or risk indicators in literatures, which indicator can be measured and used universally particularly as components of a vulnerability index? Secondly, how does the combination of these vulnerability indicators affect severity of building damage based on past tsunami events with different triggering mechanisms? The first question is answerable by identifying and defining vulnerability indicators and their associated proxies that are relevant to buildings or structures from past research and tsunami field survey. On the other hand, the second research question can be addressed by reviewing field survey data from previous tsunami events, through which a measurement of tsunami hazard intensity is correlated with observed damages via a proposed, moderating building vulnerability index.

2 Tsunami vulnerability indicators

While hazardous events may have been traditionally believed to be the sole or primary factor for disaster risks, the current state of knowledge requires two other determinants as components that make up the concept of risk, viz. vulnerability and exposure [14]. Exposure can be defined as elements potentially experiencing damages and/or losses in at-risk areas where hazard event may occur [15]. These elements can usually be identified based on satellite images, city plan or aerial photos, and may be categorised based on different degree of exposure to predicted or historical events. Meanwhile, as a term that has been more common in disaster research since the 1980s, vulnerability was generally used in reference to systems in a geographical region or locality including socio-political, socioeconomic and industrial practices [16]. Presently, vulnerability can be more accurately identified in reference to characteristics of exposed elements or attributes in their proximity that may influence their susceptibility to adverse effects following interactions with hazard events [14].

In relation to earlier views, vulnerability characteristics may indeed be derived from physical, environmental, economic, social, and political settings. From a macro-perspective lens, vulnerability can be identified in the form of population growth, rate of urbanisation, governance, economic disparity and environmental degradation, among others [17]. However, in the context of this paper where buildings are concerned, various tsunami vulnerability indicators, parameters and/or their proxies have in fact been identified and widely established at micro level in various studies previously. Commonly used vulnerability indicators include building typologies, construction materials, number of stories, building design code, building age, and building geometry [9, 18-21]. Among presently available methodologies, the Papathoma Tsunami Vulnerability Assessment (PTVA) is one of the more comprehensive method with inclusion of various building vulnerability and building surrounding attributes as part of its structural vulnerability component, S. [11]. Aside from aforementioned indicators, PTVA via its building vulnerability attributes (BV) proposed inclusion of ground floor hydrodynamics, foundation type, and preservation condition. For building surroundings attributes (denoted as *Surr*), building row, natural barriers, seawall height and shape, brick walls around the building, and sources of movable objects were proposed for inclusion as well. In addition to these parameters, level of water intrusion and exposure to inundation also make up the Relative Vulnerability Index (RVI) equation. For assessment, the vulnerability index is then calculated based on weightages assigned subjectively to each of the attributes. Two distinct group of vulnerabilities are described further in the following sub-sections in connection to the topic of this paper.

2.1 Building characteristics

Based on past studies and impact assessment methodologies such as the Hazards U.S. Multi-Hazard (HAZUS-MH) and RISK-UE, assessment of a building's vulnerability against tsunami via its characteristics may include one or more of the following factors:

a. Building materials, also identifiable as construction materials or building frame: usually classified based on the primary materials that make up the structure to distinguish different strength to withstand tsunami loading, viz. timber, masonry, concrete (reinforced or non-reinforced), and steel.
b. Number of storeys: multi-storey buildings are typically designed to be more resistant to external forces compared to single-story buildings. Classification is typically done based on literal number of storeys, or in some cases the number of storeys may be grouped in ranges and buildings are then categorised as low-rise, medium-rise or high-rise.
c. Foundation type: used as an indicator of foundation strength and in turn the building's resistance against damages resulting from scouring effects by tsunami inundation. A structure may resist the initial punching force caused by tsunami flow inland, but evidence from past events showed that scouring at the base often lead to the structure being collapsed, broken or washed away especially where foundation is shallow [22, 23].
d. Geometry: may refers to the shape of the structure or its elements, span lengths as well as orientation relative to tsunami wave direction [24].
e. Design code: as a parameter that can be used to differentiate design specifications or standards for the buildings assessed. Particularly in developed

countries prone to disasters, most buildings tend to be designed and constructed to be resistant to disaster hazards and induced lateral loads. While this may be specific towards seismic hazard, the classification has been applied in tsunami studies as well [18].

f. Building age and preservation condition: similar to other physical parameters that make up a building typological features, building age can be used as reference to characterise its vulnerability to hazard events with aged buildings typically being more vulnerable. Additionally, preservation condition could help further distinguish the state of deterioration of aged buildings and thus variations in its level of vulnerability.

2.2 Building proximity

Environmental characteristics, structure and conditions in the relatively nearby vicinity of the building can directly or indirectly affect its vulnerability. Identified and commonly used characteristics associated with tsunami vulnerability and/or fragility include:

a. Protective structures, also referred to as sea defence: this parameter refers to identifiable structural mitigation (engineered or otherwise) against coastal hazards, with seawalls being the common coastal defence structure in tsunami-prone areas. In areas where tsunami is much less frequent, coastal protection structures such as levees, revetments and embankments may still protect exposed buildings and reduce their vulnerabilities particularly for tsunami with lower inundation depths [25].
b. Coastal topology, also identifiable as coastal topography: the type of the coastal area evidently has major influence on tsunami inundation and flow velocity, particularly for buildings closer to the shoreline. Buildings along ria-type coasts are more vulnerable and have higher damage probabilities (at similar inundation depths) than those in low-lying plains [4, 26].
c. Coastal vegetation: various studies have found that coastal vegetation and forests are able to dampen tsunami effects and reduce its runup through energy dissipation and wave-breaking capabilities [27]. Parameters related to coastal vegetation affecting tsunami vulnerability include band width (between shoreline and exposed elements), density, species composition, forest structure, age and height, among others [28].

3 Methods

Three approaches are considered to appropriately address questions posed earlier in this paper. The aim as framed based the questions is to formulate a building vulnerability index that can be utilised to categorise buildings (as exposed elements in at-risk areas), and subsequently identify potential correlations between hazard intensity, vulnerability levels and impact intensity. The first approach is the identification of vulnerability indicators and measurement parameters that are most suitable for a universally applicable assessment tool. Following this, the Analytical Hierarchy Process (AHP) is utilised to express the basis of the vulnerability index with appropriate priority weightages assigned for each vulnerability component. Thirdly, the vulnerability index is applied to two tsunami events (with different triggering mechanisms) for the purpose of identifying the relationship that can be observed between all risk determinants involved.

3.1 Vulnerability indicators and measurement parameters

In comparison to the methodologies presented through other comprehensive and more complex methodologies such as PTVA or Building Vulnerability Model (BTV) [29], this paper contends that a simplified assessment tool with generalisable components and measurement may be more practical and applicable especially for a wider risk communication purpose, as well as both ex-ante and post-disaster assessment utilisations. In addition, data limitations and discrepancies in data recording between different tsunami events may also pose challenges for effective application of more complex tools. Hence, in order to attain this generalisability and practicality, established indicators and/or measurement parameters as briefly described in 2.1 and 2.2 are screened for inclusion in a newly proposed index. The variables, measurement parameters and justification for selection are listed in Table 1.

3.2 Building vulnerability index

For assessment of building vulnerability, the Analytical Hierarchy Process or AHP was employed to formulate the index equation. As a mathematical process, AHP is a heuristic algorithm that involves arranging factors in a hierarchy of importance [33], whereby for this paper the context is the indicators' influence and effectiveness in reducing tsunami damage. Levels of importance are normally represented numerically between 1 (least important) to 9 (most important), depending on the quantity of indicators being considered. The indicators are then compared pairwise, and the priority vectors (PV) as calculated based on this comparison are used as the weightage for each indicator. Being the first line of defence, protective structures are designated as the most important indicator, followed by coastal vegetation, building materials and lastly the building's number of storeys. This subjectively assigned hierarchy are shown in Table 2, while the pairwise comparison and the resulting priority vectors are shown in Table 3.

Table 1. Selected vulnerability indicators and measurement parameters.

Indicator	Measurement parameter	Justification
Building materials (BM)	Four categories: i. Timber ii. Masonry iii. Steel iv. Reinforced concrete	This indicator and its parameters are among the most commonly used, and well-established in their relationship and influence towards observable damages.
Number of storeys (St)	Three categories: i. 1 storey ii. 2 storeys iii. More than 2 storeys	While there are variations in the measurement parameters in past studies, the number of storeys considered in this paper is limited to the three categories. This categorisation is also supported by evidence from analysis of building damages during the 2011 tsunami in Japan [4].
Protective structures (PS)	Four categories: i. None ii. Breakwater only iii. Presence of at least one hard protection (seawall, revetment etc.) iv. Combination of more than one coastal protection structure (breakwater, seawall, revetments etc.)	Past studies have found that in areas where low seawalls and/or revetments are present, buildings landward were protected to some degree particularly in non-extreme tsunami events [23]. Tsunami inundation and its impacts have also been observed as much lesser in such area compared to non-protected areas [25]. On the other hand, another study concluded that breakwater does not offer significant protection to coastal buildings mainly due to openings between sections of the breakwater [30].
Coastal vegetation (CV)	Three categories: i. None ii. Band width < 200 meters iii. Band width ≥ 20 meters	Despite the availability of multitude of factors affecting wave attenuation in literature, the three categories were selected upon considering practicality, ease of data collection and/or limitations in data availability. The measurement parameters were also determined based on past studies on the different effectiveness for different coastal forest band width towards tsunami reduction [31, 32].

Table 2. Preference for each vulnerability indicator based on level of importance to reduce tsunami damage.

Value	Level of importance	Vulnerability indicator
1	Equally important	Number of storeys, St
3	Moderately important	Building materials, BM
5	Highly important	Coastal vegetation, CV
7	Extremely important	Protective structures, PS

Table 3. Matrix of pairwise comparison and the calculated priority vector, PV for the vulnerability indicators.

Indi-cator	PS	CV	BM	St	PV
PS	1	3	5	7	0.564
CV	0.333	1	3	5	0.263
BM	0.200	0.333	1	3	0.118
St	0.143	0.200	0.333	1	0.055

Based on the above priority vector values, the building vulnerability index (VI) equation is expressed as:

$$VI = (0.055 \times St) + (0.118 \times BM) + (0.263 \times CV) + (0.564 \times PS) \quad (1)$$

In equation (1), the variable *St, BM, CV* and *PS* refer to the ratings for each vulnerability indicator. For this paper, these ratings as shown in Table 4 are assigned according to the parameters' effectiveness in reducing tsunami damage as suggested or evident in literature, as well as the interval of values between 0 (least effective) and 1 (most effectiveness). The resulting VI values for each building are further segregated into quartiles as shown in Table 5.

Table 4. Measurement parameters and assigned ratings for vulnerability assessment.

Parameter Ratings	BM	St	PS	CV
0			None	None
0.25	Timber			
0.33		1 storey	Break-water	
0.50	Masonry			Band width < 200 m
0.67		2 storeys	Single protective structure	
0.75	Steel			
1.00	Reinforced concrete	More than 2 storeys	Multiple protective structures	Band width ≥ 200 m

Table 5. Vulnerability index values and respective vulnerability levels.

VI values	Designation and quartiles	Description
$0 \leq VI \leq 0.25$	V4	Critically vulnerable
$0.25 < VI \leq 0.50$	V3	Highly vulnerable
$0.50 < VI \leq 0.75$	V2	Vulnerable
$0.75 < VI \leq 1.0$	V1	Less vulnerable

3.3 Application for the 2018 tsunami events

Two major tsunami events that occurred in 2018 were selected to test this vulnerability index equation. The first of these is the 28 September Sulawesi tsunami which was triggered following a magnitude 7.5 earthquake 70 km north of Palu, Indonesia [34]. Despite this identified triggering event, the magnitude and intensity of the resulting tsunami highly suggested that submarine landslides at multiple locations following the earthquake, in addition to high tide during the compound disaster event and liquefied gravity flow further contributed towards the devastating damages observed [34, 35]. Tsunami waves exceeding 10 m high took more than 5,000 lives including those missing, left 10,679 others injured and tens of thousands more displaced as over 40,000 houses were either destroyed or damaged [1].

The second event chosen for this paper occurred merely three months afterwards in the Sunda Strait region, making it the second major tsunami event in Indonesia in 2018. It is also one of three volcanogenic tsunami from the 271 tsunami events in the past 20 years. The Anak Krakatau volcano which was formed following the Krakatau Islands collapse in 1883 had been showing signs of active Strombolian activities from June to October 2018 with series of explosions and lava flows [36]. These activities reached a climax on 22nd of December 2018 with series of eruptions and showers of incandescent ballistic blocks which led to a moderate collapse in the western and southern section of the volcano [37, 38]. Tsunami waves up to approximately 25 m were recorded on the surrounding islands, and runup in coastal cities and town near the region reached up to 15 m high [39]. This event recorded relatively lower deaths at 437, but the population injured were significantly higher at over 31,000 people. Also in contrast, 2,844 houses or buildings were damaged in this event [1].

Additional information required for the vulnerability assessment that is not provided in the field survey data such as coastal vegetation cover, protective structures and distance from shoreline are identified and gathered visually from available satellite image with recording date preceding the tsunami events. For Palu city, satellite images used for reference were recorded on August 2018, while for Banten and Lampung area, the satellite images dated back to October 2018. Measurements for buildings' distance in particular are done perpendicular to the shoreline, with visible vegetation cluster in its direct path considered and measured for the CV indicator. Polygons are drawn for ease of measurement due to irregularly shaped shorelines. Sample measurement is shown in the figure below.

Fig 1. Marking buildings' locations based on coordinates provided in field survey data sets, measuring distances and identifying surrounding features.

4 Results

Data set from two field surveys were used as the source for inundation depth measurements, building typologies' information and building damaged data. The field survey for Sulawesi tsunami in Palu was conducted from 13 to 17 November 2018, while the field survey for the Sunda Strait tsunami was conducted between 24 December 2018 to 3 January 2019 at Banten and Lampung [40, 41]. Based on the data set provided by both studies, inundation depth profile as shown in Figure 2 can be observed. The Sulawesi tsunami disaster recorded inundation depths for buildings ranging between 0.1 – 3.65 m, with a median of 0.98 m. On the other hand, the Sunda Strait tsunami recorded inundation depths ranging from 0.2 – 6.6 m, as well as a median depth of 2.72 m.

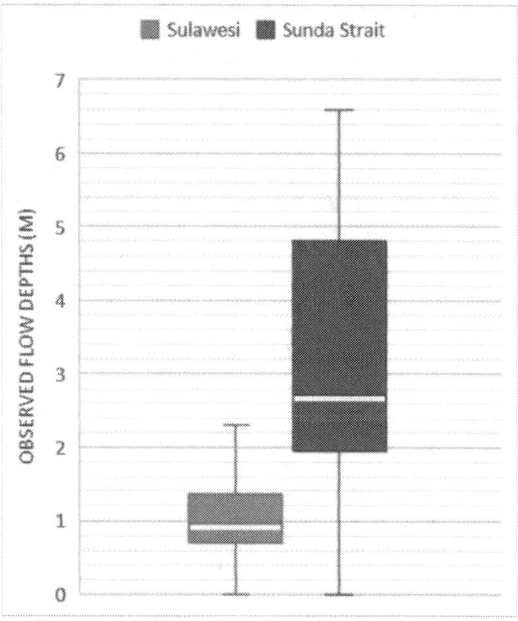

Fig 2. Inundation depth ranges for both tsunami events used as reference in this paper.

Considering the two data sets did not employ the same scale to categorise tsunami damage on buildings, the following re-categorisation is also done for all data gathered to harmonise the data and for ease of comparison. The designation for each grade as listed in Table 6 are adapted based on descriptions provided in both field surveys, as well as common classification in literatures [42].

4.1 Vulnerability profile

Application of the proposed vulnerability index equation for buildings affected by the two tsunami events yielded distribution as shown in Figure 3. In total 468 buildings were assessed, 371 of which were affected by the Sulawesi tsunami event, while 97 others were damaged by the Sunda Strait event. No building in either event could be identified as being "less vulnerable", likely contributed by the fact that only 4% of buildings in Palu were engineered reinforced concrete buildings, while majority in both events were of confined masonry construction [40, 41]. Across affected regions, majority of the buildings were categorised as V3 or highly vulnerable owing to the lack of protective features in between the buildings and the sea, particularly coastal vegetation. It can be noted that for the most part, road and buildings at both locations were built on elevated ground, and/or separated from the sandy shores by rocky levees and low seawalls.

Table 6. Damage state categorisation for buildings damaged in both tsunami events and the equivalent used in this paper.

Paulik et al. [40]	Syamsidik et al. [41]	Re-categorised grades
DL0 (No damage)	DC0 (No damage)	DS1 (Flooded only; No damages sustained)
	DC1 (Minor damage)	DS2 (Minor damages including external scouring, damaged roof)
DL1 (Partial damage, repairable)	DC2 (Moderate damage)	DS3 (Moderate damages; damaged walls, door, windows. Repairable)
DL2 (Partial damage, unrepairable)	DC3 (Major damage)	DS4 (Major damages; damages primary members e.g. columns. Requires major repairs)
DL3 (Complete damage)	DC4 (Completely or washed away)	DS5 (Collapse or washed away; Total or irreparable damages)

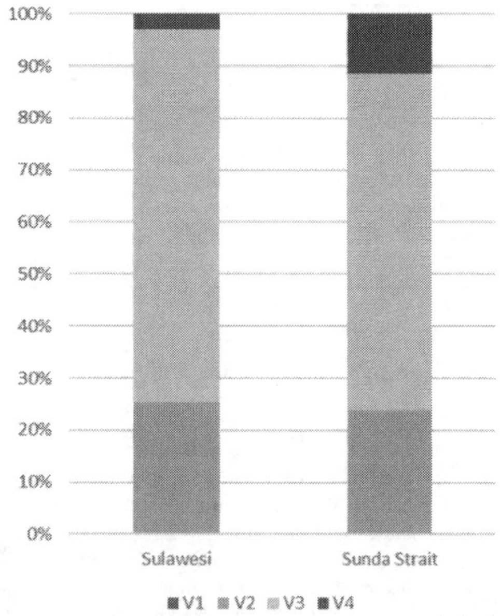

Fig 3. Vulnerability distribution for buildings affected by the 2018 Sulawesi and Sunda Strait tsunami events. V1 – V4 refer to vulnerability levels as described in Table 5.

While these may have been designed to protect landward area from coastal flooding or high tide, in this paper they are counted towards reducing tsunami vulnerability. Additionally, majority of the buildings were low-rise residential buildings made of timber or masonry, hence the relatively high vulnerability ratings. Over 10% of buildings in the Banten and Lampung region were observed as being critically vulnerable, while only 3% could be categorised similarly in Palu.

4.2 Correlation to hazard intensity and damages observed

Similar to the vulnerability distribution, a frequency heatmap in Figure 4 is used to visually represent the relationship between hazard intensity (inundation depth), vulnerability levels, and damages recorded. Inundation depths were grouped into eight grades with the intervals shown in the figure. From 0 to 2 m depth, the intervals were 0.5 m each to allow for a better observation of damage distribution at low flow depths and as deemed appropriate considering the data available. Otherwise, for flow depths above 2 m, the interval for each grade was 1 m. These intervals were categorised in such a way since building damages have been found to be certain (DS2 and above) when the inundation depth is above 2 m regardless of the building typology [4]. Furthermore, at inundation depth of 5 m and above, there is significantly higher probability for complete damage or for buildings to be washed away particularly for timber and masonry buildings [43, 44].

In this case for Palu, Banten and Lampung, most buildings with vulnerability classes 2, 3 and 4 suffered damages ranging from DS1 (flooded only) to DS3 (moderate damages) when inundation depth is below 1.0 m, except for a number of cases in Palu where the buildings were severely damaged or destroyed. As flow depth increases beyond 1 m, the proportion of buildings sustaining major damages markedly increased, as expected based on various tsunami intensity models previously. While the findings presented in Figure 4 cannot be claimed as conclusive and uncertainties remain, such as the degree of damages or vulnerabilities induced by event(s) preceding the tsunami, a clear trend can be observed in the figure whereby damage severity increases quiet linearly with vulnerability and hazard intensity. DS4 (major damages) dominates at flow depths of 1.5 m to 4 m, while above that majority of the buildings across all vulnerability classes collapsed or were washed away. To further demonstrate the discrepancies caused by the preceding tsunami triggering event, we take the example of inundation depths grade 1 and 2 (below 1.0 m), whereby buildings in Palu appeared to have sustained damages of DS4 and DS5. In the Sunda Strait region, only DS1 (flooded only) were recorded for buildings inundated at these lower depths. This is despite the fact that the buildings in Palu were mostly located within 100 m to 180 m from the shoreline, while in Lampung and Banten the buildings were mostly less than 100 m away. It is thus important to note that the compounding effects of earthquake, soil liquefaction and the tsunami during the 2018 event in Palu significantly intensified the damages observed.

Comparing damage probabilities presented in fragility curves for the two events have found that the distribution of damages in Figure 4 reflected the estimated probabilities fairly well [41, 45]. While direct comparison between fragility curves and observations from Figure 4 below may not provide the most accurate or desirable validation for the vulnerability index, interesting observations can still be made which may warrant further exploration.

Based on the fragility curves for confined masonry buildings presented in [41] for the Sunda Strait event, at under 1 m flow depth the probability distribution is partially reflected since moderate damage (DS3)

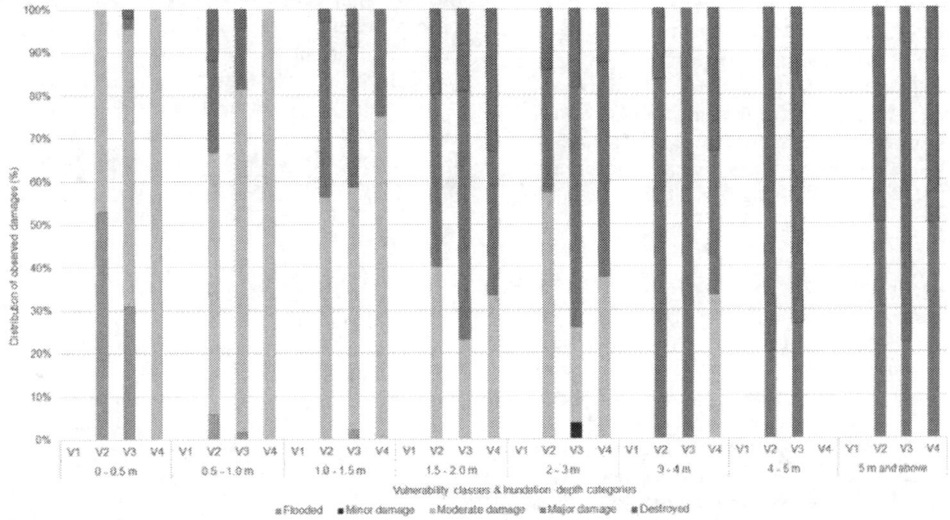

Fig 4. Distribution of building damages for 2018 Palu and Sunda Strait tsunami categorised by flow depth and vulnerability classes.

 https://doi.org/10.1051/e3sconf/202234004002

accounted for the majority of the buildings even at vulnerability class V2, with DS1 being prominent at under 0.5 m, while DS4 and DS5 are accountable at a lesser proportion as well. Meanwhile, for the Sulawesi tsunami event, the fragility curve available for buildings with similar typology showed better correlation particularly for minor to moderate, as well as major damages distribution for class V2 and V3 buildings [45]. This is most likely due to higher number of field survey data being sampled and utilised in the analysis, as mentioned in section 4.1.

As for timber buildings damaged by the Sunda Strait tsunami, probability distribution presented in [45] for flow depths under 2 m were reflected quite well especially in reference to buildings with vulnerability class V4 in the figure. For flow depths of 2 m – 3 m, significant differences were noticeable however with moderate damages still being significant for class V4. However, as the flow depth increases, the probable damages for timber buildings were again reflected quite accurately in the distribution for class V4 due to the fact that inundation during the Sulawesi tsunami event did not exceed 3 m in depth.

Another interesting observation in relation to the topic of this paper is the potential effectiveness of coastal protection, in particular coastal vegetation. In both events, coastal vegetation that may function as effective protective are relatively scarce. However, in areas where vegetation is available, some degree of protection can be observed. Figure 5(a) shows the before and after photo of two masonry residential buildings both with vulnerability class V2. Building B68 sustained moderate damage (DS3) despite being inundated by over 2.5 m of seawater and being closer to the shoreline, while building B67 suffered major damage, even though the inundation depth recorded at its location was only 1.44 m high and the building was twice as far from the shores. The only other difference in this case is the width and density of vegetation in the direct wave path between the buildings and the shoreline. The second snapshot in Figure 5(b) shows another building in Kabupaten, South Lampung. The building identified as B97 (in the field survey data set) is located approximately 80 m from the shoreline. However, inundation depth recorded at its location is low at 0.5 m, and no damage was recorded for the vulnerability class 2 building. Based on the slight difference in vegetation cover, it is highly probable that the band of trees effectively protected the building from the incoming waves.

Fig 5. (a) Before and after snapshot for building B67 and B68. (b) Snapshot of before and after for building B97. Both cases are from the Sunda Strait event. Visible changes in vegetation are highlighted in yellow. Building damage is also visible for building B67 in figure (a).

5 Conclusion

This paper provides a discussion for the potential, development, and application of a tool to assess vulnerability level of buildings in at-risk areas towards tsunami-induced building damages via a set of generalisable vulnerability indicators. Subsequently, the links that can be observed between hazard intensity, vulnerability class and impacts from the tsunami were discussed. The proposed index is developed with the aim of improving usability of presently available tsunami risk assessment tools and models, while also maintaining simplicity and relative accuracy in both ex-ante and ex-post usage. However, much like fragility curves, the accuracy of damage distribution as a result of the interaction between tsunami hazard intensity and vulnerability levels is influenced greatly by diversity, quantity and quality of field data. For this paper, the probable damages estimated in some cases based on the damage distribution rely heavily on data from a particular event, at a particular location. Due to this, addition of data from different localities and tsunami events may be able to address this inaccuracy. Furthermore, the index in its current state has potential for improvements which may be achievable via:

i. application of expert opinions' survey similar to PTVA's revision for the purpose of reviewing and refining vulnerability indicators used in the index as well as the weightages assigned to each.
ii. inclusion of data sets from the 2011 Japan tsunami where abundance of data are available. Comparison can be done with cities where all vulnerability indicators used in this study can be accounted for, and thus may also offer better insight in terms of drawing conclusive correlation between all independent, moderating and dependent variables as used in this paper.

iii. the significance of each vulnerability indicator as a moderating variable component may also be statistically tested and proven to reduce uncertainties in the index.

iv. consideration can be given to other indicators that are reasonably generalisable to be included in the index such as building row and distance from shoreline, where possible.

Additionally, in future work, this vulnerability index can be harmonised with, and incorporated into a descriptive tsunami intensity scale to facilitate better tsunami risk understanding, assessment and communication at all levels particularly among community stakeholders.

References

1. National Geophysical Data Center / World Data Service: NCEI/WDS Global Historical Tsunami Database. NOAA National Centers for Environmental Information. doi:10.7289/V5PN93H7 (Accessed on 20 July 2021)
2. CRED, D. Guha-Sapir, EM-DAT: The Emergency Events Database. https://public.emdat.be/data (Accessed on 20 July 2021)
3. UNISDR, *Words into Action guidelines: National Disaster Risk Assessment* (2017)
4. A. Suppasri, E. Mas, I. Charvet, R. Gunasekera, K. Imai, Y. Fukutani, Y. Abe and F. Imamura, Nat. Hazards **66**, 319–341 (2013)
5. I. Charvet, I. Ioannou, T. Rossetto, A. Suppasri and F. Imamura, Nat. Hazards **73**, 951–973 (2014)
6. R. de Risi, K. Goda, N. Mori and T. Yasuda, Stoch. Environ. Res. Risk Assess. **31**, 1253–1269 (2016)
7. Y. Xiong, Q. Liang, H. Park, D. Cox and G. Wang, Coast. Eng. **144**, 1–14 (2019)
8. J. Bonacho and C.S. Oliveira, Int. J. Disaster Risk Reduct. **31**, 275–280 (2018)
9. H. Park, M.S. Alam, D.T. Cox, A. R. Barbosa and J.W. van de Lindt, Int. J. Disaster Risk Reduct. **35**, 101076 (2019)
10. M. Mück, H. Taubenböck, J. Post, S. Wegscheider, G. Strunz, S. Sumaryono and F.A. Ismail, Nat. Hazards **68**, 97–114 (2013)
11. F. Dall'Osso, D. Dominey-Howes, C. Tarbotton, S. Summerhayes and G. Withycombe, Nat. Hazards **83**, 1229–1256 (2016)
12. P. Srivihok, K. Honda, A. Ruangrassamee, V. Muangsin, P. Naparat, P. Foytong, N. Promdumrong, P. Aphimaeteethomrong, A. Intavee, J.E. Layung and T. Kosin, Cont. Shelf Res. **79**, 3–15 (2014)
13. K. Goda and J. Song, Stoch. Environ. Res. Risk Assess. **30(8)**, 2271–2285 (2015)
14. O.D. Cardona, M.K. van Aalst, J. Birkmann, M. Fordham, G. McGregor, R. Perez, R.S. Pulwarty, E.L.F. Schipper, and B.T. Sinh, *Determinants of risk: exposure and vulnerability,* in Managing the Risks of Extreme Events and Disasters to Advance Climate Change Adaptation [C.B. Field, V. Barros, T.F. Stocker, D. Qin, D.J. Dokken, K.L. Ebi, M.D. Mastrandrea, K.J. Mach, G.-K. Plattner, S.K. Allen, M. Tignor, and P.M. Midgley (eds.)] (Cambridge University Press, Cambridge, UK, and New York, NY, USA, pp. 65-108, 2012)
15. GFDRR/World Bank, *Understanding risk: review of open source and open access software packages available to quantify risk from natural hazards* (World Bank Group, Washington, DC, 2014)
16. B. Wisner and H.R. Luce, GeoJournal **30(2)**, 127–140 (1993)
17. T. Cannon, *Vulnerability analysis, livelihoods and disasters*, in Risk 21: Coping with Risks Due to Natural Hazards in the 21st Century [W.J. Ammann, S. Dannenmann, and L. Vulliet (eds.)] (Taylor and Francis Group, London, UK, pp. 41-49, 2006)
18. H. Park, D.T. Cox and A.R. Barbosa, Coast. Eng. **122**, 10–26 (2017)
19. K. Goda and R. de Risi, Int. J. Disaster Risk Reduct. 28, 539–554 (2018)
20. S. Karafagka, S. Fotopoulou and K. Pitilakis, Soil Dyn. Earthq. Eng. **112**, 118–137 (2018)
21. S. Belliazzi, G.P. Lignola, M. di Ludovico and A. Prota, Structures **31**, 68–79 (2021)
22. G.S. Leonard, N. Evans, G. Prasetya, W.S.A. Saunders, L. Pearse, D. Monastra, and S. Fraser, *Scoping study for evaluating the tsunami vulnerability of New Zealand buildings for use as evacuation structures*, GNS Science Report 2011/36 39 p. (2011)
23. R.A. Darlymple and D.L. Kriebel, *Lessons in engineering from the tsunami in Thailand* in The Bridge Vol. 35, No. 2, G. Bugliarello (ed.) (National Academy of Sciences, Washington, DC, 2005) pp. 4–13
24. S. Medina, J. Lizarazo-Marriaga, M. Estrada, S. Koshimura, E. Mas and B. Adriano, Eng. Struct. **196**, 109309 (2019)
25. R.S.M. Samarasekara, J. Sasaki, M. Esteban and H. Matsuda, Int. J. Disaster Risk Reduct. **23**, 80–92 (2017)
26. R. de Risi, K. Goda, T. Yasuda and N. Mori, Earth Sci. Rev. **166**, 64–82 (2017)
27. J.C.L. Bayas, C. Marohn, G. Dercon, S. Dewi, H.P. Piepho, L. Joshi, M. van Noordwijk and G. Cadisch, *Influence of coastal vegetation on the 2004 tsunami wave impact in west Aceh* in Proceedings of the National Academy of Sciences of the United States of America **108(46)**, 18612-18617 (2011)
28. A.M. Hashim, SM.P. Catherine and H. Takaijudin, Res. J. Appl. Sci. Eng. Technol. **5(18)**, 4483-4488 (2013)
29. R. Omira, M.A. Baptista, J.M. Miranda, E. Toto, V. Catita and J. Catalao, Nat. Hazards **54,** 75–95 (2010)
30. H. Takagi and J.D. Bricker, Coast. Eng. **56(4),** 1450018 (2014)
31. K. Harada and F. Imamura, *Effects of coastal forest on tsunami hazard mitigation – a preliminary investigation*, in: K. Satake (eds) Tsunamis. Advances in Natural and Technological Hazards Research, Vol. 23, Springer, Dordrecht (2005)
32. H. Ismail, A.K. Abd Wahab and N.E. Alias, Nat. Hazards **63**, 939–963 (2012)
33. T.L. Saaty, Eur. J. Oper. Res. **48**, 9–26 (1990)

34. K. Nakata, A. Katsumata and A. Muhari, Earth Planets and Space **72**:44 (2020)
35. A.R. Gusman, P. Supendi, A.D. Nugraha, W. Power, H. Latief, H. Sunendar, S. Widiyantoro, Daryono, S.H. Wiyono, A. Hakim, A. Muhari, X. Wang, D. Burbidge, K. Palgunadi, I. Hamling and M. R. Daryono, Geophys. Res. Lett. **46(15)**, doi: 10.1029/2019GL082717 (2019)
36. Global Volcanism Program, *Report on Krakatau (Indonesia)* (Crafford, A.E., and Venzke, E., eds.). Bulletin of the Global Volcanism Network, 43:10, Smithsonian Institution (2018)
37. Global Volcanism Program, Report on Krakatau (Indonesia) (Krippner, J.B., and Venzke, E., eds.). Bulletin of the Global Volcanism Network, 44:3, Smithsonian Institution (2019)
38. L. Ye, H. Kanamori, L. Rivera, T. Lay, Y. Zhou, D. Sianipar and K. Satake, Sci. Adv. **6(3)**, eaaz1377 (2020)
39. T. Takabatake, T. Shibayama, M. Esteban, H. Achiari, N. Nurisman, M. Gelfi, T.A. Tarigan, E.R. Kencana, M.A.R. Fauzi, S. Panalaran, A.S. Harnantyari, T.O. Kyaw, Coast. Eng. J. **61**, 423–443 (2019)
40. R. Paulik, A. Gusman, J.H. Williams, G.M. Pratama, S.L. Lin, A. Prawirabhakti, K. Sulendra, M.Y. Zachari, Z.E.D. Fortuna, N.B.P. Layuk and N.W.I. Suwarni, Pure Appl. Geophys. **176**, 3305–3321 (2019)
41. Syamsidik, Benazir, M. Luthfi, A. Suppasri and L.K. Comfort, Nat. Hazards Earth Syst. Sci. **20**, 549–565 (2020)
42. A. Suppasri, I. Charvet, J. Macabuag, T. Rossetto, N. Leelawat, P. Latcharote and F. Imamura, *Building Damage Assessment and Implications for Future Tsunami Fragility Estimations* in M. Esteban, H. Takagi and T. Shibayama (eds.) Handbook of Coastal Disaster Mitigation for Engineers and Planners (Butterworth-Heinemann, Oxford, 2015)
43. S. Reese, B.A. Bradley, J. Bind, G. Smart, W. Power and J. Sturman, Earth Sci. Rev. **107**, 156–173 (2011)
44. K.I.U. Nanayakkara and W.P.S. Dias, Nat. Hazards **80**, 471–486 (2016)
45. E. Lahcene, I. Ioannou, A. Suppasri, K. Pakoksung, R. Paulik, Syamsidik, F. Bouchette and F. Imamura, Nat. Hazards Earth Syst. Sci. **21**, 2313–2344 (2021)

Community strategies in the development of Merapi permanent settlements

Bagus Ramadhan[1], *Gardyas* Bidari Adninda

Urban and Regional Planning. University of Amikom Yogyakarta, Jl. Ring Road Utara, Ngringin, Condongcatur, Depok, Sleman, Daerah Istimewa Yogyakarta 55281, Indonesia

Abstract. Merapi relocated settlements has been inhabited since 2012, two years after the Merapi Volcano eruption. Despite the adequate facilities, the land and house are not enough to accommodate the additional family members. Adapting to the condition, the community made several improvements in the physical condition of the permanent settlement. This study aims to determine the community's strategies in the Karangkendal and Plosokerep resettlement area in developing their dwellings. The observation includes the community's methods in enhancing the existing housing and its funding strategies in developing houses in general. Employing a case study approach, observation and in-depth interviews conducted with the head of the family and the head of the local hamlet. Based on the analysis, most of the community in the Karangkendal and Plosokerep resettlement area prefer to develop their existing houses. Although several residents consider leaving the resettlement area, most of them remain to stay and develop the existing house because building a new house requires a lot of money. They spend their private savings that are soon considered as the community's financial strategy. To fulfill that demand, the community needs external support. For instance, the government could provide revolving funds or build a coop that effectively supports housing improvement.

1 Introduction

The eruption of Merapi Volcano in 2010 caused severe damage to the hamlets, which included disaster-prone areas of Merapi Volcano. There are 28 hamlets in the Cangkringan district with 2,411 houses damaged [1]. In Cangkringan Subdistrict, three villages were severely affected: Umbulharjo Village, Kepuharjo Village, and Glagaharjo Village. Of these three villages, nine hamlets cannot be occupied because they are directly affected by the eruption of Merapi Volcano.

After the eruption, determination of the disaster-prone area of Merapi Volcano was made considering the damaged area of the disaster. The classification of the disaster-prone area is divided into KRB (*Kawasan Rawan Bencana*) I, II, and III, with different consequences. The most dangerous area is KRB III, or the area directly affected by the eruption of Merapi Volcano in 2010. The government's decision not to allow the area as residential. This policy led to the permanent relocation of residents' dwellings provided by the government or the community independently.

After almost eight years of living in permanent settlements, the residents currently face new problems. Those whose family members increase either through marriage or by adding children or grandchildren need a new space in their house.

The land area of 100 m^2 with a house area of 36 m^2 is insufficient to accommodate the growing number of their family members. While the initial land is in a disaster-prone area, it cannot be occupied according to the agreement when they got a permanent settlement. The people who live in the permanent settlement have agreed that the land left in the disaster-prone area remains their property, but they promise not to build in the area by being compensated in the form of permanent housing with a certificate of ownership.

This study examines the community strategies in the Karangkendal and the Plosokerep permanent settlement in developing their residential area. The observation including the strategy of developing the existing permanent settlement conducted by the community. In addition, this research also focuses on funding strategies carried out by the community in developing the house.

1.1 Housing development

An area or building can develop physically by three models [2]:

a. Horizontal development
   This development grows outward while maintaining the height of the building. Horizontal development can occur in areas where there is vacant land around the building.
b. Vertical development
   This development is carried out by increasing the height of the building and usually occurs in downtown areas with expensive land prices.
c. Intercystal development

[1] Corresponding author: bagusramadhan@amikom.ac.id

This development means an inward development. In other words, this development is carried out by compacting an area or using the space between buildings and not increasing the height of the building[2].

A house that develops physically can be referred to as a growing house. A growing house has grown from a house with a small area to a larger one [3]. The growing area occurs because it is adjusted to the financial capacity of each household. Self-growing houses can be developed vertically and horizontally and can be developed without the need to dismantle the initial building, only by adding new spaces. However, this will be more difficult if the initial house design does not adhere to the principle of a house that will grow[3].

A study on growing houses developed by Perumnas, stated that it usually increases its occupants' space requirements [4]. The most added room is the bedroom because of the incompatibility of the number of family members with the facilities offered apart from other rooms [4]. The increasing number of occupants is one of the solid factors for residential development.

This growing house is one of the efforts in the provision of homes to meet the needs of its residents. In general, the housing supply scheme, consists of:

a. Industrialized housing, where there is a process of production, distribution, and consumption of products in the form of houses.
b. Petty commodity, where small-scale housing developments are organized by architects/foremen with a few workers. One part of this petty commodity is a self-help housing process where producers and consumers are the same parties. Self-help housing is usually carried out by low-income people who do not have enough funds to buy a house as a commodity or employ a supervisor and several handypersons. Several household members can do self-help housing according to the knowledge known to the household [5].

In Indonesia, there are types of self-help housing. Self-help housing in Indonesia depends on the age of household members, development process, and demographic profile [6]. Regarding the post-disaster settlement, the provision of post-disaster permanent shelter requires adequate funding. Even the United States federal government has not been able to provide proper funding schemes for the victim of Hurricane Katrina in 2005 [7]. These conditions impact the natural disasters refugees returning to their initial living places. They assume that the financial and resources are adequate in the initial place. However, the place is not safe from the next disaster.

Self help housing means personal fulfillment in the community where people want to express their aspirations or needs without relying on bureaucracy[8]. Self-help is a gradual process regularly, starting with the basic structure then adding parts of the house with higher standards and material selection [8].

1.2 Post disaster settlements

The destructive natural disaster of the Merapi Volcano eruption in 2010 caused material losses and claimed lives. The main impact is that the hamlets directly affected by the eruption of Merapi Volcano are designated as uninhabitable areas as a form of protection against eruptions that may occur in the future. The consequence is post-disaster housing needs to be relocated. The first response to post-disaster housing damage is to organize emergency sheltering to accommodate temporary refugees [9]. After the conditions are gradually safe, permanent settlements can accommodate refugees with more decent conditions. Damage to dwellings that are usually inhabited makes the need for permanent settlement while the initial house is repaired or the option to relocate to permanent housing. The fourth phase is permanent housing, where people can return to their old house that has been repaired or relocated to a new, safer house, as in the case of permanent housing due to the 2010 eruption of Merapi Volcano.

Post-disaster settlement is one of the most important things that must be provided [10]. The house is the core of routine household activities, and without a house, the household cannot run properly and will impact the daily activities of household members. The provision of post-disaster permanent settlement in the United States often depends on market mechanisms [10]. However, this impacts low-income communities because they cannot access quality housing, living in low-quality houses. These low-income communities can live on marginal lands at low prices, are not licensed, and can return to disaster-prone areas so that when a disaster strikes again, they are a community that is very vulnerable to disasters.

The process of developing permanent housing ranges from 2 years to indefinitely if the relocation program is poorly designed or poorly implemented [11]. Meanwhile, several stages of relocation [12] consists of:

a. Emergency response, consisting of identifying victims, saving lives in the event of a disaster, distributing basic needs for victims, and providing settlement;
b. The transitional phase, consisting of clearing debris and dirt from affected areas, relocating victims from refugee camps to temporary shelters by providing basic needs, reconstructing and repairing shelters and public facilities.
c. The development phase, this long-term phase seeks to provide permanent shelter/safe relocation housing for affected communities, facilitate affected communities in relocation shelters with basic infrastructure, improve the living standards of affected communities in new locations, and develop an ideal relocation residential environment to be inherited to the next generation [12].

In this case, the community's development of permanent housing is in the development phase. The community prepares for the long term by trying a strategy so that the residents' children and grandchildren can live properly with on-site residential development and off-site strategies. The relocation process is building shelter and introducing new living spaces for affected communities to reduce danger in the future [13]. In this process, community activities and the built space will adjust to each other. However, this adaptation does not run in a short time without the availability of the community's basic needs.

In developing the permanent settlement, physically, the existing permanent settlement can be developed in several ways. However, the limited land area where the house has an area of 36 m^2 and each household's land, which is only 100 m^2, also becomes an obstacle in carrying out on-site development. In addition, if many houses need to develop, the three growth schemes above can occur in one shelter area. This condition could cause quality degradation of the settlement area if no policies regulate it.

The criteria for livable houses according to Permenpera Number 22 of 2008 concerning Minimum Service Standards in the Public Housing Sector at the Provincial and Regency/City Levels states that one of the parameters for the feasibility of a house is to meet a minimum area of 7.2 m^2/person to 12 m^2/person [14]. This standard means that a permanent residence with an area of 36 m^2 can only be filled by three to a maximum of five people. If the family member increases, the house needs several adjustments, or it becomes less livable. One of the first settlement adaptations from the area was the shape adaptation of the house that was adapted to the behavior and daily culture of the community[15]. One of these behavioral patterns is adding walls to increase private space since the relocation house has limited space.

The physical condition of Merapi's permanent settlements is quite different from the initial living area [16]. The permanent settlements does not have a front and back yard and with an area of only 100 m^2 for each household. However, the residents are satisfied with the conditions of the shelter because there are adequate facilities, both in the form of community halls, communal cages, children's playrooms, mosques, and so on.

2 Methodology

This study employs a case study approach. The data was collected through observation and in-depth interviews with the head of the family or local hamlet. A case study is a method that thoroughly investigates a program, event, activity, process, or several individuals [17]. This study explores the strategies adopted by households in two permanent settlements in developing their dwellings.

The two permanent settlements chosen are Karangkendal and Plosokerep. Karangkendal permanent settlement was chosen because the relocation process of housing is carried out with the full awareness of all residents where residents find their relocation land. The development process is carried out with the assistance of the Rekompak program. Meanwhile, the Plosokerep permanent settlement is a residential development process that relies more on the process driven by Rekompak with the involvement of residents who are not as much as the development of the Karangkendal.

The data collection method in this study is observation and in-depth interviews with the head of the family or the head of the local hamlet to determine the need for residential development. The sampling technique used purposive sampling by asking community leaders and village heads first to find out the family with the following criteria:

a. A family who was relocated after the eruption of Merapi Volcano in 2010;
b. Having four or more family members;
c. Permanent dweller.

Based on the in-depth interviews with community leaders and village heads, respondents (n=14) met these criteria. The next step is to observe the physical changes of permanent settlement modification to adapt to the increasing number of family members. The observations were done by looking at developments that might be carried out outside the shelter area. In addition, in-depth interviews were conducted to deeply determine the residential development process carried out by residents, from funding, construction, and adaptation to post-construction development.

The data obtained in the observation process was analyzed using descriptive analysis to determine the existing resettlement improvement strategy, housing preferences outside the resettlement area, and financial strategy. The existing improvement strategies are considered as the pattern of resident's adaptation to develop the area since 2012. Meanwhile, the housing location preference outside the resettlement area could be analyzed to find the desired location for the community. Moreover, financial strategies are explored to determine community effort to obtain funding for housing reconstruction.

3 Results and discussions

3.1 Relocation process

The Karangkendal and Plosokerep permanent settlements were built for the residents of two hamlets in Umbulharjo Village, whose places were directly affected by the Merapi Volcano eruption in 2010. However, in both places, there are differences in the relocation process.

The residents of Pelemsari Hamlet voluntarily relocated to the Karangkendal permanent settlement. The main reason is that their living place was damaged heavily and most of the house entirely collapsed. They decided to move in to the saver place [18].

The residents of Pelemsari even searched for their relocation land and bought it together. However, the planning and construction process was assisted by the government and Rekompak, the organization responsible for the reconstruction of the post-disaster. After six months of the construction, in mid-2012, 81 families of Dusun Pelemsari started living in the permanent settlement

In opposite to residents of Pelemsari Hamlet, Pangukrejo residents are not interested in relocating. The reason is that not all of the houses in Pangukrejo Hamlet were heavily damaged. Moreover, after the eruption, there is a new tourist attraction around Pangukrejo Hamlet. This Merapi lava tour invites many visitors, making the residents prefer to stay in their old dwellings because of the promising economic potential. Due to several factors above, only 35 out of 114 families of Dusun Pangukrejo participated in the relocation, which is now becoming Plosokerep's permanent settlement. However, the other left shortly relocated to the Plosokerep permanent settlement because they were increasingly aware of the dangers of occupying Pangukrejo Hamlet.

3.2 The condition of Karangkendal and Plosokerep permanent settlements

Since occupied in 2012 and 2013, the residents have improved their housing conditions. These changes are based on the dissatisfaction or need of each household. Since 2012 at least the residences in both permanent settlements have made at least one renovation. Starting from cleaning the walls to adding floors depends on the needs of each occupant. According to residents, the current quality of the existing shelters is as follows:

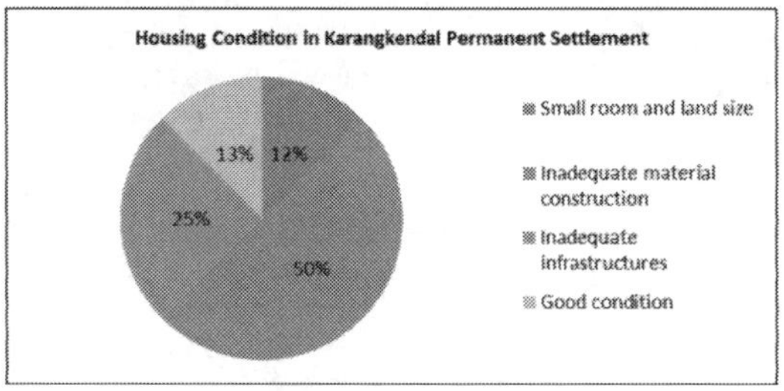

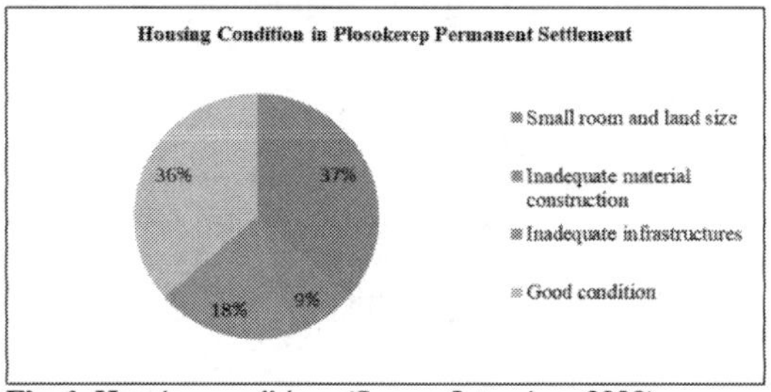

Fig. 1. Housing conditions (Source: Interview, 2020).

The results show that there are differences in the residents' perspectives. Residents' perception is based on their experience living in their old dwellings. They can compare the permanent settlements reach their ideal expectations or not. Almost half of the Karangkendal residents assume that the existing building materials are inadequate. Meanwhile, the Plosokerep residents are not too concerned about this. The respondent in Plosokerep (one-third of the respondents) complained about the small size of the area. Although at the same time, they also accepted the current permanent settlement's conditions.

On the other hand, a small part of the community complains that the settlements and daily living infrastructure are inadequate in both permanent settlements. Although some respondents already felt that the shelters were good, some wanted to improve their houses. Several reasons are illustrated in the following graphs.

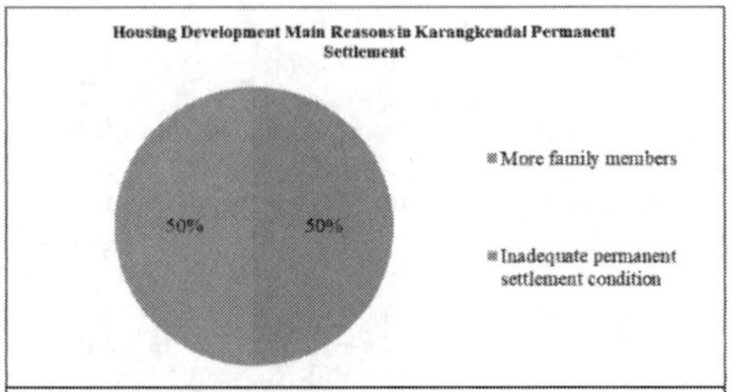

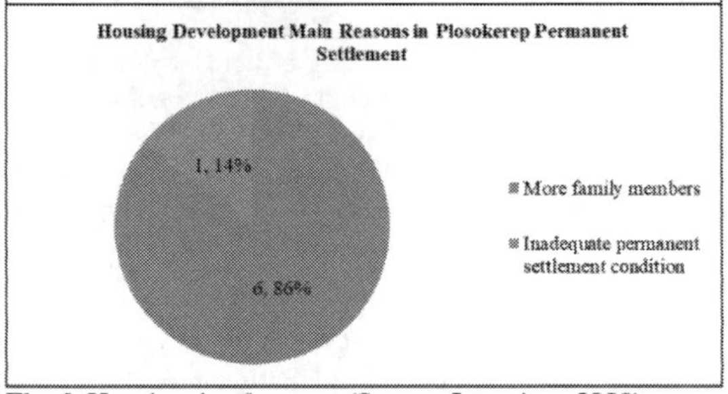

Fig. 2. Housing development (Source: Interview, 2020).

Based on the interviews, the housing development occurs due to the increasing number of family members. Additional family members included children or children who are married. In consequence, the need for space in the house increases as well.

3.3 The housing development strategies

The community's residential development strategy is divided into developing existing permanent settlements, development outside the houses, and community financial strategy.

3.3.1 The existing housing development strategy

The development of existing housing is the primary choice for the community in both the Karangkendal Huntap and Plosokerep permanent settlements. It could be a sign that the community accepted the shelter as their living place. The residents prefer developing the shelter rather than developing outside the shelter. In addition, the economic factor is one of the main reasons because most of them work in the tourism and agricultural sectors. Based on the interviews, the preferences for existing residential development are as follows:

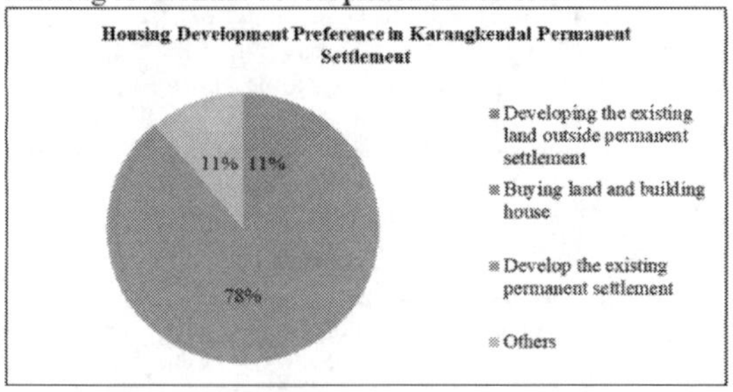

https://doi.org/10.1051/e3sconf/202234004003

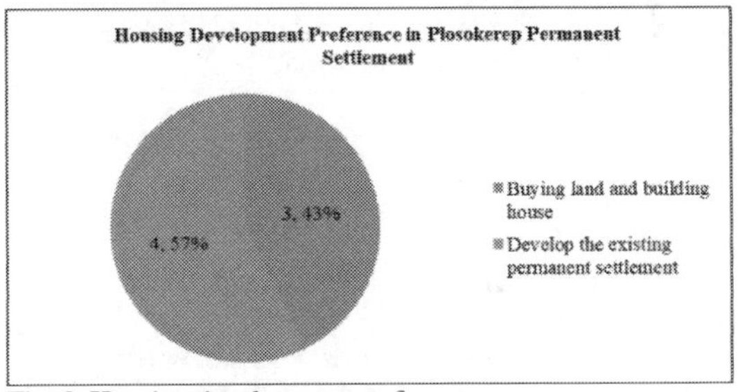

Fig. 3. Housing development preference.
(Source: Interview, 2020)

Based on the interviews, most residents want to live in the existing permanent settlement. The community's main reason is limited funds and does not own land outside the temporary shelter, so they prefer to develop the permanent settlement they currently have.

All residents of Karangkendal permanent settlement have the ownership certificate of the initial land, but the land cannot be re-inhabited after the disaster. In addition, after more than five years of living, the community has gained a sense of belonging to the place. Adaptation to daily activities and livelihoods successfully carried out, and the kinship between communities in the permanent settlement growing more robust so that people prefer to develop their existing permanent settlement.

One community even has a strategy to buy a neighbor's land as an addition to their house. Meanwhile, several residents want to buy land outside the shelter and develop new housing because they want to be safe from the threat of the Merapi Volcano.

Since 2012, the community has made many modifications to its permanent settlement. Respondents have made at least one modification to their permanent occupancy. The pattern of community development in the early days of inhabiting is modifying the walls and the house's appearance. The permanent settlement wall was initially an unpainted brick wall so that the community renovates it by repainting or painting the house.

Table 1. The residents room renovation.

No	Renovated Rooms	Total Number
1	Bedroom	13
2	Kitchen	12
3	Terrace	12
4	Toilet	1
5	Garage	2
6	Storage	1
7	Adding floor	2

(Source: Interviews, 2020)

Furthermore, the community develops housing according to the needs of each household. Some respondents develop bedrooms, kitchens, terraces, toilets, garages, and warehouses. Moreover, some of them add more floors in the dwelling. The terrace was developed for a communal space that can be shared with the surrounding neighbors. The increasing number of the family makes the residents develop adequate toilets and bedrooms. Kitchens are needed so that there is adequate and proper cooking space. The residents need to improve garages and warehouses to store household items and various private vehicles.

Fig. 4. Terrace renovation, adding rooms and adding floor (Source: Field Survey, 2020).

The development process of the permanent settlements illustrates in three stages. In the early stage, almost all respondents modified the housing appearance by painting the walls and install ceilings. Over time, several modifications were made, such as adding terraces and rooms, modifying the kitchen, adding warehouses, and adding floors according to the abilities and needs of the community. So that the building modifications carried out by the community are not carried out at one time, but gradually, this is due to financial factors because people do not have much money at one time to develop their dwellings. Because of the financial factor, each household has a different time developing their house. Residents who have an excellent financial level can develop their house immediately, but those who have limited finances have to save their money first.

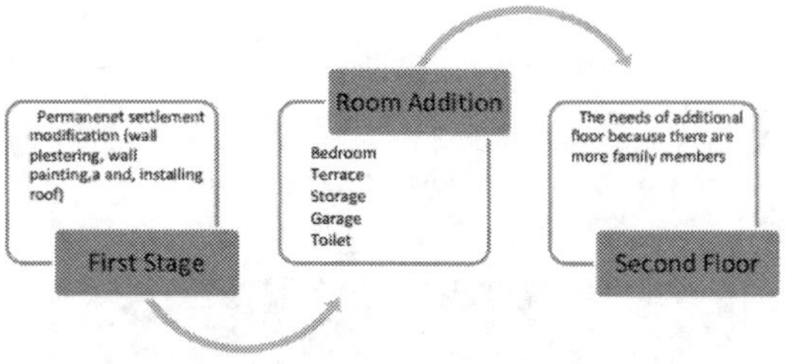

Fig. 5. Stages of permanent settlement development. (Source: Analysis, 2020)

In the future, several residents want to enhance spaces in the settlement such as toilets, terraces, rooms and add second floors. Some other want to renovate the floor and roof. The next stage shows that the community will add the second floor after the first floor is fully used.

3.3.2 Housing development preferences outside the permanent settlement

Residential development outside the permanent settlement is not the community's primary choice. However, several families develop their house in the permanent settlement area, and several others are moving back to the land in their old residences located in disaster-prone areas.

Community strategy facing the development options outside the permanent settlement area is still developing itself. Culturally, the community is more accessible and familiar with developing its houses. Buying a house requires much money.

Some residents wanted to stay in the village because of their social attachment and livelihood. They choose to be in the scope of one subdistrict but different villages or outside the sub-district. In addition, the community also wants to be safer from the threat of disasters by staying outside the affected areas of the eruption of Mount Merapi.

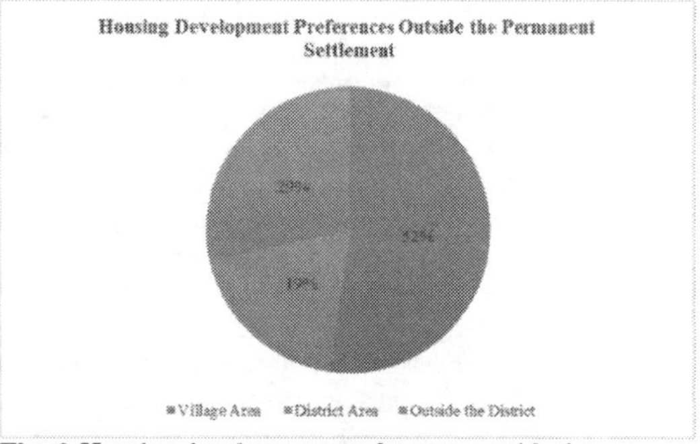

Fig. 6. Housing development preferences outside the permanent settlement (Source: Interview, 2020).

Based on these preferences, people are not entirely tied to the old locations or the locations they currently inhabit. Financial limitations are the main factor of society who stayed in the permanent settlement. If the government develops more housing to extend the living place for residents, the government can find alternative locations around Cangkringan District or outside Cangkringan District. Some people who choose to live outside the district are willing to feel safe from the threat of the eruption of Merapi Volcano.

3.3.3 Aspects of community financial strategy

In developing houses, it is essential to have resources, specifically funding. In low-income communities, self-help construction could become the most suitable solution. If the community has enough money, they could build and stop the construction whenever the money is available. In addition, the other family member could help by becoming the substitution of the primary worker.

The same goes for communities in both Karangkendal and Plosokerep permanent settlements. They have the same funding method, which is saving some of their income. In addition to having more funding, a small number of residents sell their assets, such as cattle bred. The other residents take additional loans to banks or financial institutions. This option is generally taken by residents who have more assets or residents who have fixed incomes.

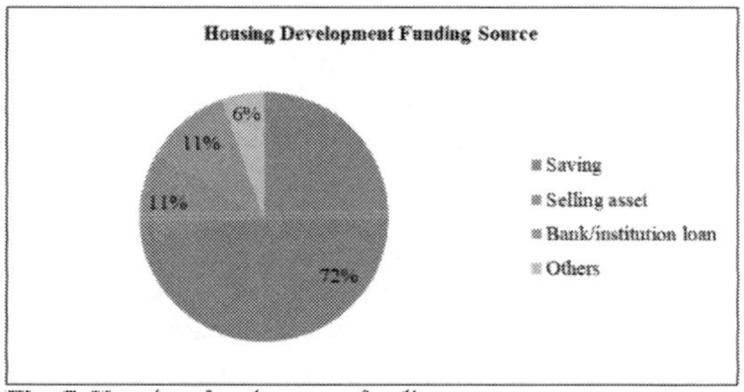

Fig. 7. Housing development funding source. (Source: Interview, 2020)

Looking at the strategy of most of the residents who save their money to develop housing, the government could provide special revolving funds for those who need to enhance their houses. Considering the next 5-10 years, the increasing number of families means more accommodation. The revolving fund could become an appreciation for residents living in the shelters and not returning to their initial village. Furthermore, the sustainability of the people who live in the shelters can be maintained.

4 Conclusion

Based on the analysis, most of the community in the Karangkendal and Plosokerep resettlement area prefer to develop their existing dwellings. Although several residents consider leaving the resettlement area, most of them remain to choose to develop the existing place.

The development of existing dwellings is also carried out gradually, starting from improving the appearance, adding space, and floor addition. The improvement has a different timeline for each family because of the financial capabilities and needs.

Some families who prefer to develop housing outside permanent settlement areas have various opinions. Some choose to be in the village or within one district because it is close to their old village. At the same time, some choose outside the district for safety reasons. Development outside of permanent settlement requires a high cost, so that this community needs a better financial strategy.

The community's financial strategy in developing housing is limited to their saving. This condition needs to be considered because the community demands external support. For instance, the government provides revolving funds or building a coop that effectively supports housing improvement.

Acknowledgment. This research is funded by Internal Research Grant of Lembaga Penelitian dan Pengabdian Masyarakat (LPPM) Universitas Amikom Yogyakarta for year 2020.

References

1. P. Bawole, Tesa Arsitektur **13**, 114 (2015)
2. M. Zahnd and H. Frick, *Perancangan Kota Secara Terpadu* (Kanisius, Yogyakarta, 1999)
3. N. Agusniansyah and K. Widiastuti, Modul **16**, 1 (2016)
4. S. M. Aryani, I. E. S. Wahyuningsih, and M. Mulyadi, Tesa Arsitektur **14**, 64 (2017)
5. P. M. WARD and G. C. MACOLOO, International Journal of Urban and Regional Research **16**, 60 (1992)
6. D. Tunas and A. Peresthu, Habitat International **34**, 315 (2010)
7. J. N. Levine, A. M. Esnard, and A. Sapat, Journal of Planning Literature **22**, 3 (2007)
8. C. Pugh, Third World Planning Review **19**, 91 (1997)
9. E. L. Quarantelli, Disaster Prevention and Management: An International Journal **4**, 43 (1995)
10. W. G. Peacock, N. Dash, and Y. Zhang, in *Handbook of Disaster Research* (Springer, 2007), pp. 258–274
11. A. Oliver-Smith, Disasters **15**, 12 (1991)
12. K. Keraminiyage and P. Piyatadsananon, International Journal of Disaster Resilience in the Built Environment **4**, 352 (2013)
13. P. Sridarran, K. Keraminiyage, and D. Amaratunga, in *Procedia Engineering* (Elsevier Ltd, 2018), pp. 125–132
14. Kementerian Perumahan Rakyat, *Permenpera Nomor 22 Tahun 2008 Tentang Standar Pelayanan Minimal Bidang Perumahan Rakyat Di Tingkat Provinsi Dan Kabupaten/Kota* (2008)
15. J. E. Duyne Barenstein, International Journal of Disaster Resilience in the Built Environment **6**, 140 (2015)
16. E. T. W. Mei, A. Fajarwati, S. Hasanati, and I. M. Sari, Procedia - Social and Behavioral Sciences **227**, 361 (2016)
17. J. Creswell, *Research Design: Pendekatan Kualitatif, Kuantitatif, Dan Mixed* (Balai Pustaka, Yogyakarta, 2012)
18. B. Ramadhan, Proses Konsensus Dalam Relokasi Hunian Korban Erupsi Gunung Merapi 2010 Dan Faktor Yang Mempengaruhi, Studi Kasus Dusun Pelemsari Dan Dusun Pangukrejo, Desa Umbulharjo, Kecamatan Cangkringan, Kabupaten Sleman, 2013

Salt attack and rising damp on house buildings at the tsunami-affected areas

Ella Meilianda[1,2,*], *Dedy* Alfian[2], *Yunita* Idris[1,2]

[1]Civil Engineering Department, Engineering Faculty, Universitas Syiah Kuala, Jl. Tgk. Abdur Rauf, Kopelma Darussalam, Banda Aceh, 23111, Indonesia
[2]Tsunami and Disaster Mitigation Research Center (TDMRC), Universitas Syiah Kuala, Jl. Hamzah Fansuri, Kopelma Darussalam, Banda Aceh, 23111, Indonesia

Abstract. Post tsunami 26 December 2004, rising damp and salt attack are two underestimated phenomena emerging on building construction on the tsunami-affected areas in Banda Aceh city, Aceh Province of Indonesia. The severity of building quality deterioration, particularly on the typical masonry walls, has never been observed by local inhabitants to have been escalated before the tsunami event. Such phenomena persistently present even after approaching the second decade of the post-tsunami event. The present study is the first attempt to explore the evidence of rising damp and salt attacks of the houses around the Banda Aceh city. Forty-five houses were purposively sampled based on their visual appearances being affected by the salt attack and rising damp, the heights of the rising damp, and also groundwater salinity in the vicinity of the sampled houses. The results show that the rising damp heights on the walls are relatively high in any location within the tsunami inundation boundary 500 m from the shoreline. However, a turning point where rising damp height reduced remarkably to less than 1 meter was identified as the distance of houses increased beyond 500 m from the shoreline. Overall, the high level of water salinity brought by tsunami inundation during the early post-tsunami has been an important controlling factor contributing to the salt attack and rising damp, regardless of the houses' distance to the shoreline, but may have been indirectly influenced those phenomena in the long run.

1 Introduction

Coastal properties exposed to both the ocean breeze and the shallow groundwater containing salt substance commonly result in salt attack and rising damp, particularly on concrete and masonry walls. Paint discoloration, peeling off paint and plaster, and also the presence of moss, the crystallization of salt along with the plaster to crumble on walls are the most common appearances of the walls infested with combined salt attack and rising damp. Factors that influence the height of capillary impact or rising damp on masonry walls depend on material properties (wall thickness and porous material composition). Evaporation is also an important factor in rising damp which include the factor that may control it, including the temperature, humidity, air movement and surface condition of the wall. If left untreated, they can cause extreme damage to the structure of buildings, particularly to both brick and mortar [1], [2]. Several studies on the effect of rising damp and salt attack have been conducted to investigate the strength decay of building structures, mostly for historic buildings located at the humid coastal areas [2], [3], [4].

Post tsunami 26 December 2004, salt attack and rising damp are two underestimate phenomena emerging on building construction on the tsunami-affected areas in Banda Aceh city, Aceh Province of Indonesia, being one of the most devastated coastal areas affected by the tsunami. The severity of building quality deterioration, particularly on the typical masonry walls have never been observed before by the coastal inhabitants in the pre-tsunami era at the coastal areas of Banda Aceh city. Such phenomena persistently present even after approaching the second decade of the post-tsunami event.

The present study is the first attempt to explore the evidence of rising damp and salt attacks of the houses around the Banda Aceh city. The heights of rising damp and salt attack visually appear on the exterior walls of the sampled houses, and the distance of those houses from the shoreline are the two main factors being analyzed to capture the spatial distribution of the rising damp and salt attack phenomena affecting the houses at the coastal area. In addition, the correlation with the level of groundwater salinity was also examined. The results of the study ultimately contribute to a better understanding of the potential long-term and persistent problems for housing construction around the tsunami-affected coastal areas.

*Corresponding Author: ella_meilianda@unsyiah.ac.id

1.1 Salt attack and rising damp

Salt attack and rising damp are two separate but interrelated processes [3],[5]. Rising damp is a phenomenon of groundwater pressure vertically transferred to building walls through capillary pressure. The capillaries draw water from the soils beneath a building against the force of gravity, leading to damp zones at the base of walls. Factors influencing the height of capillary impact, or rising damp, on wall is depending on material properties, i.e., wall thickness and porous material composition [6], [7].

When the groundwater or soil underlying a building foundation contains little or no salt concentration, the physical property of the water may induce damp walls in houses affected by rising damp. However, in most cases, the dampness will have salt associated with it, resulting in salt attack. The salt attack is the decay of masonry materials, such as stone, brick and mortar, by soluble salts forming crystals within the masonry pores [5]. The salt commonly comes from the soils beneath and is carried up into walls by rising damp. Evaporation is also an essential factor in rising damp which includes the factors that may control it, i.e., temperature, humidity, air movement and surface condition of the wall [7]. When the dampness evaporates from the walls, the salts are left behind and slowly accumulate to the point where they can cause damage. Repeated wetting and drying with seasonal changes lead to the cyclic precipitation of salts and the progressive decay of the masonry. If left untreated, the salt attack can cause extreme damage to the structure of properties.

1.2 Water salinity

A coastal aquifer is susceptible to anthropogenic activities, such as industry, household, and agriculture activities. On the other hand, the groundwater trapped within the aquifer layer is vulnerable to reduced quality and quantity by saline water intrusion, land subsidence, and water pollution [8], [9]. In general, topography, hydrostratigraphy, and the associated groundwater flow system determine the spatial distribution of saline land and water. Moreover, the redistribution of soluble salts accumulated in a catchment is evident mainly in topographically lower areas by terminal salt lakes, dry salinas, and regions of saline seeps and scalds [10].

In an extreme case, when a tsunami wave hits a coastal area, the waves inundated the low-lying coastal region. They deposited sediments and debris containing water-soluble salts in the soil, and there is no difference in chemical properties between tsunami deposits and the tsunami-inundated ground [11], [12]. Standing water, the thickness of tsunami sediment deposition and debris, and poor irrigation resulting salt traps in some locations. Moreover, the impact of tsunami inundation on groundwater salinity in the unconfined coastal aquifer not only increased acutely right after the period of inundation but is likely to persist over the years [8]. Such a condition causes a significant impact on the agricultural industry and requires special soil remediation treatment [13], [14]. In coastal aquifers, salinity could cause premature concrete deterioration due to seawater intrusion and sulfate concentration [15].

2 Methods

Walkthrough fieldwork to identify houses affected by rising damp and salt attack was conducted in March 2021 over several villages in the region of Meuraxa and Kuta Raja subdistricts to identify and observe the visual appearance of houses infested with rising damp and salt attack over the coastal area of Banda Aceh.

Forty-five houses have been purposively sampled based on their visual appearances being affected by the rising damp and salt attack. Figure 1 reveals the typical wall conditions infested with rising damp and salt attack observed around the coastal area.

Fig. 1. Typical rising damp and salt attack appearances on houses' walls at Banda Aceh coastal region.

Apart from measuring the rising damp heights on the exterior walls of those houses, we conducted a water salinity survey from the dug-wells at each house or those found in the neighboring area. The salinity level of each sample was obtained from a simple refractometer that reads the salinity concentration of the sampled water. For the data analysis, we compared the results from the salinity concentration reading with the water salinity measurement obtained in the same coastal region in the early post-tsunami in 2005, performed by the Tsunarisque team of Laboratoire de Géographie Physique, Centre National de la Recherche Scientifique (LGP-UMR 8591 CNRS) in Meudon, France in collaboration with Gadjah Mada University. The purpose is to know the state of salinity level of the groundwater as one of the leading indicators of potential salt concentration contributes to the underlying soil of the sampled house buildings.

Another parameter being considered as necessary is the distance between the houses location and the shoreline. Here, the shoreline is defined as the landward-most relatively permanent water bodies at the seashore at present (post-tsunami), as a result of severe coastal erosion by the 2004's tsunami.

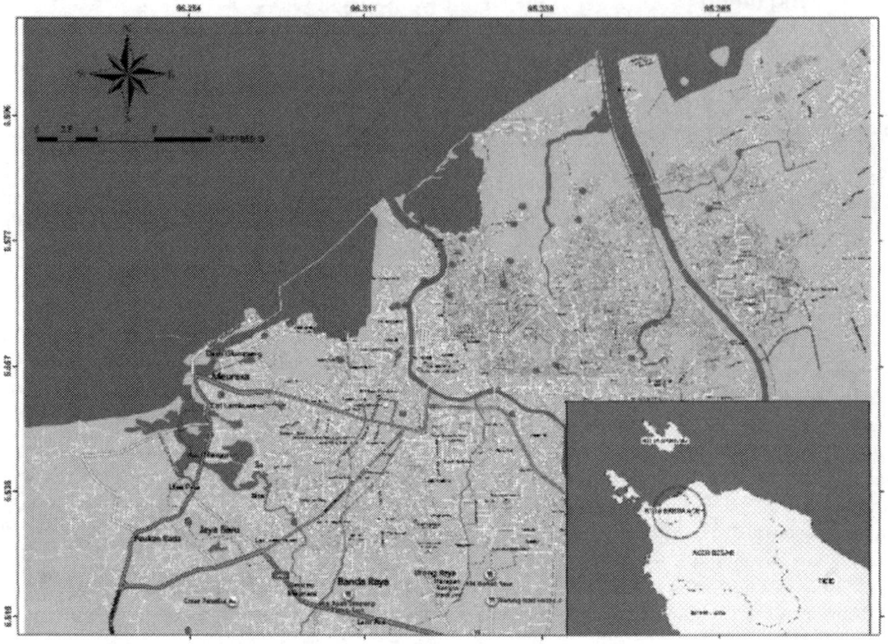

Fig. 2. Spatial distribution of houses sampling over Meuraxa and Kuta Raja Subdistrict of Banda Aceh city.

All the attribute data associated with the sampled houses are stored in a geodatabase in the GIS. We examined the spatial distribution of the house locations, the salinity concentration of the groundwater, the variability of rising damp heights and measured the distance of the house location from the post-tsunami shoreline. We subsequently examined the correlation between the height of the rising damp and salt attack of each house and their distances to the shoreline which are described in the Pearson's correlations.

3 Results

Salt attack and rising damp are problems related to both brick and mortar, particularly in the saline-prone locations such as coastal areas. All masonry materials are porous, including bricks. Bricks consist of voids or pores due to the nature of brick-making techniques. Bricks are permeable; water can pass through the pores via capillary action or a wicking effect. Therefore, salt attack and rising damp need to be tackled effectively and resolved by experts.

Most buildings in Indonesia are constructed by using reinforced concrete material and masonry structures. Bricks and mortar are the main materials for walls. This type of material is porous and it may increase the rate of water capillary and raising the moisture content of the wall. The durability of the building will be affected; thus, the building lifetime will be decreased [16]. It is therefore important to know how to control the rising damp, particularly the one instigating salt attack, to improve the sustainable building development and to increase the resistant capacity of building at the coastal areas prone to earthquake and tsunami.

The sampled houses of the present study are well-distributed within the tsunami-inundated zone over up to 5 km from the coastline of Banda Aceh (Fig. 2). The results of Pearson correlation show that there are strong positive correlations between the rising damp infested houses locations and their distance from the shoreline (0.60), and moderate correlation between the early post-tsunami water salinity level and the height of rising damp (0.49). On the other hand, weak positive correlation found between the rising damp infested houses with the 2021's water salinity (0.19).

We examined the distribution and variability of rising damp heights of the sampled houses relative to their distances from the shoreline, which depicted in Fig. 3. It shows high variability of rising damp heights on houses located at the nearest and up to 500 m from the shoreline (Fig. 3a and 3b). All the appearance of rising damp was associated with visible salt attack on the mortar; exposing bricks or corroded steel as a result (Fig. 1).

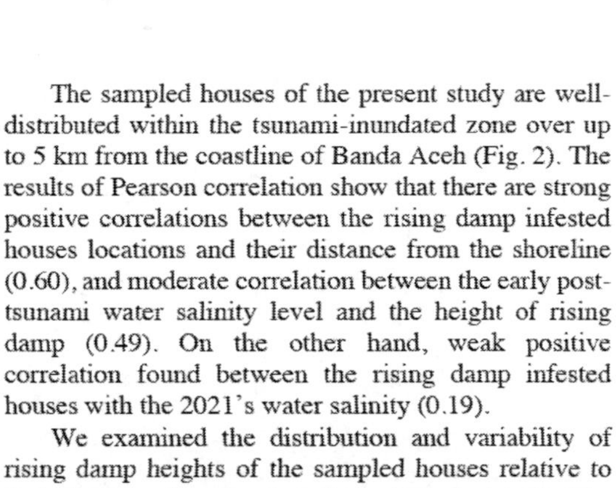

(a)

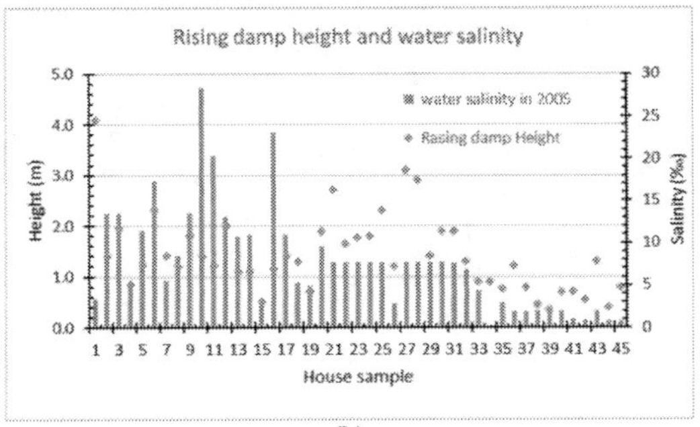

(b)

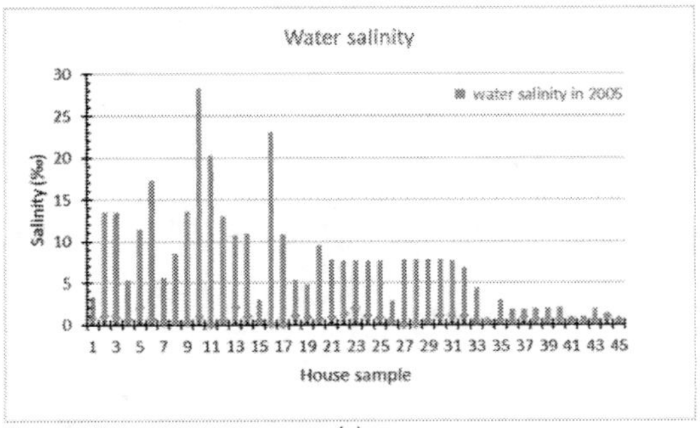

(c)

Fig. 3. Results of data analysis; (a) Distribution of rising damp heights at sampled houses relative to the distance of houses from the shoreline; (b) Distribution of rising damp height and water salinity early after the 2004 tsunami.; (c) Comparison between the early post-tsunami and the most recent water salinity condition.

It is noteworthy that out of 33 houses sampled within the area of less than 500 m from the shoreline, only three of those houses were built before the 2004 tsunami. On the other hand, out of 11 houses sampled on the area beyond 500 m from the shoreline, there were only 3 houses were built before the tsunami. This suggests that the salt attack induced by the rising damp has been persistent over the last 17 years since the tsunami event, regardless whether the building was constructed before or after the tsunami event.

Figure 3b reveals the distribution of the rising damp heights and the associated initial water salinity after the tsunami event. The rising damp heights averagely above 2 meters occurred at around the middle of the inundation area, around 400 m from the shoreline, while lower heights, but mostly well above 1 meter, were observed at those houses closer to the shoreline. This suggests that the rising damp heights on the walls are relatively high in any location within the tsunami inundation boundary within 500 m from the shoreline. However, a clear turning point where rising damp height reduced remarkably to less than 1 meter was identified as the distance of houses increased beyond 500 m from the shoreline (Fig.3a). Despite this, the effect of rising damp remains persistent until the present.

On the comparison between the water salinity levels of the early post-tsunami in 2005 and that of 2021 (Fig. 3c), it reveals that the water in 2021 has a significantly lower level of salinity, with an average of merely 0.69 ‰, compared to that of staggeringly high salinity level in 2005. This suggests that the variability of rising damp on houses built post-tsunami remains persistent to affect the building quality, while the quality of groundwater has been recovered over the years.

Overall, the high-water salinity brought by tsunami inundation during the early post-tsunami has been an important controlling factor contributing to the salt attack and rising damp, regardless of the houses' distance to the shoreline, but may have been indirectly influencing the persistence of salt attack and rising damp in the long run. The high variability but persistent appearance of the rising damp on those houses closest to the shoreline may have been contributed by several other factors. The tsunami inundation and inland deposition are inherently brought about a high concentration of salt to the soil [11,16], in particular at the area where the tsunami deposit may have been trapped, such as on a depression, aquaculture ponds, or paddy fields with poor irrigation to release the salt by rainfall [13,14]. The enhanced capillary pressure of the salinated soil from the tsunami inundation deposition or the residual salinity contained in the entrapment of tsunami deposits underlying the buildings may have a direct influence on the persistence of salt attack and rising damp, which deserves an outlook for a further study. Some other factors that needed to be considered include the houses' age, the soil condition underlying the houses' foundation, and the quality of the construction material.

4 Conclusions

Rising damp and salt attack phenomena are typical problems to the coastal properties and have been studied in many coastal settings worldwide. However, such phenomena have never been thoroughly investigated for the case of tsunami-affected coastal areas. This study investigates the controlling factors causing the persistent and escalating rising damp and salt attack to the houses exist in the tsunami-inundated coastal regions. The tsunami inundation areal coverage and salt-substance inundation materials may have been the two main controlling factors causing the rising damp and salt attack phenomena in the tsunami-inundated coastal areas. Soil salinity distribution over both preserved tsunami deposits and non-tsunami deposits should be further investigated to support these preliminary findings of the rising damp and salt attack phenomena on house buildings in the post-tsunami coastal settings.

References

1. M. Karoglou, A. Bakolas, A. Moropoulou, and A. Papapostolou, Construction and Building Materials **48**, (2013)
2. A. G. Ahmad and F. A. Rahman, Journal of Construction in Developing Countries **15**, 93 (2010)
3. A. Al-Adili and S. M. Sabih, in *International Journal of Civil Engineering and Technology (IJCIET)* (2019), pp. 2556–2565
4. E. Aghimien, C. Aigbavboa, and D. Aghimien, in *Proceedings of the International Conference*

on Industrial Engineering and Operations Management (Pretoria / Johannesburg, South Africa, 2018), pp. 1238–1246
5. D. Young, *Salt Attack and Rising Damp: A Guide to Salt Damp in Historic and Older Buildings* (Red Rover, Adelaide, 2008)
6. E. Rirsch and Z. Zhang, Construction and Building Materials **24**, (2010)
7. C. Hall and W. D. Hoff, Proceedings of the Royal Society A: Mathematical, Physical and Engineering Sciences **463**, (2007)
8. S. Violette, G. Boulicot, and S. M. Gorelick, Comptes Rendus Geoscience **341**, (2009)
9. C. A. J. Appelo and D. Postma, *Geochemistry, Groundwater and Pollution* (CRC Press, 2004)
10. R. B. Salama, C. J. Otto, and R. W. Fitzpatrick, Hydrogeology Journal **7**, (1999)
11. T. Yoshii, M. Imamura, M. Matsuyama, S. Koshimura, M. Matsuoka, E. Mas, and C. Jimenez, Pure and Applied Geophysics **170**, (2013)
12. K. J. Lee, S. Yoo, S. W. Park, S. J. Lee, V. Suriyapananont, and L. O. Namuco, Forest Science and Technology **4**, (2008)
13. H. Basri, S. Syakur, and A. Rusdi, Jurnal Floratek **6**, 144 (2011)
14. S. Syakur, H. Basri, S. Sufardi, and M. Hatta, Jurnal Floratek **7**, 1 (2012)
15. M. Mancuso, M. Deluchi, E. Kruse, C. Jorge, and A. Goncalves, in *AFTES - Congress International - Espaces Souterrainse de Demain* (Lyon, France, 2011), pp. 1–7
16. E. Franzoni, Construction and Building Materials **54**, (2014)

Coastal management status of Banda Aceh city and Aceh Besar and the future trends

Muhammad Irham[1,2,3,*], *Muhammad* Authar[4], *Edy* Miswar[3,5], *Anwar* Deli[3,6]

[1]Geographical Information Systems Laboratory, Faculty of Marine and Fisheries, Universitas Syiah Kuala
[2]Marine and Fisheries Science Study Program, Universitas Syiah Kuala
[3]Center for Environmental and Natural Resources Research (PPLH-SDA), Universitas Syiah Kuala
[4]Program Studi Agribisnis, Fakultas Pertanian, Universitas Malikussaleh, Lhokseumawe, Indonesia
[5]Center for Marine and Fisheries Research (PRKP), Universitas Syiah Kuala
[6]Department of Agribusiness, Faculty of Agriculture, Universitas Syiah Kuala

Abstract. The study aims to identify how to manage coastal zones, the scope, rationality of coastal development, challenges, frameworks, environmental impacts, and future trends in Banda Aceh City and Aceh Besar. This research is based on information from the scientific literature and secondary data. The results of the study inform that currently integrated coastal management in Banda Aceh City and Aceh Besar is not mainly used for coastal development strategies. Therefore, this coastal area requires integrated management to build reciprocal interactions between coastal communities, policies, the environment to achieve sustainable development. This coastal zone has a diversity of natural resources, minerals and tourism potential and much more to explore, therefore Banda Aceh City and Aceh Besar need to take roles and initiatives in marine governance. Judging from the results of the trend that the status of coastal management in Banda Aceh City and Aceh Besar has development opportunities that can reduce coastal poverty and contribute to local economic development.

1 Introduction

The coastal zones of Banda Aceh City and Aceh Besar are geomorphologically and hydrologically dominated by the Krueng Aceh river system and the Malacca Strait in the north and the Indian Ocean in the west. Banda Ache City and Aceh Besar coastal zone (Figure 1) covers an area of 3,030 km² or about 5% of the total area of Aceh Province. About 70% of the population lives in the coastal zone out of 695,536 residents [1]. The coastal zones of Banda Aceh City and Aceh Besar consist of 11 sub-districts that are directly adjacent to the sea and only 21 sub-districts that are not directly connected to the sea. Geographical features, the coastal zones of these two areas consist of two parts, namely the western zone and the northern zone [2]. The western region is known as the hilly coast consisting of rocky areas and sandy beaches. The northern region is a swamp and mangrove area, the most active accretion and erosion processes occurring in this area [3, 4]. The estuary of Krueng Aceh river is located in this zone. The region is also covered by stable hilly areas and sandy beaches [5]. Many coastal residents are still poor, and vulnerable to natural disasters and anthropogenic hazards [6]. Climate change-driven events such as sea level rise, storm surges, coastal inundation, salinity intrusion and soil erosion are major natural disasters [7, 8, 9].

The coast of Banda Aceh City and Aceh Besar are very dynamic and naturally always changing due to the interaction between land, sea, run off, and the atmosphere [10]. This interaction will be increasingly complex affecting the coastal environment as a result of weather changes such as storms and waves which are further exacerbated by sea level rise [11]. This vast coastal area requires holistic management and framework for sustainable development, conservation of coastal ecosystems and safety of coastal communities [12]. The coastal management and development program in this area has been going on for decades through a number of initiatives such as coastal embankment construction, coastal green belt development, and land reclamation, community-based management [13]. Therefore, this coastal area requires integrated holistic management.

Given the importance of the coastal conditions of Banda Aceh City and Aceh Besar holistically in order coastal planning management is more sustainable, it is necessary to look at the status of coastal management that has been carried out by the current government. Furthermore, in detail, it can be seen how the future trend of the management status in question is. Therefore, the purpose of this study is to identify how the management of coastal areas, scope, rationality of coastal development, challenges, frameworks, environmental impacts, and future trends in Aceh Besar and Banda Aceh City are identified.

* Corresponding author: irham@unsyiah.ac.id

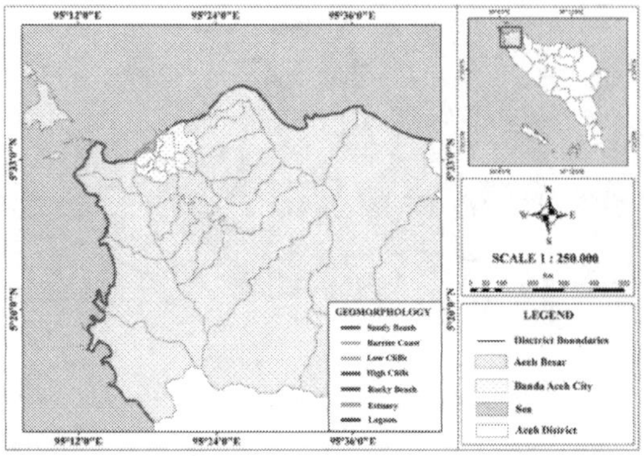

Fig. 1. Map of study area.

2 Method

This research is based on information from the scientific literature and secondary data. To collect secondary data, intensive literature review related to coastal demographic statistics, coastal development status, coastal management approaches, coastal resources, anthropogenic impacts related to coastal zones, coastal government surveys. In addition, several relevant government policy documents and reports were also collected from government agencies both online and offline.

Especially for management status, the method used is the Integrated Coastal Zone Management (ICZM) approach which is widely accepted as a means to achieve sustainable development in coastal areas. ICZM refers to a continuous and iterative dynamic process that includes a full cycle of information gathering, planning, decision making, management and implementation monitoring.

3 Result and discussion

3.1 Coastal demographic

The results of the study on the demographics of the Banda Aceh City and Aceh Besar residents can be seen in Table 1. The results show that the population of Banda Aceh City is growing at around 1.9% per year while for the Aceh Besar area, population growth is around 1.02% per year. Population growth in these two areas is almost the same but the number of immigrants continues to increase with an increasing trend of around 2-5% per year.

Table 1. The statistic of population and economic growth of Banca Aceh City and Aceh Besar.

Social		Banda Aceh City			Aceh Besar		
		2017	2018	2019	2017	2018	2019
1	Population	2,581,000	265,110	270,320	409,109	417,302	425,216
2	Population growth	1.96	2.00	1.97	1.02	1.02	1.02
3	Poor people	19,230	19,130	19,420	62,710	60,080	58,900
4	% of poor people	7.44	7.25	7.22	15.41	14.47	13.92
5	Human development index	83.95	84.37	85.07	72.00	72.73	73.55
Economic							
1	Gross Regional Domestic Bru	16,685.21	17,570.75	18,617.56	9,213.40	9561.5	9977.7
2	Economic growth	3.39	4.40	4.18	3.19	3.6	3.27
3	Per capita of GRDP (mil. Rp.	62.94	65.52	67.38	28.23	29.52	30.12

Source: BPS 2021 [1]

If we look at the economic growth in this area, it can be seen that economic growth is linear with population growth (Table 1). This indicates that the increase in the economy has triggered an increase in the urbanization population due to population mobility [14]. Migration is a complicated problem for every development of a coastal area because it is one of the factors in accelerating the rate of population growth in addition to the natural population growth factor.

Since the early 1980s, these two districts have experienced a demographic transition marked by a decrease in the death rate and birth rate as a consequence of increased welfare from economic development [15]. Apart from that, population growth also occurs due to urbanization and population mobility as a consequence of economic growth. Infrastructure development and implementation of programs that support coastal development are considered to have contributed significantly to the population growth of the study area. As a result of this population growth, the need for land for settlements, industry, infrastructure and services has also increased [16]. The use of limited resources by many people can result in exceeding the carrying capacity of the area. Therefore, the utilization of swamp, mangrove and coastal areas will have an impact on changing the landscape and trigger negative environmental impacts [17, 18]. This impact will be broad and long term.

Coastal population growth, unfortunately, is not matched by good coastal zone management policies in the study area. This is indicated by the lack of policies to protect vulnerable and sensitive areas, the existence of several damaged ecosystem areas, the disappearance of some mangrove land cover and others [19]. Pollution handling policies, for example, which so far, some people still think that coastal areas are places for waste disposal, must be eliminated immediately [20, 21, 22]. The technology used in the management of coastal areas has not supported conducive conditions for coastal areas, both environmental management and handling of existing ecosystems and plants, especially mangroves.

3.2 Coastal resources

The results of the survey conducted there are four major groups of potential coastal resources in the study area, namely: (1) renewable resources, (2) non-renewable resources, (3) marine energy and (4) environmental services. The four groups are in the area of the most valuable resources both in terms of socio-economic importance and natural-cultural heritage. These resource areas include coastal transition areas, lagoons, embankments, salt marshes, mangrove forests, sea grass, coral reef, fisheries and marine offshore [23, 24].

The renewable resources consist of various types of fish, shrimp, seaweed, seagrass beds, mangroves, coral reefs including coastal aquaculture activities and marine culture. Non-recoverable resources include minerals, mining/excavation materials and hydrocarbon. Energy resources consist of OTEC (Ocean Thermal Energy Conservation), tides, waves and so on. Meanwhile,

https://doi.org/10.1051/e3sconf/202234004005

marine environmental services include tourism and sea transportation.

Table 2. Coastal community behavior in managing coastal resources.

No	Behavioral Elements	Coastal community behavior	Expected behavior
1.	Knowledge	• Lack of understanding of the potential of Coastal Resources that can be utilized • The principle that Coastal Resources can be exploited continuously without being accompanied by recovery • There is an understanding that activities on land have no effect on Coastal Resources	• Understanding the potential of natural resources and access to their optimal use • The principle that the preservation of Coastal Resources needs to be maintained • Knowing the relationship between activities on land and at sea
2.	Response to conservation	• A priori to coastal and marine damage • Oriented to the past and difficult to accept change • Don't take risks	• Actively seeking breakthroughs in environmentally friendly utilization technologies • Future oriented and open to change • Perform calculations against risk and uncertainty
3.	Skills & Ability to utilize Resources	• Too dependent on one type of business due to limited skills • Using fishing gear without caring about the impact • environment Not able to maintain the condition of Coastal Resources • There are limitations in processing and marketing the results	• Implementing business diversification • Using environmentally friendly response tools (supporting environmental preservation) • Carry out conservation efforts in the coastal and marine environment • Able to process and market the results or market availability for product distribution

The utilization of resources that are not optimal until now, among others, because there are still obstacles, namely the average business scale is small below the economic scale, the quality of human resources is still low, the lack of mastery of technology and the level of utilization of resources is uneven and does not pay attention to the carrying capacity of the environment.

The results of the survey on community knowledge on coastal and marine resource management can be seen in Table 2. Table 2 shows that the behavior of coastal communities in understanding and managing marine and coastal resources is minimal and does not seem to understand the importance of marine and coastal resources for the future. This is marked by the non-optimal use of the existing resources and the absence of the principle of sustainable use of existing resources.

3.3 Anthropogenic impact related to coastal zones

The massive impact of human intervention on the coast has resulted in water pollution. The results of the survey and analysis in the study area show that more than 80% of coastal pollution comes from land due to urban waste, the rest comes from marine transportation and marine discharge activities. As a result of this pollution, many marine habitats are polluted and die.

Another pressure experienced by coastal areas is land use (Table 3) and human populations (Table 1). The population growth of Banda Aceh and Aceh Besar, which averages 2.3% per year, has caused dependence on coastal land to increase [1]. Furthermore, human populations will indirectly depend on coastal and marine ecosystems, habitats and resources for food, building materials, building sites, agricultural and recreational areas. This limited capacity is also used as a waste disposal area, garbage and toxic materials. This pressure will continue to increase along with the increase in urbanization, transportation and industrialization due to increasing human demography and increasing use of coastal areas [25].

Table 3. Land land cover of Banda Aceh City and Aceh Besar District.

Land Use/ Land Cover	Banda Aceh			Aceh Besar		
	2009	2014	2021	2009	2014	2021
Bare Land	16.4	15.3	14.2	1.8	1.6	1.4
Mangrove	7.1	6.8	5.6	12.9	11.8	11.4
Settlement	21.4	23.7	27.3	16.1	18.7	22.1
Agriculture	22.3	21.1	19.5	56.3	54.3	50.8
Pond	15.8	16.4	17.2	11.1	11.7	12.2
Water	17	16.7	16.2	1.8	1.9	2.1

Economic improvement in the study area also encourages industrial growth that will alter, disrupt and destroy coastal ecosystems, including sensitive habitats. The main industrial activities that affect coastal areas include port activities, fishery processing, ship building and other associated industries related to coastal and marine activities. Construction engineering activities often cause permanent habitat destruction or habitat degradation and fragmentation, due to land claims, coastal protection, extraction of basic materials, dumping and disposal [1].

3.4 Coastal management approach

The coastal area development policies that have been implemented so far are often partial and have a "top-down" pattern, so that they often do not reflect the interests and needs of the coastal area. Local communities, not having a "bottom-up" pattern, while in its implementation it does not utilize the existing potential optimally, including the values of local wisdom [24].

Community participation from the beginning in determining the problem will further ensure that the program to be implemented truly meets the needs of the local community. Such an approach also makes the community feel ownership and responsibility for the program so that it is easier to invite the community to participate in managing the resources they have. In turn, this approach method will increase public confidence and reduce community dependence on the government.

3.5 Strategic issues

As regions that share special autonomy, the cities of Banda Aceh and Aceh Besar have implemented a new paradigm of sustainable marine development [25, 26]. The existence of a new paradigm shows that several serious problems have become strategic issues in the management of coastal and marine resources in the cities of Banda Aceh and Aceh Besar. The results of the analysis show that there are five issues that need to be resolved immediately. The following is a description of the five issues based on the data [1].

The first issue is the condition of coastal and marine resources that are common property with quasi-open access. The existence of resources that are quasi-open access, then the actions of one party that harm the other party cannot be corrected by the market (market failure). This creates economic inefficiency because all parties will try to exploit resources as much as possible. This is the situation that is currently happening. Supported by financial resources and technology, parties who are stronger and able to exploit resources excessively will cause natural production resources to be disrupted. The degradation of the coastal and marine environment will be massive if this is not immediately anticipated. In the early 80's, many were surprised to see that development policies that only pursued economic growth and productivity had in fact caused serious damage to the environment. For example, the fisheries modernization program, which aims to increase the production of fishermen's catches using increasingly modern fishing technology, is not accompanied by socialization of a good understanding of the marine environment. This has a fatal impact on environmental sustainability because there is maximum exploitation of resources without regard to the existing sustainable potential.

The second issue is the degradation of the coastal and marine environment which poses a threat to the survival of coastal communities and fishermen such as pollution of the aquatic environment due to industrial and household waste. In addition to destroying the potential of aquatic resources, environmental degradation is also bad for the health and survival of fishermen, especially coastal communities. The degradation of the coastal and marine environment occurs due to several interrelated and very complex factors. Some experts state that this factor is more due to the level of education or human resources, poverty, facilities and infrastructure and government policies.

The third issue is poverty and social inequality. Poverty, social inequality, and the pressures of life do not allow family members to be actively involved in social responsibilities outside of the problems of life that are substantial for them. Such factors are often the reason for other parties to negatively assess the social behavior of coastal communities, especially fishing communities. Socio-economic changes in coastal villages or fishing villages have clarified the line of social stratification of the community. This perception only preserves the gap in social relations in political relations between the government and society.

The fourth issue is access to the use of limited technology. The increasing competition in the use of marine and coastal resources requires the community to maximize their production. One way that is used is the use of technology. The limited knowledge and ability in using this technology is one of the obstacles and triggers for the exploitation of resources that damage the sustainable potential and have a negative impact on the environment. One of them is the gap in the use of technology between large and traditional fishermen which results in the increasing urgency of traditional fishermen in competing for the use of marine resources.

The last issue is regulations and policies that are not conducive. Local governments can manage coastal and marine resources more rationally considering the availability of resources will determine the level of community prosperity. Marine and fisheries development policies that are implemented should not only pursue economic interests, but also be balanced proportionally with the commitment to preserve the existing marine fisheries resources. In addition, there must also be a high and consistent commitment in enforcing applicable legal regulations in order to avoid social and economic conflicts. Local wisdom must be accommodated as one of the legal institutions that can minimize conflicts between fishermen.

3.6 Coastal management status

The results of the study inform that currently integrated coastal management in Aceh Besar and Banda Aceh is mainly used for coastal development strategies. Therefore, this coastal area requires integrated management to build reciprocal interactions between coastal communities, policies, the environment to achieve sustainable development. Judging from the results of the trend that the status of coastal management in Aceh Besar and Banda Aceh has development opportunities that can reduce coastal poverty and contribute to local economic development. This coastal zone has a diversity of natural resources, minerals and tourism potential and much more to explore, therefore Aceh Besar and Banda Aceh need to take roles and initiatives in marine governance.

The coastal management approach relies on available information on various aspects of coastal ecosystems, processes, resources, natural hazards and their impacts, effective response by governments. Coastal zone management is very important for implementing sustainable development strategies in Aceh Besar and Banda Aceh City. The existence of linkages between the national level, sub-national level and local communities in each approach taken must be able to ensure sustainability both socio-economically and naturally.

3.7 Future trends in coastal zone management

Recently the management, development and planning of coastal areas in Banda Aceh and Aceh City must receive attention by the government. The ICZM approach should be seriously implemented which not only aims to help reduce the impact of disasters, but also provides opportunities for sustainable use of resources. ICZM will provide ecological environmental benefits which in turn will also provide survival strategies for millions of people who live at the mercy of nature. The Coastal Zone policy and strategy is an important step towards implementing ICZM. The results of the study conclude that there are several scopes and issues that need to be anticipated by policy holders as listed in **Table 4**.

Table 4. Scope, work, challenges, opportunities and threats of ICZM in Banda Aceh City and Aceh Besar.

Scopes of ICZM	Issue of ICZM
1. Coastal community management 2. Coastal resources management 3. Environmental management 4. Coastal economic management	1. Population growth 2. Infrastructure 3. Supply and demand analysis 4. Opportunity analysis 5. Coastal risk and challenge assessment
Challenges	**Opportunity**
1. Weakness in policy 2. Lack of coordination 3. Human resources 4. Inadequate public participation 5. Inadequate stake holder inclusion	1. Long, large and beautiful coastal area 2. Enriched with resources coastal zone 3. Biodiversity of coastal ecosystem 4. Mangrove area 5. Coral reefs
Major treat	
1. Population 2. Over exploitation of coastal resources 3. Coastal pollution 4. Rapid environment changes 5. Constrain of law and policy	

The coastal zone management policy (CZMP) in the study area covers a wide range of issues from economic development to ecosystem protection and conservation. Unfortunately, there is no specific plan for dealing with natural disasters or reducing vulnerability. There are no clear guidelines on how local development agencies should work with government agencies. There are several reasons why this is happening due to: (1) non-inclusion of professionals in Coastal Zone Management, (ii) unreasonable views of coastal zones, (iii) failure to create policies that are beneficial to coastal and marine areas.

4 Conclusion

Coastal zone management is very complex and multidimensional because it must address all socio-economic and natural related problems. However, coastal zone management is very important on the coast of Aceh Besar and Banda Aceh City because almost 70% of the population lives in these areas. The analysis results of the status of coastal management are still lacking so that the sustainable development strategy in this area has not been able to guarantee the safety, security and sustainability of coastal communities.

Acknowledgment. The authors gratefully acknowledge Pusat Penelitian dan Pengabdian Kepada Masyarakat (LPPM) Universitas Syiah Kuala for their financial support.

References

1. BPS Provinsi Aceh. *Provinsi Aceh dalam Angka (Aceh Province in Figures)*, Aceh (2021).
2. M. Irham, I. Rusydi, H. A. Haridhi, I. Setiawan, Y. Ilhamsyah, A. Deli, M. Rusdi, A. M. Siregar. J. Mar. Sci. Eng. **9**, 8 (2021).
3. I. Adela, S. S. Nugraha, M. Irham, S. Syahreza. IOP Conf. Series: Earth and Environmental Science **273** (2019).
4. Rais, M. Irham, F. Adrian, A. Deli. IOP Conference Series: Materials Science and Engineering **1087** (2021).
5. E. Meilianda, C. M. Dohmen-Janssen, B. H. P. Maathuis, S. J. M. H. Hulscher, J. P. M. Mulderb. Marine Geology, **275**, 1-4 (2008).
6. S. El-Masri, G. Tipple. International Planning Studies **7**, 2 (2002).
7. M. R. Islam, N. A. Khan. Journal of the Asia Pacific Economy **23**, 2 (2018).
8. M. Mousavi, *Relationship Between Climate Change, Natural Disaster, and Resilience in Rural and Urban Societies*, Handbook of Climate Change Resilience, **1** (2019)
9. W. J. W. Botzen, J. C. J. M. van Den Bergh. Environmental Hazard **8**, 3 (2011).
10. M. Irham, R. Suri, I. Setiawan, A. Fuadi. IOP Conference Series: Earth and Environmental Science **674** (2021).
11. H. Ahmad. J. Coast. Zone Manag, **22** (2019).
12. Syamsidik, Tursina, A. Suppasri, M. Al'Ala, M. Luthfi. Nat. Hazards Earth Syst. Sci. **9**, (2019).
13. J. Schmiedel, *Environmental values and threats to coastal areas and the Baltic Green Belt*. Coastline Reports. (2012).

14. A. Marques, I. S. Martins, T. Kastner, C. Plutzar, M. C. Theurl, N. Eisenmenger, M. A. Huijbregts, R. Wood, K. Stadler, M. Bruckner, J. Canelas. Nature ecology & evolution **3**, 4 (2019).
15. Oxfam. *Green Coast for Nature and People after the Tsunami*, Dokumen anlisis kebijakan. (2007).
16. M. Rusdi, M. Irham, S. Sugianto, R. Roosli, MSS Ahamad, Y. Haditiar. Depik. **48**, 1 (2021.
17. I. P. Sari, M. A. Sarong, M. Rusdi, A. Deli, M. Irham. IOP Conference Series: Earth and Environmental Science **674** (2021)
18. R. Azhar, M. Rusdi, M. Irham, A. Fuad. IOP Conference Series: Earth and Environmental Science 674 (2021).
19. Z. Nasution, M. T. D. Sunarno. J. Kebijakan Perikanan. Ind. **1**, 1 (2009.
20. B. Neumann, A. T. Vafeidis, J. Zimmermann, R. J. Nicholls. PloS one **10**, 3 (2015).
21. J. L. Merkens, D. Lincke, J. Hinkel, S. Brown, A. T. Vafeidis. Climatic Change **151**, 3 (2018).
22. T. T. Jiang, J. F. Pan, X. M. Pu, B. Wang, J. J. Pan. Estuarine, Coastal and Shelf Science **164** (2015).
23. D. Dirhamsyah. Ocean & coastal management. **49**, 1-2 (2006).
24. E. M. Ferrer, C. M. Nozawa. *Community-based coastal resources management in the Philippines: Key Concepts, Methods and Lessons Learned*. (1997).
25. PEMDA Kota Banda Aceh. Rencana Pembangunan Jangka Menengah Daerah (RPJMD) Kota Banda Aceh, (2017).
26. PEMDA Aceh Besar. Rencana Pembangunan Jangka Menengah Daerah (RPJMD) Kabupaten Aceh Besar, (2017).

Modeling search and rescue, medical disaster team response and transportation of patients in Ishinomaki city after tsunami disaster

Erick Mas[1,*], *Shinichi* Egawa M.D.[1], *Hiroyuki* Sasaki M.D.[1], and *Shunichi* Koshimura[1]

[1]International Research Institute of Disaster Science, Tohoku University, Sendai, Japan

Abstract. The 2011 Great East Japan Earthquake caused massive devastation and numerous affected people. In particular, the Ishinomaki area of the Miyagi Prefecture was one of the hardest hit areas during the event. In addition, due to tsunami inundation, several medical facilities were destroyed or inactive at the time. The Ishinomaki Red Cross Hospital served as the last bastion to respond and attend injuries and illnesses in the disaster. We aim to build an agent-based model of triage and transportation of patients to the hospital. The individual and joint activities of response teams were considered to build the agent architecture and they were incorporated into the model. We model the first three days of response activities within the central Ishinomaki area and evaluated the number of patients transported to the hospital including their evaluated health status. We discuss here the case closest to the 2011 event and evaluate the strategies for search and rescue and patient transportation to speed up medical response in the aftermath of a disaster.

1 Introduction

In the aftermath of the 2011 Great East Japan Earthquake and Tsunami (GEJET), search and rescue (SAR) operations were massive and complex. SAR activities started locally from firefighters and the police. Similarly, sea search and rescue actions were leaded by the Japan Coast Guard (JCG). In special cases, such as this one, the Japan Self Defense Forces (JSDF) are also involved in the operations. The Japan's Fire and Disaster Management Agency (FDMA) has conducted reviews of rescue activities in response to the GEJET [6] where they summarized the lessons for SAR focusing on the activities at tsunami-hit areas during the "golden 72 hours". The activities conducted by first responders in the fire services include shutting down water gates, delivering tsunami warning to local residents and guiding evacuation. Unfortunately, member volunteers of fire corps died closing tsunami gates in the area [2]. On the following hours of the tsunami impact, local fire departments were aided by less affected inland departments in operations of SAR, fire suppression and injury transportation. SAR activities conducted by fire service personnel, police, JCG and JSDF saved approximately 27,000 people. Koresawa (2013)[6] details the number of rescued people by local fire departments in tsunami-affected areas of Aomori, Iwate, Miyagi, Fukushima and Ibaraki prefectures to be over 3,500 people during the first 48 hours, and later consistently around 200 people every day from the third day towards the end of the month.

The objective of this study is to develop an agent-based model of disaster response incorporating search and rescue, medical assistance evaluation through the Disaster Medical Assistance Team (DMAT) and transportation of patients within the first three days after the event. A case study on the central Ishinomaki area in Japan is simulated following the data from the 2011 Great East Japan Earthquake and Tsunami.

*Corresponding author e-mail: mas@irides.tohoku.ac.jp

2 Study Area

Ishinomaki City (Figure 1) in the Miyagi Prefecture had the largest number of death and missing bodies in the 2011 event (3,972). Also, medical centers were inundated and out of service due to the damage caused to them. Within the Ishinomaki area, one of the main hospitals out of the inundation zone was the Ishinomaki Red Cross Hospital (IRCH) which remained active. At the IRCH the total number of patients in the first month was at least five times the regular number of patients before the earthquake [1]. The IRCH coordinated all medical teams from the Japanese Red Cross Society and other agencies at evacuation centers throughout Ishinomaki city [4]. Several predisaster preparedness activities conducted by the IRCH were key to a better response during the 2011 GEJET [3] On a 2015 report by the IRCH, they reported the situation of first days of transported patients and their condition classified in minor, medium, severe and serious injuries as shown in Table 1

The target area for analysis is shown in Figure 1 within the inundation area. It is bounded by the Old-Kitakami river in the east and north near the Fire Department Headquarters, the channel connecting Old-Kitakami and Jo rivers in the west side and the inundated areas below the 2m inundation depth. We focused the study below the 2 m inundation depth since the areas over that level of inun-

 https://doi.org/10.1051/e3sconf/202234005001

Table 1: Number of patients transported to the Ishinomaki Red Cross Hospital within the first three days after the event classified by injury level. In () are the number corresponding to the target area for this study.

Day	Minor	Medium	Severe	Serious	Total
3/12	47 (24)	33 (16)	17 (8)	2 (1)	99 (49)
3/13	552 (271)	142 (70)	74 (36)	11 (5)	779 (382)
3/14	1,037 (509)	152 (74)	39 (19)	23 (11)	1,251 (613)
Total	1,636 (804)	327 (160)	130 (63)	36 (17)	2,129 (1,044)

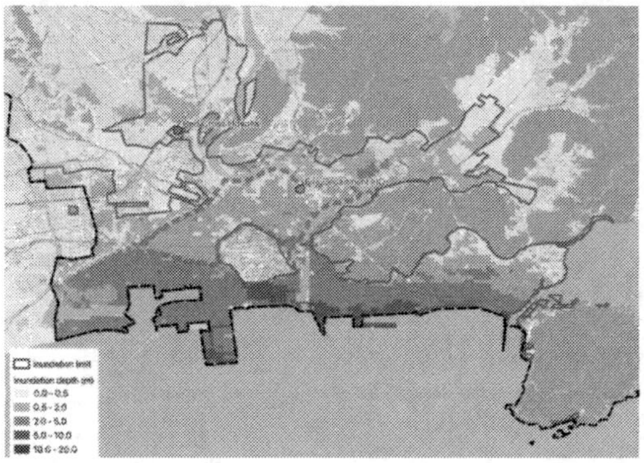

Figure 1: The area of study at Ishinomaki. The inundation limit was provided by the Geospatial Information Authority of Japan. Inundation depths were provided by the Ministry of Land, Infrastructure, Transport and Tourism of Japan. The background is an Open Street Map provided within the QGIS software. The Ishinomaki Red Cross Hospital and the Ishinomaki District Wide-Area Administrative Affairs Union Fire Department HQ are the two points shown in the map. The polygon inset at the center corresponds to the target area for analysis in this study.

dation were heavily damaged and immediate access was reported to be limited due to water presence [6].

Serikawa et al. (2020)[5] showed that the Miyagi Prefectural Police Headquarters reported 3,058 corpses within the whole Ishinomaki area (a larger area that includes Oshika Peninsula and the areas near the Kitakami river). On the other hand, the activity records from the Ishinomaki Wide-Area Administrative Office of the Fire Department reported 2,954 corpses. In addition, within the inundation area shown in Figure 1, 1,448 bodies were found. The rate of bodies within the target area versus the whole Ishinomaki medical area ($1,448/2,954 = 0.49$) is used as the rate to calculate the number of patients to be transported to the Ishinomaki Red Cross Hospital from the area in study. The values to be used for verification of our model results are shown in parenthesis in Table 1

3 Materials and Methods

We compiled several sources detailing the activities following the 2011 GEJET in the Ishinomaki area. In particular, sources related to the IRCH and the triage and transportation of patients [4] (see Table 1). In addition, for the agent architecture and interactions among them we also followed the descriptions and architecture described in [7–10].

With these, we developed an agent-based model with the aid of the AnyLogic™ software. AnyLogic is a multimethod simulation tool based on Java language. Some of the simulation paradigms possible to combine within the software are: discrete event or process-centric simulations, system dynamics and agent-based modeling. In this study we focus on the use of agent-based approach. AnyLogic 8.7.5 University edition was used to create and simulate our models.

The model contains four types of agents whose behaviors are summarized in Table 2 and described hereafter.

The first type of agent is called "patient" agent. Within this type a major attribute corresponding to the level of injury makes a distinction on the agent behavior. For instance, a "serious" injury is represented as a black agent and corresponds to a direct casualty from the event. Upon finding this type of agent, SAR team agents will not request for transportation to hospital and leave that agent. In our model assumption, police is in charge of corpses and they will be requested to take care of such agents. Since including such type of agent and activity is out of the scope of our study we did not include a police agent type and the actions taken after finding a serious injury patient is none.

Next, "severe" injury patients are identified as red colored agents in the model. These agents cannot walk but their injuries are threatening to life. Thus, immediate assistance and transportation to medical facilities is needed. When SAR teams find this type of agent, they will request to the Ishinomaki Firefighters Department (IFD) for transportation assistance. A transport team is dispatch from IFD or from nearby locations to carry the patient to the IRCH. Similarly, a "medium" injury patient, a yellow colored agent in the model, has less urgency but similar needs. Then, these type of agent is also carried to the hospital.

Then, within the "patient" agent, those with "minor" injury are represented with a green color in the model and since there is no life-threatening injury, those agents are assumed to be able to walk by themselves and find assistance elsewhere. Within the model once found by the SAR team the agent walks towards the hospital looking for assistance.

Finally, a "DMAT unit" agent is initially located at the IRCH and is in charge of a secondary triage. In this case a reassessment of the urgency of attention is conducted. Thus, a modelled set "severe" agent can be wrongfully triagged as "medium" or viceversa by SAR teams in the field, however, once at the hospital the correct triage level is revealed. Such initial triage inaccuracy can be adjusted in the model in order to reflect and evaluate the effect of low-capacity and training of in-field first responders.

A snapshot of the ABM model is shown in Fig 2. On the right, the Ishinomaki Fire Department marks the start point of SAR teams. The colored dots represent the population and their level of injury or illness based on random selection and statistical information from triage records at the IRCH. SAR activities yield to a first triage when a pa-

Table 2: Agents' characteristics and initial location for simulation

	Agent		Initial Location	Description (i.e. triage, task, etc.)
1	Patients	Serious	Random	[Black] Casualty
		Severe	Random	[Red] Cannot walk. Immediate assistance required.
		Medium	Random	[Yellow] Cannot walk. Not urgent but requires assistance.
		Minor	Random	[Green] Can walk. Not a life-threatening injury.
2	Firefighter (SAR team)		IFD	Search and Rescue of patients. Request for transportation.
3	Firefighter (Transport team)		IFD	Pick up and transport to hospital.
4	DMAT unit		IRCH	Triage at hospital. Keeps track of patients.

IFD: Ishinomaki Firefighters Department
IRCH: Ishinomaki Red Cross Hospital

tient is found and then send to the hospital to a second triage by DMAT personnel. Based on the level of injury or illness, the transportation to IRCH is prioritized.

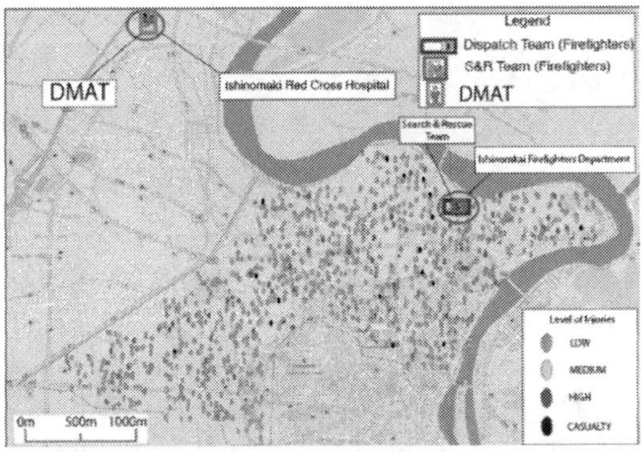

Figure 2: Screen-shot of the developed agent-based model

Regarding the rules for interaction among agents, first, patients are found by SAR teams who conduct a first triage and only severe and medium injury patients are candidates for transportation to the hospital by the Transport team. Requests are stacked and from the resource pool of available vehicles, a Transport team moves to pick up the patient and move together to the IRCH where a DMAT unit is waiting to conduct a second triage. The case of serious injured people is supposed to be taken by another agent (i.e. the police) and their dynamics is out of the scope of this research. Thus, no action is taken in the case of serious injured agents. On the other hand, minor injury patients can move and they are instructed to go to the hospital by their own means.

A particular feature to explore here is the strategy for SAR. This is, the technique employed to explore the area and find survivors. In cases of missing persons hasty searches are common, however in large disaster stricken areas grid search techniques or organized search is necessary. A particular way to organized the SAR operation is through a "roller search", which divides the affected area in subareas for search. In this study we divided 26 subareas where survivors are searched area by area. We have assumed a random time for each SAR operation before a new patient is found. The set time is between 15 to 45 min. In addition, when activities are conducted during night time we doubled the time needed for SAR operations to find the next patient. The sweeping order of subareas can vary. Here we explore 3 cases (see sec. 4).

In addition, the model considers some fix parameters related to the time spend in certain activities. For instance, triage task to one person is set to a random time between 1 or 2 minutes. Also the average time for ambulance dispatch and carry-on of patients into the ambulance has been reported to an average of 2.5 min and 8.6 min, respectively[11], here we used 3 min and 9 min.

4 Simulation cases

We have simulated three days of disaster response activities including SAR and transportation of patients to the IRCH. The reference event corresponds to the 2011 GEJET and the time of the event was 2:46 pm (JST), then, we assumed a period of one hour and minutes for headquarters to be established and then SAR and transportation activities to get started. Thus, our simulation starts from 4:00 pm until the end of a third day after the event, this is 56 hours of simulation. We tested three cases of SAR patterns to evaluate the rapidity of finding and dispatching survivors from the affected area to the IRCH.

1. Case A.- In this case the order of search starts from the areas closest to the gathering point of SAR teams (i.e. the Firefighters Department) towards the west south of the study area (Fig 3a). Here, the expected advantage would be the fastest exploration of easy-to-reach zones to then increase difficulty due to deeper water and expected larger damage. One may argue the suitability of this sweeping pattern if expectations are that a larger number of serious and severe injury patients would be found towards the coast where deeper tsunami inundation and larger damage occurred. However, on an earthquake and tsunami event, survivors due to collapsed structures in shallow inundated zones or survivors swept inland by tsunami can also be considered, and the pattern in this case would make more sense. Unfortunately very few studies have looked into the spatial distribution of survivors or patients found during post disaster response, possibly due to the difficulties posed in the midst of a disaster to collect such information.

2. Case B.- In this case SAR operations begin near to the hospital location and in shallower areas into

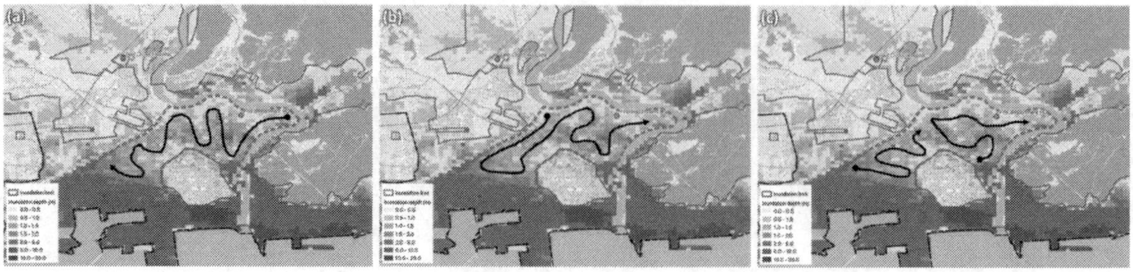

Figure 3: Three search patterns for SAR teams. (a) Case A: easy access prioritization; (b) Case B: closest transportation prioritization; (c) Case C: deeper inundation prioritization.

deeper inundation zones and zigzagging north-south towards the east (Fig 3b). Here, the main criteria for search is to speed up initial patient transport tasks by prioritizing areas near to the hospital. Similarly, this pattern aims to go as fast as possible to highly affected areas where possibly more injuries are gathered. In addition, alternating between deeper and shallow inundated areas where differences in easiness of mobility may contribute to less fatigue to SAR teams and higher efficiency in the task.

3. Case C.- In this case the SAR operation focuses on deeper inundated zones first, then the action moves inland towards shallower inundated areas (Fig 3c). This means that SAR activities will start at the farthest point from the dispatching point (i.e Firefighters Department) and the hospital where patients are transported.

5 Results and Discussion

Table 3 shows the average of patients' searched and rescued from the tsunami affected area. Slight differences among cases are shown. For instance, Case A appears to be the pattern giving better results with a higher average value. Apparently the three cases of SAR operations does not present significant differences on the timing of patients' transportation.

Table 3: Number of patients with severe or medium level injuries transported within 3 days in each case of SAR pattern.

	Case A	Case B	Case C
Average	210	206	204
Maximum	223	223	223
Minimum	191	184	184
Std. Dev.	5.19	4.38	4.94

Table 4 shows the number of patients transported each day on the three cases in study, as well as the IRCH records of the event. The model shows underestimation on the first days of activity and overestimation in the last day. Overall day by day discrepancies are balanced out in the total sum where the model shows good agreement. Some of the reasons for such differences can be explained by the limitations of the model to represent other activities that contribute to the transportation of patients. For instance, the JSDF transported patients via helicopter and several patients where transported by private vehicles without the assistance of emergency responders.

Table 4: Number of patients transported to the IRCH according to triage level and the days after the event. Three cases were compared to the IRCH record of patients and their medical level of need.

Number of patients transported to the hospital								
	Case A		Case B		Case C		IRCH records	
	Medium	Severe	Medium	Severe	Medium	Severe	Medium	Severe
Day 1	2	1	2	1	2	1	16	8
Day 2	54	21	51	19	49	20	70	36
Day 3	94	37	95	38	96	36	74	19
Total	150	59	148	58	147	57	160	63
(*) the values are the average of 40 repetitions								

In addition, outcomes of Cases A, B and C differ little due to the reduced size of the area covered in SAR and transportation. Results at larger areas might be different for the three cases since patients' location would be more scattered and distances and time for finding and move would be longer. The overall outcome for patients transported within 56 hours are similar to the IRCH records. This is expected due to the nature of the input in the model, however, reaching such ceiling within 3 days simulation verifies the model scheduling activities (agent behavior) within a reasonable marginal error.

It is possible that in the real situation, when limited resources where available at the beginning self-response and other means of transportation contributed to higher number of patients being transported compared to the model. Similarly, on the third day, when resources are plenty, the model compensates for the number of patients, while in the real event, by that time plenty of resources would have been more than enough to move the remaining survivors.

Conclusions

This study has developed an agent-based model to simulate search and rescue operations together with transportation of patients to the hospital. A portion of Ishinomaki city was taken as a case study to compared the outcomes of the model to the actual data of patients that were received at the Ishinomaki Red Cross Hospital.

We averaged the outcome values of 40 simulation runs showing agreement to the actual records. At a finer level, the model presents underestimated results during the first two days of relief activities, while overestimation is found on the third day. Limitations on the model, as well as a possible number of self-response via private means of transportation might explain such discrepancies. Further improvement in the model to incorporate other means of transportation is necessary, however data for verification might be scarce in this case.

References

[1] World Health Organization. The Great East Japan Earthquake. A story of a devastating natural disaster, a tale of human compassion. 2012

[2] Ranghieri, Federica and Ishiwatari, Mikio. Learning from megadisasters: lessons learnt from the Great East Japan earthquake and tsunami. Economics and Policy of Energy and the Environment, 3(3), 5–17, 2015.

[3] Egawa, Shinichi and Suda, Tomomi and Jones-Konneh, Tracey Elizabeth Claire and Murakami, Aya and Sasaki, Hiroyuki. Nation-Wide Implementation of Disaster Medical Coordinators in Japan. The Tohoku Journal of Experimental Medicine, 243(1), 1–9, 2017.

[4] Ishii, Tadashi. Medical response to the Great East Japan Earthquake in Ishinomaki City. Western Pacific Surveillance and Response Journal, 2(4), 2011.

[5] Serikawa, Tomoki and Seto, Shuji and Suppasri, Anawat and Imamura, Fumihiko. Spatial Distribution of Causes of Death in the 2011 Tohoku Tsunami at Ishinomaki City, Miyagi Prefecture. Journal of Disaster Research, 15(7), 943–958, 2020.

[6] Koresawa, Atsushi. Evidence-Based Analysis of Search and Rescue Operations Following the Great East Japan Earthquake. Journal of Disaster Research, 8, 746–755, 2013.

[7] Araki, Yasuhiro. A study for modeling transportation activities of casualties by Multi Agent Simulation in consideration of the Niigataken Chuetsu-oki Earthquake in 2007. Journal of Social Safety Science, 10(11), 1–9, 2008. (in Japanese).

[8] Ishinomaki District Wide Area Administrative Office Fire Department. Great East Japan Earthquake 3.11 Ishinomaki Wide Area Firefighting Activity Record. 2012. (in Japanese).

[9] Japanese Red Cross Society Planning and Public Relations Office. Great East Japan Earthquake-All records from relief activities to reconstruction assistance. 2013. (in Japanese).

[10] Ishinomaki City. Great East Japan Earthquake: History of Ishinomaki City. 2017 (in Japanese).

[11] Kataoka, Motomune and Yoshii, Toshio and Futagami, Toru and Oguchi, Takashi. Development of a model of the transportation time sent to a hospital by ambulance. Seisan Kenkyu 67(2), 137–142, 2015. (in Japanese).

https://doi.org/10.1051/e3sconf/202234005002

Do social restriction policies responding to COVID-19 pandemic associated with economic growth?

Heru Fahlevi[1*], *Teuku* Rizky Saevic[2], *Jalaluddin* Jalaluddin[1] and *Muslim* A Djalil[3]

[1]Accounting Department, Faculty of Economics and Business, Syiah Kuala University, 23111 Banda Aceh, Indonesia
[2]Economics Department, Faculty of Economics and Business, Syiah Kuala University, 23111 Banda Aceh, Indonesia
[3]Management Department, Faculty of Economics and Business, Syiah Kuala University, 23111 Banda Aceh, Indonesia

Abstract. The purpose of the study is to examine the statistical relationship between restriction and containment policies in respond to COVID-19 pandemic and the countries' economies. This study also maps countries based on their economic growth and the public mobilization restriction policies. The GDP growth was extracted from the economy prognosis from the International Monetary Fund. To determine the restriction policy level, the stringency index (SI) produced by the Oxford COVID-19 Government Response Tracker (OxCGRT) was used. The SI is between 0 – 100 in a range that is computed based on containment and closure policies, for instance, stay at home requirement and closure of public facilities, and public events. The SI used in this study is the average score of the first 3 months after the WHO declared COVID-19 as a pandemic (11 March 2020). Using the Spearman correlation test to investigate the relationship between GDP growth and restriction level in 166 countries, this study demonstrates a negative association between GDP growth and the SI. This means that countries that imposed stricter policies in the early COVID-19 pandemic tend to have higher GDP contraction. However, the correlation coefficient is relatively weak (-0.200), and thus, every government has the opportunity to reduce the unintended outcomes of restriction policies by providing economy and financial support packages for businesses, people, and other affected groups.

1 Introduction

The spread of the coronavirus disease of 2019 (COVID-19) has reached almost every country in the world. Its effect is multisectoral and affects governments, industries, and the people, particularly due to the slow down economic activity resulting [1]. The COVID-19 requires governments to take immediate actions to prevent further outbreaks for a relatively unknown period of time.

Responding to the COVID-19 pandemic and rapid increase of infected cases and death, most countries across the globe have adopted containment, social distancing, and closure policies. The policies include the closure of public facilities (school, offices, and restaurants), working from home, quarantines and even lockdown [2]. These policies often vary in terms of duration, and level of restrictions [3]. Governments tend to strengthen the policy when the case number has risen and relax the restriction when the case starts to decline. This reflects a national dilemma whether to save lives or the country's economy.

It is indeed that restriction policy may reduce COVID-19 spread (see for example [1]). However, this policy is widely questioned and even rejected by some groups of citizens [4]. Some countries, for example, China, Australia, and Vietnam imposed a national lockdown, while the other countries, for example, Indonesia adopt softer restriction rules for the sake of economic stability ([5] [6]). Despite its effectiveness in containing the spread of COVID-19, experts believe that the restriction policies cost the country economy [4] [7] [8]. [9] and [10] argue that the policies lead to economic crisis and recession as the social distancing contributes to economic and financial shutdowns.

Research on the impact of COVID-19 and the related government policies for economy and finance are still limited and demonstrated mixed findings. [11] investigated the impact of duration of national lockdown on public health and economy of 6 countries in Europe. The study revealed that the length of national lockdown has a negative influence on economic growth. In the same vein, [12] investigated the impact of lockdown policy on the African countries' economies. They revealed that the global pandemic and the subsequent lockdown led to a GDP decline of 34 percent, a reduction of wage earnings especially lower-educated workers.

In contrast, [13] studied the influence of the pandemic and its subsequent lockdown on the daily stock returns of Vietnamese companies. This study uncovered unexpected results that show a positive influence of imposed lockdown on the stock market. [13] believes that the positive influence mirrors enhanced investors' confidence and trust in the

* Corresponding author: hfahlevi@unsyiah.ac.id

government's strategies and capacity dealing with the pandemic.

As there is no consensus in terms of the results, this study aims to examine the statistical relationship between public mobilization policies and countries' economies. Besides, cross countries-based study that examined the correlation between social restriction policies and the country economy is still very scanty.

This study is required to understand how the COVID-19 policies affect the country economy and to determine the optimum level of restriction to minimize its unintended impact on the economy. Finally, this study also maps countries based on their economic growth and public mobilization restriction policy amid the COVID-19 pandemic.

2 Research method

Data were collected from published reports and databases (secondary data). The annual GDP growth refers to a change in GDP between 2019 and 2020 (prediction). The GDP growth was extracted from the economic prognosis of the International Monetary Fund [14]. To determine the restriction policy level, the stringency index (SI) produced by the Oxford COVID-19 Government Response Tracker (OxCGRT) [3] was used.

SI is between 0 – 100 in a range that is computed based on containment and closure policies. There are 8 dimensions to measure the stringent level of restriction policies. These include the closure of school and office, cancelation of public events, gathering restrictions, transportation arrangement, stay at home policy, restriction of internal movement, and international travel ban. The SI used in this study is the average score of the first 3 months after the WHO declared COVID-19 as a pandemic (11 March 2020).

This study used the Spearman correlation test to investigate the correlation between GDP growth and restriction level in 166 countries. The Spearman correlation coefficient (ρS) was adopted because our data do not meet the assumption of the bivariate normal distribution [15]. This data analysis method has been widely used to compare two unrelated datasets (see for instance, [16], [17]).

3 Results and discussion

The results consist of descriptive statistic results and correlation test results. Both provide corroborating evidence on the relationship between economic performance and restriction policies.

3.1 Descriptive statistic results

GDP growth has declined during the first year of pandemic. Table 1 exhibits that the average GDP growth in 2020 is -6.23. It indicates that in general, most countries around the world experienced a negative economic growth in 2020 that is related with the pandemic. Some countries gained positive GDP growth, while the other experienced economic turmoil.

In addition, the data shows that the average SI is 70 (medium to very stringent). Some countries have adopted a relatively very light restriction policy during the first three months of the COVID-19 pandemic (14).

Table 1. Results of descriptive statistics.

Variables	N	Mean	Maximal value	Minimum value
GDP Growth	166	-6.23	31.54	-64.33
Stringency index	166	70	95	14

Table 2 exhibits divergent SI but a similar economic performance across continents. The economy of South American and Oceania countries has been affected more severed compared to other countries in Africa and Europe. This may be linked to a more stringent polices adopted by South American countries than Oceania countries.

Table 2. Results of descriptive statistics.

Countries	GDP growth	SI
Africa	-4.84%	67
Asia	-5.75%	70
Europe	-5.35%	71
North America	-8.99%	75
Oceania	-7.06%	55
South America	-11.51%	80

Furthermore, some countries' economies were more severely affected by COVID-19. Table 3 shows 10 countries with the smallest GDP growth in 2020. Lebanon and Libya adopted relatively less stringent SI policies and their economic performance was the lowest. Most of the countries adopted medium to high restriction policies.

Table 3. Top 10 countries with the lowest 2020 GDP growth

Countries	GDP growth	SI
Lebanon	-64.33%	78
Libya	-45.26%	88
Suriname	-31.35%	73
Seychelles	-27.44%	59
Fiji	-27.28%	74
Brazil	-25.85%	74
Zimbabwe	-25.12%	75
Venezuela	-24.00%	80
Iraq	-22.61%	89
Zambia	-21.72%	46

Moreover, Table 4 shows countries that adopted the highest level of restriction policy to combat COVID-19. All of the countries experienced negative GDP growth. Compared to other countries on the list, Libya and Iraq have relatively lower GDP growth (-45.26%

and -22.61%) as they imposed strictest restriction policies. Meanwhile, other countries experienced better economic performance with milder restriction policies.

Table 4. 10 countries with the most stringent restriction policy.

Countries	GDP growth	SI
Honduras	-3.76%	95
Philippines	-2.50%	92
Peru	-15.16%	91
Guatemala	-0.66%	91
El Salvador	-8.29%	90
Bolivia	-5.47%	90
Iraq	-22.61%	89
Puerto Rico	-9.00%	89
Libya	-45.26%	88
Kuwait	-19.74%	88

3.2 Correlation test results

Spearman correlation test was performed to examine the association between GDP growth and SI. Table 5 shows a significant level is 0.01, smaller than 0,05. Therefore, this study demonstrated a significant association between GDP growth and SI.

Table 5. Results of Spearman correlation test.

Variables	Correlation coefficient	Sig. (2 tailed)
GDP_growth and SI	-0.265	0.01

Moreover, the results of this study revealed a negative association between GDP growth and SI. This means that countries that implemented stricter policies in the early COVID-19 pandemic tend to have higher GDP contraction. However, the correlation coefficient is relatively weak (-0.265) and thus, the economic damage can be mitigated using economic and fiscal incentives and optimizing the duration and level of restrictions. Before imposing lockdown, for instance, governments should have effective scenarios that minimizes the economic impact of the lockdown. The scenarios may include a different of lockdown intensity and duration across the country region and people group. For instance, a longer lockdown plan is imposed for elderly and children and for region with the highest case increase and the highest population density.

3.3 Discussion

Most governments are in a dilemmatic situation as they need to impose restriction policy to contain the rising number of infected cases and death while preventing the economy shutdown and its consequences. There is been a long debate on the impact of the restriction policies on a country's economy [18], and in some countries leads to people's resistance and demonstration.

The results of this study are consistent with the prior studies, for instance, [18] and [19]. The former found that restriction policy in 42 countries is an important driver of GDP growth [18]. The latter demonstrated that social distancing and other restriction policies explained divergent across 46 countries in the first semester of 2020 [19].

Moreover, the results of this study contribute to the debate on restriction policies and their impact on the economy. The negative correlation between GDP growth and SI indicates the need for periodic assessment. Governments are required to save both economy and people live by determining equilibrium through lift restriction policies when the economy suffered at a certain point.

Finally, the restriction policies should be accompanied by financial assistance policies to lighten the burden faced by industries and people's economies. The policies may include tax support, credit access financial support for workers and companies, and interest cuts [20].

Few studies demonstrate early evidence on the effectiveness of some government interventions during the lockdown. For instance, [21] revealed that the payroll protection program (a bridge loan program) and Main Street Lending Program can prevent corporate bankruptcies. Meanwhile, [22] evaluated the effectiveness of government support packages for small-medium enterprises in Russia in the COVID-19 pandemic. They found that the support policy is more effective if the government provides a comprehensive package rather than one type of government support. For instance, SMEs should receive tax, administrative and financial support collectively to survive in this pandemic era.

4 Conclusions

This study confirms that the restriction of public mobilization and other related strategies adapted by governments amidst the COVID-19 pandemic was negatively correlated with economic growth. However, the correlation is relatively weak. This suggests that the containment and closure policies do not always lead to economic decline especially if governments adopt economic policies that ease the burden faced by the industry, companies, and the people.

Each government, thus, should prepare anticipation policies to mitigate the COVID-19 related policies on the country and people's economy. Moreover, any support economy and financial package not only is required to save people and the country economy but also to ensure that people adherend to the restriction policies. Further studies are imperative to facilitate better understanding and empirical evidence on the trade-off between economy and health in the current difficult situation. More importantly, research on finding the most effective combination of economic and social supports is required as the pandemic can be longer than we expect and it can occur again in the future.

References

1. M. Chaudhary, P. R. Sodani, and S. Das, J. Health Manag. **22**, 169 (2020)
2. A. Atalan, Ann. Med. Surg. **56**, 38 (2020)
3. T. Hale, N. Angrist, R. Goldszmidt, B. Kira, A. Petherick, T. Phillips, S. Webster, E. Cameron-blake, L. Hallas, S. Majumdar, and H. Tatlow, Nat. Hum. Behav. **5**, 529 (2021)
4. R. K. Biswas, S. Huq, and A. Afiaz, Int. J. Heal. Policy Manag. **9**, 488 (2020)
5. N. Nurcahyono, A. N. Hanum, and F. Sukesti, J. Din. Akunt. Dan Bisnis **8**, 47 (2021)
6. R. Kuswanto, J. Din. Akunt. Dan Bisnis **8**, 105 (2021)
7. M. Mckee and D. Stuckler, Nat. Med. **26**, 640 (2020)
8. P. Boettke and B. Powell, Sourthern Econ. J. 1 (2021)
9. P. Ozili and A. Thankom, *Spillover of COVID-19: Impact on the Global Economy* (2020)
10. M. C. Howard, Pers. Individ. Dif. **168**, 110335 (2021)
11. M. Coccia, J. Econ. Libr. **3**, 100 (2016)
12. C. Arndt, R. Davies, S. Gabriel, L. Harris, K. Makrelov, S. Robinson, W. Simbanegavi, D. Van Seventer, and L. Anderson, *Impact of Covid-19 on the South African Economy An Initial Analysis* (2020)
13. D. L. T. Anh and C. Gan, J. Econ. Stud. **48**, 836 (2020)
14. International Monetery Fund, (2021)
15. R. Artusi, P. Verderio, and E. Marubini, Int. J. Biol. Markers **17**, 148 (2002)
16. E. Fallan, J. Account. Organ. Chang. **11**, 247 (2015)
17. H. Fahlevi, M. Indriani, and R. S. Oktari, Jamba J. Disaster Risk Stud. **11**, 1 (2019)
18. M. König and A. Winkler, Intereconomics **56**, 32 (2021)
19. M. Koenig and A. Winkler, CEPR Covid Econ. 132 (2020)
20. X. Cirera, M. Cruz, E. Davies, A. Grover, L. Iacovone, J. E. L. Cordova, D. Medvedev, F. O. Maduko, G. Nayyar, S. Reyes Ortega, and J. Torres, World Bank Res. Obs. **36**, 41 (2021)
21. V. Elenev, T. Landvoigt, and S. Van Nieuwerburgh, SSRN Electron. J. (2020)
22. E. Razumovskaia, L. Yuzvovich, E. Kniazeva, M. Klimenko, and V. Shelyakin, J. Open Innov. Technol. Mark. Complex. **6**, (2020)

https://doi.org/10.1051/e3sconf/202234005003

The impact of widespread covid-19 on small business venture in Aceh province: a recuperation strategy from entrepreneurs' point of view

Syafruddin Chan[1*], *Kurnia* Asni[2].

[1]Management Department, Syiah Kuala University, USK Darussalam Campus, Banda Aceh, 23111, Indonesia
[2]Master in Management Department, Abulyatama University, Unaya Lampoh Keude Campus, Aceh Besar, 23272, Indonesia

Abstract. This study aims to analyse the real impact of the COVID-19 pandemic for small businesses and what recuperation strategy needs to be implemented from the perspective of business owners including opportunities for using digital business technology and what the business owner expects from government to help the recover. This research was conducted using a quantitative descriptive method. Data collected using online questionnaires, through various MSBE communities in Aceh. 150 questionnaires distributed using non-probability sampling. In addition, several interviews with respondents were conducted to gain insight into the impact of covid-19 on their business. The spread of Covid-19 did have a negative impact on the small businesses that were respondents in this study. The biggest impact is the decline in sales because the demand for its products and services is decreasing. As a result, employers did the downsizing of by decreasing their business scale, and laying off employees. Entrepreneurs also ask for debt restructuring in banking institutions to reduce pressure on the declining cash in the company. At the same time, companies are starting to use technology to enter digital businesses, including joining online applications and selling in e-markets more intensively.

1 Introduction

Micro, Small and Medium Enterprises (MSMEs) are one of the sectors most affected by the Covid-19 pandemic, including MSMEs in Aceh Indonesia. Economic growth in Aceh before covid-19 grew by an average of 2.90 percent per year. However, after the existence of covid-19, the growth shrank by 0.37 percent. One of the reason is because of government policies that limit people's movement as well as changes in people's own behavior which tends to limit activities outside their house (1). The sectors most affected by this outbreak are food and beverage, accommodation and retail businesses. The impact on MSMEs is the decline in sales, the difficulty of raw materials, hampered distribution, as well as finance and production process (2). This phenomenon is supported by Bank Indonesia research which states that as many as 87.5 percent of MSMEs are affected by the Covid-19 pandemic Of this amount, around 93.2 percent of them were negatively affected on the sales side which ultimately put pressure on profits and cash flow (3). The COVID-19 crisis is causing major problems and its impact will be felt in the long term on the global economy (4). Although vaccination programs that have been carried out throughout the world have generated optimism, the economic outlook in the long term is always uncertain. In general, global economic growth declined by 3.4% in 2020 (5). Economic growth is expected to recover by 5.6% in 2021, but recovery will depend largely on the spread of virus variants and the effective distribution of vaccines, which the government is undertaking to prevent its development (6).

With the COVID-19 pandemic, its impact is clearly felt in every region and every industry. On the other hand, when lockdowns or social distancing forced consumers to stay at home, the e-commerce industry experienced fairly rapid growth which was reflected in a surge in online sales volume, as well as online business ratings and reviews that increased from 40 to 80 percent higher during this pandemic. Old business strategies like brand equity, more advertising spend, and big promotions are simply less relevant in an age when consumers have access to so much word of mouth (7).

This pandemic additionally affects the development of small enterprises. Micro, Small and Medium Enterprises (MSMEs) are regularly more influenced than large companies undertakings by the COVID-19 emergency particularly for the most weak sectors, for example, discount and retail exchange, transportation, convenience store, food industries, proficient and other individual services. Also, small organizations think that it is harder to exploit various wellsprings of money, including from the market. Then again, small

* Correspondent author: syafruddin.chan@unsyiah.ac.id

organizations frequently depend intensely on held income and bank obligation. Small organizations have more fragile production network capacities than their bigger partners. MSMEs for the most part have more modest inventories and provider networks making them more powerless against store network disturbances and cost expands. Essentially, they have less dealing ability to authorize alluring instalment conditions. As indicated by the review among European MSMEs directed among February and May 2020, 51% of respondents announced that late instalments put focus on their liquidity during the COVID-19 emergency, contrasted with 39% in 2019. Also, there has been a striking increment in the level of MSMEs that needed to acknowledge longer instalment terms than they anticipated (6). Small organizations are falling behind in engrossing advanced instruments and innovations that can assist with building flexibility in the current pandemic emergency. Reviews show that the pandemic has expanded the utilization of advanced innovation by MSMEs, notwithstanding generous contrasts between nations. Simultaneously, the contrasts between MSMEs – and specifically smaller firms – and big companies keep on being critical, with MSMEs take-up of advanced innovation generally a small portion of that by bigger companies (7).

The weakness of these small companies converted into a sharp decrease in income since the beginning of the pandemic at a rate quicker than their capacity to reduce working expenses, compromising an expected liquidity emergency among MSMEs. Between January 1 and April 1, 2020, US MSMEs income diminished by 40% to half (11) with income in January 2021 still down 31% contrasted with the earlier year.

The effect of the COVID-19 emergency on the appearance of MSMEs is significant. Of the MSMEs studied, 80% announced steady or expanding revenue for the year before the pandemic started. Right now, 80% of MSMEs say their revenue is declining. They likewise announced a few different impacts for instance worries about credit defaults (25%), worries about their capacity to hold the staffs (24%) and questions in their capacity to keep up with their inventory chains (28%), any expectations of decreasing the quantity of workers after the pandemic (28%) and deferring speculation for business extension (36%). Generally, miserable assumptions for the general economy and the negative pattern in SME profit paint a stressing picture. Under current conditions, very nearly one out of five MSMEs reviewed accept they will be bankrupt by August 2021. That figure ascends to 60 percent for April 2022. A decrease in SME revenue will excessively affect the practicality of MSMEs: almost three of every four say they will be bankrupt by January 2022 if their pay falls by 10 to 30 percent (6). The monetary effect of the COVID-19 pandemic has likewise been felt by the MSMEs in Indonesia. The effect brought about by this pandemic incorporates 5 viewpoints, specifically the decrease of sales is by 61%, the decline in gross profit is by 61%, the quantity of MSMEs encountering capital issues expanded to 71.4%, decreased the quantity of workers by 22%, and reducing the capacity to pay bank loan. Practically all MSMEs experience issues in doing their commitments to the banks.

1.1 For what reason do MSMEs need assistance?

Protecting small businesses from widespread permanent closures is important because of the many roles they play in the economy, as employers, engines of entrepreneurship, economic multipliers, and community centers (12). Preceding the COVID-19 pandemic, independent ventures represented portion of private sectors occupations two third of the net new positions made from 2000 to 2017. Little and medium-sized ventures (MSMEs) represented 50% of the total revenue produced by organizations and 44 percent of labor force. Extremely durable conclusion of private companies is thought to bring about longer joblessness than leaves and impermanent cutbacks. Indeed, even as the economy recuperates, current representatives should look for some kind of employment at various organizations, conceivably diminishing their pay. Investigations of past downturns showed that individuals who lost their positions acquired 17.5 percent less in their new positions (12).

1.2 Government support

The COVID-19 emergency seriously affects SME incomes, and government support is key for these organizations (13). Policymakers can expect take-up in different help projects to keep on expanding for the time being. Governments all throughout the globe are reacting rapidly and unequivocally to the extraordinary difficulties MSMEs face because of the COVID-19 pandemic, through different improvement and backing measures. These actions zeroed in on crisis liquidity support in different structures, yet were progressively being joined by underlying scaffolding and a more extensive recuperation bundle. Both focal and nearby and provincial governments participated in the arrangement exertion (6).

General government support measures can be mentioned for example Job retention schemes in the form of wage subsidies. This approach targets forestalling a sharp ascent in joblessness just as to lift customer interest (14). Then, at that point conceded instalments incorporate deferral of instalment of pay and corporate assessments, esteem added charge, government managed retirement and benefits instalments, a ban on obligation instalments and waivers on lease instalments, waivers or premium cost decreases focused on keep up with liquidity and diminish the weight of MSMEs working expenses.

Other monetary help incorporates giving advance ensures, improved on layaway loaning methods, and backing through awards and sponsorships, or through value or semi values, including convertible credits (15). What's more, the public authority has additionally upheld the parts of SME business association instruments that plan to assist MSMEs with adjusting the changing business climate, for instance support including for teleworking and E-deals, support for innovation advancement and improvement, support for abilities redesigning, support for new companies, and for discovering new elective business sectors. From the table above, it can be seen that the most interventions carried out by the government to keep MSME companies operating and prevent them from going bankrupt are wage subsidies, loan guarantees, and income tax relaxation.

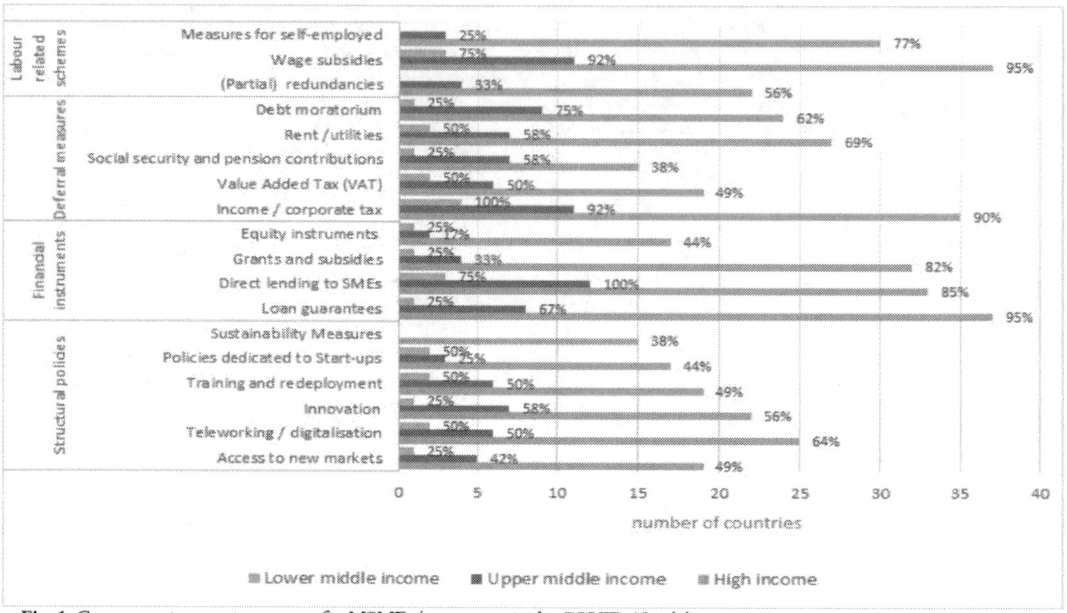

Fig. 1. Government support measures for MSMEs in response to the COVID-19 crisis.
Source: Annex A and OECD (2021), A one-year in-depth analysis of SME and entrepreneurship policy responses to COVID-19: lessons learned for the road to recovery.

2 Opportunity during pandemic

The condition of the COVID-19 corona pandemic, which was followed by the implementation of social distancing, gave rise to new consumer behavior, but on the one hand it opened up opportunities for banking, financial, and financial services businesses to spur marketing. The pandemic situation seems to have given rise to new consumer behavior, which is different in every Southeast Asian country. Now is a good time for banking, financial, and financial services businesses to do marketing. The players in the industry can take advantage of the digital space to promote loyal users, and even reach new users. At times like this, the majority of people tend to choose cashless transactions in order to maintain their health. The COVID-19 pandemic has indeed caused panic in financial markets. Many industries are feeling the impact of this situation, for example travel, hospitality, food and beverage, automotive, and entertainment (2,16). In this situation, many companies and brands have temporarily suspended their marketing activities. Some even withhold these activities until the situation returns to normal and is under control. This results in a general decreasing in marketing activity. This is an opportunity for businesses to be able to take advantage of this situation to form new habits, and change their communication and sales channels to the digital space. This new habit is believed to persist even though the situation returns to normal. The shopping interest of the Indonesian people has not disappeared. Especially for online shopping, this interest actually grew rapidly during this crisis situation (17). With the closure of the majority of shopping centres, online shopping has become the main choice in meeting the needs, from daily to hobby. Seeing the interest in spending that does not subside, this opens up opportunities for banking, financial, and other financial services businesses. Moreover, several online buying and selling platforms encourage buyers and sellers to make cashless transactions by utilizing payment services such as credit cards, transfers, or e-wallet. So, he continued, this is the right time for banking, financial, and financial services businesses to boost marketing.

The urge to practice social distancing reduces community activities. In addition to decreasing the number of people visiting business activities such as shopping centres, the number of individuals traveling outside the city has also decreased drastically. The average decline in visits at several malls in Indonesia was reduced by more than 50 percent compared to the beginning of 2020 (18). COVID-19 is changing our lives. In this situation, we prefer to stay at home instead of traveling. Social distancing also makes us spend a lot of time in the digital space either for work, communicating, shopping, or just looking for entertainment. In Indonesia, the pandemic situation and social distancing gave rise to several new behaviors, for example, the adaptive shopper and working-from-home professionals. Regarding the adaptive shopper, since social distancing was announced, the use of shopping applications has increased by up to 300 percent. Applications that are widely used are shopping applications that sell various kinds of daily needs as well as special applications for buying and selling used goods. Meanwhile, for most workers in Indonesia, working-from-home is the same as working in a normal situation. They continue to do work, collaboration, communication, and meetings as usual. However, all the

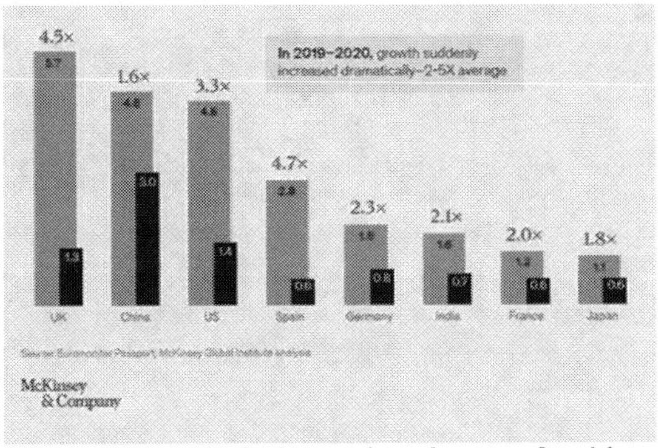

Fig. 2. Increase the business volume because of social distancing.

This phenomenon is certainly become a unique opportunity for small businesses affected by the pandemic to start taking advantage of this situation with changes in people's behavior, especially to enter the digital space and leave physical business activities as happened before the COVID-19 pandemic.

3 Research method

The following section discusses the recovery strategy carried out by small and medium enterprises in Aceh. The data presented in this section are primary data collected through questionnaires and direct interviews with business owners. This research was conducted using a quantitative descriptive method. Data collected using online questionnaires, through various MSME communities in Aceh. 150 questionnaires distributed using non-probability sampling. In addition, several interviews with respondents were conducted to gain insight into the impact of covid-19 on their business and of which 50 are small businesses which are assisted by the Syiah Kuala University Business Incubator, most of which are located in Banda Aceh and Aceh Besar.

4 Recovery strategy

The types of business are also quite diverse, although most of them are food producers. And then there is also personal care, diet food, herbs and dish cleaner.

work is done at home with the help of productivity applications. The use of productivity applications rose by more than 400 percent in mid-March 2020 (18).

The opinion above is strengthened by the results of research conducted by (17) in various countries around the world and found an increase in the volume of e-commerce business up to 4.5 times compared to before the pandemic period as shown in the following figure.

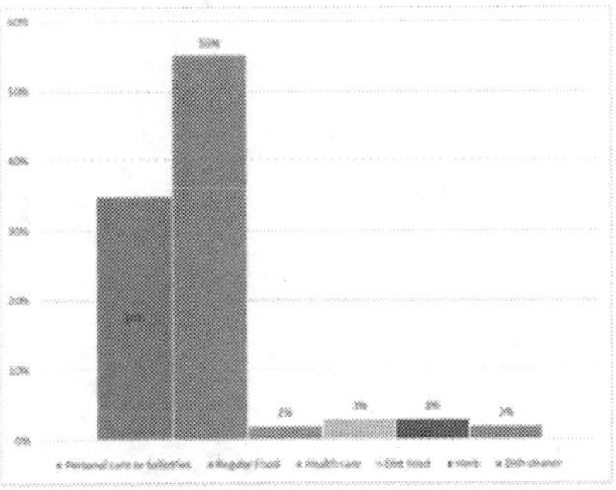

Fig. 3. Respondent's business profiles.

From the description of the respondent's business, it can be seen that the business of producing MSME products based on food is very dominant. Generally using raw materials that are in the area where the business stands. Such as tuna fish for Abon Ikan products produced by the Aroma brand, sambal Ijo which uses cayenne pepper as its main raw material or Cilet Chocolate which uses cocoa as raw material which is widely spread throughout Aceh. The second category is personal care, also generally produced by this MSME company using local raw materials such as Minyeuk Pret, a perfume product that uses local patchouli oil as a fragrance binder, then there is also Gayuen, which is a deodorant product that uses the main raw materials of whiting and lime and Yagi Natural is a cosmetic product that uses cocoa fat as its main raw material.

Of the several examples of MSMEs mentioned above, the majority do use local raw materials as their main raw materials. The use of specific local resources can lead to a comparative advantage. Comparative advantage can be turned into a competitive advantage if in utilizing local raw materials the company can save on raw material costs, which do not have to be imported from other regions. According to (19). When a company can do something or have something that competitors want, then the company represents a competitive advantage. According to the theory of competitive advantage, it is known that a company to be superior to other companies must have certain strategies and pay attention to the company's performance in order to improve every day, every month, and every year, if this is achieved then the company has a competitive advantage. In the context of companies in Aceh, they are actually in a better off position to win the competition.

In relation to the COVID-19 pandemic, the regional and central governments have actually made various efforts to help MSME companies to continue to exist during this pandemic crisis (20). However, not all of

https://doi.org/10.1051/e3sconf/202234005003

these MSME companies enjoy the convenience facilities provided by the government, either because the numbers are limited or the types are not in accordance with the needs of the MSMEs themselves, which are very specific and differ from one company to another. When asked what help do they hope for the most if there will be more assistance in the future related to this pandemic? So the answer is very diverse from one MSME to another. The largest number or 25% of the total respondents stated that working capital assistance was very much needed. Meanwhile, 20% of other MSMEs stated that what they need most at the moment is wage subsidies for their employees. For those who still have loans at the bank, they expect a moratorium and scheduling of instalment payments as well as relaxation of bank interest that must be borne (15%). The same number (15%) of them expect support for conversion to digital businesses and improvements to their digital business models (15%), so that not only can they get out of crisis situations, but also take advantage of opportunities that arise as a result of the implementation of social distancing during this pandemic.

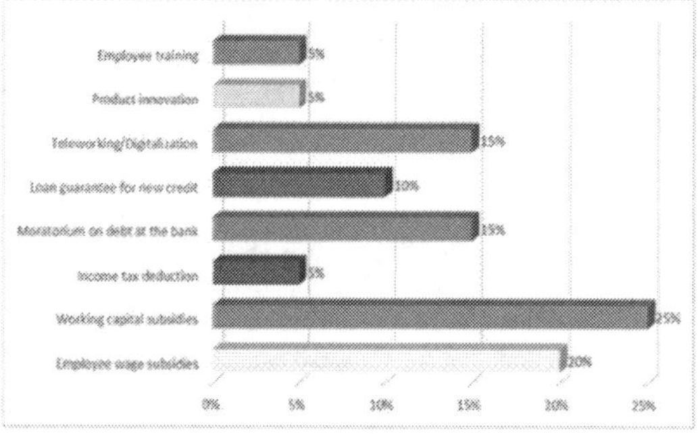

Fig. 4. Assistance expected from the government.

Some MSMEs also find it difficult to get bank loans, because they do not have enough guarantees to back up these bank loans (21). Therefore, 10% of the respondents hope that there will be guarantee assistance from the government so that they can increase their debt to the bank in the form of credit loans. While the remaining 5% each expect new product development assistance, employee training and corporate income tax reductions.

The Indonesian government through the financial services authority (OJK) has taken various actions to assist MSMEs, including the interest subsidy program for micro, small and medium enterprises (MSMEs), a fund placement program to ensure banking liquidity which will be channelled to the banking sector, real estate loans, working capital credit guarantee programs for MSMEs, and programs for providing working capital to corporations, especially in the labour-intensive sector (Fathiyah Wardah, 2020)

Regarding what has been done by MSMEs entrepreneurs in the Aceh region to keep their companies operating during this pandemic period, most of them have reduced production volume due to declining market demand (25%). Efforts made by other MSME entrepreneurs are increasing their product marketing channels by utilizing online business or digital marketing as much as 18%. Many were also forced to reduce employees (18%) due to the decreasing number of activities in the company in the hope that they would be rehired when the situation improved. In addition, there are those who diversify their businesses by building fully digital businesses to respond to the increasing demand in online businesses as a result of social distancing (14%).

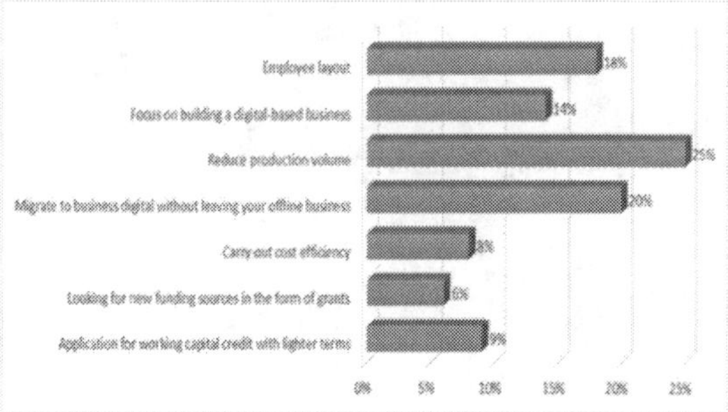

Fig. 5. What have been done to recover the business.

When asked to companies that use digital platforms as their recovery strategy, what do they do by migrating to digital businesses, then the answer is opening an online store in one or more e-market places in Indonesia such as Lazada, Shoppe, Bukalapak and Tokopedia (80%). Some of them said they did promotions through social media such as Instagram shops, Facebook shops and took advantage of the Google shopping facilities. A small number of MKMs such as Yagi and Minyeuk Pret built their own websites, of course combining them with various social media application platforms to drive traffic to their websites which functioned as landing pages (20%). Some of them outsource their financial and marketing activities to third parties (20%). Most of the respondents who trusted the most as their suppliers in this study were the MOKA application. Moka is an associate company of Gojek that provides various solutions for offline & online sales needs. Moka has a very sophisticated POS system that can help offline sales and allows MSMEs to build personal websites through GoStore. Moka POS is a cloud-based Point of Sale system that can help offline sales activities. This application provides services with complete features, namely daily transaction reports, promotional activities, storage & inventory management, as well as access to dashboards and real-time reports. And what's interesting is that some have tried to implement a business auto pilot scheme in their business (10%). Moka has a very sophisticated POS system that can help offline sales and allows MSMEs to build personal websites through GoStore. Moka POS is a cloud-based Point of Sale system that can help offline sales activities. This application provides services with complete features, namely daily transaction reports, promotional activities, storage & inventory management, as well as access to dashboards and real-time reports. And what's interesting is that some have tried to implement a business auto pilot scheme in their business (10%). Moka has a very sophisticated POS system that can help offline sales and allows MSMEs to build personal websites through GoStore. Moka POS is a cloud-based Point of Sale system that can help offline sales activities. This application provides services with complete features, namely daily transaction reports, promotional activities, storage & inventory management, as well as access to dashboards and real-time reports. And what's interesting is that some have tried to implement a business auto pilot scheme in their business (10%). as well as access to real-time dashboards and reports. And what's interesting is that some have tried to implement a business auto pilot scheme in their business (10%). as well as access to real-time dashboards and reports. And what's interesting is that some have tried to implement a business auto pilot scheme in their business (10%).

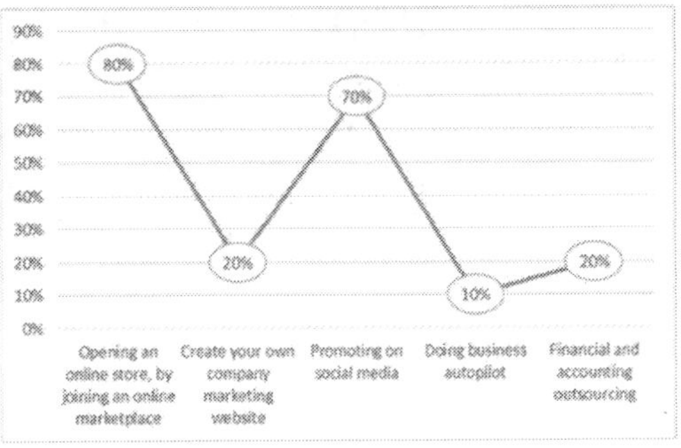

Fig. 6. Digital Business Strategies.

The Auto-Pilot business is a condition where the business can run on its own without the owner always being actively involved in it. That is, the business can run alone, the business owner can go for a walk. This means that apart from financial freedom, the business owner also has time freedom. Of course, this takes advantage of various available applications to help business owners continue to monitor their companies even remotely.

The last question asked is what exactly you are doing so that the effectiveness of your digital marketing efforts can be run well, and the traffic that comes to your landing page will increase significantly. Then the answer is automate my email marketing (7%), use website builder (5%), Integrated my application (5%), use a scheduling tool for social media (5%), Run Automated Survey Responses (5%) and none of the above (70%).

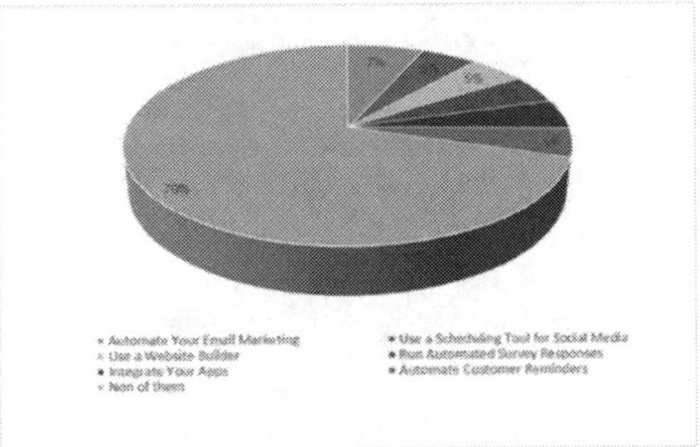

Fig. 7. Intention to have auto-pilot business.

This means that even though they have used digital platforms to maximize their revenue, not many have used strategies to maximize the traffic that comes to their stores. Plus, their penetration into online businesses is also that the majority of them only join the e-market place. Whereas optimal results will be obtained if they can sell themselves through their own website using various existing and available strategies to maximize their sales. Thus, he does not need to share profits with e-marketplaces which can make his ability to maintain the level of liquidity and solvency of the company to be lower to keep the company operating.

5 Conclusion

From the results of the research conducted, it appears that SMEs in Aceh actually have a comparative advantage because they can utilize raw materials obtained from Aceh's nature without having to bring in from outside. Its comparative advantage can be converted into a competitive advantage because it can sell at a lower price. In the era of the pandemic, the average MSME in Aceh experienced financial difficulties, so they hoped that financial assistance from the government through existing credit channelling banks would make it easier for them to get out of the financial crisis caused by the COVID-19 pandemic. The assistance in question can be in the form of grants or ease of requirements in lending business loans. The Singapore government has also done the same for its MSMEs. In addition to providing grants for the digitization of its MSME companies, the Singapore Government also helps MSMEs with credit risk from financial institutions of up to 90 percent. This proves the government's seriousness is the same as in Indonesia to help their SMEs survive this pandemic.

In addition to hoping for a helping hand from the government, MSME business actors in Aceh have also made various efforts that they consider to be able to ensure the survival of their businesses during this pandemic, for example by increasing cost efficiency, including by reducing temporary employees, reducing the amount of production, Application for working capital, credit with lighter terms, looking for new funding sources in the form of grants and migrate to business digital without leaving the existing offline business.

References

1. A. Dua, K. Ellingrud, D. Mahajan D, J. Silberg McKinsey Co. (2020)
2. N.F. Fabeil, K.H. Pazim, J. Langgat. T. J Econ Bus. **3,** 2. (2020)
3. K.F. Sandy. Survei BI (2021)
4. R. Fairlie. National Bureau of Economic Research (2020)
5. L. Demmou, OECD Working Papers. (2021).
6. OECD. OECD Policy Responses to COVID-19. (2021).
7. OECD. OECD SME and Entrepreneurship Papers. (2020)
8. E. Kiran, Gaziantep Univ J Soc Sci 752–66. (2020)
9. D. Graeber, A.S. Kritikos, J. Seebauer. J Popul Econ. **34(4),** 1141–87 (2021).
10. J.Z. Ayanian. JAMA Heal Forum **1(6),** e200689. (2020)
11. R. Zhang, Cold Spring Harbour Laboratory (2021).
12. M. Albert. Australia's Economic Measures and Employment Laws Changes in Response to Covid-19 (2020).
13. S. Kalemli-Ozcan, P-O. Gourinchas, V. Penciakova, N. Sander. IMF Work Pap **20**, 207. (2020)
14. M. Brülhart, R. Lalive, T. Lehmann, M. Siegenthaler, Swiss J Econ Stat **156,** 1 (2020)
15. F. Eckert, H. Mikosch, Swiss J Econ Stat (2020).
16. S. Kalogiannidis. Int J Soc Sci Econ Invent. **6(12),** 387-to. (2020)
17. S. Lund, A. Madgavkar, J. Manyika, S. Smit, K. Ellingrud. McKinsey (2020).
18. M. Syahran. Ekonomi dan Bisnis (2021)
19. R.D. Fred. Manajemen Strategi dan Konsep. 12th ed. Jakarta: Salemba Empat; (2011).
20. W. Fathiyah, Voice of Indonesia. (2021)
21. S. Hollander, A. Verriest. J finance Econ (2016)

Do pandemic COVID 19 and business cycle influence the Indonesia composite index?

Chenny Seftarita[1*], *Ferayanti*[1], *Fitriyani*[1], and *Asri* Diana[1]

[1]Development Economics Department, Universitas Syiah Kuala, Aceh, 23111, Indonesia.

Abstract. The impact of the Covid 19 pandemic on the Indonesian economy was worse than the impact of the global financial crisis in 2008. We can see the business cycle from the contraction of economic growth, volatility of the exchange rate, and inflation. The shocks to these business cycles could create systemic risks and influence financial system stability. Financial system stability is more sensitive to shocks from changes in one of the macroprudential indicators, namely the Indonesia Composite Index (ICI). This study will examine the effect of the Covid 19 pandemic and the business cycle on ICI. This research was conducted in Indonesia using quarterly secondary data from 2008-2021. The data was obtained from Bank Indonesia. The model used is the Auto Regressive Distributed Lag (ARDL) model. The purpose of using this model is to see the long-term and short-term relationships between the variables studied. The results show that the Covid 19 pandemic and business cycle variables such as; economic growth, exchange rate, and inflation have a relationship on the ICI both in the short and long term. The decline in economic growth during a crisis is difficult to avoid. However, the government must continue to accelerate the economic growth so it does not decline sharply during the crisis. Monetary authorities must also maintain exchange rate stability and inflation so it does not fluctuate unsteadily.

1 Introduction

Economic shock is characterized by a significant decline in economic growth. The indicators of economic shocks can be seen from fluctuations in economic growth [1]. Economic fluctuations will usually be followed by inflation and unemployment as a result of the cycle [2]. In the Indonesian case, the economic recession in 2020 is because of the Covid-19 pandemic. It caused a deep contraction shown by negative growth in the second and third quarters of 2020. This can be seen in Figure 1, where economic growth was recorded at -5.32% in the second quarter and -3.49% in the third quarter. This was the lowest recorded compared to pre-pandemic crises such as the 2008/2009 global crisis and the 2015 European crisis. During these two periods, economic growth is declined but continued to grow positively.

In 2020, the Indonesian economy experienced a sharp contraction due to the COVID-19 pandemic. This pandemic was firstly discovered in China [3] COVID-19 and caused a severe acute respiratory illness that not only spreads fast around the world but also causes a lot of fatalities. It was confirmed to have spread to Indonesia in March 2020. The enormous number of deaths and sicknesses caused by the pandemic has caused extraordinary shocks to public health and the global economy. Stock markets, global supply networks, labor markets, and consumer behavior have all been impacted negatively by the pandemic.

Economic recession shown by negative growth did not only affect Indonesia but also became a threat to the global economy. The Covid 19 pandemic has infected approximately 50.27 million people in the world and nearly 1.26 million people died [4]. The pandemic is expected to cause the world economy to grow negatively related to mass movements (social and physical distancing), such as the results of studies by [5,6]. Even in developed countries such as the United States and the Euro Area, the predicted crisis due to this pandemic is greater than the 2008/2009 Global crisis and almost equal to the Great Depression period of 1930 [7].

Other indicators of the economic shock are fluctuations in inflation and the rupiah exchange rate. Figure 2 shows the growth of inflation as a variable that is strongly influenced by the economic cycle. The decline in economic performance due to economic shocks has complex impacts, such as; declining household consumption and declining business performance. Weakening purchasing power causes deflation. This condition is shown, Inflation has fallen and deflation occurred. Moreover, in the financial sector, capital outflows due to the shock caused the rupiah to depreciate. Despite being shaken by the pandemic, the rupiah exchange rate against the dollar in the second quarter became weakened to Rp16,310/USD.

* Corresponding author: chenny@unsyiah.ac.id

This depreciation was higher than the 1997/1998 monetary crisis period.

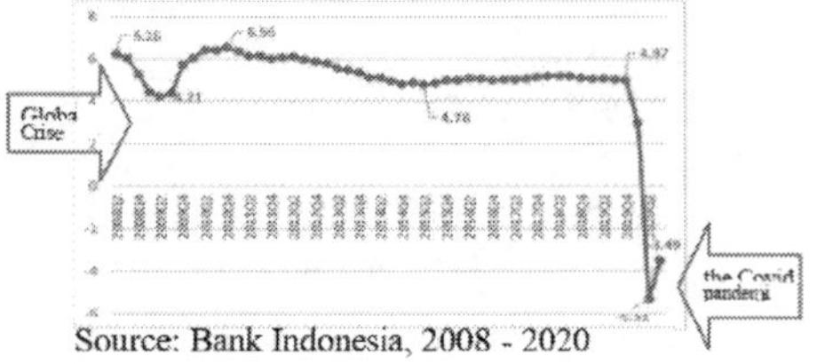

Source: Bank Indonesia, 2008 - 2020

Fig. 1. Economic Growth Before and After the Covid 19 Pandemic.

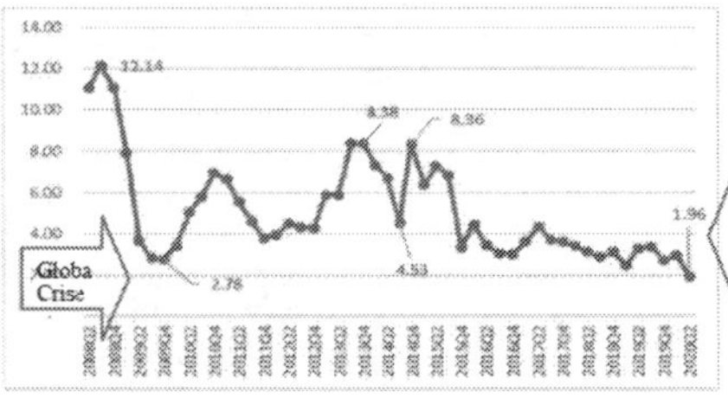

Source: Bank Indonesia, 2008-2020

Fig. 2. Inflation Before and After the Covid 19 Pandemic.

Furthermore, economic shocks have a major impact on Financial System Stability. The existence of a potential system makes the financial system vulnerable and leads to a deeper crisis. The instability of financial system causes several things; first, the monetary policy transmission mechanism becomes ineffective, the intermediation function is hampered, public distrust of financial institutions, and high-cost restructuring [8].

The financial system is divided into two, namely; financial market or capital market, financial intermediary institutions such as banking [9]. There are several steps in the macroprudential policy, including; financial institutions, macroeconomic conditions, financial markets and infrastructure, corporations, and households [10]. Several macroprudential variables are used as financial system measurements, one of which is the Composite Stock Price Index (ICI). A stable capital market reflects good economic conditions, on the other hand, unfavourable and unstable economic conditions will cause capital market performance to be vulnerable to decline. This can be seen from the decline in the ICI.

Economic cycle fluctuations will greatly affect macroprudential indicators, including the Jakarta Composite Index. Several studies analyse how the impact of economic shocks on the stock market. [11,12], analysed the effect of shock on the economic cycle in the stock market using the Vector Autoregressive (VAR) approach. The findings found that economic turmoil is a condition of uncertainty and affects the decline in stock market performance. [13] examines the distribution of the economic crisis in global finance and the real sector in several regions, such as Asia and Europe. A larger regional shock to the real sector can be seen from the collapse of financial markets due to the crisis. [14] looks at the synchronization between business cycles in the United States and the rest of the world. The crisis in America triggered a global crisis and financial crises in various countries. The related equity markets across countries experienced a sharp decline during the crisis period.

Research on the effect of economic shocks and their comparison with the period before and after the occurrence of Covid 19 using dummy variables is relatively new in this study. Financial System Stability is also newly being studied in depth by monetary authorities. Based on this background, the researcher will examine the effect of the business cycle and the shocks of the Covid 19 pandemic on the Composite Stock Price Index in the long and short term.

2 Research methodology

This study analyses the effect of the economic cycle and the Covid 19 pandemic on the Indonesian Composite Stock Price Index as one of the macroprudential variables in Indonesia. The data used in this study is secondary data sourced from Bank Indonesia, Federal Reserve Economic Data (FRED), and the Financial Services Authority (OJK). The data used is quarterly data from 2008-2020. This study uses a dynamic model and analyses the long-term and short-term relationships between variables. This research uses ARDL (Autoregressive Distributed Lag) model. The following equation will be regressed with the ARDL model:

$$\Delta L\,ICI_t = \varepsilon_0 + \sum_{i=1}^{p} \emptyset_i \Delta L\,ICI_{t-1} + \sum_{i=1}^{p} \varphi_i \Delta L\,GDP\_Growth_{t-1} + \sum_{i=1}^{p} \gamma_i \Delta EXR_{t-1} + \sum_{i=1}^{p} \theta_i \Delta INF_{t-1} + \sum_{i=1}^{p} \omega_i \Delta LD\_COVID19_{t-1} + \lambda_1 L\,ICI_{t-1} + \lambda_2 L\,GDP\_GROWTH_{t-1} + \lambda_3 EXR_{t-1} + \lambda_4 INF_{t-1} + \lambda_5 LD\_COVID19_{t-1} + u_t \quad (1)$$

The dependent variable used is one of the macroprudential indicators, namely ICI or the Composite Stock Price Index as the dependent variable. Independent variables as proxies for economic shocks include economic growth, the exchange rate of EXR or Rupiah against the dollar, INF or inflation, and D_Covid19 or Dummy before and before the Covid 19 pandemic. $\emptyset$, φ, γ, θ, ω shows a short-term relationship, λ_1, λ_2... λ_5 shows long term relationship, t is time, ut is error term. All variables are in log form, except those in the form of proportions whose purpose is to see the sensitivity of the dependent variable to the independent variable (economic shocks). In the ARDL model, several tests were carried out, including (1) Stationarity Test, (2) Optimum Lag Test, and (3) Model Stability Test (classical assumption test, such as autocorrelation, multicollinearity, and heteroscedasticity).

3 Results and discussion

Before entering the regression stage, a unit root test is firstly performed on each variable used. Based on the test results with the Phillips-Perron test statistic, all variables are not stationary at the level. The test is then continued at the first difference and the results show that all variables are stationary at the first difference. After the stability test, the next step is the optimal lag test using Akaike Information Criterion (AIC)). Based on these tests, the optimal lag used in this study is the ARDL lag (1,1,1,2,3). This lag will be used in the estimation of ARDL in this study.

Tabel 1. Unit root test results unit root test results.

Variables	At Level	At First Difference
ICI	Not Stationary	Stationary
GDP_GROWTH	Not Stationary	Stationary
INF	Not Stationary	Stationary
EXR	Not Stationary	Stationary
D_COVID19	Not Stationary	Stationary

The next stage is regression using the ARDL model, followed by model testing and classical assumption tests, namely autocorrelation, multicollinearity and heteroscedasticity tests. Table 2 shows the Autocorrelation Test Results. Based on the test results, the estimation model is free from autocorrelation problems, as seen from the insignificant probability value at the 5% confidence level.

Table 2. Autocorrelation test.

Q-statistic probabilities adjusted for 1 dynamic regressor

Autocorrelation	Partial Correlation		AC	PAC	Q-Stat	Prob*
		1	-0.114	-0.114	0.6787	0.410
		2	-0.023	-0.037	0.7069	0.702
		3	0.012	0.005	0.7147	0.870
		4	-0.170	-0.171	2.3137	0.678
		5	0.138	0.103	3.3921	0.640
		6	-0.110	-0.100	4.0944	0.664
		7	-0.057	-0.073	4.2866	0.746
		8	0.001	-0.050	4.2867	0.830
		9	0.068	0.102	4.5749	0.870
		10	-0.063	-0.106	4.8300	0.902
		11	-0.037	-0.046	4.9224	0.935
		12	0.090	0.081	5.4748	0.940
		13	-0.088	-0.064	6.0169	0.946
		14	0.018	-0.054	6.0389	0.966
		15	-0.044	-0.029	6.1785	0.977
		16	0.094	0.131	6.8488	0.976
		17	0.143	0.102	8.4453	0.956
		18	-0.138	-0.108	9.9713	0.933
		19	0.012	0.004	9.9837	0.953
		20	-0.172	-0.158	12.539	0.896

The test was carried out with a multicollinearity test and heteroscedasticity test. The multicollinearity test uses the Breusch-Godfrey Serial Correlation LM Test method. The result shows that this model is free from multicollinearity problems, where the probability of Chi-Square is not significant. Likewise, the heteroscedasticity test using Breusch-Pagan-Godfrey, the model also rejects the heteroscedasticity problems as seen from the insignificant Chi Square probability value. The last stage is testing the value of the model category as shown in Figure 3. Based on the results of the CUSUM and CUSUM OF SQUARE tests, it can be seen that the most stable model (as seen from Cusum and Cusum from the boxes in a significant area). After going through all the testing stages, the ARDL model can be used for interpretation.

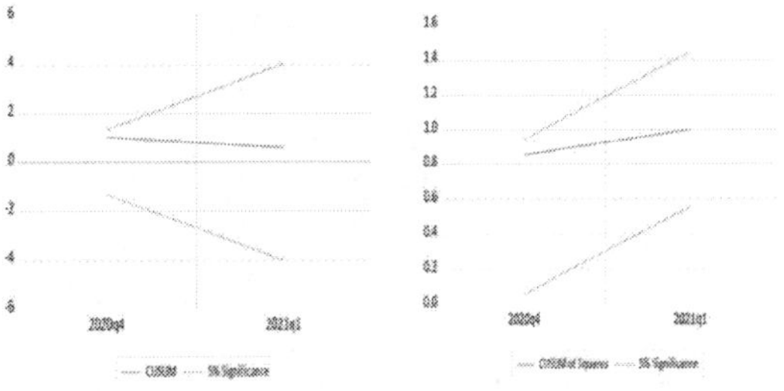

Fig. 3. Model stability test.

Table 3 is the result of short-term estimation. In general, the economic cycle and the COVID-19 pandemic significantly affect the Jakarta Composite Index (JCI) in the short term. This can be seen from the value of CointEq (-1) known as the Error Correction Term (ECT) which is negative and significant, and <1. Partially, almost all variables are significant at 5% alpha. The Covid 19 pandemic in the t_{-1} and t_{-2} periods greatly affected the decline in ICI. The pandemic caused stress for investors and demand for portfolio investments such as stocks and bonds to decline. These conditions cause a general decline in stock prices. The economic cycle variables, namely economic growth, exchange rates, and inflation also show a significant effect in the short term. The decline in economic growth caused the ICI to be sharply corrected in period t (current year). The weakening of the exchange rate in period t also has a negative effect on ICI. The inflation variable previously showed a positive effect on the ICI. This condition is probably caused by targeting inflation which ensure the inflation to be under control and thus, its percentage is still low. Even a short-term increase raises positive expectations among investors so that public demand or purchasing power is still quite high.

Table 3. ARDL Error Correction Regression
Case 2: Restricted Constant and No Trend.

Variable	Coefficient	Std. Error	t-Statistic	Prob.
D(GDP_GROWTH)	-388.5826	107.8668	-3.602431	0.0009
D(EXR)	-0.447629	0.067427	-6.638738	0.0000
D(INF)	2.416004	24.17954	0.099919	0.9210
D(INF(-1))	84.99678	27.19477	3.125483	0.0035
D(D_COVID19)	-148.9885	219.6205	-0.678391	0.5019
D(D_COVID19(-1))	-1502.042	341.1534	-4.402833	0.0001
D(D_COVID19(-2))	-4186.599	999.9504	-4.186807	0.0002
CointEq(-1)*	-0.083294	0.014031	-5.936448	0.0000

R-Squared : 77,45 %
Adj R_Squared: 73,59 %
DW Stat: 2.223

https://doi.org/10.1051/e3sconf/202234005004

Table 4. Long-term estimation with bound tests.

Test Statistic	Value	Signif.	I(0)	I(1)
F Statistic	5.16	10 %	2.2	3.09
K	4	5%	2.56	3.49
		2.5%	2.88	3.87
		1%	3.29	4.37

EC= ICI- (-1093,8 * GDP_GROWTH – 0,089 * EXR – 538.5 * INF – 209,2 * D_COVID19 + 16005.1 (2)

Based on the results of the long-term test using the Bound Test in Table 4, it can be seen that all independent variables have a long-term relationship with the dependent variable. This can be seen from the values of I(0) and I(1) which are less than the value of F Statistics, at a confidence level of 1-10 percent. This means that in the long term, the economic cycle and the Covid-19 pandemic will greatly affect ICI. Partially, in the long term, all variables do not show a significant effect on ICI. This can be seen in equation 2 above. The relationship between the variables above is in line with the results of research by Tsai (2017) and Caldara et al (2016). Therefore, the slowdown in economic growth in the event of an economic shock will have a negative impact on the composite stock price index, especially in the short term.

4 Conclusion

The decline in economic growth during a crisis is difficult to avoid. However, the government must continue to accelerate the economic growth so it does not decline sharply during the crisis. Monetary authorities must also maintain exchange rate stability and inflation to not fluctuate. The weakening of the exchange rate and rising inflation will further exacerbate the crisis. Regarding the Covid 19 pandemic, the government must try harder to reduce the spread of the coronavirus in various regions in Indonesia, considering that the root cause of the economic downturn that occurred in the 2020-2021 period is the Covid 19 pandemic. In addition, it is essential to weaken the profound impact of the pandemic on the financial system and economy.

References

1. R. Dornbusch and S. Fischer, *Macroeconomics*, Eleventh E (McGraw-Hill, 1994, 2011)
2. O. Blanchard, A. Amighini, and F. Giavazzi, *Macroeconomics: A European Perspective* (Pearson Education Limited, England, 2010)
3. Kementerian Kesehatan, Kementeri. Kesehat. RI **2**, 1 (2020)
4. Kementerian Kesehatan, (2021)
5. N. Fernandes, IESE Bus. Sch. Work. Pap. **No. WP-124**, 0 (2020)
6. Centre for Economic Policy Research, *Economic in the Time of Covid-19* (CEPR Press, London, 2020)
7. R. Barichello, Can. J. Agric. Econ. **68**, 219 (2020)
8. Bank Indonesia, *Macroprudential Policy* (2018)
9. F. S. Mishkin, *The Economics of Money, Banking, and Financial Markets*, Tenth Edit (Pearson, 2013)
10. G. J. Schinasi, *Understanding Financial Stability: Towards a Practical Framework* (2007)
11. D. Caldara, C. Fuentes-Albero, S. Gilchrist, and E. Zakrajšek, Eur. Econ. Rev. **88**, 185 (2016)
12. K. Sugimoto, T. Matsuki, and Y. Yoshida, Emerg. Mark. Rev. **21**, 201 (2014)
13. D. Kenourgios and D. Dimitriou, Econ. Model. **44**, 283 (2015)
14. J. H. Pyun and J. An, J. Int. Money Financ. **67**, 172 (2016)

Economic fluctuations and monetary policy on the stock market during the COVID-19 pandemic in Indonesia

Suriani[1*], *Muhammad* Nasir[1], and *Ernawati*[1]

[1]Faculty of Economics and Business, Universitas Syiah Kuala, Indonesia

Abstract. The purpose of this study is to examine the stock market's resiliency in the face of economic fluctuations and a crisis caused by the COVID-19 outbreak. using monthly data from 2011 to 2020 and an error correction model with the Composite Stock Price Index (conventional stock market system) and the Jakarta Islamic Index (Islamic stock market system) as dependent variables and policy interest rates, exchange rates, inflation, and crises as independent variables. In both stock market models, the regression results reveal that there is a short-term to long-term equilibrium relationship. Interest rates set by policymakers and crisis variables have a detrimental impact. In the long run, the exchange rate has a beneficial effect, and inflation has no influence on the two stock markets. In the short run, the findings in both markets are nearly identical, meaning that inflation and policy interest rates have little influence, whereas exchange rates have a negative impact. The crisis, on the other hand, had no impact on the JCI stock market. The crisis had a positive impact on the JII stock market, unlike the JII stock market. These findings are significant because they reveal a distinction that leads to a significant increase in the number of investors who believe it is safe to invest in the Islamic stock market in the short term, notwithstanding the crisis.

1 Introduction

Every developing country desires sustained economic growth conditions. However, in the event of an unavoidable recession caused by an economic crisis, optimal preparation and strategy are required to resolve the problem. A recession is an economic condition characterized by a decline and sluggishness in aggregate economic activity. It can be stated from the state of gross domestic product (GDP), which has decreased, or it can demonstrate that real economic growth has been negative for two consecutive quarters or for more than a year. Prolonged recessions can lead to economic depression and volatile price levels. According to reference [1], significant internal and external shocks can cause the occurrence of an economic recession, i.e. fluctuations in output (1).

Indonesia, as a developing country, was not invulnerable to the crises of 1997/1998 and 2008/2009. Currently, a recession could occur again if the data in Figure 1 show that Indonesia's economic growth was negative in the first two quarters of 2020 (BPS, 2020). At the end of 2018, economic growth was 5.18 percent in the fourth quarter. It continued to fall until the first quarter of 2020 (2.97 percent), then dropped precipitously to -5.32 percent in the second quarter of 2020.

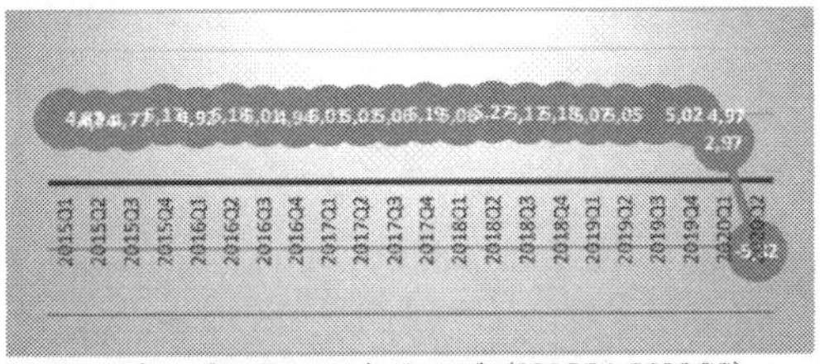

Fig. 1. Indonesia's Economic Growth (2015Q1-2020Q2).

It is known that the Covid-19 virus created a global pandemic in early 2020, which claimed many victims worldwide due to the virus's aggressive transmission. In terms of economics, the Covid-19 outbreak has had a significant impact on the fall in national income in a number of nations. Following the data presented in Figure 1, reference [2] claimed that the impact on the economy is predicted to be quite unfavorable, cutting Indonesia's estimated economic growth rate for 2020 from about 5 percent to between 4.2 and -3.5 percent. In the first two quarters of 2020, Indonesia's economic growth continued to deteriorate.

Indonesia has a dual financing system, namely conventional and sharia systems. The stock market is one of the essential sectoral of investment for Indonesia. The condition of the Covid-19 Pandemic affects all-economy segmentation, especially for productivity to get national output. Covid-19 affect many people to produce goods and service. It disturbs the health caused the people not to work, income decreases, poverty increases, and the investment sector is not interest for investors.

* Corresponding author: suriani@unsyiah.ac.id

Overcome to Covid-19 pandemic, Indonesia never stops trying to get the optimal strategies. In this case, the central bank of Indonesia (BI) used monetary instruments to control the unstable economic condition in Indonesia.

The goal of this study is to examine the state of economic fluctuations and monetary policy tools in sustaining financial system stability in the Indonesian stock market through interest rate policies. This study of the composite stock price index was conducted by the stock market, and the Jakarta Islamic Index reflects each system. During the Covid-19 pandemic, the study examined into the impact of inflation, currency rates, and policy interest rates on the composite stock price index (conventional stock market system) and the Jakarta Islamic Index (Islamic stock market system).

This research is necessary since the state of the Covid-19 pandemic has an impact on economic circumstances in all real sectors and has the potential to induce a recession. If the recession is not addressed quickly, it could lead to an economic depression. Both domestically and globally, bad things can have an impact on Indonesia's political and economic conditions. As a result, a solid strategy is required to provide the best answer for minimizing unfavorable economic conditions and avoiding a recession.

2 Literature review

2.1 Macroeconomic determinants and economic fluctuation

The rise in national output over time is a useful indicator of a country's development process. However, current situations illustrate that global economic development in both emerging and developed countries can fluctuate. Internal and external shocks, according to reference [1], might induce output volatility. It is confirmed by the findings of reference [3], who discovered that financial shocks were the primary driver of cyclical fluctuations, particularly during the Great Recession.

During a recession, macroeconomic and industrial shocks can exacerbate the situation. Positive shocks have a short-term and unfavorable effect on output growth or price rebound in the long run, according to reference [4] research. The macroeconomic impact of cyclical fluctuations in the energy industry has a stronger price effect than an output effect and a greater impact on producers than on consumers.

According to reference [5], recession circumstances can also be caused by macroeconomic policy. Their findings are that I two countries are in a long-term recession, (ii) macroeconomic policies in the most prominent Arab countries do not generate dynamic gains from trade, (iii) inflationary pressures stimulate economic growth through domestic investment, (iv) financial integration and financial development suggest that structural reforms in the banking sector and financial markets are necessary, and (v) short-term stabilization is warranted.

However, it is clear that the relationship between recessionary conditions and macroeconomic sectors, such as the labor market in the trade sector, would change more than the relationship between recessionary conditions and non-trade labor sectors. Following to reference [6] findings, the recession and subsequent financial crisis have sparked a lot of interest in the causes of unemployment variations. The tradable sector provides employment over three times quite so many employees as the non-trade sector, and wages are approximately twice as volatile.

2.2 Economic growth and pandemic recession

The results of reference [7] research explained that the Covid-19 outbreak greatly affected the decline in national income in various countries. The International Monetary Fund (IMF) predicts that China's economy will slow down by -0.4 percent. Meanwhile, global growth will slow down by -0.1 percent. In addition, reference [8] estimates that world economic growth will be hit in the range of 3-6 percent, depending on the country. In another scenario, GDP could fall by more than 10 to 15 percent.

While reference [9] estimates that the real GDP contraction in the United States (US) year-on-year fell by almost 11 percent in the fourth quarter of 2020, with the 90 percent confidence interval extending to a contraction of nearly 20 percent. Prediction for Indonesia itself from research by reference [2] estimates that the economic impact is expected to be very bad, reducing Indonesia's projected economic growth rate for 2020 from around 5 percent to between 4.2 percent -3.5 percent.

The manufacturing sector has also been affected by the outbreak of the Covid-19 pandemic. First, supply disruptions will hamper production. Second, supply chain disruptions will make it difficult for countries less affected by Covid-19 to obtain the necessary imported industrial inputs from countries hit by the Covid-19 outbreak. Third, it will be demand disruptions due to a downturn in macroeconomic conditions or a recession. The public and other economic actors tend to "wait and see" and tend to postpone the purchase of consumer goods (especially durable consumption goods) and delay investment spending by companies [10]. This research focuses on the determinants of macroeconomic variables on recessionary conditions by paying attention to macro policy—namely monetary policy through interest rates and analyzing its effect on the conventional and sharia stock markets.

3 Method

This study employs a quantitative research method based on secondary data. The monthly data for this study's sample ranges from 2011 to 2020. As a result, the total amount of data used is 120. The information comes from the Indonesian central bank (BI) and the Financial Services Authority (OJK). Ordinary least squares regression with an error correction model (ECM) approach was used as the analytical method. The ECM model has the advantage of being able to predict both long-term and short-term effects.

In this study, two equations of models are used: the first for the regression of the JCI stock market equation and the second for the regression of the JII stock market equation. All research variables must be stationary at first difference and cointegrated residuals in order for the ECM model to be chosen. The fact that the error correction term is negative is a distinguishing feature of ECM. To ensure that the regression results were valid and unbiased, the Classical assumption test was used.

This study focuses on economics, with the variables studied being economic growth and macroeconomic variables, as well as the study of Indonesian monetary policy. Using dummy variables, the analysis will be performed on economic fluctuations (recession) in Indonesia from 2011 to 2020 (monthly data) and crisis conditions during the pandemic.

3.1 Model of analysis

Multiple linear regression was used as an analytical method to establish the error correction model to be used. The model was chosen because it can predict the independent variables' short- and long-term effects on the two dependent variables over time. In this model, the stages of time series analysis are stationary data, cointegration, and the classical assumption.

To see the unit root test results, which are stationary in the same order, the Unit Roots test used in this study employs two methods, namely ADF (Augmented Dickey-Fuller) and PP (Phillips Perron) [11]:

$$y_t = \rho\, y_{t-1} + \varepsilon_t \qquad -1 \leq \rho \leq 1 \quad \ldots\ldots\ldots\ldots \quad (1)$$

In equation (3), the variable *t* represents the stochastic error term as defined by the classical assumption, implying that the difference (variance) constant is zero, and y represents the time series. The data for unit roots is not stationary at level or I(0), if the *p*-value is greater than one, the data should be tested at the first difference level, or I(1) level. The first difference is that if the y_t data is not stationary, it will regress the lag value of one period, y_{t-1}. The data is stationary if the first difference value of p smaller than 1. The optimal lag test is then accomplished.

Cointegration is a useful notion for two or more non-stationary time series variables that will be cointegrated if the combination is also linear over time. It is possible, however, that each variable is not stationary. Starting with a long-term partnership, reference [12] illustrate the cointegration:

$$y_t = a + b1\, x_t + \varepsilon_t \quad \ldots\ldots\ldots\ldots\ldots\ldots\ldots\ldots \quad (2)$$

Cointegration is included in the ECM model, according to Granger's Representation Theorem, which means that if xt ~ I(1), yt ~ I(1), t ~ I(0), the ECM equation can be written as follows [13]:

$$\Delta y_t = \alpha + \textstyle\sum_{i=1}^{k} \beta i \Delta y_{t-1} + \sum_{i=0}^{k} \gamma \Delta x_{t-1} + \phi \varepsilon_{t-1} + \varphi t \; \ldots (3)$$

Where < 0 must have a negative, or a valid ECM must be negative and statistically significant. It means that the cointegration is convergent. The ECM model is created by incorporating the ECT value into the regression equation, as shown below:

$$Y_t = \mu_i + \textstyle\sum_{i=1}^{n} A_i \Delta Y_{t-1} + \sum_{i=1}^{n} \lambda\, \Theta_{t-1} + v_t \; \ldots\ldots\ldots\ldots (4)$$

The two equation models studied in this study can be written as follows:

Model 1:

$$\Delta CSPI_t = \alpha_1 + \gamma_1 \Delta ER_t + \gamma_2 \Delta INF_t + \gamma_3 \Delta BIr_t + \gamma_4 \Delta D_t + \gamma_5 ECTcspi_{t-1} + \varepsilon_t \; \ldots\ldots\ldots (5)$$

Model 2:

$$\Delta JII_t = \alpha_1 + \gamma_1 \Delta ER_t + \gamma_2 \Delta INF_t + \gamma_3 \Delta BIr_t + \gamma_4 \Delta D_t + \gamma_5 ECTjii_{t-1} + \varepsilon_t \ldots\ldots\ldots (6)$$

Where model 1 refers to a study of the sensitivity of changes in the conventional stock market as represented by the CSPI. Model 2 depicts the sensitivity of changes in the Islamic stock market via JII as a result of economic fluctuations and crises brought on by the Covid-19 outbreak in Indonesia. The validity of the regression results was then double-checked using the classical assumption test.

4 Results and discussion

Statistic descriptive of the data (Tabel 1) shows that the maximum value of exchange rate (ER) is IDR 16367, with an Inflation rate of 8.7%, BIR 7.7, conventional stock of 6605, and JII 2376039. While the minimum value of the exchange rate (ER) is IDR 8532, with an Inflation rate of 1.32%, BIR 3.75, conventional stock 3409, and JII 1016725. The fluctuations of the data shown from the changes of the maximum and minimum value of the data reach until 50% from 2011 to 2020 (monthly).

Table 1. Statistic descriptive of data.

	ER	INF	BIR	CSPI	JII
Mean	12463.38	4.5	5.9	5073.7	1869297.
Median	13293.00	4.0	5.7	5077.5	1906429.
Max	16367.00	8.7	7.7	6605.6	2376039.
Minimum	8532.000	1.3	3.7	3409.1	1016725.
Std. Dev.	2186.104	1.8	1.2	834.30	309674.7
Obs	120	120	120	120	120

Whereas Model 1 describes the analysis conducted to determine the sensitivity of changes in the conventional stock market, Model 2 describes the sensitivity of changes in the Islamic stock market as a result of economic fluctuations and crises caused by the Covid-19 pandemic in Indonesia. This study relied on classic assumptions such as normality, multicollinearity, heteroscedasticity, and autocorrelation tests. Table 2 shows the results of the data stationarity test.

At the level of either the ADF-Fisher or PP-Fisher tests, all study variables are not stationary, as shown in Table 1. The ADF-Fisher and PP-Fisher tests, on the other hand, show that all are stationary at the first difference level. As a result, using these data as a guide, choose the correct ECM model, which is supported by the ECT value of Model 1 and Model 2, which both exhibit a substantial negative value on the stationary test at level or I(0). Because the residual variable (ECT) is cointegrated, it confirms that ECM regression may be done. Then, by analyzing the residuals and between the data's independent variables, utilize the classic assumption test to verify that the data is legitimate.

Table 2. Data stationarity test results.

Variables	ADF-Fisher		PP-Fisher	
	I(0)	I(1)	I(0)	I(1)
CSPI	0.1578	0.0000***	0.2760	0.0000***
JII	0.1510	0.0000***	0.1471	0.0000***
INF	0.1624	0.0000***	0.3513	0.0000***
ER	0.6125	0.0000***	0.7501	0.0000***
BIR	0.8573	0.0000***	0.7651	0.0000***
ECT_model1	0.0033***	-	0.0025***	-
ECT_model2	0.0003***	-	0.0002***	-

Description: (.) P-Value. Level of significance ***(1%), **(5%), *(10%)

Table 3. Classic assumption test results.

Tests	The Name of Tests	Model 1	Model 2
Normality	Histogram	0.1804	0.5011
Multicollinearity	VIF	< 10	< 10
Autocorrelation	Correlation LM test	0.1250	0.4583
Heteroscedasticity	Heteroscedasticity test	0.0731	0.2862

Description: (.) P-Value. Level of significance α di atas (5%)

Table 3 shows the results of the classical assumption test, which show that the significance of all test findings is greater than 5%. As a result, it may be stated that the data utilized in both models pass the classical assumption test. Then, as shown in Tables 3 and 4, the Model 1 and Model 2 regression results can be computed.

Table 4 shows that the regression findings in Models 1 and 2 (Conventional Stock Market (CSPI) and Sharia index (JII)) indicate no difference. The stock market's composite index is unaffected by inflation. On both stock markets, the exchange rate has a favorable impact. It indicates that if the rupiah exchange rate rises, the asset value of the two stock markets rises as well, resulting in a higher share offering. It is based on the Tobins'q theory [14] and earlier study (King & Kusairi, 2019), which describes the expectation of asset value on investment [15]. The JCI and JII stock markets, on the other hand, were negatively affected by policy interest rates and crisis conditions. It indicates that if policy interest rates rise and a long-term crisis emerges, the value of assets in the CSPI and JII stock markets will be reduced.

Table 4. Long-run regression test results.

Variables	Model 1 (CSPI)		Model 2 (JII)	
	Coefficient	Prob	Coefficient	Prob
C	3126.218	0.0000***	998514.7	0.0000***
Inf	-20.45521	0.5750	4566.425	0.7635
Er	0.2883	0.0000***	107.178	0.0000***
BIr	-244.3646	0.0000***	-79961.2	0.0008***
D	-0.969.6137	0.0000***	-336087.5	0.0000***

Description: (.) P-Value. Level of significance ***(1%), **(5%), *(10%)

Table 5's regression results show a short-term to long-term equilibrium relationship in both stock market models, as evidenced by the negative and significant ECT value [16]. Long-term regression results for the two stock market models partially explain why policy interest rates have a negative effect on the stock market index. This is supported by previous research, which shows that interest rates have a significant negative relationship with stock prices [17], and interest rate growth has a significant negative relationship with stock price growth as well [18]. Crisis conditions have a negative effect on the stock market. t is consistent with previous research by reference [19] which stated that the stock market would suffer as a result of the Covid-19 pandemic. Their findings, however, do not match the results of this regression for the exchange rate, which has a positive impact, and inflation, which has a negative impact.

Table 5. Short-run regression test results.

Variables	Model 1(CSPI)		Model 2 (JCI)	
	Coefficient	Prob	Coefficient	Prob
C	26.0007	0.1092	9219.619	0.2167
dInf	-29.2988	0.3205	-1067.14	0.4375
dEr	-0.2530	0.0000***	-86.7332	0.0001***
dBIr	-109.0788	0.1607	-33040.94	-244.3646
dD	237.6544	0.1714	216970.3	0.0075***
Ect	-0.0911	0.0205	-0.1322	0.0025

Description: (.) P-Value. Level of significance ***(1%), **(5%), *(10%)

In both markets, the conclusions are nearly identical in the short term, meaning that inflation and policy interest rates have no influence. The exchange rate has a negative impact on the stock market; this finding is consistent with reference [20] and reference [17] research findings. However, there are short-term changes in the findings on the crisis variable. The crisis had little impact on the CSPI stock market. The crisis variable, unlike the JII stock market, has a positive effect. This conclusion is significant because it reveals disparities in the behaviour of investors who believe that, despite the crisis, it is safe to continue investing in the Islamic stock market in the short term.

5 Conclusion

The cycle of economic activity is indicated by the conditions of economic fluctuations. As a result, monetary policy is required to overcome unfavourable financial conditions. For the long term, the findings are the same in both stock market models (conventional and Islamic stock markets). For the time period under consideration, it appears that inflation has no effect on the two stock markets. It implies that it has the ability to control the level of economic fluctuations in the event of price increases. The stock market benefits from changes in exchange rates. It demonstrates that the exchange rate is closely related to the value of stock market assets. The stock market suffers as a result of interest rate policies and crisis conditions.

In this case, the Covid-19 pandemic could cause a decline in the stock market's stock price index. Long-term investors prefer not to invest in the event of a crisis. In an increase in interest rates, monetary policy through interest rates can also reduce demand in the stock investment market. In order to maintain price stability, interest rates are raised when inflation occurs.

Another important short-term finding is that exchange rates have a negative impact on the two stock markets. The crisis variable, on the other hand, has a positive impact on the JII stock market. It reinforces the difference in investor behaviour when deciding whether or not to invest in times of crisis in the short term. Investors will choose to invest if they are confident in the safety of their assets. Policy recommendations based on these findings are required in order to develop effective strategies.

References

1. Rasaki MG, Malikane C. Macroeconomic shocks and fluctuations in African economies. Econ Syst [Internet]. 39(4):675–96 (2015)
2. Suryahadi A, Al Izzati R, Suryadarma D. The Impact of COVID-19 outbreak on poverty: An Estimation for Indonesia (Draft). SMERU Work Pap [Internet] (2020)
3. Altinoglu L. The origins of aggregate fluctuations in a credit network economy. J Monet Econ [Internet] (2020)
4. He Y, Lin B. Time-varying effects of cyclical fluctuations in China ' s Energy. Energy (2018)
5. Ramzi K, Asma M, Chebbi A. Growth, fluctuations and macroeconomic policies: Evidence from Arab open economies. Econ Anal Policy [Internet] (2017)
6. Zhang Y. Unemployment fluctuations in a small open-economy model with segmented labour markets: The case of Canada. Q Rev Econ Financ [Internet] (2018)
7. McKibbin WJ, Fernando R. The global macroeconomic impacts of COVID-19: Seven Scenarios. SSRN Electron J (2020)
8. Fernandes N. Economic effects of coronavirus outbreak (COVID-19) on the world economy Nuno Fernandes Full Professor of Finance IESE Business School Spain. SSRN Electron Journal, ISSN 1556-5068, Elsevier BV,. (2020)

9. Baker S, Bloom N, Davis S, Terry S. COVID-induced economic uncertainty. Natl Bur Econ Res, (2020)
10. Baldwin R, Mauro BW di. The Economic Effects of a Pandemic [Internet]. Economics in the Time of COVID-19. CEPR Press; (2020)
11. Gujarati DN. Basic Econometrics. McGrow Hill; (2003)
12. Moosa IA, Vaz JJ. Financial markets, institutions & money cointegration, error correction, and exchange rate forecasting q. J Int Financ Mark Institutions Money [Internet]. (2016)
13. Granger CWJ. Investigating Causal Relations by Econometric Models and Cross-Spectral Methods. Econometrica (1969)
14. Chappell HW, Cheng DC. Expectations, Tobin's q, and investment: A Note. J Finance (1982)
15. Qing YK, Kusairi S. The Effect of money supply, exchange rate, and interest spread toward the performance of stock market in Malaysia. Widyakala J (2019)
16. Miller SM. Monetary dynamics : An Application of cointegration and error-correction modeling. Journal of Money, Credit and Banking, (1991)
17. Khan MK. Impact of exchange rate on stock returns in Shenzhen Stock Exchange: Analysis through ARDL Approach. Int J Econ Manag (2019)
18. Uddin G, Alam MM. The Impacts of interest rate on stock market: Empirical evidence from Dhaka Stock Exchange. South Asian J Manag Sci Vol 4(1), pp 21-30, (2010)
19. Rahmayani D, Oktavilia S. Does the Covid-19 pandemic affect the stock market in Indonesia? J Ilmu Sos dan Ilmu Polit, (2020)
20. Suriani, Majid MSA, Masbar R, Wahid NA. Macroeconomic determinants of the capital market in Indonesia: A Comparative analysis between sukuk and bonds markets. Int J Acad Res Econ Manag Sci, (2018)

The Covid-19 pandemic impact on community mental health

Rosalia Putri[1*], *Nurul* Fajri[1], *Safrida* Hanum[1], *Misra* Hanum[1], *Sofia* Sofia[2], *Husnah* Husnah[2], and *Marty* Mawarpury[3]

[1] Master of Public Health Study Program, Faculty of Medicine, Universitas Syiah Kuala, Banda Aceh, Aceh, 23111, Indonesia
[2] Faculty of Medicine, Universitas Syiah Kuala, Banda Aceh, Aceh, 23111, Indonesia
[3] Psychology Department, Faculty of Medicine, Universitas Syiah Kuala, Banda Aceh, Aceh, 23111, Indonesia

Abstract. The impact of the pandemic is felt by people all over the world. This is caused by stigmatization, discrimination, positive experiences of Covid, and disruption of social support. The purpose of this systematic literature review was to find out the psychological impact during the pandemic on the community. This systematic literature review was compiled by searching for articles from 2020 to 2021 contained in health databases from several journals such as Elsevier, BMC, Index, Cambridge University Press, Dove Press, and Wiley which was performed electronically. The inclusion criteria included in this study were focusing on the impact of the Covid-19 pandemic and mental health. From 20 articles, it was shown that psychological factors also played an important role in adherence to preventive measures in public health (such as vaccination) and how people deal with the threat of this virus infection with good health protocols. A person with a low level of education had a risk of being prone to depression and anxiety. This was because if someone with a high level of education had extensive knowledge about virus transmission so that they were better able to protect and cope with the anxiety and depression. Theoretically, the implication of this study is to enrich insight related to the impact of the Covid-19 pandemic on community mental health. While practically, this study might empower the community to maintain a healthy lifestyle during the pandemic that can prevent mental disorders. We hope this study will also empower stakeholders to create a policy that works in facilitating mental health assistant during the pandemic, especially for the Covid-19 patients in the community.

1 Introduction

The Covid-19 pandemic that has lasted for more than a year leaves a huge impact on all aspects of life. One of the impacts is the mental health of the community. The impact of the pandemic is felt by people all over the world. This is caused by stigmatization, discrimination, positive experiences of Covid, and disruption of social support, these are included in short-term mental health disorders. In addition, losing a job due to the pandemic also causes long-term mental health problems. Mental health is influenced by environmental factors and the current situation, where the pandemic has caused the situation to change drastically from usual, today's society is forced to be able to deal with pressures that occur from the social, economic, educational, and cultural sides [1].

Health is a state of physical, mental, and social well-being. This understanding implies that physical, mental, and social health are in balance. The pandemic, which is still ongoing and not sure when it will end, has had a huge impact on people's lives, both physically and mentally. The massive issue of Covid-19 in all media at the beginning of the pandemic has caused confusion, fear, anxiety, anxiety, and panic among the public. The number of deaths due to Covid-19 is increasing anxiety. The act of self-isolation, work from home, online learning systems cause obstacles in social interaction in society which in turn causes increasing frustration and stress for students, college students, and office workers [2].

The pandemic also has a negative impact on the economy. Many companies have closed and laid off their employees. The number of unemployed increased in all regions, the feeling of anxiety about how the family's financial fate was felt by workers who became the backbone of the family. These workers feel confusion, anxiety, frustration, fear, feelings of helplessness and can even trigger insomnia [3]. Starting to feel inferior because they have no income.

*corresponding: putri.rosalia96@gmail.com

Teenagers and children who have to follow the online learning system feel bored and constrained because they cannot interact with their friends. Children and adolescents who are in the process of growing and developing need self-appreciation through a series of interactive activities with their friends, tutors, school teachers, dance coaches in the studio, or coaches of the sports they are involved in. The pandemic that requires restrictions on activities outside the house has changed the interaction patterns of these children and adolescents [3]. This will cause feelings of laziness, boredom, and constraint which in turn will trigger stress. Women who play an important role in every household also feel the impact of the pandemic on their mental health. These mothers feel anxious and afraid of contracting the virus, worried about the financial fate of the family when their husbands are also affected by layoffs. How to manage family finances, how to still provide nutritious food for family members in the middle of an economic problem, how to keep family health guaranteed. Not to mention the fear of thinking that if you catch the virus you have to self-isolate or worse if you have to be treated in a hospital and when you return home you will be ostracized by friends, relatives, and neighbors. This confusion, anxiety, fear, feeling of helplessness, boredom, feeling of restraint and low self-esteem are pieces of evidence that the pandemic has caused mental health problems in all levels of society. However, not everyone can manage the level of stress and anxiety they are facing. For those who cannot manage anxiety, it will result in mental disorders that will greatly interfere with daily activities. Therefore, special attention is needed so that all people can go through this Pandemic without mental health problems [4]. The purpose of this systematic literature review was to find out the psychological impact during the pandemic on the community, especially the occurrence of mental health disorders in the community so that this impact could be anticipated as early as possible.

2 Methods

This systematic literature review was compiled by searching for articles from 2020 to 2021 contained in health databases from several sources such as Elsevier, BioMed Central (BMC), Cambridge University Press, Dove Press, and Wiley. The inclusion criteria included in this study were focused on the impact of the Covid-19 pandemic and mental health, which were divided into three groups: children, adults, and Covid-19 patients. Articles searching was done by using keywords: Covid-19, mental health, mental illness, societal level effect, Covid-19 patients, insomnia, anxiety, stigmatization, and fear. While for the Indonesian article we used databased from E-resources Perpusnas and LIPI. The keyword used was: Covid-19, stres psikologis, remaja, dewasa, anak-anak, dan kecemasan, dan gangguan mental. Each article was screened by its title and abstract.

Based on the references obtained, the research or studies included in this paper meet the criteria as followed:

(1) Studies focusing on general public health (e.g. Children, adults, and Covid-19 patients).
(2) Articles focusing on the effect of Covid-19 towards mental health disorders, social and government support for this mental health issue during the pandemic.
(3) Including studies from countries across the world.
(4) The article is written in English and Indonesian

From the total of 60 articles that were obtained, only 20 articles met the inclusion criteria, and the other 40 articles were excluded, including 5 articles with abstracts only without providing full papers.

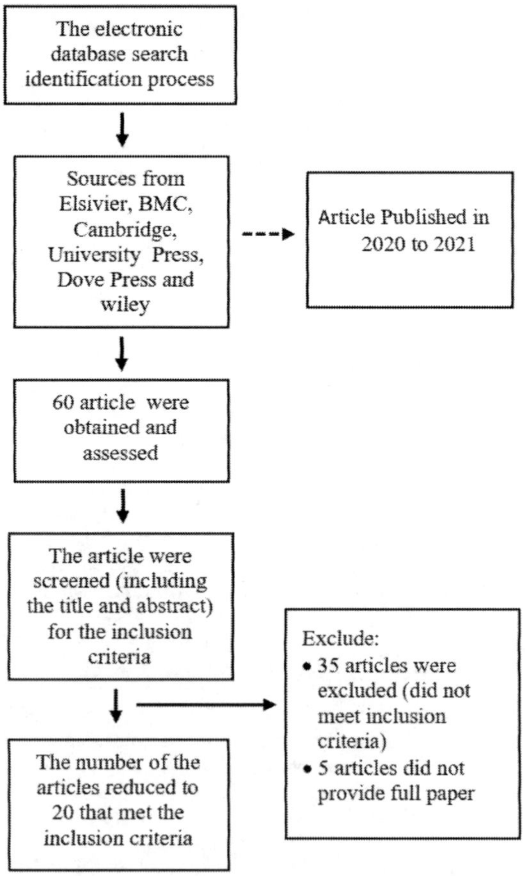

Fig. 1. Flow diagram of the systematic review procedure

3 Results

A systematic literature review study of 20 articles showed that the current state of the pandemic could have an impact on people's mental health. Among them appear the impact of post-traumatic stress symptoms including anxiety, confusion, and anger. Broadly speaking, the psychological impact felt by Covid-19 patients and the wider community in the form of feelings of depression, stress, and anxiety. The impact of this community's mental health mostly begins with stress, worries and fears, then causes conflicts in the family and the environment. The pandemic also has an impact on the incidence of insomnia in the community and worsens the symptoms of patients who have been diagnosed with insomnia. This happened because of people's fear of the Covid-19 disease and there were many changes in their daily routines during the pandemic. There is a lot of pressure that is felt during a pandemic, where fear and anxiety occur not only for yourself but for those closest to you; there are those who feel changes in sleeping patterns and changes in eating patterns, feeling depressed and having difficulty concentrating, this situation arises because of limited mobilization and being constantly at home, especially children, resulting in psychotic disorders.

From several studies, it was stated that women and the younger generation were at higher risk of anxiety and depression compared to men and older people, this was based on research in China during the pandemic. However, according to the World Health Organization (2020) survey in 2020, the number of depression increased in the elderly because it was accompanied by various comorbidities. Psychological factors also played an important role in adherence to preventive measures in public health (such as vaccination) and how people deal with the threat of this virus infection with good health protocols. A person with a low level of education had a risk of being prone to depression and anxiety. This was because if someone with a high level of education had extensive knowledge about virus transmission so that they were better able to protect and cope with the anxiety and depression, so that depression and anxiety were at lower levels.

In addition, someone who had received information about the virus from articles and journals or even from social media sources would reduce the level of depression and anxiety of course by presenting readers who can be trusted in conveying information. The peak of psychological reactions included maladaptive behavior, emotional distress, and defensive responses. Kristen.R Choi predicted that there was a risk that the outbreak would create a "second wave pandemic" of mental health crisis in the health system and society. Thus, the public health response must be comprehensive to the pandemic, which includes (a) attention to the psychological aspects of hospitalization for patients, families, and staff affected by Covid-19; (b) emergency psychiatric patient care planning if the hospital is overwhelmed with Covid-19 patients; and (c) innovation to provide mental health care in the community while social distancing is required and health system resources are being strained. According to Agus Kusnayat et al, in terms of the world of education, it had quite an impact on the mental health of students, where learning conditions suddenly had to change from face-to-face to an online system resulting in students objecting to the tasks given by the teacher which results in the stress level of students increasing by 60%. If this was allowed to continue, it would be fatal in the mental development of students [9].

4 Discussions

During the last twenty years, diseases caused by viruses that continue to emerge cannot be denied this has become a serious problem for health. The latest disease caused by a virus, namely the covid-19 virus, which initially appeared in the city of Wuhan that is the largest metropolitan area in the Hubei Province of China, and spread rapidly to other countries became a pandemic. This virus that attack the respiratory tract was originally classified as pneumonia of unknown etiology, this case was first reported to WHO on 31 December 2019 [6]. The impact of the pandemic has certainly caused anxiety. This condition is a normal reaction toward a threatening and unexpected situation such as the pandemic. Stress during the pandemic can occur due to changes in concentration, irritability, lack of productivity. To suppress the spread of the virus, various countries also implement a system quickly. Due to these policies many people experience psychological effects that increase stress symptoms, the severity of stress symptoms also differs depending on the duration and period of negativity, feelings of loneliness, excessive fear and stigmatization [6].

The impact of the pandemic on mental health in each population is different depending on how to prevent and treat the problems they experienced. In the United States, it is stated that 45% of adults experience anxiety and stress, this number continues to increase, followed by exposure to the virus. Also in the UK, about 33% of people experience high levels of anxiety during the pandemic. Meanwhile, in Italy, based on survey results, the percentage of post-traumatic syndrome reached 37%, 21% stress, 20% severe anxiety, 17% depressive symptoms and 7% insomnia [2]. The current pandemic gives a significant threat to public health including the impact of mental health on society.

4.1 The impact on childrens

This pandemic condition gives children the feeling of boredom and stress because they are constantly at home, where children's daily routines usually play outside the house with their friends [4]. Millions of children have been adversely affected by the pandemic, the greatest impact being felt in low socio-economic populations. Of course the impact is felt in terms of positive impacts and negative impacts. The positive impact on the stage of child development where children can learn new skills, develop self-ability, increase self-confidence. Cultivating great awareness regarding unknown diseases and improve health. Parental support is very important in providing a good understanding of behavior that can maintain personal safety and the safety of others. Developing relationships and a sense of empathy in family interactions to become closer, form bonds of love and affection and develop a sense of loyalty toward surrounding people [7].

The negative impacts felt by children include the impact on education, during the pandemic, face-to-face learning at school has been shifted to online. This gave a serious impact on children's education such as decreasing quality of knowledge, learning does not take place effectively, interference with internet network access, a fear of slow learning to the next stage, related to health problems, lack of physical activity outside the room, lack of nutritional intake, susceptible to various health disorders. Changes in aggressive behavior more easily such as irritability, frustration, and anxiety. Addiction to social media and internet access increases the risk of child exploitation [7]. Kusnayat et al. [8] in their research found a close relationship between online learning system and the mental attitude of the students participating in the class. Around 60.5% of students from universities who are the object of research were ready to adapt to online lectures, although there were some who found it difficult to use the applications used as many as 32.5% but around 47.5% were not ready to adapt. This condition would create psychological pressure for students. Survey data showed 59.5% objected toward assignments given by lecturers which resulted in increased stress levels for students so that they had difficulty to sleep. If this continues, it would be fatal in the psychological development of students, the impact would be study failure or drop out and other things that are more severe [9].

In teenagers, which is a period of transition from children to adults where this period is marked by the occurrence of emotional and behavioral development as well as biological, psychological and social changes [18]. In this group of age, interaction with peers doing various activities outside the home is very popular and plays an important role in the process of self-actualization and is a characteristic of teenagers identity. During this pandemic period where large-scale social restrictions (LSSR) are imposed, it is certainly very disturbing for teenagers because they cannot interact with their peers anymore. The implementation of LSSR, social distancing, online school system causes teenagers to feel isolated, restricted, bored, anxious, and sad. Sudden loss of activities outside the house with peers such as school, sports activities, art activities can result in the emergence of depressive symptoms in teenagers which of course will result in disruption of mental health [18].

4.2 The impact on adults

Sudarsana et al. [1] in his book mentioned that there were positive and negative psychological impacts on society during this pandemic. The positive psychological impacts of some individuals include having positive experiences such as feeling proud when they find ways to overcome pressure and survive during the pandemic, the community also shows an act of sacrifice and cooperation with one another in facing this pandemic condition so that they can feel the goods that comes from helping others. While the negative psychological effects can include post-traumatic stress symptoms including anxiety, confusion, and anger [1]. According to Brook et al (2020) there are several impacts of the pandemic on community psychology in the form of post traumatic stress disorder (PTSD) such as anxiety, confusion, fear, frustration, feeling helpless, insomnia [11]. The high mortality rate in the community, the implementation of social restrictions in an area has triggered depression, anxiety, excessive fear to changes in people's sleep patterns. All these things that happen will worsen the physical and mental state of the community [16].

Rifani et al 2020 in their research on 50 samples related to emotional and mood instability during the pandemic, this is the main factor causing individual mental disorders during the pandemic. A healthy mental state is closely related to physical health conditions. During the pandemic in Indonesia, there was a 20% increase in the number of mental disorders (Tugurejo Hospital Semarang from Ratih Widayati Sp.KJ) quoted from JAWA POS Radar Semarang, 2020: 10. The increase in the number of patients was the majority of the causative factors, among others, were the layoffs and people with mental disorders due to the pressure of too long to undergo a quarantine period at home then lead to conflicts in the family such as domestic violence or even divorce. Based on the results of the questionnaire, many respondents felt fear, either fear of the news that appears in the media or fear of social conflicts that occured during the pandemic. News related to the pandemic certainly has an impact on people's emotions and moods, one of which is feelings of worry which are then followed by feelings of fear. However, with many opportunities to gather with family during quarantine at home, it is enough to have a positive impact on mental health, namely feelings of pleasure, because we have the opportunity to support each other in the difficult times they are going through [2]. According to Setyaningrum & Amildha (2020) the massive layoffs that occurred in Malang City were the biggest factor in the occurrence of anxiety and stress in the people of Malang City which resulted in disturbances both to psychological conditions and to the mental health of the people of Malang City [19]. A Survey of Family Resilience in the Pandemic Period conducted by the Bogor Agricultural University (IPB) showed that for married women, the most common psychological disorders experienced during the pandemic were anxiety and restlessness, sadness and difficulty in concentrating [20].

Yulia (2020) in a review article Insomnia during the Covid-19 Pandemic not only had an impact on physical health, but also had an impact on mental health, which would affect the quality of individual sleep pattern and even cause insomnia or worsen existing insomnia symptoms. Fear and anxiety about being infected with Covid-19 can affect people's sleep quality and sleep patterns. Isolation at home during the pandemic has also changed people's daily routine habits. This irregular schedule of activity and rest will affect the circadian rhythm of society. Therefore, it is necessary to intervene in preventing and overcoming insomnia and reducing insomnia symptoms in the community during the pandemic [3]. Ilpaj et al [4] in a literature review Analysis of the effect of the Covid-19 death rate on the mental health of people in Indonesia stated that the large number of people who were infected and people die because of the Covid-19 will have a major impact on mental health. This is illustrated because the majority of people in isolation in their own condition can cause a person to be in an uncomfortable situation. Government regulations regarding social distancing obligations also create emotional distance between family, friends, friends, or people in places of worship who can support each other. Furthermore, with the feelings of loneliness and isolation can increase the suicide rate. In addition to excessive fear and anxiety disorders and changes in sleeping patterns, eating patterns, feeling depressed and having difficulty concentrating, there is also an increased risk of drug and alcohol abuse. People who are in a state of excessive stress and anxiety can stimulate the individual to consume alcohol and drugs, this bad habit must be stopped immediately, because it will interfere with the physical and mental health of the individual [4].

Rosyanti et al (2020) stated that quarantine taken by individuals and communities would have negative psychological impacts such as stress symptoms. The severity of symptoms depends on the duration and extent of quarantine, feelings of loneliness, fear to be infected, inadequate information, and stigma on more vulnerable groups including psychiatric disorders, health workers, and people with low socio-economic status [7]. Cullen et al (2021) in Commentary Mental health in the Covid-19 pandemic, It is known that psychology is a factor that plays an important role in community compliance with public health efforts (such as vaccinations) and health protocols, this is an important issue in efforts to control any infectious disease, including Covid-19. Psychological symptoms that have emerged in the middle of the pandemic include maladaptive behaviors, emotional distress, and defensive responses. People who are prone to psychological problems are especially at risk. A study on 1210 respondents from 194 cities in China in January and February 2020 noted that 54% of respondents reported the psychological impact of Covid-19 with moderate or severe symptoms; 29% reported moderate to severe anxiety symptoms; and 17% reported moderate to severe depressive symptoms. Despite the possibility of bias, it was a very high proportion that some people were likely to be at greater risk [9].

The elderly group is also vulnerable to the negative impacts of the pandemic. The elderly are prone to be infected by the Covid-19 due to the degenerative process that occurs in the elderly and causes a decrease in body resistance. In general, the elderly already have comorbidities such as hypertension, diabetes mellitus, heart disease so that the elderly are very vulnerable to be infected. Anxiety about the spread of the Covid-19 has a bad influence on the strength and physical health of these elderly people [20].

4.3 The impact on Covid-19 patients

Sudarsana et al. [1] in his book stated that the impact of mental health on Covid-19 patients could be in the form of psychological disorders such as feelings of depression, stress, and anxiety. This impact was referred to as a stressor. These stressors are divided into two, namely [1]:

1. Stressor during quarantine

Quarantine duration for 14 days can lead to increased mental disorders, especially PTSD (Post Traumatic Stress Disorder), avoidant behavior, and anxiety. Like the other people, they also experience fear where there is a fear that the infection they experience will infect others. This fear mostly occurs in pregnant women and those with children. In addition, there is also frustration and boredom due to them stopping their daily routines and reducing physical and social contact with other people, as well as feeling isolated from the outside world. Covid-19 patients also feel worried about the supply of their basic needs such as food, drinks, clothing, or other accommodation, including medicines to personal protective equipment, plus related information from the government which is inadequate both in quantity and quality. The information includes guidelines for steps in efforts to overcome Covid-19, the level of risk, and the purpose of quarantine. This inadequate information is one of the predictors of the onset of PTSD symptoms.

2. Post-quarantine stressor

After the quarantine is completed, the next post-quarantine stressor will occur including financial stressors, namely a decrease or loss of income sources, especially for workers with low income levels. This can be a trigger factor for the emergence of mental health disorders such as the risk of anger and anxiety as well as the development of socio-economic distress in the months after quarantine. In addition, stigmatization also occurs, which is a negative characteristic attached to a person's personality due to the influence of his environment according to Ministry of National Education (2008). Stigma refers to people who are quarantined, so that people are treated differently, avoided, not invited to social events, considered scary and dangerous, infectious, and criticized.

Rosyanti et al (2020) stated that quarantine could exacerbate pre-existing conditions such as depression, anxiety, and post-traumatic stress disorder. In addition, the risk of disease transmission might increase the fear of contamination in patients with obsessive-compulsive disorder and hypochondria, or individuals with a history of paranoia. Although quarantine protect against the spread of the Covid-19, their isolation process creates psychosocial stress that can trigger or worsen mental health disorders [5]. In addition, stigma has begun to emerge in the community towards Covid-19, such as discrimination against sufferers and families of Covid-19 sufferers and rejection. People are afraid to be infected by the Covid-19 so many people stay away from them and isolate the sufferers and their families. This certainly adds a mental burden to the sufferer where the sufferer feels sad, feels alone, feels afraid of transmitting the disease to others [16]. Stigma that is formed in society can cause people to hide illness to avoid discrimination, prevent people from seeking immediate health care and prevent people from adopting healthy behaviors. Such a huge effect of the stigma on the sufferers. Covid-19 sufferers, who should receive moral support from their family and community in order to recover quickly, actually get an additional mental burden that can worsen the patient's condition [16].

4.4 Strategies for maintaining mental health during the pandemic

Maintaining mental health must actually involve efforts that can be made to survive stress and can be used as a reference in maintaining mental health during the Covid-19 pandemic as follows:[1,4,8,9]

1. Taking selfcare to maintain immune system and emotional state.
2. Seeking social support such as communicating online with family and friends can provide a feeling of social support for each other during isolation.
3. Helping others by giving goods according to the needs of others and praying for them.
4. Limiting yourself to access to social media, it is not uncommon for news on social media to become a source of anxiety. Therefore, we must be wise in receiving information, choose official and trusted sources of information from the government.
5. Focus on things that can be controlled.
6. Doing light exercise at home such as jogging or push-ups and so on. Cleaning the house can make the body produce endorphins that can relieve stress, reduce worry and improve mood. Sun bathing in the morning to boost the immune system.
7. Consuming nutritious food because adequate nutritional intake can maintain mental health. Such as foods that contain protein, healthy fats, carbohydrates, vitamins, minerals, and fiber.
8. Building good relationships with family and friends, by working from home, we can have more free time to communicate with family, friends and co-workers via telephone, video calls and virtual. By telling stories about conditions and joking with each other, the stress and anxiety felt during this global pandemic can be reduced.
9. Doing meditation to control anxiety. This is a simple technique to train the center of the mind and increase a calm feeling, regulate breathing slowly and regularly for about 15-20 minutes, reduce negative suggestions, train to control stress and see things from the positive side.
10. Special attention should be paid to government workers, health care administrators, support staffs, and food delivery workers. They consistently had a higher risk of chronic PTSD than other groups. This group mostly has no experience of disaster response and they continue to do their job well, often not by choice but because of economic necessity.
11. Recommendations for the provision of psychological intervention services intended for communities affected by Covid-19. Special support is needed for people at high risk of psychological morbidity. Raising awareness and early diagnosis of mental disorders as well as increasing access to psychological interventions can help reduce and prevent mental health disorders in the future.

Kristen et al (2020) stated that the threat of a second wave pandemic of mental health crises must be addressed as part of a comprehensive public health response to Covid-19. Evidence suggested that survivors of ICU care were at high risk for posttraumatic stress disorder, depression, sleep disturbances, poor quality of life, and cognitive dysfunction. Collaboration among health system leaders from hospitals and community mental health systems must be done to ensure the safety and continuity of care for individuals experiencing mental health crises. In addition, the health system should offer support mechanisms for mental health services including interventions for patients and families (e.g., providing trauma-information ICU care, providing mental health and education resources after discharge, supporting virtual visits) and for staff (e.g., work mental occupational health resources, support for illness and family leave, ensuring adequate staff, and a safe practice environment with adequate personal protective equipment. Various innovations should be developed in the community such as the Telehealth model (including insurance coverage for telehealth), adequate drug supply, increased mental health training for officers [5]. Stuijfzand et al. [10] mentioned that social predictors in the context of a pandemic, namely organizational support and family/friend support could serve as protective factors from mental health disorders, social isolation could have an impact on poorer mental health.

5 Conclusion

The current Covid-19 pandemic has a huge impact on mental health of the community and health providers. This was related to stress reactions that include changes in concentration, anxiety, insomnia, reduced productivity, and interpersonal conflicts. Health providers has also experienced more severe mental disorders due to separation from family, abnormal situations, increased exposure, fear of Covid-19 transmission, feelings of failure to handle a poor prognosis, inadequate technical facilities such as personal protective equipments and other equipments used to treat the patients. In addition, women, children, and the elderly were also at high risk for mental health disorders during the pandemic. To prevent mental health disorders in the long term, appropriate interventions and approaches are needed to restore concerns about personal health, excessive fear of being infected with covid-19 and anxiety about transmitting the virus to family or others. Strategies for maintaining mental health during the pandemic are needed by applying many positive activities in daily routine. In addition, comperehensive cooperation between public health and other stakeholders are important in creating innovative strategies to prevent worsen mental health disorder among public during the pandemic.

6 References

1. I. Sudarsana, et al, Yayasan Kita Menulis, **1**, 27-32 (2020)
2. D.A. Rifani, Rahadi, D.R., J Manaj Bisnis. **18**, 22–34 (2021)
3. D.Y.P. Haryanti, J Ilm Kesehat Sandi Husada, **9**, 1111–6 (2021)
4. S.M Ilpaj, Nurwati N., Focus J Pekerj Sos. **3**, 16 (2020)
5. K.R. Choi, Heilemann MS V., Fauer A., Mead M., J Am Psychiatr Nurses Assoc, **26**, 340–3 (2020)
6. R. Lilin, Hadi I., Heal Inf J Penelit, **12**, 107–30 (2020)
7. S. Gupta, Kaur M., Acta Paediatr, **109**, 2181–3 (2020).
8. A. Kusnayat, Sumarni N., Mansyur A.S, Zaqiah Q.Y., EduTeach J Edukasi dan Teknol Pembelajaran, **1**, 153–65 (2020)
9. W. Cullen, Gulati G, Kelly B.D, Qjm, **113**, 311–2 (2020)
10. S. Stuijfzand, Deforges C., BMC Public Health, **20**, 1-18 (2020)
11. B. Semo, Frissa S.M., Psychol Res Behav Manag, **13**, 713–20 (2020)
12. L. García-fernández, Romero-ferreiro V., López-roldán P.D., Padilla S., Calero-sierra I., Monzó-garcía M., et al. Psychol Med, **3**, 8–10 (2021)
13. R. Ransing, et al. Asian J Psychiatry J, **51**, (2020)
14. S.A.M. Rauch, Simon N.M., Rothbaum B.O., Depress Anxiety, **37**, 505–9 (2020)
15. R. Shahriarirad, et al, Int J Ment Health Syst, **15**, 1–13 (2020)
16. S. Aslamiyah, Jurnal Riset dan Pengabdian Masyarakat, **1**, 56–69 (2021)
17. F. Kaligis, Indraswari M.T., Ismail R.I., Medical Journal of Indonesia, **29**, 436–441 (2020)
18. Oktaviany, R, OSF Preprint (2021)
19. W. Setyaningrum, Yanuarita, H.A., JISIP, **4**, **550-57** (2020)
20. D. Vibriyanti, Jurnal Kependudukan Indonesia, **Juli**, 69–74 (2020)

Preventing COVID-19 pandemic outbreak: physical distancing and the socio-economic wisdom of local community in Aceh, Indonesia

Gunawan[1], *Hafas* Furqani[2], *Ratna* Mulyany[3*] and *Ernita* Dewi[4]

[1]Faculty of Tarbiyah and Education, Universitas Islam Negeri Ar-Raniry, Banda Aceh, Indonesia
[2]Faculty of Islamic Economics and Business, Universitas Islam Negeri Ar-Raniry, Banda Aceh, Indonesia
[3]Faculty of Economics and Business, Universitas Syiah Kuala, Banda Aceh, Indonesia
[4]Faculty of Social and Political Science, Universitas Islam Negeri Ar-Raniry Banda Aceh, Indonesia

Abstract. Human as a social being have an instinct to live together and interact with one another. Nevertheless, COVID-19 has significantly limited face to face interaction. In a local community such as Aceh, Indonesia, such situation makes the community somewhat indifferent as their traditions carried out so far are actually interacting closely with each other as a form of intimacy. This research aims to explore the problem of physical distancing based on local wisdom in Acehnese society. Qualitative approach was utilized with data from field study and library research. This study found that there still exist various local wisdoms in Aceh community which are parallel in the current recommended mechanism in avoiding the spread of COVID-19. The socio-economic ties are also still strong among the local population who believe in tradition which prevent the COVID-19 outbreak. This study implies that local wisdom which have been applied by Acehnese for years could be one of solutions to mitigate the COVID-19 outbreak.

1 Introduction

In the beginning 2020, the world was shaken by the presence of a deadly virus named severe acute respiratory syndrome coronavirus-2 (SARS-Co V-2). Its spread was uncontrollable, crossing national borders quickly as the incubation period for this virus to be transmitted from human to human is within 14 days [1]. Since this virus is very difficult to predict, even its existence is the scariest killer, it only took five months for this virus to infect millions of people around the world. Hundreds of thousands of people lost their lives, and there are still millions of people in care, although it cannot be denied, many of the infected patients have been declared cured. The World Health Organization (WHO) has declared the new terminology for this pandemic and become well known in today society as COVID-19 [2].

COVID-19 pandemic has become a global problem and WHO has declared this to be a pandemic, and requires all countries to work together to overcome this pandemic. Certainly it is not easy to explain how to break the chain of transmission of this virus, especially for people who still cannot fully accept the outbreak. Various attempts were made so that the virus did not spread more and people who were infected did not spread the virus to other parties quickly.

Various bodies such as US Centers for Disease Control and Prevention (CDC) and including the WHO have issued the preventing advice for the further spread of this virus [3], and one of the efforts made is to forbid people from crowding and carrying out activities involving many people. People are encouraged to do social distancing and physical distancing, as WHO has declared regarding to suppress the further spread of this virus, maintaining the distance among other people or more known as physical distancing is crucial to be done [4]. Besides that the cardinal features of the response to COVID-19 is that people are required to remain enhanced the practice of hygienic such as frequent handwashing and avoiding touching the face, and also to wear masks. All of these are important part of breaking the chain of virus spread [5].

However, these appeals made some people confused as some have no proper understanding of the rationale for such new normal. This is due to the lack of information and education. In addition, the government policy in managing COVID-19 rarely take into account the local wisdom or tradition. Hence, there are some resistance to follow the instruction of the government. This paper aims at elaborating the Acehnese local wisdom in mitigating the impact of pandemic COVID-19 through various local wisdoms.

2 Literature review

2.1 The emergence of physical distancing norm

Prior to the introduction of the term "physical distancing" by WHO, Ibn Sina (980-1037 AD) an

* Corresponding author: ummuhaziq@unsyiah.ac.id

Islamic philosopher, known in the Western world by the name Avicenna, had printed his exemplary history as the "Father of Modern Medicine" and was able to fight the virus at that time, where conditions were almost the same as the condition of COVID-19 pandemic. Ibn Sina ordered the people not to gather in the streets, markets and places of worship as he asserted that a person infected with a virus can spread to hundreds of other people. Ibn Sina who was born in Afsana (Uzbekistan) could be the first person to practice social distancing, psychological distancing and no social panic in fighting the virus that caused the death of "*al-maut al-aswad*" in the past ten centuries. To anticipate the virus, Ibn Sina advised people to use vinegar and gargle with water mixed with apsitus /wormwood (*artemisia absinthium*) leaves [6].

According to the authors of the book *al-Syifa* (Healing) and *Qanun fi al-Tib* (Qonun Medicine), the small size of the virus cannot be seen with the naked eye if it is spread will cause people to contract various symptoms such as coughing, shortness of breath and can even lead to death sudden. The virus can spread through hands (shaking hands), face, hair and even through clothing worn by people exposed to viruses such as corona today. Because the distance recommended by Ibn Sina was also conveyed by the World Health Organization (WHO) and all world governments in line with what Ibn Sina had done. At this time, especially in Indonesian society, it is not uncommon to have suspicions among people caused by the corona virus. In fact, the fear of being close to anyone and trying in all activities to stay away from others as experienced by the author himself.

Physical distancing or physical distance limitation is a substitute term for social distancing, which both aims to prevent the corona virus. Physical distancing refers to making changes in one's everyday routines, minimizing close contact with others which includes avoiding crowded places and non-essential gatherings, common greetings (hand-shakes), and limiting contact with people at higher risk such as older adults and those in poor health. Not only when outside the home, has the government even recommended that this method is also done when inside the house. When undergoing physical distancing, a person is asked not to travel to crowded places, such as markets, places of worship, schools, marriages, and other crowded places. As much as possible avoid too crowded public transportation.

In practice, physical distancing can also be done in several ways namely do not leave the house, except for important matters, such as buying basic necessities or getting treatment when sick, greeting others with a wave of a hand, not by shaking hands, working or learning from home, using the telephone handheld or video call to stay connected with relatives and coworkers, buying necessities through online platform when ordering food or other items.

Previously, efforts to limit the distance with others was known as social distancing. However, some time ago, WHO recommended replacing the term with physical distancing. The reason, the use of the term social distancing was feared to be misinterpreted by cutting off communication or social interaction with family and relatives. In fact, social interaction also has an important role in efforts to deal with the COVID-19 pandemic. By maintaining communication with others, people can give each other news and enthusiasm, so they do not feel lonely, sad, or alienated. These negative feelings can trigger stress and depression, and weaken the immune system [7].

WHO in their press briefing said "We started saying physical distance because we want people to stay connected. So, find ways to stay socially connected, through various social media to stay connected with others because one's mental health is just as important with their physical health". The government and WHO hope that this change of terms can make it easier for the public to understand that the effort needed to prevent the spread of the Corona virus is to maintain a safe physical distance, not to cut off social contact [8].

Physical distancing is essentially effective enough to prevent the spread of the Corona virus. However, this certainly needs to be accompanied by other preventive measures, such as wearing a mask when leaving the house, diligently washing hands, cleaning the house and eating utensils properly, and strengthening the body's resistance by consuming foods that contain protein and vitamins. More importantly everyone is expected to avoid the crowd, if you have to be there because it is very urgent, then keep a minimum distance of 1 meter to 2 meters.

2.2 Physical distancing and socio-economics changing patterns of community life

Humans are social creatures (*zoon politicon*) that live in community life and interact with each other in the form of social framework of a society. Human being as a social creatures does not mean to show the nature of humans to socialize with each other, but the idea of social beings is primarily intended to point directly to the perfection of human identity which means sociality is human nature and identity because humans cannot live alone [9].

Humans need each other for their live and they will grow to develop tradition and civilization. By nature, humans are creatures who have a tendency to live together with others, to learn to live as humans. Humans are creatures who seek perfection in the order of living together. Humans are born, grow and become adults because of and with other humans. Then the definition of humans as social beings directly intends to emphasize that only in the sphere of shared living order will human perfection find its perfection. Human life and development, even what is called the meaning and value of human life is only possible in the context of being together with other humans. The meaning and value of life will be manifested when humans agree and acknowledge the existence of their fellow human beings. Also the division of a personality will reach its perfection if humans are able to accept the presence of their neighbor [10].

 https://doi.org/10.1051/e3sconf/202234005007

Humans as social beings have an instinct to help one another, to be loyal, friends and tolerance, as well as to have sympathy and empathy for each other. This condition can make a good and harmonious society, so that there arises norms, ethics, and traditions adopted by the community. If the things above are violated or ignored, then there is what is called social deviation.

A survey was conducted by Snap cart to see how much impact the corona virus has had on Indonesian lifestyles. Conducted from 17-28 March 2020, a survey involving 2000 men and women aged 15-50 years in 8 major cities in Indonesia (Jakarta, Bandung, Semarang, Surabaya, Medan, Palembang, Makassar, and Manado) showed that a virus pandemic corona has the greatest impact on people's social lives. Followed by career or work factors and changes in travel plans or vacations due to the corona co-19 virus. The most disturbing impact of COVID-19 is social life, ranking highest at 48%, followed by career and work concerns 44%, then vacation and travel plans 39%, furthermore concern about the limited religious activities 31%, and not being able to do shopping habits 24% [11].

As in the current condition, where the impact caused by COVID Pandemic 19 has a considerable influence on the rate of social change. Social change caused by the COVID Pandemic, is one of the changes that are not desired (unintended change). This concept of undesirable change includes changes that are not expected by the community itself. For example, much local wisdom in Acehnese society must be replaced because of the presence of COVID-19, maintaining physical distance and not shaking hands, two things must be done, but changes along with the presence of adaptation of new habits (new normal).

In addition, the influence of the media today is very determining one's mindset in responding to the corona virus outbreak. With various issues from the news on social media and television related to this corona virus made people more confused. Actually the virus is really dangerous as it was given before or just a strategy of world politics to frighten people for the needs of certain groups. Sometimes afraid of a lot of sudden death, but there is a feeling that the virus is like a common cold virus that does not need to be afraid to be free to move and hang out with friends.

Thus the pattern of one's life depends on the mind set of seeing and digesting a situation for lifestyle changes. Through the optimal use of reason, humans can become more sensitive, smart, and able to think and act positively. These conditions can be said that a person already has common sense, good, and positive. Continuing education really needs to be given by the government, leaders of community, teachers, academics, so that people really understand and want to apply this physical distancing.

3 Research method

Due to the nature of this study, it is deemed that qualitative design based upon conceptual and interpretive approach is most suitable [23]. . Prior literature were learned in depth to identify the root of new normal measures such as physical distancing and hygiene propagation. Interview were conducted with local religious scholar (*ulama*), leader, as well local community members during May-June 2020.

4 Findings and discussion

4.1 Physical distancing and socio-economic changes in acehnese community

Local Wisdom is all forms of knowledge, beliefs, understandings, or insights as well as customs or ethics that guide human behavior in life in the community. Local wisdom is also defined as truth that has been directed or fixed in an area. From both definitions then local wisdom can be interpreted as a value that is considered good and right that lasts for generations and is carried out by the community concerned as a result of the interaction between humans and their environment [12].

Relationships or social interactions between individuals with one another, of course, be hampered in this condition of widespread pandemic. Individuals will be eventually feel their habitual activities have been changed. Social interaction is limited as a result of the friendship relationship decreasing. The notion of social change is the changes that occur in social relations or changes in the equilibrium of social relations [13]. This then explains that COVID's Pandemic actually has an influence or impact on social change itself. In this case, the reality that occurs shows that social conditions are declining. Various aspects of life became contaminated and suffered a setback due to the COVID Pandemic.

Changing a tradition that is ingrained in a society is not easy. Habits that grow from a child even become a character cannot be changed in a short time [14]. The traditions of the Acehnese people, which are mostly colored by crowd activities called *kenduri*, clearly cannot be abandoned, even though this event is very unfriendly with the COVID-19 pandemic.

After the government officially banned the wedding party during co-19, the community only obeyed to follow the instruction of wearing mask, washing hand, and physical distancing, but the activities of commemorating death (*takziyah* and *samadiyah*), wedding celebration (*kanduri walimah*), *khanduri blang* (festivity in rice filed), festivity of *nisfu sha'ban*, and others, seems that it is still often done, but with less number of people. Likewise, the *meugang* tradition, where people buy meat to celebrate *puasa ramadhan* and *uro raya* (*'eid* festive), seems to keep going. It seems this is outside the control of the local government, and the community keeps carrying out this tradition by mutual agreement at the village level.

This is because according to Interviewee 1 (*Ulama* in Pidie) [15]:

> *The Acehnese community highly respect their tradition. If this tradition is banned there will certainly be upheaval and unwillingness from the community. For the people of Aceh who embrace Islam, their religion, culture and local wisdom are an inseparable part of their daily lives. The motto "Adat ngon hukom lagee zat ngon sifeuet (tradition*

> *and law are like the body and soul)" is a reflection that for Acehnese people, local wisdom and Islamic Shariah law are one, like substance and nature, inseparable and apply back and forth. In fact, there is also a belief in Acehnese society that making visits to neighboring places or to family homes (silaturrahmi) including attending various khanduri (festivities) gives benefits such as lengthened age and ease of fortune* (interview in June 20, 2020).

Therefore, when the prohibition of visiting each other comes, it must be accompanied by an explanation that is acceptable to the community that for the time being these activities cannot be carried out, when the pandemic ends, the activity can be done again.

The corona virus has restricted all community activities as usual. In order to prevent transmission of corona virus infection, many parties are calling for a social distance or social distancing appeal. However, the term is prone to misperceptions in the community. Social ties are actually needed, but not physically close together. Interacting with other people can create a feeling of security and happiness. However, with this corona virus, everyone must maintain distance or self-distancing and self-isolating at home. Although hospitality as a basic need for the people of Aceh, in these conditions can still be done by telephone, video calls, and using social media. If you meet you can do it by waving your hands and smiling without having to shake hands. But, for the people of Aceh, this new habit is not yet grounded. No matter how big the news about the dangers of shaking hands and hugging, there are still many people who do this activity, even when there are people who refuse to do these acts for security reasons and keep from transmitting COVID-19, this attitude is actually considered as being polite, lacking respect other people or relatives, even considered to have been arrogant and arrogant, which ended with social sanctions by the omission of other communities.

Therefore, interviewee 2 (one of the local community leaders in Pidie, Aceh) in our interview says there is a need to cultivate consciousness in Aceh society toward the effect of COVID-19 in order to prevent it. He advices his community by saying [16]:

> *For people who are not busy with outside activities, it's better to just stay at home or don't leave the village because it's very vulnerable to the virus. To prevent this virus, we have to take care our self and family and always pray. The villagers also must remain cautious of foreigners coming from outside the city. The term new normal, as popularized by the government, does not mean that the community is free to move anywhere like before the emergence of this pandemic, the space is still tightened by always keeping themselves safe, keep a distance, wear masks, diligently wash hands and may not gather (crowding). If this is violated, it will be very dangerous, it could be like a time bomb when the virus is spread throughout the community. It should wait until the virus disappears 100% of this world* (Interview in May 17, 2020).

This corona virus has spread to almost all regions of Indonesia, including Aceh and therefore must be cautious and obedient to the regulations set by the government. Deputy MPU (*Majelis Permusyawaratan Ulama*) Aceh Province, Tgk. Faishal Ali (2021) told "even though people in Aceh can have a friendly relationship directly during *eid al-fitr* or *eid al-adha*, even in the midst of the outbreak of the corona virus, they are advised not to shake hands. Visiting each other during the festive can only be carried out by fellow citizens who are not exposed to the corona virus or are not undergoing independent quarantine" [17].

The Acehnese call the disease outbreak as "*taeun*" or "*taeut*" which was adopted from Arabic "*thaun*". Some of Aceh's local wisdom is relevant in dealing with the "*taeun*" outbreak [18]. There are several Acehnese local wisdoms that are relevant in dealing with the "*Taeun*" outbreak, namely:

- Using salt in various cooking and before eating. There is a tradition of there is a tradition of providing salt in a small plate at a feast at a feast in Aceh. It is believed a good tradition because salt besides functioning as a flavor enhancer, is also believed to be an antidote to various viruses for the food served.
- The tradition of "*manoe rabu abeh*" or "*manoe safa*" which means sea bathing to soak the body with salt water. This practice of bathing in the sea is believed to function as an effort to eliminate various germs and diseases that exist in the human body.
- The tradition of washing feet through "*ie lam guci*". Another local wisdom in Acehnese society in dealing with disease is to always maintain cleanliness. Every house of the ancient Acehnese would provide a jar (jug, urn) filled with clean water complete with a dipper in front of the house. This water is used to wash feet, hands, and even face by anyone who wants to enter the house. One must ensure that it is clean and sterile before stepping into the house.
- The tradition of "*pakek gaca*" or using henna using henna (gaca) on the fingers alternately, not on all fingers. This symbolizes that we should keep our distance when there is a *tauen* epidemic.
- The tradition of "*pajoh ranup*" The Acehnese community is also having tradition of having *oen ranup* (betel leaves) to be tasted by visiting guests so that they are free from viruses when chatting. Betel leaves are recognized by health experts as an antiseptic and anti-microbial capable of killing the sars-cov-2 virus, the cause of COVID-19 disease.
- The tradition of "*toet leumang*". Another custom that the ancients did when a disease plagued the country was to burn *leumang* and then distribute it to neighbors. The philosophy of this tradition are, *first*, burning *leumang* from glutinous rice indicates that the person no longer has rice stock as the main food in his house, this indicates that the food stock in the family is running low. The message is that people who have the ability must immediately provide basic food assistance to the poor in their neighborhood. Second, behind the action of burning leumang there is a message to get used to sharing (alms) even in difficult conditions. The burnt lemang is not eaten by a family alone, but with the community.

- The tradition of "*manok keunong taeun*" whereby chickens that show signs of contracting the virus (such as shaking their heads or salivating) are immediately confined (isolated) in a cage so that they are separated from other chickens. After that, animal health officers (*mantri*) were invited to give vaccines to "*manok keunong taeun*" and also to chickens that had not been infected with the virus as a preventive measure. This is a symbol of isolation for victims of the pandemic and given drugs or vaccines to recover.
- The tradition of "*do'a let tauen*" or prayer to reject and stop pandemic to spread by reading a verse from the Qur'an: "*wa qul jaa-al haqqul wa zahaqal baathil. Innal baathila kaana zahuuqa*". This is then followed by reading the Qur'an, especially Surah al-Kahf, Surah Yaasin, and others. Prayer is an effort to surrender oneself to the Creator. The Acehnese believe that humans have limitations so that sometimes they are unable to control the epidemic even though they have tried their best.

4.2 The role of women in facing Covid-19 pandemic in Acehnese society

Women play an important role in the process of applying physical distancing through education and modeling provided in married life in the traditional system of Aceh local community.

Ibu Suryani, in Pidie region for example, continuously warn her children so they could keep physical distancing in the relationship with others beside giving good nutrition. During recent festival of *Eid al-Fitr*, usually Acehnese will visit their extended family, but, Ibu Suryani did not do that for preventing the spread of corona. She attempts to give logical reason for doing that and usually Acehnese will listen to the voice of mother. This shows the role of women is very significant in contributing to break the chain of distribution of corona in Aceh [19].

Women actually have a great power to be able to provide coercion so that their families or communities are compliant with a government policy especially on urgent matter such as health. Moreover, these women have the ability to speak, sit as community leaders, have more general and religious knowledge than the average woman, and of course have a high economy compared to other communities. This advantage is a bargaining value for women so that their opinions are heard and can be implemented. If all women in this category have the same voice, then certain physical distancing can be carried out properly.

The figure of women in the family has a very important role. The most important education is in the family environment which is played by a mother. The mother gives birth to intelligent children as the next generation of the nation. A housewife in Pidie, says [20]:

> *To prevent the virus, we have to continiuesly maintain health by living cleanly, teaching children to stay alert to this virus. This "tauen" corona is like a disease in poultry, while corona in humans.*

Religious Education and Approach in Mitigating the Impact of Pandemic

Religion in Acehnese society plays an important role in eliminating excessive panic and worries. There is a wisdom in Acehnese tradition to return back anything happened in human life to Allah SWT (*pu wo bak Po*). It is believed that if someone can accept this situation and surrender all to Allah, then that person will get the strength and patient to deal with everything without any fear. However, if someone is restless, panics and worries because they are afraid of getting infected by the virus, it will certainly make their immunity weak, and eventually they will get infected.

This phenomenon has indeed become an inseparable part of the life of the religious community. Religion plays a role in providing an opportunity for the community that there is a source of strength and hope that is greater than human strength, as a helpless human being. The only hope is to depend on the best scenario outlined by God

The role of religion can be maximized in encouraging people to be able to prevent the transmission of this virus. The way is to encourage their respective adherents to be obedient to health experts, public security experts, as well as to the authorities authorized to maintain the security, comfort and safety of their citizens.

Moreover, the Acehnese has been known as a religious society that strongly upheld the principles of Syariah Islam and follow the instructions issued by their religious leaders. The religious leader (*teungku imum/Ulama*) in Pidie said [21]:

> *to overcome the emergence of the corona virus, we have to follow the advice of the government by keeping the physical distancing and be aware of outsiders who enter the settlement. Regarding the procedure of worship is not a problem if there is no conflict with religious teachings, sufficient conditions and harmony. This is because, Islam has taught his people to always maintain good health both physical and spiritual health as in the teaching of thaharah (purification), which is the beginning of acts of worship even the first chapter in studying Islamic law, which must be done at any time not just when praying. This proves that Muslims should be clean because they maintain their ablution, therefore all diseases are not easy to attack. This has become clear evidence for previous scholars who rarely got sick. In handling the virus now related to the problem of worship in congregation there is little criticism with the protocol distance of one meter left, right, front and back. According to him it was wrong because it was not an alternative event to be able to pray in congregation in the mosque because it could not be seen by praying far apart. But it's better to pray alone at home to protect yourself from the corona virus.*

Another religious leader in Aceh also said that [22]:

> *A Muslim life in the world does follow religious advice, but living in a country must first follow government regulations. In dealing with the*

corona virus, it must follow the direction of the government such as keeping a distance, wearing a mask, and limiting the usual business, but it must also be strengthened with religious advice such as increasing worship (ibadah), zikr, and praying (du'a). Good government regulations will not collide with religious regulations, they will strengthen each other to provide prosperity and peace.

5 Conclusion

The COVID-19 pandemic has been significantly affecting all lines of life without exception in transforming people's lifestyle throughout the world, including small province like Aceh, Indonesia. The most striking change can be seen from the way people exercising their social role. With the restriction on social interactions, the society have been under pressure as their habitual activities have been limited unlike in the normal life situation. Many people have even objected when some preventive measures are taken such as on physical distancing policy.

Nevertheless, local wisdom that Acehnese have applied for years could be one of solutions to prevent the pandemic COVID-19 outbreak. The society have been following this tradition wisely in the past and the provision of a thorough education will be effective in ensuring the compliance towards new normal measures. This participation of the community from various elements is needed so that the implementation of a new life structure that is still very foreign to the people of Aceh in particular and the people of Indonesia in general can be carried out voluntarily without coercion. All people are expected to understand that implementing new rules such as physical distancing is only temporary, and people must be patient with this situation, when the pandemic ends and vaccines are given, then all these traditions can be re-established.

References

1. D. Wang, B. Hu, C. Hu, F. Zhu, X. Liu, J. Zhang, Wang, B., Xiang, Hui., Z. Cheng, Y. Xiong, Y. Zhao, Y. Li, X. Wang, and Z. Peng, Clinical characteristics of 138 hospitalized patients with 2019 novel coronavirus–infected pneumonia in Wuhan, China. *JAMA*. Vol. 323 **11**, (2020).
2. C. Lai, T. Shih, C. Ko, H. Tang, and P Hsueh, Severe acute respiratory syndrome coronavirus 2 (SARS-CoV-2) and corona virus disease-2019 (COVID-19): the epidemic and the challenges. *International Journal of Antimicrobial Agents*, **55** (2020).
3. C Sohrabi, Z. Alsafi, N. O'Neill, M. Khan, A. Kerwan, A. Al-Jabir, and R. Agha, World Health Organization declares global emergency: A review of the 2019 novel coronavirus (COVID-19). *International Journal of Surgery*, (2020).
4. K. Siste, E. Hanafi, L. Sen, H. Christian, Adrian, Siswidiani, and C. Suwartono, The Impact of Physical Distancing and Associated Factors Towards Internet Addiction Among Adults in Indonesia During COVID-19 Pandemic: A Nationwide Web-Based Study. *Frontiers in Psychiatry*, **11** (2020).
5. D. Bausch,. Precision Physical Distancing for COVID-19: An Important Tool in Unlocking the Lockdown, The American Journal of Tropical Medicine and Hygiene **103** (2020).
6. A. Maulana, Pandemi dalam Worldview Islam; Dari Konsepsi ke Konspirasi. *Tribakti: Jurnal Pemikiran Keislaman*, **31**, 2 (2020).
7. "Pemerintah ubah istilah Social Financing menjadi Physical Distancing" Kompas, 23 Maret 2020, available at https://nasional.kompas.com/read/2020/03/23/14332461/pemerintah-ubah-istilah-social-distancing-jadi-physical-distancing
8. WHO, COVID-19, *Press Briefing* (2020).
9. H. Gintis, C. Schaik, and C. Boehm, Zoon politikon: The evolutionary origins of human socio-political systems. *Behav Processes,* **April 161** (2019).
10. Judson R. Landis, *Sociology, Concepts and Characteristics* (California: Wadsworth Publishing Company, 1986).
11. "Riset tunjukkan gaya hidup orang Indonesia berubah karena virus Corona COVID-19", available at https://today.line.me/id/v2/article/eMNBDg (13 April 2020)
12. Jundiani. Local Wisdom in the Environmental Protection and Management. *IOP Conf. Series: Earth and Environmental Science* **175** (2018).
13. Soekanto, S, *Pengantar Sosiologi* (Jakarta: Raja Grafindo Persada, 1990).
14. E. Ismawati, *Ilmu Social Budaya Dasar* (Yogyakarta: Ombak, 2002).
15. Interview with Teungku Syarwan, Ulama in Pidie, on June 20, 2020.
16. Interview with Yusuf, community leader in Pidie on May 17, 2020.
17. "Tausyiah MPU Aceh perbolehkan masyarakat Shalat Idul Adha" available at https://aceh.antaranews.com/berita/225146/tausyiah-mpu-aceh-perbolehkan-masyarakat-shalat-idul-adha (20 July 2021).
18. A. Panton and M. Hasan Basri. *Taeun Corona dan 7 Local Wisdom di Aceh, Mulai dari Sira, Ie Lam Guci, Toet Leumang, Hingga On Ranup*, available at https://aceh.tribunnews.com/2020/04/09/taeun-corona-dan-7-local-wisdom-di-aceh-mulai-dari-sira-ie-lam-guci-toet-leumang-hingga-on-ranup?page=3 (9 April 2020).
19. Interview with Suryani, a housewife in Pidie on 27 May 2020.
20. Interview with Maryam, a housewife in Pidie on 21 June 2020.
21. Interview with Teungku Syarwan, Ulama in Pidie, on June 20, 2020.
22. Interview with Abu Yusri, Ulama in Pidie, on May 17, 2020.

23. D. Gioia, A systematic methodology for doing qualitative research. *The Journal of Applied Behavioral Science*, Vol. 57 **1,** 2020.

The fading of disaster memory in Pulau Sebesi: A historical construction

Devi Riskianingrum[1,*] and *Herry* Yogaswara[2]

[1] Department of History, Faculty of Humanity, University of Indonesia
[2] Research Centre for Demography, Indonesian Institute of Sciences

Abstract. The island has been impacted by volcanic eruptions which led to tsunami in a different scale, where in 1883 claimed 36000 lives, subsequently in 2018 took one victim. Some island communities succeed in pertaining memory of tsunami through oral tradition, namely Simeulue, differently Sebesi communities fail to maintain its memory on tsunami. The gap of 138 years seems to bury the memory of tsunami in Sebesi island. This paper aims to explore why the Sebesi communities fail to maintain the disaster memory. To build an understanding the way in which the Sebesi forgetting their past disaster, this paper uses longue durée approach, oral history framework and archival studies to analyses the structures—both environmental structures and socio-politic and cultural structures—that play roles in the disappearance of disaster memory. The study revealed that none of survivor during its catastrophic tsunami in 1883 and repopulation of this island occurred only after 1940s. This resulted to the formation of community without disaster memory. Only after the 2018 tsunami, the community of Sebesi Island began to aware that there are hazards among their environments. Uncovering the fading of disaster memory in Pulau Sebesi elucidates lessons to pursue resilient development trajectory on the island.

1 Introduction

"...every full moon I feel frightened Remember the high waves chasing our vessel. If only it rolled over our vessel, I may be death by then ..." (Bapak Jefry—eyewitness and survivor of 2018 tsunami/ a staff of BKSDA Lampung). The tsunami disaster that occurred on December 22, 2018, at around 21:27 WIB, exhibited an important value for the people of Sebesi Island. The disaster that stricken during the full moon killed 430 people in the Pandeglang area and the coast of South Lampung [1]. It became a milestone for a disaster awareness of the people in Sebesi on the dangers of their surrounding environment.

PVMBG (Volcanology and Geological Disaster Mitigation Center) has been monitoring this mountain since early December 2018. The agency detected an increase in eruption activity of Anak Krakatau on Friday, December 21, 2018, with a throw height reaching 738 meters above sea level. Therefore, Anak Krakatau received an *alert* status from this institution. In line with this, BMKG (Meteorological, Climatological, and Geophysical Agency) also monitored the threat of high waves on December 22 and gave warnings to the public [2]. Unwittingly, continuous eruption since early December triggered landslides on the slopes of Anak Krakatau. According to research, an area of 64 hectares of the southwest slope of Anak Krakatau experienced a landslide on Saturday at 20:56 WIB, hence generated tsunami [3].

Sebesi Island located only 20 KM or 10.8 nautical miles from Anak Krakatau also hit by the tsunami. This led the island dealt with communication paralyze and isolation, hence caused panic from around 2700 residents living in Sebesi. The help from rescue team came a day after the tsunami due to high waves. The government also helped evacuate the residents of Sebesi to Kalianda, South Lampung. Based on observations, there were about 50 houses in Regahan Lada Hamlet that were destroyed. However, it claimed only one victim, a child of a seasonal worker from Pandeglang who at that time stayed in Gubuk Seng area in Sebesi island. The 5-year-old toddler was separated from his parents' grip and swept by the waves [4]. Instead of evacuated to Kalianda, some residents prefer to stay on the island to protect their property and they belief that Anak Krakatau will soon get better. Until now, the Sebesians opt to stay in the island with various reasons, from assets to livelihoods that are easier to find on Sebesi Island [5]. It seems that they have anchored their life in Sebesi, where they spend their live and die with the island.

* Corresponding author: devichudori@gmail.com

For the people of Sebesi, the 2018 tsunami disaster appeared to be a new experience, although history recorded a similar disaster with larger scale had hit this island. The eruption of Mount Krakatau on August 27, 1883 was so intense that it generated a tsunami and devastated the coasts of Pandeglang, South Lampung, and Sebesi Island, and even affected the world [6]. The span of 138 years has succeeded in fading the disaster memory in the minds of Sebesians. Although, on other affected area like in Labuan sub-district, Pandenglang, the communities continue to commemorate the calamity by convening *haul kalembak*, a tradition where people gathered in Labuhan Great Mosque to pray for the victims of Krakatau 1883. In addition, the people of Simeulue island were able to preserve the experience of the tsunami disaster that had befallen them in 1907. Through oral tradition, namely lullaby and storytelling, the experience was recorded and continued to be conveyed to the younger generation of this island community. The tradition, known as *smong*, proved to save the people of Simeulue island from the great tsunami in 2004 [7].From a historical perspective, Bankoff [8] explains that disasters can be seen in two trajectories, namely 'natural'—in this case the forces of nature—on the one hand, and society on the other. When these two trajectories meet at a certain place and time, a catastrophic event occurs. At that time a new experience is formed for the survivors. The experience of individual will form a shared memory in a particular community. This is in line with Halbwach's explanation which states that collective experiences could form into 'collective memory.' He explains that memory refers to individual processes, but it has a role in constructing "collective memory" within the community by forming specific behaviour limited to specific communities, location, certain area, and time [9]. He concluded that a shared framework for a memory is a result or combination of individual's memory from people in the same community [10]. Based on this concept, the paper tries to reconstruct the failure of the Sebesi community in preserving the memory of tsunami Krakatau in 1883. Furthermore, new knowledge that emerged after the 2018 tsunami as a form of new behaviour in the Sebesi community will also be analysed.

2 Methodology and research sites

This paper aims to reconstruct the fading of tsunami 1883 memory on the people of Sebesi island. Further, it also analyses the new knowledge arise as an impact of the recent tsunami. To have broader understanding about collective memories and how people maintained it, this paper applies limited comparative case study with local knowledge of Smong from Simeulue of Aceh [7], [11], [10]. It is the knowledge that attracted many scholars to study and translated into policy implementation.

This research employed longue durée approach with archival study and oral history. Longue durée is a structural analysis of historical reality without leaving the aspect of time in it, which presupposes structural change itself, although it runs very slowly. It is not always referring to temporalities, but rather a way of seeing reality that is formed slowly over a long period, which then becomes a precondition for the problems that arise on the surface. This approach enables to find the relationship between agency and environment over the long period of time. The slow-moving time that running in a long process becomes the precondition of fast-moving time as *conjuncture*, which later comes to the surface as an *event* [12]. Accordingly, the change of community structure as the result of natural structure recorded in the memory could be investigated throughout the approach.Primary data and secondary information presented in this paper were gathered during irregular visits to Sebesi island for several weeks in 2019, 2020, and 2021. Data were gathered through archival studies, literature reviews, and interviews with older individuals and tsunami survivors that they remember, in search of stories about disasters in this island. Secondary data were gathered from the Internet and published reports. Academic papers, as well as relevant documents of governmental and non-governmental organizations, were studied and evaluated. Some relevant reviews of selected literature and published stories about earthquakes and tsunamis were also analysed and elaborated in this paper. In addition, a field notes of a project conducted by Indonesian Institute of Sciences in some coastal areas of Banten in 2015 related to Risk Culture is using in this paper to enrich knowledge about people and disaster in coastal area.

Sebesi Island is in Lampung Bay with coordinates at the position of 05°055'37.43"-05°058'44.48" South Latitude and 105°027'30.50"-105°030'47.54" East Longitude. It has an area of 2620 hectares, currently there are 787 families or about 2795 people recorded living on the island. For livelihood, most of the people in Sebesi work as farmers as much as 75%, and fishermen as much as 20%, and 5% consists of government employees, traders, and tourism actors. Meanwhile, the dominant ethnic group on this island is Banten (Javanese-Serang or Jaseng), while the second largest ethnic group is South Lampung, and several other ethnic groups such as Sundanese, Batak, Nusatenggara, Bugis, and Padang [13].

Sebesi island is bordered by Lampung Bay and Sebuku Island in the north, the Indian Ocean in the west, the Krakatau Island complex—including Mount Anak Krakatau—in the south, and the Sunda Strait in the east. Based on its administration, Sebesi island is included in the administrative area of Tejang Village, Rajabasa District, South Lampung Regency. In Tejang Village, there are four hamlets, namely Hamlet I Bangunan, Hamlet II Inpres,

Hamlet III Regahan Lada, and Hamlet IV Segenom [14].

Fig.1 The map of Sebesi Island.
Source: Biologi LIPI 2019

3 Result and discussion

3. 1 The Dynamics of post-tsunami 1883 in Sebesi Island: from Plantation to Settlement and Tourism

Although it is obscure in the Dutch archives, the Sebesi island's fertility has lured the people to settle, both those who came from around the Rajabasa and Kaimbang areas, as well as those who came from Banten. Long before Krakatau erupted, this island had become a stopover for Dutch ships before entering the Banten waters. If the English ships with their weapons and merchandise were to land on Legoendi Island, then the Dutch ships, one of which was commanded by Pieter de Carpentier in 1624, had a place on Sebesi Island to fill water and supplies, repair the ship, and wait for permission to enter the Port of Banten [15].

After the eruption of Krakatau, Sebesi island became uninhabited. The first visit to Sebesi was 14 days after the eruption. It was conducted by Berg from *Javabode* in 1883 described that this island buried with 6 meters thick mudbank and giant pumice stone along the shore [16], [17], [18]. Another visit on May 21, 1884, explained that living creature was only consisted of a few banana treetops, tall weed, and bushes and full of desolation. The atmosphere of the island was only a stretch of white sand and bones—which was unknown whether they were animal or human bones [19].

After the eruption of Krakatau, the ownership of the Sebesi island was on dispute due to the death of all the descendants of Pangeran Singa Brata and all existing evidence of ownership. Pangeran Minak Poetra claimed to be the younger brother of the remaining Pangeran Singa Brata and appealed for these two islands. According to the customary law, since all the descendants of Singa Brata have become victims of Krakatau, hence Minak Poetra, as a younger brother, can be appointed as head of the Raja Basa clan or *penyimbang*. He also appointed himself as Bandar, that traditionally with this position has power over these islands. The appointment of Minak Poetra's also received the blessing from the Sultan of Banten in the 1890s [20].

Furthermore, the Dutch also approved Minak Poetra as a *penyimbang* as well as *bandar* of Raja Basa. This is known from his action in giving assistance to the Dutch to cease the rebellion in Cilegon-Banten in 1888. Furthermore, the Dutch approval was also seen as an effort to cut off the descendants of Pangeran Singa Brata. He was known as important ally of Radin Inten II's in struggling against the Dutch. In the hands of Minak Poetra, Sebuku island was initially leased to a Lanberg firm in 1889 to grow pepper and timber, while Sebesi was left to remain. Subsequently, Sebesi and Sebuku Islands were sold to the trader named Hadji Djamaloedin in 1896. He and Pangeran Minak Poetra were awarded a gold medal from the Dutch government in 1888 for helping to stop the riots in Cilegon. Sebesi island was purchased for f.7000 while Sebuku was priced at f.3000. under Djamaloedin, these two islands obtained the status of individueel bezight recht (private property/goods) through the Governor General's besluit in 1906 [21], [22].

Plantation

As if a *blessing in disguise* from Krakatau eruption 1883, the land on Sebesi became very fertile so that the agricultural products of this island were abundant. A report on a trip to Sebesi island in the 1920s stated that the island was filled with thriving coconut trees planted by Hj Djamaloedin and his workers. An archive states that a ship loaded with chickens and goats was sent from Kalianda port to Sebesi island by Hj. Djamaloedin in the early 1900s and was followed by the dispatch of workers from Banten to Sebesi [23].

Entering the 1940s, the world economic conditions declined and were overshadowed by the threat of the second world war. This influenced the economic condition on Sebesi. Coconut prices have fallen quite badly. In the era of the 1950s, coconut prices improved again and gave decent results. This evidenced by several supervisor who went to hajj from the sales of coconuts at that time [24].

As the coconut continues to support Sebesi's economy, bananas have also become an important commodity on this island. Bananas became staples food for the community since it always available in the island. In line with this, this island once became a shot of clove cultivation in 1990s. Some farmers change their gardens by planting cloves. This commodity was destroyed during the 1997 monetary storm, and residents began destroying clove trees and replacing them again with coconut and bananas.

In 2004, the government came up with a cocoa planting program. Unfortunately, without market clarity, the results of chocolate in Sebesi are not as sweet as expected. Today, society continues to focus on coconut—which is sold in the form of granules for one thousand rupiah per grain—and bananas, as the main commodities to support their lives [25].

Settlement

The development of settlements on this island is closely related to plantation activities. The first wave was in the 1920s, where most of them were farmers from the Canti, Way Muli and surrounding areas also came to ask Hadji Djamaloedin's permission to open agricultural land. Later, the waves of workers that brought by the land owner also opted to settle in the island during 1930s. They were allowed to farm with a profit-sharing system, where 2/3 is owned by the farmer and 1/3 is given to the owner of the island. The results that are shared were not in the form of commodities, in this case coconut, but 1/3 of the total trees grown by the farmer [26].

Like their fellow farmers, these workers also applied for approval to work on a piece of land with profit-sharing system. The requests were granted by Hadji Djamaloedin and this system was continued by his son, Saleh Ali. Since 1937, Saleh Ali issued a letter of agreement with several laborers who asked to become cultivators with a profit-sharing system. Entering the 1940s, these workers brought their families with them, and began to build an umbul (like a small hamlet) consisting of several houses from fellow workers. In this era, the island of Sebesi returned to a settlement for the workers community [27].

In other part, the waters where Mount Krakatau used to stand showed its seismic and volcanic activities in 1927. The roar and large bubbles rise to the surface, erupted and release ash and sulfuric gas. These bubbles were an indication that it was an effort of a new mountain to build itself. On January 26, 1928, a pile of ash and solid rock began to emerge from the sea in a shape of a curve. This was followed by explosive and earthquake activities which continued until 1929. The solid rock began to grow into a new black plain and continued to expand to form an island. It was a Russian geophysicist, W.A Petroeschevsky, who first noticed this condition and observed it closely from Panjang Island. He named this nascent plain as 'Child of Krakatau.' The island began to show its stability on August 11, 1930, and further solid growth in the 1931 [6]. In response to this, the colonial government carried out detailed observation on the growth of this newly born mountain. A monitoring post in Kalianda was also obliged to report to the central government in Buitenzorg about daily condition of the mountain. The report lasted from 1928 to 1931 [28].

The era of the 1950s to the 1960s was a time of immigrant influx to Sebesi. The first migrant began to invite their relatives and neighbours migrating to Sebesi. Umbul began to turn into a hamlet in the 1960s, as new houses grew. During this period, there were three main hamlets, namely Tejang hamlet, Segenom hamlet, and Regahan Lada hamlet with a population of around 500 people.

In 1980, a village administrative was initiated in Sebesi and named as Desa Tejang, which included in the Kalianda sub-district. In 1985, with the help of the military, roads were built and connected from one hamlet to another. Furthermore, the port was repaired and enlarged as well as a scheduled public transport ship was formed. The local government also built supporting facilities, such as elementary school building and office for village staff. To fulfil the need of health services to its 800 inhabitants, the head village asked local government to send health workers to the island. It was only in 1992 that the Health Center was built. Subsequently, one midwife and one *mantri* were stationed there [29].

The existence of Anak Krakatau in 1930, which is included as the most active volcano in Indonesia and has erupted 40 times over 85 years, does not seem to be a threat to the community. They live side by side in peace with nature. The volcano activities of Anak Krakatau, such as rumbling sounds, small eruptions that emit smoke, the smell of sulfuric, and volcanic ash that splashed Sebesi, did not distract them. The routine activities of Sebesians take part in fading their memory of past disaster [30]. The people of Sebesi consider the volcanic activities of Anak Krakatau as harmless and routine phenomena hence they are courageous to keep on living there. The risk of explosion does not reflect as hazard that could create disaster.

Tourism

With an area of less than 3000 Hectare, Sebesi Island is included in the category of a small island. The significant characteristics of an island is its smallness and remoteness, which triggers vulnerability to natural hazards [31]. Furthermore, economically the main characteristics of islands are market, human and non-human resources limitation, hence their economic activities are less diversified and closely related to water, such as fisheries, inter-island trade, and tourism. Similarly, Sebesi Island also rely on agriculture, fishing, and tourism [32].

Render to its natural environments, Sebesi island appointed as one of the leading marine tourism destinations by the South Lampung Regency. White sandy beach with its beautiful scenery and Mount Anak Krakatau on its direct opposite become main reason why Sebesi is promoted as tourism destination. The existence of Anak Krakatau proved to be a magnet for the development of tourism activities in Sebesi island.

https://doi.org/10.1051/e3sconf/202234005008

In 2008, the provincial government designated Sebesi Island as one of the Tourism Destination Objects (ODTW) in South Lampung Regency by offering marine ecotourism and adventure nature ecotourism. To support its tourism revenue, the Lampung government is also promoting an annual event entitled the *Krakatau Festival*. Confirmed through the Decree of the Governor of Lampung Number G/126/Diparda/1991, the *Krakatau Festival* was inaugurated as annual tourism activity for the Lampung Region. Various cultural events, fairs, and parades invigorate this activity, with the main pursuit was visiting Mount Anak Krakatau. This involved Sebesi island since the participants would stay overnight there and visit Mount Anak Krakatau in the next morning [33].

Fig 2. The Map of Festival Krakatau routes in 2016 onward.
Source: Evadianti, 2017 [34]

As a part of their lives, Anak Krakatau is also considered a blessing from God. According to Ibu Rumanah, Ibu Jamilah and Ibu Maemunah, at a grocery shop where they usually spend their afternoons, explained that Sebesi was their hope to fulfil their necessitates because Sebesi is a 'home' for them. Entering the year 2008 onwards, a wave of tourists both from local and foreign flooded Sebesi, especially when provincial government conducted Festival Krakatau. Accordingly, visitors who stayed at Sebesi could reach to a thousand people. Even at the time of the tsunami, several houses in Dusun I *Bangunan* were receiving about 30 guests from Jakarta [35].

Based on its historical background, the decision to select Sebesi Island as a tourist spot requires careful attention and planning from the government. Exposing Sebesi into tourist destination would lead people to gather round, hence preceded to the increasing vulnerability when a disaster occurs. Consequently, it requires awareness and preparedness in management both from local people, the government, and relevant stakeholders.

3.2. The 2018 tsunami: a memory remind

It is not easy for Jefrey (35 years) to live life as prior the tsunami since he was witnessing the enormity of the tsunami waves caused by the flank collapse of Anak Krakatau. He still clearly remembered that evening, he and 7 colleagues had been in the post for a week, assigned to control Anak Krakatau. He and four of his colleagues patrolled around the island, while two others stationed at the safeguard post on the island. The aim was to prevent the entry of fishermen to the island because of its *alert* status due to its rising volcanic activities. It seems that the common people are still not aware of the dangers of this mountain, so they still saw three fishing boats landing on the island. Indeed, this island is rich in crabs accordingly entice many fishermen try their luck hunting crabs on this island despite the dangers that lurk due to the lava and sulfuric fumes [36].

As the night was getting late, they saw 4 fishing boats docked on the island. They tried to chase these boats away by shining flashlights at the fishermen. From the distance, Jefry could see that these fishermen were lighting their stove. The fishermen looked relaxed and not at all frightened although Anak Krakatau erupted continually at that night. It was a full moon, and the sky was so clear that the sparks from Anak Krakatau could clearly be seen. According to Jeffry, one unusual view from Anak Krakatau that night was the black smoke from the fire that didn't leave the peak, where usually after the smoke came out it would immediately evaporate away from the peak. Furthermore, they saw the peak was in reddish and a flash like a path of fire on the seabed. Finally, the BKSDA patrol boat approached, not only dispel the fishermen but also to summon their colleagues who were on guard at the post. At that time, according to Jefry, the Captain felt something was wrong with Anak Krakatau, so he ordered all his crew to leave the guard post immediately and return to Sebesi although they supposed to return on the following [37].

It was at about 9 PM when the BKSDA ship slowly left the island, after driving away the fishermen. Unfortunately, of the four ships that docked, only one ship rushed to leave the island, while three ships stayed behind with the excuse of having dinner and would leave the island soon. The ship was just move further for no more than ten minutes, about 500 meters away from the island, when there was a loud bang, followed by rumbling and lightning that erupted from the peak of Anak Krakatau and debris that almost reached their ship. The debris rolled up three fishing boats that were docked and created very high waves. The ship's engine did not turn off but for some reason it could not run faster. The ship was running very slowly although there were high waves behind them, also to their right and left. Somehow it happened, the ship was moving very slowly but the waves were slowly disappeared without hitting the ship. Accompanied by the continuous *azan* from one of their comrades, the ship slowly made its way towards Sebesi. At that time, there was no signal so that they failed to contact their colleagues on Sebesi Island to inform the condition of Anak Krakatau and ask for help to pick them up.

After an hour away from Anak Krakatau headed to Sebesi, they started getting signals. One of Jefry's colleagues' telephones rang and informed them that there was a tsunami in Sebesi and asked for help from the BKSDA, whereas they were just about to ask for help from Sebesi's people. Hearing this news, Jefry's mind became increasingly frantic, remembering his wife and child who lived in Sebesi. Jefri is a native born in Sebesi, whose father is from West Nusa Tenggara and a mother is native born in Sebesi. They finally arrived safely in Sebesi. The trip, which usually only took two hours from Anak Krakatau to Sebesi, took nearly four hours. Arriving at Sebesi, they saw the Port was slightly destroyed by the tsunami and the atmosphere of the village was deserted. Upon information from his brother, Jefry immediately headed to Mount Sebesi, followed his wife and children and gathered with his extended family. The horror was not over yet because on Mount Sebesi, the roar sound of Anak Krakatau along with heavy rain continuously until morning came. The sound and rain are finally ceased as the morning came.

It is not easy for Jefri to forget this disastrous experience. Horrors often come every time he received assignment to monitor Anak Krakatau. In addition, every full moon after the tsunami, he still feels the fear of high waves. However, life must go on, and giving up his job at BKSDA is also unrealistic hence he decided to fight his fear by surrender to God.

Regarding the tsunami, when referring to Krakatau in 1883, Jefry only remembers it vaguely. He never paid any attention to the story. Honestly, Jefry admits that he forgot about the incident, although he may have heard the story either from his parents or grandfather or even from the media or at school. For Jefry, the 2018 tsunami was the new disaster knowledge that was formed from personal experience [38].

While Jefry witnessed the eruption of Anak Krakatau, whose avalanche triggered a tsunami, people in several areas on Sebesi Island also experienced the horror of being hit by the tsunami waves. One of them is Nenek Kani (65 years old), a resident of Regahan Lada. When people screamed for the tsunami, nenek was still in her house, trying to get some clothes and some basic food. At that time, the first wave began to arrive which was more like a tidal wave that entered her house. Noticing water that enter the house, Nenek Kani, who lived alone, immediately left the house and walked towards Mount Sebesi, following the advice of the village head and the people around her [39].

On the way to the mountain, Nenek Kani saw the second wave that rolled up high and hauling 3 boats that were docked to the mainland and hit her house and her neighbours. With horror, Nenek Kani walk immediately toward the mountain. Due to her age, nenek Kani could not reach further to the top, hence she decided to stop at a high enough place with those who were not able to walk again. Nenek Kani subservient to God if indeed the waves would reach and rolled her in her shelter. Fortunately, the third wave was not as powerful as the second one, so her shelter was safe. It was the second wave that caused many houses destroyed in Regahan Lada hamlet.

In the following morning, Nenek Kani returned to her house. At that time, she saw so much garbage that entered the house and the right-side wall of her house was collapsed by the tsunami. Several houses to the right of her house were badly destroyed so could no longer be lived in. She preferred to return to the shelter for another night as she was afraid of another tsunami. Nenek Kani, who came to Sebesi in 1955, remembered that so long as she lives on Sebesi, this was her first experience, where it was the most terrifying disaster. Initially, she had never heard the word *tsunami*, just followed the words that people often said without understanding it. However, after the incident, she realized that the wave was called as tsunami. For nenek Kani, she never had experienced or heard any stories about the catastrophic disaster in Sebesi, including stories from her parents or grandparents. The tragedy of Krakatau in 1883 appears to have never recorded in the memory of this 65-year-old grandmother. Furthermore, Anak Krakatau to nenek Kani was nothing more than a volcano near the island of Sebesi.

Recalled to the 2018 tsunami, the people of Sebesi unlikely to believe that their island would strike by such a severe disaster. Throughout his life on Sebesi Island, Mr. Mochtar (currently around 70 years old) never imagined that he would suffer from the tsunami disaster. For him, Mount Anak Krakatau has become like God's creatures who live side by side with them. However, Mr. Mochtar is aware that the island they occupy has indeed experienced total destruction after the eruption of Krakatau in 1883. He still remembers the story told by his *dato* (grandfather) regarding the enormity of the eruption [40].

Andi (21 years old) and Hasmy (31 years old), like most other Sebesian youths, were relaxing at Tejang Pier. It was a full moon at that time, since it was Saturday night, many people were still relaxing around the pier, including the Head of the Dusun *Bangunan*, Mr. Achmad Kurtubi (40 years old), who was looking at the pier while relaxing. It was around 9PM, and suddenly the sea water receded for about ten meters, many small ships that were docked were immediately sunk because of the dry water. At that time, many fish were floundering on the dry sand, but the people on the pier did not dare to go down. The people in pier found this as a strange phenomenon and began to mumble about the tsunami in Aceh which began with low tide. It did not clear who was first shouted *tsunami,* but the people on the pier started screaming and ran to their hamlets to save themselves. Some people immediately fled to Mount Sebesi, but many also ran to their homes to save their families while telling other people. Instantly people panicked and flocked to Mount Sebesi to save themselves from the tsunami [41].

At that time, a group of tourists from Jakarta, about 30 people, were staying at Mr Hayun's homestay. They were very enthusiastic to see Anak Krakatau who was active because its reddish feature was an exciting phenomenon. After dinner, they plan to visit Gubuk Seng, an area where people can see the mountain clearly and where a monitoring tower belong to the South Lampung Regional Government stood. Luckily, these group had not yet left for the area due to the delay of dinner schedule, because the Gubuk Seng area was the worst place hit by the tsunami. These tourists together with local people immediately fled to Mount Sebesi and spent their night there [42].

Looking back on the tsunami disaster, Andi believes that it was the worst he has ever experienced, while he does not retain the story of Krakatau 1883 at all. As he remembers, neither his parents nor his grandparents ever told stories related to the 1883 tsunami disaster. It is rare for young Sebesi children to hear stories from their parents because their parents are busy with their plantation in the middle of the forest. In line with Andi, Hasmy also stated that he did not remember the story of Krakatau in 1883, except it was vaguely obtained through television or from social media. As far as Hasmy remembers, Hasmy's parents and grandparents never share stories to him, including about the 1883 Krakatau disaster. Differently, Kurtubi remembered the 1883 Krakatau story which he got from conversations with the elders in Sebesi, but the story was blurred due to the absence of a major disaster befall Sebesi.

Memory is an individual process whose abilities are limited. As an individual, a person is capable to remember the information obtained as well as to forget something he got earlier. Forgetting is the process of failing to recall existing information. Forgetting is also described as the loss of the ability to recall or reproduce previously learned knowledge. When the stored information is getting weaker, it will be difficult to retrieve it from storage, and it tends to get lost over time if the information is not used. This is called as the *Decay theory*. This is in line with Ebbinghaus' research which explains that there is a close relationship between forgetting and time. Important factors in forgetting are how the information obtained and how often the information repeatedly used. The less it is used, the more the memory fades and the longer the span of time, the more likely it is to be forgotten [43].

This is reflected to people in Sebesi where the 2018 tsunami believed as their first experience, although history records that a same type of disaster in a much larger scale have stricken. Some people, especially the older generation, still remember the story of the 1883 Krakatau eruption that triggered the tsunami, especially from the story of their *dato* or grandmother. However, the disaster gap, the rare occurrence of severe hazards during more than a century, is significant element in explaining the loss of disaster memory. Most of the other people forgot whether they had received stories or information related to the incident or not. However, for the younger generation of Sebesi, the information about the 1883 Krakatau disaster is mostly received either from schools or the media. Unfortunately, this information appeared easily replace by other information due to the infrequent repetition hence it was fading away.

Another reason why this devastated disaster was blurred among people in Sebesi is the fading of oral tradition practice from older to the current generation, either in the form of storytelling, lullaby, recite a poetry, or narrating legend. There are some efforts from tourism activists to create tell-tale tradition about Krakatau, but only to attract incoming tourists. Usually at night when these travellers get together, the guide would start to recount the story where this received positive response from them. Regrettably, this practice ceased to perform to local communities resulting in a total lack of civilian disaster awareness and preparedness, also diminishing cultural value in society.

The 2018 tsunami played important role for the Sebesians in recurring disaster memory. The disaster experience for the Sebesians led to the formation of specific attitude or behaviour in looking at their natural environment. The result of and combination on individual experience on 2018 tsunami initiate a collective memory for the communities since it confines and binds their most intimate remembrances to each other. It reminds them that they live in a hazardous environment. Therefore, this memory encourages the awareness attitude within the community. Currently, the Sebesian arranged themselves when they smell a strong sulfuric smell, or a sudden rain of ash, or a roar sound, as a sign that something is happening to Anak Krakatau. Furthermore, they also pay attention to the waves, its environs and believed that the safest place for evacuation is Mount Sebesi. Accordingly, what occurred in Sebesi is correspond to the concept of Halbwach's collective memory. Subsequently, preserving this shared memory, either in the form of storytelling, poetry recite, or stories, as well as children's songs are important ways to transmitting messages, to bring communities together, and maintain the historical ties. Thus, memories of the 2018 tsunami can last for several generations.

Fig. 3 the debris of houses in Regahan Lada and Teluk Baru due to 2018 tsunami.
Source: private collection taken on August 3, 2019

3.3 The *Cerita* Smong as comparative case studies

The Smong story (Cerita Smong) is a local knowledge that origins from Simeulue Island of Aceh. The story is about a very devastated event of Tsunami on 4 January 1907, which hit some areas of the island. The report from colonial archive mentioned that some areas in West Simeulue was damaged and many victims reported. However, due to the traumatic event of tsunami, survivors of tsunami were telling stories over generation and put messages that if there is a strong earthquake and retreating water on the shore, people should run to hills for evacuation purposes, and avoid the shore to collect fish, "*...bila ada gempa kuat, laut surut maka larilah ke bukit dan jangan ke pantai untuk mengambil ikan...*". That story has been transferred over generations, not only among native people of Simeulue, but also to migrants. During the Indian Ocean Tsunami of December 26, 2004, the tsunami was hit Simeulue with the similar preceeding like 4 January 1907. There was evidence that many people warned again the smong stories and they successfully evacuated to the safe place. There were 7 people killed on the processes of evacuation, mostly because of panic and riding motorbike to evacuation areas. This event has made many peoples believed that knowledge of Smong safe people of Simeulue from tsunami.

However, there are different version related to how smong story transferred from one generation to other. The study sponsored by UNESCO conducted by LIPI in 2005 stated that smong stories transferred through story telling from older generation to their children, the grandchildren and so on. Usually, the time of storytelling related to its situation, mainly while there is natural hazard such as earthquake, floods and other. The elders would tell stories that hazard is not so dangerous compared to tsunami 1907. Another version says that elders or parents just telling stories without any particular situation, just tell story whenever they have chance to meet.

Other version about transfer story is through local folklore namely singing *buai-buai* (traditional lullaby) and narrating *nandong* (poem) [10], [11]. There are varies of lyrics depend on dialect of local dialect in Simeulue. However, those all shown that *smong* is part of local knowledge in Simuelue.

Report by UNESCO (Yogaswara and Yualianto, 2005)) stated that why smong stories has been existed over generation until the 2004's tsunami happened, because of several factors, namely (1) frequently of earthquake and other natural hazard that remind people about tsunami (smong) 1907 (2) survivors of tsunami 1907 that willing to telling stories and (3) kinship system and cultural bound among people of Simeulue and (4) material culture (natural and human made) that become 'living monument' remembering people about tsunami 1907. In addition, geographical features of Simeulue island that dominated of coastal and hilly areas provide people to monitor water situation after earthquake and easy access for evacuation route.

The tsunami of December 2004 that hit Simeulue island creating new knowledge about *smong* as local knowledge that safe people. The smong story has been researched by many agencies, including academic community, government agencies, international funding, and the like. The story of Smong transformed to different medium of expression, including digital tools and purposes for millennial generation. Indeed, the smong story be part of policy making of *qanun* related to disaster education, The story of Smong will always in mind of the people in Simeulue and Aceh.

4 Conclusion

History records the rise and fall of the Sebesi island due to a catastrophic disaster. The eruption of Krakatau on August 27th,1883 has led to enormous tsunami and took life of all the people in Sebesi. Based on the report from the officers who visited Sebesi after the eruption, it is unlikely to have survivors of this disaster. Therefore, the people living today are not those who witnessed or survived from the severe disaster. The current society is a migrant community who arrived in Sebesi to seek their chance from the development of agriculture. The fertile plantations cannot be separated from Krakatau's "blessing" because the volcanic ash that buried Sebesi has proven to make the land on this island very fertile.

The birth of Anak Krakatau since 1930 were a new challenge for newly established settlement in Sebesi. However, the economic reason was their main motivation to stay, greater than the threat of an unknown danger. Over time, Sebesi has become home to more than two thousand people. The growing of Anak Krakatau brought piece of good fortune to them, namely the development of tourism, which is economically profitable. The threat of danger viewed as just a figment since Anak Krakatau is so calm and pampers the Sebesians with the beauty of its evolution and its abundant marine life.

The rare occurrence of severe disaster from post Krakatau in 1883 to 2017 has caused people in Sebesi unaware to their 'risky' environment. The continuous eruption of Anak Krakatau, from booms, rain of ash, to the strong smell of sulfuric acid are believed as normal and harmless by the Sebesians. The disaster gap, different perceptions in looking at Anak Krakatau, as well as social, economic, and cultural life, are significant element in explaining the loss of disaster information they learnt from elders. Additionally, the absence of oral tradition practices, in the form of storytelling, lullaby, poetry and others led to the loss of disasters memories in current generations. Even though local government is regularly celebrating annual commemoration of Krakatau eruption *(Festival Krakatau)*, but this does not generate awareness and disaster risk reduction of the communities in Sebesi increase. It seems there is memory distance between Sebesi people with people in mainland Lampung area and Banten coast area.

However, tsunami-related knowledge among young people in Sebesi mostly been formed by learning at school and from the media. It was only after the Aceh Tsunami in 2004 that the public became familiar with the word *tsunami* due to the very intensive media coverage. In the past, people were more familiar with *rajuh* or known as rising tide into the land. Subsequently, the condition in Sebesi confirmed with the concept of cultural memory where the media plays an important role in shaping memories and further regulating people's behaviour.

The power of nature once again revealed when Anak Krakatau initiate a tsunami that strike Sebesi island, the coast of South Lampung, as well as Anyer and Pandeglang on December 22, 2018. The disastrous has changes the way in which the Sebesians perceived Anak Krakatau, from a blessing to a thread. The Sebesians are currently responsive to their surrounding environment, hence the hidden blessing of this disaster is the re-establishment of disaster memories and the revival of disaster awareness and preparedness. Lesson to learn from the Smong story of Simeulue is that a cultural preservation needs to be consider through appreciation for local people who maintain their knowledge. Local knowledge is always changing due to socio-cultural change in the community. However, recognize the local knowledge is the best way to acknowledge that local people owned their knowledge, not only scientist or the government.

References

1. "Tsunami Selat Sunda: Korban Tewas 430 orang, Krakatau jadi 'siaga', Hujan Abu di beberapa tempat", 27-12-2018. Diakses pada https://www.bbc.com/indonesia/live/indonesia-46663949
2. "BMKG Paparkan Kronologi Tsunami Selat Sunda", 23-12-2018 diakses dari https://www.cnnindonesia.com/teknologi/20181231002758-199-357400/bmkg-paparkan-kronologi-tsunami-selat-sunda
3. Zengefinnen, T., Lovhort, F., Pedersen, G.K., Muhari, Abdul. "Modelling 2018 Anak Krakatoa Flank Collapse and Tsunami: Effect of Landslide Failure Mechanism and Dynamics on Tsunami Generation." *Pure and Applied Geophysics* (2020): 2493-2516
4. Interview by phone with Bapak Hayun, a member of Tourism activist in Sebesi island, on 30 June 2020, at 19.30 WIB.
5. Interview with Bapak Andi (staff of Desa Tejang) on 3 August 2019 in the Office of Tejang Village in Sebesi island.
6. Simon Winchester, S. *Krakatau: Ketika Dunia Meledak 27 Agustus 1883* (Terjemahan) (Jakarta: Serambi Ilmu Semesta, 2006).

7. Yogaswara, H., Yulianto, E. *Smong—Local Knowledge and Strategies on Tsunami preparedness in Simeulue Island, Nanggroe Aceh Darussalam,* (LIPI and UNESCO, 2005).
8. Bankoff, G. "Comparing Vulnerabilities: Toward Charting an Historical Trajectory of Disasters," in *Historical Social Research.* Vol. 32, No. 3 (212), (2007) pp. 103-114.
9. Halbwachs, M. *On Collective Memory.* Chicago: The University of Chicago Press, (1992).
10. Tulius, J. (2020). "Lesson from The Past, Knowledge for The Future: Roles of Human Memories in Earthquake and Tsunami Narratives in Mentawai, Indonesia." *Paradigma Jurnal Kajian Budaya,*, Vol 10 (2), (2020).
11. Rahman. A., Sakurai.A., Munadi, K. 'Indigenous Knowledge management to Enhance Community Resilience to Tsunami Risk: Lesson Learned from Smong Traditions in Simeulue Island, Indonesia.' in *IOP Conference Series Earth and Environmental Science.* 56((1): 012018. IOP Publishing: 2017.
12. Gani A, Jaelani, "Seberapa Panjang Longue Durée? Catatan tentang Longue Durée dalam Praktik Penulisan Sejarah," in *Metahumaniora,* Volume 9 Nomor 3, Desember 2019, 313 – 327
13. Interview with Bapak Rozali—secretary of Tejang Village term 2020—on 15 October 2020 and interview with Bapak Tb. Firdaus—secretary of Tejang Village term 2021-2023—on 14 April 202
14. Wiryawan, B., D.G. Bengen, I. Yulianto, H.A. Susanto, A.K. Mahi, M. Ahmad. *Sumberdaya Pulau Sebesi, Kecamatan Rajabasa Kabupaten Lampung Selatan.* (Penerbitan Khusus Proyek Pesisir, Coastal Resources Centre - University of Rhode Island. Narraganset, Rhode Island, 2002)
15. Nieuwe Hoornsche Courant, Friday 29 January 1937.
16. Soerabaijasch Handelsblad, 22 October 1833;
17. Algeemen Handelsblad, 22 November 1833;
18. De Standaard, 4 December 1883.
19. *"Een Bezoek aan Krakatau,"* De Locomotief, 17 Maart 1885.
20. J.W Van Royen, *"Nota Over de Lampongsche Merga's"* (Landsdrukerij, 1930, pp. 49)
21. Twee Eilanden in Straat Soenda Te Koop," *De Locomotief,* Zaterdag 22 Augustus 1936, Zesde Blad, p. 2.
22. *De Sumatra Post,* 27 Augustus 1936, "Adat Rechten en Legenden", Derde Blad.
23. Ian Thorton, *Island Colonization. The Origin and Development of Island Communities.* Cambridge: Cambridge University Press, 2002
24. Interview with Bapak Busry on 15 October 2020 at Dusun Segenom. He was a supervisor for Segenom area. He was born in 1947 in Sebesi island and became 4th generation of his families that stayed in the island. In 1955, his father, who also a supervpervisor in Segenom, and his fellow supervisor from Regahan Lada, Mad Husein, were able to conduct Hajj pilgrim because of coconut sales during that period. Bapak Busry is a son of Hadji Abdoellah, a first supervisor appointed by Hadji Djamaloedin in 1925. He was among the first settler in Sebesi.
25. Interview with Bapak Busry on 15 October 2020; interview with Ibu Rohani on 16 April 2021; interview with Dato Mohamad Saleh (*Saleh Salui*) on 15 April 2021; Interview with nenek Kani on 25 May 2021
26. Interview with Bapak Hayun during field research in 2020 and 2021.
27. Interview with Bapak Mochtar on 17 April 2021 and 25 May 2021. His Dato (kakek) was the first batch workers that granted permission from Saleh Ali as profit-sharing cultivators in 1937. This was proved by a copy of an agreement letter between his Dato and Saleh Ali.
28. ANRI, 8552 TZG Ag 1931/29051
29. Interview with Bapak Mochtar on 23/5/2021, 17/4/2021, and 14/10/2020. He was appointed as Head Village for the period from 1988-2000.
30. Syamsidik, Benazir, Mumtaz Luthfi, Anawat Suppasri, and Louise K. Comfort, "The 22 December 2018 Mount Anak Krakatau volcanogenic tsunami on Sunda Strait coasts, Indonesia: tsunami and damage characteristics" in *Natural Hazards and Earth System Sciences,* (20, 549-565, 2020).
31. Lewis, James. (2009). "An Island Characteristic: Derivative Vulnerabilities to indigenous and exogenous Hazards." *Shima: The International Journal of Research into Island Cultures,* (Vol 3: 1).
32. Jedrusik, "Island Studies. Island Geography. But What is an Island?" in *Miscellania Geographica,* (Volume 15(1):201-212, 2011).
33. Yuli Evadiyanti, "Reposisi Brand dalam Festival Krakatau oleh Dinas Kebudayaan dan Pariwisata Kota Lampung." *PRofesi Humas: Jurnal Ilmiah Ilmu Hubungan Masyarakat,* (Vol 2: No. 1, 2017).
34. Yuli Evadianti, *ibid*
35. Interview with Ibu Rumanah, Ibu Jamilah, and Ibu Maemunah at the grocery store belong to Ibu Rumanah on 13 October 2020 at 14.00 PM
36. Interview with Bapak Jefry on 22 May 2021. He is a staf in BKSDA since 2008.
37. Interview with Bapak Jefry on 22 May 2021

38. Different with Pak Jefry, Bapak Herman, his elder brother from different mother, still clearly remember the catastrophic tsunami Krakatau because his grandmother from his mother's side continuously told the story to him. Bapak Herman works as a ship crew in KM Dewi Putri, a ship in Canti-Sebesi line. The experience of Bapak Herman explain that some Sebesians might know and still remember the previous tsunami, although most of them fairly admitted that they forgot about a devastated disaster in 1883. Bapak Herman, personal communication, 21 May 2021.
39. Interview with nenek Kani on 25 May 2021. She was among the first settler who moved in to Sebesi in 1955 at the age of 10 years old, the age where she can clearly remember her journey to Sebesi and the development of her *umbul* into hamlet. She was a wife of Mad Huein, a supervisor of Regahan Lada hamlet, so that she knew a development of this hamlet from 1960s to 1990s. Until recently, nenek Kani is an honourable figure as healer and traditional mid-wives.
40. Interview with Bapak Mochtar on 14 October 2020.
41. Interview with Hasmy, Andi, and Bapak Achmad Khurtubi di separate place during field research on 13-16 October 2020.
42. [Interview with pak Hayun during field research on 13-16 October 2020.
43. Nirupma Rath, "Remembering and Forgetting," in *Psychology of Education* (New Delhi: Global Vision Publishing House, 2008)

Islamic psycho-immunological approaches in increasing immunity during the COVID-19 pandemic

Imam Maulana[1], *Raisha* Fathima[1], *Haiyun* Nisa[2], and *Rina* Suryani Oktari[1,3*]

[1]Faculty of Medicine, Universitas Syiah Kuala, 23111 Jl. Tgk. Syech Abdul Rauf, Darussalam, Banda Aceh, Indonesia
[2]Department of Psychology, Faculty of Medicine, Universitas Syiah Kuala, 23111 Jl. Tgk. Syech Abdul Rauf, Darussalam, Banda Aceh, Indonesia
[3]Tsunami & Disaster Mitigation Research Centre (TDMRC), Department of Family Medicine, Faculty of Medicine, Universitas Syiah Kuala, Banda Aceh, Indonesia

Abstract. The COVID-19 pandemic has a significant impact on mental health because it can increase the risk of panic, anxiety, and fear in society. New normal conditions with various limitations require people to be adaptive in order to cope with stressors that arise during the pandemic. This condition could interfere with mental health and adversely affect the body's immunity. Stress, through the nervous system, functions to help individuals move actively. In individuals who experience stress, the sympathetic system will activate various organs of the body causing the increment of blood sugar, blood pressure, muscle activities and metabolism which can reduce the immunity of individuals. Besides that, various efforts have been made as a form of controlling the spread of infection such as the application of health protocols, the concept of *Iman* (Faith), *Imun* (Immune), *Aman* (Secure), and vaccination. In addition, there are other efforts done by the Islamic religious people that hold approaches in the form of beliefs and daily life rituals, namely the concept of *Dua, Ikhtiar, Tawakkul,* and *Tawadhu* in facing the COVID-19 pandemic disaster. This concept is related to the psychoneuroimmunology aspect which, if applied correctly, can have an impact in finding mental resilience.

1 Introduction

According to the United Nations, individuals in the United States of America (45%), Iran (60 %) and Republic of China (35%) got distress during COVID-19. The Lancet Commission on global mental health and sustainable development reported that people are situated on a continuum from mild, time-limited distress to severe mental health conditions in COVID-19 pandemic. Many people are now less able to cope than before because of the stressors generated by the pandemic. Society tends to be in increasing anxiety and distress especially those who had less experience of these. Some get worsening conditions if once affected previously and even develop a mental health condition [1].

In response to the increasing number of COVID-19 cases in Indonesia, The National Task Force informs a program called *iman, aman, imun*, as the addition to "3M" program that stands for "*menggunakan masker*" (wearing mask), "*mencuci tangan*" (washing hands), and "*menjaga jarak*" (social-distancing). *Iman* means faith, *aman* means safety and *imun* means people encouragement to maintain their immunity [2]. This concept is reflecting the values of faith in conjunction with spirituality as one of the main factors to pursue comprehensive protection against COVID-19.

As the pandemic continues and significantly affects the economy and social life, it's a challenge to prepare society, not only in terms of disease prevention, but also mental attentiveness. Mentally prepared state means a normal life to run along with the period of crisis caused of COVID-19. As distress is the ultimate risk of every individual during the pandemic, immunity is on the edge of its impact. Human body responds to stressors by activating the hypothalamus-pituitary-adrenal axis (HPA axis) that elevates cortisol level, suppressing the immune system. The vicious cycle of stressors and immune suppression carries harms towards psychoneuroimmunology.

The needed approaches are supposed to connect the spiritual aspect to the physical aspect, as implemented in Islamic point of view. all which are *dua, ikhtiar, Tawakkul, tawadhu*. *Dua*, such as a prayer to be avoided by disasters; *Ikhtiar*, such as complying with health protocols correctly, sorting information, and

* Corresponding author: okta@unsyiah.ac.id

https://doi.org/10.1051/e3sconf/202234005009

maintaining a lifestyle, as well as vaccinations; *Tawakkul*, such as accepting the situation and surrendering to God after doing the best we could; and *Tawadhu*, which is humbling so that we always stay alert of our conditions. This concept is in line with the mechanism of psychoneuroimmunology which explains the reciprocal effect of mind-body on immunity.[3] Islamic religious rituals have become a part of life, including depicting disasters as a test instead of a punishment that can definitely be faced so that we are out of despair.

The purpose of this study is to examine the approaches of the Islamic concept *(Dua, Ikhtiar, Tawakkul, Tawadhu)* in facing the COVID-19 pandemic regarding psychoneuroimmunology. This study is expected to be an insight for the society so they can be educated better through life rituals, related to surrounding characteristics and cultures to be easier accepted.

2 Method

This study used a systematic literature review with a descriptive method using search engines from databases such as Scopus, Science Direct, DOAJ, Google Scholar, PubMed, SpringerLink, and Sage. The following research questions directed the search strategy, including: i) What is the psychoneuroimmunology concept in Islam? ii) How does it influence immunity? and iii) How does the Islamic psychoneuroimmunology approach increase mental health resilience during the COVID-19 pandemic. To enable for a broader spectrum of literature to emerge from the searches, the databases were searched using the following key terms: *"Islam"*, *"mental health"*, *"spirituality"*, *"psychology"*, *"immunology"*, *"COVID-19"*, *"psychoneuroimmunology"* and *"HPA axis"* by adding Boolean Operator "AND" and "OR". As an inclusion criterion, this review only considered articles published in English and Indonesian, then discussing the Islamic concept of psychoneuroimmunology. Articles that do not have full access were excluded from this review. The literature search was conducted from April 10 to August 10, 2021. The total minimum of 30 articles that met the inclusion and exclusion criteria, were included for further analysis.

3 Result and discussion

3.1 Stress and COVID-19 pandemic

During COVID-19 pandemic, people are constrained to adapt to regulations of social distancing, physical distancing, work from home, and school from home. However, this condition raises new problems, namely limitations in daily activities or living in quarantine. Recent evidence suggests that people who are kept in isolation and quarantine experience significant levels of anxiety, anger, confusion, and stress [4]. Research has also shown that frequent media exposure may cause distress.[5] A systematic review and meta-analysis conducted by Salari et al. 2020 conclude that COVID-19 not only causes physical health concerns but also results in a number of psychological disorders (Figure 1) [6].

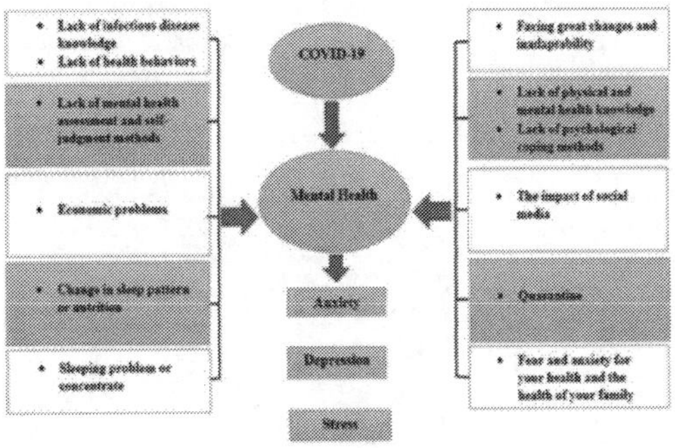

Fig 1. Impacts of COVID-19 pandemic on mental health [6].

Stress defined by Selye (1956) as the effects of anything that fatally threatens homeostasis [7]. Dhabhar-McEwen said that stressor (source of stress) will be responded by the brain (stress perception) then responded by other systems (stress response) [8]. Since COVID-19 has been stated as a pandemic, it unveiled some stressors due to unclarity and changes for all of sudden in certain vital sectors like health, economy and social life. Although the concept of "new normal" has been socialized, the demands to adapt to these conditions have become a challenge.

Physiologically, stress is divided into 3 stages known as the General Adaptation Syndrome (GAS): 1) Alarm Reaction Stages, refers to the initial symptoms of the body under acute stress and the "fight or flight" response. After the initial shock of the stressful event, the body begins to repair itself by lowering cortisol levels and normalizing the physiologic responses (i.e., blood pressure and heart rate). During this recovery phase, the body remains on alert until the stressful event is no longer an issue. However, if the stressful event persists for extended periods of time, the body will adapt to cope with the higher level of stress. The body will continue to secrete stress hormones which keep the body's physical response to stress elevated. This induces the second stage; 2) Resistance stage includes symptoms of poor concentration, irritability, and frustration. If the stressful event continues to persist, the body will enter the third stage; 3) Exhaustion Stage, symptoms of this stage include burnout, fatigue, depression, anxiety, and reduced stress tolerance. As the stressful event persists, the body's immune system will continue to weaken. This is due to the suppressive effects of stress hormones on cells of the immune system [9].

Although it seems that stress has only negative effects on the body, it actually also has advantageous effects. Stress which gives disadvantage on the body known as distress, and the one which has an advantageous effect known as eustress. Eustress energizes and motivates us to make a change. It gives us a positive outlook and makes us capable of overcoming

https://doi.org/10.1051/e3sconf/202234005009

obstacles and sickness [10]. Distress or eustress depends on how the individual perceives the stress. If they feel that their coping skills are inadequate, they will experience distress. On the other hand, if an individual perceives their coping skills as adequate, they will experience eustress. Transactional Approach outlines that an individual's experience of stress is dependent on their appraisal of their ability to cope with the stressor (Figure 2) [11].

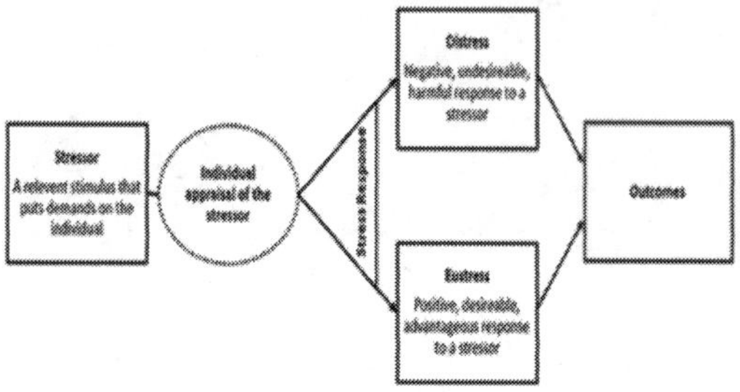

Fig. 1. A visual description of the partial-consensus definition of the stress process [11].

Distresses increase the risk of anxiety, anger, disappointment, frustration, financial problems, and stigma [4]. The most common stressors during the pandemic are fear and anxiety about the self-health and people surrounding [12]. The human body responds to stressors by activating molecular and cellular changes in the brain mediated by activation of the hypothalamus-pituitary-adrenal axis (HPA axis) which can elevate cortisol level. Cortisol suppresses the immune system [13].

Psychological stress and distress have been associated with higher levels of oxidative damage [14]. The mechanism remains unclear, and is likely mediated in part by stress-related hormones (e.g., cortisol) as well as health behaviors [15]. The first evidence that low-to-moderate doses of stress may have *beneficial* effects ("eustress") emerged over a century ago as an "inverted U" relationship between arousal and performance (Figure 3) [16].

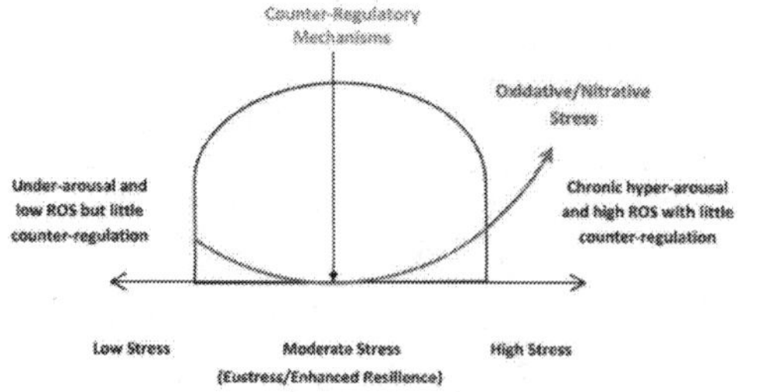

Fig. 3. Moderate Stress can enhance resilience [17].

Research by Aschbacher et al. shows that moderate level of stress perceived were associated with reduced levels of oxidative damage compared to low and high which supports the emerging model that chronic stress exposure promotes oxidative damage through frequent and sustained activation of the HPA axis [17].

This proves that even COVID-19 is a real ordeal, but the wise reaction, *Dua, Ikhtiar, Tawakkul* and *Tawadhu*, towards the emerging stress, can maintain the level of stress under its high level.

3.2 The concept of *Dua, Ikhtiar, Tawakkul* and *Tawadhu* towards psychoneuroimmunology

In Islam, the stress management concept towards stressor is comprehensive. In disaster for example, there are three different point of views: 1) Disaster as *musibah*, which means something that befalls human beings in the form of something either pleasant or unpleasant. It is usually caused by human sins, including being ignorance of fact; 2) Disaster as *bala*, which means test (human promotion). *Bala* is God's will without human involvement. *Bala* aims to raise the human standing, forgive his sins and purify his soul. it's used in *"...And We tried them with good (favors) and bad (disasters) in order to return (to the truth)." Surah al-A'raf [7]: Verse 168; 3)* Disaster as *azab,* which means punishment (punishment of God). In the Qur'an, *azab* is interpreted as torture or an excruciating punishment inflicted by God only on those who are ungodly and do not believe. The religious leaders interpret the disaster that befell the people of the previous prophets such as the people of Prophet Shuaib, the people of Noah, the people of Prophet Lut, the People of Prophet Saleh and the tribe 'Ad people of Prophet Hud as a punishment for not believing [18].

The comprehensive concepts of stress management in Islam consist of *Dua, Ikhtiar, Tawakkul, and Tawadhu*. The concept is inseparable from the lives of Muslims can be a supporting modality in dealing with existing distress. *Dua, Ikhtiar, Tawakkul and Tawadhu* were the cultivated values in the perspective of Islam, including the people of Aceh, depicted from the daily religious life rituals. These spiritual activities give tranquility upon the individuals psychologically.

Dua is defined as an expression of the birth of conscience or feeling of need to ask for help or help from Allah (SWT). *Dua* is essentially an expression of submission of faith to God and of one's neediness. Two types of Dua: 1) *Du'ā al-mas'alah* (دُعَاءُ الْمَسْأَلَة *du'ā'u 'l-mas'alah*), or the 'dua of asking.' This type of dua is when one asks for the fulfillment of a need, or that some harm be removed from him/her. For instances, *Wakulja alhaqqu wazaqal batil, Innal batilaka nazahuka, meaning: "And say, the truth has come and the falsehood has disappeared, and indeed the falsehood will be destroyed." (Qs: Al-Isra' [17]: 81) which used by Acehnese as part of ritual Tulak Bala (translate: Rejections of disaster);* 2) *Du'ā al-'ibadah* (دُعَاءُ الْعِبَادَة *du'ā'u 'l-'ibādah*), or the 'dua of worship.' This type of *dua* includes every single act of worship. Examples would include when a Muslim prays or gives *zakāt* or fasts. Pajevic et al (2017) in his research towards 100 war veterans divided into two groups (practicing and not practicing daily praying) shows that practice groups have a reduction towards risk, impulsiveness, and aggression than their peers who did not practice religion [19].

Besides supplicating through *dua*, Islam's view in guaranteeing someone's attainment is not separated from God's instruction to humans to make certain efforts, or so called *ikhtiar*. *Ikhtiar* means to endeavor to receive better results [20]. It is stated in *Quran* that emphasized the importance of *ikhtiar*, as follows, "God does not change the condition of a people unless they change what is in themselves", stated in surah 13 (Ar-Ra'd), verse 11. This indicates that no matter how truly someone does *dua*, it must be accompanied by effort. During COVID-19, *ikhtiar* might mean complying with some rules such as wearing a mask, washing hands and keeping a distance for each individual to hamper the transmission of the virus.

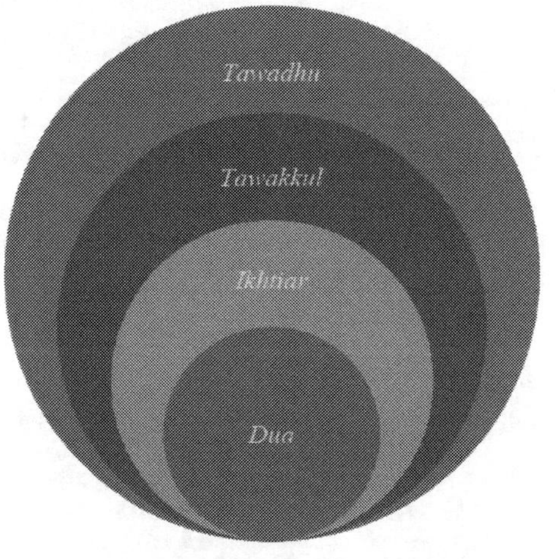

Fig. 3. The configuration of *Dua, Ikhtiar, Tawakkul* and *Tawadhu*.

After *dua* and *ikhtiar*, Islam also concerned about human's mind serenity by surrendering to Allah (SWT) following all prayers and efforts done. This is named *tawakkul*. *Tawakkul* is an Arabic word emphasizing the reliance on God or "trusting in God's plan" [21]. Being *tawakkul* also leads someone to do the efforts as best as they could and feel no regrets because they know Allah (SWT) is the best helper and no burdens He gives upon a soul more than it can bear, as stated in *Quran*, "God does not burden any soul with more than it can bear", surah 2 (*Al-Baqara*), verse 286. Peace upon someone's feeling after accomplishing hard work is a result after Tawakkul, as strengthen in Quran that said, "And whoever puts all his trust in Allah, He will be enough for him", surah 65 (*At-Talaq*), verse 3.

Islam concept in shaping someone wise in managing stress is also directing humans to be humble and keep sustaining good deeds, or named *tawadhu*. *Tawadhu* or humility indicates humans to stay alert of the recent conditions and do not be too satisfied as if no more efforts are needed. The Messenger of Allah (Peace Be Upon Him) has said, *"Allah has commanded me as follows: Show such humility that none of you boasts to others and none of you oppresses another"* cited by Muslim, Jannah, 64 [22]. That statement is highlighting the importance of having sincere intentions wholeheartedly to sustainably follow the COVID-19 health protocols amid the pandemic, even in healthy condition or negative from COVID-19 test.

3.3 Psychoneuroimmunology in Islamic concept

Conceptualizing Islamic values as the approach in psychoneuroimmunology, in the form of religious-derived mindfulness practices, is considered as mind/body medicine. The mind/body medicine emphasizes the mental state of humans that is able to connect to the psychology and the physical aspect.[23] The mental state is maintained through meditation-based spiritual activities [24].

Several studies found out the improved neural-plasticity and the increased cerebral blood flow (CBF) in the transcendental meditation [25], the increment of abrineurin or brain-derived neurotrophic factor (BDNF) and the decrement of cortisol production in persistent Buddhist meditation [12], and the upgraded brain connectivity in long-term meditators [26]. Interestingly, brain imaging researchers have discerned brain anatomy changes in long-term meditators, like the increased brain cortical thickness in the areas processing sensory and attention [27, 28], and the improved grey matter density in the cerebellum, the temporo-parietal junction, the posterior cingulate cortex [29], and also the brain stem [30]. In addition, short term meditation, like a few hours of integrated mind-body training, is advantageous in enhancing white matter integrity of the superior and anterior corona radiata and corpus callosum [31]. It is also claimed that religious-derived mindfulness practices affect positively upon the psychological and physical sector and even become a part of stress-related disorders treatment [32-35].

There are two types of Islamic approach in mind/body medicine, which are *Al-tibb al-jismani* (physical medicine) and *Al-Tibb al-Ruhani* (psychological/spiritual medicine). Both approaches are related conjointly since in Islam point of view, human is having two main aspects, spirit (*ruh*) and body (*jism*). While it is different from modern medicine that focuses only on the physical aspect [36]. This explanation creates the balance between body and mind to reach overall peace and health of humankind. The implementation of this concept as shown in Table 1.

Table 1. The instances of *Dua, Ikhtiar, Tawakkul* and *Tawadhu* implementation along COVID-19.

	Pre COVID-19	During COVID-19	Post COVID-19
Dua	Asking for being avoided from illness	Asking for the sickness to be lifted	Ask for the healthiness
Ikhtiar	Following Health Protocols Getting vaccinated Filtering information and hoax	Getting COVID-19 test Report to healthcare facilities Following the doctors order	Following Health Protocols Filtering information and hoax
Tawakkul	Not being panicked Believe it could a *bala*, instead of *azab*	Believe that the sickness is a test from God	Being grateful for being healthy again
Tawadhu	Stay alert of the recent conditions not be too satisfied as if no more efforts are needed Keep following health protocols	Not blaming anyone (including ourselves) for our sickness	Stay alert of the recent conditions Keep following health protocols

In terms of COVID-19 that causes grief for all of the sudden adaptation and significant health and economic loss, the Kübler-Ross cycle presents five stages of grief. It consists of denial, anger, bargaining, depression and acceptance [37]. The concept of *Dua, Ikhtiar, Tawakkul* and *Tawadhu* is expected to speed up or cut the step as it is said that each individual does not always experience the whole stages. This approach also builds up a belief that any stress is supposed to be eustress instead of distress. This belief is created by several key points, which includes: i) *Dua*, it creates a belief that there is a higher power that will help us, it makes us feel not alone and be able to overcome the situations; ii) *Ikhtiar*, by doing everything at best creates a positive value which also provide us with solution and solving the problem; iii) *Tawakkul*, after the best action comes to a perceive that the result will be in line with our action; and iv) *Tawadhu*, which makes us keep on track and not overconfidence of our deeds so we will continue to do *dua*, *ikhtiar*, and *tawakkul*.

4. Conclusion

The concept of *Dua, Ikhtiar, Tawakkul, and Tawadhu* during the COVID-19 pandemic is useful in coping with distress conditions. Spirituality and belief, as well as following health protocols appropriately, can suppress the body's response to stressors so that the body's immunity is maintained. It is expected that by implementing this concept, people will have a more resilient psychological state that provides a more robust immunity system.

Conflict of interest. The authors declare that there is no conflict of interest regarding the manuscript.

Author contribution. All authors are contributed equally to the content of the study.

References

1. Nations, U., *Policy Brief: COVID-19 and the Need for Action on Mental Health*. 2021, United Nations.
2. (WHO), W.H.O., *Coronavirus Disease 2019 (COVID-19) Situation Report - 28*. 2020, World Health Organization (WHO): Indonesia. p. 1-23.
3. Francisco T Ausk, I.E., Jan Moynihan, *Psychoneuroimmunology*. Dermatol Ther, 2008. **21** p. 22–31.
4. Brooks SK, W.R., Smith LE, Woodland L, Wessely S, Greenberg N, et al., *The psychological impact of quarantine and how to reduce it: rapid review of the evidence*. Lancet, 2020. **395**(10227): p. 912–20.
5. Neria Y, S.G., *Understanding the mental health effects of indirectexposure to mass trauma through the media*. JAMA, 2011. **306**(12): p. 1374–5.
6. Nader Salari, A.H.-F., Rostam Jalali, Aliakbar Vaisi-Raygani, Shna Rasoulpoor, Masoud Mohammadi, Shabnam Rasoulpoor, Behnam Khaledi-Paveh, *Prevalence of stress, anxiety, depression among the general population during the COVID-19 pandemic: a systematic review and meta-analysis*. Globalization and Health, 2020. **16**(57).
7. H, S., *The Stress of Life*. 1956, New York: McGraw-Hill.
8. Dhabhar FS, M.B., *Acute stress enhances while chronic stress suppresses cell-mediated immunity in vivo: a potential role for leukocyte trafficking*. Brain Behav Immun, 1997. **11**(4): p. 286-306.
9. Brianna Chu, K.M., Terrence Sanvictores, Derek Ayers. *Physiology, Stress Reaction*. StatPearls 2021 2021 Sep 18 [cited 2021 01 August].
10. Ross, D., *Distress vs. Eustress*. Drake University. p. 1-3.
11. Victoria Branson, M.J.D., Edward Palmer, Deborah Turnbull, *The Adolescent Distress-Eustress Scale:Development and Validation*. SAGE Open, 2019. **9**(3): p. 1-14.
12. Thaddeus W.W. Pace, L.T.N., Daniel D. Adame, Steven P. Cole, Teresa I. Sivilli, Timothy D. Brown, Michael J. Issa, Charles L. Raison, *Effect of Compassion Meditation on Neuroendocrine, Innate Immune and Behavioral Responses to Psychosocial Stress*. Psychoneuroendocrinology, 2009. **34**(1): p. 87-98.
13. Dedy Syahrizal, R.F., Imam Maulana. *Psychoneuroimmunology: The Approach for*

Increasing the Immunity of Disaster Victims. in *2017 Conference on Health Management in Post Disaster Recovery*. 2017.
14. Gidron Y, R.K., Tissarchondou H, Warner J, *The relation between psychological factors and DNA-damage: a critical review*. Biol Psychol, 2006 **72**(3): p. 291-304.
15. Ballal K, W.C., Harmancey R, Taegtmeyer H, *Obesogenic high fat western diet induces oxidative stress and apoptosis in rat heart*. Mol Cell Biochem, 2010. **344**(1-2): p. 221-30.
16. Yerkes RM, D.J., *The relation of strength of stimulus to rapidity of habit-formation*. J Comp Neurol Psychol, 1908. **18**: p. 459–482.
17. Aschbacher K, O.D.A., Wolkowitz OM, Dhabhar FS, Su Y, Epel E *Good stress, bad stress and oxidative stress: insights from anticipatory cortisol reactivity*. Psychoneuroendocrinology 2013. **38**(9): p. 1698-1708.
18. Aksa, F.I., *Islamic perspectives in disaster: An alternative to changing fatalistic attitudes*. Jàmbá: Journal of Disaster Risk Studies, 2020. **12**(1): p. a942
19. Izet Pajević, O.S., Mevludin Hasanović, *Association of Islamic Prayer with Psychological Stability in Bosnian War Veterans*. Journal of Religion and Health, 2017. **56**(1): p. 2317–2329.
20. Azizunisak Hidayati Wahyuna, S.F., *The Concept of Hope in The Western and Eastern Perspective*. International Journal of Islamic Educational Psychology (IJIEP), 2020. **1**(1): p. 1-12.
21. Librande, L., *Studia Islamica*. Vol. 100. 2005: Maisonneuve & Larose.
22. Topbas, O.N. *Tawadhu' (Humility)*. [cited 2021 20 June]; Available from: https://en.osmannuritopbas.com/tawadhu-humility.html.
23. Saniotis, A., *Understanding Mind/Body Medicine from Muslim Religious Practices of Salat and Dhikr*. J Relig Health, 2018. **57**(3): p. 849-857.
24. Chittaranjan Andrade, R.R., *Prayer and healing: A medical and scientific perspective on randomized controlled trials*. Indian J Psychiatry, 2009. **51**(4): p. 247-253.
25. R Jevning, R.A., M Biedebach, G Fernando, *Effects on regional cerebral blood flow of transcendental meditation*. Physiol Behav, 1996. **59**(3): p. 399-402.
26. Eileen Luders, K.C., Katherine L. Narr, Arthur W. Toga, *Enhanced Brain Connectivity in Long-term Meditation Practitioners*. Neuroimage, 2011. **57**(4): p. 1308–1316.
27. Britta K. Hölzel, J.C., Mark Vangel, Christina Congleton, Sita M. Yerramsetti, Tim Gard, Sara W. Lazara, *Mindfulness practice leads to increases in regional brain gray matter density*. Psychiatry Res, 2011. **191**(1): p. 36-43.
28. Amishi P. Jha, J.K., Michael J. Baime, *Mindfulness training modifies subsystems of attention*. Cognitive, Affective, & Behavioral Neuroscience, 2007. **7**: p. 109-119.
29. Britta K Hölzel, S.W.L., Tim Gard, Zev Schuman-Olivier, David R Vago, Ulrich Ott, *How Does Mindfulness Meditation Work? Proposing Mechanisms of Action From a Conceptual and Neural Perspective*. Perspect Psychol Sci, 2011. **6**(6): p. 537-59.
30. Peter Vestergaard-Poulsen, M.v.B., Joshua Skewes, Carsten R Bjarkam, Michael Stubberup, Jes Bertelsen, Andreas Roepstorff, *Long-term meditation is associated with increased gray matter density in the brain stem*. Neuroreport, 2009. **20**(2): p. 170-4.
31. Yi-Yuan Tang, Q.L., Xiujuan Geng, Elliot A. Stein, Yihong Yang, Michael I. Posner, *Short-term meditation induces white matter changes in the anterior cingulate*. Proc Natl Acad Sci U S A, 2010. **107**(35): p. 15649–15652.
32. J Kabat-Zinn, A.O.M., J Kristeller, L G Peterson, K E Fletcher, L Pbert, W R Lenderking, S F Santorelli, *Effectiveness of a meditation-based stress reduction program in the treatment of anxiety disorders*. Am J Psychiatry, 1992. **149**(7): p. 936-43.
33. J J Miller, K.F., J Kabat-Zinn, *Three-year follow-up and clinical implications of a mindfulness meditation-based stress reduction intervention in the treatment of anxiety disorders*. Gen Hosp Psychiatry, 1995. **17**(3): p. 192-200.
34. D Goldberg, A.H., A Furomoto-Dawson, H Nelson-Johnson, *Mindfulness-based stress reduction and its effects on well-being*. Journal of Investigative Medicine, 1998. **46**: p. 278A.
35. Grossman P, N.L., Schmidt S, Walach H, *Mindfulness-based stress reduction and health benefits. A meta-analysis*. J Psychosom Res, 2004. **57**: p. 35-43.
36. N Deuraseh, M.A.T., *Mental health in Islamic medical tradition*. The International Medical Journal 2005. **4**(2): p. 76–79.
37. Ann Scheck McAlearney, J.L.H., Cynthia J Sieck, Timothy R Huerta, *The Journey through Grief: Insights from a Qualitative Study of Electronic Health Record Implementation*. Health Serv Res, 2015. **50**(2): p. 462–488.

E3S Web of Conferences **340**, 05010 (2022) https://doi.org/10.1051/e3sconf/202234005010
The 13th AIWEST-DR 2021

Psychological preparedness of coastal communities in Surabaya: a preliminary finding

Listyati Setyo Palupi[1*2345]

[1]Psychology Department, Universitas Airlangga, Jl. Airlangga 4-6, Surabaya, East Java, Indonesia
[2]Addiction Study Centre, Jl. Airlangga 4-6, Surabaya, East Java, Indonesia
[3]Health and Environmental Sustainability Research Group, Jl. Airlangga 4-6, Surabaya, East Java, Indonesia
[4]Technology and Psychology of Social Change Research Group, Jl. Airlangga 4-6, Surabaya, East Java, Indonesia
[5]Direktorat Riset dan Pengabdian Masyarakat (DRPM) RISTEKDIKTI Indonesia Penelitian Dasar (PD) research funding scheme 2021 (No. 528/UN3.15/PT/2021)

Abstract. Indonesia has been acknowledged as one of countries that has the largest coastline in the world after Canada and Norway. Situated between Australia plate, Pacific plate, and Eurasia plate has made the country prone to disaster. Tsunami is one of the disasters that struck coastal communities in Indonesia. Tsunami is one type of disaster that bring devastated impact economically, socially and psychologically especially for the coastal communities including for Surabaya community that live along the coast. Studies found that psychologically preparing individual to face disaster could help to reduce the psychological impact of the disaster. Therefore, identifying the psychological preparedness is pertinent to reduce the risk of disaster especially for the coastal communities in Surabaya that were prone to tsunamis and other natural hazards. The purpose of the study is to describe the psychological preparedness for disaster of coastal communities in Surabaya. The result shows that the psychological preparedness for disaster of the participants was mostly in average level. Therefore, psychological preparedness for disaster needs to be improved especially among female with age between 12-35 years old population in order to reduce the risk of psychological impact of disaster.

1 Introduction

Indonesia is surrounded by Australia plate, Pacific plate, and Eurasia plate. This situation has made Indonesia prone to disaster. As an archipelagic country, Indonesia is known as country that has the largest coastline in the world after Canada and Norway. Therefore, Indonesia is included as an archipelagic country that are prone to disaster.

According The National Disaster Management Agency of Indonesia record, there are 28.100 disaster that happen in Indonesia in 10-year period from 2011 until 2021 [1]. Record of disaster composition that happen in Indonesia from January until August 2021 could be seen in figure 1 below.

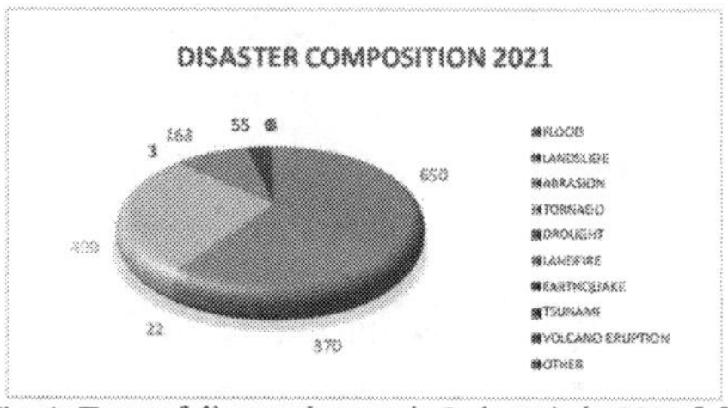

Fig. 1. Type of disaster happen in Indonesia in 2021 [1].

Disaster that happens from January to August 2021 has happen in almost every province in Indonesia from Aceh to Papua. Several provinces in Indonesia that struck by disaster in 2021 has been record by Indonesian Disaster Data and Information Database (DIBI) below.

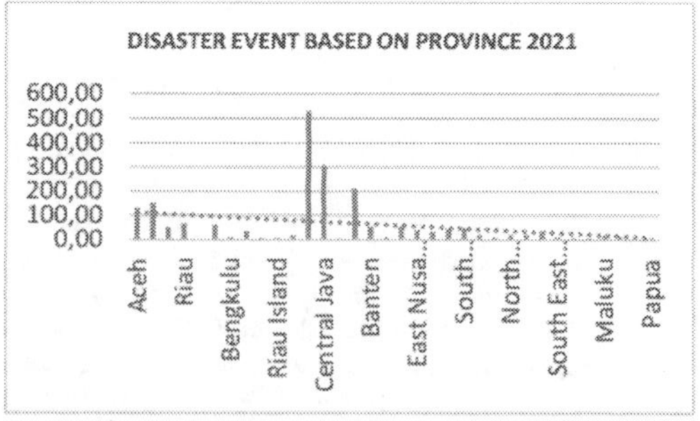

Fig. 2. Disaster occurrence based on province 2021 [1].

East Java Province is on the third position among 34 provinces in Indonesia that experience disaster in 2021. Disaster that often happen in East Java in 2021 are flood, cyclone, landslide, and eruption [2]. There are several cities in East Java that are identified as disaster-prone area including Surabaya.

Surabaya is one of region in Indonesia that reside along the coastline. Geographical location of Surabaya

* Corresponding author: listyati.palupi@psikologi.unair.ac.id

made the region prone to hydrological hazard such as flood, drought, and tsunami.

Disaster could incur devastated impact economically, socially and psychologically to the society. Infrastructure damage, settlement damage and also industrial damage has affected the economic condition of the community affected by the disaster. Beside of that, being separated from other family member, have to live in a shelter, and loss of privacy is among other social impact of disaster [3]. Disaster victim could also experience mental health issue as a result of being exposed to a disaster event. Research that focuses on psychological impact of disasters has found that disasters experience has induces several psychological problems to the survivor. The survivor could develop acute stress disorder, post-traumatic stress disorder (PTSD) and also other psychological disorders [4–8].

There have been several interventions that are develop to reduce the psychological impact of disaster. However, the interventions were only focus in technical preparedness and neglecting the psychological aspects of disaster preparedness. Studies found that psychologically preparing individual to face disaster could help to reduce the psychological impact of the disaster [9–11].

Several authors have defined psychological preparedness. According to Roudini et al (2017), psychological preparedness is "… a state of awareness and expectation of an individual's psychological reactions to the disaster warning … mental health preparedness will significantly support the ability to identify particular stress-related emotions and thoughts … this also contains the perception, appraisal, and understanding of threatening events and the risk communication" [12]. Another definitifon of psychological preparedness is propose by Zulch (2015) which is "a heightened state of awareness, anticipation, and readiness for: (1) the uncertainty and emotional arousal in expectation of the possible occurrence of the threat, (2) one's own psychological response to the unfolding threat situation, and (3) the ability to manage the demands of the situation" [13]. Boyland () extend the definition provide by Zulch and define psychological preparedness as multidimensional construct … the multiple attributes are: 1) perceived control over one's coping ability; 2) perceived control over the environmental outcomes; 3) a general understanding of the risk context; and 4) proactive planning and preparing for the threat" [14]. Various definition has been proposed nonetheless psychological preparedness for disaster is still understudied. Study found that the available research that investigate psychological preparedness for disaster are limited [15].

Research found that, psychologically preparing individual to successfully manage a disaster situation or disaster could also influence their resilience [16]. Thus, psychological preparedness for disaster is an important aspect to reduce the risk of disaster especially for coastal communities such as in Surabaya that were prone to tsunamis and other natural hazards. The purpose of this study is to describe the psychological preparedness for disaster of coastal communities in Surabaya.

2 Method

This study was conducted using online survey research with 583 people participants involved in this research. The invitation to the general population to be involve in this study was send in social media such as Facebook, Instagram and online communication platform such as WhatsApp.

All respondents should complete Psychological Preparedness for Disaster scale that are developed by the author.

2.1 Measurements

The Psychological Preparedness for Disaster (PPD) scale consists of three parts which are informed consent, identity, and the question or statement. The participants

First part of this scale was provided to record the participant's informed consent. The second part is a part where participants fill their identity which includes: age, gender, and socioeconomic status. The third part is the Psychological Preparedness for Disaster scale 1.1. This scale consists of 34 items version which include 17 items of affective dimension, 8 item of behaviour dimension, and 9 item of cognitive dimension. The internal consistency of the scale is a = 0.84. Data were analysed by using descriptive statistical method.

The total score of PPD scale is acquire by summing up the score on each item. The total Scores can range from 34 to 68. The mean of the PPD score is Higher score indicates higher psychological preparedness for disaster.

The participants total score then classified into three group which is low, average, and high. The classification was made based on the equal interval classification [17]. Participant with score below 46 is considered low, score between 46 up to 57 is average, and score above 57 is high.

3 Results

Based on age, data shows that the majority of the participant or 99% participant in young adulthood ((15-24 years old) category. The majority of the participant that have average level of psychological preparedness for disaster in this study is 67%. High and low level of psychological preparedness are found in 23% and 10% respectively.

Participant socioeconomic status is mostly at low-income category with 63%. While, middle-, and high-income group were 22%, and 15% respectively.

3.1 Demographic profile

The participants involve in this study were 583 people Surabaya resident. The gender composition in this study are 73% female and 27 % male. The participants age group range between 15 years old until 33 years old.

The participants were mostly from low-income socio-economic status (57%). While the middle-income group and high-income group were only 22% and 15% respectively.

There are only three age categories that count in his study which are youth (11-14yo), young adulthood (15-24 years old) and middle adulthood (25-44 years old) with 0.2%, 99% and 0.8% respectively. Participants demographic profile are depict in table 1 below.

Table 1. Demographic profile.

Characteristic	f	%
Age group (years)		
Youth (1-14yo)	1	0.2
Young adulthood (15-24yo)	585	99
Middle adulthood (25-44yo)	5	0.8
Gender		
Female	431	73
Male	160	27
Socioeconomic status		
Low income	375	63
Middle income	128	22
High Income	88	15

3.2 Psychological preparedness level

Most of the population in this study has average level of psychological preparedness for disaster (66%). People with average level of psychological preparedness were dominated by female with 50% of total population. They mostly come from low-income group with 43% out of the total respondents in this study. The participant level of psychological preparedness for disaster summarized in Table 2.

Table 2. Psychological preparedness level

Psychological preparedness level	f	%
Low	58	10
Average	395	67
High	138	23

3.3 Psychological preparedness level based on socioeconomic status

The participants of this study were come from low-, middle-, and high-income socioeconomic status. The participants from low-income level were 375 participants or 63%. While, people from middle and high-income group are 22% and 15% respectively.

The majority of the respondent from low-income group were have average level of psychological preparedness for disaster with 43%. The number of participants in the low-income status that have low and high level of psychological preparedness were only 5% and 15% respectively.

Respondents in middle income group mostly have average level of psychological preparedness with 14% of the total population. Whereas the second biggest number of middle-income participants were in the high level of psychological preparedness with 3% from the total population. The last was participants with low level of psychological preparedness for disaster with only 3%.

The greatest number of people with high income status have average level of psychological preparedness for disaster. The high-income status with low and high psychological preparedness were both only 2% from the total population.

The summary of psychological preparedness for disaster level based on socioeconomic status could be found in table 3.

Table 3. Psychological preparedness level based on socioeconomic status.

Psychological Preparedness level	Socioeconomic	f	%
Low	Low income	32	5
	Middle income	15	3
	High income	11	2
Average	Low income	252	43
	Middle income	80	14
	High income	63	10
High	Low income	91	15
	Middle income	33	6
	High income	14	2

3.4 Psychological preparedness level based on age

The vast majority of the population were in young adulthood category follow by youth, young adulthood, and middle adulthood category with 0.2%, 99% and 0.8% respectively.

Table 4. Psychological preparedness level based on age

Psychological Preparedness level	Age	f	%
Low	Youth	1	0.2
	Young adulthood	57	10
	Middle adulthood	0	0
Average	Youth	0	0
	Young adulthood	391	66
	Middle adulthood	4	0.6
High	Youth	0	0
	Young adulthood	137	23
	Middle adulthood	1	0.2

Most of the participants in the youth category have low level of psychological preparedness for disasters with 0.2% from the total population.

Participants in the age between 15-24 years old were classified in the young adulthood category. Participant in this age category is mostly have average level of psychological preparedness for disaster (67%). Whereas the low-level psychological preparedness for disaster in this age category were 10% and 23% of them were have a high level of psychological preparedness for disaster.

The distribution of level of psychological preparedness across different age are provided in table 4.

4 Discussions

This study found that mainly population that participate in this study has average level of psychological preparedness for disaster. Beside of that, the descriptive statistical analysis shown that there are differences of psychological preparedness for disaster level between socioeconomic status and age group.

The data shows that low-income status participants largely have average level of psychological preparedness for disaster followed by the group with high level of psychological preparedness for disaster and the last was group with low level of psychological preparedness for disaster. This pattern is also applied for participants in middle-income and high-income group. Different pattern of psychological preparedness for disaster could be seen between people in high income group. In this group, the percentage of people with high level of psychological preparedness for disaster is the same as with the percentage of people that have low level of psychological preparedness for disaster. The data shows that majority of the population with low level of psychological preparedness for disaster were coming from low-income status.

Study on the relationship between natural disasters and poverty in Fiji found that there is intercorrelation between the two variables [18]. In this study, poverty increase the Fiji people vulnerability to disaster. the same conclusion also found in the study of socioeconomic vulnerability of Vietnamese toward flooding [19]. The study found that poor household are the most vulnerable group that will experience a severe impact of flood compare to their counterpart. This is also true in the context of Sri Lanka population [20]. The financial condition influences people's sensitivity toward disaster and its impact. It is also found on several disaster preparedness research that the number of available resources affect the disaster preparedness reciprocally [21,22]. Low socioeconomic people have the lowest level of psychological preparedness since they could only have a limited resources an access to disaster risk reduction effort (e.g., training, post disaster aid and other related support for survivor after disaster). Hence, lack of access to disaster related aid support is one of the reasons that made and people in the low-income group have low level of psychological preparedness for disaster compare to others group. Thus, economic vulnerability influences individual disaster preparedness capacity.

Data analysis also show that level of psychological preparedness for disaster also differ between age. Low level of psychological preparedness for disaster were found among youth and young adulthood age group. While high level of psychological preparedness for disaster were found between young and middle adulthood. Study on the disaster preparedness has found that youngster have higher level of fear toward disaster compare to the others age group [23]. According to this study, young people have several sources of fear that related to disaster. first fear of their parent health condition. Secondly, afraid of their own health, and afraid of material losses as a result of disaster. Although there has been found that psychological preparedness for disaster of coastal communities is found difference between socioeconomic status and age. Nevertheless, future research should further investigate the role of fear toward psychological preparedness for disaster.

5 Conclusions

The result shows that the psychological preparedness for disaster of the participants was mostly in average level. Therefore, there is a need to improve the condition especially among people in youth and young adulthood category with age between 1-24 years old in order to mitigate and lower the risk of psychological impact of disaster.

Acknowledgement. We would like to thank you Indonesia Ministry of Education, Culture, Research and Technology which has fund this research as part of the Penelitian Dasar (PD) research grant scheme.

References

1. BNPB, (2021)
2. CNN Indonesia, CNN Indonesia (2021)
3. P. B. S. Putro, Social and Economic Impacts of the Sidoarjo Mud Flow: Community Resettlement After Disaster, Tohoku Unversity of Japan, 2012
4. M. Filipski, L. Jin, X. Zhang, and K. Z. Chen, European Economic Review **116**, 107 (2019)
5. G. Wong-Parodi and I. Feygina, Weather, Climate, and Society **10**, 747 (2018)
6. L. S. King, J. D. Osofsky, H. J. Osofsky, C. F. Weems, T. C. Hansel, and G. M. Fassnacht, Current Psychology **34**, 524 (2015)
7. M. B. Jacobs and E. W. Harville, Child and Youth Care Forum **44**, 511 (2015)
8. N. Makwana, Journal of Family Medicine and Primary Care **8**, 1 (2019)
9. H. Zulch, *Psychological Preparedness for Natural Hazards – Improving Disaster Preparedness Policy and Practice* (Brisbane, 2019)
10. H. Zulch, J. Reser, and P. Creed, 1 (2012)
11. S. A. Morrissey and J. P. Reser, The Australian Journal of Emergency Management **18**, 46 (2003)
12. J. Roudini, H. R. Khankeh, and E. Witruk, Health Psychology Open **4**, 1 (2017)
13. H. Zulch, Psychological Preparedness for Natural Disasters – Improving Disaster Management Theory and Practice, 2015
14. J. L. Boylan, The Development and Validation of The Bushfire Psychological Preparedness Scale (BPPS), The University of Western Australia, 2016

15. L. S. Palupi, IOP Conference Series: Earth and Environmental Science **649**, (2021)
16. L. S. Palupi and M. Noor Rahman Himawan, in *E3S Web of Conferences*, edited by B. W. Sudarno and T. T. Putranto (E3S Web Conference, 2020), pp. 0–5
17. B. Stern, L. Hurni, S. Wiesmann, and Y. Ysakowski, (2011)
18. P. N. Lal, R. Singh, and P. Holland, *Relationship Between Natural Disasters and Poverty: A Fiji Case Study* (2009)
19. R. Brouwer, A. Admiraal, K. Dusseldorp, T. Bui, and T. H. Tran, 1 (2010)
20. M. M. G. T. De Silva and A. Kawasaki, Ecological Economics **152**, 131 (2018)
21. D. P. Eisenman, D. Glik, R. Maranon, L. Gonzales, and S. Asch, Journal of Health Care for the Poor and Underserved **20**, 330 (2009)
22. B. M. Reininger, M. H. Rahbar, M. Lee, Z. Chen, S. A. Raja, J. Pope, and B. Adams, Social Science and Medicine **83**, 50 (2013)
23. V. M. Cvetković, A. Öcal, and A. Ivanov, International Journal of Disaster Risk Reduction **35**, (2019)

Community's knowledge, attitude, and misperception towards COVID-19: a future demand to prevent advanced wave of pandemic

Novi Maulilna[1,2,5], *Nova* Fajri[3,5], *Rosaria* Indah[2,5*], *Shylvana* Adella[4,5]

[1]Doctoral Student, Medical Faculty, Syiah Kuala University, Aceh, Indonesia, 23111
[2]Medical Faculty, Syiah Kuala University, Aceh, Indonesia, 23111
[3]Nursing Faculty, Syiah Kuala University, Aceh, Indonesia, 23111
[4]Public Health Faculty, Serambi Mekkah University, Aceh, Indonesia, 23123
[5] IMANI-PROKAMI (Islamic Medical Association and Network of Indonesia-Perhimpunan Profesi Kesehatan Muslim Indonesia), Aceh, Indonesia

Abstract. Coronavirus disease 19 is currently a global health threat. Viral spread prevention is essential as a feasible intervention strategy. The aim of this study was to determine the level of community knowledge, attitude, and perception towards Covid-19 in Aceh province. An electronic questionnaire was shared with 1,058 participants in the Covid-19 seminars organized by IMANI-PROKAMI Aceh Chapter (Islamic Medical Association and Network of Indonesia - Perhimpunan Profesi Kesehatan Muslim Indonesia), which were held between August 30th to Sept 20th, 2020. The questionnaire collected demographic data as well as data related to knowledge, perception, and attitude towards Covid-19 prevention. There were 1,058 survey participants, 841 (79.5%) of them were women, 366 (34.6%) were between 26 and 35 of age, 705 (66.6%) had a college degree and 540 (51.0%) were government employees. Most of participants had good knowledge and attitude about Covid-19 and its prevention. However, 44.1% of participants believed that the virus was created in research laboratories, as a result of conspiration (35.4%), and related to political issues (20%). They doubted the accuracy of Covid-19 diagnostic test (52.5%) and effectiveness of vaccines (28.2%). The results show multiple cases of misinformation around Covid19. Therefore, stakeholders need to counteract the community misperception towards disease by a transparent and strategic collaboration between society, government, and media, to increase positive perception, attitude, and practice of the community toward Covid19's prevention.

1 Introduction

The unprecedented pandemic of Covid-19 [1] has altered the way public live. The total confirmed cases and deaths worldwide increased at an alarming rate [2]. After 17 months since the first case of Covid-19 in Hubei Province, China, Indonesia bears a heavy burden, counting the highest case prevalence and fatality rate in Southeast Asia. Catastrophic damages in public health, economic and social stability has been severely affected the country.[3,4]

In order to develop the herd immunity, vaccination, social distancing, and good critical care services remain best strategies to decrease mortality. [5,6] However, as the number of people infected with COVID-19 grows, Indonesia and many other countries are, or will be, overburdening their health-care systems. Strengthening the non-pharmaceutical intervention including 5M (Wearing mask, washing hands with soap, social distancing, avoiding the crowd and mobility restriction) in Indonesia is an essential effort to decrease casualties and flatten the curve. [6–8] Nevertheless, one of the key factors influencing these strategies is the public active engagement and compliance with the measures. [8–10]

Public compliance depends on how people act toward risk-seeking or risk-avoiding actions, called public risk perception. Individual risk perception was influenced toward the threat. [11,12] Besides, individual knowledge, attitude, and practices will shape their readiness to fight against the disease, in line with societal conformity of the proposed regulations. [11–13]

As the most devastating effects of Covid-19 spread globally, it causes confusion, anxiety and fear among the public.[13,14] In some places, the existence of the virus is doubted, since the virus is invisible, abstract, and some people cannot really accept that having flu will threaten their lives. [14,15] They think that this pandemic benefits health care workers (HCWs). But, absurdly, in some places, they attacked health care workers for not allowing them to take care of the deceased body of their family after the death. [14,15]

Social media played a main role in escalating covid-19 to vaccine-related fake news. [16] Misinformation regarding vaccines in social media generated a fear and

* Corresponding author: rosaria.indah@unsyiah.ac.id

hesitancy toward vaccine administration. [17] Vaccine safety and effectivity were misperceived daily in the media, augmentate the hesitancy and activate the anti-vaccine movement in society. [18,19]

As time and research evolve, covid-19 related facts are growing. However, fake news is also increasing in Indonesian communities. Parallel with frequent social media use, this fake news is outspreading quickly. [20] Yet, there is scant research on what kind of misinformation is influencing people's perception toward Covid19 situation. This study aimed to evaluate knowledge, attitude, and perception of Aceh community during the coronavirus pandemic. The result of this study may inform future strategies focusing on public preparedness toward pandemic situation.

2 Methods

2.1 Study design and sample

This cross-sectional study was conducted involving 1,058 respondents participating in virtual seminar about Covid-19 provided by IMANI-PROKAMI Aceh Chapter (Islamic Medical Association and Network of Indonesia - *Perhimpunan Profesi Kesehatan Muslim Indonesia*). The seminars were held weekly for four weeks, from August 30th until September 20th, 2020. A link to the survey was distributed via Zoom chat box, right before the seminars started.

2.2 Measurement tool and data analysis

The self-reported questionnaire was developed, according to the COVID-19 prevention guidelines at a community level. The questionnaire was designed in Bahasa Indonesia, inviting respondents living in Aceh province, aged 18 years or older, and agree to participate completing the questionnaire.

The questionnaire consisted of four primary sections. The first section gathered information on respondents' sociodemographic characteristics, including age, gender, education level, work status, and region of residence. The second section assessed participants' knowledge of Covid-19. This section included 12 items on modes of transmission, clinical symptoms, disease risks, prevention measures, and treatments. The third section assessed participants' attitudes by asking their agreement toward statements about Covid-19. The final section of the questionnaire assessed respondents' misperception towards Covid-19.

2.3 Independent variables

For sociodemographic variables, gender was coded as one for men, and zero for women. The age variable was divided into categories: 18–25, 26-35, 36-45, 46-55, 55-65, and >65.

Educational levels were categorized into high school or below, college/university degree, and postgraduate degree.

Occupational data were grouped into students, government employee, non-government employee, self-employed, and un-employed.

We also obtained the data on the location of participants, which were residing in 23 districts in Aceh province: Banda Aceh, Aceh Besar, Sabang, Pidie, Pidie Jaya, Bireun, Aceh Utara, Lhokseumawe, Aceh Timur, Langsa, Aceh Tamiang, Aceh Tengah, Bener Meriah, Gayo Lues, Aceh Tenggara, Aceh Jaya, Aceh Barat, Nagan Raya, Aceh Barat Daya, Aceh Selatan, Simeulue, Subulussalam, and Singkil.

2.4 Dependent variables

Respondents were asked to respond to knowledge questions as either true or false, with an additional "I have no idea" option. The incorrect and additional response were given a score of zero, and correct answers were assigned a score of one. The total score for knowledge ranged from zero to 12, with high scores indicating better knowledge of COVID-19. Scores of 7 and above were categorized as good, and scores of 6 and below were categorized as poor.

In the section on attitudes, scores were calculated based on the respondents' answers. Total scores ranged from zero to four, with high scores indicating positive attitudes.

In the section on perception related covid-19 statements, respondents were asked to respond "yes" or "no" to the items. A score of one was given to answers that reflected good perception, and a score of zero was given for answers that reflected false statement. The total score ranged from zero to eleven, with high scores indicating better perception.

2.5 Analysis methods

Univariate analysis was used to analyze demographic data. Chi-Square tests was used to differentiate between good and poor knowledge, good and poor attitude and positive and negative perception related Covid-19. All analyses utilized SPSS Software.

3 Results

A total of 1,058 responses were recorded in the study. One participant could only give one response. All the participants were above 18 years old and Aceh origin. The total sample consist of 841 (79.5%) female participants and 217 (20.5%) were male. A large portion of the participants (34.6%) aged between 26 and 35. More than half of the sample (66.6%) had a college or university degree. In terms of work status, 540 (51.0%) were government employees and 50 (4.7%) were self-employed. Participants came from different regions in Aceh and the highest portion, 334 of them (31.5%) were residing in Banda Aceh.

Table 1. Demographic Characteristic of Participants.

Variable	N	%
Gender		
Male	217	20.5
Female	841	79.5
Age		
18-25	280	26.5
26-35	366	34.6
36-45	320	30.2
46-55	76	7.1
56-65	14	1.3
>65	2	0.1
Education		
High School	172	16.2
Bachelor's Degree	705	66.6
Post Graduate Degree	181	17.1
Occupation		
Students	240	22.7
Government Employee	540	51.0
Non-Government Employee	222	20.9
Self-Employed		
Un-employed	50	4.7
	102	9.6
Region		
Banda Aceh	334	31.5
Aceh Besar	230	21.7
Sabang	8	0.7
Pidie	37	3.5
Pidie Jaya	0	0
Bireun	74	6.9
Aceh Utara	15	1.4
Lhokseumawe	41	3.8
Aceh Timur	10	0.9
Langsa	30	2.8
Aceh Tamiang	20	1.8
Aceh Tengah	57	5.3
Bener Meriah	3	0.2
Gayo Lues	2	0.1
Aceh Tenggara	11	1.0
Aceh Jaya	11	1.0
Aceh Barat	48	4.53
Nagan Raya	17	1.6
Aceh Barat Daya	48	4.5
Aceh Selatan	15	1.4
Simeulue	6	0.5
Subulussalam	53	5.0
Singkil	5	0.4

As shown in table 2, the average of COVID-19 knowledge score was 11.67 (SD = 0.76, range: 4–12), indicating good knowledge. The average score in attitude toward COVID-19 was 3.73 (SD = 0.51, range: 1-4), indicating positive attitudes, and the average score for perception was 7.91 (SD = 2.91, range: 0–11).

Table 2. Knowledge, Attitude, and Perception towards Covid-19 among participants.

Variable	Mean	SD	Min	Max
Knowledge Score	11.67	±0.76	4	12
Attitude Score	3.73	±0.51	1	4
Perception Score	7.91	±2.91	0	11

Table 3 depicts our findings about relation between community knowledge, attitude and perception towards Covid-19. There was a statistically significant correlation between knowledge and attitude. However, there was no correlation between knowledge and perception and between perception and attitude. In other words, good knowledge and attitude do not guarantee high risk perception of Covid-19.

Table 3. Relation between knowledge, attitude and perception.

		Knowledge		P-value	Perception		P-value
		Good	**Poor**		**Pos**	**Neg**	
Attitude	Good	811	2	<0.0001	410	401	0.104
	Poor	237	8		139	106	
Perception	Pos	542	7	0.345			
	Neg	504	3				

Table 4 shows participant's responses to misinformation towards Covid-19. Almost half of participants (44.1%) believe that the virus was created in research laboratory, as a conspiration power (35.4%), and a political issue (20%). More than half of participants (52.7%) doubted the accuracy of Covid-19 diagnostic test. They believe that positivity test of Covid-19 was aimed to benefit medical staffs and hospitals financially (25.6%). The vaccine production was a multi-profit project was also believed by 28.2% participants.

Table 4. Community Perception towards Covid-19.

Statements	Yes (%)
Covid-19 virus was deliberately created in a research laboratory	467 (44.1%)
Covid-19 is a conspiracy created to reduce the world's population	375 (35.4%)
Covid-19 is a political fabrication. Medical personnel and lab examining Covid-19 tests get a lot of benefits from this pandemic	218 (20.6%)
Covid-19 is a self-limiting disease. It's no need for special drugs and isolation	164 (15.5%)
The funerals of Covid-19 patients can be carried out as usual without special protocols	159 (15.0%)
In-patient hospital, either because of respiratory symptoms (fever, cough and runny nose) or other diseases will be called positive for Covid-19, to increase the number of cases and the aid fund	271 (25.6%)
The creation of a Covid-19 vaccine is a multi-profit project	299 (28.2%)
The results of diagnosis test of Covid-19 are changing, so the accuracy can be doubted	558 (52.5%)

Table 5 shows the responses to items related to knowledge, attitude and perception towards COVID-19. As shown in the table, there was a statistically significant difference in attitudes, across gender and occupation. In sum, there is a significant corelation between attitude in gender and occupation. Perception towards Covid-19 was also has a statistically significant difference across occupations.

Table 5. Comparison of demographic characteristics and knowledge, attitude and perception towards COVID-19.

Variable		Knowledge		Attitude		Perception	
		Good/ Poor	P-value	Good/ Poor	P-value	Pos/Neg	P-Value
Gender	Men	213/4	0.254	149/68	0.002	117/100	0.574
	Women	835/6		664/177		432/407	
Age	18-25	276/4	0.847	210/70	0.416	148/132	0.574
	26-35	362/4		284/82		192/172	
	36-45	318/2		250/70		156/164	
	46-55	76/0		54/22		43/33	
	56-65	14/0		13/1		8/6	
	>65	2/0		2/0		2/0	
Education Level	High School	171/1	0.666	132/40	0.745	90/82	0.981
	Under-graduate	697/8		538/167		364/339	
	Graduate	180/1		143/38		95/86	
Occupation	Students	238/2	>0.005	183/57	0.002	127/113	0.029
	Govt Employee	444/0		355/89		220/224	
	Non-Govt Employee	217/5		171/51		106/115	
	Self-Employee	47/3		27/23		31/19	
	Un-employed	102/0		77/25		65/36	
Region	Banda Aceh	329/5	0.485	258/76	0.896	175/158	0.485
	A. Besar	229/1		178/52		127/103	
	Sabang	8/0		7/1		4/4	
	Pidie	31/0		28/3		14/16	
	Pidie Jaya	5/1		5/1		6/0	
	Bireun	73/1		54/20		41/33	
	A. Utara	15/0		13/2		7/8	
	Lhokseumawe	41/0		32/9		18/23	
	A. Timur	10/0		6/4		6/4	
	Langsa	30/0		23/7		18/12	
	A. Tamiang	20/0		17/3		10/10	
	A. Tengah	57/0		43/14		23/34	
	Bener Meriah	3/0		2/1		2/1	
	Gayo Lues	2/0		2/0		2/0	
	A. Tenggara	11/0		8/3		4/7	
	A. Jaya	11/0		8/3		5/6	
	A. Barat	35/1		27/9		19/17	
	Nagan Raya	17/0		14/3		6/11	
	A. Barat Daya	12/0		7/5		6/6	
	A. Selatan	15/0		12/3		8/7	
	Simeulue	6/0		4/2		3/3	
	Subulussalam	52/1		37/16		25/28	
	Singkil	5/0		5/0		3/2	

4 Discussion

This study aimed to evaluate knowledge, attitude, and perception of Aceh communities toward Covid-19. Due to the disease's serious threats, preventive measures are crucial in decreasing infection rates and controlling the disease spread. These measures are impossible to conduct without public engagement and support, which is affected by their knowledge, attitudes, perception, and practices.[21,22]

Located in the western part of Indonesia with 5.371.532 population, Aceh had a significant increase of Covid-19 prevalence in May 2021.[23] On August 27th 2021, this region had the highest positivity rate of 51.55% nationally followed by a high bed occupancy rate (BOR) for Covid-19 of 59%, while it had been struggling with low vaccination coverage.[24,25]

The result of this study shows that most of the participants showed a good knowledge and optimistic attitude toward COVID-19. Most of them concur the virus transmission can be avoided by complying with preventive measures, applying cough etiquette, and offering social support to other community members during outbreak (table 2 and 3).

However, in the same time, they misperceived Covid-19 facts. Although most of the participants understand that the disease caused by virus which could be transmitted airborne by cough and sneeze, some of them believed that the virus was a conspiracy of certain parties. They believed that the virus was deliberately created in laboratories to reduce the population (35.4% & 44.1%). Twenty percent of the participants thought that healthcare providers, including laboratory staff conducting Covid-19 tests obtained financial benefits from this pandemic. Furthermore, they doubted the accuracy of Covid-19 tests as they produce different results overtime. There were also misperceptions of vaccine-related facts reported by most of the participants (52.7% and 28,2%, respectively). The result concurs with findings in other studies as the belief on the myth of laboratory-produced virus was also reported elsewhere. [26–29] This interpretation raises another question: do these people believe that Covid-19 is real? Will they comply with the preventive measures and accept vaccine?

A key factor contributes to the commitment to disease prevention during outbreaks of pandemic is risk perception.[30,31] The type of information someone received may mediate the risk perception of Covid-19, whether it is an accurate information or fake one. Misinformation or lack of information may challenge risk of infection. [30–32] The interesting part to consider is that people make judgments based on their own risk perception, not on the actual risk. [31,32]

Risk perception is also related to individual background such as socio-economy status. [28] This may explain the discrepancy between knowledge and perception of Covid-19 in our findings. Most of the participants had good knowledge but good portion of them believed in misinformation around Covid-19. The latter may lead to low risk perception.

It appears that the low risk perception may lead to practices that disregard early Covid-19 symptoms. Considering the strong infectivity of Covid-19 and the symptoms of the disease [29], some of practices based on low-risk perception may lead to recurrent episode of case outbreaks. Furthermore, the rejection toward vaccination is related to the myth of the laboratory-produced virus and this may lead to low vaccination coverage. [33,34] Therefore, health campaigns should be well designed to address the misinformation (a.k.a. hoaxes) related to Covid-19. [30]

This study calls for stronger efforts to improving public risk perception in disease prevention, which may include stronger collaboration between society, government, and the media. [35] To increase public risk perception, public health institutions may utilize massive promotion in social media in an transparent, strategic, and proactive manner. In addition, larger investments should also involve efforts in increasing media education and digital literacy in order to create a critical understanding of how media and digital technologies are utilized. [36,37] Finally, it is essential that public health organizations, religious leaders, and other stakeholders to continue to educate, via offline tools and conventional media, as part of multi-channels approach, so that no segment of the community is left behind. [38,39]

5 Limitation of the study

This study is a subject of limitations. First, the study obtained data from participants of the seminars, but it did not compare the obtained data to the data from general population in Aceh. Second, the majority of participants (31.5%) came from the capital city of the province (Banda Aceh), and less from other regions. Therefore, a larger survey using proportional regional sampling may provide a more accurate result. Third, this study does not aim to find causation. Future research may utilize regression analysis to assess whether there is a correlation between Covid-19 knowledge and mortality, which may concur with causation.

6 Conclusion

Studying knowledge, attitude, and practice of a community toward Covid-19 may provide explanation of community perception on the disease. In addition to those three data, data on misperception are also crucial to assess population risk perception. It seems that low risk perception in this study might lead to population disengagement with public health measures. Therefore, the government, media, and other community stakeholders should collaborate to design health promotion approaches counteracting the hoaxes/misinformation and delivering quality information to increase positive perception, attitude and practices of the community.

Acknowledgement. The head and all members of PROKAMI Aceh for their great encouragement and hard work.

References

1. WHO, WHO (2020)
2. Reliefweb JHU, (2020)
3. K. K. RI, (2020)
4. R. Djalante, J. Lassa, D. Setiamarga, A. Sudjatma, M. Indrawan, B. Haryanto, C. Mahfud, M. S. Sinapoy, S. Djalante, I. Rafliana, L. A. Gunawan, G. A. K. Surtiari, and H. Warsilah, Prog. Disaster Sci. 6, (2020)
5. CDC, (2020)
6. E. J. Rubin, D. Ph, L. R. Baden, S. Morrissey, and D. Ph, 2021 (2020)
7. K. K. RI, (2020)
8. A. M. Tri Sakti, E. Mohamad, and A. A. Azlan, J. Med. Internet Res. 23, (2021)
9. O. Al-Wutayd, A. E. Mansour, A. H. Aldosary, H. Z. Hamdan, and M. A. Al-Batanony, Sci. Reports | 11, 16769 (123AD)
10. D. E. Alagili and M. Bamashmous, J. Infect. Public Health 14, 1398 (2021)
11. F. R, Curr. Opin. Psychol. 1, 85 (2015)
12. E. Chavarría, F. Diba, M. E. Marcus, Marthoenis, A. Reuter, L. Rogge, and S. Vollmer, J. Dev. Stud. 57, 1245 (2021)
13. M. A. Fardin, Arch. Clin. Infect. Dis. 15, 15 (2020)
14. Dawn, (2020)
15. SAMAA, (2020)
16. A. Smith, J. Yarwood, and D. M. Salisbury, Vaccine 25, 3996 (2007)
17. H. J. Larson, L. Z. Cooper, J. Eskola, S. L. Katz, and S. Ratzan, Lancet 378, 526 (2011)
18. E. J. Gangarosa, A. M. Galazka, C. R. Wolfe, L. M. Phillips, R. E. Gangarosa, E. Miller, and R. T. Chen, Lancet 351, 356 (1998)
19. E. Dubé, C. Laberge, M. Guay, P. Bramadat, R. Roy, and J. Bettinger, Hum. Vaccin. Immunother. 9, 1763 (2013)
20. J. Cifuentes-Faura, Online Inf. Rev. (2020)

21. F. D. Algahtani, M. A. Alzain, N. Haouas, K. Angawi, B. Alsaif, A. Kadri, M. A. Dkhil, M. Snoussi, and R. Zrieq, Int. J. Environ. Res. Public Health 18, 1 (2021)
22. D. Roy, S. Tripathy, S. Kumar, and N. Sharma, Asian J. Psychiatry J. (2020)
23. BPS Prov Aceh, (2020)
24. DetikHealth, (2021)
25. Dinas Kesehatan Aceh, (2021)
26. SabahD, (2020)
27. ConversationT, (2020)
28. Saleem A, (2020)
29. Wire T, (2020)
30. T. C. Hansel, L. Y. Saltzman, and P. S. Bordnick, Disaster Med. Public Health Prep. 14, 670 (2020)
31. L. Cori, F. Bianchi, E. Cadum, and C. Anthonj, Int. J. Environ. Res. Public Health 17, (2020)
32. E. Samadipour, F. Ghardashi, and N. Aghaei, Disaster Med. Public Health Prep. (2020)
33. A. Bhagavathula, W. A. Aldhaleei, J. R. Rahmani, M. A. Mahabadi, and D. K. Bandari, JMIR Public Heal. Surveill. (2020)
34. B. L. Zhong, W. Luo, H. M. Li, Q. Q. Zhang, X. G. Liu, W. T. Li, and Y. Li, Int. J. Biol. Sci. 16, 1745 (2020)
35. Rohrmann B, (2008)
36. Rohrmann B, (2000)
37. A. Bridgman, E. Merkley, P. J. Loewen, T. Owen, D. Ruths, L. Teichmann, and O. Zhilin, Harvard Kennedy Sch. Misinformation Rev. 1, 1 (2020)
38. M. Yu, Z. Li, Z. Yu, J. He, and J. Zhou, Curr. Issues Tour. 24, 2699 (2021)
39. L. Liu, W. Wu, and D. A. Mcentire, Int. J. Disaster Risk Reduct. (2021)

Medical doctors and community engagement in the aftermath of a disaster

Rosaria Indah*[1,2]

[1]Fakultas Kedokteran, Universitas Syiah Kuala, Banda Aceh, Indonesia
[2]Pusat Riset Ilmu Sosial dan Budaya, Universitas Syiah Kuala, Indonesia

Abstract. Medical doctors conduct many health care and promotion activities after a disaster. However, there are limited studies on how doctors engage disaster-affected community members during the activities. This paper aims to examine medical doctors' experiences in engaging their community during the rehabilitation period after a big disaster. The study is based on a more extensive ethnographic study at disaster-affected health centers and communities in Aceh. The researcher conducted interviews with eleven community doctors and 45 hours of participant observations involving three doctors. Findings show different levels of community engagement in the doctor participants' experiences. Most of them only 'informed' rather than 'engaged' the disaster-affected patient and community. This paper calls for the enactment of community engagement in healthcare practices and education.

1 Introduction

Disasters may cause significant disruptions in healthcare services for several reasons: healthcare facilities may be severely damaged, deaths of healthcare workers. and patients might be displaced to areas afar. These disruptions provide significant challenges for medical doctors to cope. Therefore, medical doctors working in disaster areas should be prepared to face the challenges together with the community they serve. Medical doctors are not well-positioned to respond to disasters without proper training and link to public health officials and community elements [1]. Yet, there is scant research on how doctors engage disaster-affected communities in healthcare.

This paper aims to explore the life experiences of medical doctors working in post-disaster areas. It also examines how they engaged community members to cope with challenges in the aftermath of a huge disaster. It starts with a literature review on the medical doctor's role in disaster management, concepts of community engagement, and elaboration on data collection and data analysis methods. Subsequently, the findings will be explained and discussed further. The paper concludes and suggests some recommendations for further research.

1.1 Medical doctor's role in disaster management

Disaster may expose medical doctors to many uncertainties, both personally and in the community. These uncertainties hinder medical doctors from working synergically with community members to manage the disaster's effects [2].

Disaster management works in different phases connected in a cycle: Emergency responses, reconstruction, rehabilitation (post-disaster phases), and these efforts are designed to prevent, mitigate and prepare (pre-disaster stages) for the next possible disaster. Medical doctors play essential roles depending on the stages of disaster management, both before the disaster and its aftermath [3].

Within the stages of disaster management, doctor roles slightly transform from conventional roles [2]. Before a disaster, the ultimate task of medical doctors is to increase the population's disaster resilience by improving their health status. In the aftermath, doctors' primary mission is to rebuild the health status and reduce health disparity between the disaster-affected and non-affected communities. To do so, doctors need to engage the targeted post-disaster community in close collaboration to assure health programs achieve an optimum result [4].

1.2 Community engagement: concept, principles, and frameworks

There are several concepts of community engagement in healthcare. The first one was inspired by Paul Farmer [5], who applied 'liberation theology' to address public health issues, such as tuberculosis (TB) in low-income and developing countries (LIDCs). Liberation theology argues that authentic transformation is rooted in marginalized communities.

* Corresponding author: rosariaindah@unsyiah.ac.id

To activate the transformation, one needs to consider a simple methodology consisting of three linear steps: (i) observe, (ii) judge, and (iii) act[5]. *Observing* includes various elements, such as seeking the causes of the problem, followed by eliciting the experiences and views of a marginalized community and then incorporating the result of *judgment* and *actions* [5]. The second step, *judgment*, is produced by careful observation that should be directed to raise the consciousness of people and explain the processes of coming to an understanding of how social structures can cause injustice [5]. The *act* linear step is more than just reporting one's findings because it also requires accompanying a marginalized community in its struggle for liberation.

Applying the idea of liberation theology, Farmer's project in Haiti incorporated a program of funding food vouchers, offering travel expenses, and supplying Tuberculosis (TB) drugs for TB patients. Initially, Farmer observed that failures in previous TB treatments were due to a lack of community engagement [5]. When he asked community members why the last treatment of TB failed, they told him they could not afford to buy food or travel to health centers to collect their TB drugs. Based on this finding, it was concluded that food vouchers were to be incorporated because the root of the cause was malnutrition and structural poverty [5]. By increasing the participation and role of community members in TB treatment, Farmer had addressed the problem of managing TB cases effectively.

There are principles in community engagement in healthcare [6]. First, healthcare practitioners have to map existing networks and resources. If the healthcare program uses infrastructures, it is essential to make the community take ownership of the structures and maintain them. Third, cultural competence [4], [7], fourth, ensure two-way communication, and fifth, engage vulnerable groups, build trust and acknowledge existing culture.

Several organizations developed several frameworks of community engagement, such as Health Canada [8], US Clinical and Translational Science Award [9], and Association for Public Participation [10]. These three frameworks were quite similar, underlining gradual steps to increase the level of power-sharing (see Fig. 1.).

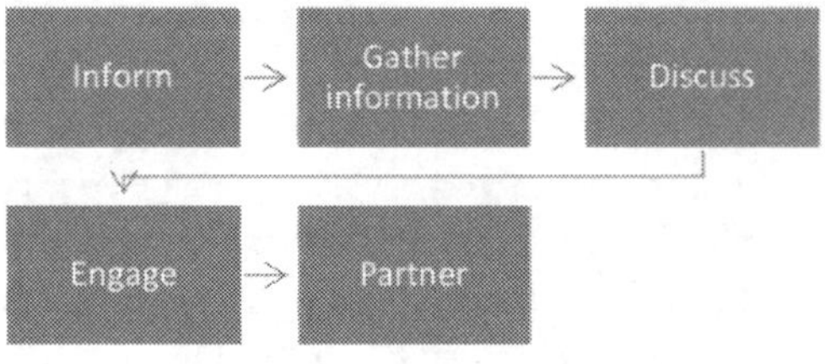

Fig. 1. Public involvement continuum, by Health Canada [8].

However, different terms were used. They described the lowest level of engagement as informing or educating, followed by listening, consulting, involving/engaging, and collaborating. The highest level of engagement was partnering with shared decision-making and leadership. A comparative study of the approaches concluded that there was a need to provide robust studies on the success of implementation and the extent of community engagement because current publications only explain these approaches at the theoretical level without adequate empirical data [11].

These models help provide a general understanding of different levels of engagements in community-based healthcare projects to align with the spirit of power-sharing. The lower levels of engagement, such as 'consulting,' appreciate the situated knowledge possessed by the community. In comparison, the moderate to higher levels of engagement (involve, collaborate, and shared-leadership) seek to recognize the community's ability to speak and therefore act. However, one must be aware that the lowest level of engagement (educate or inform) potentially perpetuates the effects of asymmetrical power relations between healthcare providers and patients.

2 Method

This study was a part of a larger ethnographic research. It used ethnographic tools to explore medical doctors' experiences engaging disaster-affected patients and communities after the 2004 Indian Ocean tsunami in Aceh, Indonesia. The province of Aceh was the devastated province in Indonesia, in which most of its coastlines were hit by the 2004 Indian Ocean tsunami. The official death toll was 166,760, while 635,384 survived the disaster but were internally displaced. Among the survivors, there were 1,736 people injured, and 61% of them lost one or more immediate family members that exposed them to a greater risk of mental illnesses such as depression, anxiety, and post-traumatic stress disorder (PTSD). Some of the participants also experienced the loss of core family members and assets.

The inquiry was set in natural settings, post-disaster healthcare facilities. Ethnographic tools allowed the researcher to explore cultural patterns and perspectives of medical doctors in natural environments. Although this methodology required a prolonged contact to locate interpretations about those occurrences into profound scripts [12], ethnographic tools may help depict the participants' meanings [13]. Ethnographic tools were also helpful to focus on social processes that facilitate daily life [14], especially on those related to community engagement. To collect data, the researcher conducted interviews and participant observations.

2.1 Interview

This data collection method aimed to understand medical doctors' perspectives upon their interactions with communities affected by the 2004 Indian Ocean tsunami in Aceh. The author chose to conduct interviews with experienced medical doctors working in post-tsunami areas, which have been there for at least ten years.

There were some challenges in conducting the interviews. First, it was difficult to locate medical doctors who had been working in the disaster-affected area for such an extended period (more than ten years). Most of the doctors working in the aftermath of the tsunami have moved to other places outside the affected areas after their two years of new graduate mandatory service had been completed. Only some of them decided to continue their works in those underserved areas. By conducting the snow-ball method, some medical doctors can be located. They were willing to spend some of their precious time for one or more interviews.

Second, when the researcher was trying to obtain their consent, some doctors were reluctant due to the time constraint. The researcher's identity as a medical doctor did not necessarily help her attain consent. They rejected because they assumed that the researcher could discern the circumstances of a high intensity of emotion and workload, with great responsibilities and stakes involved. Hence, the researcher decided to treat them as 'busy informants' [15] She waited until they offered her a time to meet.

Other challenges include locating and setting the interview. The researcher must travel more than 600 km from her main area to reach some participant doctors. However, it was worth spending time since most of them were willing to raise the issues that were central to her research topic.

Third, although most doctor participants agreed to be interviewed; however, one was reluctant to recall 'horrible memories' related to her personal experiences in the 2004 Indian Ocean tsunami that may trigger their PTSD. To prevent coerciveness, the researcher reminded the participant's right. She may decide to cease or continue with the interview at any point in time. The doctor participant, who was initially reluctant, chose to continue.

Other doctors have asked the researcher to wait and provide some time to deal with a potential emotional discomfort caused by remembering their experiences after the traumatic event. Talking about their experiences regarding disasters, death, and suffering is almost always categorized as a 'difficult conversation' in the medical world that needs careful mental preparedness [16]. The researcher expressed her understanding and patiently waited for their consent. She realized that gaining access to disaster-affected participants is one of the challenges that disaster-focused researchers encounter. However, most of the participants eventually agreed when they discussed the information statements and understood the risk and benefit associated with their involvement within the research.

Another way to obtain consent was by asking other medical colleagues to be a gatekeeper. They facilitated the interviews with targeted informants. An ethnographer should seek a gatekeeper's help in ethnographic research, such as what 'Doc' did in the classic ethnography piece 'Street Corner Society.' The gatekeeper provided a more accessible entry to the field. In this case, the participants admitted that they conducted 'a research' on the researcher's identity and credibility before granting their approval on setting up the interviews.

This 'research on the researcher's credibility is partly due to the assumption that the researcher might be obtaining financial gain from the research or might be a 'spy' investigating their 'good practice.' Fortunately, the researcher's experience as a teacher provided her with networking with former students. They acted as gatekeepers and thus verified researcher's identity and credibility. Eventually, she succeeded in obtaining consent from most of her planned participants. Yet, some of them have refused to be audio-recorded.

Individual semi-structured interviews were conducted with 11 community doctor participants. In the beginning, the researcher explained the aim of this study and information written on the participant information sheet. Participants may ask any questions regarding the research being answered.

Most interviews lasted 45 to 90 minutes. The interviews were flexible, interactive, dialogic, and reflective. An interview guide was used to support and organize the responses conveyed in Bahasa Indonesia. All doctors chose their office as the location for their interview. The characteristics of doctor participants are described in Table 1.

Table 1. Doctor participants characteristics.

No	Font	Sex	Age	Experience (years)
1	Dr. Alia	F	40	10
2	Dr. Jalal	M	45	12
3	Dr. Endang	F	40	17
4	Dr. Aisha	F	55	20
5	Dr. Fajar	M	54	25
6	Dr. Eva	F	55	20
7	Dr. Julie	F	54	21
8	Dr. Mahmud	M	55	12
9	Dr. Anna	F	50	15
10	Dr. Sarah	F	53	12
11	Dr. Rita	F	54	20

2.2 Participant observation

To further understand how the doctor participants engaged their community in healthcare, the researcher chose three out of 11 community medical doctors interviewed. Before the participant observation was conducted, the objectives were explained in detail, and consent was sought since the observation was a more intrusive method of data collection. During the observation process, sometimes the researcher attended clinical encounters as a full observer and sometimes helped them as their administrative assistant or aid.

Initially, the researcher felt uncertain about how to conduct the observation, how to dress or participate in the different settings around health centers in villages. The researcher was unsure about what to expect because the setting was unfamiliar to her. Therefore, the researcher allowed doctors to define the boundaries within clinical encounters. The researcher asked them about her attire, where she should sit, when she should join in a conversation, ask or answer questions, or whether she should not. The researcher tried not to observe merely. She listened actively to their doctor-patient conversations to capture any means of engagement processes. The community doctors introduced the researcher to their patients and informed them that she was there to observe doctor-patient interactions. The doctors or the researcher asked for each patient's consent to be observed. If any patient refused permission, the researcher exited the examination room. For example, one male patient who did not want to be observed asked the researcher to stop her observation. The community medical doctor participants allowed the researcher to participate with community members inside and outside the health center, such as in health promotion campaigns, home visits, and non-clinical events, such as luncheons and wedding ceremonies of family members, or just enjoying coffee with community chiefs. As a result, rich descriptions of data were produced from these observations. The data from participant observations were documented in a logbook, field notes, and research memos.

2.3 Data analysis

Data analysis included carefully reading and rereading the transcripts of interviews, field notes, writing research memos, and generating codes. Some codes were generated deductively from the literature on the first round of data reading. Others were inductively produced from the data corpus at the second and third rounds of reading the entire data. Then the researcher made a coding scheme and categorized data into relevant categories based on these codes to generate preliminary themes. The author then did the fourth round of data reading before conducting the next round of thematic analysis to finalize research themes. The data analysis process was facilitated by NVivo version 10 software and guided by the study's research question, which included life experiences of medical doctors in engaging their communities in the aftermath of the 2004 Indian Ocean tsunami. In the writing up of the analysis, the names of all medical doctors were given pseudonyms to protect their anonymity.

3 Findings

As mentioned previously, the 2004 Indian Ocean tsunami devastated, destroyed, and damaged houses and displaced survivors from their homes [17]. Survivors were forced to flee and find new dwellings, often recovering from injury. Challenges related to health issues and internal displacement were intertwined, creating a complicated situation that needed to be understood by community medical doctors working in the aftermath of a disaster. Hence, patients in this context required special attention.

One day the researcher was allowed to observe the practice of Dr. Mahmud, a camp doctor, for a week. He had been working in a subdistrict with a high prevalence of tuberculosis, one of the effects of the overcrowded displacement situation. He used local languages, such as Achehnese or Aneuk Jamee, to converse with patients accordingly. He explained the reason:

> *Language could be a big problem for doctors. I want to engage patients in conversation about therapy. Therefore, they need to understand the details of the treatment offered. Unfortunately, I see most new doctors do not have the skill to speak the local languages, maybe because they came from different cultural backgrounds. (Dr. Mahmud, male, 55 years old)*

It seems that Dr. Mahmud had successfully engaged his patients and their family members in the course of their treatment. During the observed conversations, Dr. Mahmud asked questions, responded to patients' questions, used jokes to establish a good rapport. He also made sure patients understood their essential roles in the treatment processes. However, Dr. Mahmud had never mentioned or suggested patients utilize resources in their communities, such as nutrition and sanitation programs in their community health centers. He engaged his patients at individual and family levels.

Other medical doctors expanded their engagement at the community level. To improve access to healthcare provisions for disaster-affected patients, community medical doctors often conducted outreach programs, as explained by Dr. Rita.

> *I went to mosques or people's houses to provide healthcare services and health promotions, mostly in religious community gatherings where people were gathered, even when they happened outside working hours. Even though I was not paid to do home visits, I was happy to come because I prefer seeing health messages arrive in the community. (Dr. Rita, female, 54 years old)*

Community doctors, such as Dr. Rita, were committed to serving disaster-affected patients displaced in their relatives' houses for several years, helping patients resume their works until they moved to reconstructed sites. However, many tsunami survivors were forced to live in overcrowded displacement camps, which were also often located at a great distance from health centers and clinics. Some patients had previously been displaced due to armed conflict. Therefore, they had been displaced several times, as Dr. Jalal observed.

> *Most of them were affected by double disasters. Initially, they resided on the mountainside, but they moved to the shore due to armed conflict. But then they were attacked by the tsunami, and they had to return to the mountainside. While living in the shelters, some of them got panic attacks by loud sounds like gunshots. It was a rough time for them. (Dr. Jalal, male, 40 years old)*

Displaced patients chose to stay at the camps for different reasons. For example, some patients wanted better access to healthcare and aids, some decided to live closer to their previous settlement, or they did not want to be a 'burden' to their relatives. Living in camps had its challenges and advantages. They had to deal with problems, such as overcrowding, a lack of clean water, supplies, and inadequate sanitation facilities [18]. An advantage of living in the camps was access to relatively better healthcare facilities since every camp received visits from mobile clinics. Doctors working in mobile clinics and health centers were required to assist displaced patients in wrestling with the context of the camp.

Community medical doctors working in displacement periods after the tsunami also wrestled with infectious diseases, including malaria. In the following quote, Dr. Anna describes how she employed a community-based approach to providing healthcare for disaster-affected patients with severe malaria infection.

I had several patients that we found unconscious during their displacement periods in a camp far up the mountainside. Previously, their family brought them to a traditional healer to conduct an exorcism to take the 'bad spirit' away from the patients. When it did not work, then the family members contacted me. I visited these patients and took their blood for the rapid test. The result revealed that they had falciparum malaria that had been severely affecting their brain. It was a kind of cerebral malaria. I tried to inject them with Arsucam, a new medicine supplied by an international aid organization. Alhamdulillah, they responded very well. Then I ran a malaria prevention campaign. I was glad that we could distribute mosquito nets; we gave them the 'real things,' so there was no reason to reject the idea anymore. We also explained how to use the Acehnese language to increase people's understanding. (Dr. Anna, female, 50 years old)

During the interview, Dr. Anna mentioned that malaria-associated symptoms were familiar to her. Before the tsunami hit the district, she had encountered many patients with malaria. However, displacement after the tsunami rapidly increased the number of malaria cases due to overcrowding and suboptimal living situations in displacement camps.

Dr. Anna was also familiar with the healthcare-seeking behavior of people in her community; they inevitably go to a traditional healer before seeking a medical doctor's help. By that time, the disease may have deteriorated to a severe stage where, in this case, malaria parasites had created inflammation of the brain, causing decreased consciousness, often leading to death. This culture of medical pluralism can have severe consequences for the patient and their family. However, Dr Anna did not blame them for consulting a traditional healer, because she understood that it is their culture, and they had limited choices as their village was located far from healthcare facilities. Therefore, she arranged mobile clinics as soon as possible to provide such a service to improve patient access to better healthcare provision.

The use of an incompatible language in medical encounters was a common problem in communication. Dr. Anna, who originally came from the area, understood the challenges faced by her community. Therefore, she used the local language to promote healthcare and increase people's understanding of malaria prevention measures. With full support from multiple international organizations, Dr. Anna gathered resources to provide disadvantaged people in the community with 'real things' (pre-medicated mosquito nets). Before the tsunami, she could not do this due to limited funds. Like other local doctors in rural areas in Indonesia, Dr. Anna had to wait for the appropriate moment where good intention and facilitation came together. However, she did not use other means of malaria prevention that were aligned with local wisdom, such as stocking ponds and wells with fish that consume mosquito larvae or using particular plants that repel mosquitoes. Her engagement was different from Dr. Julie's, who had more experience with community engagement programs.

In the aftermath of the 2004 Indian Ocean tsunami, Dr. Julie found some survivors from her village displaced in a soccer stadium on higher ground. Initially, her district was located on the shore near the harbor of Banda Aceh. While only approximately 5,000 people in the subdistrict survived the tsunami from 27,000 total occupants, the survivors decided to live together as a community to support each other and improve access to daily logistics at the stadium. Dr. Julie visited them at the stadium from time to time before they moved back to their village several months later. When an international non-government organization from Kuwait reconstructed her health center a year later, Dr. Julie worked there in the district to support the survivors returning to their village. However, the village was not as it was previously because lingering complications of the tsunami overshadowed the culture of the community. Dr. Julie observed how those patients coped with the disaster through a unique reproductive behavior involving herself in many informal conversations with her community members.

I talked with the mothers in informal meetings and understood that the survivors wanted to regain their family by remarrying, and they decided to have children as soon as possible. The birth control program failed [laughing]. There was a local 'baby boom' period two years after the tsunami. Therefore, instead of conducting a birth control program, our 'mother and child health' team promoted exclusive breastfeeding. We successfully organized a massive event for an exclusive breastfeeding campaign that was attended by 10,000 mothers from all around Aceh from many displacement camps with support from various NGOs. It was a national record! (Dr. Julie, female, 54 years old)

Dr. Julie attempted to learn and understand her community's needs by becoming involved in health programs. Instead of merely conducting centralized health programs assigned by the local health office (birth control program), Dr. Julie adjusted them to the local community's needs. To come to this understanding, she observed the phenomena in the post-tsunami community, that is, the urge to remarry and have children. Therefore, she encouraged healthier reproduction behavior by promoting exclusive breastfeeding. The action was an opportunity for better health for babies as the exclusive breastfeeding rate before the tsunami was lower than 50%. Dr. Julie and her staff reached out beyond their work area by inviting participants from adjacent districts. After the success of this campaign, Dr. Julie was offered a promotion to work at the local health office. However, she turned it down and continued to work at her community healthcare center.

The above quotes from doctors describe how medical doctors interacted with their patients and disaster-affected community members during the displacement period. Some explained how they came to understand the community's needs by reflecting on their previous experiences. Several doctors applied similar strategies from before the disaster (such as what Dr. Mahmud did), some changed their approaches (similar to Drs. Anna and Julie). Some doctors limited their interventions to an individual level, and others expanded their intervention to the family and community level.

The doctors also tried to apply their cultural competence to increase patient participation. Drs Rita and Anna used the local culture. Dr. Anna attempted to understand medical pluralism cultures and conducted outreach programs to improve the community's engagement with malaria prevention programs. However, her action of informing communities about the importance of mosquito nets is considered a low-level of community engagement and may not be wholly capable. Dr. Julie's approach was an example of an engagement encompassing multiple levels, informing patients at personal, family, and community levels. However, despite noble intentions to help patients, every medical interaction has dynamic power relations, especially when the interaction involves marginalized community members, such as disaster-affected patients who are less privileged than ordinary patients. Several cases presented above showed that asymmetrical power relations existed, and further analysis of these cases applying a postcolonial lens may provide valuable insight applicable to improving future medical practice and education.

4 Discussion

The findings highlight community engagement's importance in stimulating mutual power-sharing in doctor-patient interactions. The findings complemented what has been described in limited literature of how community engagement works in low-income and developing countries. Building on the research findings, a community engagement model can be developed to inform practitioners working in post-disaster areas in Indonesia and perhaps beyond (Fig. 1.). Deceptively simple, Fig. 1. demonstrates trust-building among community members and other stakeholders, encourages long-term collaboration, and ensures the ownership of healthcare programs belongs to the community rather than NGOs or the government.

Instead of a linear process similar to those produced and used by Health Canada, CTSA, and IAP, as reviewed in the literature review section, this community engagement model utilizes a cyclic, never-ending process that emphasizes continuous improvement. Instead of starting from an informed or educated step, it begins with 'listening' to community demands. This step aims to map community vulnerabilities and capacities and locate people's perspectives toward their health problems. It also aims to find indigenous knowledge that has been in place for years as each community's uniqueness.

Fig. 2. Proposed model of community engagement suggested by findings of this study.

The community engagement processes work in a cycle. First, listening to the community's needs is essential to identifying health problems and people's perspectives on existing issues and potential solutions. For example, Dr. Julie found the issue of exclusive breastfeeding was more specific and essential to promoting infant health rather than focusing on birth control by carefully listening to the community's demands. Second, health practitioners should gather information on potential accessible resources to resolve health problems, as discussed in the 'listening' step. Dr. Julie gathered information by contacting health stakeholders, including her staff, heads of villages, community representatives, the government, NGOs, and INGOs. Third, collaborating with community members and stakeholders in the planning and executing of health programs. The fourth process is implementing the partnership into action. The partnership requires equal sharing of resources and responsibilities. For instance, Dr. Julie shared duties with neighbors to support older women patients in the community who had mental health issues. This process is reiterated, which involves listening to community responses to the program to improve its implementation.

5 Conclusion

Medical doctors are encouraged to continuously pay attention to several aspects of doctor-patient interactions to engage their communities. These aspects include using local languages, participating in cultural activities, and carefully listening to people's voices. These approaches may create more balanced power relations, which can help doctors to value each patient's lived experiences as valuable knowledge and an essential resource in healing processes.

The experiences of medical doctors working in post-disaster areas suggest various steps in engaging communities in their health programs. These steps may be described in a cyclic model, comprising listen- gather information-collaborate-partnership- and back to listening to communities' voices to expand the engagement. Instead of linear models established previously, this cyclic model suggests continuing improvement in doctors and community engagement. Based on the experiences of community medical doctor participants, these approaches may help them overcome many challenges in providing integrated post-disaster healthcare services.

This study also calls for medical schools to incorporate community engagement approaches in their curriculum. Hopefully, future doctors will better understand their community needs and improve their willingness and ability to work hand in hand with their communities to overcome more severe challenges in the future.

Acknowledgments. This work was funded by Indonesian Endowment Fund for Education/ Lembaga Pengelola Dana Pendidikan (LPDP) (Grant No. PRJ-1125/LPDP/2015). Dr. Matthew A.M. Thomas and Dr. Ruth Phillips are greatly thanked for inspiring and constructive comments.

References

1. P. Burns, K. Douglas, W. Hu, P. Aitken, and B. Raphael, "When Disaster Strikes what is the Role of the Local Primary Healthcare Doctor?," *Prehospital Disaster Med.*, vol. 32, no. S1, pp. S70–S71, 2017.
2. S. Abeysinghe, C. Leppold, A. Ozaki, and M. Morita, "Risk, uncertainty and medical practice: changes in the medical professions following disaster," *Evid. Policy J. Res. Debate Pract.*, vol. 16, no. 2, pp. 285–303, 2020.
3. W. Kusumasari, "Technical guidelines for health crisis responses on disaster," in *Guidelines for health workers involved in health crisis responses on disaster in Indonesia*, The Ministry of Health of Republic of Indonesia Jakarta, 2011, p. 228.
4. J. C. Erves *et al.*, "Needs, priorities, and recommendations for engaging underrepresented populations in clinical research: a community perspective," *J. Community Health*, vol. 42, no. 3, pp. 472–480, 2017.
5. P. Farmer, *Pathologies of power*. University of California Press, 2004.
6. A. Ramsbottom, E. O'Brien, L. Ciotti, and J. Takacs, "Enablers and barriers to community engagement in public health emergency preparedness: a literature review," *J. Community Health*, vol. 43, no. 2, pp. 412–420, 2018.
7. M. D. Morris, S. T. Popper, T. C. Rodwell, S. K. Brodine, and K. C. Brouwer, "Healthcare barriers of refugees post-resettlement," *J. Community Health*, vol. 34, no. 6, pp. 529–538, 2009.
8. Health Canada, *Health Canada Policy Toolkit for Public Involvement in Decision Making*. Author Ottawa, ON, Canada, 2000.
9. Clinical and Translational Science Award, "Principles of Community Engagement," Jun. 25, 2015. https://www.atsdr.cdc.gov/communityengagement/pce_ctsa.html (accessed Apr. 29, 2019).
10. International Association for Public Participation, "IAP2 Public Participation Spectrum," 2014. https://www.iap2.org.au/Resources/IAP2-Published-Resources (accessed Apr. 29, 2019).
11. E. Manafò, L. Petermann, V. Vandall-Walker, and P. Mason-Lai, "Patient and public engagement in priority setting: a systematic rapid review of the literature," *PloS One*, vol. 13, no. 3, p. e0193579, 2018.
12. M. Hammersley, "Ethnography: problems and prospects," *Ethnogr. Educ.*, vol. 1, no. 1, pp. 3–14, 2006.
13. R. C. Bogdan and S. K. Biklen, "Research for education: An introduction to theories and methods," 2007.
14. M. D. LeCompte, J. Preissle, and R. Tesch, *Ethnography and qualitative design in educational research*. Academic Press, 1993.
15. R. Indah, "Insider Dilemmas: An Ethnographic Study on Community-Based Medical Education in Aceh, Indonesia," *Explorations*, vol. 15, pp. 86–92, 2019.
16. R. Indah, "Probing problems: Dilemmas of conducting an ethnographic study in a disaster-affected area," *Int. J. Disaster Risk Reduct.*, vol. 31, pp. 799–805, 2018, doi: https://doi.org/10.1016/j.ijdrr.2018.07.025.
17. H. Arie, B. Syahbudin, and Indonesia. Badan Rehabilitasi & Rekonstruksi, *BRR Book Series: Education, health, women empowerment, preparing quality generation*. Executing Agency of Rehabilitation and Reconstruction for Aceh and Nias, 2009. [Online]. Available: https://books.google.co.id/books?id=C5nJxAEACAAJ
18. E. K. Noji, "Public health issues in disasters," *Crit. Care Med.*, vol. 33, no. 1, pp. S29–S33, 2005.

Investigating the midwives' knowledge on infant and young child feeding during the COVID-19 pandemic

Nuril Annissa Niswanto[1*], *Nasaruddin* Nasaruddin[2], and *Marthoenis* Marthoenis[3]

[1]Department of Disaster Science, Post Graduate School, Universitas Syiah Kuala, Banda Aceh 23111, Indonesia
[2]Department of Electrical & Computer Engineering, Engineering Faculty, Universitas Syiah Kuala, Banda Aceh 23111, Indonesia
[3]Department of Psychiatry and Mental Health Nursing, Nursing Faculty, Universitas Syiah Kuala, Banda Aceh 23111, Indonesia

Abstract. Implementing the WHO gold standards for Infant & Young Child Feeding (IYCF) may help in preventing stunting, which may worsen during Covid-19 Pandemic. Midwives can help educating the mother in proper child feeding. This research aimed to assess the knowledge of midwives working in the community on IYCF during Covid-19. A total of 88 midwives working in various healthcare facilities participated in this study. Using online survey methods, they were invited to answers questions that measure their knowledge on basic practices of IYCF, Covid-19 transmission risk in breastfeeding dyads, breastfeeding practices, and complementary feeding during the pandemic. It was found that the rate of good, moderate and poor knowledge on IYCF during Covid-19 Pandemic were 23%, 67% and 19%, respectively. In conclusion, less than one third of the respondents showed good knowledge – something requiring investigation on the strategies to build stronger capacity of midwives in terms of IYCF during Covid-19 Pandemic.

1 Introduction

A community incapable of understanding – thus ensuring the optimal practices of – IYCF before a crisis/emergency takes place would only accelerate "the secondary crisis" during the phase of the early emergency response of a disaster [1]. Aceh is at stake as the pre-Pandemic data showed that the prevalence of stunting in Aceh had shown an alarming number of 37.9% compared to the national average of 29.9% [2]. Aceh is a province with the highest prevalence of stunting in Indonesia in infants under two years of age, "conquering" the provinces of Papua, Maluku and other Eastern regions [3].

The prevalence of stunting in Aceh, which was still above 30% in the 2007-2018 period fell into the very high and high categories. Even without an emergency or disaster, Aceh already faced nutrition problems – a dangerous predisposition to a worsening malnutrition problem during a crisis we are all facing today; the Covid-19 Pandemic. Children who faced and survived events of malnutrition might lose certain cognitive traits so permanent it could be inherited to the next generations genetically [4]. Hence, the proper practices of IYCF during the Pandemic, among multiple other factors, may avert the stunting cases in the future.

The proper practices of Infant & Young Child Feeding (IYCF) in Emergencies (or often referred to as IYCFE) could save the infant and young children from infectious illnesses or even death threats [5]. IYCF (early breastfeeding initiation, exclusive breastfeeding 0-6 months old, complimentary solid feeding since the age of 6 months, while continuing the breastfeeding up until 2 years of age or more) needs to be addressed in any emergencies – including Covid-19 Pandemic – for infants and young children. They are amongst the vulnerable groups to a following "secondary crisis" in the form of acute malnutrition, which may end with death when not treated properly [6].

1.1 IYCF during Covid-19 pandemic

The National Health Ministry has stated that there has been no change in IYCF recommendations during the Covid-19 Pandemic, which is aligned to the WHO [7]. Mothers with confirmed cases of Covid-19 should continue to breastfeed and or provide complimentary home-cooked foods while maintaining the correct health protocols; by washing hands before and after contact with their babies or the food preparation processes, wearing masks all the time when being present with the babies/children, and sanitizing all surfaces periodically [8]. The risk – and fear – of children contracting Covid-19 is outweighed by the abundant benefits of ensuring the IYCF practices during this pandemic. The correct IYCFE could actually help increase the children's immunity [9] and prevent malnutrition from happening during and after the pandemic ends [10].

1.2 Midwives, IYCF, and Covid-19 pandemic

Midwives possess accessibilities in educating the communities about disaster/emergency preparedness

* Corresponding author: nuril.an@unsyiah.ac.id

and other disaster risk reduction actions. Midwives are encouraged by The International Confederation of Midwives (ICM) to contribute in managing a disaster – including a pandemic – particularly regarding mothers and children's health issues [11].

Midwives are the closest health workers working amongst "end-users" in society, including those stationed in Local Community Health Center or as known as Puskesmas. Every sub district (or more known as "kecamatan") usually is equipped with 1 or 2 Puskesmas, depending on the area width and the population in it.

Midwives stationed in Puskesmas generally serve longer duration of working in the assigned area, thus more frequent interactions with the society. They are amongst the most effective health educators in the grassroots [11], [12]. Thus, their knowledge on IYCF is crucial in running their services and for the stakeholder to have a better foundation for trainings needed to improve their capacity, especially during the on-going pandemic. However, data regarding the implementation of IYCFE in the midst of the Covid-19 pandemic is not yet available. Hence this study aimed to get the picture on where the midwives were with IYCF during Covid-19 Pandemic.

Identifying the current knowledge status will benefit the related stakeholders for better capacity building on IYCF practices. This, hopefully, will clear the way for the stakeholders to build stronger capacity of the midwives as one of the many ways to decrease the stunting prevalence in Aceh during or after the pandemic ends.

2 Methods

2.1 Study setting

The study was conducted in the city of Banda Aceh City, the capital of Aceh Province. The city has been stated as red zones due to high number of Covid-19 cases in Aceh for several times. The infection had been widely spread within the population, including breastfeeding mothers or those taking care of a child aged 0–2-year-old. Hence, correct understanding on IYCF plays a significant role for the healthcare workers – including midwives – to protect the IYCF practices during this challenging time.

Banda Aceh has 11 Puskesmas [13] where more than 100 midwives are in contact with women and children within Banda Aceh.

2.2 Study design

This cross-sectional study was conducted among midwives with close contact to mothers and children under 2-year-old in all Puskesmas in the city of Banda Aceh. A Google form was developed to collect data on the demographic information and their knowledge on the IYCF. The survey questions were constructed based on the IYCF guidelines both the international and national ones on IYCF Practices during Covid-19 Pandemic [9]–[15]. The knowledge obtained was aimed at several aspects related to the IYCF during Covid-19 Pandemic consisting of their knowledge on basic practices of IYCF, Covid-19 transmission risk in breastfeeding dyads, breastfeeding practices – in general, breast milk expression, and breast milk substitution – and complementary feeding during the pandemic.

The surveys consist of 54 true-or-false statements in total that were given to the subjects in randomized order. The questions were designed so that every question needed to be filled for the survey to conclude. The researcher asked for participants' written consent beforehand the surveys were done during a non-working day simultaneously in limited time to prevent information-sharing during the surveys.

2.3 Data collection & management

Data were collected directly from the electronic forms filled by the participants via Google Form platforms. Data were anonymized, and the original data were stored separately on an encrypted hard storage device. The results of the study were informed as a compiled and processed data for the respondents.

3 Results & discussion

3.1 Characteristics of respondents

All 11-community health centres or Puskesmas in Banda Aceh City were planned to be involved in this study. However, in the course of the research, several respondents from one of the Puskesmas turned out to have just started self-isolation due to Covid-19. In order to respect the healing time of the respondents infected with Covid-19 with mild-moderate symptoms, the researchers excluded 1 Puskesmas thus leaving only 10 Puskesmas to be included in the study. All respondents were 100% all-female with the following characteristics:

Table 1. Respondents' characteristics of the Study.

Characteristic	Sub Characteristic	% (n=88)
Age Groups	30–40-year-old	55%
	> 40-year-old	45%
Education	Diploma	76%
	Graduate/ Postgrad	24%
Working Period	1-5 years	3%
	5-10 years	15%
	> 10 years	82%

Most of the respondents were senior in terms of working years and age, yet the results showed that only 23% of the respondents possessed good knowledge of IYCF during Covid-19 Pandemic while the remaining showed moderate and poor knowledge.

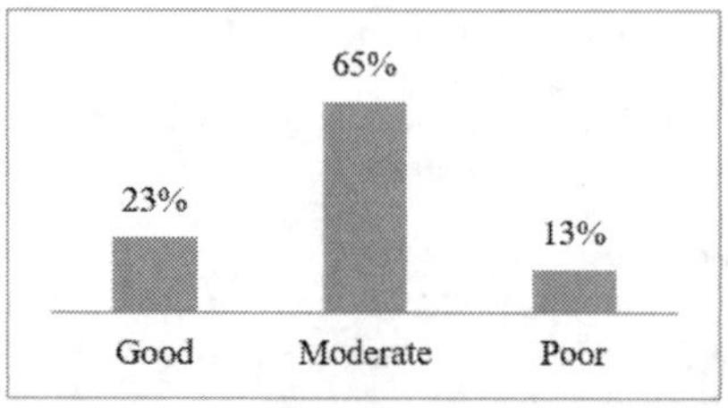

Fig. 1. Respondents' knowledge on IYCF during Covid-19 pandemic in general.

3.2 Deciphering Midwives' knowledge on IYCF during Covid-19 pandemic

Table 2. Midwives' Knowledge on IYCF during Covid-19 Pandemic.

Knowledge Level	Good	Moderate	Poor	Very Poor
IYCF Basic Principles	17%	77%	6%	0%
Covid-19 Transmission Risk in Breastfeeding Dyads	41%	31%	17%	11%
Health Protocols of A Mother with Covid-19	38%	40%	18%	5%
Breast Milk Substitutes when Breastfeeding Not Possible	15%	50%	18%	17%
Managing Complimentary Foods during Covid-19 Pandemic	72%	26%	2%	0%

The results showed that there were less than one fifth or only 17% were within good knowledge on basic principles of IYCF in general. This might be due to more than half of them (68%) never received the needed-training in IYCF before. Only one third or 32% of the respondents stated that they had the training once by UNICEF in the past before the Pandemic started. The results might also be caused by more work was emphasized into infection control and prevention of Covid-19, and not as much onto the other supporting factors such as nutrition or IYCF. Having stakeholders – including health workers such as midwives – with a non-adequate capacity of understanding issues related to nutrition such as IYCF during non-disaster contexts could lead to even worse nutritional statuses when in an emergency situation [1].

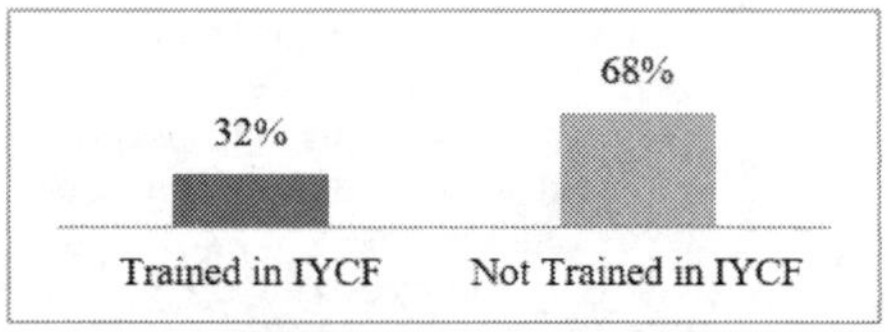

Fig. 2. Respondents' training experiences on IYCF.

Almost half or 41% of the respondents were having good knowledge on Covid-19 Transmission Risk in Breastfeeding Dyads. The remaining majority showed moderate knowledge (31%) while the rest were still showing poor and very poor scores. This might be caused by the few or no contacts some of them had been having with such scenarios in their real work life during the Pandemic. Only 6% of the respondents encountered pregnant/breastfeeding mothers and infant with Covid-19 from period March 2020 up until June 2021.

During this Covid-19 pandemic in which the virus easily transmitted via air droplets, breastfeeding is still highly supported to be continued by the WHO anytime an infected mother is fit enough to nurse [16]. This is based upon weighing the abundant benefits of continuing breastfeeding over the risk of Covid-19 transmission in which experts stated that mother-child dyads separation is not the answer to it [17].

Fig. 3. Respondents' Encounter with Mothers & Infants with Covid-19 (March 2020 – June 2021).

Midwives are among health workers that are often asked for their opinion on maternal and child health, especially during this Pandemic. Having knowledge about the health protocol of breastfeeding mothers suspected of or confirmed positive for Covid-19 is an important capacity for the midwives on advising/counselling mothers with the condition.

There were less than half or only 38% of the subjects showed good knowledge of the health protocols recommendations for when a mother with Covid-19 infection wills to breastfeed. This is a critical point in which the 6% of the respondents who stated that they had encountered mothers with Covid-19 might have had counselled their patients poorly on IYCF.

Breastfeeding mothers with COVID-19 in a stable condition are to be counselled to continue breastfeeding their babies [8,9,14]. However, infection control efforts should still be maintained through health protocols in accordance with existing recommendations [15,16]. These precautions need to be taken into accounts in order to minimize the risk of transmission while still promoting breastfeeding for the breastfeeding dyads with Covid-19 [10].

When breastfeeding is not possible, alternatives or substitutions should still be given, especially for babies under 6 months of age who are not suitable of solid foods. However, giving alternative or inappropriate breast milk substitutes such as the incorrect preparation for formula feeding will only yield "new disasters" in the form of threats of infection or even death [5]. Expressed breast milk remains the first priority if the mother cannot breastfeed directly. Expressed breast milk is proven not to transmit Covid-19 [17]. Even more beneficial for the infants consuming them, expressed breast milks from mothers with Covid-19 are proven to contain antibodies against SARS Cov-2 that will protect the infants – which is not found in formula milk [18]. It is important for health workers, including respondents of this study, to have this knowledge in order to be able to properly inform the public on the matter.

Unfortunately, only 15% of the respondents obtained good scores of knowledges on handling breast milk substitution when the direct breastfeeding is not possible. Improper handling of breast milk substitute such as improper preparation of formula feeding could cause malnutrition or worse – infection or even death – when not handled correctly [19]. This might be due to the incomplete understanding of IYCF practices. Highlights of IYCF generally are placed upon breastfeeding and less on how to handle the unimaginable non-ideal situations previously. Today, situations such as when a mother died of Covid-19 and no other breast milks are available to feed the child with, has been "normal news" during Pandemic. Incomplete understanding of IYCF, including on when and how to prepare breast milk substitutes, could lead to possible cases of infection and worsening nutritional status.

For example, there were 8% of the respondents who had experienced suggesting breast milk substitutes to their patients without necessarily assisting them with the proper dosages, methods and risks attached to the product consumption. The listed reasons were merely converging into one excuse of "the patients complained of not having enough breast milks". Detailed indications for the breastfeeding could actually take place. The indications are usually diagnosed by medical specialists for medical conditions preventing mothers to breastfeed such as several cases of mothers' poor mental health status or proof of baby's dehydration or the malnutrition status in need of fast catching up and a bucket of other legitimate causes.

In addition to breastfeeding, appropriate complementary feeding according to WHO and Ministry of Health guidelines needs to be continued. The Covid-19 pandemic itself has made conditions for many families economically difficult which has the potential to contribute to the risk of malnutrition cases [20]. Thus ensuring the continuation of proper complimentary food management in an emergency – including this pandemic – as much as possible is one of the keys to preventing the decline in children's nutritional status [21].

Almost two thirds or 72% of them showed good scores of knowledge in preparing complimentary foods safely with the suitable infection control precautions even when the caretakers are infected by Covid-19. These results are "fresh air" at the end of the surveys, where possible good practices of safe complementary feeding were already established pre-pandemic thus continued to reside as good knowledge on the matter.

4 Conclusion

In conclusion, there are less than one third of the respondents that showed good knowledge on IYCF during Covid-19 Pandemic in general. More than half of them possessed moderate and poor knowledge – something that needs a further investigation on the underlying causes and the strategies needed to build even stronger capacity of midwives in terms of IYCF during Covid-19 Pandemic. There were clearly rooms for improvement for these not-yet-trained midwives that can be filled from other sources of information upgrading and capacity building on the matter.

This pandemic seems to not cease in near future. Thus, giving more attention on the correct knowledge on IYCF practices particularly during Covid-19 Pandemic not only has the potential to strengthen their immunities to survive the pandemic, but also help them grow well even after the pandemic is contained or even better, ended. These study results may serve as the ground to better understanding on how to develop programs to prevent the stunting cases from re-emerging or worse, sky-rocketing during or after this Covid-19 Pandemic ends.

References

1. D. de Rosari and A. Saepudin, **17**, (2013)
2. Pemerintah Aceh, (2019)
3. Kementerian kesehatan Repubelik Indonesia, (2018)
4. C. M. Jacob, D. D. Briana, G. C. Di Renzo, N. Modi, F. Bustreo, G. Conti, A. Malamitsi-Puchner, and M. Hanson, The Lancet Public Health **2667**, 19 (2020)
5. T. L. Creek, A. Kim, L. Lu, A. Bowen, J. Masunge, W. Arvelo, M. Smit, O. MacH, K. Legwaila, C. Motswere, L. Zaks, T. Finkbeiner, L. Povinelli, M. Maruping, G. Ngwaru, G. Tebele, C. Bopp, N. Puhr, S. P. Johnston, A. J. Dasilva, C. Bern, R. S. Beard, and M. K. Davis, Journal of Acquired Immune Deficiency Syndromes **53**, 14 (2010)
6. K. Gribble, Journal of Human Lactation **34**, 40 (2018)
7. Kemenkes RI and United Nations Indonesia, Kesmas: National Public Health Journal **1**, 1 (2020)
8. WHO, 1 (2020)
9. C. Tomori, K. Gribble, A. E. L. Palmquist, M. T. Ververs, and M. S. Gross, Maternal and Child Nutrition **16**, 1 (2020)
10. W. Lubbe, E. Botha, H. Niela-Vilen, and P. Reimers, International Breastfeeding Journal **15**, 82 (2020)
11. N. Hesti, H. Yetti, and E. Erwani, Jurnal Kesehatan Andalas **8**, 338 (2019)
12. N. Syamiyah and H. Helda, Jurnal Penelitian Dan Pengembangan Pelayanan Kesehatan **2**, 29 (2018)

13. Kementerian Kesehatan RI, *Profil Kesehatan Indonesia Tahun 2019* (2020)
14. D. Vu Hoang, J. Cashin, K. Gribble, K. Marinelli, and R. Mathisen, BMJ Nutrition, Prevention and Health (2020)
15. GKIA and WHO, 1 (2020)
16. Satgas Covid RI, (2020)
17. M. Á. M. Gabriel, A. M. M. Martínez, M. E. M. Martínez, and J. A. Pedroche, Breastfeeding Medicine **15**, 492 (2020)
18. G. G. A. Fouda, J. J. Kwiek, and M. Yotebieng, **147**, (2021)
19. K. Gribble and C. Fernandes, World Nutrition **9**, 261 (2018)
20. D. Headey, R. Heidkamp, S. Osendarp, M. Ruel, N. Scott, R. Black, M. Shekar, H. Bouis, A. Flory, L. Haddad, and N. Walker, The Lancet **396**, 519 (2020)
21. P. Codjia, M. Volege, M. T. Le, A. Donnelly, F. F. Sesay, J. V. Senesie, and L. Kiige, Field Exchange 35 (2018)

AUTHOR INDEX

AUTHOR INDEX